FORD

FORD MUSTANG
1989-92 REPAIR MANUAL

CHILTON'S

CHILTON BOOK COMPANY

ONE OF THE ***DIVERSIFIED PUBLISHING COMPANIES,***
A PART OF ***CAPITAL CITIES/ABC, INC.***

Manufactured in USA

Chilton Way, Radnor, PA 19089
ISBN 0–8019–8253–7
Library of Congress Catalog Card No. 91–058819
2345678901 3210987654

Contents

Contents

SAFETY NOTICE

Proper service and repair procedures are vital to the safe, reliable operation of all motor vehicles, as well as the personal safety of those performing repairs. This manual outlines procedures for servicing and repairing vehicles using safe, effective methods. The procedures contain many NOTES, CAUTIONS and WARNINGS which should be followed along with standard safety procedures to eliminate the possibility of personal injury or improper service which could damage the vehicle or compromise its safety.

It is important to note that the repair procedures and techniques, tools and parts for servicing motor vehicles, as well as the skill and experience of the individual performing the work vary widely. It is not possible to anticipate all of the conceivable ways or conditions under which vehicles may be serviced, or to provide cautions as to all of the possible hazards that may result. Standard and accepted safety precautions and equipment should be used when handling toxic or flammable fluids, and safety goggles or other protection should be used during cutting, grinding, chiseling, prying, or any other process that can cause material removal or projectiles.

Some procedures require the use of tools specially designed for a specific purpose. Before substituting another tool or procedure, you must be completely satisfied that neither your personal safety, nor the performance of the vehicle will be endangered.

Although information in this manual is based on industry sources and is complete as possible at the time of publication, the possibility exists that some car manufacturers made later changes which could not be included here. While striving for total accuracy, Chilton Book Company cannot assume responsibility for any errors, changes or omissions that may occur in the compilation of this data.

PART NUMBERS

Part numbers listed in this reference are not recommendations by Chilton for any product by brand name. They are references that can be used with interchange manuals and aftermarket supplier catalogs to locate each brand supplier's discrete part number.

SPECIAL TOOLS

Special tools are recommended by the vehicle manufacturer to perform their specific job. Use has been kept to a minimum, but where absolutely necessary, they are referred to in the text by the part number of the tool manufacturer. These tools can be purchased, under the appropriate part number, from your Ford dealer or regional distributor, or an equivalent tool can be purchased locally from a tool supplier or parts outlet. Before substituting any tool for the one recommended, read the SAFETY NOTICE at the top of this page.

ACKNOWLEDGMENTS

The Chilton Book Company expresses appreciation to Ford Motor Co.; Ford Parts and Service Division, Service Technical Communications Department, Dearborn, Michigan for their generous assistance.

1

GENERAL INFORMATION AND MAINTENANCE

HOW TO USE THIS BOOK

Chilton's Total Car Care Manual for 1989–92 Ford Mustang is intended to help you learn more about the inner workings of your vehicle and save you money on its upkeep and operation.

The first two sections will be the most used, since they contain maintenance and tune-up information and procedures. Studies have shown that a properly tuned and maintained car can get at least 10% better gas mileage than an out-of-tune car. The other sections deal with the more complex systems of your car. Operating systems from engine through brakes are covered to the extent that the average do-it-yourselfer becomes mechanically involved.

A secondary purpose of this book is a reference for owners who want to understand their car and/or their mechanics better. In this case, no tools at all are required.

Before removing any bolts, read through the entire procedure. This will give you the overall view of what tools and supplies will be required. There is nothing more frustrating than having to walk to the bus stop on Monday morning because you were short one bolt on Sunday afternoon. So read ahead and plan ahead. Each operation should be approached logically and all procedures thoroughly understood before attempting any work.

All sections contain adjustments, maintenance, removal and installation procedures, and repair or overhaul procedures. When repair is not considered practical, we tell you how to remove the part and then how to install the new or rebuilt replacement. In this way, you at least save the labor costs. Backyard repair of such components as the alternator is just not practical.

Two basic mechanic's rules should be mentioned here. One, whenever the left side of the car or engine is referred to, it is meant to specify the driver's side of the car. Conversely, the right side of the car means the passenger's side. Secondly, most screws and bolt are removed by turning counterclockwise, and tightened by turning clockwise.

Safety is always the most important rule. Constantly be aware of the dangers involved in working on an automobile and take the proper precautions. (See the procedure in this section Servicing Your Vehicle Safely and the SAFETY NOTICE on the acknowledgement page.)

Pay attention to the instructions provided. There are 3 common mistakes in mechanical work:

1. Incorrect order of assembly, disassembly or adjustment. When taking something apart or putting it together, doing things in the wrong order usually cost you extra time; however, it CAN break something. Read the entire procedure before beginning disassembly. Do everything in the order in which the instructions say you should do it, even if you can't immediately see a reason for it. When you're taking apart something that is very intricate (for example, a carburetor), you might want to draw a picture of how it looks when assembled at one point in order to make sure you get everything back in its proper position. (We will supply exploded view whenever possible). When making adjustments, especially tune-up adjustments, do them in order; often, one adjustment affects another, and you cannot expect even satisfactory results unless each adjustment is made only when it cannot be changed by any order.

2. Overtorquing (or undertorquing). While it is more common for over-torquing to cause damage, undertorquing can cause a fastener to vibrate loose causing serious damage. Especially when dealing with aluminum parts, pay attention to torque specifications and utilize a torque wrench in assembly. If a torque figure is not available, remember that if you are using the right tool to do the job, you will probably not have to strain yourself to get a fastener tight enough. The pitch of most threads is so slight that the tension you put on the wrench will be multiplied many, many times in actual force on what you are tightening. A good example of how critical torque is can be seen in the case of spark plug installation, especially where you are putting the plug into an aluminum cylinder head. Too little torque can fail to crush the gasket, causing leakage of combustion gases and consequent overheating of the plug and engine parts. Too much torque can damage the threads, or distort the plug which changes the spark gap.

There are many commercial products available for ensuring that fasteners won't come loose, even if they are not torqued just right (a very common brand is Loctite®). If you're worried about getting something together tight enough to hold, but loose enough to avoid mechanical damage during assembly, one of these products might offer substantial insurance. Read the label on the package and make sure the products is compatible with the materials, fluids, etc. involved before choosing one.

3. Crossthreading. This occurs when a part such as a bolt is screwed into a nut or casting at the wrong angle and forced. Cross threading is more likely to occur if access is difficult. It helps to clean and lubricate fasteners, and to start threading with the part to be installed going straight in. Then, start the bolt, spark plug, etc. with your fingers. If you encounter resistance, unscrew the part and start over again at a different angle until it can be inserted and turned several turns without much effort. Keep in mind that many parts, especially spark plugs, used tapered threads so that gentle turning will automatically bring the part you're threading to the proper angle if you don't force it or resist a change in angle. Don't put a wrench on the part until its's been turned a couple of turns by hand. If you suddenly encounter resistance, and the part has not seated fully, don't force it. Pull it back out and make sure it's clean and threading properly.

Always take your time and be patient; once you have some experience, working on your car will become an enjoyable hobby.

TOOLS AND EQUIPMENT

➧ SEE FIGS. 1-3

Naturally, without the proper tools and equipment it is impossible to properly service you vehicle. It would be impossible to catalog each tool that you would need to perform each or any operation in this book. It would also be unwise for the amateur to rush out and buy an expensive set of tool on the theory that he may need on or more of them at sometime.

The best approach is to proceed slowly gathering together a good quality set of those tools that are used most frequently. Don't be misled by the low cost of bargain tools. It is far better to spend a little more for better quality. Forged wrenches, 6- or 12-point sockets and fine tooth ratchets are by far preferable to their less expensive counterparts. As any good mechanic can tell you, there are few worse

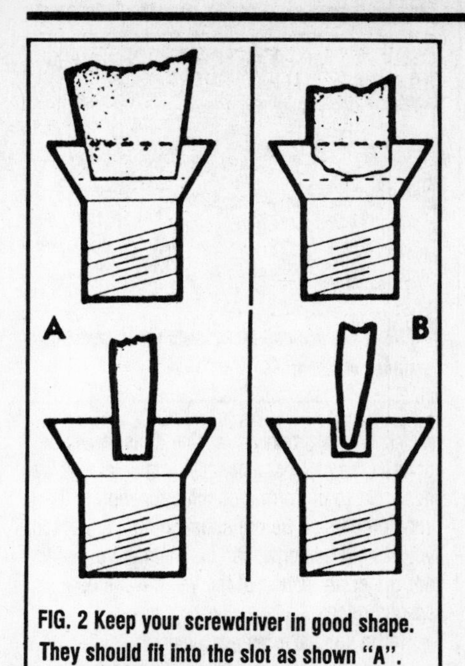

FIG. 2 Keep your screwdriver in good shape. They should fit into the slot as shown "A". If they look like those in "B", they need grinding or replacing.

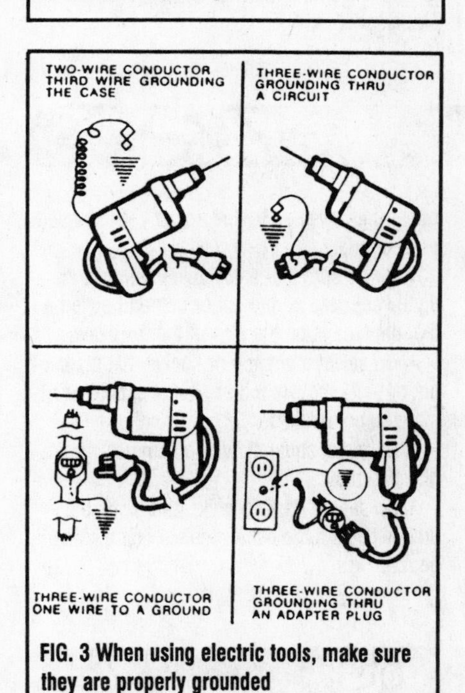

FIG. 3 When using electric tools, make sure they are properly grounded

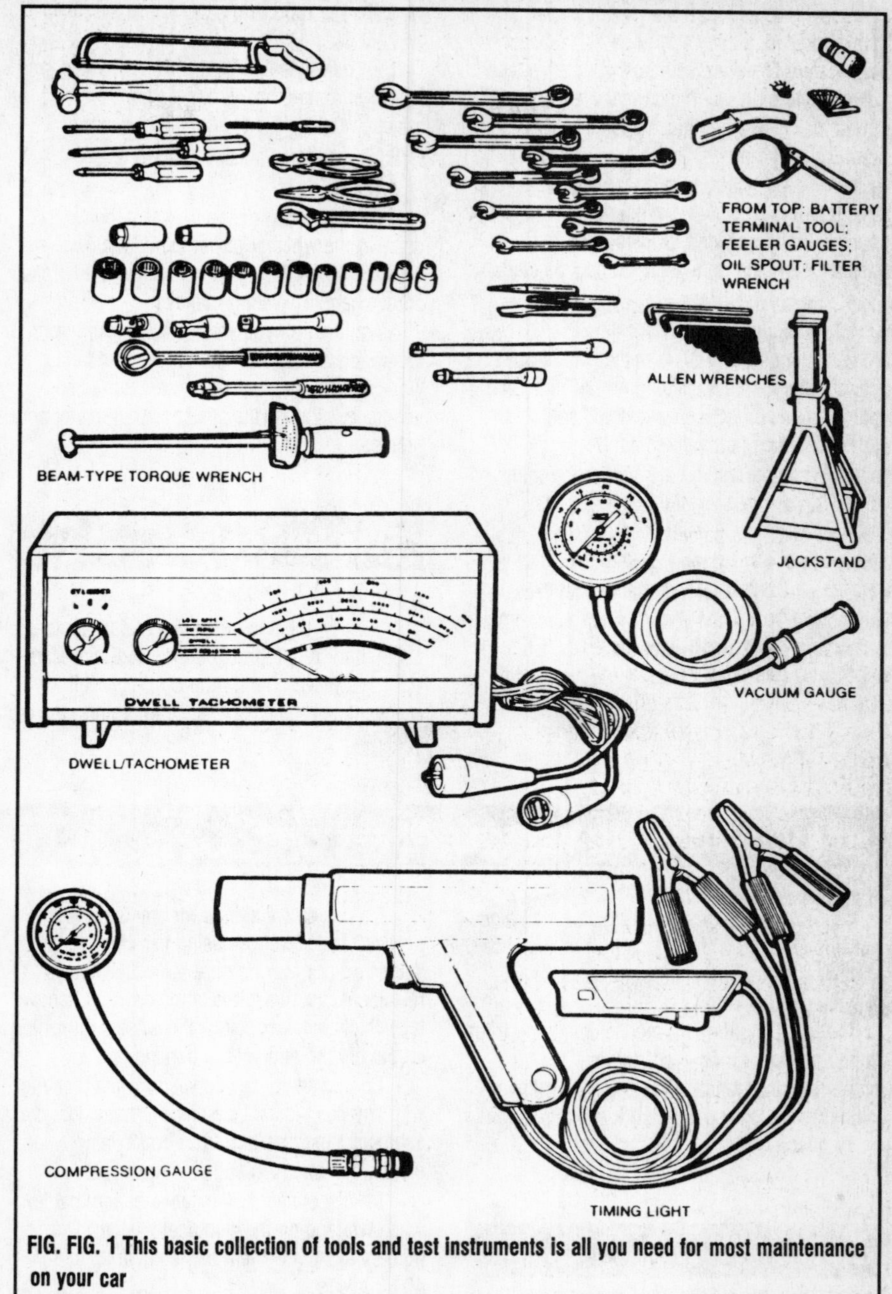

FIG. FIG. 1 This basic collection of tools and test instruments is all you need for most maintenance on your car

experiences than trying to work on a car with bad tools. Your monetary savings will be far outweighed by frustration and mangled knuckles.

Begin accumulating those tools that are used most frequently; those associated with routine maintenance and tune-up.

In addition to the normal assortment of screwdrivers and pliers you should have the following tools for routine maintenance jobs:

1. SAE (or Metric) or SAE/Metric wrenches-sockets and combination open end-box end wrenches in sizes from 1/8 in. (3mm) to 3/4 in. (19mm) and a spark plug socket (13/16 in. or 5/8 in. depending on plug type).

If possible, buy various length socket drive extensions. One break in this department is that the metric sockets available in the U.S. will all fit the ratchet handles and extensions you may already have (1/4 in., 3/8 in., and 1/2 in. drive).

2. Jackstands for support.
3. Oil filter wrench.
4. Oil filler spout for pouring oil.
5. Grease gun for chassis lubrication.
6. Hydrometer for checking the battery.
7. A container for draining oil.
8. Many rags for wiping up the inevitable mess.

In addition to the above items there are several others that are not absolutely necessary, but handy to have around. these include oil dry, a transmission funnel and the usual supply of lubricants, antifreeze and fluids, although these can be purchased as needed. This is a basic list for routine maintenance, but only your personal needs and desire can accurately determine you list of tools.

The second list of tools is for tune-ups. While

the tools involved here are slightly more sophisticated, they need not be outrageously expensive. There are several inexpensive tach/dwell meters on the market that are every bit as good for the average mechanic as a $100.00 professional model. Just be sure that it goes to a least 1,200–1,500 rpm on the tach scale and that it works on 4, 6, 8 cylinder engines. (A special tach is needed for diesel engines). A basic list of tune-up equipment could include:

1. Tach/dwell meter.
2. Spark plug wrench.
3. Timing light (a DC light that works from the car's battery is best, although an AC light that plugs into 110V house current will suffice at some sacrifice in brightness).
4. Wire spark plug gauge/adjusting tools.
5. Set of feeler blades.

Here again, be guided by your own needs. A feeler blade will set the points as easily as a dwell meter will read well, but slightly less accurately. And since you will need a tachometer anyway. . . well, make your own decision.

In addition to these basic tools, there are several other tools and gauges you may find useful. These include:

1. A compression gauge. The screw-in type is slower to use, but eliminates the possibility of a faulty reading due to escaping pressure.
2. A manifold vacuum gauge.
3. A test light.
4. An induction meter. This is used for determining whether or not there is current in a wire. These are handy for use if a wire is broken somewhere in a wiring harness.

As a final not, you will probably find a torque wrench necessary for all but the most basic work. The beam type models are perfectly adequate, although the newer click type are more precise.

Special Tools

➧ SEE FIG. 4

Normally, the use of special factory tools is avoided for repair procedures, since these are not readily available for the do-it-yourself mechanic. When it is possible to preform the job with more commonly available tools, it will be pointed out, but occasionally, a special tool was designed to perform a specific function and should be used. Before substituting another tool, you should be convinced that neither your safety nor the performance of the vehicle will be compromised.

FIG. 4 Always use jackstands when working under your car

Some special tools are available commercially from major tool manufacturers. Others can be purchased from your Ford Dealer or from the Owatonna Tool Company, Owatonna, Minnesota 55060.

SERVICING YOUR VEHICLE SAFELY

It is virtually impossible to anticipate all of the hazards involved with automotive maintenance and service but care and common sense will prevent most accidents.

The rules of safety for mechanics range from "don't smoke around gasoline" to "use the proper tool for the job." The trick to avoiding injuries is to develop safe work habits and take every possible precaution.

Do's

- Do keep a fire extinguisher and first aid kit within easy reach.
- Do wear safety glasses or goggles when cutting, drilling, grinding, or prying, even if you have 20/20 vision. If you wear glasses for the sake of vision, then they should be made of hardened glass that can serve also as safety glasses, or wear safety glasses over your regular glasses.
- Do shield your eyes whenever you work around the battery. Batteries contain sulphuric acid; in case of contact with the eyes or skin, flush the area with water or a mixture of water and baking soda and get medical attention immediately.
- Do use safety stands for any under-car service. Jacks are for raising vehicles; safety stands are for making sure the vehicle stays raised until you want it to come down. Whenever the vehicle is raised, block the wheels remaining on the ground and set the parking brake.
- Do use adequate ventilation when working with any chemicals. Like carbon monoxide, the asbestos dust resulting from brake lining wear can be poisonous in sufficient quantities.
- Do disconnect the negative battery cable when working on the electrical system. The primary ignition system can contain up to 40,000 volts.
- Do follow manufacturer's directions whenever working with potentially hazardous materials. Both brake fluid and antifreeze are poisonous if taken internally.
- Do properly maintain your tools. Loose hammerheads, mushroomed punches and chisels, frayed or poorly grounded electrical cords, excessively worn screwdrivers, spread wrenches (open end), cracked sockets, slipping ratchets, or faulty droplight sockets can cause accidents.
- Do use the proper size and type of tool for the job being done.
- Do when possible, pull on a wrench handle rather than push on it, and adjust your stance to prevent a fall.
- Do be sure that adjustable wrenches are tightly adjusted on the nut or bolt and pulled so that the face is on the side of the fixed jaw.
- Do select a wrench or socket that fits the nut or bolt. The wrench or socket should sit straight, not cocked.
- Do strike squarely with a hammer. Avoid glancing blows.
- Do set the parking brake and block the drive wheels if the work requires that the engine be running.

Don't's

- Don't run an engine in a garage or anywhere else without proper ventilation — EVER! Carbon monoxide is poisonous; it takes a long time to leave the human body and you can build up a deadly supply of it in your system by simply breathing in a little every day. You may not realize you are slowly poisoning yourself. Always use proper vents, window, fans or open the garage door.
- Don't work around moving parts while wearing a necktie or other loose clothing. Short

sleeves are much safer than long, loose sleeves and hard-toed shoes with neoprene soles protect your toes and give a better grip on slippery surfaces. Jewelry such as watches, fancy belt buckles, beads or body adornment of any kind is not safe working around a car. Long hair should be hidden under a hat or cap.

• Don't use pockets for toolboxes. A fall or bump can drive a screwdriver deep into your body. Even a wiping cloth hanging from the back pocket can wrap around a spinning shaft or fan.

• Don't smoke when working around gasoline, cleaning solvent or other flammable material.

• Don't smoke when working around the battery. When the battery is being charged, it gives off explosive hydrogen gas.

• Don't use gasoline to wash your hands; there are excellent soaps available. Gasoline may contain lead, and lead can enter the body through a cut, accumulating in the body until you are very ill. Gasoline also removes all the natural oils from the skin so that bone dry hands will such up oil and grease.

• Don't service the air conditioning system unless you are equipped with the necessary tools and training. The refrigerant, R-12, is extremely cold and when exposed to the air, will instantly freeze any surface it comes in contact with, including your eyes. Although the refrigerant is normally non-toxic, R-12 becomes a deadly poisonous gas in the presence of an open flame. One good whiff of the vapors from burning refrigerant can be fatal.

• Don't ever use a bumper jack (the jack that comes with the vehicle) for anything other than changing tires! If you are serious about maintaining your car yourself, invest in a hydraulic floor jack of at least 1½ ton capacity. It will pay for itself many times over through the years.

SERIAL NUMBER IDENTIFICATION

Vehicle

➧ SEE FIG. 5-6

The vehicle identification number is located on the left side of the dash panel behind the windshield.

A seventeen digit combination of numbers and letters forms the Vehicle Identification Number (VIN). Refer to the illustration for VIN details.

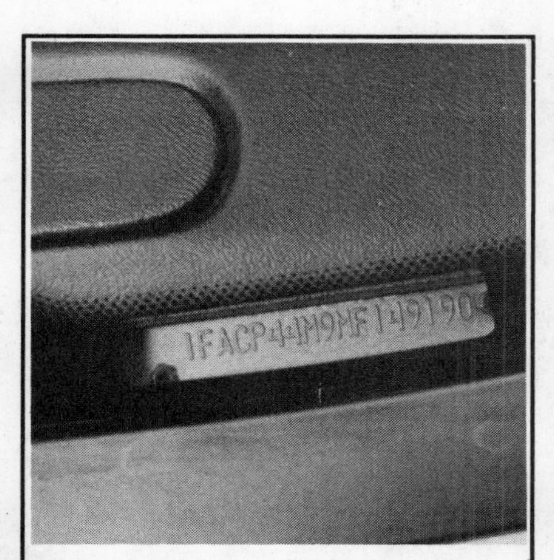

FIG. 5 The VIN plate is located on the driver's side of the dash

1FABP43M2 [N] X100001

VIN Code	Year
H	1987
J	1988
K	1989
L	1990
M	1991
N	1992

FIG. 6 The vehicle model year position is VIN position 10

Vehicle Safety Compliance Certification Label

➧ SEE FIG. 7

The label is attached to the driver's door lock pillar. The label contains the name of the manufacturer, the month and year of the vehicle, certification statement and VIN. The label also contains gross vehicle weight and tire data.

Engine

The engine identification tag identifies the cubic inch displacement of the engine, the model year, the year and month in which the engine was built, where it was built and the change level number. The change level is usually the number one (1), unless there are parts on the engine that will not be completely interchangeable and will require minor modification.

The engine identification code is located in the VIN at the eighth digit. The VIN can be found in the safety certification decal and the VIN plate at the upper left side of the dash panel. Refer to the "Engine Identification" chart for engine VIN codes.

Transmission

➧ SEE FIGS. 8-10

The transmission identification letter is located on a metal tag or plate attached to the case or it is stamped directly on the transmission case. Also, the transmission code is located on the Safety Certification Decal. Refer to the "Transmission Application" chart in this section.

Drive Axle

➧ SEE FIG. 11

The drive axle code is found stamped on a tag secured by one of the differential housing cover bolts.

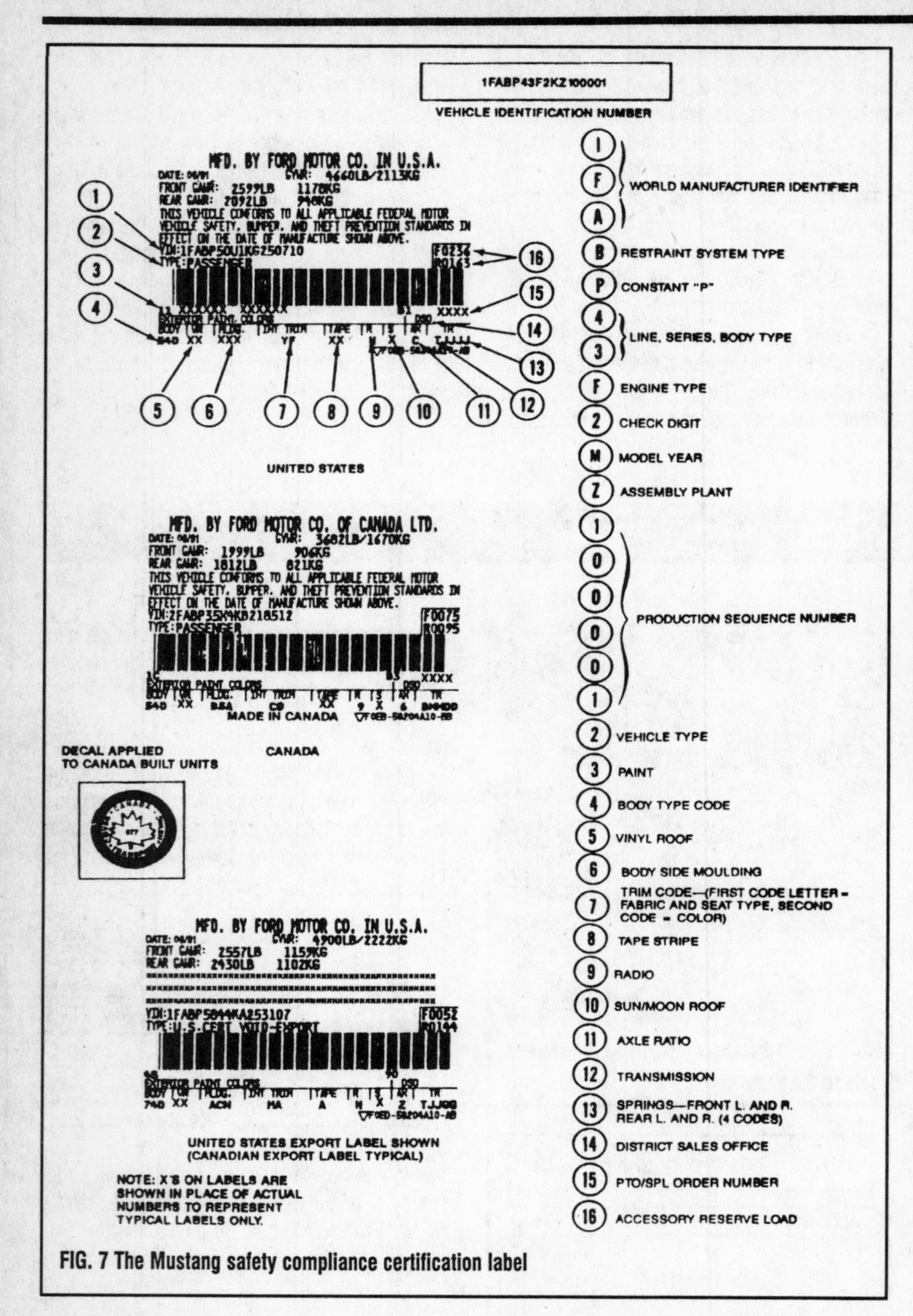

FIG. 7 The Mustang safety compliance certification label

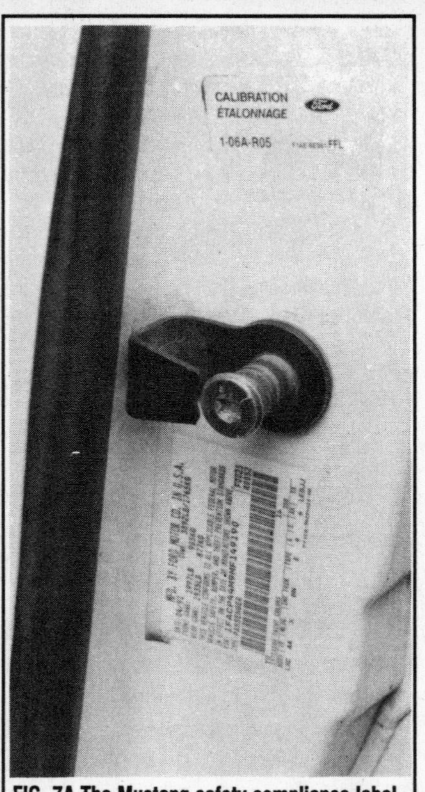

FIG. 7A The Mustang safety compliance label location

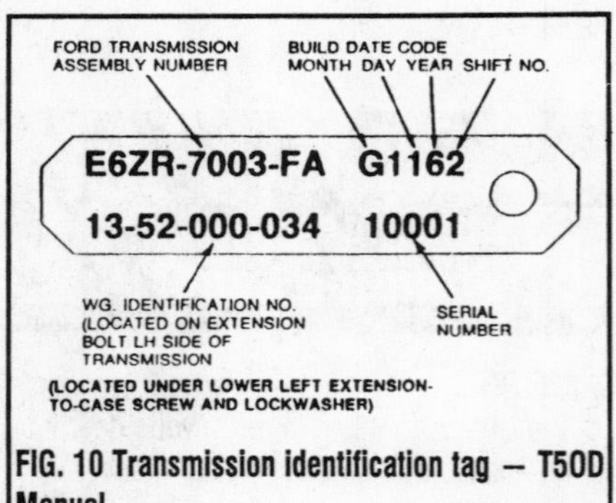

FIG. 10 Transmission identification tag — T50D Manual

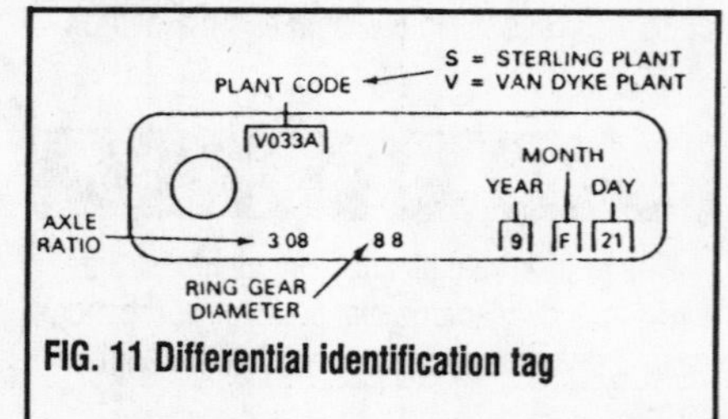

FIG. 11 Differential identification tag

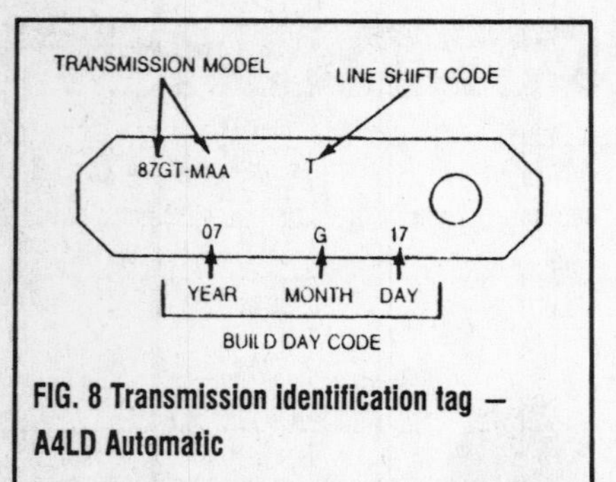

FIG. 8 Transmission identification tag — A4LD Automatic

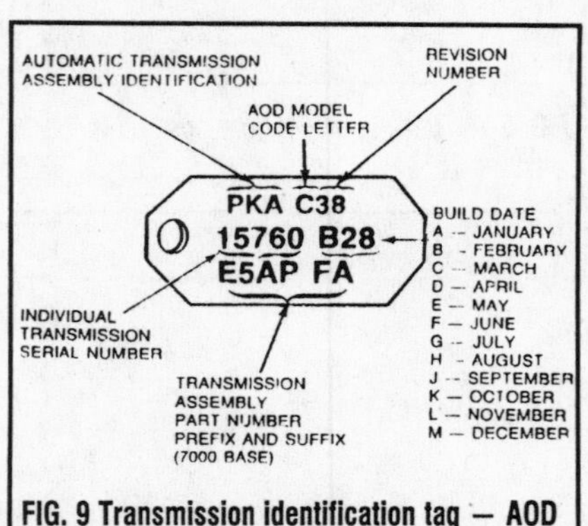

FIG. 9 Transmission identification tag — AOD Automatic

VEHICLE IDENTIFICATION CHART

It is important for servicing and ordering parts to be certain of the vehicle and engine identification. The VIN (vehicle identification number) is a 17 digit number visible through the windshield on the driver's side of the dash and contains the vehicle and engine identification codes. The tenth digit indicates model year and the eighth digit indicates engine code. It can be interpreted as follows:

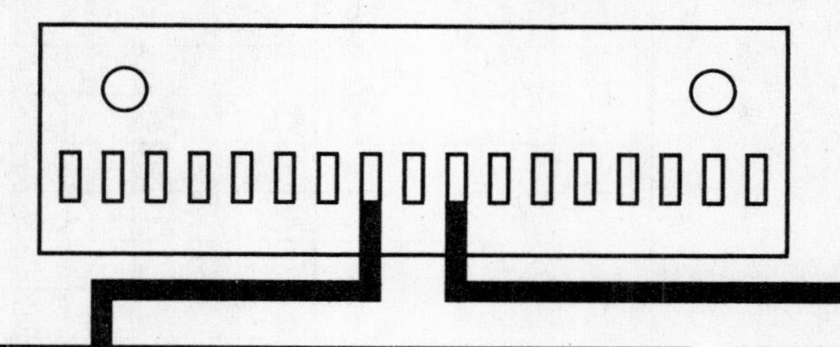

Engine Code

Code	Liter	Cu. In. (cc)	Cyl.	Fuel Sys.	Eng. Mfg.
A	2.3	140 (2300)	4	EFI	Ford
M	2.3	140 (2300)	4	EFI	Ford
E	5.0 (HO)	302 (5000)	8	EFI	Ford

EFI—Electronic Fuel Injection
HO—High Output

Model Year

Code	Year
K	1989
L	1990
M	1991
N	1992

ENGINE IDENTIFICATION

Year	Model	Engine Displacement Liter (cc)	Engine Series (ID/VIN)	Fuel System	No. of Cylinders	Engine Type
1989	Mustang	2.3 (2300)	A	EFI	4	OHC
	Mustang	5.0 HO (5000)	E	EFI	8	OHV
1990	Mustang	2.3 (2300)	A	EFI	4	OHC
	Mustang	5.0 HO (5000)	E	EFI	8	OHV
1991	Mustang	2.3 (2300)	M	EFI	4	OHC
	Mustang	5.0 HO (5000)	E	EFI	8	OHV
1992	Mustang	2.3 (2300)	M	EFI	4	OHC
	Mustang	5.0 HO (5000)	E	EFI	8	OHV

OHC—Overhead Camshaft
OHV—Overhead Valve
HO—High Output

TRANSMISSION APPLICATION CHART

Year	Model	Transmission Identification	Transmission Type
1989	Mustang	2	5-Speed OD M5 Manual
	Mustang	L	A4LD Automatic
	Mustang	T	AOD Automatic
1990	Mustang	2	5-Speed OD M5 Manual
	Mustang	L	A4LD Automatic
	Mustang	T	AOD Automatic
1991	Mustang	2	5-Speed OD M5 Manual
	Mustang	L	A4LD Automatic
	Mustang	T	AOD Automatic
1992	Mustang	2	5-Speed OD M5 Manual
	Mustang	L	A4LD Automatic
	Mustang	T	AOD Automatic

DRIVE AXLE APPLICATION CHART

Year	Model	Axle Identification	Axle Type
1989	Mustang	8	2.73 Axle Ratio Conventional
		M	2.73 Axle Ratio Limited-Slip
		Y	3.08 Axle Ratio Conventional
		Z	3.08 Axle Ratio Limited-Slip
		F	3.45 Axle Ratio Conventional
		R	3.45 Axle Ratio Limited-Slip
		5	3.27 Axle Ratio Conventional
		E	3.27 Axle Ratio Limited-Slip
		6	3.73 Axle Ratio Conventional
		W	3.73 Axle Ratio Limited-Slip
		2	3.55 Axle Ratio Conventional
		K	3.55 Axle Ratio Limited-Slip
1990	Mustang	8	2.73 Axle Ratio Conventional
		M	2.73 Axle Ratio Limited-Slip
		Y	3.08 Axle Ratio Conventional
		Z	3.08 Axle Ratio Limited-Slip
		F	3.45 Axle Ratio Conventional
		R	3.45 Axle Ratio Limited-Slip
		5	3.27 Axle Ratio Conventional
		E	3.27 Axle Ratio Limited-Slip
		6	3.73 Axle Ratio Conventional
		W	3.73 Axle Ratio Limited-Slip
		2	3.55 Axle Ratio Conventional
		K	3.55 Axle Ratio Limited-Slip
1991	Mustang	8	2.73 Axle Ratio Conventional
		M	2.73 Axle Ratio Limited-Slip
		Y	3.08 Axle Ratio Conventional
		Z	3.08 Axle Ratio Limited-Slip
		F	3.45 Axle Ratio Conventional
		R	3.45 Axle Ratio Limited-Slip
		5	3.27 Axle Ratio Conventional
		E	3.27 Axle Ratio Limited-Slip
		6	3.73 Axle Ratio Conventional
		W	3.73 Axle Ratio Limited-Slip
		2	3.55 Axle Ratio Conventional
		K	3.55 Axle Ratio Limited-Slip
1992	Mustang	8	2.73 Axle Ratio Conventional
		M	2.73 Axle Ratio Limited-Slip
		7	3.07 Axle Ratio Conventional
		Y	3.08 Axle Ratio Conventional
		Z	3.08 Axle Ratio Limited-Slip
		F	3.45 Axle Ratio Conventional
		R	3.45 Axle Ratio Limited-Slip
		5	3.27 Axle Ratio Conventional
		E	3.27 Axle Ratio Limited-Slip
		6	3.73 Axle Ratio Conventional

DRIVE AXLE APPLICATION CHART

Year	Model	Axle Identification	Axle Type
1992	Mustang	W	3.73 Axle Ratio Limited-Slip
		2	3.55 Axle Ratio Conventional
		K	3.55 Axle Ratio Limited-Slip

ROUTINE MAINTENANCE

Air Cleaner

➧ SEE FIGS. 12-13

The air cleaner is a paper element type.

The paper cartridge should be replaced according to the Preventive Maintenance Schedule at the end of this Section.

➡ Check the air filter more often if the vehicle is operated under severe dusty conditions and replace or clean it as necessary.

REPLACEMENT

1. Label, then disconnect all hoses and tubes connecting to the air filter assembly.
2. Remove the screws (2.3L),or clamps (5.0L) that attach the air cleaner lid to the housing.
3. Remove the air filter element.

To install:

1. Clean all the inside surfaces of the air cleaner body and cover.
2. Install the air filter element and position the air cleaner cover on the air cleaner body.
3. Install the screws (2.3L),or clamps (5.0L) that attach the air cleaner lid to the housing.
4. Connect all hoses and tubes connecting to the air filter assembly.

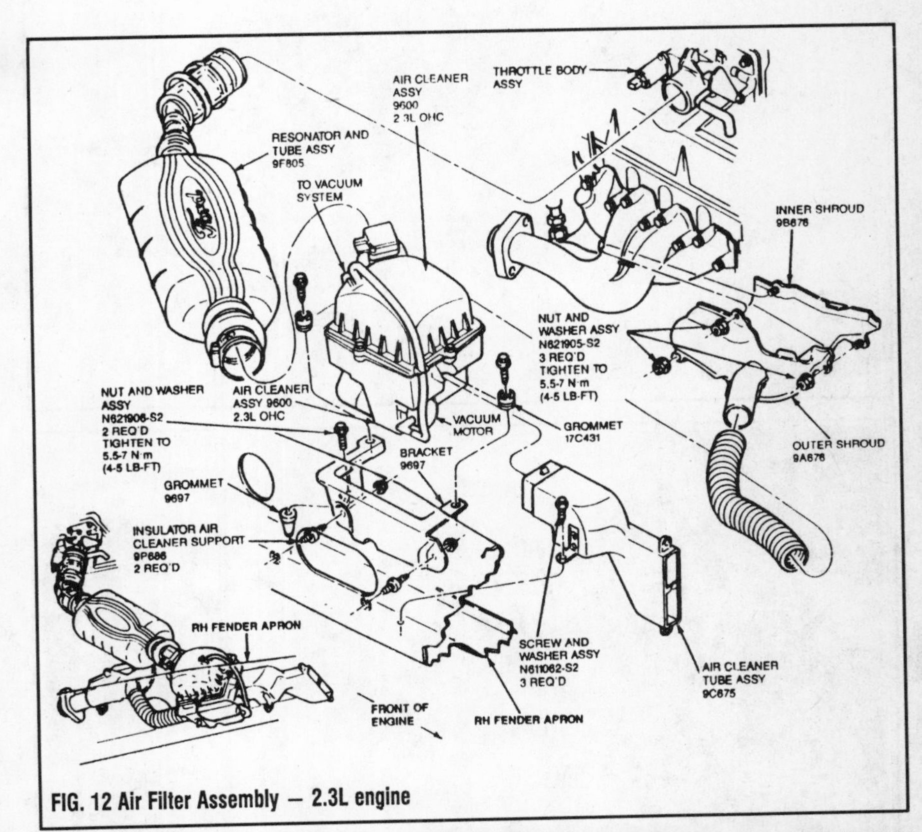

FIG. 12 Air Filter Assembly — 2.3L engine

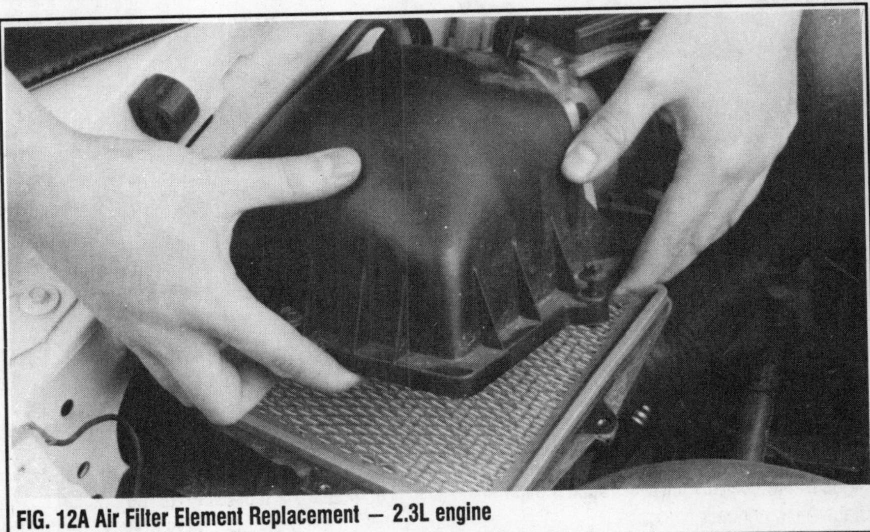

FIG. 12A Air Filter Element Replacement — 2.3L engine

Fuel Filter

➧ SEE FIG. 14

REPLACEMENT

CAUTION

NEVER SMOKE WHEN WORKING AROUND OR NEAR GASOLINE! MAKE SURE THAT THERE IS NO IGNITION SOURCE NEAR YOUR WORK AREA!

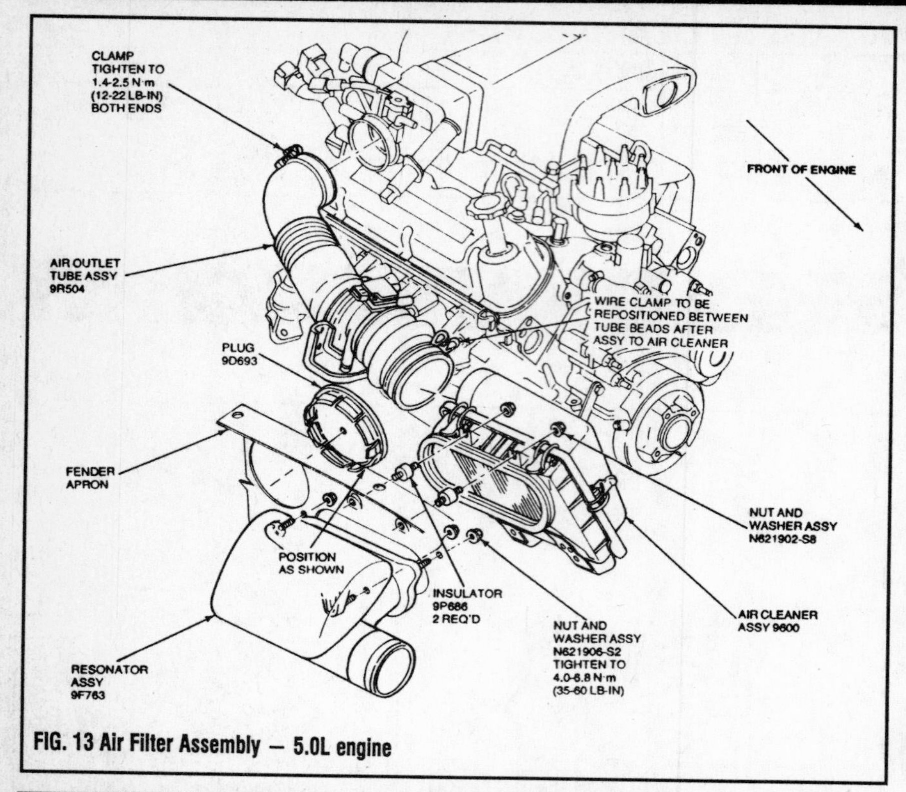

FIG. 13 Air Filter Assembly — 5.0L engine

FIG. 14 Fuel Filter Assembly mounted to frame rail

IN-LINE FILTER

➡ To prevent siphoning of fuel from the tank when the filter is removed, slightly raise the front of the vehicle above the level of the tank.

1. Disconnect the negative battery cable and relieve the fuel system pressure.
2. Raise and safely support the vehicle.
3. Remove the push connect fittings at both ends of the filter. Install new retainer clips in each push connect fitting.
4. Remove the fuel filter from the bracket by loosening the worm gear clamp. Note the direction of the flow arrow as installed in the bracket to ensure proper direction of fuel flow through the replacement filter.
5. Remove the filter from the retainer. Note that the direction of the flow arrow points to the open end of the retainer. Remove the rubber insulator rings.

To install:

6. Install the fuel filter into the bracket, ensuring the proper direction of flow. Tighten the worm gear clamp to 15–25 inch lbs. (1.7–2.8 Nm).
7. Install the push connect fittings onto the filter ends. Start the engine and check for leaks.
8. Lower the vehicle.

PCV Valve

➧ SEE FIG. 15

Check the PCV valve according to the Preventive Maintenance Schedule at the end of this Section to see if it is free and not gummed up, stuck or blocked. To check the valve, remove it from the engine and work the valve by sticking a screwdriver in the crankcase side of the valve. It should move. It is possible to clean the PCV valve by soaking it in a solvent and blowing it out with compressed air. This can restore the valve to some level of operating order. This should be used only as an emergency measure. Otherwise the valve should be replaced.

Evaporative Canister

➧ SEE FIGS. 16-17

The fuel evaporative emission control canister should be inspected for damage or leaks at the hose fittings every 24,000 miles. Repair or replace any old or cracked hoses. Replace the canister if it is damaged in any way. The canister is located under the hood, to the right of the engine.

Battery

➧ SEE FIGS. 18-24

Loose, dirty, or corroded battery terminals are a major cause of "no-start." Every 3 months or so, remove the battery terminals and clean them, giving them a light coating of petroleum jelly when you are finished. This will help to retard corrosion.

Check the battery cables for signs of wear or chafing and replace any cable or terminal that looks marginal. Battery terminals can be easily cleaned and inexpensive terminal cleaning tools are an excellent investment that will pay for themselves many times over. They can usually be purchased from any well-equipped auto store or parts department. Side terminal batteries require a different tool to clean the threads in the battery case. The accumulated white powder and corrosion can be cleaned from the top of the battery with an old toothbrush and a solution of baking soda and water.

Unless you have a maintenance-free battery, check the electrolyte level (see Battery under

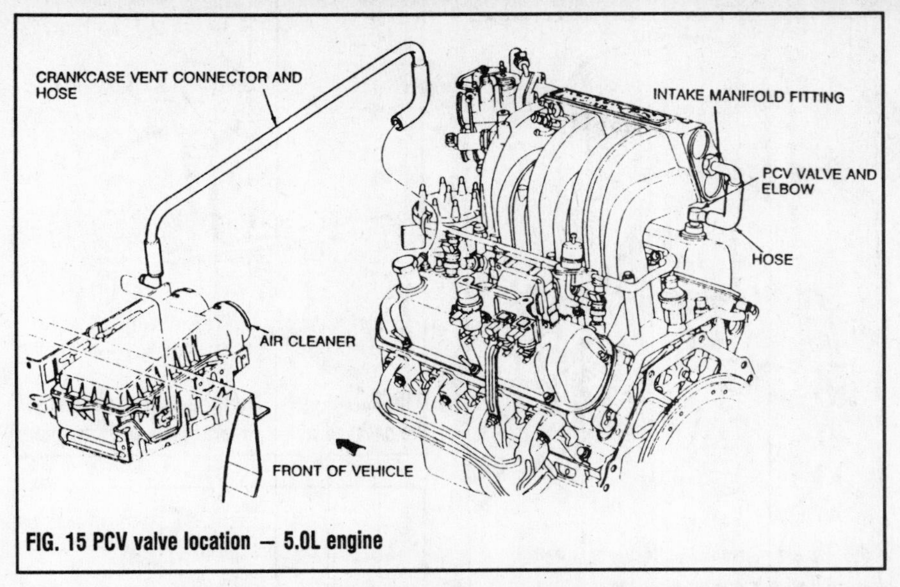

FIG. 15 PCV valve location – 5.0L engine

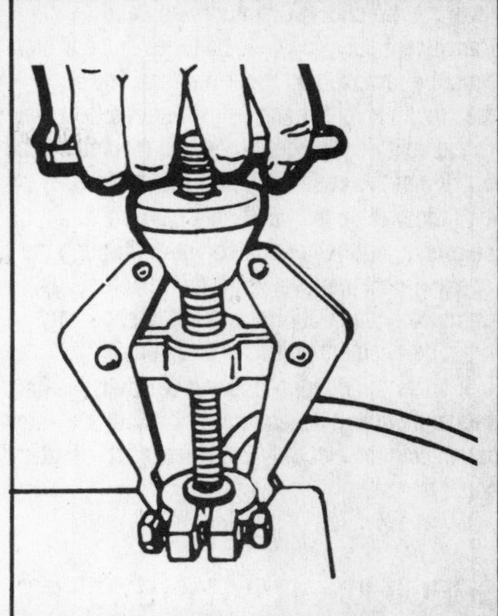
FIG. 18 Top terminal cables are easily removed with this inexpensive puller

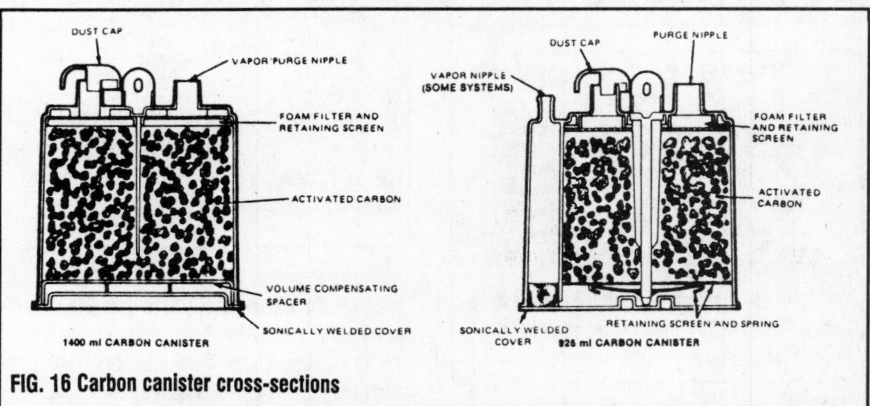

FIG. 16 Carbon canister cross-sections

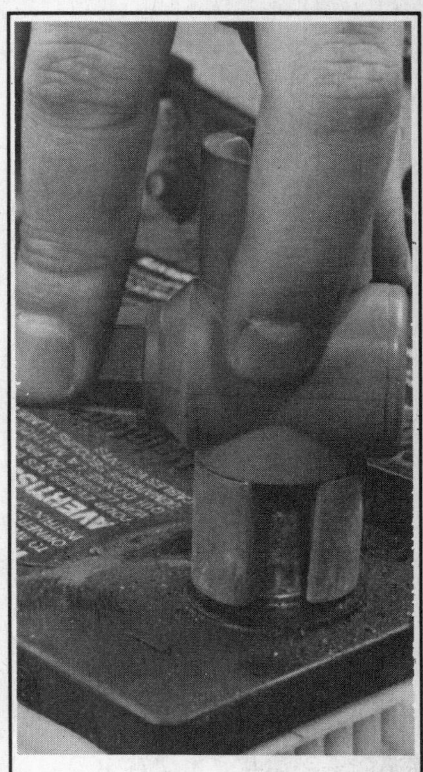
FIG. 19 Clean the battery posts with a wire terminal cleaner

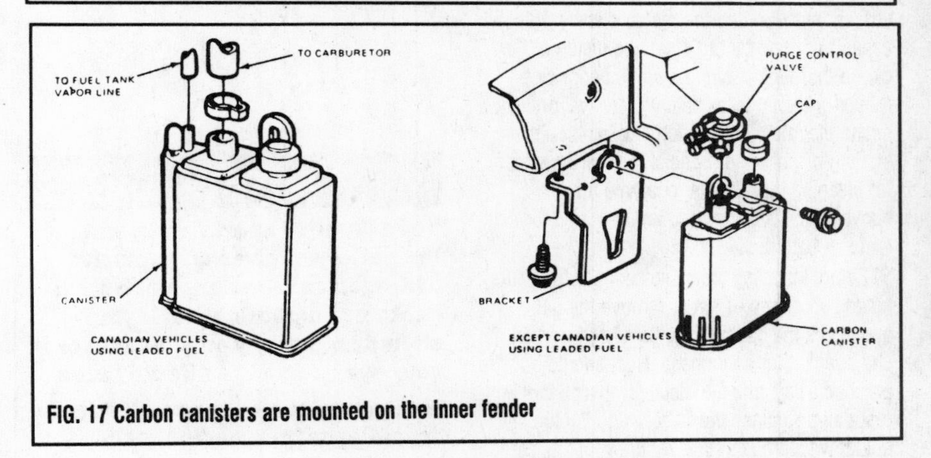

FIG. 17 Carbon canisters are mounted on the inner fender

Fluid Level Checks in this Section) and check the specific gravity of each cell. Be sure that the vent holes in each cell cap are not blocked by grease or dirt. The vent holes allow hydrogen gas, formed by the chemical reaction in the battery, to escape safely.

REPLACEMENT BATTERIES

The cold power rating of a battery measures battery starting performance and provides an approximate relationship between battery size and engine size. The cold power rating of a replacement battery should match or exceed your engine size in cubic inches.

FLUID LEVEL (EXCEPT MAINTENANCE FREE BATTERIES)

Check the battery electrolyte level at least once a month, or more often in hot weather or during periods of extended car operation. The level can be checked through the case on translucent polypropylene batteries; the cell caps must be removed on other models. The electrolyte level in each cell should be kept filled to the split ring inside, or the line marked on the outside of the case.

If the level is low, add only distilled water, or colorless, odorless drinking water, through the opening until the level is correct. Each cell is completely separate from the others, so each must be checked and filled individually.

If water is added in freezing weather, the car should be driven several miles to allow the water to mix with the electrolyte. Otherwise, the battery could freeze.

SPECIFIC GRAVITY (EXCEPT MAINTENANCE FREE BATTERIES)

At least once a year, check the specific gravity of the battery. It should be between 1.20 in.Hg and 1.26 in.Hg at room temperature.

The specific gravity can be check with the use of an hydrometer, an inexpensive instrument available from many sources, including auto parts stores. The hydrometer has a squeeze bulb at one end and a nozzle at the other. Battery electrolyte is sucked into the hydrometer until the float is lifted from its seat. The specific gravity is then read by noting the position of the float. Generally, if after charging, the specific gravity between any two cells varies more than 50 points (0.50), the battery is bad and should be replaced.

It is not possible to check the specific gravity in this manner on sealed (maintenance free) batteries. Instead, the indicator built into the top of the case must be relied on to display any signs of battery deterioration. If the indicator is dark, the battery can be assumed to be OK. If the indicator is light, the specific gravity is low, and the battery should be charged or replaced.

CABLES AND CLAMPS

Once a year, the battery terminals and the cable clamps should be cleaned. Loosen the clamps and remove the cables, negative cable first. On batteries with posts on top, the use of a puller specially made for the purpose is recommended. These are inexpensive, and available in auto parts stores. Side terminal battery cables are secured with a bolt.

Clean the cable lamps and the battery terminal with a wire brush, until all corrosion, grease,

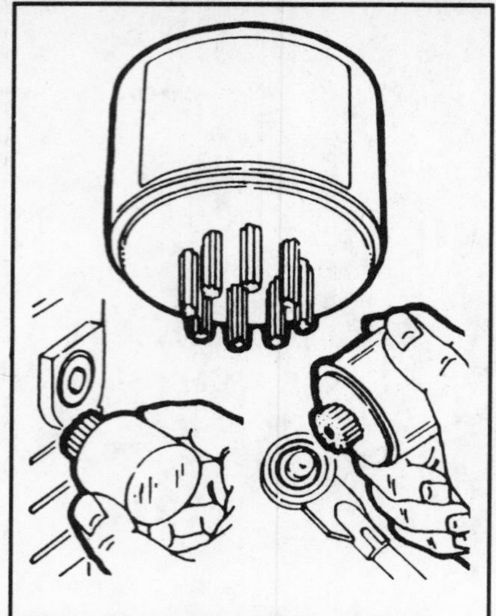

FIG. 21 Side terminal batteries require a special wire brush for cleaning

SPECIFIC GRAVITY (@ 80°F.) AND CHARGE
Specific Gravity Reading
(use the minimum figure for testing)

Minimum	Battery Charge
1.260	100% Charged
1.230	75% Charged
1.200	50% Charged
1.170	25% Charged
1.140	Very Little Power Left
1.110	Completely Discharged

FIG. 22 Battery specific gravity. Some testers have colored balls which correspond to the numerical valued in the left column

etc., is removed and the metal is shiny. It is especially important to clean the inside of the clamp thoroughly, since a small deposit of foreign material or oxidation there will prevent a sound electrical connection and inhibit either starting or charging. Special tools are available for cleaning these parts, one type for conventional batteries and another type for side terminal batteries.

Before installing the cables, loosen the battery holddown clamp or strap, remove the battery and check the battery tray. Clear it of any debris, and check it for soundness. Rust should be wire brushed away, and the metal given a coat of anti-rust paint. Replace the battery and tighten the holddown clamp or strap securely, but be careful not to overtighten, which will crack the battery case.

After the clamps and terminals are clean, reinstall the cables, negative cable last; do not hammer on the clamps to install. Tighten the clamps securely, but do not distort them. Give the clamps and terminals a thin external coat of grease after installation, to retard corrosion.

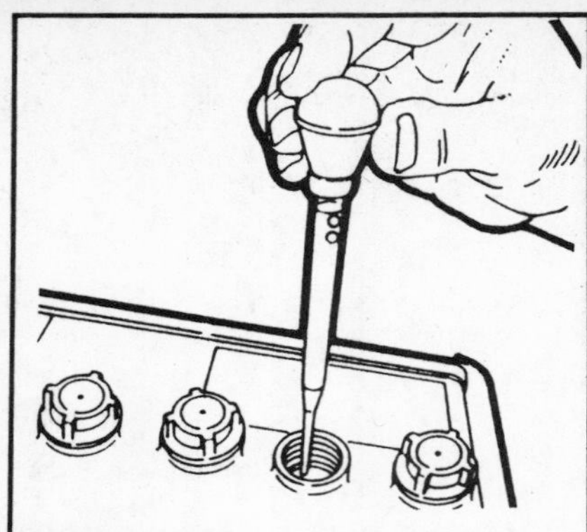

FIG. 23 The specific gravity of the battery can be checked with a simple float-type hydrometer

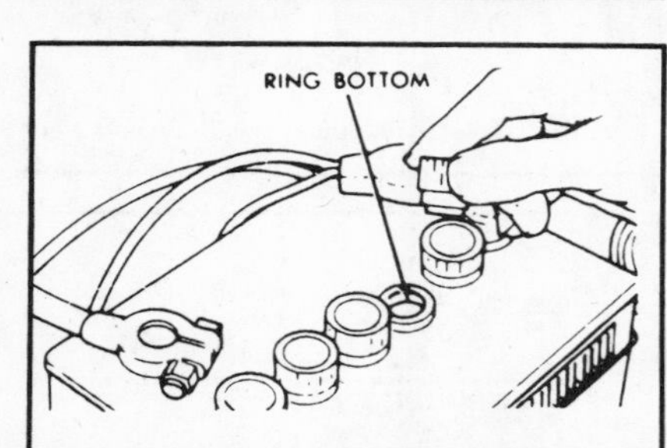

FIG. 24 Fill each battery cell to the bottom of the split ring with distilled water

Check the cables at the same time that the terminals are cleaned. If the cable insulation is cracked or broken, or if the ends are frayed, the cable should be replaced with a new cable of the same length and gauge.

CAUTION

Keep flame or sparks away from the battery; it gives off explosive hydrogen gas. Battery electrolyte contains sulphuric acid. If you should splash any on your skin or in your eyes, flush the affected area with plenty of clear water. If it lands in your eyes, get medical help immediately.

Belts

SEE FIGS. 25-29

Once a year or at 12,000 mile intervals, the belt should be checked, and, if necessary,

JUMP STARTING A DEAD BATTERY

The chemical reaction in a battery produces explosive hydrogen gas. This is the safe way to jump start a dead battery, reducing the chances of an accidental spark that could cause an explosion.

Jump Starting Precautions

1. Be sure both batteries are of the same voltage.
2. Be sure both batteries are of the same polarity (have the same grounded terminal).
3. Be sure the vehicles are not touching.
4. Be sure the vent cap holes are not obstructed.
5. Do not smoke or allow sparks around the battery.
6. In cold weather, check for frozen electrolyte in the battery. Do not jump start a frozen battery.
7. Do not allow electrolyte on your skin or clothing.
8. Be sure the electrolyte is not frozen.

CAUTION: Make certin that the ignition key, in the vehicle with the dead battery, is in the OFF position. Connecting cables to vehicles with on-board computers will result in computer destruction if the key is not in the OFF position.

Jump Starting Procedure

1. Determine voltages of the two batteries; they must be the same.
2. Bring the starting vehicle close (they must not touch) so that the batteries can be reached easily.
3. Turn off all accessories and both engines. Put both vehicles in Neutral or Park and set the handbrake.
4. Cover the cell caps with a rag—do not cover terminals.
5. If the terminals on the run-down battery are heavily corroded, clean them.
6. Identify the positive and negative posts on both batteries and connect the cables in the order shown.
7. Start the engine of the starting vehicle and run it at fast idle. Try to start the car with the dead battery. Crank it for no more than 10 seconds at a time and let it cool for 20 seconds in between tries.
8. If it doesn't start in 3 tries, there is something else wrong.
9. Disconnect the cables in the reverse order.
10. Replace the cell covers and dispose of the rags.

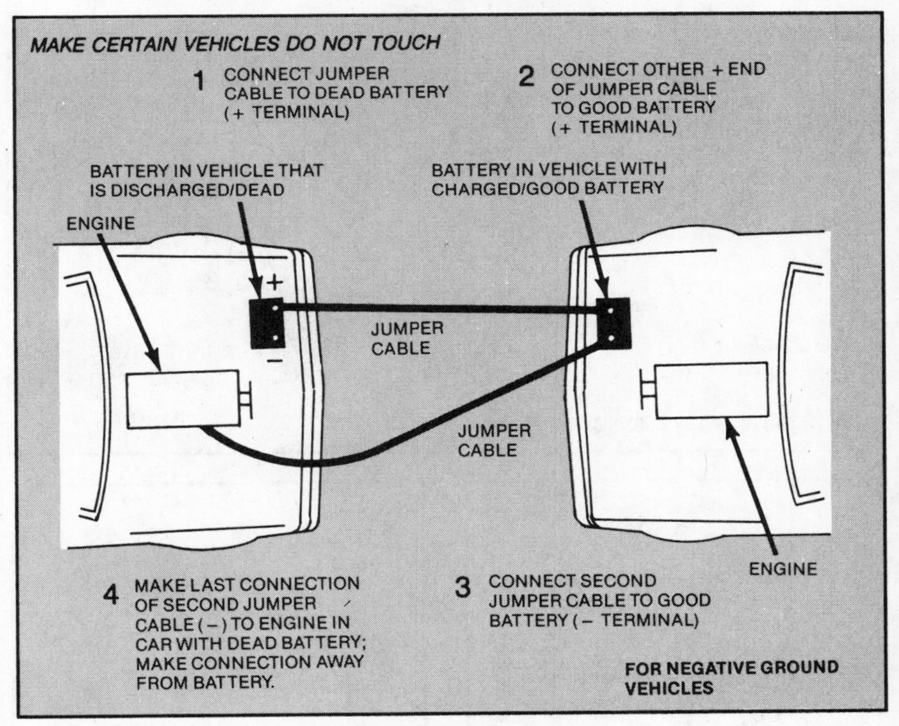

Side terminal batteries occasionally pose a problem when connecting jumper cables. There frequently isn't enough room to clamp the cables without touching sheet metal. Side terminal adaptors are available to alleviate this problem and should be removed after use

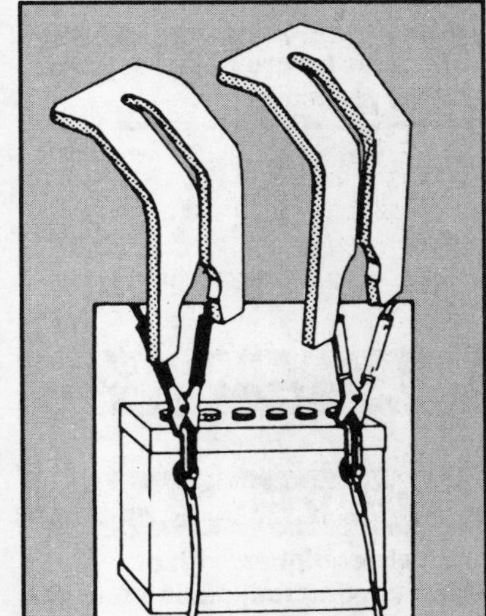

replaced. Loose accessory drive belts can lead to poor engine cooling and diminish alternator, power steering pump or air conditioning compressor output. A belt that is too tight places a severe strain on the water pump, alternator, power steering pump or compressor bearings.

Replace any belt that is so glazed, worn or stretched that it cannot be tightened sufficiently.

➡ **The material used in late model drive belts is such that the belts do not show wear. Replace belts at least every three years.**

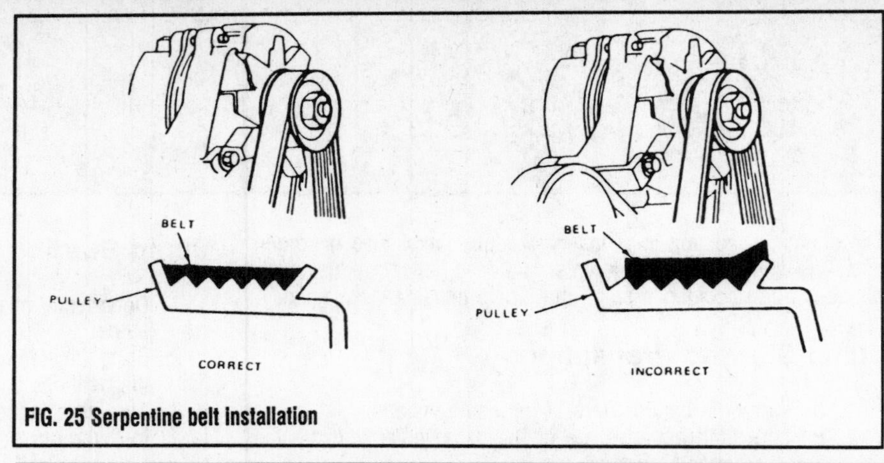

FIG. 25 Serpentine belt installation

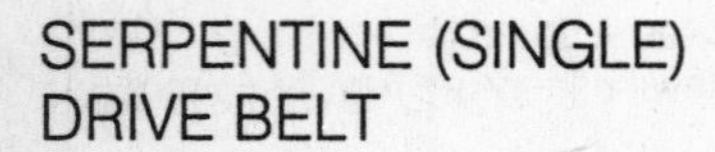

SERPENTINE (SINGLE) DRIVE BELT

Most models feature a single, wide, ribbed V-belt that drives the water pump, alternator, and (on some models) the air conditioner compressor. To install a new belt, loosen the bracket lock bolt, retract the belt tensioner with a pry bar and slide the old belt off of the pulleys. Slip on a new belt and release the tensioner and tighten the lock bolt. The spring powered tensioner eliminates the need for periodic adjustments.

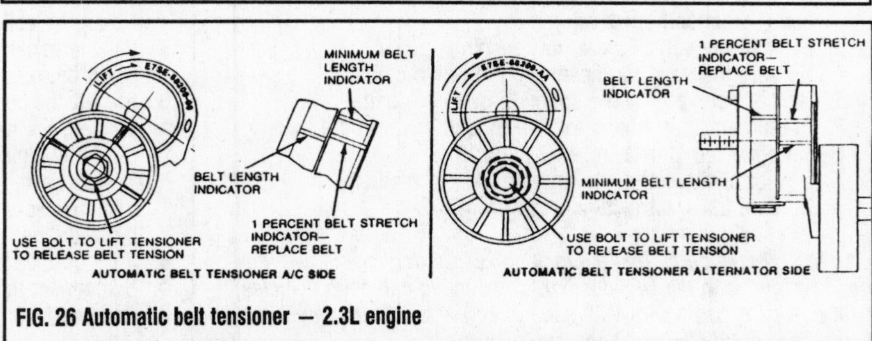

FIG. 26 Automatic belt tensioner — 2.3L engine

⁂ WARNING

Check to make sure that the V-ribbed belt is located properly in all drive pulleys before applying tensioner pressure.

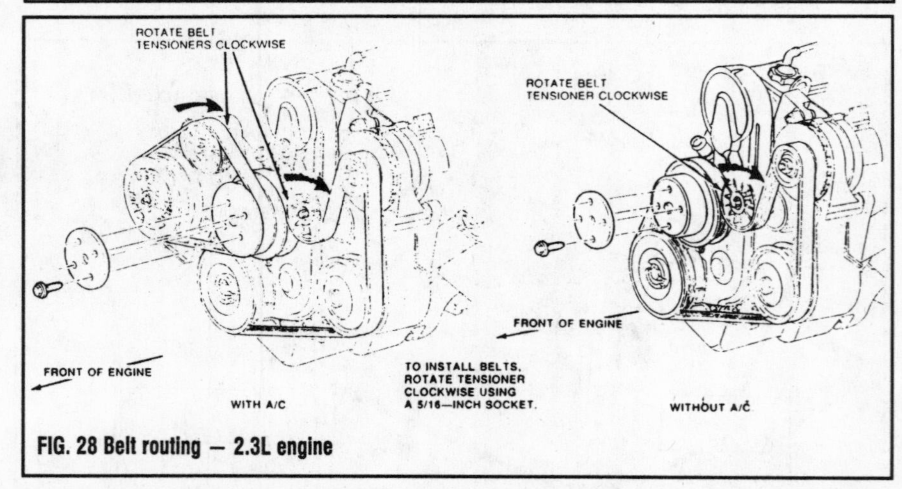

FIG. 28 Belt routing — 2.3L engine

Hoses

⁂ CAUTION

On models equipped with an electric cooling fan, disconnect the negative battery cable, or fan motor wiring harness connector before replacing any radiator/heater hose. The fan may come on, under certain circumstances, even though the ignition is Off.

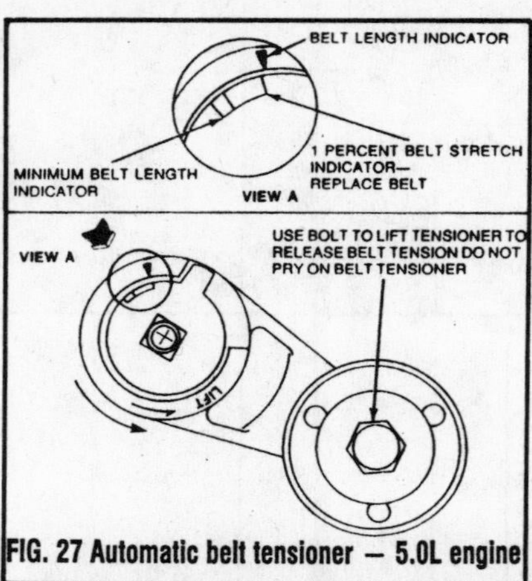

FIG. 27 Automatic belt tensioner — 5.0L engine

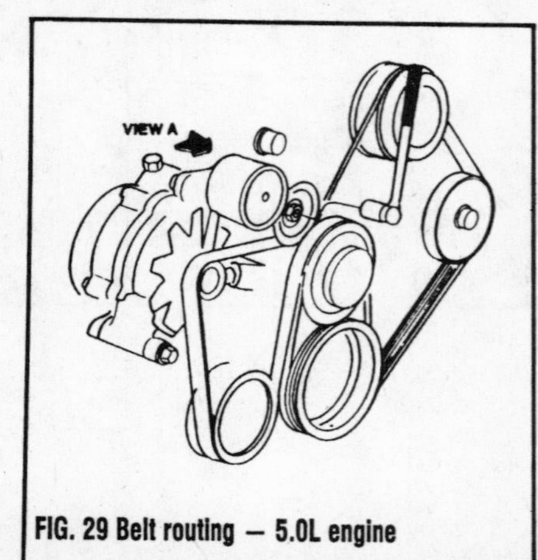

FIG. 29 Belt routing — 5.0L engine

HOW TO SPOT WORN V-BELTS

V–Belts are vital to efficient engine operation—they drive the fan, water pump and other accessories. They require little maintenance (occasional tightening) but they will not last forever. Slipping or failure of the V–belt will lead to overheating. If your V–belt looks like any of these, it should be replaced.

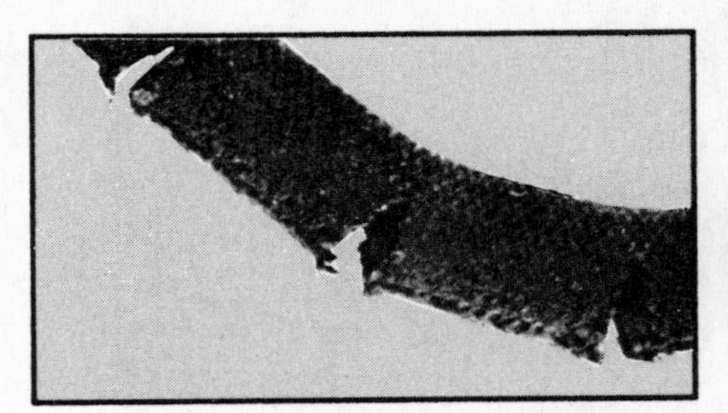

Cracking or Weathering

This belt has deep cracks, which cause it to flex. Too much flexing leads to heat build–up and premature failure. These cracks can be caused by using the belt on a pulley that is too small. Notched belts are available for small diameter pulleys.

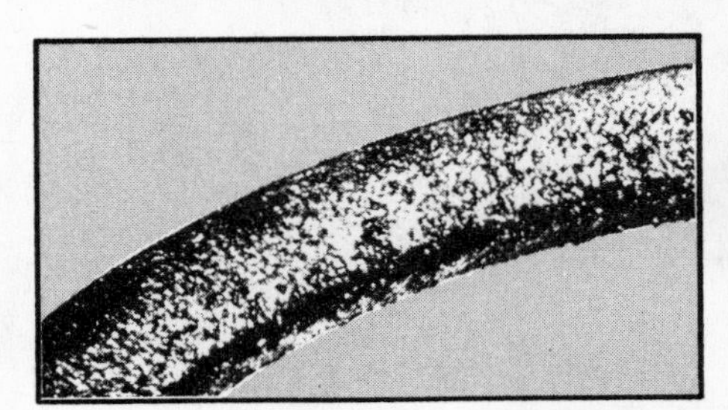

Softening (Grease and Oil)

Oil and grease on a belt can cause the belt's rubber compounds to soften and separate from the reinforcing cords that hold the belt together. The belt will first slip, then finally fail altogether.

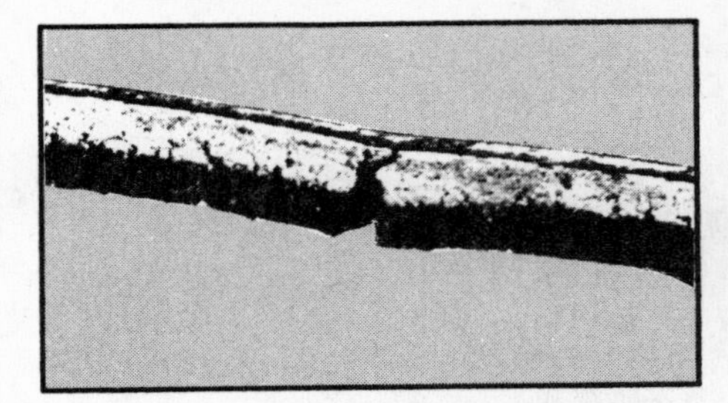

Glazing

Glazing is caused by a belt that is slipping. A slipping belt can cause a run-down battery, erratic power steering, overheating or poor accessory performance. The more the belt slips, the more glazing will be built up on the surface of the belt. The more the belt is glazed, the more it will slip. If the glazing is light, tighten the belt.

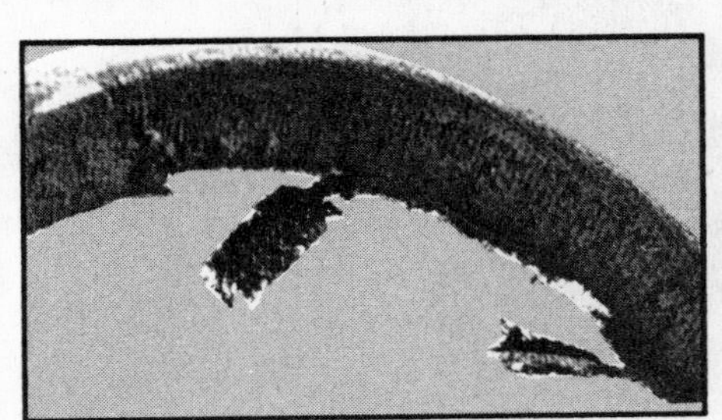

Worn Cover

The cover of this belt is worn off and is peeling away. The reinforcing cords will begin to wear and the belt will shortly break. When the belt cover wears in spots or has a rough jagged appearance, check the pulley grooves for roughness.

Separation

This belt is on the verge of breaking and leaving you stranded. The layers of the belt are separating and the reinforcing cords are exposed. It's just a matter of time before it breaks completely.

HOW TO SPOT BAD HOSES

Both the upper and lower radiator hoses are called upon to perform difficult jobs in an inhospitable environment. They are subject to nearly 18 psi at under hood temperatures often over 280°F, and must circulate nearly 7500 gallons of coolant an hour—3 good reasons to have good hoses.

Swollen Hose

A good test for any hose is to feel it for soft or spongy spots. Frequently these will appear as swollen areas of the hose. The most likely cause is oil soaking. This hose could burst at any time, when hot or under pressure.

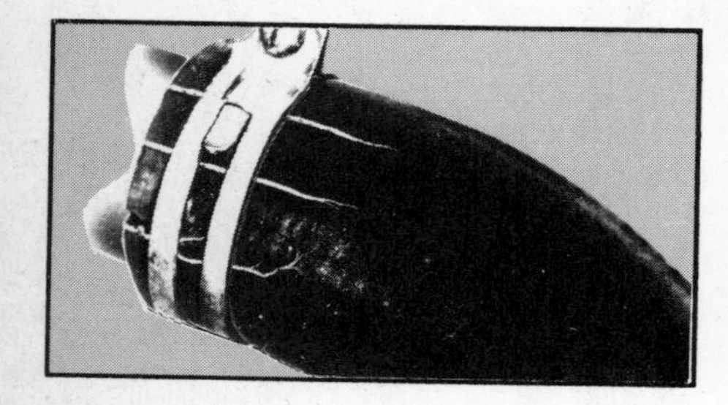

Cracked Hose

Cracked hoses can usually be seen but feel the hoses to be sure they have not hardened; a prime cause of cracking. This hose has cracked down to the reinforcing cords and could split at any of the cracks.

Frayed Hose End (Due to Weak Clamp)

Weakened clamps frequently are the cause of hose and cooling system failure. The connection between the pipe and hose has deteriorated enough to allow coolant to escape when the engine is hot.

Debris in Cooling System

Debris, rust and scale in the cooling system can cause the inside of a hose to weaken. This can usually be felt on the outside of the hose as soft or thinner areas.

REPLACEMENT

Inspect the condition of the radiator and heater hoses periodically. Early spring and at the beginning of the fall or winter, when you are performing other maintenance, are good times. Make sure the engine and cooling system are cold. Visually inspect for cracking, rotting or collapsed hoses, replace as necessary. Run your hand along the length of the hose. If a weak or swollen spot is noted when squeezing the hose wall, replace the hose.

1. Drain the cooling system into a suitable container (if the coolant is to be reused).

CAUTION

When draining the coolant, keep in mind that farm animals are attracted by the ethylene glycol antifreeze, and are quite likely to drink any that is left in an uncovered container or in puddles on the ground. This will prove fatal in sufficient quantity. Always drain the coolant into a sealable container. Coolant should be reused unless it is contaminated or several years old.

2. Loosen the hose clamps at each end of the hose that requires replacement.
3. Twist, pull and slide the hose off the radiator, water pump, thermostat or heater connection.
4. Clean the hose mounting connections. Position the hose clamps on the new hose.
5. Coat the connection surfaces with a water resistant sealer and slide the hose into position. Make sure the hose clamps are located beyond the raised bead of the connector (if equipped) and centered in the clamping area of the connection.
6. Tighten the clamps to 20–30 inch lbs. Do not overtighten.
7. Fill the cooling system.
8. Start the engine and allow it to reach normal operating temperature. Check for leaks.

Cooling System

SEE FIGS. 30-35

CAUTION

Never remove the radiator cap under any conditions while the engine is running! Failure to follow these instructions could result in damage to the cooling system or engine and/or personal injury. To avoid having scalding hot coolant or steam blow out of the radiator, use extreme care when removing the radiator cap from a hot radiator. Wait until the engine has cooled, then wrap a thick cloth around the radiator cap and turn it slowly to the first stop. Step back while the pressure is released from the cooling system. When you are sure the pressure has been released, press down on the radiator cap (still have the cloth in position) turn and remove the radiator cap.

At least once every 2 years, the engine cooling system should be inspected, flushed, and refilled with fresh coolant. If the coolant is left in the system too long, it loses its ability to prevent rust and corrosion. If the coolant has too much water, it won't protect against freezing.

The pressure cap should be looked at for signs of age or deterioration. Fan belt and other drive belts should be inspected and adjusted to the proper tension. (See checking belt tension).

Hose clamps should be tightened, and soft or cracked hoses replaced. Damp spots, or accumulations of rust or dye near hoses, water pump or other areas, indicate possible leakage, which must be corrected before filling the system with fresh coolant.

CHECK THE RADIATOR CAP

While you are checking the coolant level, check the radiator cap for a worn or cracked gasket. It the cap doesn't seal properly, fluid will be lost and the engine will overheat.

Worn caps should be replaced with a new one.

CLEAN RADIATOR OF DEBRIS

Periodically clean any debris — leaves, paper, insects, etc. — from the radiator fins. Pick the large pieces off by hand. The smaller pieces can be washed away with water pressure from a hose.

Carefully straighten any bent radiator fins with a pair of needle nose pliers. Be careful — the fins are very soft. Don't wiggle the fins back and forth too much. Straighten them once and try not to move them again.

DRAIN AND REFILL THE COOLING SYSTEM

Completely draining and refilling the cooling system every two years at least will remove accumulated rust, scale and other deposits. Coolant in late model cars is a 50/50 mixture of ethylene glycol and water for year round use. Use a good quality antifreeze with water pump lubricants, rust inhibitors and other corrosion inhibitors along with acid neutralizers.

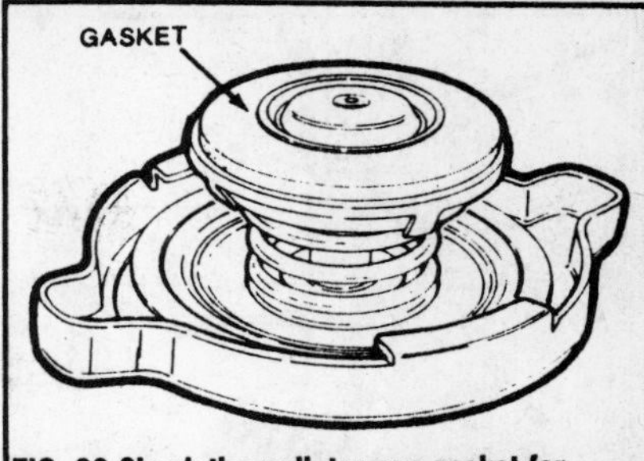

FIG. 30 Check the radiator cap gasket for cracks or wear

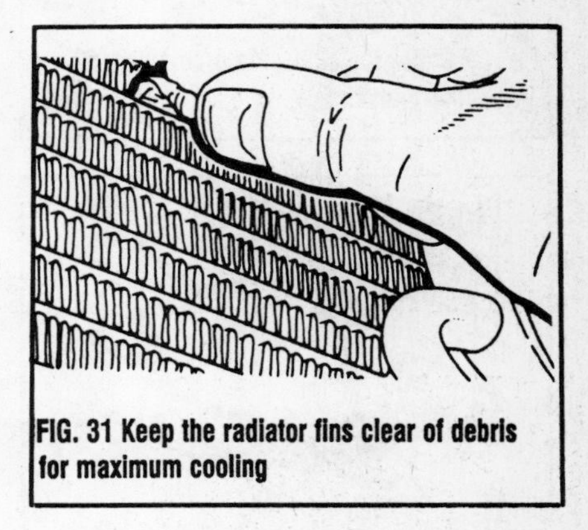

FIG. 31 Keep the radiator fins clear of debris for maximum cooling

1. Drain the existing antifreeze and coolant. Open the radiator and engine drain petcocks, or disconnect the bottom radiator hose, at the radiator outlet.

➡ Before opening the radiator petcock, spray it with some penetrating lubricant.

2. Close the petcock or reconnect the lower hose and fill the system with water.
3. Add a can of quality radiator flush.
4. Idle the engine until the upper radiator hose gets hot.
5. Drain the system again.
6. Repeat this process until the drained water is clear and free of scale.
7. Close all petcocks and connect all the hoses.
8. If equipped with a coolant recovery system, flush the reservoir with water and leave empty.
9. Determine the capacity of your coolant system (see capacities specifications). Add a 50/50 mix of quality antifreeze (ethylene glycol) and water to provide the desired protection.
10. Run the engine to operating temperature.
11. Stop the engine and check the coolant level.
12. Check the level of protection with an antifreeze tester, replace the cap and check for leaks.

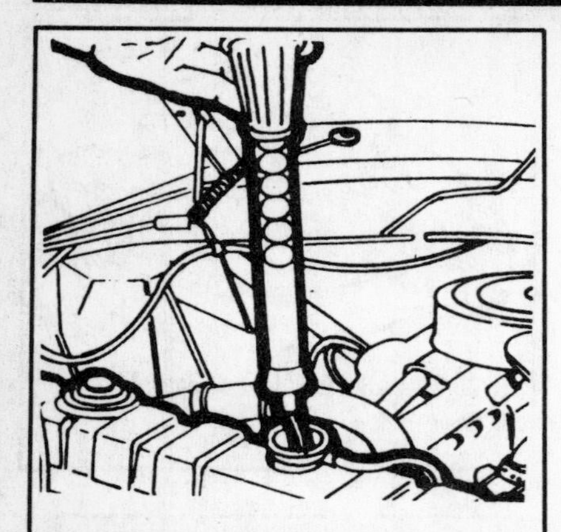

FIG. 32 Check the antifreeze protection with an inexpensive tester

FIG. 33 Open the petcock to drain the cooling system. Spray first with penetrating oil

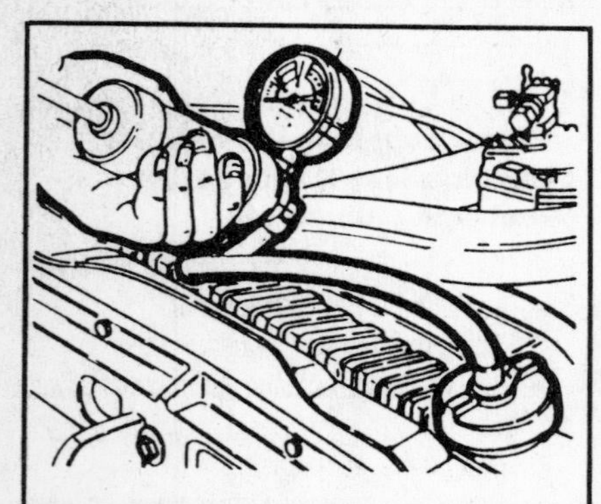

FIG. 34 The system should be pressure tested once a year

Air Conditioning

GENERAL SERVICING PROCEDURES

➧ SEE FIGS. 36-41

CAUTION

R-12 refrigerant is a chlorofluorocarbon which, when released into the atmosphere, can contribute to the depletion of the ozone layer in the upper atmosphere. Ozone filters out harmful radiation from the sun. If possible, an approved R-12 Recovery/Recycling machine that meets SAE standards should be employed when discharging the system. Follow the operating instructions provided with the approved equipment exactly to properly discharge the system.

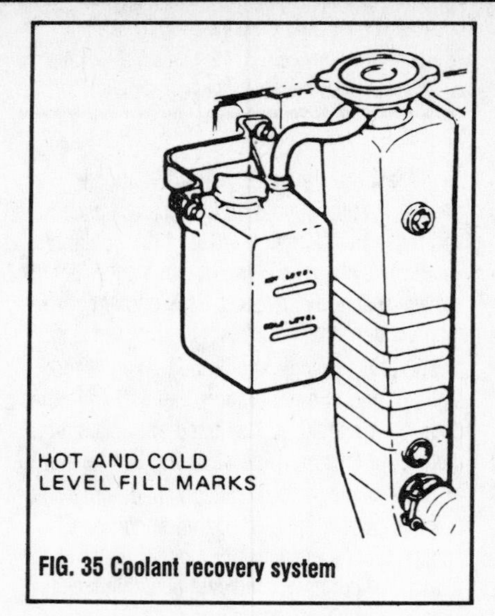

FIG. 35 Coolant recovery system

The most important aspect of air conditioning service is the maintenance of pure and adequate charge of refrigerant in the system. A refrigeration system cannot function properly if a significant percentage of the charge is lost. Leaks are common because the severe vibration encountered in an automobile can easily cause a sufficient cracking or loosening of the air conditioning fittings. As a result, the extreme operating pressures of the system force refrigerant out.

The problem can be understood by considering what happens to the system as it is operated with a continuous leak. Because the expansion valve regulates the flow of refrigerant to the evaporator, the level of refrigerant there is fairly constant. The receiver/drier stores any excess of refrigerant, and so a loss will first appear there as a reduction in the level of liquid. As this level nears the bottom of the vessel, some refrigerant vapor bubbles will begin to appear in the stream of liquid supplied to the expansion valve. This vapor decreases the capacity of the expansion valve very little as the valve opens to compensate for its presence. As the quantity of liquid in the condenser decreases, the operating pressure will drop there and throughout the high side of the system. As the R-12 continues to be expelled, the pressure available to force the liquid through the expansion valve will continue to decrease, and, eventually, the valve's orifice will prove to be too much of a restriction for adequate flow even with the needle fully withdrawn.

At this point, low side pressure will start to drop, and severe reduction in cooling capacity, marked by freeze-up of the evaporator coil, will result. Eventually, the operating pressure of the evaporator will be lower than the pressure of the atmosphere surrounding it, and air will be drawn into the system wherever there are leaks in the low side.

Because all atmospheric air contains at least some moisture, water will enter the system and mix with the R-12 and the oil. Trace amounts of moisture will cause sludging of the oil, and corrosion of the system. Saturation and clogging of the filter/drier, and freezing of the expansion valve orifice will eventually result. As air fills the system to a greater and greater extend, it will interfere more and more with the normal flows of refrigerant and heat.

A list of general precautions that should be observed while doing this follows:

1. Keep all tools as clean and dry as possible.
2. Thoroughly purge the service gauges and hoses of air and moisture before connecting them to the system. Keep them capped when not in use.
3. Thoroughly clean any refrigerant fitting before disconnecting it, in order to minimize the entrance of dirt into the system.
4. Plan any operation that requires opening the system beforehand in order to minimize the length of time it will be exposed to open air. Cap or seal the open ends to minimize the entrance of foreign material.
5. When adding oil, pour it through an extremely clean and dry tube or funnel. Keep the oil capped whenever possible. Do not use oil that has not been kept tightly sealed.
6. Use only refrigerant 12. Purchase refrigerant intended for use in only automotive air conditioning system. Avoid the use of refrigerant 12 that may be packaged for another use, such as cleaning, or powering a horn, as it is impure.
7. Completely evacuate any system that has been opened to replace a component, other than when isolating the compressor, or that has leaked sufficiently to draw in moisture and air. This requires evacuating air and moisture with a good vacuum pump for at least one hour.

If a system has been open for a considerable length of time it may be advisable to evacuate the system for up to 12 hours (overnight).

8. Use a wrench on both halves of a fitting

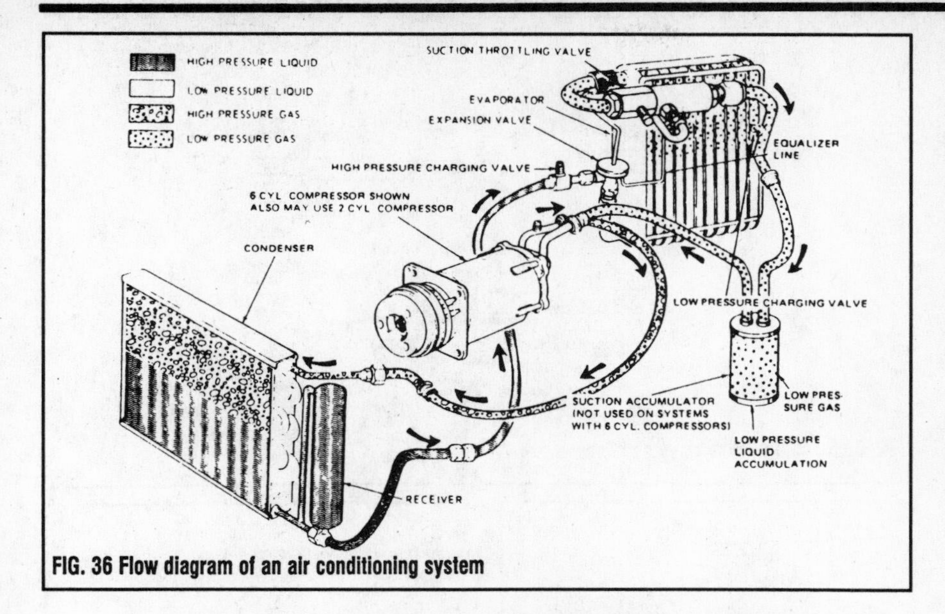

FIG. 36 Flow diagram of an air conditioning system

that is to be disconnected, so as to avoid placing torque on any of the refrigerant lines.

ADDITIONAL PREVENTIVE MAINTENANCE CHECKS

Antifreeze

In order to prevent heater core freeze-up during A/C operation, it is necessary to maintain permanent type antifreeze protection of +15°F (-9°C) or lower. A reading of -15°F (-26°C) is ideal since this protection also supplies sufficient corrosion inhibitors for the protection of the engine cooling system.

WARNING

Do not use antifreeze longer than specified by the manufacturer.

Radiator Cap

For efficient operation of an air conditioned car's cooling system, the radiator cap should have a holding pressure which meets manufacturer's specifications. A cap which fails to hold these pressure should be replaced.

Condenser

Any obstruction of or damage to the condenser configuration will restrict the air flow which is essential to its efficient operation. It is therefore, a good rule to keep this unit clean and in proper physical shape.

➡ **Bug screens are regarded as obstructions.**

Condensation Drain Tube

This single molded drain tube expels the condensation, which accumulates on the bottom of the evaporator housing, into the engine compartment.

If this tube is obstructed, the air conditioning performance can be restricted and condensation buildup can spill over onto the vehicle's floor.

SAFETY PRECAUTIONS

Because of the importance of the necessary safety precautions that must be exercised when working with air conditioning systems and R-12 refrigerant, a recap of the safety precautions are outlined.

1. Avoid contact with a charged refrigeration system, even when working on another part of the air conditioning system or vehicle. If a heavy tool comes into contact with a section of copper tubing or a heat exchanger, it can easily cause the relatively soft material to rupture.
2. When it is necessary to apply force to a fitting which contains refrigerant, as when checking that all system couplings are securely tightened, use a wrench on both parts of the fitting involved, if possible. This will avoid putting torque on the refrigerant tubing. (It is advisable, when possible, to use tube or line wrenches when tightening these flare nut fittings.)
3. Do not attempt to discharge the system by merely loosening a fitting, or removing the service valve caps and cracking these valves. Precise control is possibly only when using the service gauges. Place a rag under the open end of the center charging hose while discharging the system to catch any drops of liquid that might escape. Wear protective gloves when connecting or disconnecting service gauge hoses.
4. Discharge the system only in a well ventilated area, as high concentrations of the gas can exclude oxygen and act as an anesthetic. When leak testing or soldering this is particularly important, as toxic gas is formed when R-12 contacts any flame.
5. Never start a system without first verifying that both service valves are backseated, if equipped, and that all fittings are throughout the system are snugly connected.
6. Avoid applying heat to any refrigerant line or storage vessel. Charging may be aided by using water heated to less than 125°F (52°C) to warm the refrigerant container. Never allow a refrigerant storage container to sit out in the sun, or near any other source of heat, such as a radiator.
7. Always wear goggles when working on a system to protect the eyes. If refrigerant contacts the eye, it is advisable in all cases to see a physician as soon as possible.
8. Frostbite from liquid refrigerant should be treated by first gradually warming the area with cool water, and then gently applying petroleum jelly. A physician should be consulted.
9. Always keep refrigerant can fittings capped when not in use. Avoid sudden shock to the can which might occur from dropping it, or from banging a heavy tool against it. Never carry a refrigerant can in the passenger compartment of a car.
10. Always completely discharge the system before painting the vehicle (if the paint is to be baked on), or before welding anywhere near the refrigerant lines.

TEST GAUGES

Most of the service work performed in air conditioning requires the use of a set of two gauges, one for the high (head) pressure side of the system, the other for the low (suction) side.

The low side gauge records both pressure and vacuum. Vacuum readings are calibrated from 0 to 30 inches Hg and the pressure graduations read from 0 to no less than 60 psi.

The high side gauge measures pressure from 0 to at last 600 psi.

Both gauges are threaded into a manifold that contains two hand shut-off valves. Proper manipulation of these valves and the use of the attached test hoses allow the user to perform the following services:

1. Test high and low side pressures.
2. Remove air, moisture, and contaminated refrigerant.
3. Purge the system (of refrigerant).

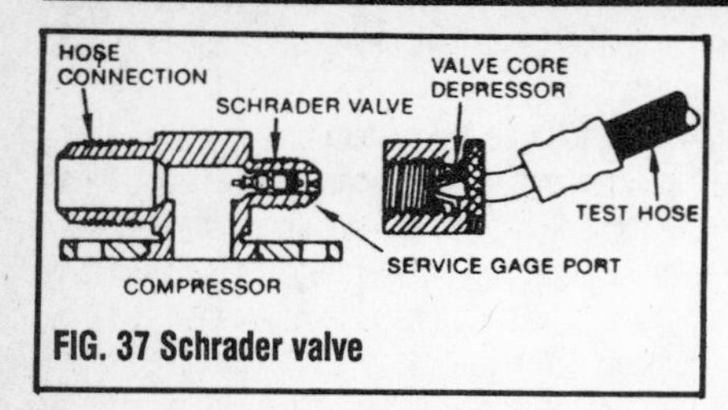

FIG. 37 Schrader valve

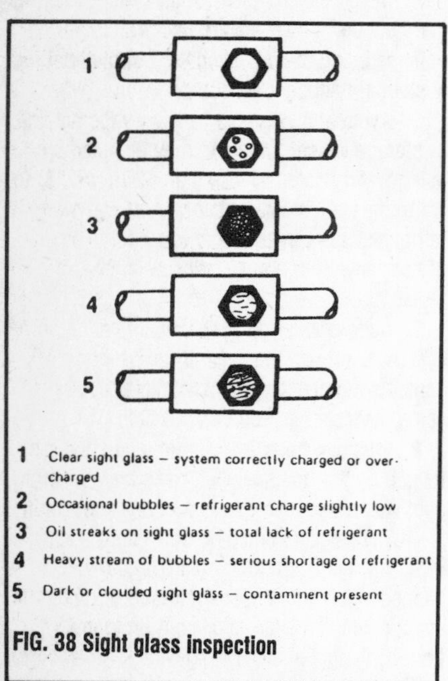

1 Clear sight glass – system correctly charged or overcharged
2 Occasional bubbles – refrigerant charge slightly low
3 Oil streaks on sight glass – total lack of refrigerant
4 Heavy stream of bubbles – serious shortage of refrigerant
5 Dark or clouded sight glass – contaminent present

FIG. 38 Sight glass inspection

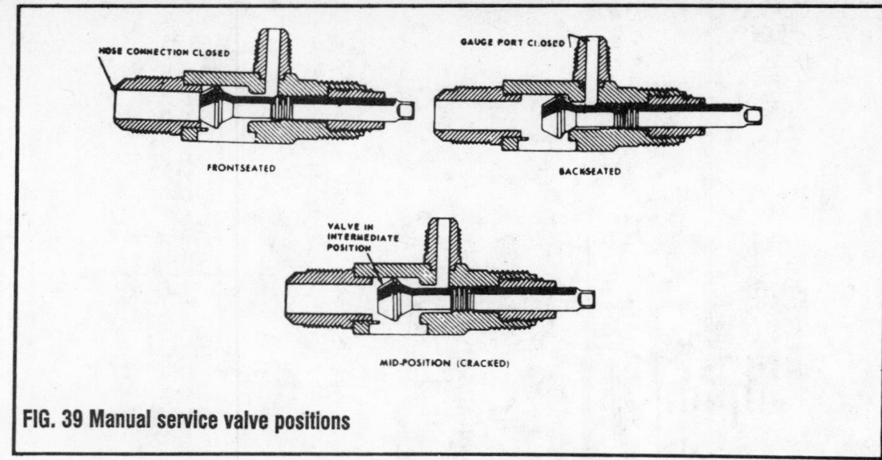

FIG. 39 Manual service valve positions

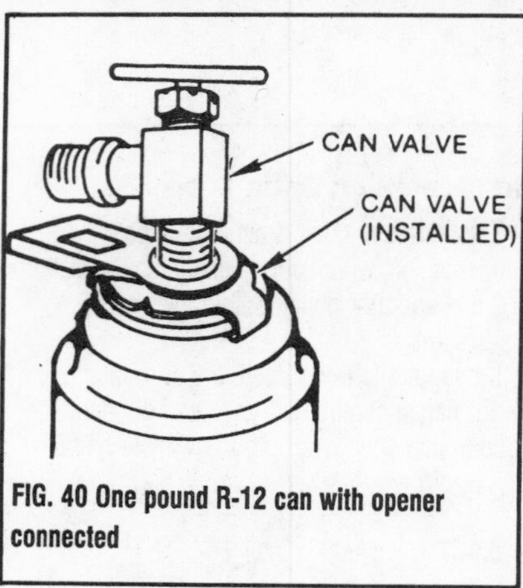

FIG. 40 One pound R-12 can with opener connected

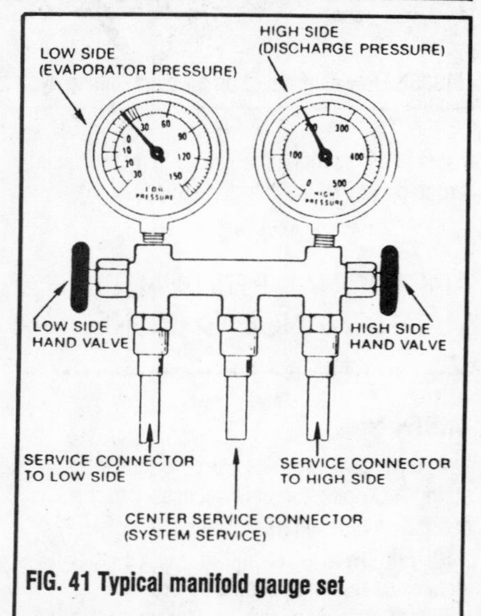

FIG. 41 Typical manifold gauge set

4. Charge the system (with refrigerant).

The manifold valves are designed so that they have no direct effect on gauge readings, but serve only to provide for, or cut off, flow of refrigerant through the manifold. During all testing and hook-up operations, the valves are kept in a close position to avoid disturbing the refrigeration system. The valves are opened only to purge the system or refrigerant or to charge it.

INSPECTION

CAUTION

The compressed refrigerant used in the air conditioning system expands into the atmosphere at a temperature of -21.7°F (-30°C) or lower. This will freeze any surface, including your eyes, that it contacts. In addition, the refrigerant decomposes into a poisonous gas in the presence of a flame. Do not open or disconnect any part of the air conditioning system.

Sight Glass Check

You can safely make a few simple checks to determine if your air conditioning system needs service. The tests work best if the temperature is warm (about 70°F [21.1°C]).

➡ If your vehicle is equipped with an aftermarket air conditioner, the following system check may not apply. You should contact the manufacturer of the unit for instructions on systems checks.

1. Place the automatic transmission in Park or the manual transmission in Neutral. Set the parking brake.
2. Run the engine at a fast idle (about 1,500 rpm) either with the help of a friend or by temporarily readjusting the idle speed screw.
3. Set the controls for maximum cold with the blower on High.
4. Locate the sight glass in one of the system lines. Usually it is on the left alongside the top of the radiator.
5. If you see bubbles, the system must be recharged. Very likely there is a leak at some point.
6. If there are no bubbles, there is either no refrigerant at all or the system is fully charged. Feel the two hoses going to the belt driven compressor. If they are both at the same temperature, the system is empty and must be recharged.
7. If one hose (high pressure) is warm and the other (low pressure) is cold, the system may be all right. However, you are probably making these tests because you think there is something wrong, so proceed to the next step.
8. Have an assistant in the car turn the fan control on and off to operate the compressor clutch. Watch the sight glass.
9. If bubbles appear when the clutch is disengaged and disappear when it is engaged, the system is properly charged.
10. If the refrigerant takes more than 45 seconds to bubble when the clutch is disengaged, the system is overcharged. This usually causes poor cooling at low speeds.

✲✲ WARNING

If it is determined that the system has a leak, it should be corrected as soon as possible. Leaks may allow moisture to enter and cause a very expensive rust problem. Exercise the air conditioner for a few minutes, every two weeks or so, during the cold months. This avoids the possibility of the compressor seals drying out from lack of lubrication.

TESTING THE SYSTEM

1. Park the car in the shade, at least 5 feet from any walls.
2. Connect a gauge set.
3. Close (clockwise) both gauge set valves.
4. Start the engine, set the parking brake, place the transmission in NEUTRAL and establish an idle of 1,100–1,300 rpm.
5. Run the air conditioning system for full cooling, in the MAX or COLD mode.
6. The low pressure gauge should read 5–20 psi; the high pressure gauge should indicate 120–180 psi.

✲✲ WARNING

These pressures are the norm for an ambient temperature of 70–80°F (21–27°C). Higher air temperatures along with high humidity will cause higher system pressures. At idle speed and an ambient temperature of 110°F (43°C), the high pressure reading can exceed 300 psi. Under these extreme conditions, you can keep the pressures down by directing a large electric floor fan through the condenser.

DISCHARGING THE SYSTEM

1. Remove the caps from the high and low pressure charging valves in the high and low pressure lines.
2. Turn both manifold gauge set hand valves to the fully closed (clockwise) position.
3. Connect the manifold gauge set.
4. If the gauge set hoses do not have the gauge port actuating pins, install fitting adapters T71P–19703–S and R on the manifold gauge set hoses. If the car does not have a service access gauge port valve, connect the gauge set low pressure hose to the evaporator service access gauge port valve. A special adapter, T77L–19703–A, is required to attach the manifold gauge set to the high pressure service access gauge port valve.
5. Place the end of the center hose away from you and the car.
6. Open the low pressure gauge valve slightly and allow the system pressure to bleed off.
7. When the system is just about empty, open the high pressure valve very slowly to avoid losing an excessive amount of refrigerant oil. Allow any remaining refrigerant to escape.

EVACUATING THE SYSTEM

➡ This procedure requires the use of a vacuum pump.

1. Connect the manifold gauge set.
2. Discharge the system.
3. Make sure that the low pressure gauge set hose is connected to the low pressure service gauge port on the top center of the accumulator/drier assembly and the high pressure hose connected to the high pressure service gauge port on the compressor discharge line.
4. Connect the center service hose to the inlet fitting of the vacuum pump.
5. Turn both gauge set valves to the wide open position.
6. Start the pump and note the low side gauge reading.
7. Operate the pump until the low pressure gauge reads 25–30 in.Hg. Continue running the vacuum pump for 10 minutes more. If you've replaced some component in the system, run the pump for an additional 20–30 minutes.
8. Leak test the system. Close both gauge set valves. Turn off the pump. The needle should remain stationary at the point at which the pump was turned off. If the needle drops to zero rapidly, there is a leak in the system which must be repaired.

LEAK TESTING

Some leak tests can be performed with a soapy water solution. There must be at least a 1/2 lb. charge in the system for a leak to be detected. The most extensive leak tests are performed with either a Halide flame type leak tester or the more preferable electronic leak tester.

In either case, the equipment is expensive, and, the use of a Halide detector can be **extremely** hazardous!

CHARGING THE SYSTEM

✲✲ CAUTION

NEVER OPEN THE HIGH PRESSURE SIDE WITH A CAN OF REFRIGERANT CONNECTED TO THE SYSTEM! OPENING THE HIGH PRESSURE SIDE WILL OVER PRESSURIZE THE CAN, CAUSING IT TO EXPLODE!

1. Connect the gauge set.
2. Close (clockwise) both gauge set valves.
3. Connect the center hose to the refrigerant can opener valve.
4. Make sure the can opener valve is closed, that is, the needle is raised, and connect the valve to the can. Open the valve, puncturing the can with the needle.
5. Loosen the center hose fitting at the pressure gauge, allowing refrigerant to purge the hose of air. When the air is bled, tighten the fitting.

✲✲ CAUTION

IF THE LOW PRESSURE GAUGE SET HOSE IS NOT CONNECTED TO THE ACCUMULATOR/DRIER, KEEP THE CAN IN AN UPRIGHT POSITION!

6. Disconnect the wire harness snap-lock connector from the clutch cycling pressure switch and install a jumper wire across the two terminals of the connector.
7. Open the low side gauge set valve and the can valve.
8. Allow refrigerant to be drawn into the system.
9. When no more refrigerant is drawn into the system, start the engine and run it at about 1,500 rpm. Turn on the system and operate it at the full high position. The compressor will operate and pull refrigerant gas into the system.

➡ To help speed the process, the can may be placed, upright, in a pan of warm water, not exceeding 125°F (52°C).

10. If more than one can of refrigerant is needed, close the can valve and gauge set low side valve when the can is empty and connect a

new can to the opener. Repeat the charging process until the sight glass indicates a full charge. The frost line on the outside of the can will indicate what portion of the can has been used.

✻✻ CAUTION

NEVER ALLOW THE HIGH PRESSURE SIDE READING TO EXCEED 240 psi.

11. When the charging process has been completed, close the gauge set valve and can valve. Remove the jumper wire and reconnect the cycling clutch wire. Run the system for at least five minutes to allow it to normalize. Low pressure side reading should be 4–25 psi; high pressure reading should be 120–210 psi at an ambient temperature of 70–90°F (21–32°C).
12. Loosen both service hoses at the gauges to allow any refrigerant to escape. Remove the gauge set and install the dust caps on the service valves.

➡ Multi-can dispensers are available which allow a simultaneous hook-up of up to four 1 lb. cans of R-12.

✻✻ CAUTION

Never exceed the recommended maximum charge for the system. The maximum charge for systems is 3 lb.

Windshield Wipers

➧ SEE FIG. 42

Intense heat from the sun, snow, and ice, road oils and the chemicals used in windshield washer solvent combine to deteriorate the rubber wiper refills. The refills should be replaced about twice a year or whenever the blades begin to streak or chatter.

WIPER REFILL REPLACEMENT

Normally, if the wipers are not cleaning the windshield properly, only the refill has to be replaced. The blade and arm usually require replacement only in the event of damage. It is not necessary (except on new Tridon® refills) to remove the arm or the blade to replace the refill (rubber part), though you may have to position the arm higher on the glass. You can do this turning the ignition switch on and operating the wipers. When they are positioned where they are accessible, turn the ignition switch off.

There are several types of refills and your vehicle could have any kind, since aftermarket blades and arms may not use exactly the same type refill as the original equipment.

Most Anco® styles use a release button that is pushed down to allow the refill to slide out of the yoke jaws. The new refill slides in and locks in place.

Some Trico® refills are removed by locating where the metal backing strip or the refill is

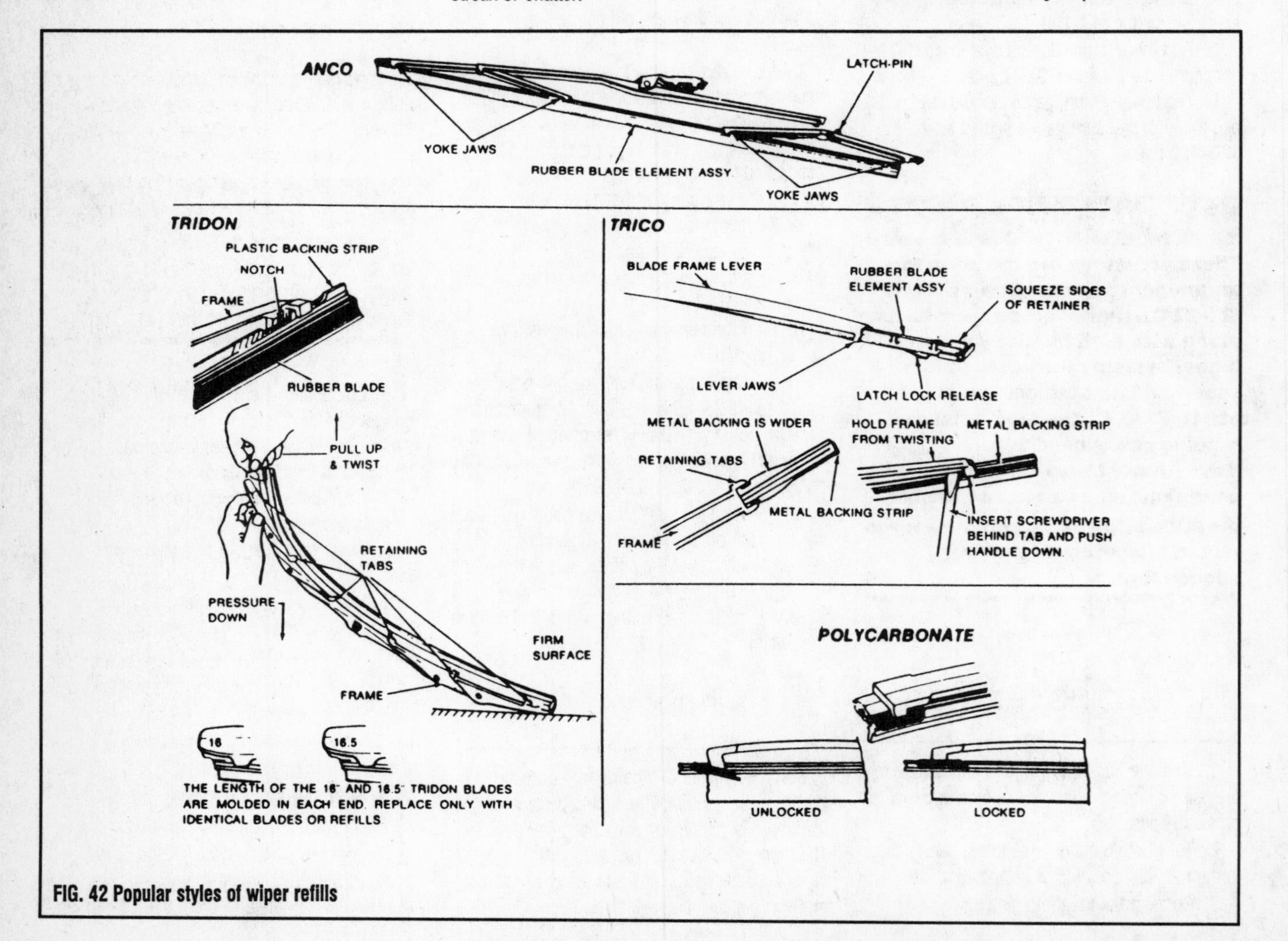

FIG. 42 Popular styles of wiper refills

wider. Insert a small screwdriver blade between the frame and metal backing strip. Press down to release the refill from the retaining tab.

Other Trico® blades are unlocked at one end by squeezing 2 metal tabs, and the refill is slid out of the frame jaws. When the new refill is installed, the tabs will click into place, locking the refill.

The polycarbonate type is held in place by a locking lever that is pushed downward out of the groove in the arm to free the refill. When the new refill is installed, it will lock in place automatically.

The Tridon® refill has a plastic backing strip with a notch about 1 in. (25mm) from the end. Hold the blade (frame) on a hard surface so that the frame is tightly bowed. Grip the tip of the backing strip and pull up while twisting counterclockwise. The backing strip will snap out of the retaining tab. Do this for the remaining tabs until the refill is free of the arm. The length of these refills is molded into the end and they should be replaced with identical types.

No matter which type of refill you use, be sure that all of the frame claws engage the refill. Before operating the wipers, be sure that no part of the metal frame is contacting the windshield.

Tires and Wheels

➧ SEE FIGS. 43-48

The tires should be rotated as specified in the Maintenance Intervals Chart. Refer to the accompanying illustrations for the recommended rotation patterns.

The tires on your car should have built-in tread wear indicators, which appear as 1/2 in. (13mm) bands when the tread depth gets as low as 1/16 in. (1.6mm). When the indicators appear in 2 or more adjacent grooves, it's time for new tires.

For optimum tire life, you should keep the tires properly inflated, rotate them often and have the wheel alignment checked periodically.

Some late models have the maximum load pressures listed in the V.I.N. plate on the left door frame. In general, pressure of 28–32 psi would be suitable for highway use with moderate loads and passenger car type tires (load range B, non-flotation) of original equipment size. Pressures should be checked before driving, since pressure can increase as much as 6 psi due to heat. It is a good idea to have an accurate gauge and to check pressures weekly. Not all gauges on service station air pumps are to be trusted. In general, car type tires require higher pressures and flotation type tires, lower pressures.

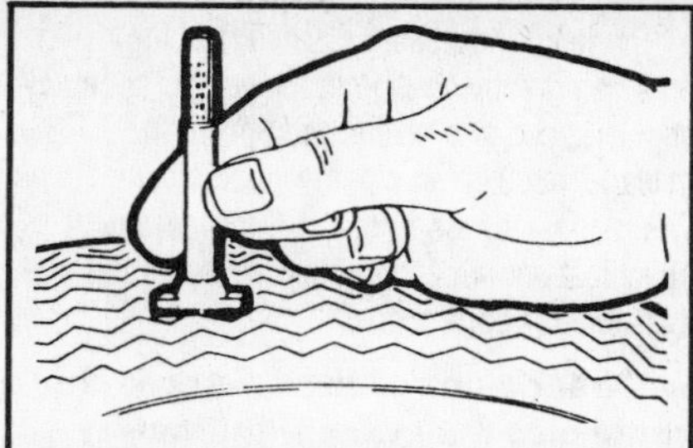

FIG. 43 Checking tread with an inexpensive depth gauge

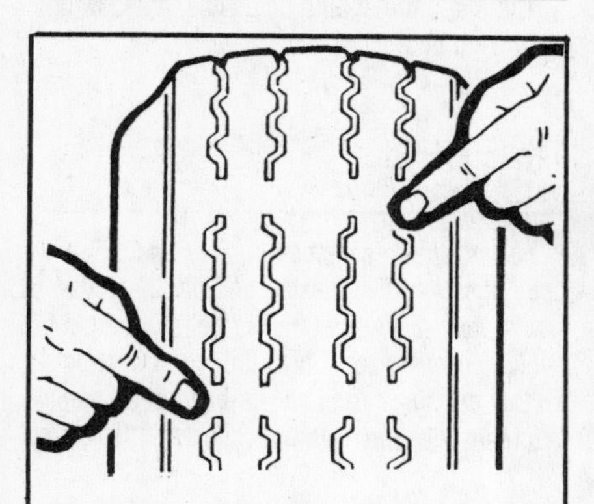

FIG. 44 Tread wear indicators are built into all new tires. When they appear, it's time to replace them

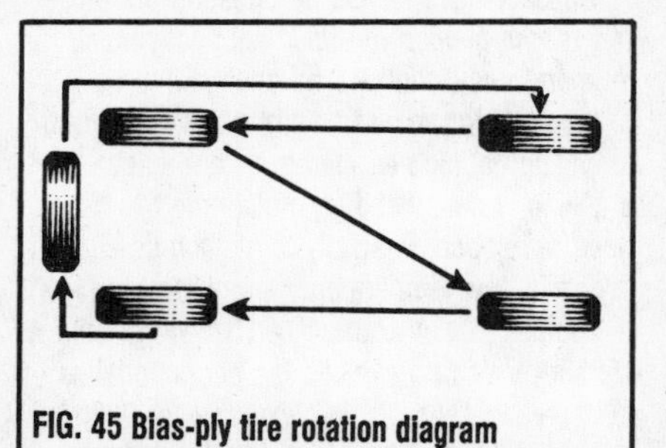

FIG. 45 Bias-ply tire rotation diagram

4 TIRE 5 TIRE

FIG. 46 Radial-ply & directional tire rotation diagram

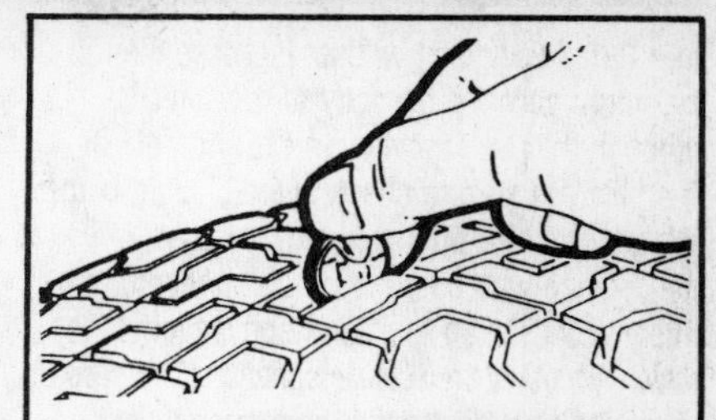

FIG. 47 Tread depth can be checked with a penny; when the top of Lincoln's head is visible, it's time for new tires

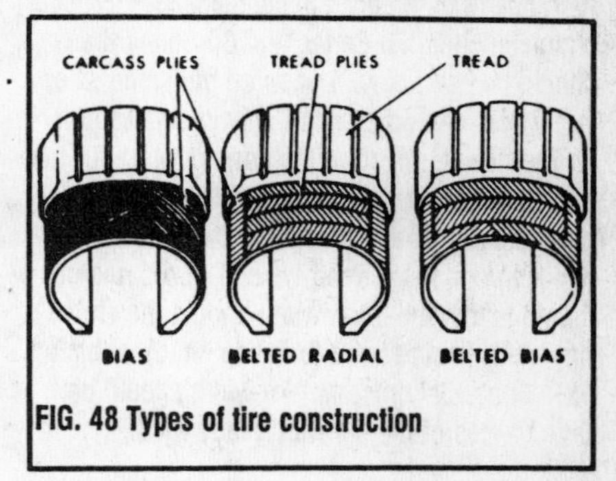

FIG. 48 Types of tire construction

TIRE ROTATION

It is recommended that you have the tires rotated every 6,000 miles. There is no way to give a tire rotation diagram for every combination of tires and vehicles, but the accompanying diagrams are a general rule to follow. Radial tires should not be cross-switched; they last longer if their direction of rotation is not changed. car tires sometimes have directional tread, indicated by arrows on the sidewalls; the arrow shows the direction of rotation. They will wear very rapidly if reversed. Studded snow tires will lose their studs if their direction of rotation is reversed.

➡ **Mark the wheel position or direction of rotation on radial tires or studded snow tires before removing them.**

If your car is equipped with tires having different load ratings on the front and the rear, the tires should not be rotated front to rear. Rotating these tires could affect tire life (the tires with the lower rating will wear faster, and could become overloaded), and upset the handling of the car.

TIRE USAGE

The tires on your car were selected to provide the best all around performance for normal operation when inflated as specified. Oversize tires (Load Range D) will not increase the maximum carrying capacity of the vehicle, although they will provide an extra margin of tread life. Be sure to check overall height before using larger size tires which may cause interference with suspension components or wheel wells. When replacing conventional tire sizes with other tire size designations, be sure to check the manufacturer's recommendations. Interchangeability is not always possible because of differences in load ratings, tire dimensions, wheel well clearances, and rim size. Also due to differences in handling characteristics, 70 Series and 60 Series tires should be used only in pairs on the same axle; radial tires should be used only in sets of four.

The wheels must be the correct width for the tire. Tire dealers have charts of tire and rim compatibility. A mismatch can cause sloppy handling and rapid tread wear. The old rule of thumb is that the tread width should match the rim width (inside bead to inside bead) within an inch. For radial tires, the rim width should be 80% or less of the tire (not tread) width.

The height (mounted diameter) of the new tires can greatly change speedometer accuracy, engine speed at a given road speed, fuel mileage, acceleration, and ground clearance. Tire manufacturers furnish full measurement specifications. Speedometer drive gears are available for correction.

➡ **Dimensions of tires marked the same size may vary significantly, even among tires from the same manufacturer.**

The spare tire should be usable, at least for low speed operation, with the new tires.

TIRE DESIGN

For maximum satisfaction, tires should be used in sets of five. Mixing or different types (radial, bias-belted, fiberglass belted) should be avoided. Conventional bias tires are constructed so that the cords run bead-to-bead at an angle. Alternate plies run at an opposite angle. This type of construction gives rigidity to both tread and sidewall. Bias-belted tires are similar in construction to conventional bias ply tires. Belts run at an angle and also at a 90° angle to the bead, as in the radial tire. Tread life is improved considerably over the conventional bias tire. The radial tire differs in construction, but instead of the carcass plies running at an angle of 90° to each other, they run at an angle of 90° to the bead. This gives the tread a great deal of rigidity and the sidewall a great deal of flexibility and accounts for the characteristic bulge associated with radial tires.

Radial tire are recommended for use on all Ford cars. If they are used, tire sizes and wheel diameters should be selected to maintain ground clearance and tire load capacity equivalent to the minimum specified tire. Radial tires should always be used in sets of five, but in an emergency radial tires can be used with caution on the rear axle only. If this is done, both tires on the rear should be of radial design.

➡ **Radial tires should never be used on only the front axle.**

FLUIDS AND LUBRICANTS

➧ SEE FIGS. 49-52

Oil and Fuel Recommendations

All 1989–92 Ford Mustangs must use lead-free gasoline.

The recommended oil viscosities for sustained temperatures ranging from below 0°F (–18°C) to above 32°F (0°C) are listed in this Section. They are broken down into multi-viscosities and single viscosities. Multi-viscosity oils are recommended because of their wider range of acceptable temperatures and driving conditions.

When adding oil to the crankcase or changing the oil or filter, it is important that oil of an equal quality to original equipment be used in your car. The use of inferior oils may void the warranty, damage your engine, or both.

The SAE (Society of Automotive Engineers) grade number of oil indicates the viscosity of the oil (its ability to lubricate at a given temperature). The lower the SAE number, the lighter the oil; the lower the viscosity, the easier it is to crank the engine in cold weather but the less the oil will lubricate and protect the engine in high temperatures. This number is marked on every oil container.

Oil viscosities should be chosen from those oils recommended for the lowest anticipated temperatures during the oil change interval. Due to the need for an oil that embodies both good lubrication at high temperatures and easy cranking in cold weather, multigrade oils have been developed. Basically, a multigrade oil is thinner at low temperatures and thicker at high temperatures. For example, a 10W–40 oil (the W stands for winter) exhibits the characteristics of a 10 weight (SAE 10) oil when the car is first started and the oil is cold. Its lighter weight allows it to travel to the lubricating surfaces quicker and offer less resistance to starter motor cranking than, say, a straight 30 weight (SAE 30) oil. But after the engine reaches operating temperature, the 10W–40 oil begins acting like straight 40 weight (SAE 40) oil, its heavier weight providing greater lubrication with less chance of foaming than a straight 30 weight oil.

The API (American Petroleum Institute) designations, also found on the oil container, indicates the classification of engine oil used under certain given operating conditions. Only oils designated for use Service SG heavy duty detergent should be used in your car. Oils of the SG type perform may functions inside the engine besides their basic lubrication. Through a balanced system of metallic detergents and polymeric dispersants, the oil prevents high and low temperature deposits and also keeps sludge and dirt particles in suspension. Acids, particularly sulphuric acid, as well as other by-products of engine combustion are neutralized by the oil. If these acids are allowed to concentrate, they can cause corrosion and rapid wear of the internal engine parts.

CAUTION

Non-detergent motor oils or straight mineral oils should not be used in your Ford gasoline engine.

Engine

OIL LEVEL CHECK

➧ SEE FIG. 53

Check the engine oil level every time you fill the gas tank. The oil level should be above the ADD mark and not above the FULL mark on the dipstick. Make sure that the dipstick is inserted into the crankcase as far as possible and that the vehicle is resting on level ground. Also, allow a

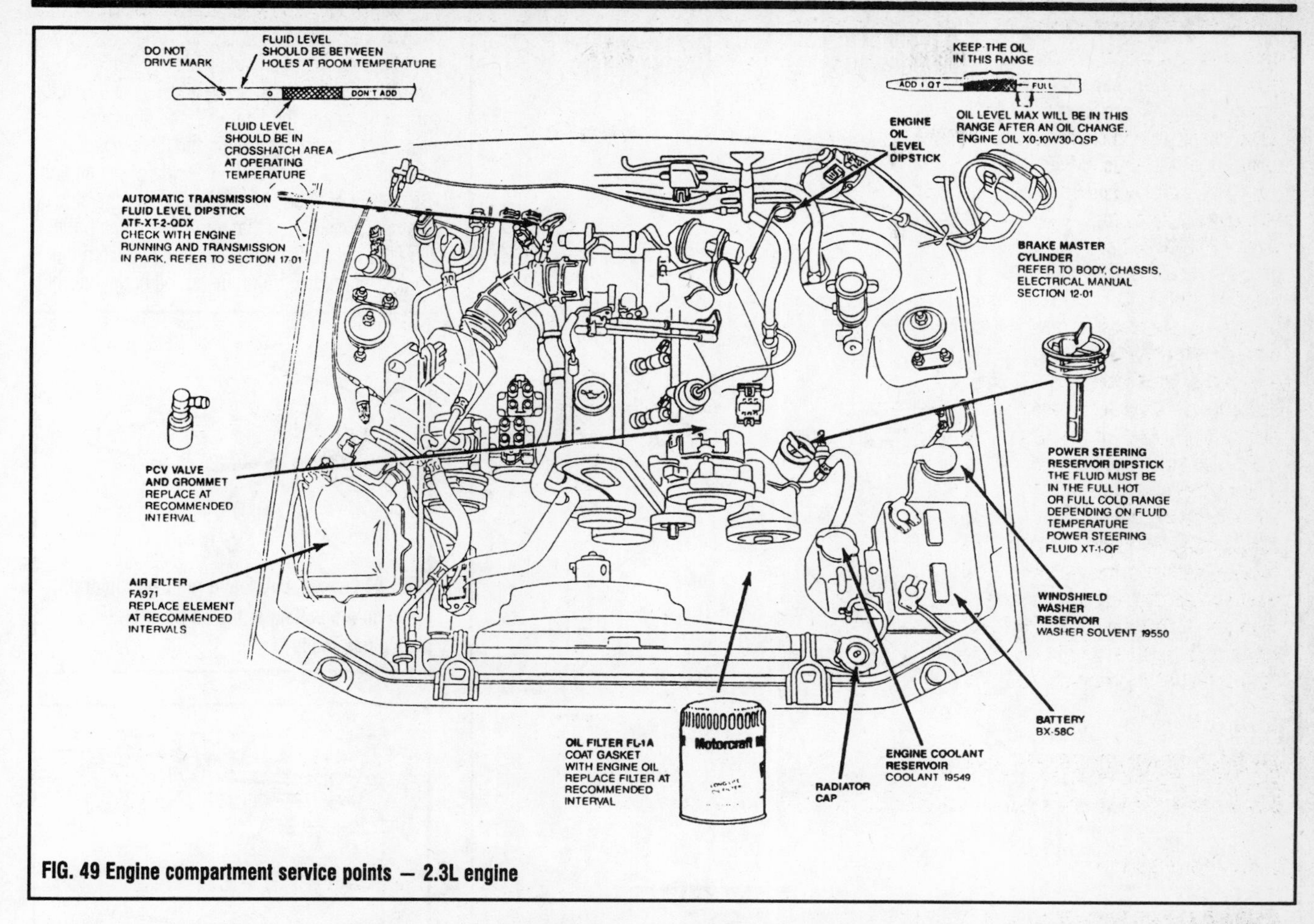

FIG. 49 Engine compartment service points — 2.3L engine

few minutes after turning off the engine for the oil to drain into the pan or an inaccurate reading will result.

1. Open the hood and remove the engine oil dipstick.
2. Wipe the dipstick with a clean, lint-free rag and reinsert it. Be sure to insert it all the way.
3. Pull out the dipstick and note the oil level. It should be between the **SAFE** (MAX) mark and the **ADD** (MIN) mark.
4. If the level is below the lower mark, replace the dipstick and add fresh oil to bring the level within the proper range. Do not overfill.
5. Recheck the oil level and close the hood.

➡ **Use a multi-grade oil with API classification SG.**

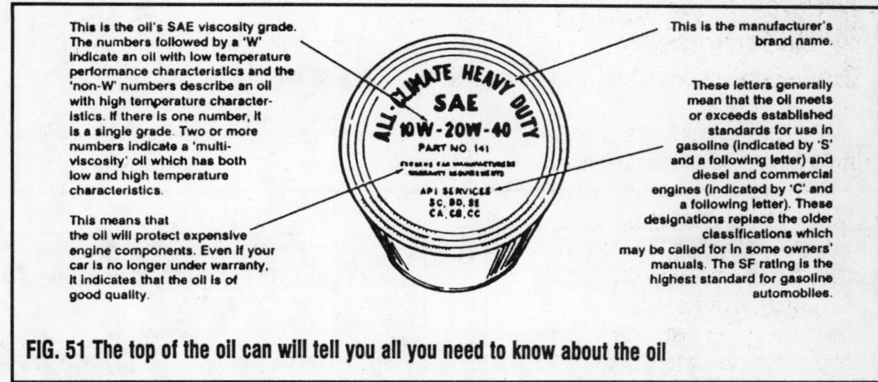

FIG. 51 The top of the oil can will tell you all you need to know about the oil

OIL AND FILTER CHANGE

▸ SEE FIGS. 54-59

➡ **The engine oil and oil filter should be changed at the same time, at the recommended intervals on the maintenance schedule chart.**

The oil should be changed more frequently if the vehicle is being operated in very dusty areas. Before draining the oil, make sure that the engine is at operating temperature. Hot oil will hold more impurities in suspension and will flow better, allowing the removal of more oil and dirt.

Loosen the drain plug with a wrench, then, unscrew the plug with your fingers, using a rag to shield your fingers from the heat. Push in on the plug as you unscrew it so you can feel when all of the screw threads are out of the hole. You can then remove the plug quickly with the minimum amount of oil running down your arm and you will also have the plug in your hand and not in the bottom of a pan of hot oil. Drain the oil into a suitable receptacle. Be careful of the oil. If it is at operating temperatures it is hot enough to burn you.

The oil filter is located on the left side of all the engines installed in Ford cars, for longest engine life, it should be changed every time the oil is changed. To remove the filter, you may need an oil filter wrench since the filter may have been fitted too tightly and the heat from the engine

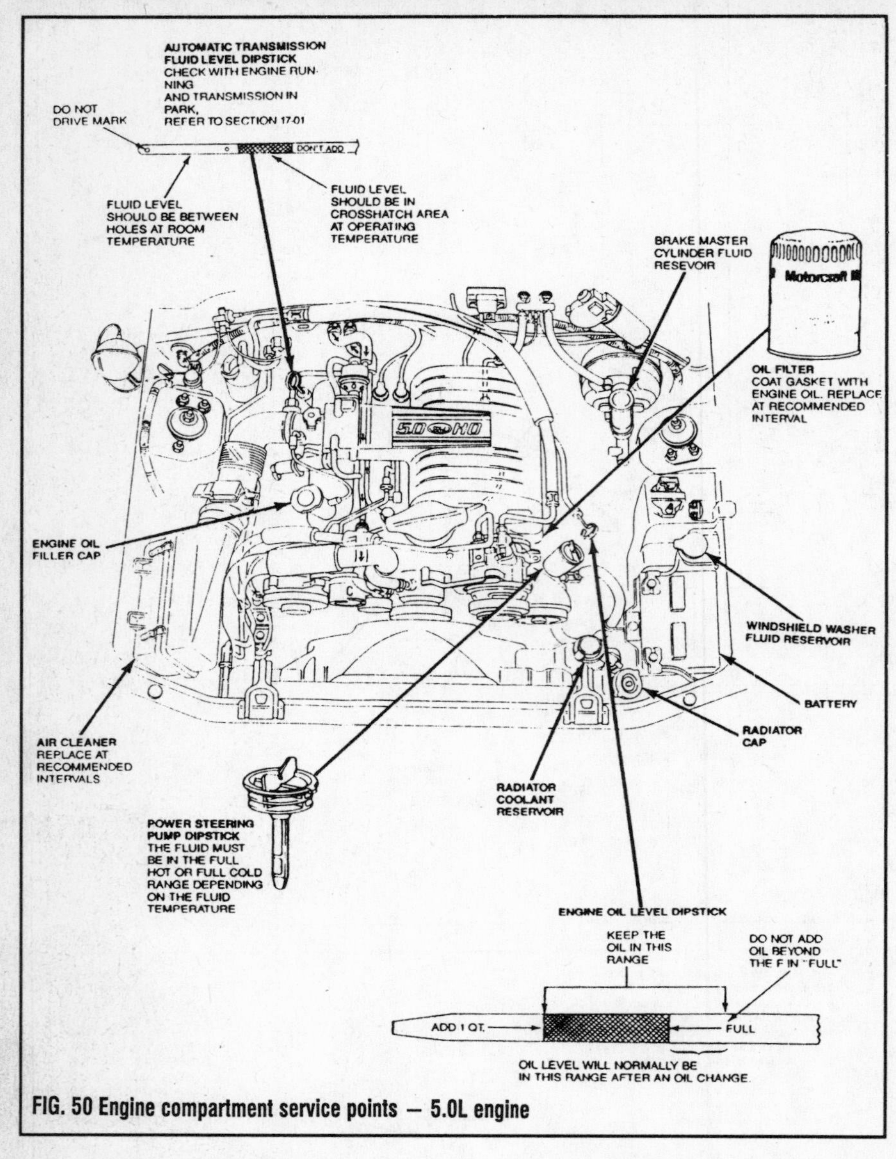

FIG. 50 Engine compartment service points — 5.0L engine

may have made it even tighter. A filter wrench can be obtained at an auto parts store and is well worth the investment, since it will save you a lot of grief. Loosen the filter with the filter wrench. With a rag wrapped around the filter, unscrew the filter from the boss on the side of the engine. Be careful of hot oil that will run down the side of the filter. Make sure that you have a pan under the filter before you start to remove it from the engine; should some of the hot oil happen to get

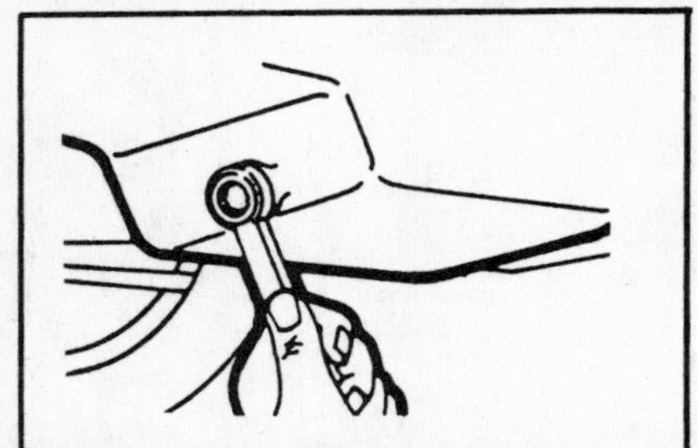
FIG. 54 Loosen, but do not remove, the drain plug on the bottom of the oil pan. Get your drain pan ready

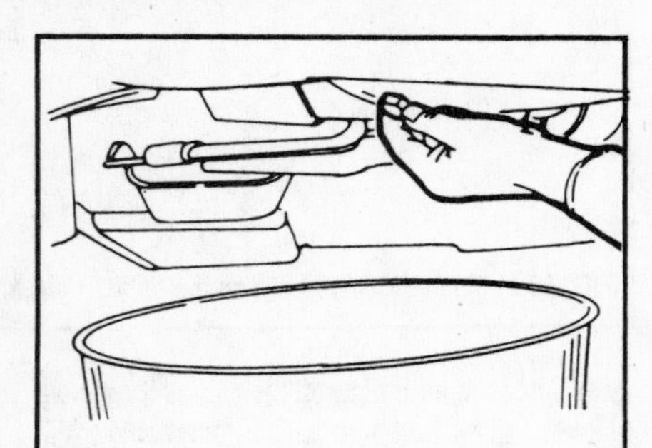
FIG. 55 Unscrew the plug by hand. Keep an inward pressure on the plug as you unscrew it, so the oil won't escape until you pull the plug away

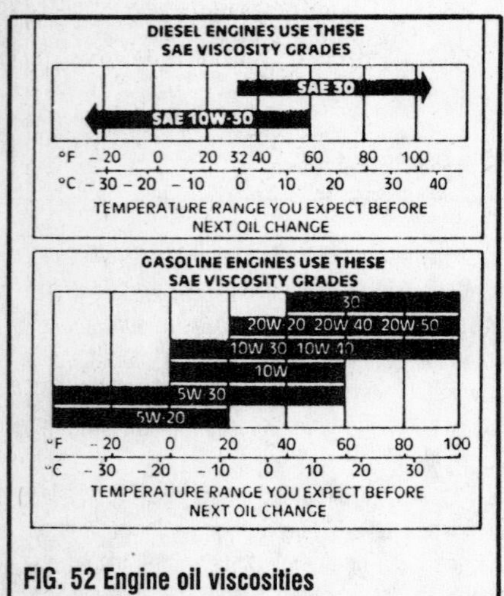

FIG. 52 Engine oil viscosities

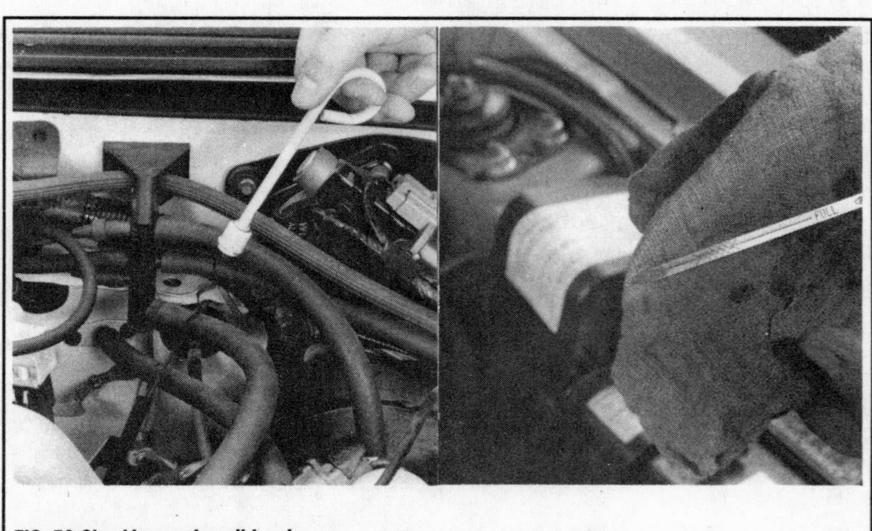

FIG. 53 Checking engine oil level

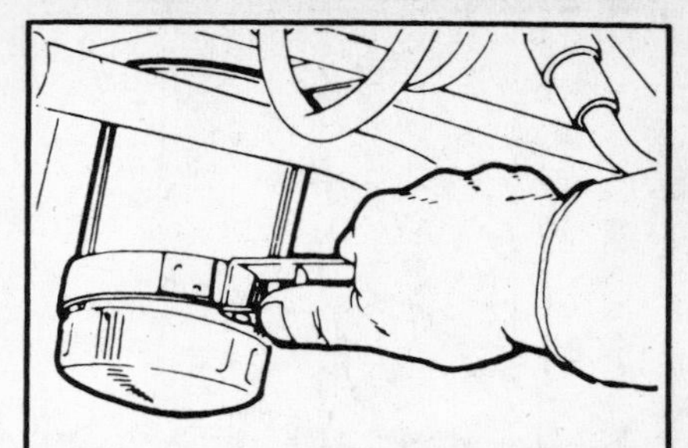
FIG. 56 Move the drain pan underneath the oil filter. Use a strap wrench to remove the filter — remember it is still filled with about a quart of hot, dirty oil

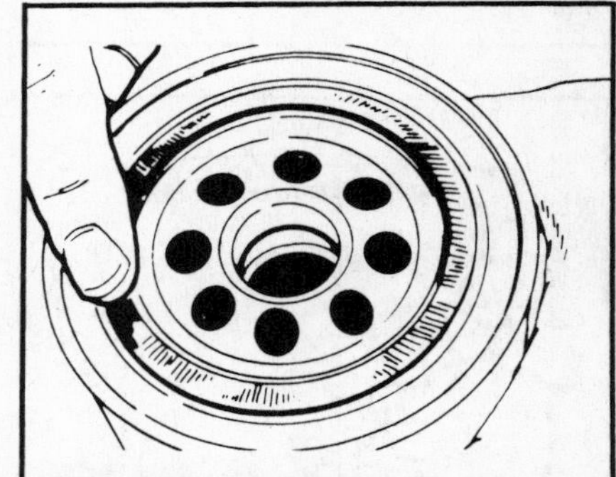
FIG. 57 Wipe clean engine oil around the rubber gasket on the new filter. This helps ensure good seal

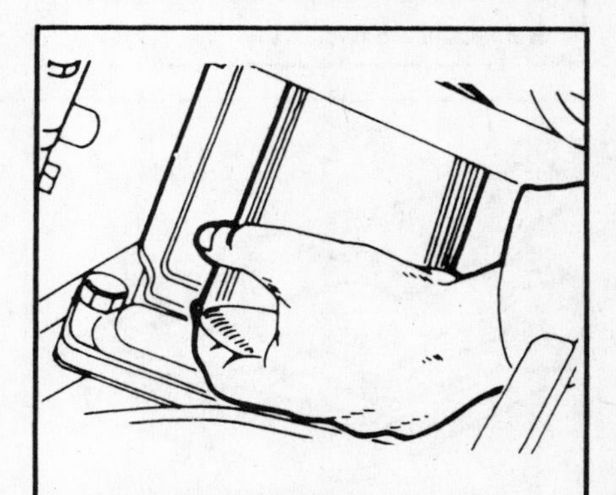
FIG. 58 Install the new filter by hand only; DO NOT use a strap wrench to install

on you, you will have a place to dump the filter in a hurry. Wipe the base of the mounting boss with a clean, dry cloth. When you install the new filter, smear a small amount of oil on the gasket with your finger, just enough to coat the entire surface, where it comes in contact with the mounting plate. When you tighten the filter, rotate if only a half turn after it comes in contact with the mounting boss.

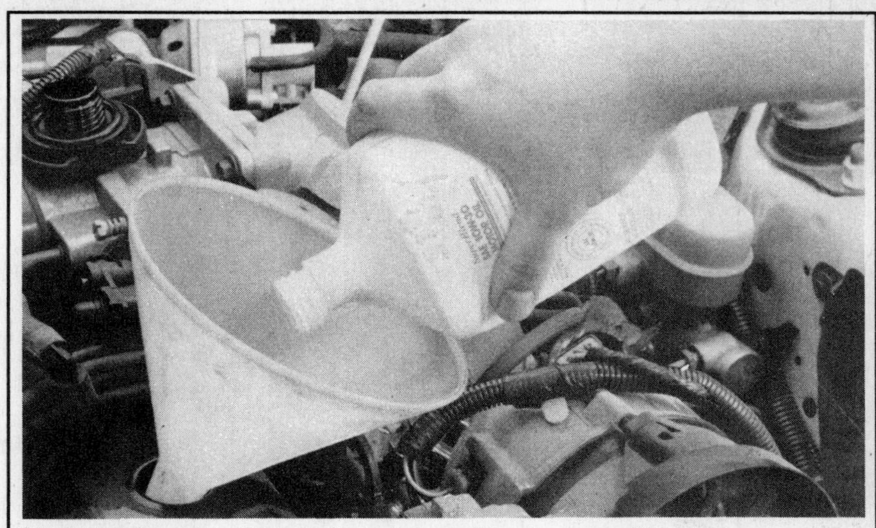
FIG. 59 Don't forget to install the drain plug before refilling the engine with fresh oil

Transmission

FLUID RECOMMENDATIONS

Manual Transmissions:

• 1989–92 5-speed models — Dexron®II ATF

Automatic Transmissions:

• All models — Dexron®II ATF

LEVEL CHECK

Automatic Transmissions

➧ SEE FIG. 60-62

It is very important to maintain the proper fluid level in an automatic transmission. If the level is either too high or too low, poor shifting operation and internal damage are likely to occur. For this reason a regular check of the fluid level is essential.

1. Drive the vehicle for 15–20 minutes to allow the transmission to reach operating temperature.
2. Park the car on a level surface, apply the parking brake and leave the engine idling. Shift the transmission and engage each gear, then place the gear selector in **P** (PARK).
3. Wipe away any dirt in the areas of the transmission dipstick to prevent it from falling into the filler tube. Withdraw the dipstick, wipe it with a clean, lint-free rag and reinsert it until it seats.
4. Withdraw the dipstick and note the fluid

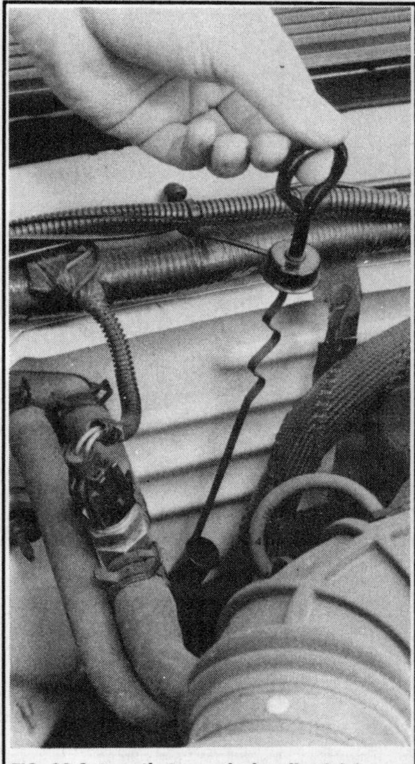
FIG. 60 Automatic transmission dipstick is found towards the rear of the engine

level. It should be between the upper (FULL) mark and the lower (ADD) mark.

5. If the level is below the lower mark, use a funnel and add fluid in small quantities through the dipstick filler neck. Keep the engine running while adding fluid and check the level after each small amount. Do not overfill.

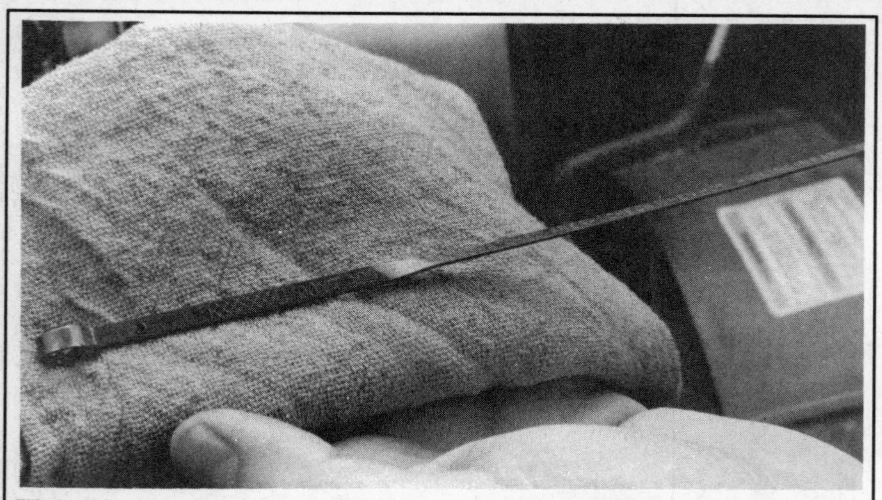

FIG. 61 Checking automatic transmission fluid level. Check transmission when it is warmed to operating temperature

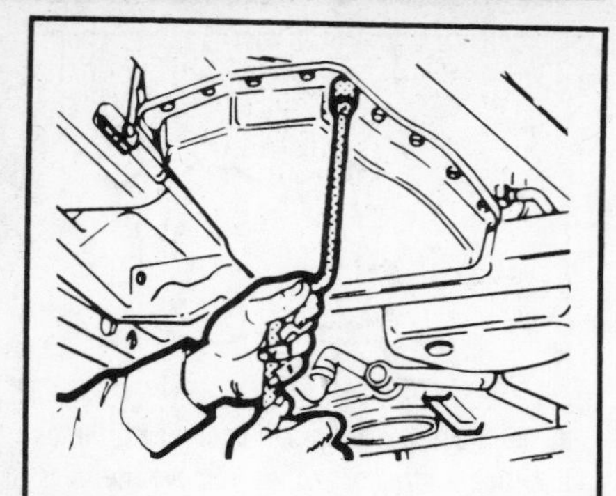

FIG. 66 Many late model vehicles have no drain plug. Loosen the pan bolts and allow one corner of the pan to hang, so that the fluid will drain out

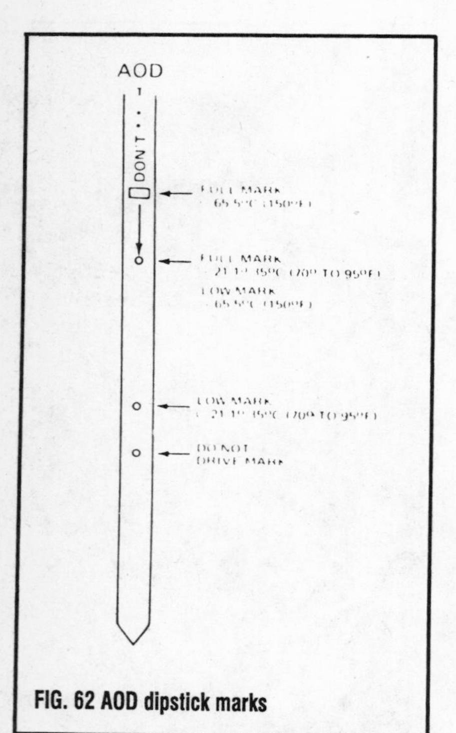

FIG. 62 AOD dipstick marks

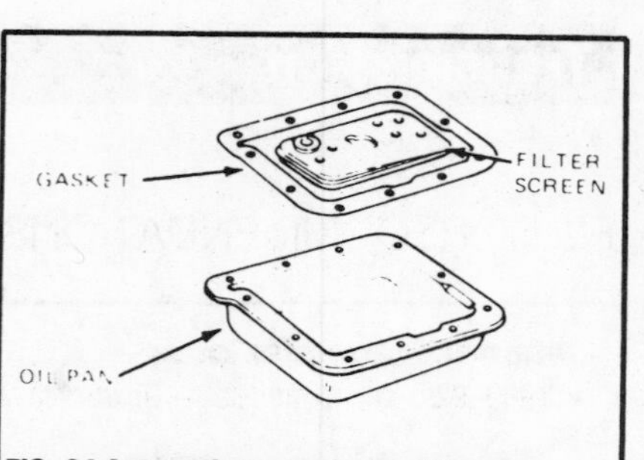

FIG. 64 Automatic transmission filters are found above the transmission oil pan

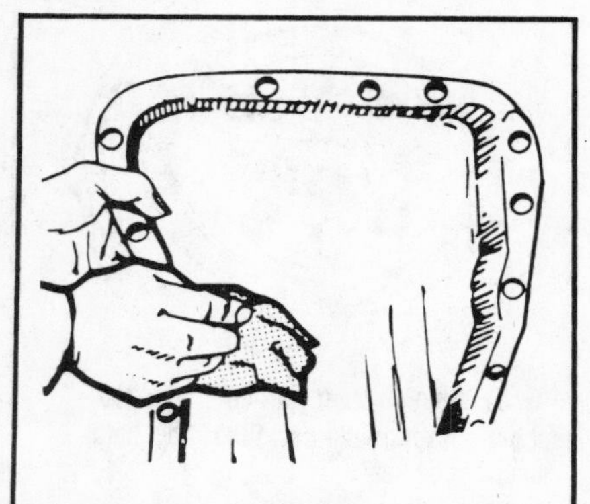

FIG. 67 Clean the pan thoroughly with a safe solvent and allow it to air dry

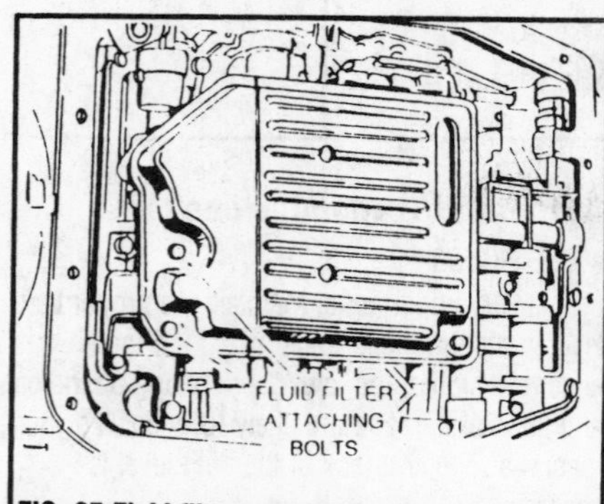

FIG. 65 Fluid filter, automatic overdrive (AOD)

FIG. 68 Install a new pan gasket

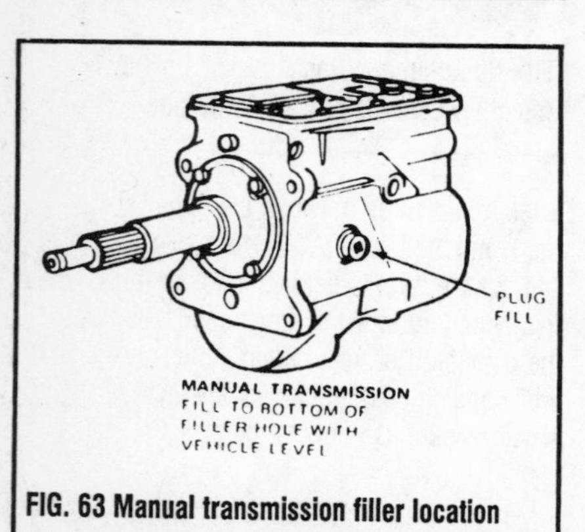

FIG. 63 Manual transmission filler location

Manual Transmission

➧ SEE FIG. 63

The fluid level should be checked every 6 months/6,000 miles, whichever comes first.

1. Park the car on a level surface, turn off the engine, apply the parking brake and block the wheels.
2. Remove the filler plug from the side of the transmission case with a proper size wrench. The fluid level should be even with the bottom of the filler hole.
3. If additional fluid is necessary, add it through the filler hole using a siphon pump or squeeze bottle.
4. Replace the filler plug; do not overtighten.

DRAIN AND REFILL

Automatic Transmission

➧ SEE FIGS. 64-69

1. Raise the car and support on jackstands.

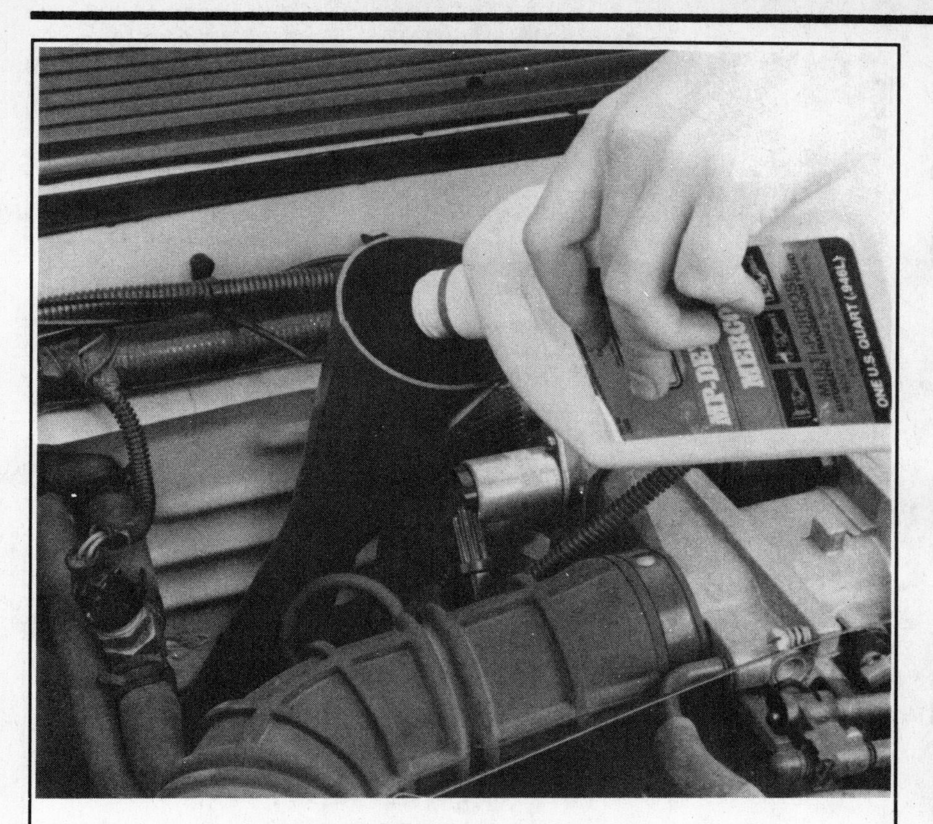
FIG. 69 Fill the transmission with required amount of fluid. Do not over fill!

2. Place a drain pan under the transmission.
3. Loosen the pan attaching bolts and drain the fluid from the transmission.
4. When the fluid has drained to the level of the pan flange, remove the remaining pan bolts working from the rear and both sides of the pan to allow it to drop and drain slowly.
5. When all of the fluid has drained, remove the pan and clean it thoroughly. Discard the pan gasket.
6. Place a new gasket on the pan, and install the pan on the transmission. Tighten the attaching bolts to 12–16 ft. lbs.
7. Add three 3 quarts of fluid to the transmission through the filler tube.
8. Lower the vehicle. Start the engine and move the gear selector through shift pattern. Allow the engine to reach normal operating temperature.
9. Check the transmission fluid. Add fluid, if necessary, to maintain correct level.

Manual Transmission

1. Place a suitable drain pan under the transmission.
2. Remove the drain plug and allow the gear lube to drain out.
3. Replace the drain plug, remove the filler plug and fill the transmission to the proper level with the required fluid.
4. Reinstall the filler plug.

Rear Axle

FLUID LEVEL CHECK

➧ SEE FIG. 70

Clean the area around the fill plug, which is located in the housing cover, before removing the plug. The lubricant level should be maintained to the bottom of the fill hole with the axle in its normal running position. If lubricant does not appear at the hole when the plug is removed, additional lubricant should be added. Use hypoid gear lubricant SAE 80 or 90.

➡ **If the differential is of the limited slip type, be sure and use special limited slip differential additive.**

DRAIN AND REFILL

Drain and refill the front and rear axle housing every 24,000 miles, or every day if the vehicle is operated in deep water. Remove the oil with a suction gun. Refill the axle housings with the proper oil. Be sure and clean the area around the drain plug before removing the plug. See the section on level checks.

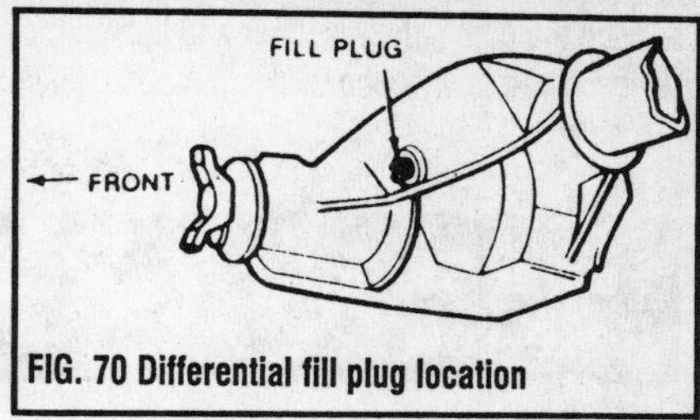

FIG. 70 Differential fill plug location

Brake Master Cylinder

➧ SEE FIG. 71

The master cylinder reservoir is located under the hood, on the left side firewall. Before removing the master cylinder reservoir cap, make sure the vehicle is resting on level ground and clean all dirt away from the top of the master cylinder. Pry off the retaining clip or unscrew the holddown bolt and remove the cap. The brake fluid level should be within $^1/_4$ in. (6mm) of the top of the reservoir.

If the level of the brake fluid is less than half the volume of the reservoir, it is advised that you check the brake system for leaks. Leaks in the hydraulic brake system most commonly occur at the wheel cylinder.

There is a rubber diaphragm in the top of the master cylinder cap. As the fluid level lowers in the reservoir due to normal brake shoe wear or leakage, the diaphragm takes up the space. This is to prevent the loss of brake fluid out the vented cap and contamination by dirt. After filling the master cylinder to the proper level with heavy duty brake fluid, but before replacing the cap, fold the rubber diaphragm up into the cap, then

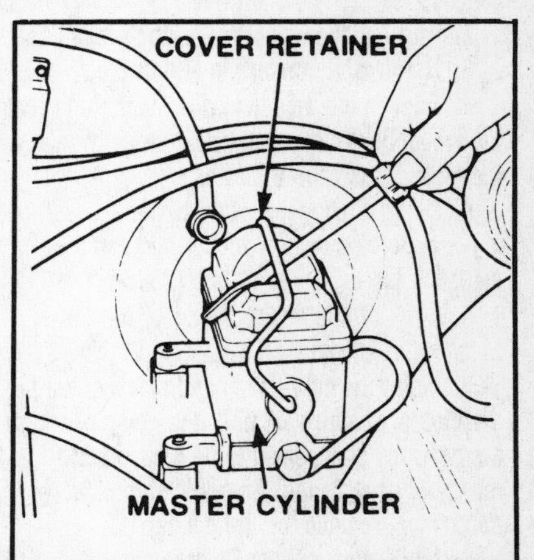

FIG. 71 Prying off the master cylinder retaining wire

replace the cap in the reservoir and tighten the retaining bolt or snap the retaining clip into place.

Power Steering Reservoir

➧ SEE FIG. 72

Position the vehicle on level ground. Run the engine until the fluid is at normal operating temperature. Turn the steering wheel all the way to the left and right several times. Position the wheels in the straight ahead position, then shut off the engine. Check the fluid level on the dipstick which is attached to the reservoir cap. The level should be between the ADD and FULL marks on the dipstick. Add fluid accordingly. Do not overfill. Use power steering fluid.

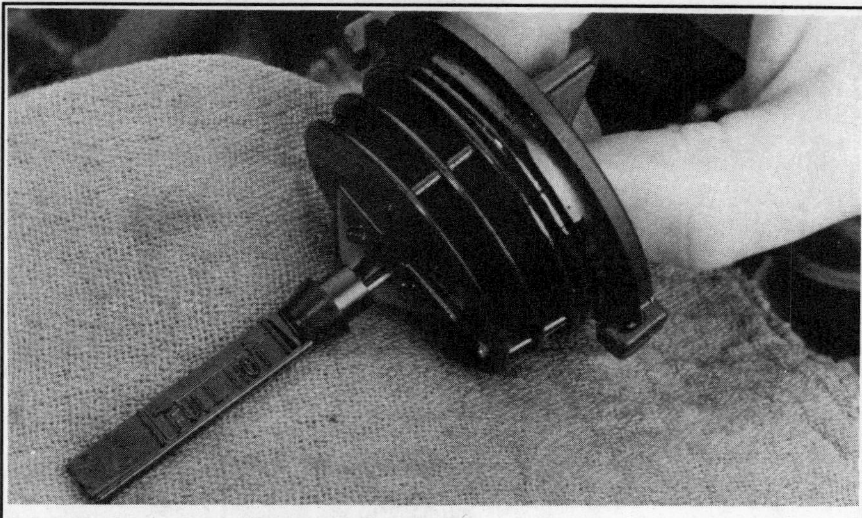

FIG. 72 Typical power steering pump reservoir dipstick

Chassis Greasing

➧ SEE FIG. 73

The lubrication chart indicates where the grease fittings are located. The vehicle should be greased according to the intervals in the Preventive Maintenance Schedule at the end of this Section.

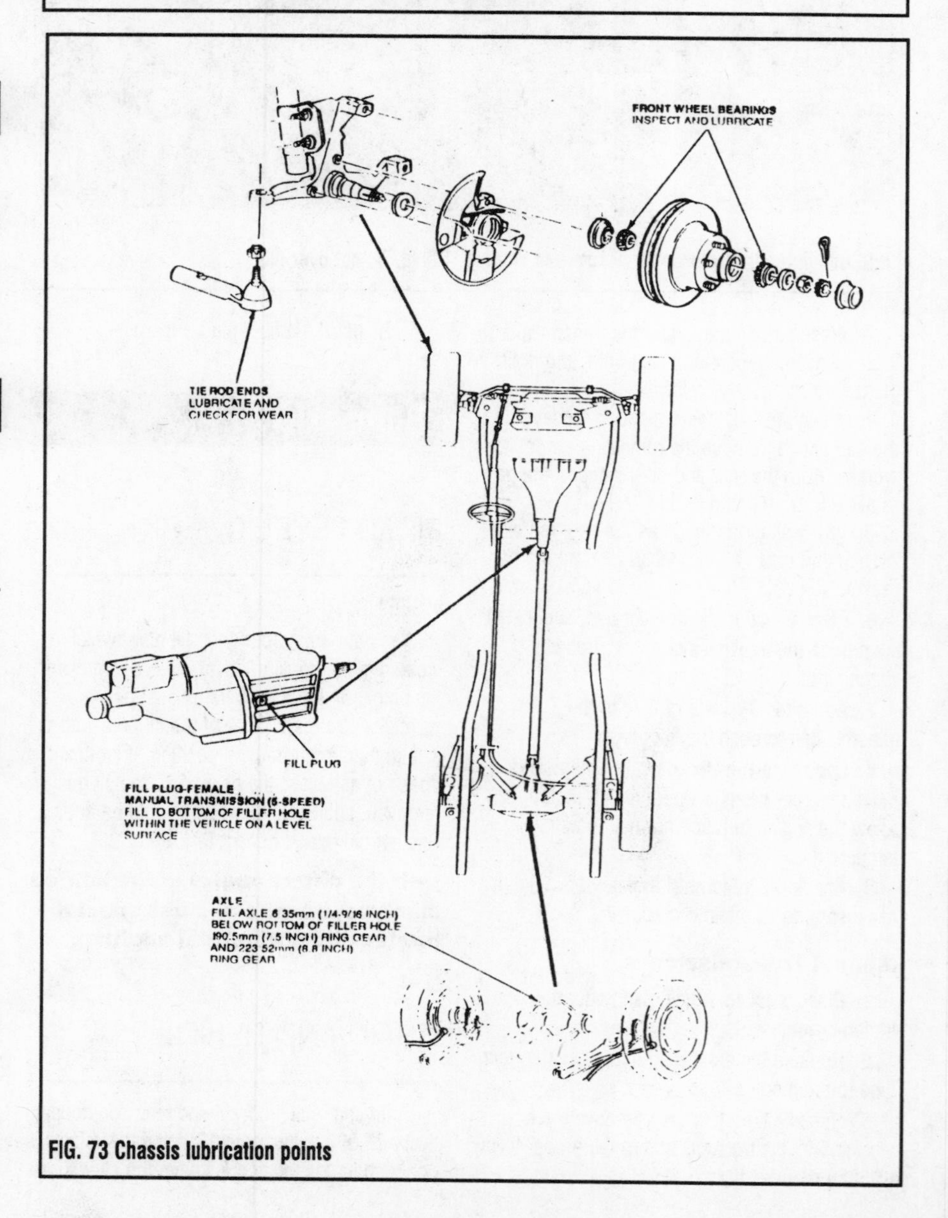

FIG. 73 Chassis lubrication points

Wheel Bearings

➧ SEE FIG. 74-75

ADJUSTMENT

1. Raise and safely support the front of the vehicle.
2. Remove the wheel cover and grease cap.
3. Remove the cotter pin and nut retainer.
4. Loosen the adjusting nut 3 turns and rock the wheel back and forth a few times to release the brake pads from the rotor.
5. While rotating the wheel and hub assembly in a counterclockwise direction, tighten the adjusting nut to 17–25 ft. lbs. (23–34 Nm).
6. Back off the adjusting nut $1/2$ turn, then retighten to 10–28 inch lbs. (1.1–3.2 Nm).
7. Install the nut retainer and a new cotter pin. Check the wheel rotation. If it is noisy or rough, the bearings either need to be cleaned and repacked or replaced. After adjustment is completed, replace the grease cap.
8. Lower the vehicle. Before driving the vehicle, pump the brake pedal several times to restore normal brake pedal travel.

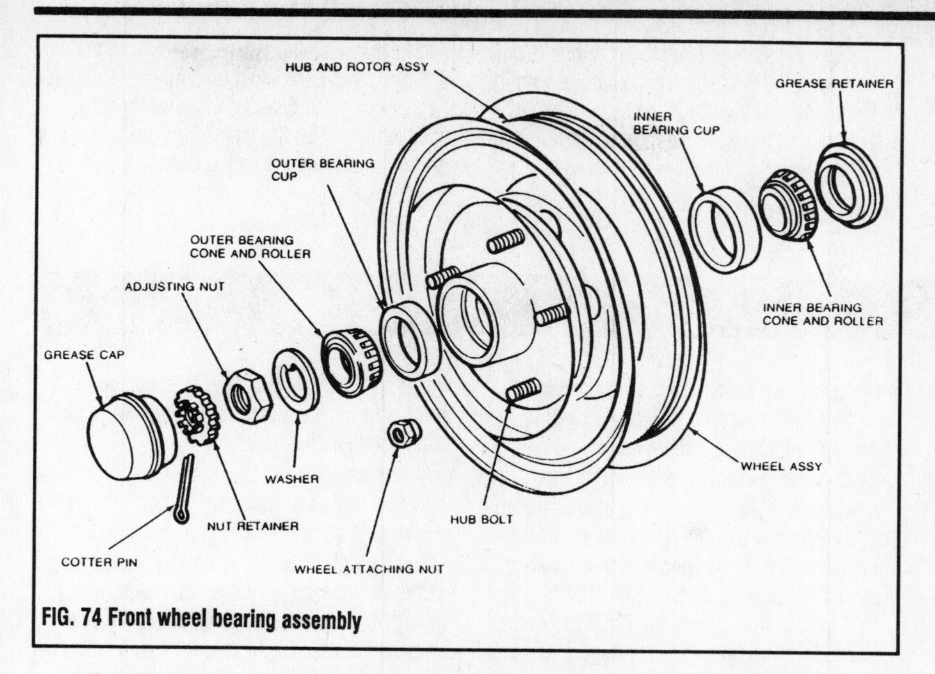

FIG. 74 Front wheel bearing assembly

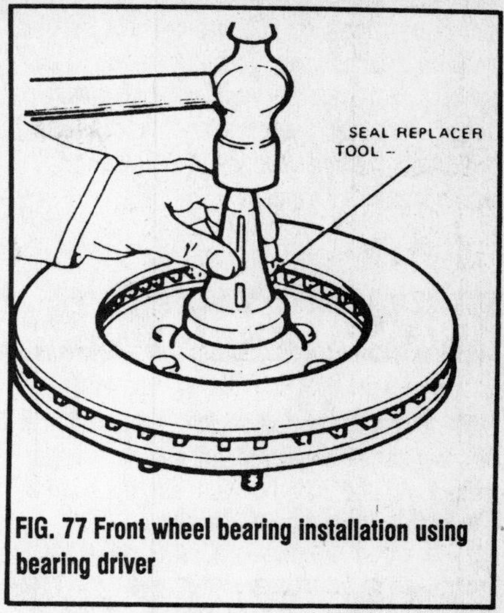

FIG. 77 Front wheel bearing installation using bearing driver

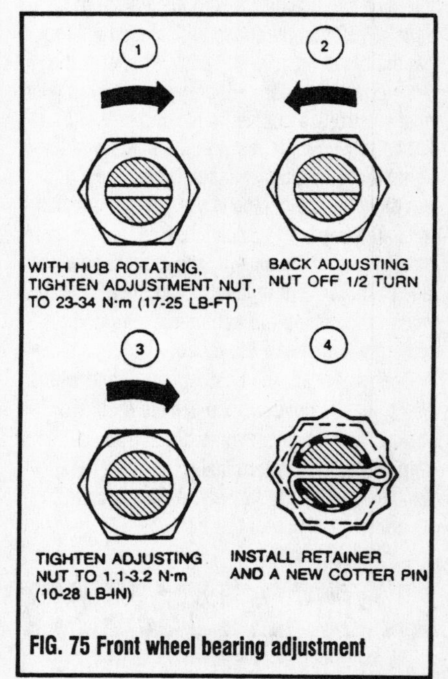

FIG. 75 Front wheel bearing adjustment

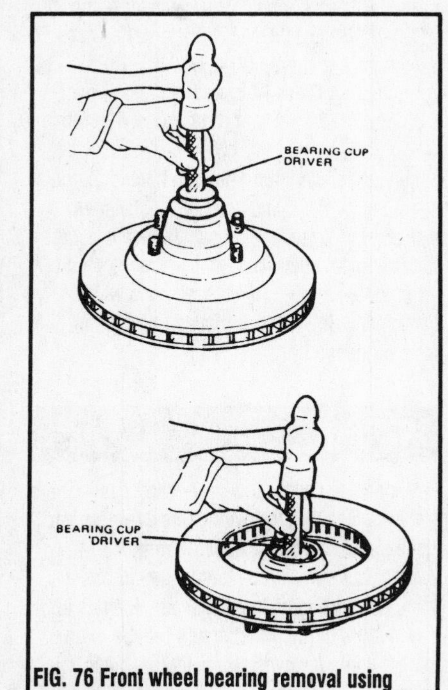

FIG. 76 Front wheel bearing removal using bearing driver

REMOVAL & INSTALLATION

➧ SEE FIG. 76-77

Before handling the bearings, there are a few things that you should remember to do and not to do.

Remember to DO the following:

• Remove all outside dirt from the housing before exposing the bearing.

• Treat a used bearing as gently as you would a new one.

• Work with clean tools in clean surroundings.

• Use clean, dry canvas gloves, or at least clean, dry hands.

• Clean solvents and flushing fluids are a must.

• Use clean paper when laying out the bearings to dry.

• Protect disassembled bearings from rust and dirt. Cover them up.

• Use clean rags to wipe bearings.

• Keep the bearings in oil-proof paper when they are to be stored or are not in use.

• Clean the inside of the housing before replacing the bearing.

Do NOT do the following:

• Don't work in dirty surroundings.

• Don't use dirty, chipped or damaged tools.

• Try not to work on wooden work benches or use wooden mallets.

• Don't handle bearings with dirty or moist hands.

• Do not use gasoline for cleaning; use a safe solvent.

• Do not spin-dry bearings with compressed air. They will be damaged.

• Do not spin dirty bearings.

• Avoid using cotton waste or dirty cloths to wipe bearings.

• Try not to scratch or nick bearing surfaces.

• Do not allow the bearing to come in contact with dirt or rust at any time.

1. Raise and support the vehicle safely. Remove the wheel and tire assembly and the caliper. Suspend the caliper with a length of wire; do not let it hang from the brake hose.
2. Pry off the dust cap. Tap out and discard the cotter pin. Remove the nut retainer.
3. Being careful not to drop the outer bearing, pull off the brake disc and wheel hub assembly.
4. Remove the inner grease seal using a prybar. Remove the inner wheel bearing.
5. Clean the wheel bearings with solvent and inspect them for pits, scratches and excessive wear. Wipe all the old grease from the hub and inspect the bearing races. If either bearings or races are damaged, the bearing races must be removed and the bearings and races replaced as an assembly.
6. If the bearings are to be replaced, drive out the races from the hub using a brass drift.
7. Make sure the spindle, hub and bearing assemblies are clean prior to installation.

To install:

8. If the bearing races were removed, install

new ones using a suitable bearing race installer. Pack the bearings with a bearing packer. If a packer is not available, work as much grease as possible between the rollers and cages.

9. Coat the inner surface of the hub and bearing races with grease.

10. Install the inner bearing in the hub. Being careful not to distort it, install the oil seal with its lip facing the bearing. Drive the seal in until its outer edge is even with the edge of the hub. Lubricate the lip of the seal with grease.

11. Install the hub/disc assembly on the spindle, being careful not to damage the oil seal.

12. Install the outer bearing, washer and spindle nut. Install the caliper and the wheel and tire assembly and adjust the bearings.

TRAILER TOWING

Factory trailer towing packages are available on most cars. However, if you are installing a trailer hitch and wiring on your car, there are a few thing that you ought to know.

Trailer Weight

Trailer weight is the first, and most important, factor in determining whether or not your vehicle is suitable for towing the trailer you have in mind. The horsepower-to-weight ratio should be calculated. The basic standard is a ratio of 35:1. That is, 35 pounds of GVW for every horsepower.

To calculate this ratio, multiply you engine's rated horsepower by 35, then subtract the weight of the vehicle, including passengers and luggage. The resulting figure is the ideal maximum trailer weight that you can tow. One point to consider: a numerically higher axle ratio can offset what appears to be a low trailer weight. If the weight of the trailer that you have in mind is somewhat higher than the weight you just calculated, you might consider changing your rear axle ratio to compensate.

Hitch Weight

There are three kinds of hitches: bumper mounted, frame mounted, and load equalizing.

Bumper mounted hitches are those which attach solely to the vehicle's bumper. Many states prohibit towing with this type of hitch, when it attaches to the vehicle's stock bumper, since it subjects the bumper to stresses for which it was not designed. Aftermarket rear step bumpers, designed for trailer towing, are acceptable for use with bumper mounted hitches.

Frame mounted hitches can be of the type which bolts to two or more points on the frame, plus the bumper, or just to several points on the frame. Frame mounted hitches can also be of the tongue type, for Class I towing, or, of the receiver type, for Classes II and III.

Load equalizing hitches are usually used for large trailers. Most equalizing hitches are welded in place and use equalizing bars and chains to level the vehicle after the trailer is hooked up.

The bolt-on hitches are the most common, since they are relatively easy to install.

Check the gross weight rating of your trailer. Tongue weight is usually figured as 10% of gross trailer weight. Therefore, a trailer with a maximum gross weight of 2,000 lb. (907 kg) will have a maximum tongue weight of 200 lb. (91 kg)Class I trailers fall into this category. Class II trailers are those with a gross weight rating of 2,000–3,500 lb., (907–1588 kg) while Class III trailers fall into the 3,500–6,000 lb. (1588–2722 kg) category. Class IV trailers are those over 6,000 lb. (2722 kg) and are for use with fifth wheel cars, only.

When you've determined the hitch that you'll need, follow the manufacturer's installation instructions, exactly, especially when it comes to fastener torques. The hitch will subjected to a lot of stress and good hitches come with hardened bolts. Never substitute an inferior bolt for a hardened bolt.

Wiring

Wiring the car for towing is fairly easy. There are a number of good wiring kits available and these should be used, rather than trying to design your own. All trailers will need brake lights and turn signals as well as tail lights and side marker lights. Most states require extra marker lights for overly wide trailers. Also, most states have recently required backup lights for trailers, and most trailer manufacturers have been building trailers with backup lights for several years.

Additionally, some Class I, most Class II and just about all Class III trailers will have electric brakes.

Add to this number an accessories wire, to operate trailer internal equipment or to charge the trailer's battery, and you can have as many as seven wires in the harness.

Determine the equipment on your trailer and buy the wiring kit necessary. The kit will contain all the wires needed, plus a plug adapter set which included the female plug, mounted on the bumper or hitch, and the male plug, wired into, or plugged into the trailer harness.

When installing the kit, follow the manufacturer's instructions. The color coding of the wires is standard throughout the industry.

One point to note, some domestic vehicles, and most imported vehicles, have separate turn signals. On most domestic vehicles, the brake lights and rear turn signals operate with the same bulb. For those vehicles with separate turn signals, you can purchase an isolation unit so that the brake lights won't blink whenever the turn signals are operated, or, you can go to your local electronics supply house and buy four diodes to wire in series with the brake and turn signal bulbs. Diodes will isolate the brake and turn signals. The choice is yours. The isolation units are simple and quick to install, but far more expensive than the diodes. The diodes, however, require more work to install properly, since they require the cutting of each bulb's wire and soldering in place of the diode.

One final point, the best kits are those with a spring loaded cover on the vehicle mounted socket. This cover prevents dirt and moisture from corroding the terminals. Never let the vehicle socket hang loosely. Always mount it securely to the bumper or hitch.

Cooling

ENGINE

One of the most common, if not THE most common, problem associated with trailer towing is engine overheating.

With factory installed trailer towing packages, a heavy duty cooling system is usually included. Heavy duty cooling systems are available as optional equipment on most cars, with or without a trailer package. If you have one of these extra-capacity systems, you shouldn't have any overheating problems.

If you have a standard cooling system,

without an expansion tank, you'll definitely need to get an aftermarket expansion tank kit, preferably one with at least a 2 quart capacity. These kits are easily installed on the radiator's overflow hose, and come with a pressure cap designed for expansion tanks.

Another helpful accessory is a Flex Fan. These fan are large diameter units are designed to provide more airflow at low speeds, with blades that have deeply cupped surfaces. The blades then flex, or flatten out, at high speed, when less cooling air is needed. These fans are far lighter in weight than stock fans, requiring less horsepower to drive them. Also, they are far quieter than stock fans.

If you do decide to replace your stock fan with a flex fan, note that if your car has a fan clutch, a spacer between the flex fan and water pump hub will be needed.

Aftermarket engine oil coolers are helpful for prolonging engine oil life and reducing overall engine temperatures. Both of these factors increase engine life.

While not absolutely necessary in towing Class I and some Class II trailers, they are recommended for heavier Class II and all Class III towing.

Engine oil cooler systems consist of an adapter, screwed on in place of the oil filter, a remote filter mounting and a multi-tube, finned heat exchanger, which is mounted in front of the radiator or air conditioning condenser.

TRANSMISSION

An automatic transmission is usually recommended for trailer towing. Modern automatics have proven reliable and, of course, easy to operate, in trailer towing.

The increased load of a trailer, however, causes an increase in the temperature of the automatic transmission fluid. Heat is the worst enemy of an automatic transmission. As the temperature of the fluid increases, the life of the fluid decreases.

It is essential, therefore, that you install an automatic transmission cooler.

The cooler, which consists of a multi-tube, finned heat exchanger, is usually installed in front of the radiator or air conditioning compressor, and hooked inline with the transmission cooler tank inlet line. Follow the cooler manufacturer's installation instructions.

Select a cooler of at least adequate capacity, based upon the combined gross weights of the car and trailer.

Cooler manufacturers recommend that you use an aftermarket cooler in addition to, and not instead of, the present cooling tank in your car's radiator. If you do want to use it in place of the radiator cooling tank, get a cooler at least two sizes larger than normally necessary.

➡ **A transmission cooler can, sometimes, cause slow or harsh shifting in the transmission during cold weather, until the fluid has a chance to come up to normal operating temperature. Some coolers can be purchased with or retrofitted with a temperature bypass valve which will allow fluid flow through the cooler only when the fluid has reached operating temperature, or above.**

PUSHING AND TOWING

To push-start your vehicle, (manual transmission only), check to make sure that bumpers of both vehicles are aligned so neither will be damaged. Be sure that all electrical system components are turned off (headlight, heater, blower, etc.). Turn on the ignition switch. Place the shift lever in Third or Fourth and push in the clutch pedal. At about 15 mph, signal the driver of the pushing vehicle to fall back, depress the accelerator pedal, and release the clutch pedal slowly. The engine should start.

When you are doing the pushing, make sure that the two bumpers match so you won't damage the vehicle you are to push. Another good idea is to put an old tire between the two vehicles. Try to keep your car right up against the other vehicle while you are pushing. If the two vehicles do separate, stop and start over again instead of trying to catch up and ramming the other vehicle. Also try, as much as possible, to avoid riding or slipping the clutch.

If your car has to be towed by a tow car, it can be towed forward for any distance with the driveshaft connected as long as it is done fairly slowly. Otherwise disconnect the driveshaft at the rear axle and tie it up.

JACKING AND HOISTING

➧ SEE FIG. 78

It is very important to be careful about running the engine, on vehicles equipped with limited slip differentials, while the vehicle is up on the jack. This is because when the drive train is engaged, power is transmitted to the wheel with the best traction and the vehicle will drive off the jack if one drive wheel is in contact with the floor, resulting in possible damage or injury.

Jack a Ford car from under the axles, radius arms, or spring hangers and the frame. Be sure and block the diagonally opposite wheel. Place jackstands under the vehicle at the points mentioned or directly under the frame when you are going to work under the vehicle.

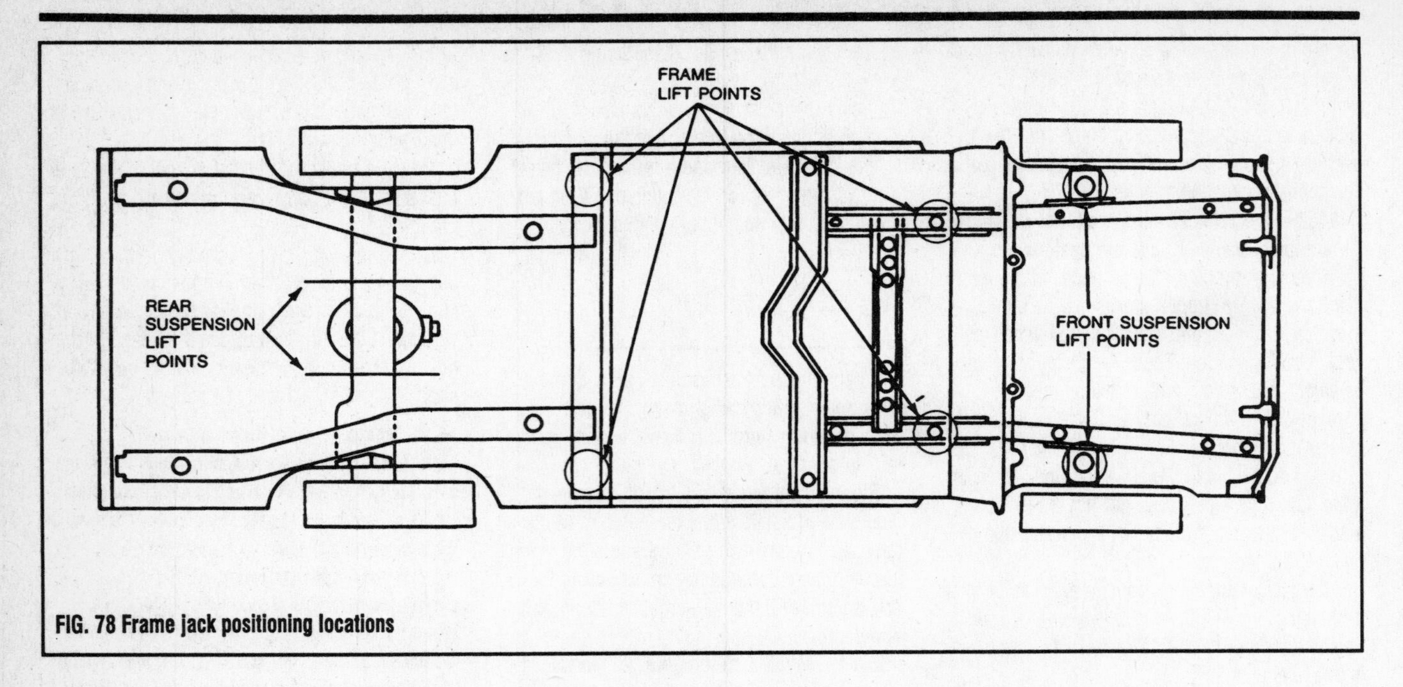

FIG. 78 Frame jack positioning locations

MAINTENANCE INTERVALS CHARTS

CUSTOMER MAINTENANCE SCHEDULE A

Follow this Schedule if your driving habits MAINLY include one or more of the following conditions:

- Short trips of less than 10 miles (16 km) when outside temperatures remain below freezing.
- Operating during HOT WEATHER
  — Driving in stop-and-go "rush hour" traffic.
- Towing a trailer or using a car-top carrier.
- Operating in severe dust conditions.
- Extensive idling, such as police, taxi or door-to-door delivery service.

SERVICE INTERVAL Perform at the months or distances shown, whichever comes first. Miles × 1000	3	6	9	12	15	18	21	24	27	30	33	36	39	42	45	48	51	54	57	60
Kilometers × 1000	4.8	9.6	14.4	19.2	24	28.8	33.6	38.4	43.2	48	52.8	57.6	62.4	67.2	72	76.8	81.6	86.4	91.2	96
EMISSION CONTROL SERVICE																				
Change Engine Oil and Oil Filter Every 3 Months OR	X	X	X	X	X	X	X	X	X	X	X	X	X	X	X	X	X	X	X	X
Replace Spark Plugs										X										X
Inspect Accessory Drive Belt(s)										X										X
Replace PCV Valve					(X)					(X)					(X)					(X)
Replace Air Cleaner Filter①										X										X
Replace Engine Coolant, EVERY 36 Months OR										X										X
Check Engine Coolant Protection, Hoses and Clamps	ANNUALLY																			
GENERAL MAINTENANCE																				
Inspect Exhaust Heat Shields										X										X
Change Automatic Transmission Fluid②										X										X
Lubricate Steering and/or Suspension Linkage										X										X
Inspect Brake Linings and Drums (Rear)③										X										X
Inspect and Repack Front Wheel Bearings										X										X

① If operating in severe dust, more frequent intervals may be required. Consult your dealer.
② Change automatic transmission fluid if your driving habits frequently include one or more of the following conditions:
- Operation during hot weather (above 32°C (90°F)) carrying heavy loads and in hilly terrain.
- Towing a trailer or using a car top carrrier.
- Police, taxi or door to door delivery service.

③ If your driving includes continuous stop-and-go driving or driving in mountainous areas, more frequent intervals may be required.
(X) This item not required to be performed, however, Ford recommends that you also perform maintenance on items designated by an (X) in order to achieve best vehicle operation. Failure to perform this recommended maintenance will not invalidate the vehicle emissions warranty or manufacturer recall liability.

CUSTOMER MAINTENANCE SCHEDULE B
Follow Maintenance Schedule B if, generally, you drive your vehicle on a daily basis for several miles and NONE OF THE UNIQUE DRIVING CONDITIONS SHOWN IN SCHEDULE A APPLY TO YOUR DRIVING HABITS.

SERVICE INTERVALS Perform at the months or distances shown, whichever comes first. Miles x 1000	7.5	15	22.5	30	37.5	45	52.5	60
Kilometers x 1000	12	24	36	48	60	72	84	96
EMISSIONS CONTROL SERVICE								
Change Engine Oil and Filter (every 6 months) or 7,500 Miles Whichever Occurs First	X	X	X	X	X	X	X	X
Replace Spark Plugs				X				X
Change Crankcase Emission Filter①				X①				X①
Inspect Accessory Drive Belt(s)				X				X
Replace Air Cleaner Filter①				X①				X①
Replace PCV Valve		(X)		(X)		(X)		(X)
Change Engine Coolant Every 36 Months or				X				X
Check Engine Coolant Protection, Hoses and Clamps	ANNUALLY							
GENERAL MAINTENANCE								
Check Exhaust Heat Shields				X				X
Lube Steering and/or Suspension		X③		X		X③		X
Inspect Brake Linings and Drums (Rear)②				X				X
Inspect and Repack Front Wheel Bearings				X				X

① If operating in severe dust, more frequent intervals may be required. Consult your dealer.
② If your driving includes continuous stop-and-go driving or driving in mountainous areas, more frequent intervals may be required.
③ All vehicles.

CAPACITIES

Year	Model	Engine ID/VIN	Engine Displacement liter (cc)	Engine Crankcase with Filter	Transmission (pts.) 4-Spd	5-Spd	Auto.	Drive Axle Rear (pts.)	Fuel Tank (gal.)	Cooling System (qts.)
1989	Mustang	A	2.3 (2300)	5	—	5.6	20	①	15.4	②
	Mustang	E	5.0 HO (5000)	5	—	5.6	24.6	①	15.4	14.1
1990	Mustang	A	2.3 (2300)	5	—	5.6	20	①	15.4	②
	Mustang	E	5.0 HO (5000)	5	—	5.6	24.6	①	15.4	14.1
1991	Mustang	M	2.3 (2300)	5	—	5.6	20	①	15.4	②
	Mustang	E	5.0 HO (5000)	5	—	5.6	24.6	①	15.4	14.1
1992	Mustang	M	2.3 (2300)	5	—	5.6	20	①	15.4	②
	Mustang	E	5.0 HO (5000)	5	—	5.6	24.6	①	15.4	14.1

HO—High Output
① With 7.5 in. axle—3.5 pts.
With 7.5 in. limited slip, 8.8 in. or 8.8 in. limited slip axles—3.75 pts.
② Equipped with manual transmission and air conditioning—9.7 qts.
Except equipped with manual transmission and air conditioning—10.0 qts.

Troubleshooting Basic Air Conditioning Problems

Problem	Cause	Solution
There's little or no air coming from the vents (and you're sure it's on)	• The A/C fuse is blown • Broken or loose wires or connections • The on/off switch is defective	• Check and/or replace fuse • Check and/or repair connections • Replace switch
The air coming from the vents is not cool enough	• Windows and air vent wings open • The compressor belt is slipping • Heater is on • Condenser is clogged with debris • Refrigerant has escaped through a leak in the system • Receiver/drier is plugged	• Close windows and vent wings • Tighten or replace compressor belt • Shut heater off • Clean the condenser • Check system • Service system
The air has an odor	• Vacuum system is disrupted • Odor producing substances on the evaporator case • Condensation has collected in the bottom of the evaporator housing	• Have the system checked/repaired • Clean the evaporator case • Clean the evaporator housing drains
System is noisy or vibrating	• Compressor belt or mountings loose • Air in the system	• Tighten or replace belt; tighten mounting bolts • Have the system serviced
Sight glass condition		
Constant bubbles, foam or oil streaks	• Undercharged system	• Charge the system
Clear sight glass, but no cold air	• No refrigerant at all	• Check and charge the system
Clear sight glass, but air is cold	• System is OK	
Clouded with milky fluid	• Receiver drier is leaking dessicant	• Have system checked
Large difference in temperature of lines	• System undercharged	• Charge and leak test the system
Compressor noise	• Broken valves • Overcharged • Incorrect oil level • Piston slap • Broken rings • Drive belt pulley bolts are loose	• Replace the valve plate • Discharge, evacuate and install the correct charge • Isolate the compressor and check the oil level. Correct as necessary. • Replace the compressor • Replace the compressor • Tighten with the correct torque specification
Excessive vibration	• Incorrect belt tension • Clutch loose • Overcharged • Pulley is misaligned	• Adjust the belt tension • Tighten the clutch • Discharge, evacuate and install the correct charge • Align the pulley
Condensation dripping in the passenger compartment	• Drain hose plugged or improperly positioned • Insulation removed or improperly installed	• Clean the drain hose and check for proper installation • Replace the insulation on the expansion valve and hoses

Troubleshooting Basic Air Conditioning Problems (cont.)

Problem	Cause	Solution
Frozen evaporator coil	• Faulty thermostat • Thermostat capillary tube improperly installed • Thermostat not adjusted properly	• Replace the thermostat • Install the capillary tube correctly • Adjust the thermostat
Low side low—high side low	• System refrigerant is low • Expansion valve is restricted	• Evacuate, leak test and charge the system • Replace the expansion valve
Low side high—high side low	• Internal leak in the compressor—worn	• Remove the compressor cylinder head and inspect the compressor. Replace the valve plate assembly if necessary. If the compressor pistons, rings or
Low side high—high side low (cont.)	 • Cylinder head gasket is leaking • Expansion valve is defective • Drive belt slipping	cylinders are excessively worn or scored replace the compressor • Install a replacement cylinder head gasket • Replace the expansion valve • Adjust the belt tension
Low side high—high side high	• Condenser fins obstructed • Air in the system • Expansion valve is defective • Loose or worn fan belts	• Clean the condenser fins • Evacuate, leak test and charge the system • Replace the expansion valve • Adjust or replace the belts as necessary
Low side low—high side high	• Expansion valve is defective • Restriction in the refrigerant hose	• Replace the expansion valve • Check the hose for kinks—replace if necessary
Low side low—high side high	• Restriction in the receiver/drier • Restriction in the condenser	• Replace the receiver/drier • Replace the condenser
Low side and high normal (inadequate cooling)	**• Air in the system** **• Moisture in the system**	**• Evacuate, leak test and charge the system** **• Evacuate, leak test and charge the system**

Troubleshooting Basic Wheel Problems

Problem	Cause	Solution
The car's front end vibrates at high speed	• The wheels are out of balance • Wheels are out of alignment	• Have wheels balanced • Have wheel alignment checked/adjusted
Car pulls to either side	• Wheels are out of alignment • Unequal tire pressure • Different size tires or wheels	• Have wheel alignment checked/adjusted • Check/adjust tire pressure • Change tires or wheels to same size

Troubleshooting Basic Wheel Problems

Problem	Cause	Solution
The car's wheel(s) wobbles	• Loose wheel lug nuts • Wheels out of balance • Damaged wheel • Wheels are out of alignment • Worn or damaged ball joint • Excessive play in the steering linkage (usually due to worn parts) • Defective shock absorber	• Tighten wheel lug nuts • Have tires balanced • Raise car and spin the wheel. If the wheel is bent, it should be replaced • Have wheel alignment checked/adjusted • Check ball joints • Check steering linkage • Check shock absorbers
Tires wear unevenly or prematurely	• Incorrect wheel size • Wheels are out of balance • Wheels are out of alignment	• Check if wheel and tire size are compatible • Have wheels balanced • Have wheel alignment checked/adjusted

Troubleshooting Basic Tire Problems

Problem	Cause	Solution
The car's front end vibrates at high speeds and the steering wheel shakes	• Wheels out of balance • Front end needs aligning	• Have wheels balanced • Have front end alignment checked
The car pulls to one side while cruising	• Unequal tire pressure (car will usually pull to the low side) • Mismatched tires • Front end needs aligning	• Check/adjust tire pressure • Be sure tires are of the same type and size • Have front end alignment checked
Abnormal, excessive or uneven tire wear See "How to Read Tire Wear"	• Infrequent tire rotation • Improper tire pressure • Sudden stops/starts or high speed on curves	• Rotate tires more frequently to equalize wear • Check/adjust pressure • Correct driving habits
Tire squeals	• Improper tire pressure • Front end needs aligning	• Check/adjust tire pressure • Have front end alignment checked

2

ENGINE PERFORMANCE AND TUNE-UP

TUNE-UP PROCEDURES

In order to extract the full measure of performance and economy from your engine it is essential that it be properly tuned at regular intervals. A regular tune-up will keep your vehicle's engine running smoothly and will prevent the annoying minor breakdowns and poor performance associated with an untuned engine.

A complete tune-up should be performed every 12,000 miles (19,300 km) or twelve months, whichever comes first. This interval should be halved if the vehicle is operated under severe conditions, such as trailer towing, prolonged idling, continual stop and start driving, or if starting or running problems are noticed. It is assumed that the routine maintenance described in Section 1 has been kept up, as this will have a decided effect on the results of a tune-up. All of the applicable steps of a tune-up should be followed in order, as the result is a cumulative one.

If the specifications on the tune-up sticker in the engine compartment disagree with the Tune-Up Specifications chart in this Section, the figures on the sticker must be used. The sticker often reflects changes made during the production run.

Spark Plugs

➧ SEE FIGS. 1-6

A typical spark plug consists of a metal shell surrounding a ceramic insulator. A metal electrode extends downward through the center of the insulator and protrudes a small distance. Located at the end of the plug and attached to the side of the outer metal shell is the side electrode. The side electrode bends in at a 90° angle so that its tip is even with, and parallel to, the tip of the center electrode. The distance between these two electrodes (measured in thousandths of an inch or hundreths of a millimeter) is called the spark plug gap. The spark plug in no way produces a spark but merely provides a gap across which the current can arc. The coil produces anywhere from 20,000 to 40,000 volts which travels to the distributor where it is distributed through the spark plug wires to the spark plugs. The current passes along the center electrode and jumps the gap to the side electrode, and, in do doing, ignites the air/fuel mixture in the combustion chamber.

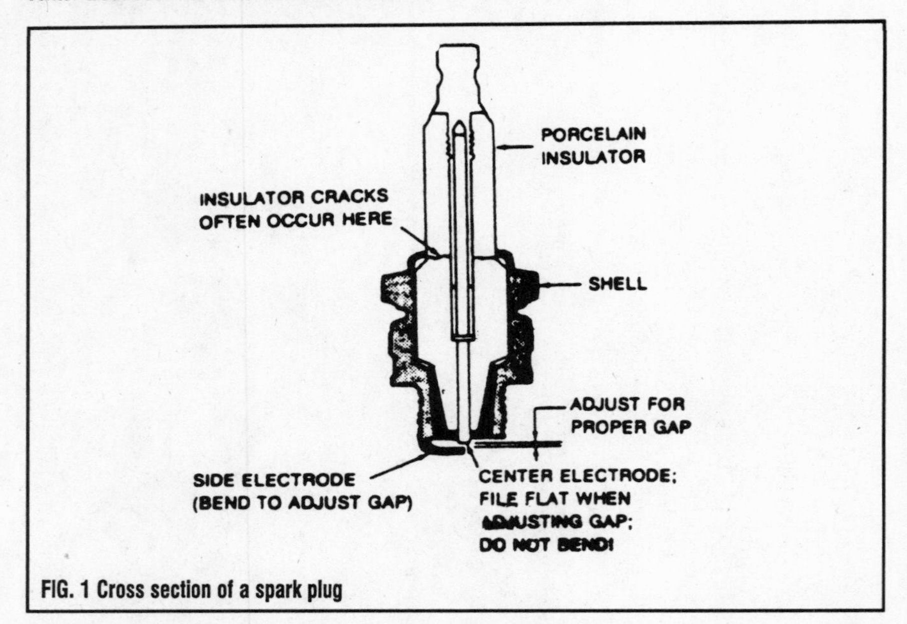

FIG. 1 Cross section of a spark plug

GASOLINE ENGINE TUNE-UP SPECIFICATIONS

Year	Engine ID/VIN	Engine Displacement liter (cc)	Spark Plugs Gap (in.)	Ignition Timing (deg.) MT	Ignition Timing (deg.) AT	Fuel Pump (psi)	Idle Speed (rpm) MT	Idle Speed (rpm) AT	Valve Clearance In.	Valve Clearance Ex.
1989	A	2.3 (2300)	0.044	①	①	35–45	①	①	Hyd.	Hyd.
	E	5.0 HO (5000)	0.054	10B	10B	35–45	①	①	Hyd.	Hyd.
1990	A	2.3 (2300)	0.044	10B	10B	35–45	①	①	Hyd.	Hyd.
	E	5.0 HO (5000)	0.054	10B	10B	35–45	①	①	Hyd.	Hyd.
1991	M	2.3 (2300)	0.032	①	①	35–45	①	①	Hyd.	Hyd.
	E	5.0 HO (5000)	0.054	10B	10B	35–45	①	①	Hyd.	Hyd.
1992	M	2.3 (2300)	0.032	①	①	35–45	①	①	Hyd.	Hyd.
	E	5.0 HO (5000)	0.054	10B	10B	35–45	①	①	Hyd.	Hyd.

NOTE: The lowest cylinder pressure should be within 75% of the highest cylinder pressure reading. For example, if the highest cylinder is 134 psi, the lowest should be 101. Engine should be at normal operating temperature with throttle valve in the wide open position.
The underhood specifications sticker often reflects tune-up specification changes in production. Sticker figures must be used if they disagree with those in this chart.
B—Before top dead center
HO—High Output
NA—Not applicable
① Calibrations vary depending upon the model; refer to the underhood sticker.

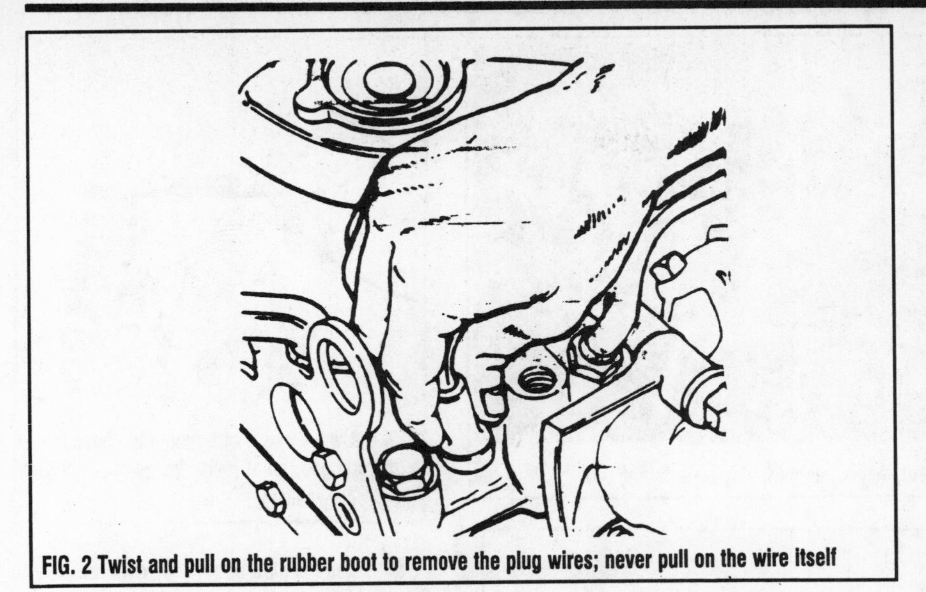

FIG. 2 Twist and pull on the rubber boot to remove the plug wires; never pull on the wire itself

FIG. 3 Removing spark plug with special plug socket and extension

SPARK PLUG HEAT RANGE

Spark plug heat range is the ability of the plug to dissipate heat. The longer the insulator (or the farther it extends into the engine), the hotter the plug will operate; the shorter the insulator the cooler it will operate. A plug that absorbs little heat and remains too cool will quickly accumulate deposits of oil and carbon since it is not hot enough to burn them off. This leads to plug fouling and consequently to misfiring. A plug that absorbs too much heat will have no deposits, but, due to the excessive heat, the electrodes will burn away quickly and in some instances, preignition may result. Preignition takes place when plug tips get so hot that they glow sufficiently to ignite the fuel/air mixture before the actual spark occurs. This early ignition will usually cause a pinging during low speeds and heavy loads.

The general rule of thumb for choosing the correct heat range when picking a spark plug is: if most of your driving is long distance, high speed travel, use a colder plug; if most of your driving is stop and go, use a hotter plug. Original equipment plugs are compromise plugs, but most people never have occasion to change their plugs from the factory-recommended heat range.

REPLACING SPARK PLUGS

A set of spark plugs usually requires replacement after about 20,000–30,000 miles (32,100–48,300 km), depending on your style of driving. In normal operation, plug gap increases about 0.001 in. (0.025mm) for every 1,000–2,500 miles (1600–4000 km). As the gap increases, the plug's voltage requirement also increases. It requires a greater voltage to jump the wider gap and about two to three times as much voltage to fire a plug at high speeds than at idle.

When you're removing spark plugs, you should work on one at a time. Don't start by removing the plug wires all at once, because unless you number them, they may become mixed up. Take a minute before you begin and number the wires with tape. The best location for numbering is near where the wires come out of the cap.

➡ Apply a small amount of silicone dielectric compound (D7AZ-19A331-A or the equivalent) to the inside of the terminal boots whenever an ignition wire is disconnected from the plug, or coil/distributor cap connection.

1. Twist the spark plug boot and remove the boot and wire from the plug. Do not pull on the wire itself as this will ruin the wire.
2. If possible, use a brush or gag to clean the area around the spark plug. Make sure that all the dirt is removed so that none will enter the cylinder after the plug is removed.
3. Remove the spark plug using the proper size socket. Truck models use either a 5/8 in. or 13/16 in. size socket depending on the engine. Turn the socket counterclockwise to remove the plug. Be sure to hold the socket straight on the plug to avoid breaking the plug, or rounding off the hex on the plug.
4. Once the plug is out, check it against the plugs shown in the Color section to determine

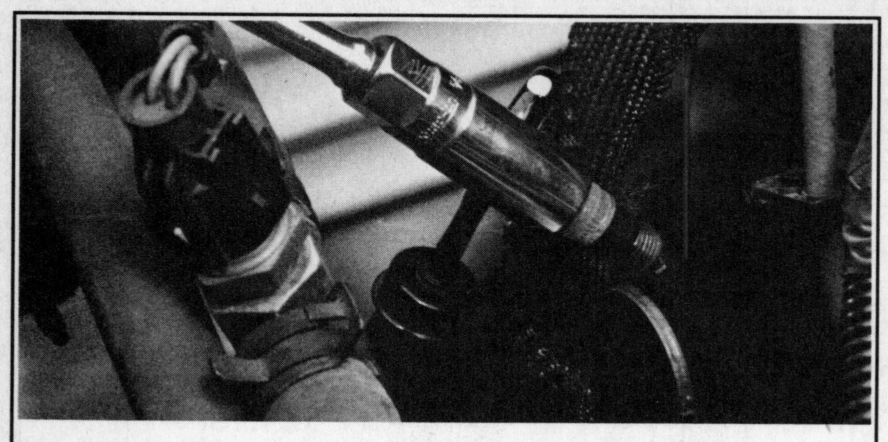

FIG. 4 Notice how the rubber insert in plug socket keeps the spark plug form falling out

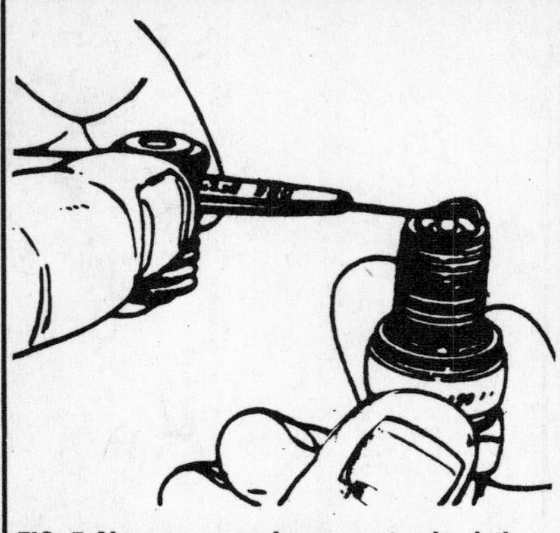

FIG. 5 Always use a wire gauge to check the electrode gap; a flat feeler gauge may not give the proper reading

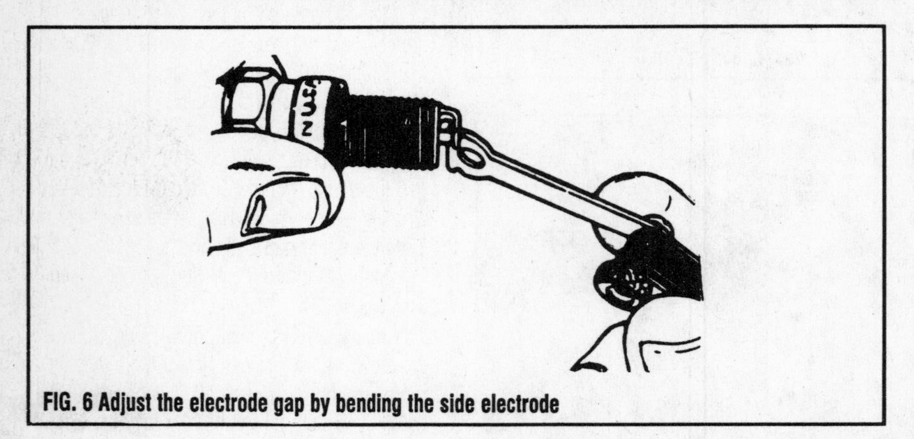

FIG. 6 Adjust the electrode gap by bending the side electrode

engine condition. This is crucial since plug readings are vital signs of engine condition.

5. Use a round wire feeler gauge to check the plug gap. The correct size gauge should pass through the electrode gap with a slight drag. If you're in doubt, try one size smaller and one larger. The smaller gauge should go through easily while the larger one shouldn't go through at all. If the gap is incorrect, use the electrode bending tool on the end of the gauge to adjust the gap. When adjusting the gap, always bend the side electrode. The center electrode is non-adjustable.

6. Squirt a drop of penetrating oil on the threads of the new plug and install it. Don't oil the threads too heavily. Turn the plug in clockwise by hand until it is snug.

7. When the plug is finger tight, tighten it with a wrench. If you don't have a torque wrench, tighten the plug as shown.

8. Install the plug boot firmly over the plug. Proceed to the next plug.

CHECKING AND REPLACING SPARK PLUG CABLES

Visually inspect the spark plug cables for burns, cuts, or breaks in the insulation. Check the spark plug boots and the nipples on the distributor cap and coil. Replace any damaged wiring. If no physical damage is obvious, the wires can be checked with an ohmmeter for excessive resistance.

When installing a new set of spark plug cables, replace the cables on at a time so there will be no mixup. Start by replacing the longest cable first. Install the boot firmly over the spark plug. Route the wire exactly the same as the original. Insert the nipple firmly into the tower on the distributor cap. Repeat the process for each cable.

FIRING ORDERS

To avoid confusion, replace spark plug wires one at a time.

➧ SEE FIGS. 7-9

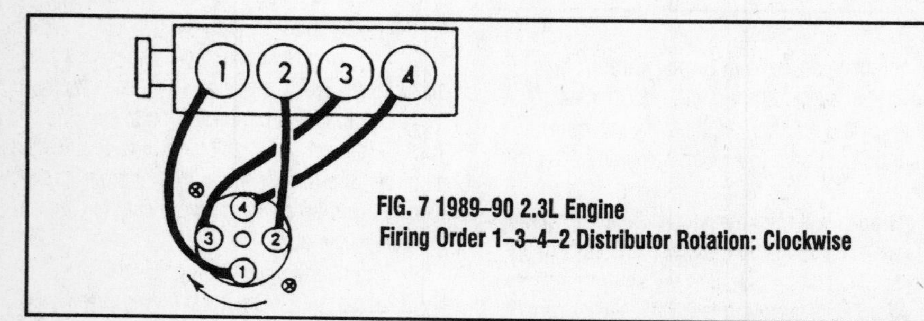

FIG. 7 1989–90 2.3L Engine
Firing Order 1–3–4–2 Distributor Rotation: Clockwise

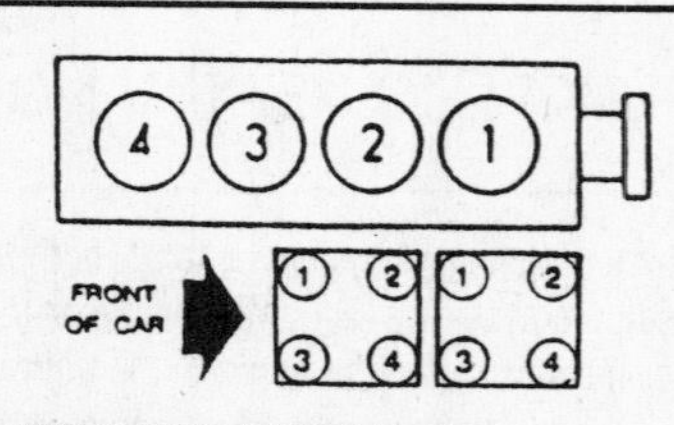

FIG. 8 1991–92 2.3L Engine

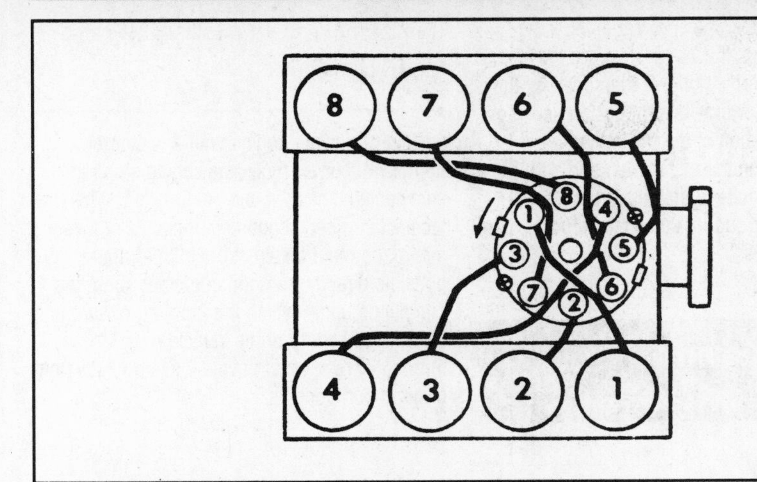

FIG. 9 5.0L HO Engine
Firing Order 1–3–7–2–6–5–4–8
Distributor Rotation: Counterclockwise

ELECTRONIC IGNITION SYSTEM APPLICATION CHART

Model	Body VIN ①	Year	Engine Liter	Engine VIN ②	Ignition Type	Fuel Type
Mustang	4	1989–90	2.3 OHC	A	TFI	EFI
		1991–92	2.3 OHC	M	DIS	EFI-MA
		1989–92	5.0 HO	E	TFI	SEFI-MA

TFI—Thick Film Integrated Ignition System
DIS—Distributorless Ignition System
① Sixth digit of VIN number
② Eighth digit of VIN number

THICK FILM INTEGRATED (TFI-IV) IGNITION SYSTEM

General Information

➧ SEE FIGS. 10-13

The Thick Film Integrated (TFI-IV) ignition system is used for all EEC-IV electronic fuel injected vehicles. The TFI-IV system module has 6 pins and uses an E-Core ignition coil, named after the shape of the laminations making up the core.

There are 2 types of TFI-IV system, and are as follow:

- PUSH START — this first TFI-IV system featured a "push start" mode which allow manual transmission vehicles to be push started. Automatic transmission vehicles must not be push started.
- COMPUTER CONTROLLED DWELL — this second TFI-IV system features an EEC-IV controlled ignition coil charge time.

The TFI-IV ignition system can be equipped with either a "Universal" or "Close Bowl" type distributor.

The TFI-IV ignition system with "Close Bowl" distributor, can be easily identified by its externally mounted TFI-IV ignition module. This module is generally mounted at the radiator support bracket or behind the engine at cowl area.

The TFI-IV ignition system with "Universal" distributor has a distributor base mounted TFI ignition module and a hall effect stator assembly. The distributor also contains a provision to change the basic distributor calibration with the use of a replaceable octane rod, from the standard of 0° to either 3° or 6° retard rods. No other calibration changes are possible.

➡ Initial timing adjustments are not required unless the distributor has been removed from the engine or moved from its initial factory setting.

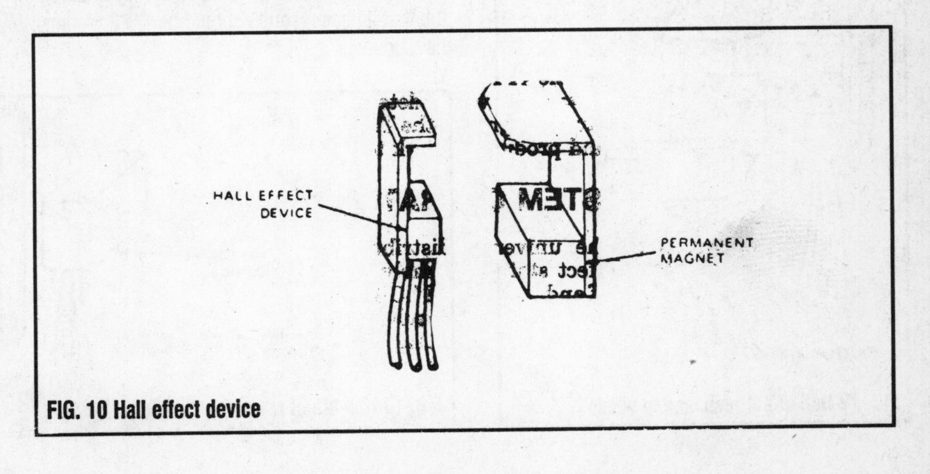

FIG. 10 Hall effect device

Both the PUSH START and COMPUTER CONTROLLED DWELL TFI-IV systems operates in the same manner. The TFI-IV module supplies voltage to the Profile Ignition Pickup (PIP) sensor, which sends the crankshaft position information to the TFI-IV module. The TFI-IV module then sends this information to the EEC-IV module, which determines the spark timing and sends an electronic signal to the TFI-IV ignition module to turn off the coil and produce a spark to fire the spark plug.

SYSTEM OPERATION

The operation of the universal distributor is accomplished through the Hall Effect stator assembly, causing the ignition coil to be switched off and on by the ECC-IV computer and TFI-IV modules. The vane switch is encapsulated package consisting of a Hall sensor on one side and a permanent magnet on the other side.

A rotary armature, made of ferrous metal, is used to trigger the Hall Effect switch. When the window of the armature is between the magnet and the Hall Effect device, a magnetic flux field is completed from the magnet through the Hall Effect device back to the magnet. As the vane passes through the opening, the flux lines are shunted through the vane and back to the magnet. A voltage is produced while the vane passes through the opening. When the vane clears the opening, the window causes the signal to go to 0 volts. The signal is then used by the EEC-IV system for crankshaft position sensing and the computation of the desired spark advance based on the engine demand and calibration. The voltage distribution is accomplished through a conventional rotor, cap and ignition wires.

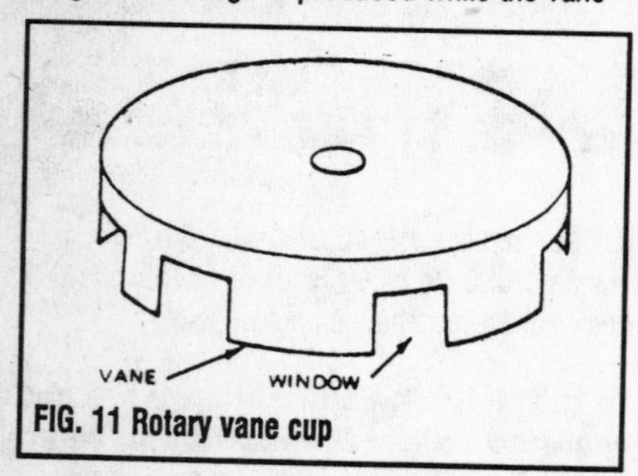

FIG. 11 Rotary vane cup

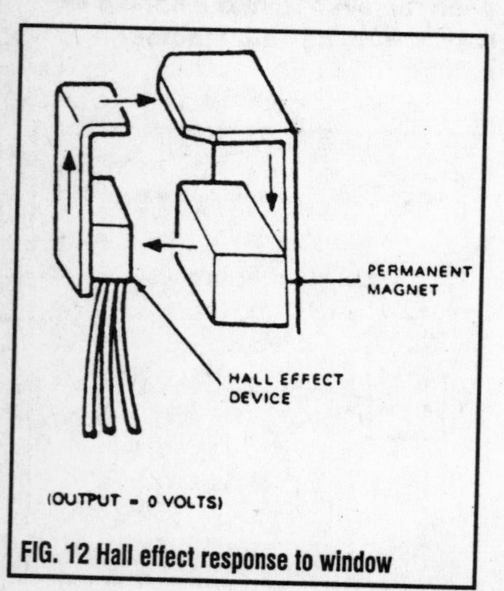

FIG. 12 Hall effect response to window

Diagnosis and Testing

SERVICE PRECAUTIONS

- Always turn the key **OFF** and isolate both ends of a circuit whenever testing for short or continuity.
- Never measure voltage voltage or resistance directly at the processor connector.
- Always disconnect solenoids and switches from the harness before measuring for continuity, resistance or energizing by way of a 12 volts source.
- When disconnecting connectors, inspect for damaged or pushed-out pins, corrosion, loose wires, etc. Service if required.

PRELIMINARY CHECKS

1. Visually inspect the engine compartment to ensure that all vacuum lines and spark plug wires are properly routed and securely connected.
2. Examine all wiring harnesses and connectors for insulation damage, burned, overheated, loose or broken conditions. Check that the TFI module is securely fasten to the side of the distributor.
3. Be certain that the battery is fully charged and that all accessories should be **OFF** during the diagnosis.

INTERMITTENT DIAGNOSIS PROCEDURE

This procedure begins with a customer complaint. For example, the engine stops at unexpected times but can be restarted. When the technician must diagnose complaints of these nature, he must obtain as much information as possible directly from the customer about the conditions under which the problem occurs and the service history of the vehicle must be thoroughly reviewed to avoid repeat replacement of good components.

TFI-IV AND TFI-IV WITH CCD

Ignition Coil Secondary Voltage Test

SEE FIG. 14

CRANK MODE

1. Connect a spark tester between the ignition coil wire and a good engine ground.
2. Crank the engine and check for spark at the tester.
3. Turn the ignition switch **OFF**.
4. If no spark occurs, check the following:
   a. Inspect the ignition coil for damage or carbon tracking.
   b. Also, check that the distributor shaft is rotating, when the engine is being cranked.
   c. If the results in Steps a, and b are okay, go to Stator Test 1.
5. If a spark did occur, check the distributor cap and rotor for damage or carbon tracking. Go to Ignition Coil Secondary Voltage (Run Mode) test.

RUN MODE

1. Fully apply the parking brake. Place the gear shift lever in **P** (automatic transmission) or **N** (manual transmission).
2. Disconnect the S terminal wire at the starter relay. Then, attach a remote starter switch.

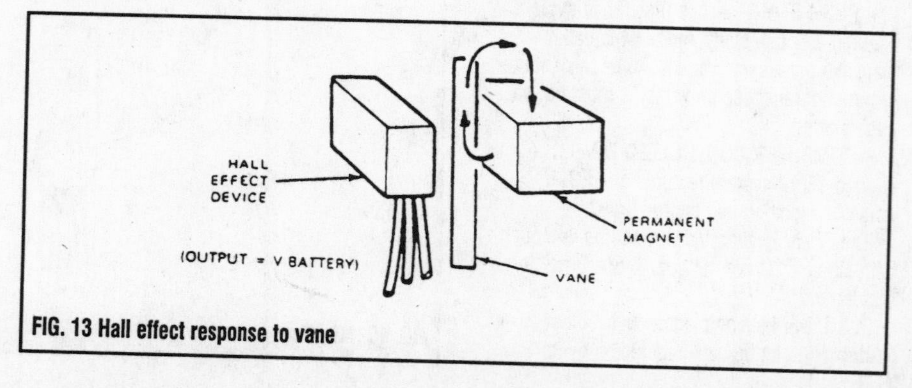

FIG. 13 Hall effect response to vane

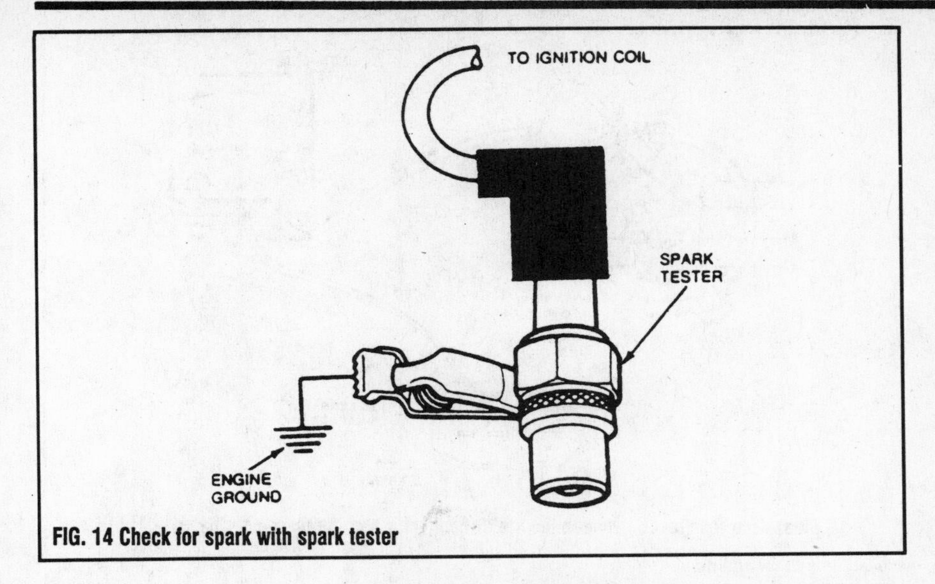

FIG. 14 Check for spark with spark tester

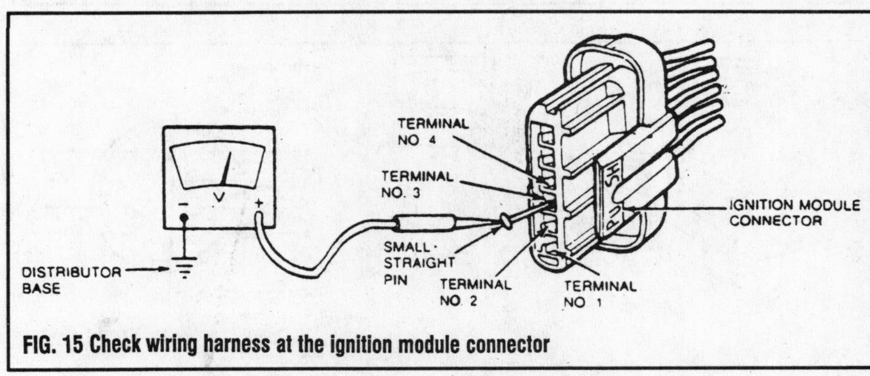

FIG. 15 Check wiring harness at the ignition module connector

3. Turn the ignition switch to the **RUN** position.

4. Using the remote starter switch, crank the engine and check for spark.

5. Turn the ignition switch **OFF**.

6. If no spark occurs, go to the Wiring Harness test.

7. If a spark did occur, the problem is not in the ignition system.

Wiring Harness Test

➧ SEE FIG. 15

1. Push the connector tabs and separate the wiring harness connector from the ignition module. Check for dirt, corrosion or damage.

2. Check that the S terminal wire at the starter relay is disconnected.

3. Measure the battery voltage.

4. Carefully insert a small straight pins in the appropriate terminal.

➡ Do not allow the straight pins to contact electrical ground, while performing this test.

5. Measure the voltage at the following points:

a. TFI Without CCD — Terminal No. 3 (Run Circuit) — with ignition switch in **RUN** and **START**.

b. TFI Without CCD — Terminal No. 4 (Start Circuit) — with ignition switch in **START**.

c. TFI With CCD — Terminal No. 3 (Run Circuit) — with ignition switch in **RUN** and **START**.

6. Turn the ignition switch **OFF** and remove the straight pin.

7. Reconnect the S terminal wire at the starter relay.

8. If the results are within 90% of battery voltage, replace the TFI module.

9. If the results are not within 90% of battery voltage, inspect the wiring harness and connectors in the faulty circuit. Also, check for a faulty ignition switch.

Stator Test 1

1. Fully apply the parking brake. Place the gear shift lever in **P** (automatic transmission) or **N** (manual transmission).

2. Disconnect the harness connector at the TFI module and connect the TFI tester.

3. Disconnect the S terminal wire at the starter relay. Then, attach a remote starter switch.

4. Using the remote starter switch, crank the engine and check the status of the 2 LED lamps.

5. Remove the tester and remote switch.

6. Reconnect the wire at the starter relay and harness to the TFI module.

7. If the PIP light blink, go to the TFI module test.

8. If the PIP light did not blink, remove the distributor cap, crank the engine and check that the distributor shaft rotates. If okay, go to the Stator Test 2.

Stator Test 2

1. Remove the distributor from the engine. Remove the TFI module from the distributor.

2. Measure the resistance between the TFI module terminals, as follow:

a. GND–PIP In — should be greater than 500Ω.

b. PIP PWR–PIP IN — should be less than 2 kΩ.

c. PIP PWR–TFI PWR — should be less than 200Ω.

c. GND–IGN GND — should be less than 2Ω.

d. PIP In–PIP — should be less than 200Ω.

3. If the readings are within the specified value, replace the stator.

4. If the readings are not as specified, replace the TFI.

Module Test

1. Use the status of the tach light from the Stator Test 1.

2. If the tach light blink, go to Ignition Coil Wire and Coil test.

3. If the tach light did not blink, replace the TFI module, then check for spark, as indicated in the Ignition Coil Secondary Voltage test. If the spark was not present, replace the coil also.

Coil Wire and Coil Test

1. Disconnect the ignition coil connector and check for dirt, corrosion or damage.

2. Substitute a known good coil and check for spark using the spark tester.

➡ The possibility of dangerous high voltage may be present when performing this test. Do not hold the coil while performing this test.

3. Crank the engine and check for spark.

4. Turn the ignition switch **OFF**.

5. If a spark did occur, measure the resistance of the ignition coil wire, replace it if the resistance is greater than 7 kΩ per foot. If the readings are within specification, replace the ignition coil.

6. If no spark occurs, the the problem is not the coil. Go to the EEC-IV — TFI-IV test.

EEC-IV — TFI-IV Test

1. Disconnect the pin-in-line connector near the distributor.
2. Crank the engine.
3. Turn the ignition switch **OFF**.
4. If a spark did occur, check the PIP and ignition ground wires for continuity. If okay, the problem is not ignition.
5. If no spark occurs, check the voltage at the positive (+) terminal of the ignition coil, with the ignition switch in **RUN**.
6. If the reading is not within battery voltage, check for a worn or damaged ignition switch.
7. If the reading is within battery voltage, check for faults in the wiring between the coil and TFI module terminal No. 2 or any additional wiring or components connected to that circuit.

FIG. 16 Measuring resistance between the stator connector and the distributor base — TFI-IV Closed Bowl distributor

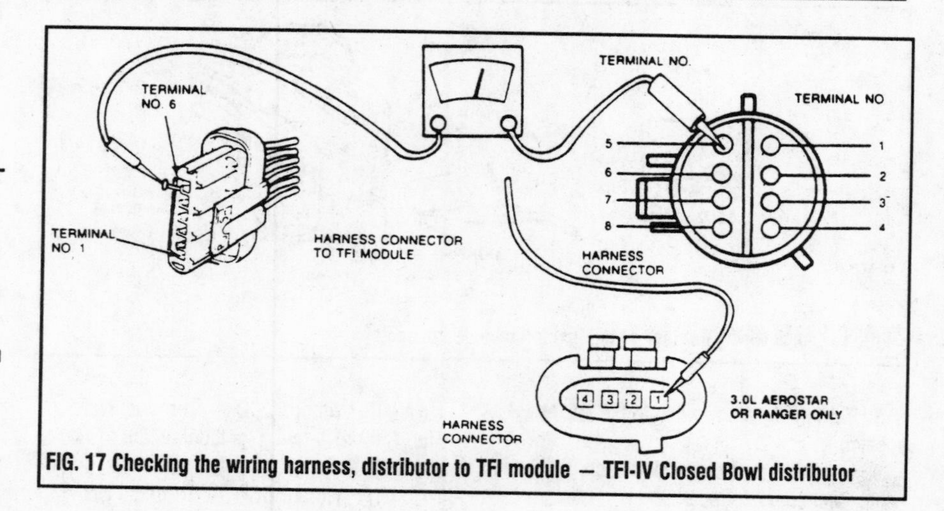

FIG. 17 Checking the wiring harness, distributor to TFI module — TFI-IV Closed Bowl distributor

TFI-IV WITH CLOSED BOWL DISTRIBUTOR

➧ SEE FIGS. 16-19

Ignition Coil Secondary Voltage Test

CRANK MODE

1. Connect a spark tester between the ignition coil wire and a good engine ground.
2. Crank the engine and check for spark at the tester.
3. Turn the ignition switch **OFF**.
4. If no spark occurs, check the following:
 a. Inspect the ignition coil for damage or carbon tracking.
 b. Also, check that the distributor shaft is rotating, when the engine is being cranked.
 c. If results in Steps a and b are okay, go to Stator Test 1.
5. If a spark did occur, check the distributor cap and rotor for damage or carbon tracking. Go to Ignition Coil Secondary Voltage (Run Mode) test.

RUN MODE

1. Fully apply the parking brake. Place the gear shift lever in **P** (automatic transmission) or **N** (manual transmission).
2. Disconnect the S terminal wire at the starter relay. Then, attach a remote starter switch.
3. Turn the ignition switch to the **RUN** position.
4. Using the remote starter switch, crank the engine and check for spark.
5. Turn the ignition switch **OFF**.
6. If no spark occurs, go to the Wiring Harness test.
7. If a spark did occur, the problem is not in the ignition system.

Wiring Harness Test

1. Push the connector tabs and separate the wiring harness connector from the ignition module. Check for dirt, corrosion or damage.
2. Check that the S terminal wire at the starter relay is disconnected.
3. Measure the battery voltage.
4. Carefully insert a small straight pins in the appropriate terminal.

➡ Do not allow the straight pins to contact electrical ground, while performing this test.

5. Measure the voltage at the following points:
 a. TFI Without CCD — Terminal No. 3 (Run Circuit) — with ignition switch in **RUN** and **START**.
 b. TFI Without CCD — Terminal No. 4 (Start Circuit) — with ignition switch in **START**.
 c. TFI With CCD — Terminal No. 3 (Run Circuit) — with ignition switch in **RUN** and **START**.
6. Turn the ignition switch **OFF** and remove the straight pin.
7. Reconnect the S terminal wire at the starter relay.
8. If the results are within 90% of battery voltage, check for faults in the wiring between the coil and TFI module terminal No. 2 or any additional wiring or components connected to that circuit.
9. If the results are not within 90% of battery voltage, inspect the wiring harness and connectors in the faulty circuit. Also, check for a faulty ignition switch.

Stator Test 1

1. Separate the wiring harness connector

from the distributor. Check for dirt, corrosion or damage.

2. Measure the resistance between the stator connector terminals 1 and 5.

3. If greater than 5Ω, replace the stator.

4. If less than 5Ω, measure the resistance between the stator connector terminal 2 and the distributor base. Also, measure the resistance between the stator connector terminal 6 and the distributor base.

5. If the resistance is greater than 1Ω in each case, inspect the stator retaining screws. If okay, replace the stator.

6. If less than 1Ω, go to Stator Test 2.

Stator Test 2

1. Fully apply the parking brake. Place the gear shift lever in **P** (automatic transmission) or **N** (manual transmission).

2. Disconnect the harness connector at the TFI module and connect the TFI tester.

3. Disconnect the S terminal wire at the starter relay. Then, attach a remote starter switch.

4. Using the remote starter switch, crank the engine and check the status of the 2 LED lamps.

5. Remove the tester and remote switch.

6. Reconnect the wire at the starter relay and harness to the TFI module.

7. If the PIP light blink, go to the TFI module test.

8. If the PIP light did not blink, replace the stator.

Module Test

1. Use the status of the tach light from the Stator test 2.

2. If the tach light blink, go to Ignition Coil Wire and Coil test.

3. If the tach light did not blink, replace the TFI module, then check for spark, as indicated in the Ignition Coil Secondary Voltage test. If the spark was not present, replace the coil also.

Coil Wire And Coil Test

1. Disconnect the ignition coil connector and check for dirt, corrosion or damage.

2. Substitute a known good coil and check for spark using the spark tester.

➡ The possibility of dangerous high voltage may be present when performing this test. Do not hold the coil while performing this test.

3. Crank the engine and check for spark.

4. Turn the ignition switch **OFF**.

5. If a spark did occur, measure the resistance of the ignition coil wire, replace it if the resistance is greater than 7 kΩ per foot. If the readings are within specification, replace the ignition coil.

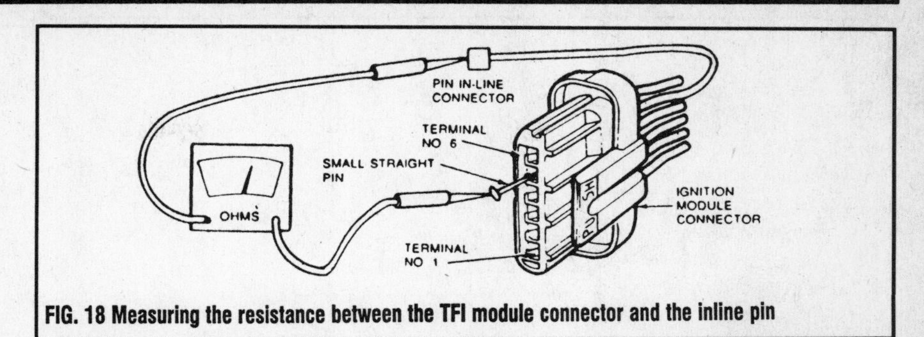

FIG. 18 Measuring the resistance between the TFI module connector and the inline pin

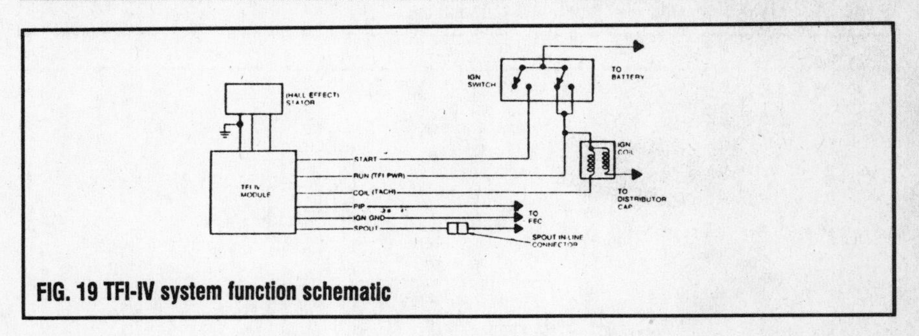

FIG. 19 TFI-IV system function schematic

PRELIMINARY NOTE

The procedure described below for setting initial timing is to be used under normal circumstances. If problems are encountered setting initial timing using this procedure, the spark timing procedure that follows should be used

Procedure	EEC-IV
① Place transmission in PARK or NEUTRAL, A/C and heater in OFF position.	X
② Remove vacuum hoses from the distributor vacuum advance connection at the distributor and plug the hoses.	
③ Connect an inductive timing light, Rotunda 059-00006 or equivalent.	X
④ Connect a tachometer, Rotunda 059-00010 or equivalent.	
⑤ Disconnect the single wire in-line spout connector or remove the shorting bar from the double wire spout connector.	X
⑥ If the vehicle is equipped with a barometric pressure switch (-12A243-) disconnect it from the ignition module and place a jumper wire across the pins at the ignition module connector (yellow and black wires).	
⑦ Start the engine and allow it to warm up to operating temperature. To set timing correctly, **a remote starter should not be used.** Use the ignition key only to start the vehicle. Disconnecting the start wire at the starter relay will cause TFI Module to revert to start mode timing after the vehicle is started. Reconnecting the start wire after the vehicle is running will not correct the timing.	X
⑧ With engine at timing rpm if specified, check/adjust initial timing to specification.	X
⑨ Reconnect single wire in-line spout connector or reinstall the shorting bar on the double wire spout connector. Check timing advance to verify distributor is advancing beyond the initial setting.	X
⑩ Remove test instruments.	X
⑪ Unplug and reconnect vacuum hoses.	
⑫ Remove jumper from ignition connector and reconnect if applicable.	

FIG. 20 Initial Set Timing Procedure

6. If no spark occurs, reconnect the coil connector and install the spark tester, then go to EEC-IV — Wiring Test.

EEC-IV — Wiring Test

1. Disconnect the pin-in-line connector near the TFI module.

2. Crank the engine.

3. If a spark did occur, check the PIP-A and ignition ground wires for continuity to the EEC system. If okay, the problem is not ignition.

4. If no spark occurs, check the wiring harness between the harness connector to the TFI module and the distributor. Service as required.

IGNITION TIMING

➧ SEE FIG. 20

➡ Do not change the ignition timing by the use of a different octane rod

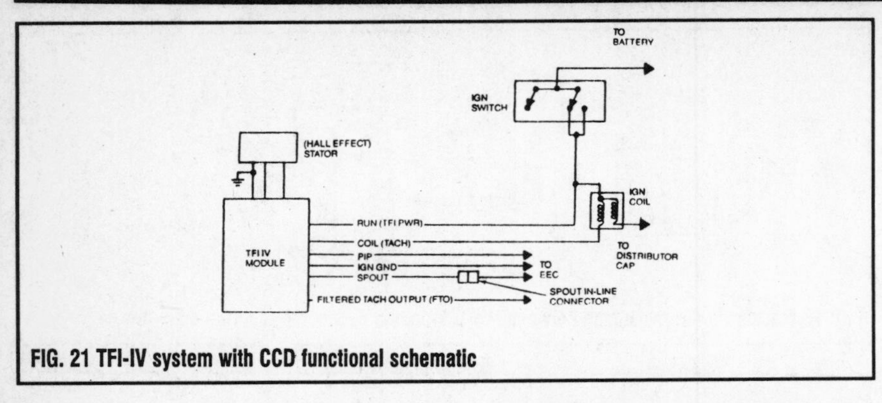

FIG. 21 TFI-IV system with CCD functional schematic

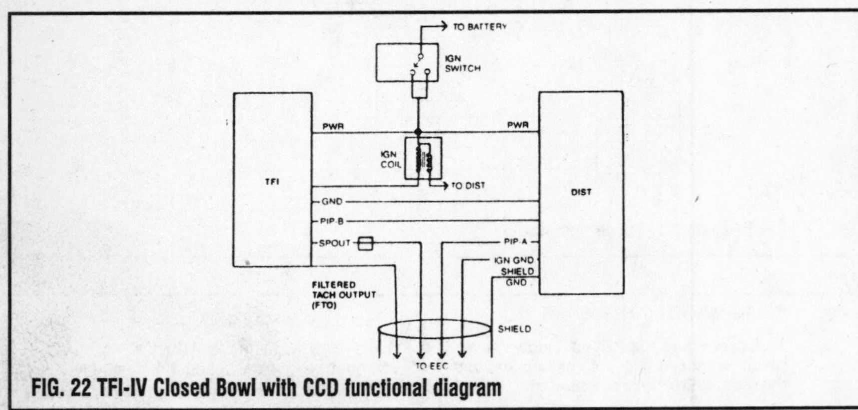

FIG. 22 TFI-IV Closed Bowl with CCD functional diagram

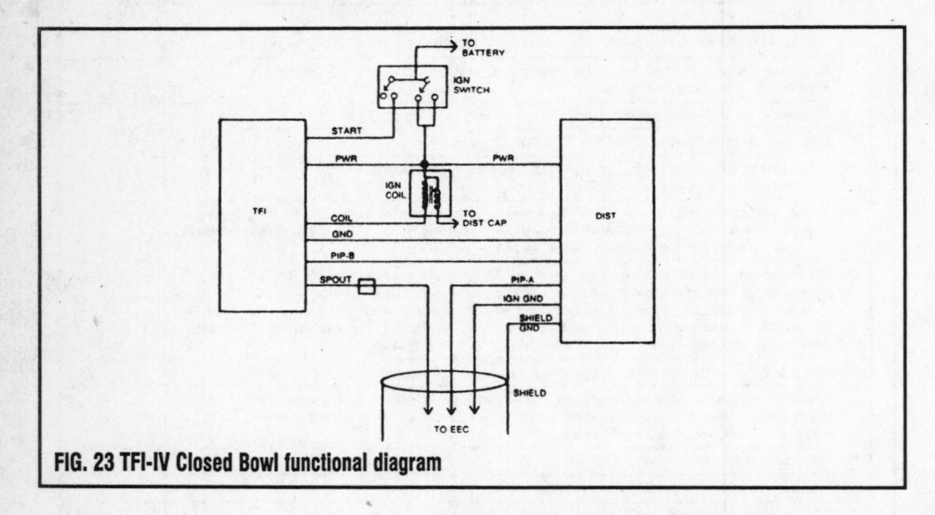

FIG. 23 TFI-IV Closed Bowl functional diagram

without having the proper authority to do so as federal emission requirements will be effected.

SPARK TIMING ADVANCE

Spark timing advance is controlled by the EEC system. This procedure checks the capability of the ignition module to receive the spark timing command from the EEC module. The use of a volt/ohmmeter is required.

1. Turn the ignition switch **OFF**.
2. Disconnect the pin inline connector (Spout connector) near the TFI module.
3. Start the engine and measure the voltage at idle. From spout connector to distributor base. Should indicate approximately battery voltage.
4. If the result is okay, the problem lies within the EEC-IV system.
5. If the result was not satisfactory, separate the wiring harness connector from the ignition module. Check for damage, corrosion or dirt. Service as necessary.
6. Measure the resistance between terminal 5 and the pin inline connector. This test is done at the ignition module connector only. Should indicate less than 5Ω.
7. If okay, replace the TFI module.
8. If the result was not satisfactory, service the wiring between the pin inline connector and the TFI connector.

THICK FILM INTEGRATED (TFI)-IV DIAGNOSTIC TEST CHARTS

➧ SEE FIGS. 24-51

Distributor

REMOVAL & INSTALLATION

1. Rotate the engine until the No. 1 piston is on top dead center of its compression stroke.
2. Disconnect the negative battery cable. Disconnect the vehicle wiring harness connector from the distributor. Before removing the distributor cap, mark the position of the No. 1 wire tower on the distributor base for future reference.
3. Loosen the distributor cap hold-down screws and remove the cap. Mark the position of the rotor to the distributor housing. Position the cap and wires out of the way.
4. Scribe a mark in the distributor body and the engine block to indicate position of the distributor in the engine.
5. Remove the distributor hold-down bolt and clamp. On 1.9L engine, remove 2 distributor hold-down bolts.

Preliminary Checkout, Equipment & Notes

CHECKOUT

- Visually inspect the engine compartment to ensure all vacuum hoses and spark plug wires are properly routed and securely connected.
- Examine all wiring harnesses and connectors for insulation damage, and burned, overheated, loose or broken conditions.
- Check that the TFI module is securely fastened to the distributor base.
- Be certain the battery is fully charged.
- All accessories should be off during diagnosis.

EQUIPMENT

Obtain the following test equipment or an equivalent:

- Spark Tester, Special Service Tool D81P-6666-A. See **NOTES.**
- Volt/Ohm Meter Rotunda 014-00407 or 007-00001.
- 12 Volt Test Lamp.
- Small straight pin.
- Remote Starter Switch.
- TFI Ignition Tester, Rotunda 105-00002.
- E-core Ignition Coil E73F-12029-AB.
- Ignition coil secondary wire E43E-12A012-AB.

NOTES

- A spark plug with a broken side electrode **is not** sufficient to check for spark and may lead to incorrect results.
- When instructed to inspect a wiring harness, both a visual inspection and a continuity test should be performed.
- When making measurements on a wiring harness or connector, it is good practice to wiggle the wires while measuring.
- References to pin-in-line connector apply to a shorting bar type connector used to set base timing.

FIG. 24 TFI-IV Diagnostic Test Charts

Functional Schematic

TFI-IV And TFI-IV With CCD

The TFI-IV system electrical schematic is shown below.

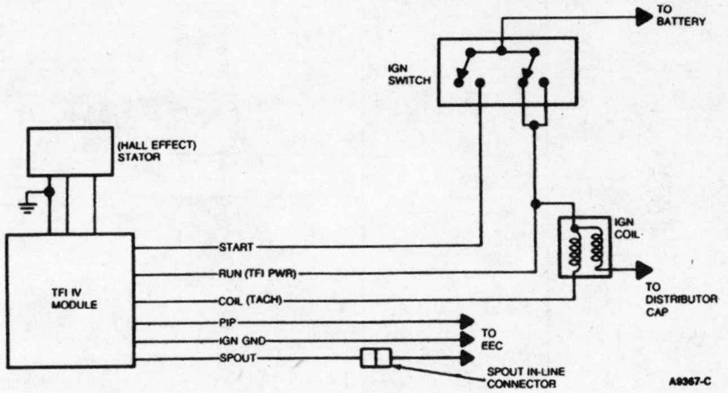

The TFI-IV with CCD system electrical schematic is shown below.

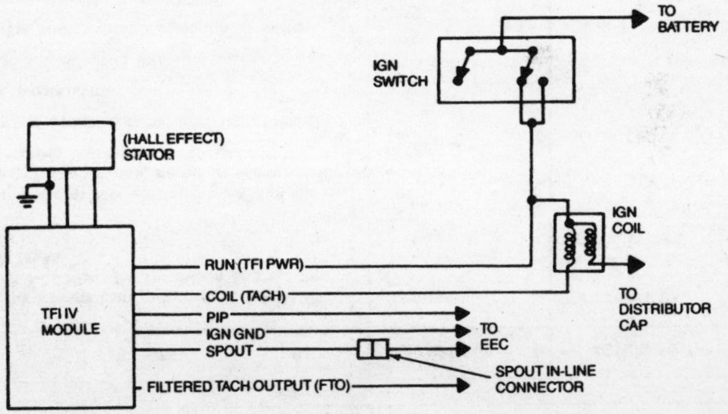

FIG. 25 TFI-IV Diagnostic Test Charts

Ignition Coil Secondary Voltage (Crank Mode) | **TFI-IV And TFI With CCD** | **Part 2 Test 1**

TEST STEP	RESULT ▶	ACTION TO TAKE
1. Connect spark tester between ignition coil wire and engine ground. 2. Crank engine. 3. Turn ignition switch to the OFF position. 4. **Was spark present?**	Yes ▶	Test result OK. INSPECT distributor cap and rotor for damage/carbon tracking. If engine starts, GO to Part 1, Test 2, otherwise GO to Test 2.
	No ▶	INSPECT ignition coil for damage/carbon tracking. CRANK engine to verify distributor rotation. GO to Test 4.

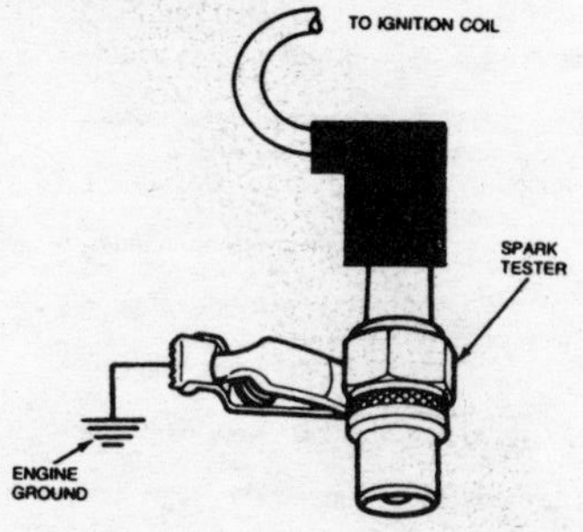

FIG. 26 TFI-IV Diagnostic Test Charts

Ignition Coil Secondary Voltage (Run Mode) | **TFI-IV And TFI With CCD** | **Part 2 Test 2**

TEST STEP	RESULT ▶	ACTION TO TAKE
1. Place the transmission shift lever in the PARK (A/T) or NEUTRAL (M/T) position and set the parking brake. CAUTION **Failure to perform this step may result in the vehicle moving when the starter is subsequently engaged during the test.**	Yes ▶	Test result OK. Problem is not in the ignition system.
2. Disconnect wire at S terminal of starter relay. 3. Attach remote starter switch. 4. Turn ignition switch to the RUN position. 5. Crank the engine using remote starter switch. 6. Turn ignition switch to the OFF position. 7. Remove remote starter switch. 8. Reconnect wire to S terminal of starter relay. 9. **Was spark present?**	No ▶	GO to Test 3.

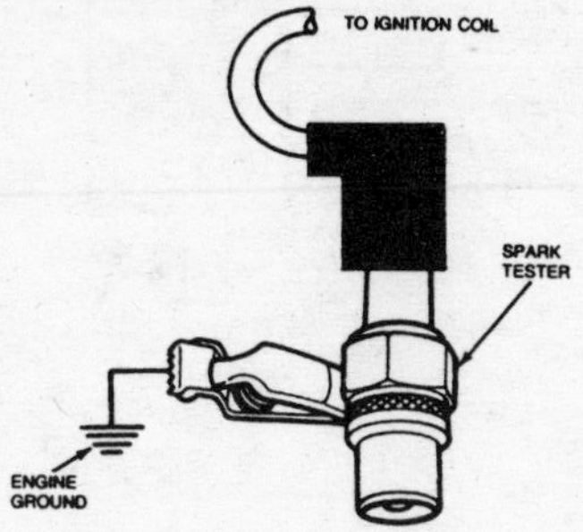

FIG. 27 TFI-IV Diagnostic Test Charts

Wiring Harness — TFI-IV And TFI With CCD — Part 2 Test 3

TEST STEP

1. Separate wiring harness connector from ignition module. Inspect for dirt, corrosion, and damage.
   NOTE: Push connector tabs to separate.
2. Verify that the wire to the S terminal of starter relay is disconnected.
3. Attach negative (−) VOM lead to distributor base.
4. Measure battery voltage.
5. Following the appropriate table below, measure connector terminal voltage by attaching VOM to small straight pin inserted into connector terminal and turning ignition switch to position shown.

CAUTION

Do not allow straight pin to contact electrical ground.

TFI without CCD

Connector Terminal	Wire/Circuit	Ignition Switch Test Position
#3	Run Circuit	Run and Start
#4	Start Circuit	Start

TFI with CCD

Connector Terminal	Wire/Circuit	Ignition Switch Test Position
#3	Run Circuit	Run and Start

6. Turn ignition switch to OFF position.
7. Remove straight pin.
8. Reconnect wire to S terminal of starter relay.
9. **Was the value at least 90 percent of battery voltage in each case?**

RESULT	ACTION TO TAKE
Yes	REPLACE TFI module.
No	INSPECT for faults in wiring harness and connectors.
	Damaged or worn ignition switch.

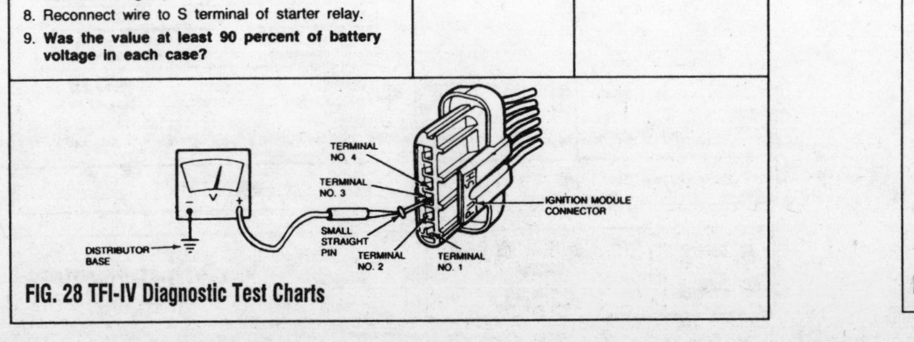

FIG. 28 TFI-IV Diagnostic Test Charts

Stator — TFI — TFI-IV And TFI With CCD — Part 2 Test 4

TEST STEP

1. Place the transmission shift lever in the PARK (A/T) or NEUTRAL (M/T) position and set the parking brake.

CAUTION

Failure to perform this step may result in the vehicle moving when the starter is subsequently engaged during the test.

2. Disconnect the harness connector from the TFI module and connect the TFI tester.
3. Connect the red lead from the tester to the positive (+) side of the battery.
4. Disconnect the wire at the S terminal of the starter relay, and attach remote starter switch.
5. Crank the engine using the remote starter switch and note the status of the two LED lamps.
6. Remove the tester and remote starter switch.
7. Reconnect the wire to the starter relay and the connector to the TFI.
8. **Did the PIP light blink?**

RESULT	ACTION TO TAKE
Yes	GO to Test 6.
No	REMOVE distributor cap and VERIFY rotation. If OK, GO to Test 5.

FIG. 29 TFI-IV Diagnostic Test Charts

Stator — TFI-IV | TFI-IV And TFI With CCD | Part 2 Test 5

TEST STEP	RESULT ▶	ACTION TO TAKE
1. Remove the distributor from the engine and the TFI module from the distributor.	Yes ▶	Replace stator.
2. Measure resistance between TFI module terminals as shown below.	No ▶	Replace TFI.

Measure Between These Terminals	Resistance Should Be
GND — PIP In	Greater than 500 Ohms
PIP PWR — PIP IN	Less than 2K Ohms
PIP PWR — TFI PWR	Less than 200 Ohms
GND — IGN GND	Less than 2 Ohms
PIP In — PIP	Less than 200 Ohms

3. **Are all these readings as specified?**

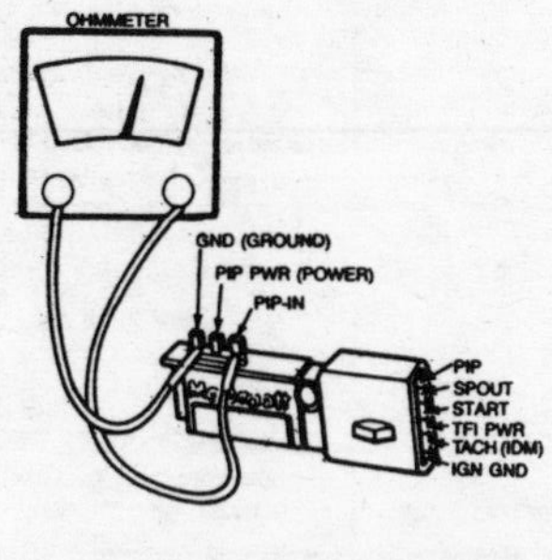

OHMMETER
GND (GROUND)
PIP PWR (POWER)
PIP-IN
PIP
SPOUT
START
TFI PWR
TACH (IDM)
IGN GND

FIG. 30 TFI-IV Diagnostic Test Charts

TFI Module | TFI-IV And TFI-IV With CCD | Part 2 Test 6

TEST STEP	RESULT ▶	ACTION TO TAKE
1. Use status of Tach light from Test 4. 2. **Did the Tach light blink?**	Yes ▶	GO to Test 7.
	No ▶	REPLACE TFI module and CHECK for spark using the method described in Test 1. If spark was not present REPLACE the coil also.

FIG. 31 TFI-IV Diagnostic Test Charts

Ignition Coil and Secondary Wire	TFI-IV And TFI With CCD	Part 2 Test 7

TEST STEP	RESULT ▶	ACTION TO TAKE
1. Disconnect ignition coil connector. Inspect for dirt, corrosion and damage. 2. Connect the ignition coil connector to a known good ignition coil. 3. Connect one end of a known good secondary wire to the spark tester. Connect the other end to the known good ignition coil. CAUTION **DO NOT HOLD THE COIL while performing this test. Dangerous voltages may be present on the metal laminations as well as the high voltage tower.** 4. Crank engine. 5. Turn ignition switch to OFF position. 6. **Was spark present?**	Yes ▶	MEASURE resistance of the ignition coil wire (from vehicle). REPLACE if greater than 7,000 ohms per foot. If OK, REPLACE ignition coil.
	No ▶	RECONNECT coil connector to the vehicle coil and spark tester to vehicle secondary wire and GO to Test 8.

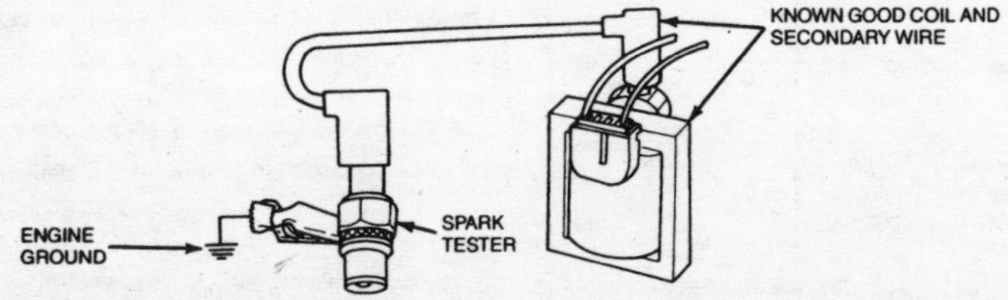

FIG. 32 TFI-IV Diagnostic Test Charts

EEC-IV — TFI-IV	TFI-IV And TFI-IV With CCD	Part 2 Test 8

TEST STEP	RESULT ▶	ACTION TO TAKE
1. Disconnect pin-in-line connector near the distributor. 2. Crank engine. 3. Turn ignition switch to OFF position. 4. **Was spark present?**	Yes ▶	CHECK PIP and Ignition ground wires for continuity. SERVICE as necessary.
	No ▶	GO to Test 9.

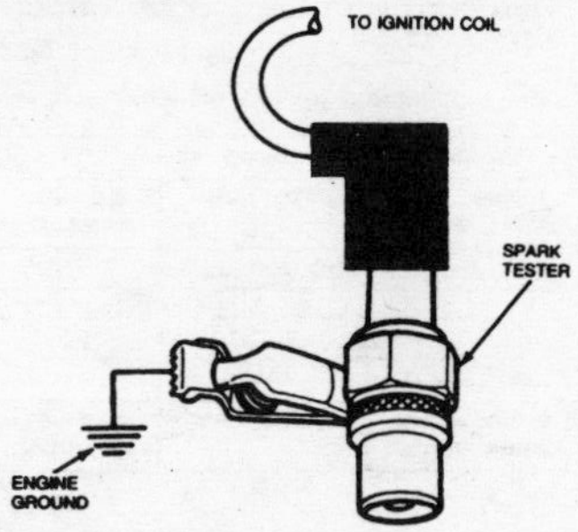

FIG. 33 TFI-IV Diagnostic Test Charts

Ignition Coil Supply Voltage — TFI-IV And TFI-IV With CCD — Part 2 Test 9

TEST STEP	RESULT ▶	ACTION TO TAKE
1. Attach negative (−) VOM lead to distributor base. 2. Measure battery voltage. 3. Turn ignition switch to RUN position. 4. Measure voltage at POSITIVE (+) terminal of ignition coil. 5. Turn ignition switch to OFF position. 6. **Was the value 90 percent of battery voltage or more?**	Yes ▶	GO to Test 10.
	No ▶	INSPECT and SERVICE wiring between ignition coil and ignition switch. Worn or damaged ignition switch.

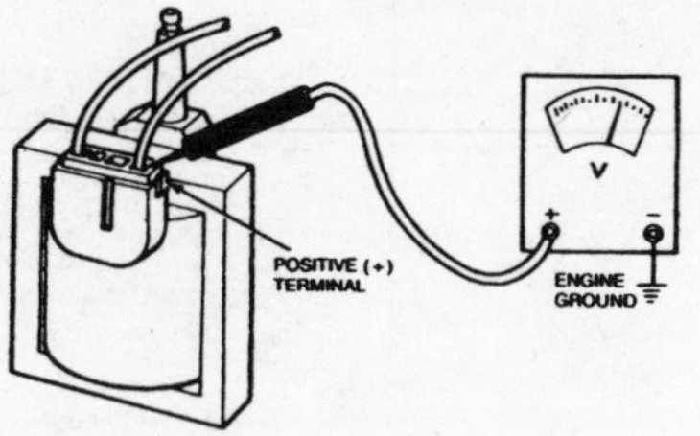

FIG. 34 TFI-IV Diagnostic Test Charts

Wiring Harness — TFI-IV — Part 2 Test 10

TEST STEP	RESULT ▶	ACTION TO TAKE
1. Separate wiring harness connector from ignition module. Inspect for dirt, corrosion, and damage. **NOTE: Push connector tabs to separate.** 2. Disconnect the wire at S terminal of starter relay. 3. Attach negative (−) VOM lead to distributor base. 4. Measure battery voltage. 5. Following the appropriate table below, measure connector terminal voltage by attaching VOM to small straight pin inserted into connector terminal and turning ignition switch to position shown.	Yes ▶	INSPECT for faults in wiring between the coil and TFI module terminal No. 2 or any additional wiring or components connected to that circuit.
CAUTION **Do not allow straight pin to contact electrical ground.**	No ▶	INSPECT for faults in wiring harness and connectors. Damaged or worn ignition switch.

TFI without CCD

Connector Terminal	Wire/Circuit	Ignition Switch Test Position
#3	Run Circuit	Run and Start
#4	Start Circuit	Start

TFI with CCD

Connector Terminal	Wire/Circuit	Ignition Switch Test Position
#3	Run Circuit	Run and Start

6. Turn ignition switch to OFF position.
7. Remove straight pin.
8. Reconnect wire to S terminal of starter relay.
9. **Was the value at least 90 percent of battery voltage in each case?**

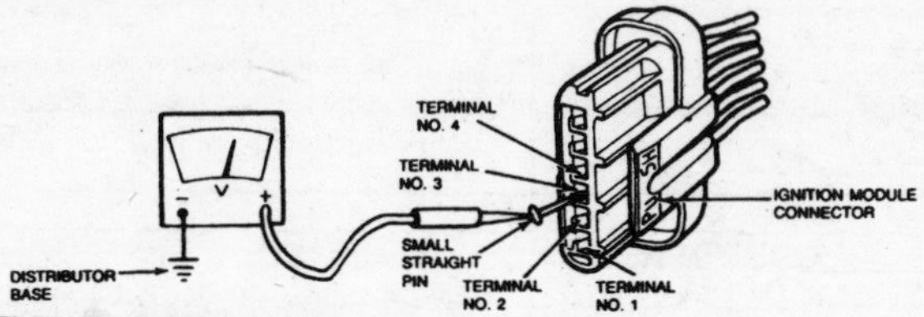

FIG. 35 TFI-IV Diagnostic Test Charts

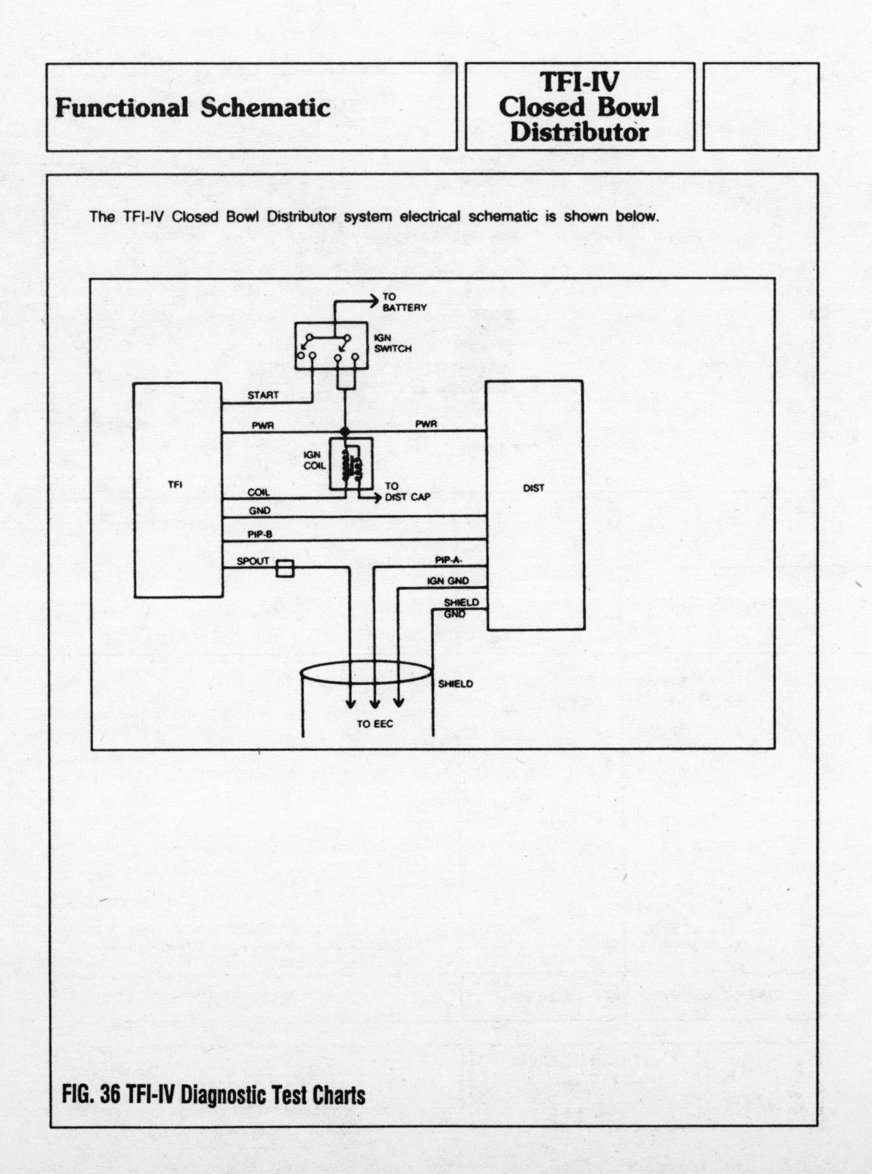

FIG. 36 TFI-IV Diagnostic Test Charts

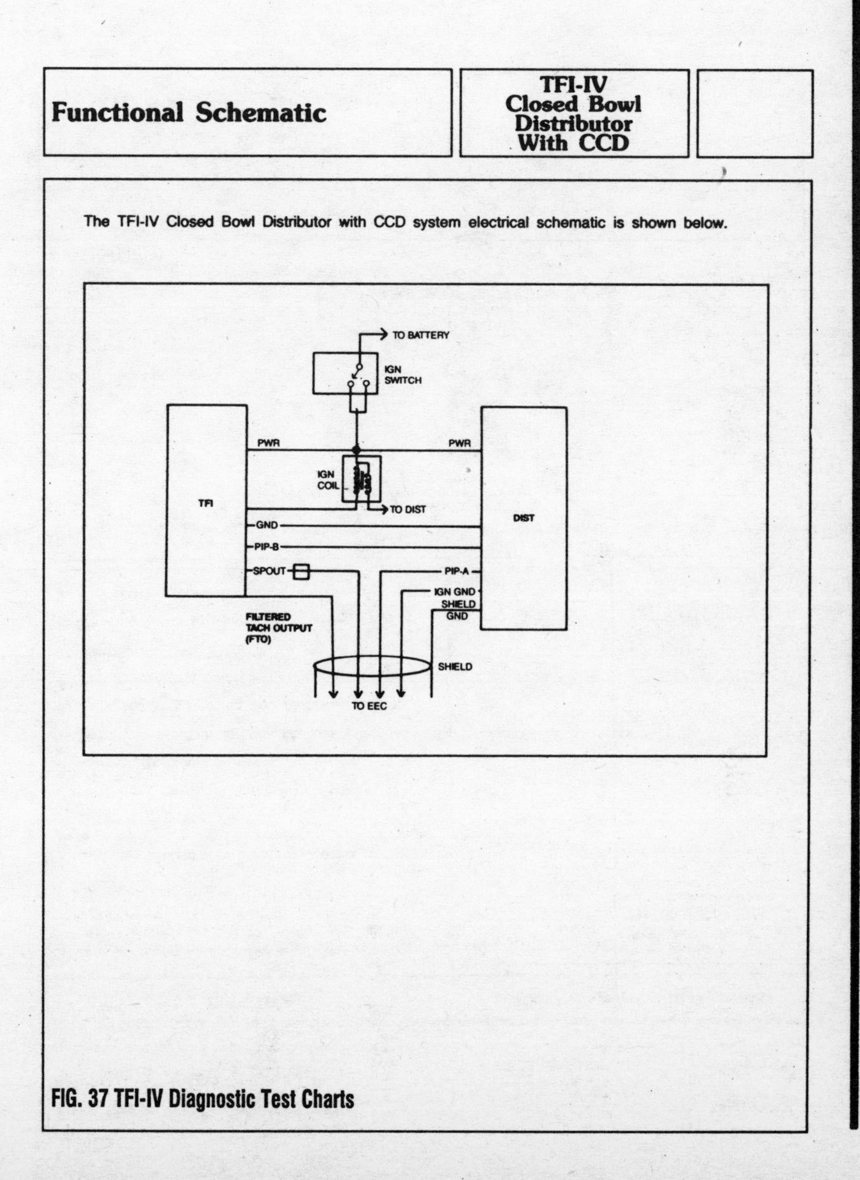

FIG. 37 TFI-IV Diagnostic Test Charts

Ignition Coil Secondary Voltage — Crank Mode | TFI-IV Closed Bowl Distributor | Part 2 Test 1

TEST STEP	RESULT ▶	ACTION TO TAKE
1. Connect spark tester between ignition coil wire and engine ground. 2. Crank engine. 3. **Was spark present?**	Yes ▶	Test result OK. INSPECT distributor cap and rotor for damage/carbon tracking.
	No ▶	INSPECT ignition coil for damage/carbon tracking. Crank engine to verify distributor rotation. GO to Test 4.

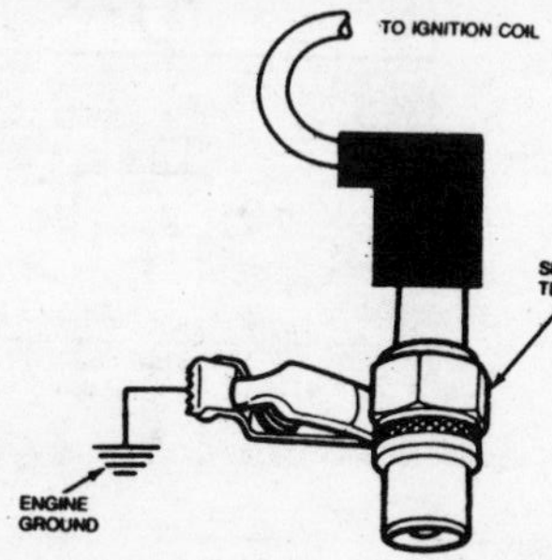

FIG. 38 TFI-IV Diagnostic Test Charts

Ignition Coil Secondary Voltage — Run Mode | TFI-IV Closed Bowl Distributor | Part 2 Test 2

TEST STEP	RESULT ▶	ACTION TO TAKE
1. Place the transmission shift lever in the PARK (A/T) or NEUTRAL (M/T) position and set the Parking Brake. CAUTION **Failure to perform this step may result in the vehicle moving when the starter is subsequently engaged during the test.**	Yes ▶	Test result OK. Problem is not in the ignition system primary circuit components.
2. Disconnect wire at S terminal of starter relay. 3. Attach remote starter switch. 4. Turn ignition switch to RUN position. 5. Crank the engine using remote starter switch. 6. Turn ignition switch to OFF position. 7. Remove the remote starter switch. 8. Reconnect wire to S terminal of starter relay. 9. **Was spark present?**	No ▶	GO to Test 3.

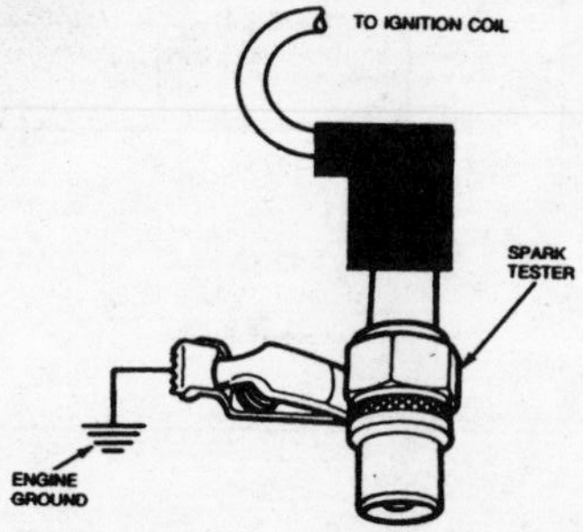

FIG. 39 TFI-IV Diagnostic Test Charts

Wiring Harness

TFI-IV Closed Bowl Distributor With CCD | **Part 2 Test 3**

TEST STEP	RESULT ▶	ACTION TO TAKE
1. Separate wiring harness connector from ignition module. Inspect for dirt, corrosion, and damage. **NOTE: Push connector tabs to separate.** 2. Disconnect the wire at S terminal of starter relay. 3. Attach negative (−) VOM lead to distributor base. 4. Measure battery voltage. 5. Following the appropriate table below, measure connector terminal voltage by attaching VOM to small straight pin inserted into connector terminal and turning ignition switch to position shown.	Yes ▶	INSPECT for faults in wiring between the coil and TFI module terminal No. 2 or any additional wiring or components connected to that circuit.
	No ▶	INSPECT for faults in wiring harness and connectors.
		Damaged or worn ignition switch.

CAUTION

Do not allow straight pin to contact electrical ground.

TFI without CCD

Connector Terminal	Wire/Circuit	Ignition Switch Test Position
#3	Run Circuit	Run and Start
#4	Start Circuit	Start

TFI with CCD

Connector Terminal	Wire/Circuit	Ignition Switch Test Position
#3	Run Circuit	Run and Start

6. Turn ignition switch to OFF position.
7. Remove straight pin.
8. Reconnect wire to S terminal of starter relay.
9. **Was the value at least 90 percent of battery voltage in each case?**

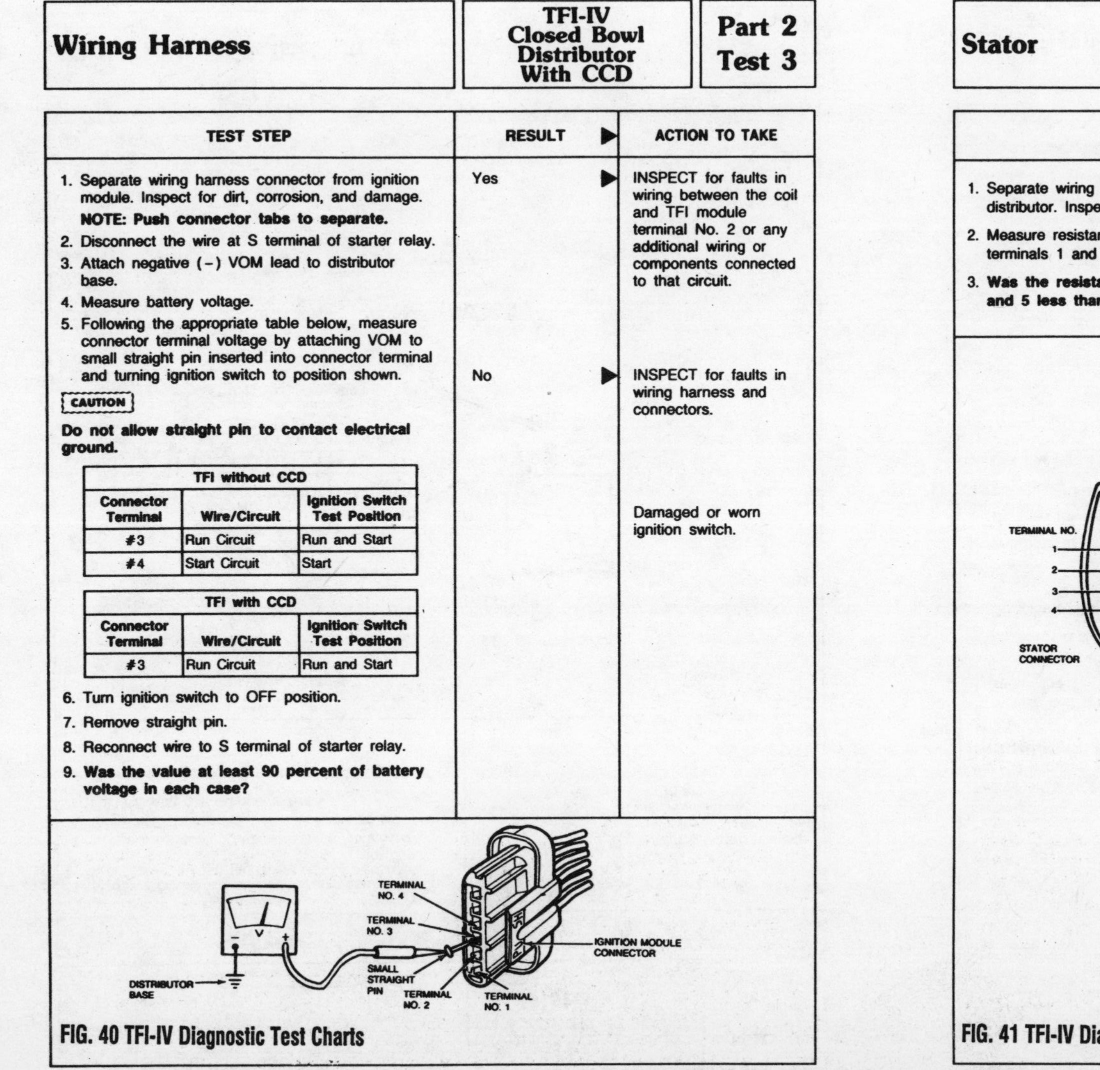

FIG. 40 TFI-IV Diagnostic Test Charts

Stator

TFI-IV Closed Bowl Distributor | **Part 2 Test 4**

TEST STEP	RESULT ▶	ACTION TO TAKE
1. Separate wiring harness connector from the distributor. Inspect for dirt, corrosion and damage. 2. Measure resistance between the stator connector terminals 1 and 5. 3. **Was the resistance between stator terminals 1 and 5 less than 5 ohms?**	Yes ▶	GO to Test 5.
	No ▶	REPLACE the stator.

TERMINAL NO.
1
2
3
4
5
6
7
8
TERMINAL NO.
STATOR CONNECTOR
OHMS

FIG. 41 TFI-IV Diagnostic Test Charts

Stator	TFI-IV Closed Bowl Distributor	Part 2 Test 5

TEST STEP	RESULT ▶	ACTION TO TAKE
1. Measure resistance between stator connector terminal 2 and the distributor base. 2. Measure resistance between stator connector terminal 6 (terminal 3 for 3.0L Aerostar or Ranger) and the distributor base. 3. **Was the resistance less than 1 ohm in each case?**	Yes ▶	GO to Test 6.
	No ▶	INSPECT the retaining screws to stator in the distributor bowl. If OK, REPLACE the stator.

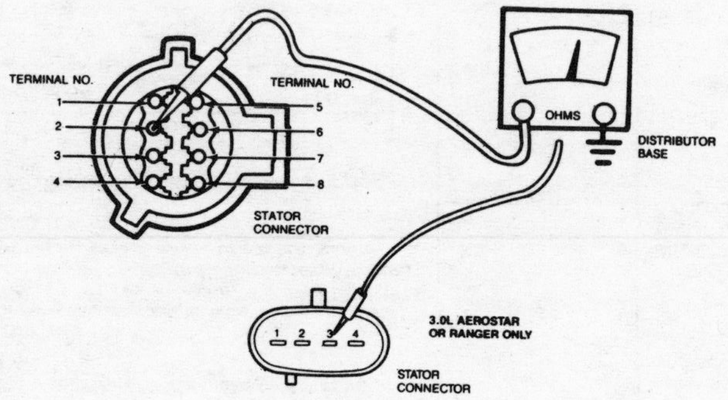

FIG. 42 TFI-IV Diagnostic Test Charts

Stator	TFI-IV Closed Bowl Distributor	Part 2 Test 6

TEST STEP	RESULT ▶	ACTION TO TAKE
1. Place the transmission shift lever in the PARK (A/T) or NEUTRAL (M/T) position and set the parking brake. CAUTION **Failure to perform this step may result in the vehicle moving when the starter is subsequently engaged during the test.** 2. Disconnect the harness connector from the TFI module and connect the TFI-IV tester to the stator and TFI module. 3. Connect the red lead from the tester to the positive (+) side of the battery. 4. Disconnect the wire at the S terminal of the starter relay, and attach remote starter switch. 5. Crank the engine using the remote starter switch and note the status of the two LED lights. 6. Remove the tester and remote starter switch. 7. Reconnect the wire to the starter relay. 8. **Did the PIP light blink?**	Yes ▶	RECONNECT harness connectors to TFI module and distributor, then GO to Test 7.
	No ▶	REPLACE the stator.

FIG. 43 TFI-IV Diagnostic Test Charts

TFI Module — TFI-IV Closed Bowl Distributor — Part 2 Test 7

TEST STEP	RESULT ▶	ACTION TO TAKE
1. Use status of Tach light from Test 6. 2. **Did the Tach light blink?**	Yes ▶	GO to Test 8.
	No ▶	REPLACE the TFI module and CHECK for spark using the method described in Test 1. If spark is not present REPLACE the coil also.

FIG. 44 TFI-IV Diagnostic Test Charts

Ignition Coil And Secondary Wire — TFI-IV Closed Bowl Distributor — Part 2 Test 8

TEST STEP	RESULT ▶	ACTION TO TAKE
1. Disconnect ignition coil connector. Inspect for dirt, corrosion and damage. 2. Connect the ignition coil connector to a known good ignition coil. 3. Connect one end of a known good secondary wire to the spark tester. Connect the other end to the known good ignition coil. CAUTION **DO NOT HOLD THE COIL while performing this test. Dangerous voltages may be present on the metal laminations as well as the high voltage tower.** 4. Crank engine. 5. Turn ignition switch to OFF position. 6. **Was spark present?**	Yes ▶	MEASURE resistance of the ignition coil wire (from vehicle). REPLACE if greater than 7,000 ohms per foot. If OK, REPLACE ignition coil.
	No ▶	RECONNECT coil connector to the vehicle coil and spark tester to vehicle secondary wire and GO to Test 9.

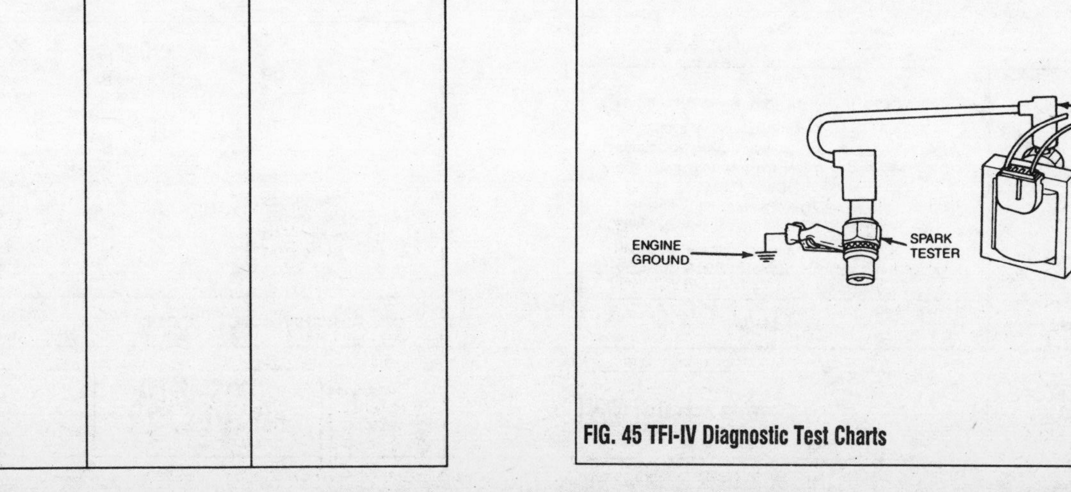

FIG. 45 TFI-IV Diagnostic Test Charts

EEC-IV — Wiring | TFI-IV Closed Bowl Distributor | Part 2 Test 9

TEST STEP	RESULT ▶	ACTION TO TAKE
1. Disconnect the in-line connector near the TFI module. 2. Crank the engine. 3. **Was spark present?**	Yes ▶	Check PIP-A and IGN GND signal wires for continuity to EEC system. SERVICE as necessary.
	No ▶	GO to Test 10.

Wiring Harness | TFI-IV Closed Bowl Distributor | Part 2 Test 10

TEST STEP	RESULT ▶	ACTION TO TAKE
1. Separate wiring harness connector from distributor and from the TFI module. 2. Measure resistance between terminal No. 5 (terminal 1 for 3.0L Aerostar or Ranger) of the harness connector which connects to the distributor and terminal No. 6 of the harness connector which connects to the TFI module by inserting a small straight pin. 3. **Is the resistance less than 5 ohms?**	Yes ▶	GO to Test 11.
	No ▶	INSPECT and SERVICE wiring between the distributor and TFI module (PIP-B circuit).

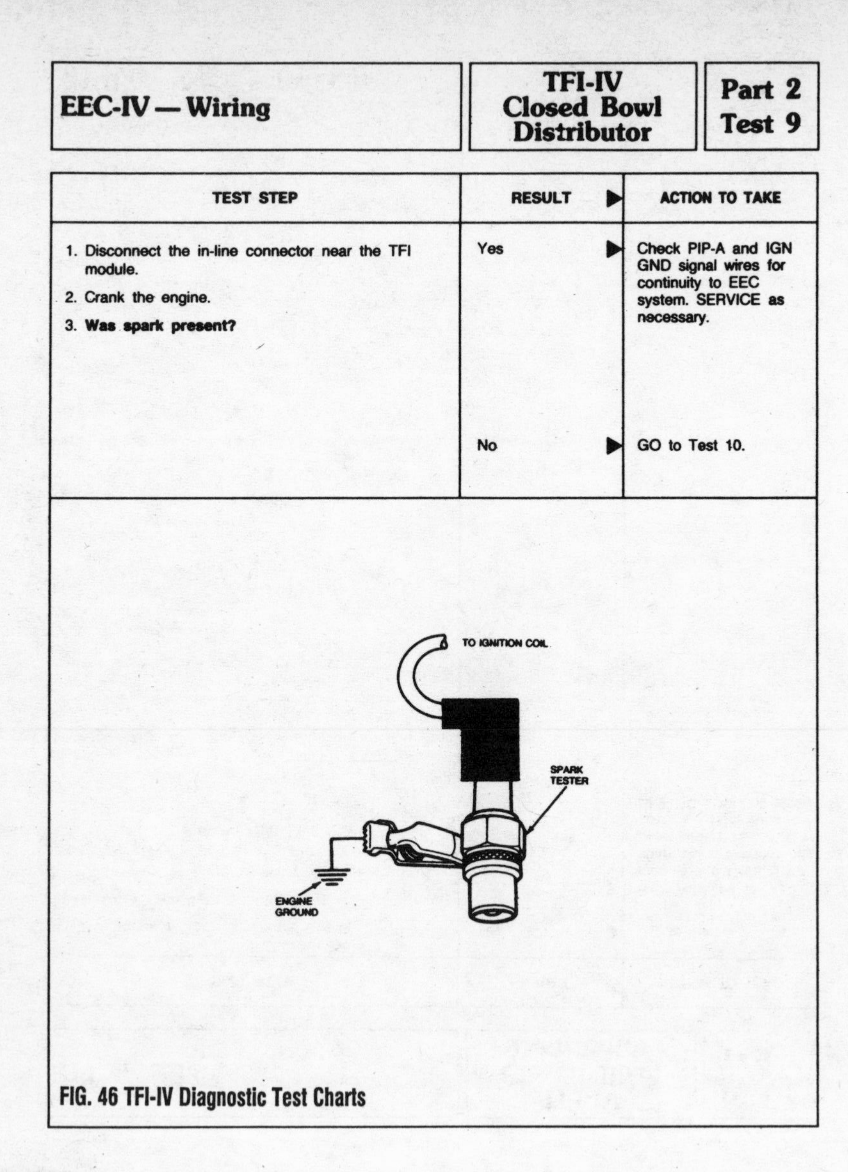

FIG. 46 TFI-IV Diagnostic Test Charts

FIG. 47 TFI-IV Diagnostic Test Charts

Ignition Coil Supply Voltage

TFI-IV Closed Bowl Distributor | **Part 2 Test 11**

TEST STEP	RESULT ▶	ACTION TO TAKE
1. Attach negative (−) VOM lead to distributor base. 2. Measure battery voltage. 3. Turn ignition switch to RUN position. 4. Measure voltage at positive (+) terminal of ignition coil. 5. Turn ignition switch to OFF position. 6. **Was the voltage at coil positive terminal at least 90 percent of battery voltage?**	Yes ▶	INSPECT ignition coil harness connector for dirt, corrosion, and damage. INSPECT ignition coil terminals for dirt, corrosion, and damage. GO to Test 12.
	No ▶	INSPECT and SERVICE wiring between ignition coil and ignition switch. Worn or damaged ignition switch.

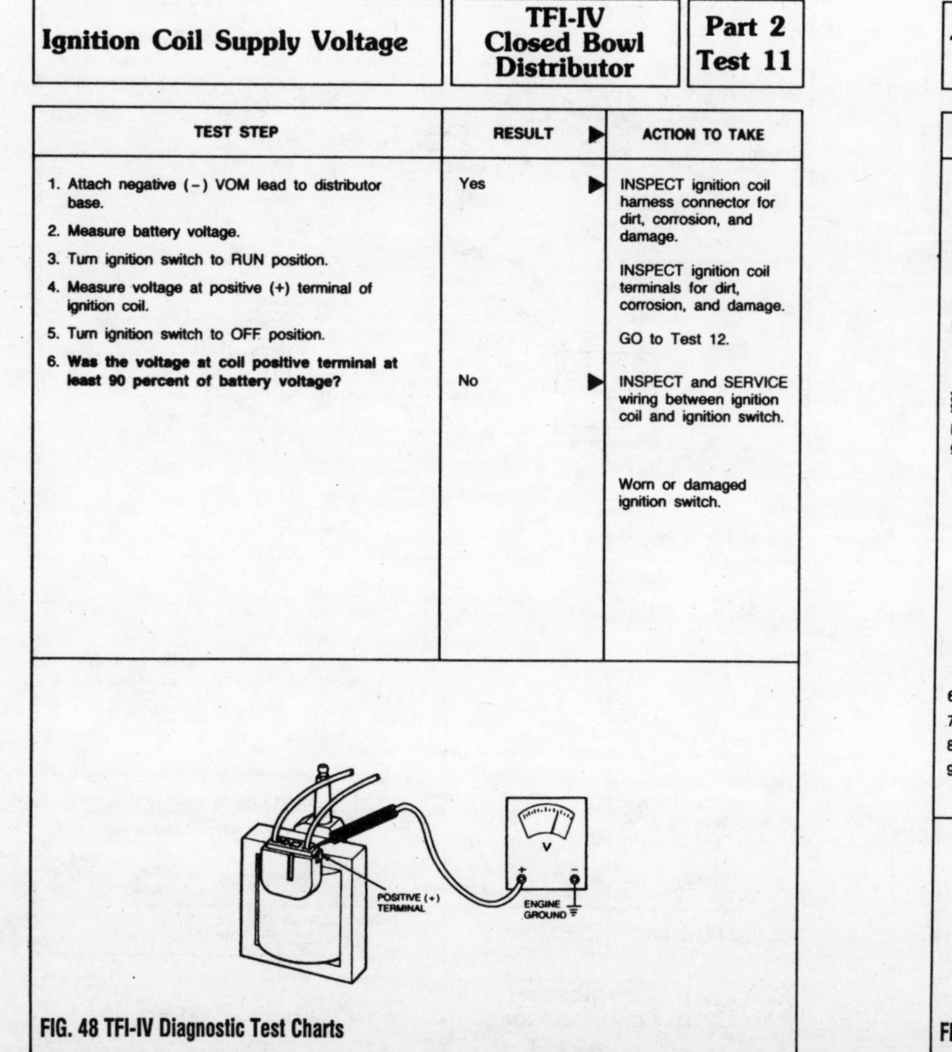

FIG. 48 TFI-IV Diagnostic Test Charts

TFI Supply Voltage

TFI-IV | **Part 2 Test 2**

TEST STEP	RESULT ▶	ACTION TO TAKE
1. Separate wiring harness connector from ignition module. Inspect for dirt, corrosion, and damage. **NOTE: Push connector tabs to separate.** 2. Disconnect the wire at S terminal of starter relay. 3. Attach negative (−) VOM lead to distributor base. 4. Measure battery voltage. 5. Following the appropriate table below, measure connector terminal voltage by attaching VOM to small straight pin inserted into connector terminal and turning ignition switch to position shown.	Yes ▶	INSPECT for faults in wiring between the coil and TFI module terminal No. 2 or any additional wiring or components connected to that circuit.
CAUTION **Do not allow straight pin to contact electrical ground.**	No ▶	INSPECT for faults in wiring harness and connectors. Damaged or worn ignition switch.

TFI without CCD

Connector Terminal	Wire/Circuit	Ignition Switch Test Position
#3	Run Circuit	Run and Start
#4	Start Circuit	Start

TFI with CCD

Connector Terminal	Wire/Circuit	Ignition Switch Test Position
#3	Run Circuit	Run and Start

6. Turn ignition switch to OFF position.
7. Remove straight pin.
8. Reconnect wire to S terminal of starter relay.
9. **Was the value at least 90 percent of battery voltage in each case?**

FIG. 49 TFI-IV Diagnostic Test Charts

Stator Supply Voltage — TFI-IV Closed Bowl Distributor — Part 2 Test 13

TEST STEP	RESULT ▶	ACTION TO TAKE
1. Attach negative (−) VOM lead to distributor base. 2. Measure battery voltage. 3. Turn ignition switch to RUN position. 4. Measure voltage at terminal 8 (terminal 4 for 3.0L Aerostar or Ranger) of the harness connector which connects to the stator. 5. Turn ignition switch to the OFF position. 6. **Was the result greater than 90 percent of battery voltage?**	Yes ▶	GO to Test 14.
	No ▶	INSPECT and SERVICE wiring between stator and ignition switch. Worn or damaged ignition switch.

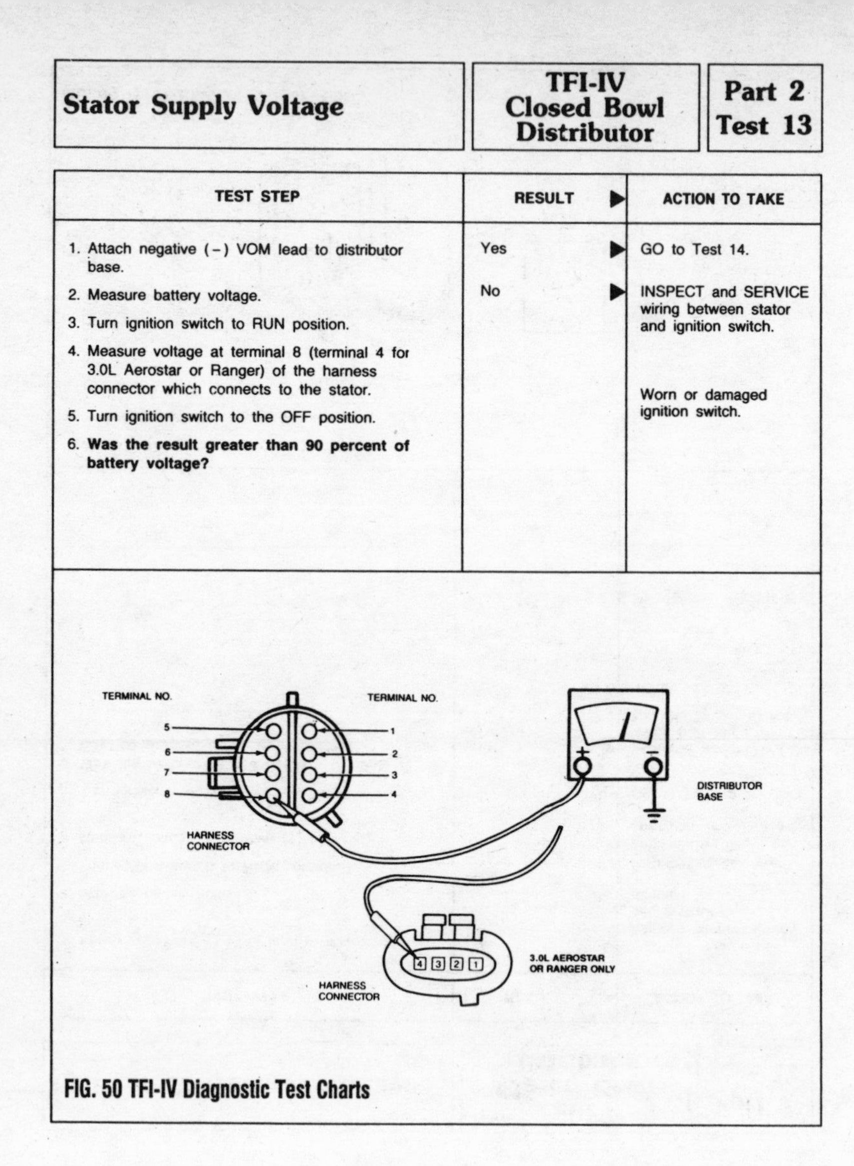

FIG. 50 TFI-IV Diagnostic Test Charts

Wiring Harness — TFI-IV Closed Bowl Distributor — Part 2 Test 14

TEST STEP	RESULT ▶	ACTION TO TAKE
1. Reconnect the wiring harness connector to the distributor. 2. Measure resistance between the distributor base and terminal No. 1 of the harness connector at the ignition module. 3. **Is the resistance less than 1 ohm?**	Yes ▶	INSPECT for faults in wiring between the coil and TFI module terminal No. 2 or any additional wiring or components connected to that circuit.
	No ▶	INSPECT and SERVICE wiring between the harness connector at the ignition module and the harness connector at the distributor (GND circuit).

FIG. 51 TFI-IV Diagnostic Test Charts

➡ Some engine may be equipped with a security type distributor hold-down bolt. If this is the case use distributor wrench T82L-12270-A or equivalent to remove the distributor retaining bolt and clamp.

6. Remove the distributor assembly from the engine. Be sure not to rotate the engine while the distributor is removed.

To install:

7. Make sure that the engine it still with the No. 1 piston up on top dead center of its compression stroke.

➡ If the engine was disturbed, while the distributor was removed, it will be necessary to remove the No. 1 spark plug and rotate the engine clockwise until No. 1 piston is on the compression stroke. Align the timing pointer with TDC on the crankshaft damper or flywheel, as required.

8. On all vehicles:
   a. Rotate the distributor shaft so the rotor points toward the mark on distributor housing, made previously.
   b. Rotate the rotor slightly so the leading edge of the vane is centered in the vane switch stator assembly.
   c. Rotate the distributor in the block to align the leading edge of the vane with the vane switch stator assembly. Make certain the rotor is pointing to the No. 1 mark on the distributor base.

➡ If the vane and vane switch stator cannot be aligned by rotating the distributor in the cylinder block, remove the distributor enough to just disengage the distributor gear from the camshaft gear. Rotate the rotor enough to engage the distributor gear on another tooth of the camshaft gear. Repeat Step 7, if necessary.

9. Install the distributor hold-down clamp and bolt(s) and tighten slightly.
10. Install the distributor cap and wires. Install the No. 1 spark plug, if removed.
11. Reconnect the negative battery cable.
12. Recheck the initial timing.
13. Tighten the hold-down clamp and recheck timing. Adjust if necessary.

Stator

➧ SEE FIGS. 52-53

REMOVAL & INSTALLATION

1. Disconnect the negative battery cable.
2. Remove the distributor assembly from the engine. Remove the rotor. Also, Remove the module from the base, if equipped.
3. Mark the armature and distributor gear for orientation during installation.
4. Hold the distributor drive gear and remove the armature retaining screws. Remove the armature.
5. Remove the distributor gear retaining pin and discard.
6. Place the distributor assembly in an arbor press. Press off the distributor gear from the shaft, using bearing removal tool D84L-950-A or equivalent.
7. Clean and polish the shaft with emery paper. Wipe clean so that the shaft slides out freely from the distributor base. Remove the shaft.
8. Remove the stator assembly retaining screws and remove the stator.

To install:

9. Position the stator assembly over the bushing and press to secure. Place the stator connector in position. The tab should fit in the notch on the base and the fastening eyelets aligned with the screw holes. Be certain the wires are positioned away from moving parts.
10. Install the stator retaining screws. Tighten to 15-35 inch lbs. (1.7-4.0 Nm).
11. Apply a light coat of engine oil to the distributor shaft, beneath the armature. Install the shaft.
12. Position a ½ in. deep well socket over the shaft and place in the arbor press.
13. Place the distributor gear on the shaft end. Make certain the mark on the armature align with the gear.

➡ The hole in the shaft and gear must be properly aligned to ensure ease of the roll pin.

14. Place a ⅝ in. deep well socket over the shaft and gear and press the gear onto the shaft. Install the roll pin.
15. Install the armature. Tighten to 25-35 inch lbs. (2.8-40Nm).
16. Rotate the distributor shaft while checking for free movement.

➡ If the armature contacts the stator, replace the entire distributor.

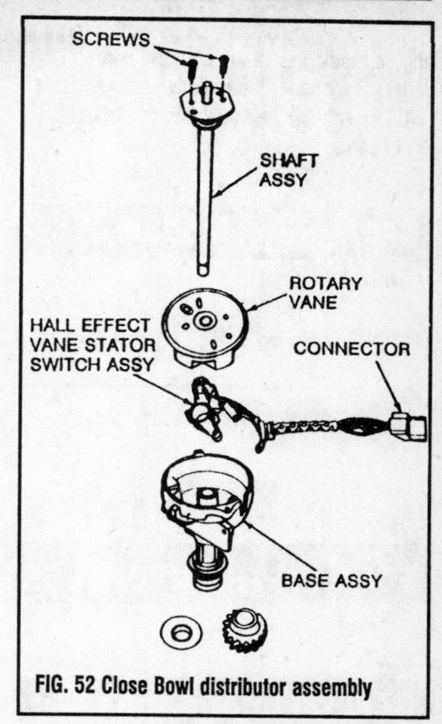

FIG. 52 Close Bowl distributor assembly

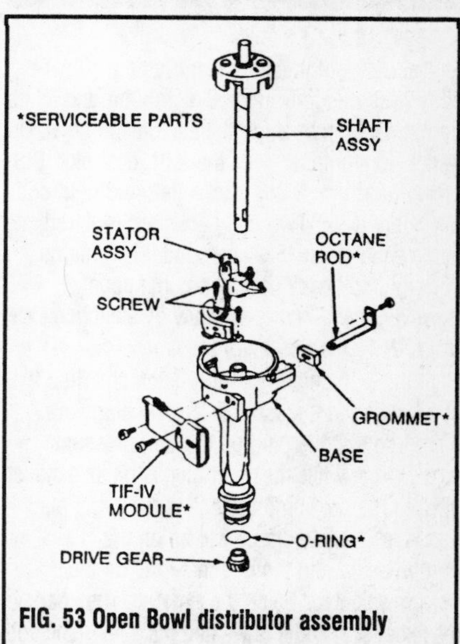

FIG. 53 Open Bowl distributor assembly

17. If equipped with TFI module, proceed as follow:
   a. Wipe the back of the module and its mounting surface in the distributor clean. Coat the base of the TFI ignition module uniformly with a $1/32$ in. (0.79mm) of silicone compound (Silicone Dieletric Compound WA-10 or equivalent).
   b. Invert the distributor base so the stator connector is in full view. Then, insert the module. Be certain the 3 module pins are inserted into the stator connector.
   c. Install the module retaining screws and tighten to 15-35 inch lbs. (1.7-4.0 Nm).

18. Install the distributor assembly into the engine block. Install the cap and wires.
19. Reconnect the negative battery cable.
20. Recheck the initial timing. Adjust if necessary.

TFI Ignition Module

REMOVAL & INSTALLATION

1. Disconnect the negative battery cable.
2. Remove the distributor assembly from the engine.
3. Place the distributor on the workbench and remove the the module retaining screws. Pull the right side of the module down the distributor mounting flange and back up to disengage the module terminals from the connector in the distributor base. The module may be pulled toward the flange and away from the distributor.

➡ **Do not attempt to lift the module from the mounting surface, except as explained in Step 3, as pins will break at the distributor module connector.**

To Install:

4. Coat the base plate of the TFI ignition module uniformly with a 1/32 in. (0.79mm) of silicone dieletric compound WA–10 or equivalent.
5. Position the module on the distributor base mounting flange. Carefully position the module toward the distributor bowl and engage the 3 connector pins securely.
6. Install the retaining screws. Tighten to 15–35 inch lbs. (1.7–4.0 Nm), starting with the upper right screw.
7. Install the distributor into the engine. Install the cap and wires.
8. Reconnect the negative battery cable.
9. Recheck the initial timing. Adjust, if necessary.

FORD DISTRIBUTORLESS IGNITION SYSTEM (DIS)

General Information

➧ SEE FIGS. 54-56

The Distributorless Ignition System (DIS) eliminates the conventional distributor and all its components, by using multiple ignition coils. On Mustangs equipped with the 2.3L dual plug DIS ignition system, 2 coil packs (left and right coil pack) are used. Two coil packs are required, since each cylinder is equipped with 2 plugs. The right coil pack operates continuously; however, the left coil pack may by switched **ON** or **OFF** by the EEC-IV processor.

In the DIS ignition system, each coil fires 2 spark plugs at the same time. The plugs are paired so that the plug on the compression stroke fires, while the other plug is on its exhaust stroke. The next time the coil is fired, the plug that was on the exhaust will be on the compression and the one that was on the compression will be on the exhaust. The spark in the exhaust cylinder is wasted but little of the coil energy is lost.

The DIS ignition system consists of the following components:

- Crankshaft timing sensor
- Camshaft sensor
- DIS ignition module
- Ignition coil pack
- Spark angle portion of the EEC-IV module
- Related wiring

The 2.3L engine uses a dual hall crankshaft sensor, containing 2 hall effect devices (PIP and CID). The Profile Ignition Pickup (PIP) cup has 2 teeth which generate 2 positive going edges each crankshaft revolution. The PIP signal provides base spark timing information. The Cylinder Identification (CID) cup has 1 tooth which generates 1 positive edge each crankshaft revolution. The CID is used by the DIS module to determine which coil should be fired. Despite the EEC-IV processor tells the DIS module when to fire the coil, its the job of the DIS module to select which of the 2 coils should be fired based on CID signal (which 2 of the 4 coils, if in the Dual Plug Inhibit (DPI) mode. The CID is also used by the EEC processor to know which bank of injectors to fire.

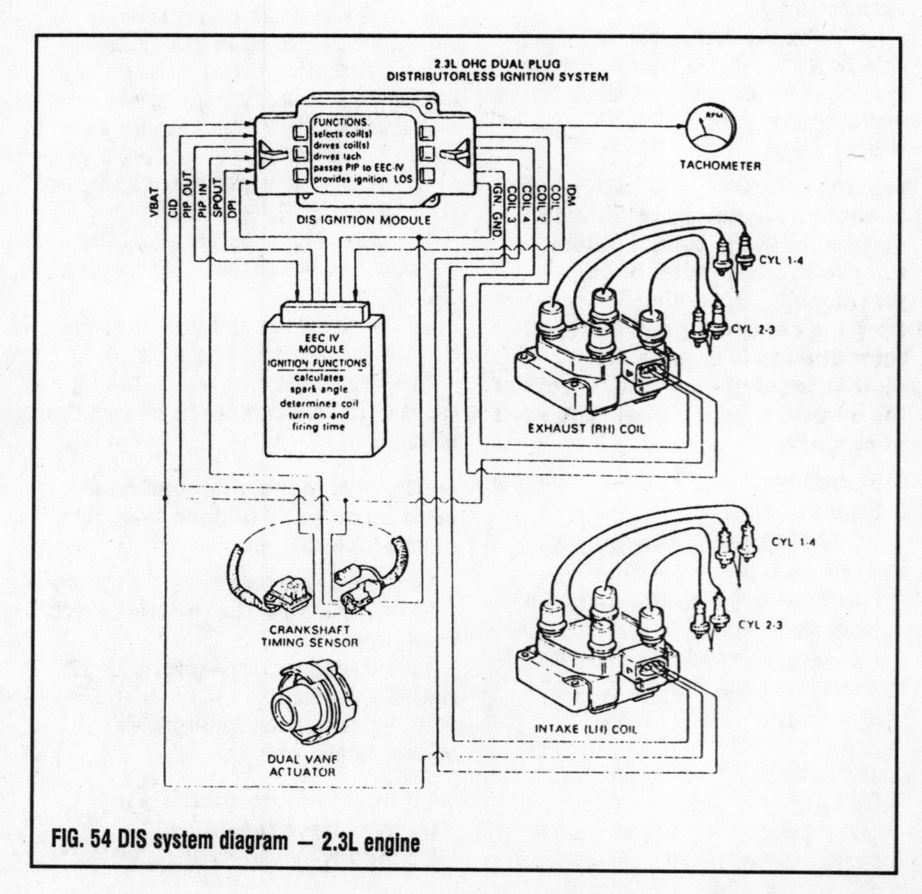

FIG. 54 DIS system diagram — 2.3L engine

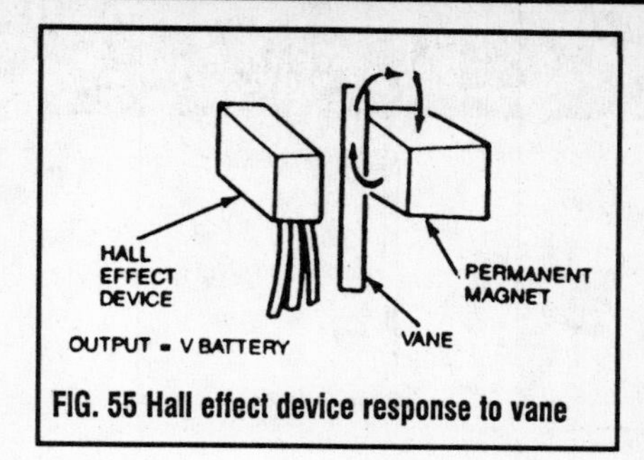

FIG. 55 Hall effect device response to vane

SYSTEM OPERATION

In the DIS ignition system, the EEC-IV processor determines the spark angle using the PIP signal to establish base timing. Spout is provided by the EEC-IV processor to the DIS module and serves 2 purposes. The leading edge fires the coil and the trailing edge controls the dwell time. This feature is referred to as Computer Control Dwell (CCD).

The DIS module incorporates an Ignition Diagnostic Monitor (IDM). This is an output signal that provides diagnostic information concerning the ignition system to the EEC-IV processor for self-test. It is also the input signal for the vehicle's tachometer.

If the CID circuit fails and an attempt to start the engine is made, the DIS module will randomly select 1 of the coils to fire. If hard starting results, turning the key **OFF** and trying to restart will result in another guess. Several attempts may be needed until the proper coil is selected, allowing the vehicle to be started and driven until repairs can be made. The Failure Mode Effects Management (FMEM) system will try to keep the vehicle driveable despite certain EEC-IV system failures that prevent the EEC-IV processor from providing spark angle or dwell commands. The EEC-IV processor opens the SPOUT line and the DIS module fires the coils directly from the PIP input. This condition will result in a fixed spark angle of 10° and a fixed dwell.

2.3L Engine

The DIS ignition module receives the PIP and the CID signals from the crankshaft timing sensor and the Spout (spark out) signal from the EEC-IV control module. During normal operation, the PIP signal is passed onto the EEC-IV control module and provides base timing and rpm information. The CID signal provides the DIS module with information required to switch between the coils for cylinders No. 1 and No. 4 and the coils for cylinders No. 2 and No. 3. DPI allows the EEC-IV processor to switch the ignition system from single to dual plug operation.

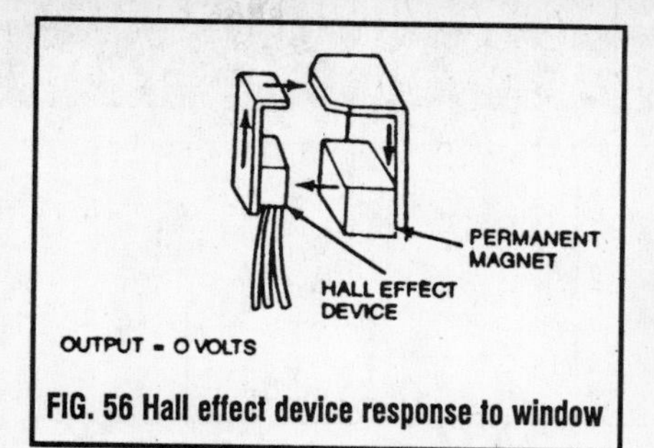

FIG. 56 Hall effect device response to window

When the window of the cup is in the air gap between the permanent magnet and hall effect device, a magnetic flux field is completed. At this time the magnetic field is allow to travel from the permanent magnet through the hall effect device and back to the magnet. This condition creates a low (0 volts) output signal. As the crankshaft rotates and the tooth on the cup move into the air gap, the magnetic field will be shunted through the tooth and back to the magnetic. The output signal, during this time, will change from a low (0 volts) to high (source volts).

SYSTEM COMPONENTS

Dual Crank Sensor

The dual hall crankshaft sensor, 2.3L engine, contains 2 hall effect devices (PIP and CID). The sensor is mounted a bracket near the crankshaft damper.

Ignition Coil Pack

➧ SEE FIG. 57

The ignition coil pack contains 2 separate coils. Each ignition coil fires 2 spark plugs simultaneously. The spark plug fired on the exhaust stroke uses very little of the ignition coil's stored energy.

DIS Ignition Module

The DIS ignition module receives the PIP signal from the crankshaft sensor. The CID signal provides the DIS ignition module with the information required to synchronize the ignition coils so that they are fired in the proper sequence.

Diagnosis and Testing

SERVICE PRECAUTIONS

- Always turn the key **OFF** and isolate both ends of a circuit whenever testing for short or continuity.
- Never measure voltage voltage or resistance directly at the processor connector.
- Always disconnect solenoids and switches from the harness before measuring for continuity, resistance or energizing by way of a 12 volts source.
- When disconnecting connectors, inspect for damaged or pushed-out pins, corrosion, loose wires, etc. Service if required.

PRELIMINARY CHECKS

1. Visually inspect the engine compartment to ensure that all vacuum lines and spark plug wires are properly routed and securely connected.
2. Examine all wiring harnesses and connectors for insulation damage, burned, overheated, loose or broken conditions.
3. Be certain that the battery is fully charged and that all accessories should be **OFF** during the diagnosis.

IGNITION TIMING

Ignition timing adjustment is not possible on the DIS system.

FIG. 57 DIS coil pack assemblies — 2.3L engine

DIS DIAGNOSTIC TEST CHARTS

2.3L Mustang (Dual Plug) DIS Pinpoint Test Diagnostics | Pinpoint Test | Index

TEST STEP	RESULT ▶	ACTION TO TAKE
These diagnostics (Pinpoint Tests A through G) are written to catch "Hard Faults"; intermittent failures will be difficult or impossible to diagnose using these procedures.		
• Engine No Start and No Ignition Service Codes.	▶	GO to A1.
• Engine No Start and 211 — PIP circuit fault.	▶	GO to A3.
• 222 and Hard Start (Engine will start normally at least once out of 5 attempts). When it fails to start, the cranking rpm is erratic (Ignition is firing out of time) — CID circuit fault.	▶	GO to B1.
• 222 and Normal Start. — IDM low circuit fault or right coil fault.	▶	GO to B12.
• 218 — IDM circuit fault high or open or left coil pack open.	▶	GO to C1.
• 224 — C1, C2, C3 or C4 circuit fault.	▶	GO to D1.
• 223 — DPI circuit fault open or high. SPOUT circuit fault high.	▶	GO to E1.
• 213 — SPOUT circuit fault open or low.	▶	GO to F1.
• No Ignition Service Code and Misfire Under Load — secondary (high voltage) short to ground.	▶	GO to G1.
• No Ignition Service Codes and intermittent miss or stall.	▶	GO to F1.

FIG. 58 DIS Diagnostic Test Charts

2.3L Mustang (Dual Plug) DIS No Start | Pinpoint Test | A

TEST STEP	RESULT ▶	ACTION TO TAKE
A1 PERFORM EEC-IV QUICK TEST		
• perform EEC-IV Quick Test. • **Are any service codes (KOEO, KOER or Continuous Memory present?**	Yes ▶	SERVICE any EEC codes first. If still no start, GO to A2.
	No ▶	GO to A2.
A2 CHECK FOR SPARK DURING CRANK		
• Using a Neon Bulb Spark Tester (D89P6666-A) check for spark at each right side spark plug wire while cranking. • **Is spark consistent on all right side plug wires (one spark per crank revolution)?**	Yes ▶	GO to A3.
	No ▶	GO to A8.
A3 CHECK PIPD — KOEC		
• Connect the DIS diagnostic cable to the EEC breakout box. • Use the 2.3L DP DIS overlay. • Install DIS diagnostic cable TEE to Pins 1 to 6 side of DIS module. • Set DVOM to the 20 volt AC range. • Connect diagnostic cable negative lead to battery. • Crank engine. • **Is the voltage greater than 1.0 volt?**	Yes ▶	GO to A4.
	No ▶	REPLACE the DIS module. No PIP output. REMOVE all test equipment. RECONNECT all components. CLEAR Continuous Memory. RERUN Quick Test.
A4 CHECK PIP TO EEC CONTINUITY — KEY OFF		
• Key off. • Disconnect EEC processor. • Set DVOM to 200 ohm range. • Connect the second EEC breakout box to the EEC vehicle harness connector. • Measure resistance between J31 (PIP EEC) and J56 of the second breakout box (PIP at EEC). • **Is the resistance less than 5 ohms?**	Yes ▶	GO to A5.
	No ▶	SERVICE harness and connectors. PIP is open. REMOVE all test equipment. RECONNECT all components. CLEAR Continuous Memory. RERUN Quick Test.

FIG. 59 DIS Diagnostic Test Charts

2.3L Mustang (Dual Plug) DIS No Start | **Pinpoint Test** | **A**

TEST STEP	RESULT ▶	ACTION TO TAKE
A5 CHECK IGND AT DIS MODULE		
• Key off. • Set DVOM to 200 ohm range. • Measure resistance between J2 (IGNDD) and J60 (BAT−). • **Is the resistance less than 5 ohms?**	Yes ▶ No ▶	GO to A7. GO to A6.
A6 IGND FAULT—CHECK DIS MOUNTING SCREWS—KEY OFF		
• Key off. • **Are the mounting screws clean and tight?**	Yes ▶	REPLACE DIS module. Ignition ground open.
	No ▶	CLEAN, TIGHTEN or REPLACE mounting screws. CLEAN the mounting area of the DIS module. Be sure to replace any heat sink grease removed. REMOVE all test equipment. RECONNECT all components. CLEAR Continuous Memory. RERUN Quick Test.
A7 CHECK IGND CONTINUITY TO EEC		
• Key off. • Set DVOM to 200 ohm range. • Remove the EEC processor and install the second EEC breakout box at the EEC vehicle harness connector. • Measure resistance between J2 (IGNDD) and J16 (IGNDE) at the second breakout box. • **Is the resistance less than 5 ohms?**	Yes ▶	**Go to fuel system diagnosis.** Ignition system checks OK. REMOVE test equipment.
	No ▶	SERVICE harness and connectors. IGND is open.

FIG. 60 DIS Diagnostic Test Charts

2.3L Mustang (Dual Plug) DIS No Start | **Pinpoint Test** | **A**

TEST STEP	RESULT ▶	ACTION TO TAKE
A8 SPARK FAULT—DETERMINE MISSING SPARK COMBINATION		
• **Is spark missing from both 1 and 4 plug wires or was spark missing from both 2 and 3 plug wires?**	Yes ▶ No ▶	GO to A10. GO to A9.
A9 SPARK FAULT—CHECK PLUGS AND WIRES—KEY OFF		
• Check plug wires for insulation damage, looseness, shorting or other damage. • Remove and check spark plugs for damage, wear, carborn deposits and proper plug gap. • **Are the plugs and wires OK?**	Yes ▶	REINSTALL plugs wires then GO to A10.
	No ▶	SERVICE or REPLACE defective spark plugs or wires.
A10 SPARK FAULT—CHECK VBAT AT DIS KOEO		
• Connect DIS diagnostic cable to the EEC breakout box. • Select the 2.3L DP DIS overlay. • INSTALL the 1-6 DIS module tee only. • Connect the diagnostic cable negative lead to battery. • Set DVOM to 20 volt DC range. • Key on. • Measure voltage between +J5 (VBATD) and −J60 (BAT−). • **Is the voltage greater than 10 volts?**	Yes ▶	GO to A11.
	No ▶	SERVICE harness and connectors. REMOVE all test equipment. RECONNECT all components. CLEAR Continuous Memory. RERUN Quick Test.
A11 SPARK FAULT—CHECK IGND AT DIS MODULE—KEY OFF		
• Key off. • Set DVOM to 200 ohm range. • Measure resistance between J2 (IGNDD) and J60 (BAT−). • **Is the resistance less than 5 ohms?**	Yes ▶ No ▶	GO to A13. GO to A12.

FIG. 61 DIS Diagnostic Test Charts

2.3L Mustang (Dual Plug) DIS No Start — Pinpoint Test A

TEST STEP	RESULT ▶	ACTION TO TAKE
A12 SPARK FAULT — CHECK DIS MOUNTING SCREWS — KEY OFF		
• Check DIS module mounting screws for corrosion or looseness. • **Are the mounting screws clean and tight?**	Yes ▶	REPLACE DIS module. REMOVE all test equipment. RECONNECT all components. CLEAR Continuous Memory. RERUN Quick Test.
	No ▶	TIGHTEN, CLEAN or REPLACE mounting screws. CLEAN mounting area on DIS module. REPLACE heat sink grease if removed. REMOVE all test equipment. RECONNECT all components. CLEAR Continuous Memory. RERUN Quick Test.
A13 SPARK FAULT — CHECK C1D/C2D — KOEC		
• Key off. • Install right coil tee. • Set DVOM to 20 volt DC range. • Connect the DVOM between the two pairs of test jacks listed below. • Measure between +J18 (RC1D) and −J60 (BAT−). • Measure between +J10 (RC2D) and −J60 (BAT−). • **Is the voltage greater than 10 volts in both tests?**	Yes ▶	GO to A17.
	No ▶	GO to A14.

FIG. 62 DIS Diagnostic Test Charts

2.3L Mustang (Dual Plug) DIS No Start — Pinpoint Test A

TEST STEP	RESULT ▶	ACTION TO TAKE
A14 C1/C2 LOW FAULT — ISOLATE DIS MODULE — KOEO		
• Key off. • Disconnect the DIS module tee 7-12 from the DIS module. DO NOT disconnect the vehicle harness from the other side of the 7-12 module tee. • Repeat Step A13. • **Is the voltage greater than 10 volts in both tests?**	Yes ▶	REPLACE DIS module. C1 or C2 short to ground. Coil pack may be damaged. REMOVE all test equipment. RECONNECT all components. CLEAR Continuous Memory. RERUN Quick Test.
	No ▶	GO to A15.
A15 C1/C2 LOW FAULT — CHECK VBAT AT RIGHT COIL — KOEO		
• SET DVOM to 20 volt DC range. • Key on. • Measure voltage between +J26 (VBATR) and −J60 (BAT−). • **Is the voltage greater than 10 volts?**	Yes ▶	GO to A16.
	No ▶	SERVICE harness and connectors. VBAT is open. REMOVE all test equipment. RECONNECT all components. CLEAR Continuous Memory. RERUN Quick Test.

FIG. 63 DIS Diagnostic Test Charts

2.3L Mustang (Dual Plug) DIS No Start — Pinpoint Test A

TEST STEP	RESULT ▶	ACTION TO TAKE
A16 C1/C2 LOW FAULT — CHECK C1C/C2C — KOEO		
• SET DVOM to 20 volt DC range. • Connect the DVOM between the two pairs of test jacks listed below. • Key on. • Measure between +J23 (RC1C) and −J60 (BAT−). • Measure between +J24 (RC2C) and −J60 (BAT−). • **Is the voltage greater than 10 volts in both tests?**	Yes ▶	SERVICE harness and connectors. C1 and/or C2 is shorted high. REMOVE all test equipment. RECONNECT all components. CLEAR Continuous Memory. RERUN Quick Test.
	No ▶	REPLACE right coil pack. C1 and/or C2 open. REMOVE all test equipment. RECONNECT all components. CLEAR Continuous Memory. RERUN Quick Test.

FIG. 64 DIS Diagnostic Test Charts

2.3L Mustang (Dual Plug) DIS No Start — Pinpoint Test A

TEST STEP	RESULT ▶	ACTION TO TAKE
A17 SPARK FAULT — CHECK PIPD — KOEC		
• Set DVOM to 20 volt AC range. • Measure the voltage between J32 (PIPD) and J60 (BAT−) while cranking engine. • **Is the voltage greater than 1.0 volt?**	Yes ▶	GO to A18.
	No ▶	GO to A22, PIP Fault.
A18 SPARK FAULT — CHECK C1D/C2D — KOEC		
• Set DVOM to 20 volt AC range. • Read the voltage between the two pair of test jacks listed below while cranking the engine. • Measure between J18 (RC1D) and J60 (BAT−). • Measure between J10 (RC2D) and J60 (BAT−). • **Were both voltage readings greater than 1.0 volt?**	Yes ▶	GO to A19.
	No ▶	REPLACE DIS module. No C1/C2 output. REMOVE all test equipment. RECONNECT all components. CLEAR Continuous Memory. RERUN Quick Test.
A19 SPARK FAULT — CHECK VBATR — KOEO		
• Set DVOM to 20 volt DC range. • Key on. • Measure voltage between +J26 (VBATR) and −J60 (BAT−). • **Is the voltage greater than 10 volts?**	Yes ▶	GO to A20.
	No ▶	SERVICE harness and connectors. VBATR is open. REMOVE all test equipment. RECONNECT all components. CLEAR Continuous Memory. RERUN Quick Test.

FIG. 65 DIS Diagnostic Test Charts

2.3L Mustang (Dual Plug) DIS No Start | **Pinpoint Test** | **A**

TEST STEP	RESULT ▶	ACTION TO TAKE
A20 SPARK FAULT — CHECK C1C/C2C — KOEO		
• Set DVOM to 20 volt DC range. • Key on. • Read the voltage between the two pair of test jacks listed below. • Measure between +J23 (RC1C) and −J60 (BAT−). • Measure between +J24 (LC2C) and −J60 (BAT−). • **Were both readings greater than 10 volts?**	Yes ▶	GO to A21.
	No ▶	SERVICE harness and connectors and REPLACE coil pack. C1/C2 open. REMOVE all test equipment. RECONNECT all components. CLEAR Continuous Memory. RERUN Quick Test.
A21 SPARK FAULT — CHECK C1C/C2C — KOEC		
• Set DVOM to 20 volt AC range. • Read the voltage between the two pair of test jacks listed below while cranking the engine. • Measure between J18 (RC1D) and J60 (BAT−). • Measure between J10 (RC2C) and J60 (BAT−). • **Is the voltage greater than 1.0 volt both times?**	Yes ▶	REPLACE coil pack. Input OK, but no high voltage output. REMOVE all test equipment. RECONNECT all components. CLEAR Continuous Memory. RERUN Quick Test.
	No ▶	SERVICE harness and connectors. C1/ C2 open. REMOVE all test equipment. RECONNECT all components. CLEAR Continuous Memory. RERUN Quick Test.

FIG. 66 DIS Diagnostic Test Charts

2.3L Mustang (Dual Plug) DIS No Start | **Pinpoint Test** | **A**

TEST STEP	RESULT ▶	ACTION TO TAKE
A22 PIPD FAULT — ISOLATE DIS MODULE CHECK PIPD — KOEC		
• Key off. • Disconnect the DIS tee 1-6 from the DIS module, do not disconnect the vehicle harness from the other end of the tee. • Set DVOM to 20 volt AC range. • Measure the voltage between J32 (PIPD) and J60 (BAT−) while cranking the engine. • **Is the voltage reading greater than 1.0 volt?**	Yes ▶	GO to A29.
	No ▶	GO to A23.

FIG. 67 DIS Diagnostic Test Charts

2.3L Mustang (Dual Plug) DIS No Start | **Pinpoint Test** | **A**

TEST STEP	RESULT ▶	ACTION TO TAKE
A23 PIP FAULT — CHECK VBATS — KOEO		
• Set DVOM to 20 volt DC range. • Key on. • Measure the voltage between +J56 (VBATS) and −J60 (BAT−). • **Is the voltage greater than 1.0 volt?**	Yes ▶	GO to A24.
	No ▶	SERVICE harness and connectors. VBATS is open. REMOVE all test equipment. RECONNECT all components. CLEAR Continuous Memory. RERUN Quick Test.
A24 PIP FAULT — CHECK IGNDS — KOEO		
• Key off. • Set DVOM to 200 ohm range. • Measure resistance between J55 (IGNDS) and J60 (BAT−). • **Is the resistance less than 5 ohms?**	Yes ▶	GO to A25.
	No ▶	SERVICE harness and connectors. IGND OPEN. REMOVE all test equipment. RECONNECT all components. CLEAR Continuous Memory. RERUN Quick Test.
A25 PIP FAULT — CHECK PIPS — KOEC		
• Key off. • Install the crankshaft sensor tee. • Set DVOM to 20 volt AC range. • Measure the voltage between J33 (PIPS) and J60 (BAT−) while cranking the engine. • **Is the voltage greater than 1.0 volt?**	Yes ▶	SERVICE harness and connectors. PIP is OPEN. REMOVE all test equipment. RECONNECT all components. CLEAR Continuous Memory. RERUN Quick Test.
	No ▶	GO to A26.

FIG. 68 DIS Diagnostic Test Charts

2.3L Mustang (Dual Plug) DIS No Start | **Pinpoint Test** | **A**

TEST STEP	RESULT ▶	ACTION TO TAKE
A26 PIP FAULT — CHECK FOR SHORT LOW		
• Key off. • Disconnect the EEC processor. • Disconnect the HEGO sensor. • Disconnect the DIS module tee 1-6 from the DIS module, do not disconnect the vehicle harness from the other side of the tee. • Set DVOM to 2000 ohm range. • Measure resistance between J33 (PIPS) and J60 (BAT−). • **Is the resistance less than 1000 ohms?**	Yes ▶	SERVICE harness and connectors. PIP short to ground. REMOVE all test equipment. RECONNECT all components. CLEAR Continuous Memory. RERUN Quick Test.
	No ▶	GO to A27.
A27 PIP FAULT — CHECK FOR SHORT HIGH — KOEO		
• Set DVOM to 20 volt DC range. • Key on. • Measure voltage between +J33 (PIPS) and −J60 (BAT−). • **Is the voltage greater than 0.5 volt?**	Yes ▶	SERVICE harness and connectors. PIP short high. REMOVE all test equipment. RECONNECT all components. CLEAR Continuous Memory. RERUN Quick Test.
	No ▶	GO to A28.
A28 PIP FAULT — CHECK VANE — KEY OFF		
• **Does the crankshaft vane move through the sensor air gap when the engine is cranked?**	Yes ▶	REPLACE crank — shaft sensor. No output. REMOVE all test equipment. RECONNECT all components. CLEAR Continuous Memory. RERUN Quick Test.
	No ▶	Vane damaged.

FIG. 69 DIS Diagnostic Test Charts

2.3L Mustang (Dual Plug) DIS No Start | **Pinpoint Test** | **A**

TEST STEP	RESULT ▶	ACTION TO TAKE
A29 PIP FAULT — ISOLATE EEC — CHECK PIPD — KOEC		
• Key off. • Reconnect the DIS tee 1-6 to the DIS module. • Disconnect the ECC processor. • Set DVOM to 20 volt AC range. • Measure the voltage between J32 (PIPD) and J60 (BAT –) while cranking the engine. • **Is the voltage greater than 1.0 volt?**	Yes ▶	REPLACE EEC processor. PIP short. REMOVE all test equipment. RECONNECT all components. CLEAR Continuous Memory. RERUN Quick Test.
	No ▶	GO to A30 .
A30 PIP FAULT — ISOLATE DIS MODULE — CHECK FOR PIPD HIGH — KOEO		
• Key off. • Disconnect the DIS module tee 1-6 from the DIS module, do not disconnect the vehicle harness from the other side of the tee. • Set DVOM to 20 volt DC range. • Key on. • Measure the voltage between +J31 (PIPE) and – J60 (BAT –). • **Is the voltage greater than 0.5 volt?**	Yes ▶	SERVICE harness and connectors. PIP short high. REMOVE all test equipment. RECONNECT all components. CLEAR Continuous Memory. RERUN Quick Test.
	No ▶	GO to A31 .

FIG. 70 DIS Diagnostic Test Charts

2.3L Mustang (Dual Plug) DIS No Start | **Pinpoint Test** | **A**

TEST STEP	RESULT ▶	ACTION TO TAKE
A31 PIP FAULT — CHECK FOR SHORT LOW — KEY OFF		
• Key off. • Set DVOM to 2000 ohm range. • Measure resistance between J31 (PIPE) and J60 (BAT –). • **Is the resistance less than 1000 ohms?**	Yes ▶	SERVICE harness and connectors. PIP short to ground. REMOVE all test equipment. RECONNECT all components. CLEAR Continuous Memory. RERUN Quick Test.
	No ▶	REPLACE DIS module. No PIP output. REMOVE all test equipment. RECONNECT all components. CLEAR Continuous Memory. RERUN Quick Test.

FIG. 71 DIS Diagnostic Test Charts

2.3L Mustang (Dual Plug) DIS—Code 222 CID Failure, IDM Low, DPI High or Right Coil Pack Failure | **Pinpoint Test** | **B**

TEST STEP	RESULT ▶	ACTION TO TAKE
B1 CHECK FOR SPARK—KOER		
• Using a Neon Spark Tester (D89P6666-A), check for spark at each of the right side spark plug wires while cranking the engine. • **Is spark consistent on one or more plug wires (one spark per crank revolution).**	Yes ▶ No ▶	GO to B2. GO to D1.
B2 VERIFY HARD CRANK—KOEC		
• Attempt to start vehicle 5 times. **NOTE: The engine will start normally at least once out of 5 attempts. When it fails to start the cranking rpm is erratic (ignition is firing out of time.—CID circuit fault).** • **Does the engine fail to start at least once?**	Yes ▶ No ▶	GO to B3. GO to B12.
B3 CID FAULT—CHECK CIDD—KOER		
• Key off. • Install DIS diagnostic cable tee to Pins 1 to 6 side of DIS module. • Connect diagnostic cable negative lead to battery. • Connect diagnostic cable to the EEC breakout box. • Use 2.3L DP DIS overlay. • Set DVOM to 20 volt AC range. • Start the engine. • Measure the voltage between J51 (CIDD) and J60 (BAT−). • **Is the settled voltage greater than 5.0 volts?**	Yes ▶ No ▶	REPLACE DIS module. Does not respond to CID input. REMOVE all test equipment. RECONNECT all components. CLEAR Continuous Memory. RERUN Quick Test. GO to B4.

FIG. 72 DIS Diagnostic Test Charts

2.3L Mustang (Dual Plug) DIS—Code 222 CID Failure, IDM Low, DPI High or Right Coil Pack Failure | **Pinpoint Test** | **B**

TEST STEP	RESULT ▶	ACTION TO TAKE
B4 CID FAULT—CHECK CIDS—KOER		
• Key off. • Connect the crankshaft sensor tee. • Set DVOM to 20 volt AC range. • Start the engine. • Measure the voltage between J35 (CIDS) and J60 (BAT−). • **Is the settled voltage greater than 5.0 volts?**	Yes ▶ No ▶	CHECK connectors, SERVICE or REPLACE the harness. CID circuit is open. REMOVE all test equipment. RECONNECT all components. CLEAR Continuous Memory. RERUN Quick Test. GO to B5.
B5 CID FAULT—CHECK VBATS—KOEO		
• Set DVOM to 20 volt DC range. • Key on. • Measure the voltage between +J56 (VBATS) and −J60 (BAT−). • **Is the voltage greater than 10 volts?**	Yes ▶ No ▶	GO to B6. CHECK connectors, SERVICE or REPLACE harness. VBATS is open. REMOVE all test equipment. RECONNECT all components. CLEAR Continuous Memory. RERUN Quick Test.
B6 CID FAULT—CHECK IGNDS—KEY OFF		
• Key off. • Set DVOM to 200 ohm range. • Measure the resistance between J55 (IGNDS) and J60 (BAT−). • **Is the resistance less than 5 ohms?**	Yes ▶ No ▶	GO to B7. CHECK connectors, SERVICE or REPLACE the harness. IGNDS is open. REMOVE all test equipment. RECONNECT all components. CLEAR Continuous Memory. RERUN Quick Test.

FIG. 73 DIS Diagnostic Test Charts

2.3L Mustang (Dual Plug) DIS—Code 222 CID Failure, IDM Low, DPI High or Right Coil Pack Failure | **Pinpoint Test** | **B**

TEST STEP	RESULT ▶	ACTION TO TAKE
B7 CID FAULT—ISOLATE DIS MODULE CHECK CIDS—KOEC		
• Key off. • Disconnect the DIS module tee 1-6 from the DIS module. Do not disconnect the vehicle harness from the other side of the tee. • Set DVOM to 20 volt range. • Measure voltage between J35 (CIDS) and J60 (BAT−) while cranking the engine. • **Is the voltage greater than 3.0 volts?**	Yes ▶	REPLACE DIS module. CID shorted low. REMOVE all test equipment. RECONNECT all components. CLEAR Continuous Memory. RERUN Quick Test.
	No ▶	GO to B8 .
B8 CID FAULT—ISOLATE EEC PROCESSOR—CHECK CIDS—KOEC		
• Key off. • Disconnect the EEC processor from the vehicle harness. • Measure voltage between J35 (CIDS) and J60 (BAT−) while cranking the engine. • **Is the settled voltage greater than 3.0 volts?**	Yes ▶	REPLACE EEC processor. CID shorted. REMOVE all test equipment. RECONNECT all components. CLEAR Continuous Memory. RERUN Quick Test.
	No ▶	GO to B9 .
B9 CID FAULT—ISOLATE SENSOR—CHECK CIDD FOR SHORT LOW—KEY OFF		
• Key off. • Disconnect the diagnostic cable CID sensor tee from the sensor. Do not disconnect the vehicle harness from the other side of the tee. • Set DVOM to 200 ohm range. • Measure resistance between J51 (CIDD) and J60 (BAT−). • **Is the resistance less than 5 ohms?**	Yes ▶	SERVICE connectors and harness. CID shorted low. REMOVE test equipment. RECONNECT all components. CLEAR Continuous Memory. RERUN Quick Test.
	No ▶	GO to B10 .

FIG. 74 DIS Diagnostic Test Charts

2.3L Mustang (Dual Plug) DIS—Code 222 CID Failure, IDM Low, DPI High or Right Coil Pack Failure | **Pinpoint Test** | **B**

TEST STEP	RESULT ▶	ACTION TO TAKE
B10 CID FAULT—CHECK FOR CID SHORT HIGH—KOEO		
• Set DVOM to 20 volt DC range. • Measure voltage between +J51 (CIDD) and −J60 (BAT−). • **Is the voltage greater than 0.5 volt?**	Yes ▶	CHECK connectors, SERVICE or REPLACE harness. CID shorted high. REMOVE test equipment. RECONNECT all components. CLEAR Continuous Memory. RERUN Quick Test.
	No ▶	GO to B11 .
B11 CID FAULT—CHECK VANE—KOEC		
• **Does the CID vane move through the sensor air gap when the engine is cranked or running?**	Yes ▶	REPLACE crank sensor. No CID output. REMOVE test equipment. RECONNECT all components. CLEAR Continuous Memory. RERUN Quick Test.
	No ▶	Service as required.
B12 CHECK DPID—KOEO		
• Key off. • Install DIS module tee to Pins 1−6 side of DIS module. • Connect DIS diagnostic cable to EEC breakout box. • Use 2.3L DP DIS overlay. • Connect the diagnostic cable negative lead to battery. • Set DVOM to 20 volt DC range. • Key on. • Measure voltage between +J54 (DPID) and −J60 (BAT−). • **Is the voltage greater than 10 volts?**	Yes ▶	GO to B15 .
	No ▶	GO to B13 .

FIG. 75 DIS Diagnostic Test Charts

2.3L Mustang (Dual Plug) DIS — Code 222 CID Failure, IDM Low, DPI High or Right Coil Pack Failure | **Pinpoint Test** | **B**

TEST STEP	RESULT ▶	ACTION TO TAKE
B13 DPI HIGH FAULT — ISOLATE DIS MODULE — CHECK DPID — KOEO		
• Key off. • Disconnect the DIS diagnostic tee to Pins 1-6 side of the DIS module. Do not disconnect the vehicle harness from the other side of the tee. • Key on. • Measure voltage between +J54 (DPID) and −J60 (BAT−). • **Is the voltage greater than 10 volts?**	Yes ▶	GO to B14.
	No ▶	REPLACE DIS module. DPI short high. REMOVE test equipment. RECONNECT all components. CLEAR Continuous Memory. RERUN Quick Test.
B14 DPI FAULT HIGH — ISOLATE EEC PROCESSOR — CHECK DPID — KOEO		
• Key off. • Disconnect EEC processor. • Set DVOM to 20 volt DC range. • Key on. • Measure voltage between +J54 (DPID) and −J60 (BAT−). • **Is the voltage greater than 10 volts?**	Yes ▶	CHECK connectors, SERVICE or REPLACE harness. DPI shorted high. REMOVE test equipment. RECONNECT all components. CLEAR Continuous Memory. RERUN Quick Test.
	No ▶	REPLACE EEC processor. DPI shorted high. REMOVE test equipment. RECONNECT all components. CLEAR Continuous Memory. RERUN Quick Test.
B15 CHECK IDMD — KOER		
• Set DVOM to 20 volt AC range. • Start engine. • Measure voltage between J4 (IDMD) and J60 (BAT−). • **Is the voltage greater than 2.0 volts?**	Yes ▶	GO to B19.
	No ▶	GO to B16.

FIG. 76 DIS Diagnostic Test Charts

2.3L Mustang (Dual Plug) DIS — Code 222 CID Failure, IDM Low, DPI High or Right Coil Pack Failure | **Pinpoint Test** | **B**

TEST STEP	RESULT ▶	ACTION TO TAKE
B16 IDM FAULT — ISOLATE EEC PROCESSOR CHECK IDMD — KOEC		
• Key off. • Disconnect EEC processor. • Set DVOM to 20 volt range. • Measure voltage between J4 (IDMD) and J60 (BAT−) while cranking the engine. • **Is the settled voltage greater than 3.0 volts?**	Yes ▶	CHECK connectors, SERVICE or REPLACE harness. IDM shorted high. REMOVE test equipment. RECONNECT all components. CLEAR Continuous Memory. RERUN Quick Test.
	No ▶	GO to B17.
B17 IDM FAULT — ISOLATE DIS MODULE CHECK IDMD FOR SHORT LOW — KEY OFF		
• Key off. • Disconnect the DIS diagnostic cable tee from Pins 7-12 side of the DIS module. Do not disconnect the vehicle harness from the other side of the tee. • Set DVOM to 2000 ohm range. • Measure resistance between J4 (IDMD) J60 (BAT−). • **Is the resistance less than 1,000 ohms?**	Yes ▶	CHECK connectors, SERVICE or REPLACE harness. IDM shorted low. REMOVE all test equipment. RECONNECT all components. CLEAR Continuous Memory. RERUN Quick Test.
	No ▶	GO to B18.

FIG. 77 DIS Diagnostic Test Charts

2.3L Mustang (Dual Plug) DIS — Code 222 CID Failure, IDM Low, DPI High or Right Coil Pack Failure | **Pinpoint Test** | **B**

TEST STEP	RESULT ▶	ACTION TO TAKE
B18 IDM FAULT — CHECK IDMD FOR SHORT HIGH — KOEO		
• Set DVOM to 20 volt DC range. • Key on. • Measure the voltage between +J4 (IDMD) and −J60 (BAT −). • **Is the voltage greater than 0.5 volt?**	Yes ▶	CHECK connectors, SERVICE or REPLACE harness. CID open. REMOVE test equipment. RECONNECT all components. CLEAR Continuous Memory. RERUN Quick Test.
	No ▶	REPLACE DIS module. No IDM output. REMOVE test equipment. RECONNECT all components. CLEAR Continuous Memory. RERUN Quick Test.
B19 IDM LOW FAULT — ISOLATE EEC PROCESSOR — CHECK IDM TO EEC CONTINUNITY — KEY OFF		
• Key off. • Disconnect the EEC processor. • Connect the second EEC breakout box to the EEC vehicle harness connector. • Measure resistance between J4 (IDMD) and J4 (IDME) of the second EEC breakout box. • **Is the resistance less than 5 ohms?**	Yes ▶	REPLACE EEC processor. Does not respond to IDM input. REMOVE test equipment. RECONNECT all components. CLEAR Continuous Memory. RERUN Quick Test.
	No ▶	CHECK connectors, SERVICE or REPLACE harness. IDM open. REMOVE test equipment. RECONNECT all components. CLEAR Continuous Memory. RERUN Quick Test.

FIG. 78 DIS Diagnostic Test Charts

2.3L Mustang (Dual Plug) DIS — Code 218 IDM Open, IDM High or Left Coil Pack Failure | **Pinpoint Test** | **C**

TEST STEP	RESULT ▶	ACTION TO TAKE
C1 CHECK FOR ANY LEFT SIDE SPARK — KOER		
• Start engine. • Using a Neon Bulb Spark Tester (D89P-6666-A) or air gap spark tester, check for spark at each left spark plug wire. • **Is spark present at any left spark plug wire?**	Yes ▶ No ▶	GO to C2. GO to D1.
C2 CHECK IDMD — KOER		
• Key off. • Install the DIS module tee to Pins 7−12 side of DIS module. • Connect the negative lead of the DIS diagnostic cable to battery. • Connect the DIS diagnostic cable to the EEC breakout box. • Use 2.3L DP DIS overlay. • Set DVOM to 20 volt AC range. • Start engine. • Measure voltage between J4 (IDMD) and J60 (BAT −). • **Is the voltage greater than 1.0 volt?**	Yes ▶ No ▶	GO to C3. GO to C4.

FIG. 79 DIS Diagnostic Test Charts

2.3L Mustang (Dual Plug) DIS — Code 218 IDM Open, IDM High or Left Coil Pack Failure | **Pinpoint Test** | **C**

TEST STEP	RESULT ▶	ACTION TO TAKE
C3 IDM OPEN FAULT — CHECK IDM TO EEC PROCESSOR CONTINUNITY — KEY OFF		
• Key off. • Disconnect EEC processor. • Connect the second EEC breakout box to the EEC vehicle harness connector. • Measure the resistance between J4 (IDMD) and J4 (IDME) of the second breakout box. • **Is the resistance less than 5 ohms?**	Yes ▶	REPLACE EEC processor. Does not respond to IDM input. REMOVE test equipment. RECONNECT all components. CLEAR Continuous Memory. RERUN Quick Test.
	No ▶	CHECK connectors, SERVICE or REPLACE harness. IDM open. REMOVE test equipment. RECONNECT all components. CLEAR Continuous Memory. RERUN Quick Test.
C4 IDM HIGH FAULT — ISOLATE EEC PROCESSOR — CHECK IDMD — KOEC		
• Key off. • Disconnect EEC processor. • Set DVOM to 20 volt range. • Measure voltage between J4 (IDMD) and J60 (BAT −) while cranking the engine. • **Is the voltage greater than 1.0 volt?**	Yes ▶	REPLACE EEC processor. IDM short high. REMOVE test equipment. RECONNECT all components. CLEAR Continuous Memory. RERUN Quick Test.
	No ▶	GO to C5.

FIG. 80 DIS Diagnostic Test Charts

2.3L Mustang (Dual Plug) DIS — Code 218 IDM Open, IDM High or Left Coil Pack Failure | **Pinpoint Test** | **C**

TEST STEP	RESULT ▶	ACTION TO TAKE
C5 IDM HIGH FAULT — ISOLATE DIS MODULE CHECK IDMD FOR SHORT HIGH — KOEO		
• Key off. • Disconnect the DIS diagnostic cable tee to Pins 7-12 side of DIS module. Do not disconnect the other side of the tee from the vehicle harness. • Set DVOM to 20 volt DC range. • Key on. • Measure voltage between +J4 (IDMD) and −J60 (BAT −). • **Is the voltage greater than 0.5 volt?**	Yes ▶	CHECK connectors, SERVICE or REPLACE harness. IDM shorted high. REMOVE test equipment. RECONNECT all components. CLEAR Continuous Memory. RERUN Quick Test.
	No ▶	GO to C6.
C6 IDM FAULT — CHECK IDMD FOR SHORT LOW — KEY OFF		
• Key off. • Set DVOM to 2000 ohm range. • Measure resistance between J4 (IDMD) and J60 (BAT −). • **Is the resistance less than 1,000 ohms?**	Yes ▶	CHECK connectors, SERVICE or REPLACE harness. IDM short low. REMOVE test equipment. RECONNECT all components. CLEAR Continuous Memory. RERUN Quick Test.
	No ▶	REPLACE DIS module. IDM shorted high. REMOVE all test equipment. RECONNECT all components. CLEAR Continuous Memory. RERUN Quick Test.

FIG. 81 DIS Diagnostic Test Charts

2.3L Mustang (Dual Plug) DIS No Start/Continuous Memory Code 224 and/or Coil Failure | **Pinpoint Test** | **D**

TEST STEP	RESULT ▶	ACTION TO TAKE
D1 CHECK FOR SPARK DURING CRANK		
• Using a Neon Bulb Spark Tester (OTC D89P-6666-A) or Air Gap Spark Tester (D81P-6666-A). Check for spark at all spark plug wires while cranking. • **Is spark consistent on all spark plug wires?**	Yes ▶ No ▶	GO to E1. GO to D2.
D2 CHECK FOR SPARK AT RIGHT PLUG WIRES DURING CRANK		
• **Is spark present on all right spark plug wires?**	Yes ▶ No ▶	GO to D19. GO to D3.
D3 CHECK RIGHT PLUGS AND WIRES — RIGHT SPARK FAULT		
• Check right spark plug wires for insulation damage, looseness, shorting or other damage. • Remove and check right spark plugs for damage, wear, carbon deposits and proper plug gap. **NOTE: The spark plugs and wires on the right side of the engine are connected to the "right coil".** • **Are spark plugs and wires OK?**	Yes ▶ No ▶	REINSTALL plugs and wires. GO to D4. SERVICE or REPLACE damaged component. REMOVE all test equipment. RECONNECT all components. CLEAR Continuous Memory. RERUN Quick Test.

FIG. 82 DIS Diagnostic Test Charts

2.3L Mustang (Dual Plug) DIS No Start/Continuous Memory Code 224 and/or Coil Failure | **Pinpoint Test** | **D**

TEST STEP	RESULT ▶	ACTION TO TAKE
D4 CHECK FOR VBAT OPEN TO RIGHT COIL — RIGHT SPARK FAULT		
WARNING **NEVER CONNECT EEC-IV PROCESSOR TO THE EEC BREAKOUT BOX WHEN PERFORMING DIS DIAGNOSTICS.** • Key off. • Connect the right coil tee to the right coil pack and the vehicle harness connector. • Install DIS diagnostic cable to the breakout box. • Connect DIS diagnostic cable BAT(+) and BAT(−) leads to battery. • Use 2.3L DP DIS overlay. • Set DIS cable box switch to "4 cylinder" position. • Set DVOM on 20 volt DC range. • Key on, engine off. • Measure voltage between +J26 (VBATR) and −J60 (BAT−) at the breakout box. • **Is the voltage greater than 10.0 volts?** **NOTE: Do not disconnect or reconnect unless directed to do so.**	Yes ▶ No ▶	GO to D5. SERVICE open circuit. VBATR is open in harness. REMOVE all test equipment. RECONNECT all components. CLEAR Continuous Memory. RERUN Quick Test.
D5 CHECK C1 AT COIL PACK — KOEO — COIL FAULT		
• Set DVOM on 20 volt DC range. • Key on, engine off. • Measure voltage between +J23 (RC1C) and −J60 (BAT−) at the breakout box. • **Is the voltage greater than 10.0 volts?**	Yes ▶ No ▶	GO to D6. GO to D11.
D6 CHECK C2 AT COIL PACK — KOEO — COIL FAILURE		
• Set DVOM on 20 volt DC range. • Key on. • Measure voltage between +J24 (RC2C) and (−) J60 (BAT−) at the breakout box. • **Is the voltage greater than 10.0 volts?**	Yes ▶ No ▶	GO to D7. GO to D13.

FIG. 83 DIS Diagnostic Test Charts

2.3L Mustang (Dual Plug) DIS No Start/Continuous Memory Code 224 and/or Coil Failure — Pinpoint Test D

TEST STEP	RESULT ▶	ACTION TO TAKE
D7 CHECK C1 AT DIS MODULE — KOEO — COIL FAULT		
• Key off. • Connect DIS diagnostic cable tee to the DIS module and vehicle harness connector (both black and grey connectors). • Set DVOM on 20 volt DC range. • Key on, engine off. • Measure voltage between +J18 (RC1D) and −J60 (BAT−) at the breakout box. • **Is the voltage greater than 10.0 volts?**	Yes ▶	GO to D8.
	No ▶	SERVICE open circuit. C1 open in harness at DIS module. REMOVE all test equipment. RECONNECT all components. CLEAR Continuous Memory. RERUN Quick Test.
D8 CHECK C2 AT DIS MODULE — KOEO — COIL FAULT		
• Set DVOM on 20 volt DC range. • Key on, engine off. • Measure voltage between +J10 (RC2D) and −J60 (BAT−) at the breakout box. • **Is the voltage greater than 10.0 volts?**	Yes ▶	GO to D9.
	No ▶	SERVICE open circuit. C2 open in harness at DIS module. REMOVE all test equipment. RECONNECT all components. CLEAR Continuous Memory. RERUN Quick Test.
D9 CHECK FOR C1 AT COIL PACK — ISOLATE COIL		
• Key off. • Disconnected coil pack from the coil tee, leave DIS diagnostic cable connected to vehicle harness connector. • Set DVOM on 20 volt DC range. • Key on, engine off. • Measure voltage between +J23 (RC1C) and −J60 (BAT−) at the breakout box. • **Is the voltage greater than 1.0 volt?**	Yes ▶	GO to D10.
	No ▶	GO to D15.

FIG. 84 DIS Diagnostic Test Charts

2.3L Mustang (Dual Plug) DIS No Start/Continuous Memory Code 224 and/or Coil Failure — Pinpoint Test D

TEST STEP	RESULT ▶	ACTION TO TAKE
D10 CHECK FOR C1 AT COIL PACK — ISOLATE DIS MODULE — COIL FAULT		
• Key off. • Disconnect DIS module from the DIS module tee, leave DIS diagnostic cable connected to vehicle harness connector (both black and grey connectors). • Set DVOM on 20 volt DC range. • Key on, engine off. • Measure voltage between +J23 (RC1C) and −J60 (BAT−) at the breakout box. • **Is the voltage less than 0.5 volt?**	Yes ▶	REPLACE DIS module. C1 shorted high. REMOVE all test equipment. RECONNECT all components. CLEAR Continuous Memory. RERUN Quick Test.
	No ▶	SERVICE short circuit. C1 shorted high in harness. REMOVE all test equipment. RECONNECT all components. CLEAR Continuous Memory. RERUN Quick Test.
D11 CHECK FOR C1 SHORT TO GROUND PACK — COIL LOW FAULT		
• Key off. • Set DVOM on 2000 ohm range. • Measure resistance between J23 (RC1C) J60 (BAT−) at the breakout box. • **Is the resistance less than 1,200 ohms?**	Yes ▶	GO to D12.
	No ▶	REPLACE right coil pack. Coil primary open, C1 or VBATR open at right coil pack. REMOVE all test equipment. RECONNECT all components. CLEAR Continuous Memory. RERUN Quick Test.

FIG. 85 DIS Diagnostic Test Charts

2.3L Mustang (Dual Plug) DIS No Start/Continuous Memory Code 224 and/or Coil Failure — Pinpoint Test D

TEST STEP	RESULT ▶	ACTION TO TAKE
D12 CHECK FOR C1 SHORT TO GROUND IN HARNESS — COIL FAULT LOW • Key off. • Disconnect DIS module from DIS module tee, leave DIS diagnostic cable connected to vehicle harness connector (both black and grey connectors). • Set DVOM on 2,000 ohm range. • Measure resistance between J3 (RC1C) and J60 (BAT −) at the breakout box. • **Is resistance less than 5 ohms?**	Yes ▶	SERVICE short circuit in harness. C1 is shorted to ground in harness. REMOVE all test equipment. RECONNECT all components. CLEAR Continuous Memory. RERUN Quick Test.
	No ▶	REPLACE DIS module. C1 is shorted to ground. REMOVE all test equipment. RECONNECT all components. CLEAR Continuous Memory. RERUN Quick Test.
D13 CHECK FOR C2 SHORT TO GROUND — COIL FAULT LOW • Key off. • DVOM on 2,000 ohm scale. • Measure resistance between J24 (RC2C) and J60 (BAT −) at the breakout box. • **Is the resistance less than 1,200 ohms?**	Yes ▶	GO to D14.
	No ▶	REPLACE right coil pack. C2 or VBATR is open in coil pack. REMOVE all test equipment. RECONNECT all components. CLEAR Continuous Memory. RERUN Quick Test.

FIG. 86 DIS Diagnostic Test Charts

2.3L Mustang (Dual Plug) DIS No Start/Continuous Memory Code 224 and/or Coil Failure — Pinpoint Test D

TEST STEP	RESULT ▶	ACTION TO TAKE
D14 CHECK FOR C2 AT COIL PACK — ISOLATE EDIS MODULE — COIL FAULT • Key off. • Disconnected DIS module from the DIS module tee, leave DIS diagnostic cable connected to vehicle harness connector (both black and grey connectors). • Set DVOM on 2000 ohm range. • Key on, engine off. • Measure resistance between J24 (RC2C) and J60 (BAT −) at the breakout box. • **Is the resistance greater than 2000 ohms?**	Yes ▶	REPLACE DIS module. C2 shorted low. REMOVE all test equipment. RECONNECT all components. CLEAR Continuous Memory. RERUN Quick Test.
	No ▶	SERVICE short circuit. C2 shorted low in harness. REMOVE all test equipment. RECONNECT all components. CLEAR Continuous Memory. RERUN Quick Test.
D15 CHECK FOR C2 HIGH AT COIL PACK — KOEO — ISOLATE COIL • Set DVOM on 20 volt DC range. • Key on, engine off. • Measure voltage between +J24 (RC2C) and −J60 (BAT −) at the breakout box. • **Is the voltage greater than 1.0 volt?**	Yes ▶	GO to D16.
	No ▶	GO to D17.

FIG. 87 DIS Diagnostic Test Charts

2.3L Mustang (Dual Plug) DIS No Start/Continuous Memory Code 224 and/or Coil Failure | **Pinpoint Test** | **D**

TEST STEP	RESULT ▶	ACTION TO TAKE
D16 CHECK FOR C2 HIGH AT COIL PACK — KOEO ISOLATE DIS MODULE — COIL FAULT		
• Key off. • Disconnected DIS module from the DIS module tee, leave DIS diagnostic cable connected to vehicle harness connector (both black and grey connectors). • Set DVOM on 20 volt DC scale. • Key on, engine off. • Measure voltage between +J24 (RC2C) and −J60 (BAT−) at the breakout box. • **Is the voltage less than 1.0 volt?**	Yes ▶	REPLACE DIS module. C2 shorted high in DIS module. REMOVE all test equipment. RECONNECT all components. CLEAR Continuous Memory. RERUN Quick Test.
	No ▶	SERVICE short circuit. C2 shorted high in harness. REMOVE all test equipment. RECONNECT all components. CLEAR Continuous Memory. RERUN Quick Test.
D17 CHECK C1 AT COIL PACK WHILE ENGINE RUNNING OR CRANKING — ISOLATE COIL		
• Connect test lamp between J57 (VBAT+) and J23 (RC1C). • Crank or start engine. • **Does the lamp blink consistently?**	Yes ▶	GO to D18.
	No ▶	REPLACE DIS module. No C1 output. C1 open in module. REMOVE all test equipment. RECONNECT all components. CLEAR Continuous Memory. RERUN Quick Test.

FIG. 88 DIS Diagnostic Test Charts

2.3L Mustang (Dual Plug) DIS No Start/Continuous Memory Code 224 and/or Coil Failure | **Pinpoint Test** | **D**

TEST STEP	RESULT ▶	ACTION TO TAKE
D18 CHECK C2 AT COIL PACK WHILE ENGINE RUNNING OR CRANKING		
• Key off. • Connect test lamp between J57 (VBAT+) and J24 (RC2C). • Crank or start engine. • **Does the lamp blink consistently?**	Yes ▶	REPLACE right coil pack. Input to coil pack is OK, but no high voltage output. REMOVE all test equipment. RECONNECT all components. CLEAR Continuous Memory. RERUN Quick Test.
	No ▶	REPLACE DIS module. REMOVE all test equipment. RECONNECT all components. CLEAR continous memory. RERUN Quick Test.
D19 CHECK LEFT PLUGS AND WIRES — LEFT SPARK FAULT		
• Check left spark plug wires for insulation damage, looseness, shorting or other damage. • Remove and check left spark plugs for damage, wear, carbon deposits and proper plug gap. NOTE: **The spark plugs and wires on the left side of the engine are connected to the "left coil".** • **Are spark plugs and wires OK?**	Yes ▶	REINSTALL plugs and wires. GO to D20.
	No ▶	SERVICE or REPLACE damaged component. REMOVE all test equipment. RECONNECT all components. CLEAR Continuous Memory. RERUN Quick Test.

FIG. 89 DIS Diagnostic Test Charts

2.3L Mustang (Dual Plug) DIS No Start/Continuous Memory Code 224 and/or Coil Failure — Pinpoint Test D

TEST STEP	RESULT ▶	ACTION TO TAKE
D20 CHECK FOR VBAT OPEN TO RIGHT COIL		
WARNING **NEVER CONNECT EEC PROCESSOR TO THE EEC-IV BREAKOUT BOX WHEN PERFORMING EDIS DIAGNOSTICS.** • Key off. • Connect left coil tee to left coil pack and vehicle harness connector. • Connect DIS diagnostic cable negative and positive leads to battery. • Set DVOM on 20 volt DC range. • Install DIS diagnostic cable to the breakout box. • Key on, engine off. • Set DIS cable box switch to "4 cylinder" position. • Measure voltage between +J15 (VBAT L) and J60 (BAT −) at the breakout box. • **Is the voltage greater than 10.0 volts?**	Yes ▶ No ▶	GO to D21. SERVICE open circuit. VBAT L open in harness. REMOVE all test equipment. RECONNECT all components. CLEAR Continuous Memory. RERUN Quick Test.
D21 CHECK C3 AT COIL PACK — KOEO — COIL FAULT		
• Set DVOM on 20 volt DC range. • Key on, engine off. • Measure voltage between +J30 (LC3C) and −J60 (BAT −) at the breakout box. • **Is the voltage greater than 10.0 volts?**	Yes ▶ No ▶	GO to D22. GO to D27.
D22 CHECK C4 AT COIL PACK — KOEO — COIL FAULT		
• Set DVOM on 20 volt DC range. • Key on, engine off. • Measure voltage between +J28 (LC4C) and −J60 (BAT −) at the breakout box. • **Is the voltage greater than 10.0 volts?**	Yes ▶ No ▶	GO to D23. GO to D29.

FIG. 90 DIS Diagnostic Test Charts

2.3L Mustang (Dual Plug) DIS No Start/Continuous Memory Code 224 and/or Coil Failure — Pinpoint Test D

TEST STEP	RESULT ▶	ACTION TO TAKE
D23 CHECK C3 AT DIS MODULE — KOEO — COIL FAULT		
• Key off. • Connect DIS diagnostic cable tee to the DIS module and vehicle harness connector. • Set DVOM on 20 volt DC range. • Key on, engine off. • Measure voltage between +J3 (LC3D) and −J60 (BAT −) at the breakout box. • **Is the voltage greater than 10.0 volts?**	Yes ▶ No ▶	GO to D24. SERVICE open circuit. C3 open in harness. REMOVE all test equipment. RECONNECT all components. CLEAR Continuous Memory. RERUN Quick Test.
D24 CHECK C4 AT DIS MODULE — KOEO — COIL FAULT		
• Set DVOM on 20 volt DC range. • Key on, engine off. • Measure voltage between +J6 (LC4D) and −J60 (BAT −) at the breakout box. • **Is the voltage greater than 10.0 volts?**	Yes ▶ No ▶	GO to D25. SERVICE open circuit. C4 open in harness. REMOVE all test equipment. RECONNECT all components. CLEAR Continuous Memory. RERUN Quick Test.
D25 CHECK C3 AT COIL PACK — KOEO — COIL FAILURE — ISOLATE COIL		
• Key off. • Disconnect the coil pack from the coil tee, leave DIS diagnostic cable connected to the vehicle harness connector. • Set DVOM on 20 volt DC range. • Key on engine off. • Measure voltage between +J30 (LC3C) and −J60 (BAT −) at the breakout box. • **Is the voltage greater than 1.0 volt?**	Yes ▶ No ▶	GO to D26. GO to D31.

FIG. 91 DIS Diagnostic Test Charts

2.3L Mustang (Dual Plug) DIS No Start/Continuous Memory Code 224 and/or Coil Failure — Pinpoint Test D

TEST STEP	RESULT ▶	ACTION TO TAKE
D26 CHECK FOR C3 AT COIL PACK — ISOLATE EDIS MODULE — COIL FAULT		
• Key off. • Disconnected DIS module from the DIS module tee, leave DIS diagnostic cable connected to vehicle harness connector (both black and grey connectors). • Set DVOM on 20 volt DC range. • Key on, engine off. • Measure voltage between +J30 (LC3C) and −J60 (BAT−) at the breakout box. • **Is the voltage less than 1.0 volt?**	Yes ▶	REPLACE DIS module. C3 shorted high. REMOVE all test equipment. RECONNECT all components. CLEAR Continuous Memory. RERUN Quick Test.
	No ▶	SERVICE short circuit. C3 shorted high in harness. REMOVE all test equipment. RECONNECT all components. CLEAR Continuous Memory. RERUN Quick Test.
D27 CHECK FOR C3 SHORT TO GROUND PACK — COIL FAULT LOW		
• Key off. • Set DVOM on 2,000 ohm range. • Measure resistance between J30 (LC3C) and J60 (BAT−) at the breakout box. • **Is the resistance less than 1,200 ohms?**	Yes ▶	GO to D28.
	No ▶	REPLACE left coil pack. Coil primary open. C3 or VBAT L open in left coil pack. REMOVE all test equipment. RECONNECT all components. CLEAR Continuous Memory. RERUN Quick Test.

FIG. 92 DIS Diagnostic Test Charts

2.3L Mustang (Dual Plug) DIS No Start/Continuous Memory Code 224 and/or Coil Failure — Pinpoint Test D

TEST STEP	RESULT ▶	ACTION TO TAKE
D28 CHECK FOR C3 SHORT TO GROUND IN HARNESS — ISOLATE DIS		
• Key off. • Disconnect DIS module from DIS module tee, leave DIS diagnostic cable connected to vehicle harness connector (both black and grey connectors). • Set DVOM on 2,000 ohm range. • Measure resistance between J30 (LC3C) and J60 (BAT−) at the breakout box. • **Is the resistance less than 5 ohms?**	Yes ▶	SERVICE short circuit in harness. C3 is shorted to ground in harness. REMOVE all test equipment. RECONNECT all components. CLEAR Continuous Memory. RERUN Quick Test.
	No ▶	REPLACE DIS module. C3 is shorted to ground. REMOVE all test equipment. RECONNECT all components. CLEAR Continuous Memory. RERUN Quick Test.
D29 CHECK FOR C4 SHORT TO GROUND — COIL FAULT LOW		
• Key off. • Set DVOM on 2,000 ohm range. • Measure resistance between J28 (LC4C) and J60 (BAT−) at the breakout box. • **Is the resistance less than 1200 ohms?**	Yes ▶	GO to D30.
	No ▶	REPLACE left coil pack. C4 or VBAT L open in left coil pack. REMOVE all test equipment. RECONNECT all components. CLEAR Continuous Memory. RERUN Quick Test.

FIG. 93 DIS Diagnostic Test Charts

2.3L Mustang (Dual Plug) DIS No Start/Continuous Memory Code 224 and/or Coil Failure — **Pinpoint Test** — **D**

TEST STEP	RESULT ▶	ACTION TO TAKE
D30 CHECK FOR C4 AT COIL PACK — ISOLATE DIS MODULE — COIL FAULT		
• Key off. • Disconnected DIS module from the DIS module tee, leave EDIS diagnostic cable connected to vehicle harness connector (both black and grey connectors). • Set DVOM on 2000 ohm range. • Key on, engine off. • Measure resistance between J28 (LC4C) and J60 (BAT –). • **Is the resistance less than 1,200 ohms?**	Yes ▶	REPLACE DIS module. C4 shorted low. REMOVE all test equipment. RECONNECT all components. CLEAR Continuous Memory. RERUN Quick Test.
	No ▶	SERVICE short circuit. C4 shorted low in harness. REMOVE all test equipment. RECONNECT all components. CLEAR Continuous Memory. RERUN Quick Test.
D31 CHECK FOR C4 AT COIL PACK — ISOLATE COIL		
• Set DVOM on 20 volt DC range. • Key on, engine off. • Measure voltage between +J28 (LC4C) and –J60 (BAT –) at the breakout box. • **Is the voltage greater than 1.0 volt?**	Yes ▶	GO to D32.
	No ▶	GO to D33.

FIG. 94 DIS Diagnostic Test Charts

2.3L Mustang (Dual Plug) DIS No Start/Continuous Memory Code 224 and/or Coil Failure — **Pinpoint Test** — **D**

TEST STEP	RESULT ▶	ACTION TO TAKE
D32 CHECK FOR C4 AT COIL PACK — KOEO ISOLATE DIS MODULE — COIL FAULT		
• Key off. • Disconnected DIS module from the DIS module tee, leave DIS diagnostic cable connected to vehicle harness connector (both black and grey connectors). • Set DVOM on 20 volt DC range. • Key on, engine off. • Measure voltage between +J28 (LC4C) and –J60 (BAT –) at the breakout box. • **Is the voltage less than 1.0 volt?**	Yes ▶	REPLACE DIS module. C4 shorted high in DIS module. REMOVE all test equipment. RECONNECT all components. CLEAR Continuous Memory. RERUN Quick Test.
	No ▶	SERVICE short circuit. C4 shorted high in harness. REMOVE all test equipment. RECONNECT all components. CLEAR Continuous Memory. RERUN Quick Test.
D33 CHECK C3 AT COIL PACK WHILE ENGINE RUNNING OR CRANKING — ISOLATE COIL		
• Key off. • Connect test lamp between J5 (VPWRD) and J30 (LC3C). • Crank or start engine. • **Does the lamp blink consistently?**	Yes ▶	GO to D34.
	No ▶	REPLACE DIS module. COIL 3 open in DIS module. REMOVE all test equipment. RECONNECT all components. CLEAR Continuous Memory. RERUN Quick Test.

FIG. 95 DIS Diagnostic Test Charts

2.3L Mustang (Dual Plug) DIS No Start/Continuous Memory Code 224 and/or Coil Failure — Pinpoint Test D

TEST STEP	RESULT	ACTION TO TAKE
D34 CHECK C4 AT COIL PACK WHILE ENGINE RUNNING OR CRANKING		
• Key off. • Connect test lamp between J5 (VBATD) and J28 (LC4C). • Crank or start engine. • **Does the lamp blink consistently?**	Yes ▶	REPLACE right coil pack. Input to coil pack is OK, but no high voltage output. REMOVE all test equipment. RECONNECT all components. CLEAR Continuous Memory. RERUN Quick Test.
	No ▶	REPLACE DIS module. C4 open in DIS module. REMOVE all test equipment. RECONNECT all components. CLEAR Continuous Memory. RERUN Quick Test.

FIG. 96 DIS Diagnostic Test Charts

2.3L Mustang (Dual Plug) DIS Continuous Memory Code 213: DPI Open, DPI High or Spout High — Pinpoint Test E

TEST STEP	RESULT	ACTION TO TAKE
E1 CHECK SPOUT AT DIS MODULE		
• Key off. • Connect DIS diagnostic cable negative and positive lead to battery. • Install DIS diagnostic cable to breakout box and the DIS module (Pin 1-6). • Use 2.3L DP DIS overlay. • Connect DVOM between + J36 (SPOUT) and – J60 (BAT –) • Set DVOM to 2.0 Volt AC range. • Start engine. • **Is the voltage greater than 2.0 volts?**	Yes ▶	GO to E2.
	No ▶	GO to E6. SPOUT fault.
E2 DP FAULT — VERIFY DUAL PLUG OPERATION		
• Using a Neon Spark Tester (D89P-6666-A), check for spark at each left side plug wire with the engine running. • **Is spark present at one or more wires during dual plug operation?**	Yes ▶	REPLACE DIS module. Module not indicating dual plug operation. REMOVE all test equipment. RECONNECT all components. CLEAR Continuous Memory. RERUN Quick Test.
	No ▶	GO to E3.
E3 DP FAULT — CHECK DPI CONTINUITY		
• Key off. • Disconnect EEC-IV processor. • Measure resistance between Test Pin 32 (DPI) at the EEC vehicle harness connector and J54 (DPI). • **Is the resistance less than 5 ohms?**	Yes ▶	GO to E4.
	No ▶	CHECK connectors, SERVICE or REPLACE the harness. DPI is open between the DIS module and the processor. REMOVE all test equipment. RECONNECT all components. CLEAR Continuous Memory. RERUN Quick Test.

FIG. 97 DIS Diagnostic Test Charts

2.3L Mustang (Dual Plug) DIS Continuous Memory Code 213: DPI Open, DPI High or Spout High — Pinpoint Test E

TEST STEP	RESULT ▶	ACTION TO TAKE
E4 DPI FAULT — CHECK FOR SHORT TO VBAT		
• Key on. • Set DVOM to 20 volt DC range. • Measure voltage between +J54 (DPI) and −J60 (BAT−). • **Is the voltage greater than 0.5 volt?**	Yes ▶ No ▶	GO to E5. GO to E6.
E5 DPI FAULT — ISOLATE DIS MODULE		
• Key off. • Disconnect DIS module (Pin 1-6) from DIS diagnostic cable tee, leave DIS diagnostic cable connected to vehicle harness connector. • Set DVOM to 20 volt DC range. • Key on, engine off. • Measure voltage between +J54 (DPI) and −J60 (BAT−). • **Is the voltage greater than 0.5 volt?**	Yes ▶	SERVICE harness. REMOVE all test equipment. RECONNECT all components. CLEAR Continuous Memory. RERUN Quick Test.
	No ▶	REPLACE DIS module. Module not responding to DPI signal. REMOVE all test equipment. RECONNECT all components. CLEAR Continuous Memory. RERUN Quick Test.
E6 DPI FAULT — FORCE DUAL PLUG COMMAND AT DIS MODULE		
• Connect a jumper between J54 (DPI) and J60 (BAT−). • Reconnect EEC processor. • Reconnect DIS TEE (1-6) to DIS module. CAUTION **Do not Jumper J54 to VBAT+ or BAT+; the DPI circuit in the EEC-IV processor will be damaged.** • Start the engine. • **Is there spark at any left plug wire?**	Yes ▶	REPLACE EEC processor. Processor not sending DPI signal. REMOVE all test equipment. RECONNECT all components. CLEAR Continuous Memory. RERUN Quick Test.
	No ▶	REPLACE DIS module. Module not responding to DPI signal. REMOVE all test equipment. RECONNECT all components. CLEAR Continuous Memory. RERUN Quick Test.

FIG. 98 DIS Diagnostic Test Charts

2.3L Mustang (Dual Plug) DIS Continuous Memory Code 213: DPI Open, DPI High or Spout High — Pinpoint Test E

TEST STEP	RESULT ▶	ACTION TO TAKE
E7 SPOUT HIGH FAULT — OPEN SPOUT CIRCUIT		
• Remove SPOUT jumper. • Set DVOM to 20 volt AC range. • Start engine. • Measure voltage between EEC side of SPOUT jumper and J60 (BAT−). • **Is the voltage greater than 3 volts?** DIS SIDE — EEC SIDE SPOUT IN-LINE VEHICLE HARNESS CONNECTOR	Yes ▶ No ▶	GO to E7. GO to E8.
E8 SPOUT HIGH FAULT — CHECK EEC SIDE OF SPOUT CONNECTOR		
• Key off. • Disconnect EEC processor. • Set DVOM to 20 volt DC range. • Key on. • Measure voltage between +DIS side of SPOUT connector and −J60. • **Is the voltage greater than 0.5 volt?** DIS SIDE — EEC SIDE SPOUT IN-LINE VEHICLE HARNESS CONNECTOR	Yes ▶	SERVICE harness and connectors. SPOUT is shorted high between SPOUT jumper and EEC harness connector. REMOVE all test equipment. RECONNECT all components. CLEAR Continuous Memory. RERUN Quick Test.
	No ▶	REPLACE DIS module. SPOUT is shorted high. REMOVE all test equipment. RECONNECT all components. CLEAR Continuous Memory. RERUN Quick Test.

FIG. 99 DIS Diagnostic Test Charts

2.3L Mustang (Dual Plug) DIS Continuous Memory Code 213: DPI Open, DPI High or Spout High — Pinpoint Test E

TEST STEP	RESULT ▶	ACTION TO TAKE
E9 SPOUT HIGH FAULT — CHECK DIS SIDE OF SPOUT CONNECTOR		
• Key off. • Disconnect EEC processor. • Set DVOM to 20 volt DC range. • Key on. • Measure voltage between +EEC side of SPOUT connector and −J60. • **Is the voltage greater than 0.5 volt?**	Yes ▶	SERVICE harness and connectors. SPOUT is shorted high between SPOUT jumper and EEC harness connector. REMOVE all test equipment. RECONNECT all components. CLEAR Continuous Memory. RERUN Quick Test.
DIS SIDE — EEC SIDE SPOUT IN-LINE VEHICLE HARNESS CONNECTOR	No ▶	REPLACE EEC processor. SPOUT is shorted high. REMOVE all test equipment. RECONNECT all components. CLEAR Continous Memory. RERUN Quick Test.

FIG. 100 DIS Diagnostic Test Charts

2.3L Mustang (Dual Plug) DIS Engine Running Code 213/ No Codes and Lack of Power — Pinpoint Test F

TEST STEP	RESULT ▶	ACTION TO TAKE
F1 CHECK TIMING AT IDLE		
• Key off. • Install timing light. • Engine running at normal operating temperature. • Transmission out of gear. • Check the timing. • **Is the timing greater than 16 degrees BTDC?**	Yes ▶ No ▶	Reset ignition timing. GO to F2 .
F2 CHECK FOR ADVANCE SPARK ANGLE		
• Key off. • Remove SPOUT harness jumper plug from spout vehicle harness connector. • **Is the engine timing 10 ± 2 degrees BDTC?**	Yes ▶ No ▶	GO to F3 . GO to F4 .
F3 SPOUT FAULT — CHECK SPOUT AT DIS MODULE		
• Key off. • Replace SPOUT harness jumper plug to SPOUT vehicle harness connector. • Install DIS diagnostic cable to the breakout box, and ONLY to the DIS module (Pin 1-6) connector. • Use 2.3L DP DIS overlay. • Connect DIS diagnostic cable negative and positive leads to the battery. • Set DVOM on 20 volt AC range. • Start engine and measure voltage between J36 (SPOUT) and J60 (BAT −) at the breakout box. • **Is the voltage greater than 2.0 volts?**	Yes ▶ No ▶	GO to F5 . GO to F8 .

FIG. 101 DIS Diagnostic Test Charts

2.3L Mustang (Dual Plug) DIS Engine Running Code 213/ No Codes and Lack of Power — Pinpoint Test F

TEST STEP	RESULT ▶	ACTION TO TAKE
F4 SPOUT FAULT — INSPECT SENSOR/TRIGGER WHEEL		
• **Is the sensor or trigger wheel damaged i.e., loose or misaligned?**	Yes ▶	REPLACE or SERVICE as required. REMOVE all test equipment. RECONNECT all components. CLEAR Continuous Memory. RERUN Quick Test.
	No ▶	REPLACE DIS module. DIS module not responding to input signals. REMOVE all test equipment. RECONNECT all components. CLEAR Continuous Memory. RERUN Quick Test.
F5 SPOUT OPEN FAULT — CHECK EEC SIDE OF SPOUT CONNECTOR.		
• Key off. • Remove SPOUT harness jumper from SPOUT vehicle harness connector. • Set DVOM to 20 volts AC range. • Start engine and measure voltage between EEC side of SPOUT connector and J60 (BAT –). • **Is the voltage greater than 2.0 volts?**	Yes ▶	GO to F6.
	No ▶	GO to F7.

FIG. 102 DIS Diagnostic Test Charts

2.3L Mustang (Dual Plug) DIS Engine Running Code 213/ No Codes and Lack of Power — Pinpoint Test F

TEST STEP	RESULT ▶	ACTION TO TAKE
F6 CHECK SPOUT CIRCUIT CONTINUITY		
• Key off. • Set DVOM to 200 ohm range. • Measure resistance between J36 (SPOUT) and the DIS side of the SPOUT connector. • **Is the resistance less than 5 ohms?**	Yes ▶	REPLACE DIS module. SPOUT open in module. REMOVE all test equipment. RECONNECT all components. CLEAR Continuous Memory. RERUN Quick Test.
	No ▶	SERVICE open circuit in hardness between DIS module and the SPOUT connector. REMOVE all test equipment. RECONNECT all components. CLEAR Continuous Memory. RERUN Quick Test.
F7 CHECK SPOUT CIRCUIT CONTINUITY		
• Key off. • Disconnect EEC processor. • Install DIS diagnostic cable to the breakout box, and DIS module (Pin 1-6) connector. • Set DVOM to 200 ohm range. • Measure resistance between J36 (SPOUT) and the EEC side of the SPOUT connector. • **Is the resistance less than 5 ohms?**	Yes ▶	REPLACE processor. SPOUT not being transmitted by the processor. REMOVE all test equipment. RECONNECT all components. CLEAR Continuous Memory. RERUN Quick Test.
	No ▶	SERVICE open circuit in harness between the processor and the SPOUT connector. REMOVE all test equipment. RECONNECT all components. CLEAR Continuous Memory. RERUN Quick Test.

FIG. 103 DIS Diagnostic Test Charts

2.3L Mustang (Dual Plug) DIS Engine Running Code 213/ No Codes and Lack of Power — Pinpoint Test F

TEST STEP	RESULT ▶	ACTION TO TAKE
F8 SPOUT FAULT — ISOLATE DIS MODULE		
• Key off. • Disconnect the DIS module (Pin 1-6) from the DIS module TEE, leave the DIS diagnostic cable connected to the vehicle harness connector. • Set DVOM to 200 ohm range. • Measure resistance between J36 (SPOUT) and J60 (BAT −). • **Is the resistance less than 10 ohms?**	Yes ▶	GO to F9.
	No ▶	REPLACE DIS module. SPOUT not being transmitted by the module. REMOVE all test equipment. RECONNECT all components. CLEAR Continuous Memory. RERUN Quick Test.
F9 SPOUT FAULT — ISOLATE PROCESSOR		
• Key off. • Connect the DIS module TEE to the DIS module and vehicle harness connector. • Disconnect the EEC processor. • Set DVOM to 200 ohm range. • Measure resistance between J36 (SPOUT) and J60 (BAT −). • **Is the resistance less than 10 ohms?**	Yes ▶	GO to F10.
	No ▶	REPLACE processor. SPOUT shorted to ground in the processor. REMOVE all test equipment. RECONNECT all components. CLEAR Continuous Memory. RERUN Quick Test.

FIG. 104 DIS Diagnostic Test Charts

2.3L Mustang (Dual Plug) DIS Engine Running Code 213/ No Codes and Lack of Power — Pinpoint Test F

TEST STEP	RESULT ▶	ACTION TO TAKE
F10 SPOUT FAULT — CHECK CONTINUITY		
• Key off. • Set DVOM to 200 ohm scale. • Measure resistance between J60 (BAT −) and the DIS side of the SPOUT connector. • **Is the resistance less than 5 ohms?**	Yes ▶	SERVICE harness and connectors. SPOUT shorted to ground between the processor and the SPOUT connector. REMOVE all test equipment. RECONNECT all components. CLEAR Continuous Memory. RERUN Quick Test.
	No ▶	SERVICE harness. SPOUT shorted to ground in harness between the DIS module and the SPOUT connector. REMOVE all test equipment. RECONNECT all components. CLEAR Continuous Memory. RERUN Quick Test.

FIG. 105 DIS Diagnostic Test Charts

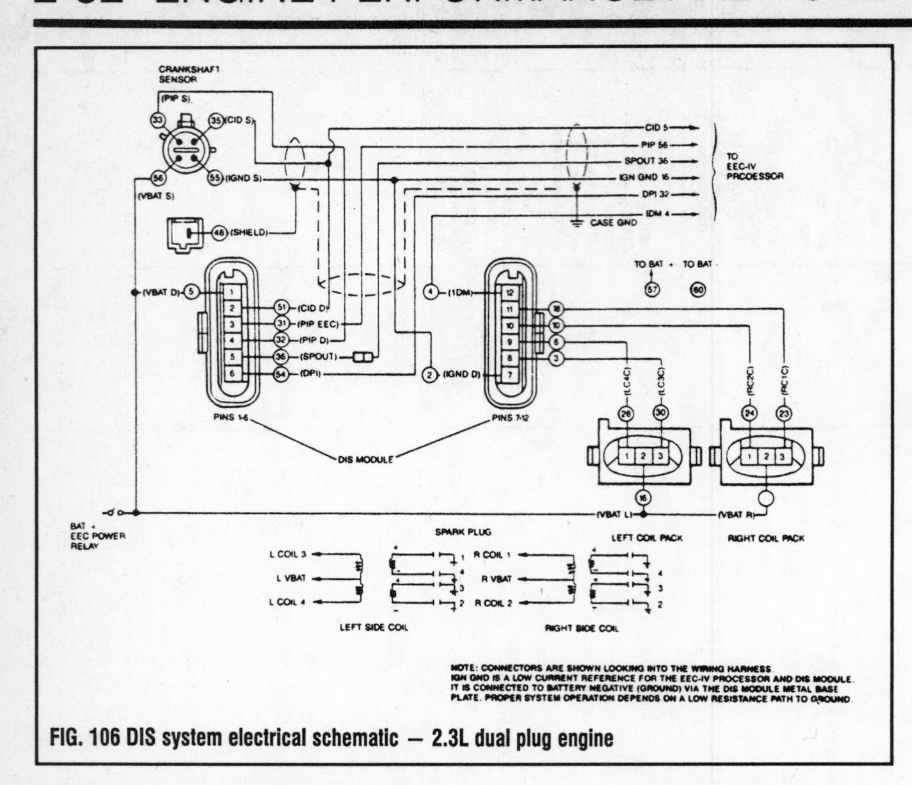

FIG. 106 DIS system electrical schematic — 2.3L dual plug engine

Crankshaft Timing Sensor

REMOVAL & INSTALLATION

➧ SEE FIG. 107

1. Turn the ignition key **OFF** and disconnect the negative battery cable.
2. Loosen the tensioner pulleys for the A/C compressor and supercharger idler pulley belts, if equipped.
3. Remove the belts from the crankshaft pulley.
4. Disconnect the sensor electrical connectors from the engine wiring harness.
5. Raise and support the vehicle safely. Remove the upper and lower damper shield assemblies.
6. Remove the crankshaft damper and pulley assembly with a suitable puller.
7. Remove the the crankshaft sensor mounting screws and remove the sensor.

To install:

8. Position the crankshaft sensor assembly on the bracket. Install the 2 sensor retaining screws and torque them to 22–31 inch lbs.
9. Install the crankshaft damper and pulley assembly using crank gear and damper replacer tool T83T–6316–B or equivalent. Torque the pulley bolt to specifications 103–132 ft. lbs. 140–180 Nm).
10. Install the upper and lower damper shields and install the retaining screws.
11. Lower the vehicle. Route the sensor wiring harness and connect both electrical connectors.
12. Install the A/C and power steering belts and adjust as necessary. Reconnect the negative battery cable and perform a vehicle road test.

Camshaft Sensor

REMOVAL & INSTALLATION

➧ SEE FIG. 108

1. Turn the ignition key **OFF** and disconnect the negative battery cable.
2. Disconnect the camshaft sensor electrical connector.
3. Remove the camshaft sensor mounting screws and remove the sensor.

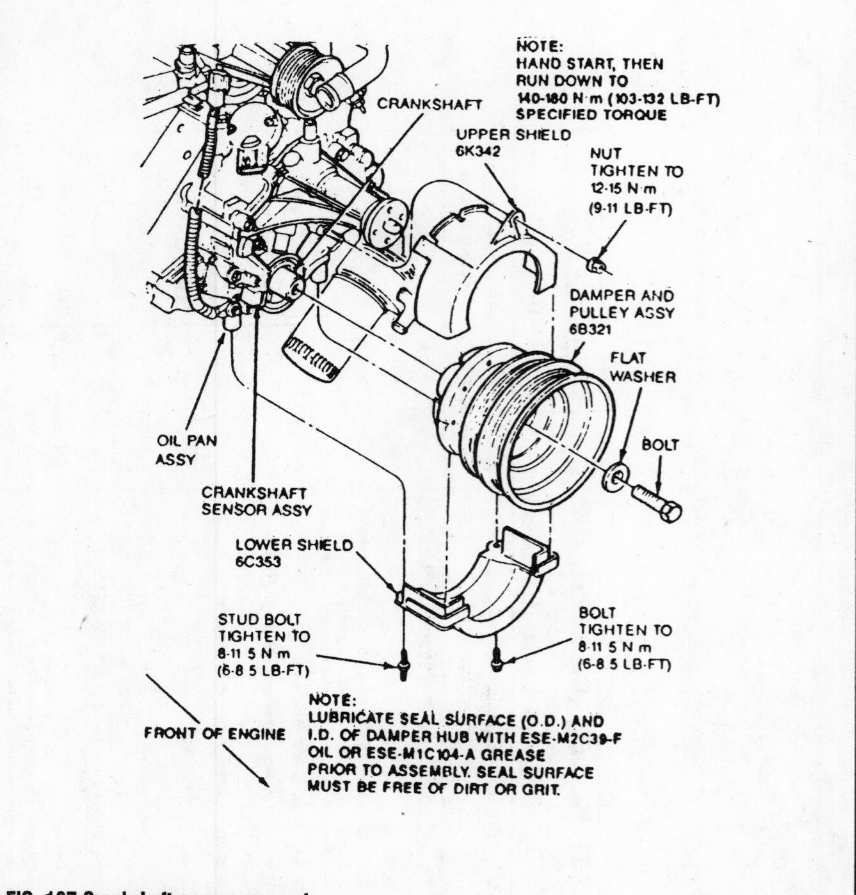

FIG. 107 Crankshaft sensor removal

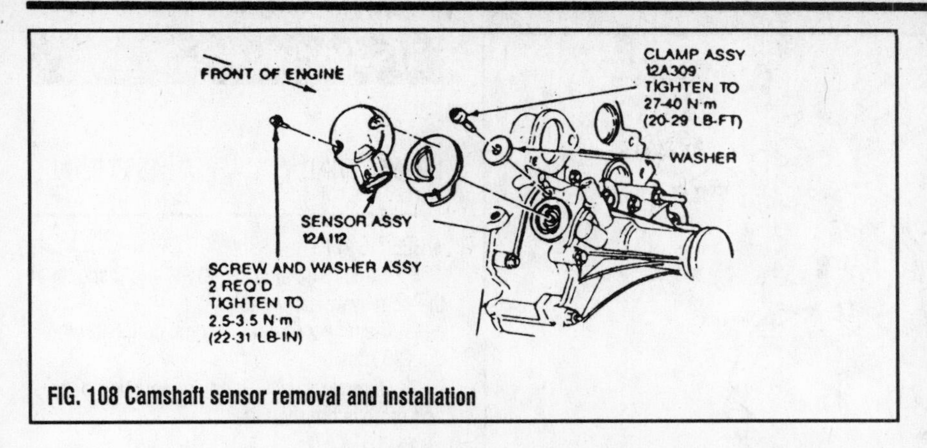

FIG. 108 Camshaft sensor removal and installation

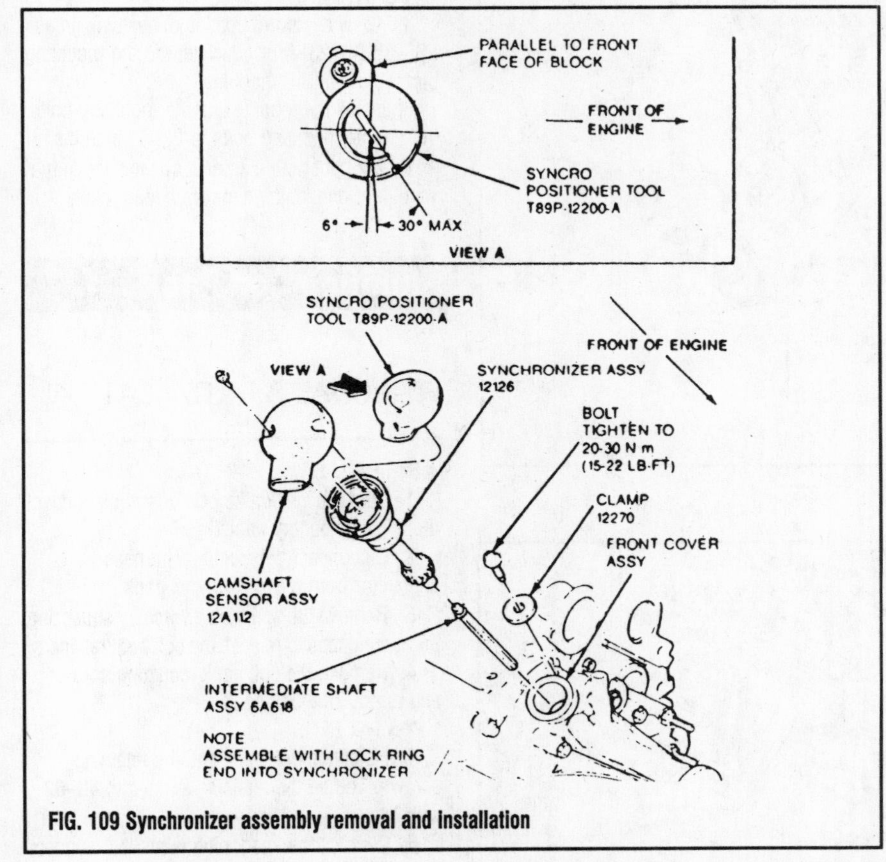

FIG. 109 Synchronizer assembly removal and installation

To Install:

4. Install the new camshaft sensor and retaining screws. Torque the retaining screws to 22–31 inch lbs.

5. Connect the sensor electrical connector and reconnect the negative battery cable.

Synchronizer Assembly

REMOVAL & INSTALLATION

➧ SEE FIG. 109

Prior to starting this procedure, set the No. 1 cylinder to 26° after top dead center of the compression stroke. Then take note of the position of the camshaft sensor electrical connector. The installation procedure requires that the connector be located in the same position.

1. Turn the ignition key **OFF** and disconnect the negative battery cable.

2. Remove the camshaft sensor as previously outlined.

3. Remove the synchronizer clamp, bolt and washer assembly.

4. Remove the synchronizer from the front of the engine cover assembly. Take note that the intermediate oil pump drive shaft should be removed with the synchronizer assembly.

To Install:

If the replacement synchronizer does not contain a plastic locator cover tool, a special service tool such as synchro positioner T89P–12200–A or equivalent must be obtained prior to installation of the replacement synchronizer. Failure to follow this procedure will result in improper synchronizer alignment. This will result in the ignition system and fuel system being out of time with the engine, possibly causing engine damage.

5. If the replacement synchronizer does not contain a plastic locator cover tool, attach the synchro positioner tool T89P–12200–A or equivalent as follows:

a. Engage the synchronizer vane into the radial slot of the tool. Rotate the tool on the synchronizer base until the tool boss engages the base notch.

b. The cover tool should be square and in contact with the entire top surface of the synchronizer base.

6. Transfer the intermediate oil pump driveshaft from the old synchronizer to the new one.

7. Install the synchronizer assembly so that the gear engagement occurs when the arrow on the locator tool is pointed approximately 30° counterclockwise from the front face of the engine block. This step will locate the camshaft sensor electrical connector in the pre-removal position.

8. Install the synchronizer base clamp, bolt and washer and torque the bolt to 15–22 ft. lbs.

9. Remove the synchro positioner tool T89P–12200–A or equivalent.

10. Install the camshaft sensor.

➡ If the camshaft sensor electrical connector is not positioned properly (i.e. contacting the A/C bracket or forward of the supercharger drive belt). Do not reposition the connector by rotating the synchronizer base. This will result in the ignition system and fuel system being out of time with the engine, possibly causing engine damage. Remove the synchronizer and repeat the installation procedure.

11. Reconnect the negative battery cable.

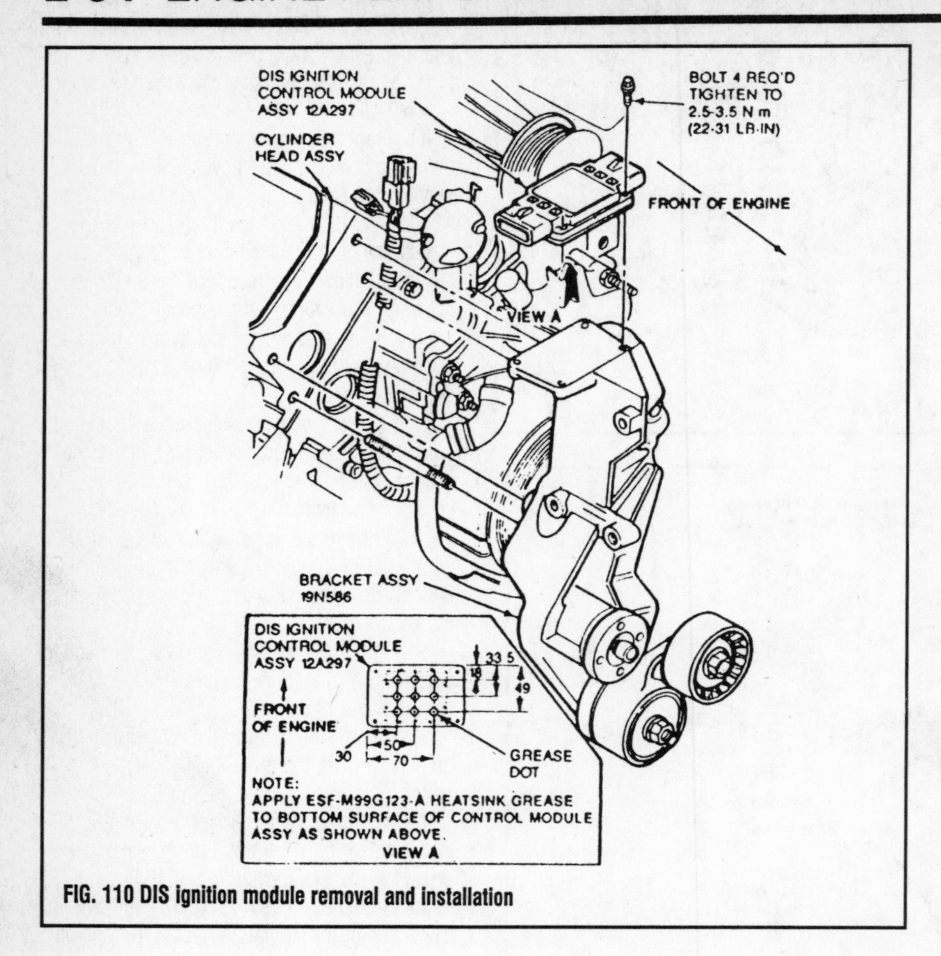

FIG. 110 DIS ignition module removal and installation

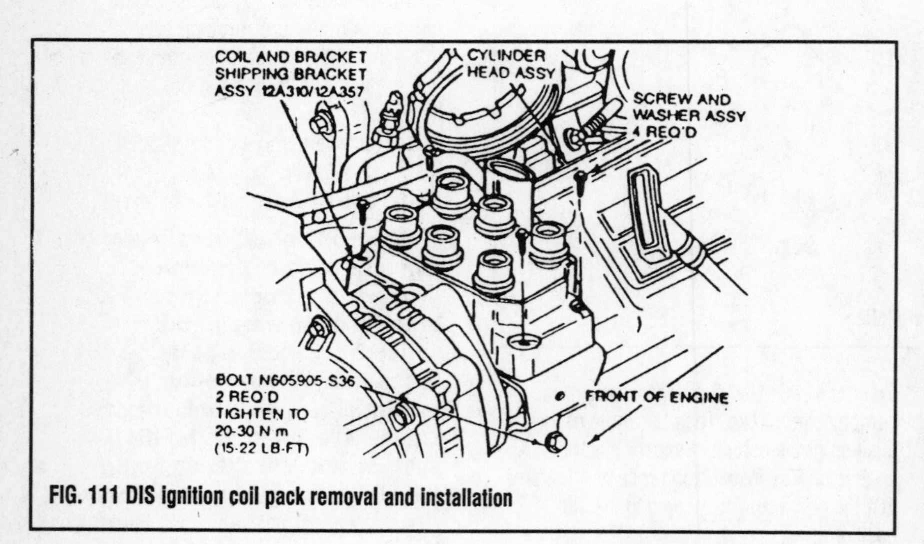

FIG. 111 DIS ignition coil pack removal and installation

DIS Module

REMOVAL & INSTALLATION

➧ SEE FIG. 110

1. Turn the ignition key **OFF** and disconnect the negative battery cable.
2. Disconnect both electrical connectors at the DIS module.
3. Remove the DIS module mounting bolts and remove the module.

To install:

4. Apply a uniform coating of heatsink grease ESF–M99G123–A or equivalent to the mounting surface of the DIS module.
5. Install the module and the mounting bolts. Torque the mounting bolts to 22–31 inch lbs.
6. Connect both electrical connectors to the module. Connect the negative battery cable.

Ignition Coil Pack

REMOVAL & INSTALLATION

➧ SEE FIG. 111

1. Turn the ignition key **OFF** and disconnect the negative battery cable.
2. Disconnect the electrical harness connector from the ignition coil pack.
3. Remove the spark plug wires by squeezing the locking tabs to release the coil boot retainers.
4. Remove the coil pack mounting screws and remove the coil pack.

To install:

5. Install the coil pack and the retaining screws. Torque the retaining screws to 40–62 inch lbs.
6. Connect the spark plug wires and connect the electrical connector to the coil pack.
7. Reconnect the negative battery cable.

➡ Be sure to place some dielectric compound into each spark plug boot prior to installation of the spark plug wire.

VALVE LASH

ADJUSTMENT

2.3L Engine

➧ SEE FIG. 112

1. Disconnect the negative battery cable.
2. Remove the valve cover assembly.
3. Position the camshaft so that the base circle of the lobe is facing the cam follower of the valve to be checked.
4. Using valve spring compressor tool T88T–6565–BH or equivalent, slowly apply pressure to the cam follower until the the lash adjuster is completely collapsed.
5. With follower collapsed, insert a feeler gauge between the base circle of the camshaft and follower. The clearance should not be more than 0.040–0.050 in. (1.0–1.3mm).
6. If the clearance is excessive, remove the cam follower and inspect for damage.
7. If the cam follower appears to be intact and not excessively worn, measure the valve spring assembled height to make sure the valve is not sticking.
8. If the valve spring assembled height is correct, check the camshaft for wear. If the camshaft dimensions are correct, replace the lash adjuster.
9. Install the valve cover and any other removed components.

5.0L HO Engine

➧ SEE FIG. 113

The valve lash is not adjustable. If the collapsed lifter clearance is found to be incorrect, there are replacement pushrods available to compensate for excessive or insufficient clearance.

1. Install an auxiliary starter switch. Crank the engine with the ignition switch off until the No. 1 piston is at TDC on the compression stroke.
2. With the crankshaft in the positions designated in Steps 4, 5 and 6, position lifter bleed down wrench tool No. T71P–6513–B or equivalent, on the rocker arm. Slowly apply pressure to bleed down the lifter until the plunger is completely bottomed. Hold the lifter in this position and check the available clearance between the rocker arm and the valve stem tip with a feeler gauge.
3. The clearance should be 0.123–0.146 in. (3.1–3.7mm) on 5.0L HO engine and 0.096–0.146 in. (2.4–3.7mm) on 5.8L engine. If the clearance is less than specification, install a shorter pushrod. If the clearance is greater than specification, install a longer pushrod.
4. The following valves can be checked with the engine in position 1, No. 1 piston at TDC on the compression stroke.
   - No. 1 intake — No. 1 exhaust
   - No. 4 intake — No. 3 exhaust
   - No. 8 intake — No. 7 exhaust
5. Rotate the engine 360° (1 revolution) from the 1st position and check the following valves:
   - No. 3 intake — No. 2 exhaust
   - No. 7 intake — No. 6 exhaust
6. Rotate the engine 90° (1/4 revolution) from the 2nd position and check the following valves:
   - No. 2 intake — No. 4 exhaust
   - No. 5 intake — No. 5 exhaust
   - No. 6 intake — No. 8 exhaust

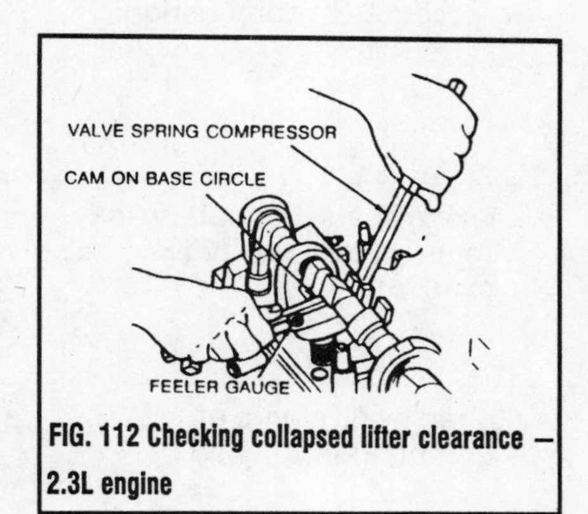

FIG. 112 Checking collapsed lifter clearance — 2.3L engine

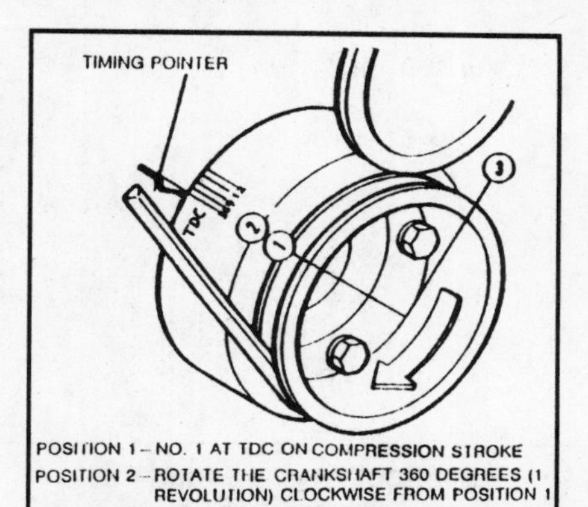

FIG. 113 Engine valve adjusting positions — 5.0L engine

Diagnosis of Spark Plugs

Problem	Possible Cause	Correction
Brown to grayish-tan deposits and slight electrode wear.	• Normal wear.	• Clean, regap, reinstall.
Dry, fluffy black carbon deposits.	• Poor ignition output.	• Check distributor to coil connections.
Wet, oily deposits with very little electrode wear.	• "Break-in" of new or recently overhauled engine. • Excessive valve stem guide clearances. • Worn intake valve seals.	• Degrease, clean and reinstall the plugs. • Refer to Section 3. • Replace the seals.

Diagnosis of Spark Plugs

Problem	Possible Cause	Correction
Red, brown, yellow and white colored coatings on the insulator. Engine misses intermittently under severe operating conditions.	• By-products of combustion.	• Clean, regap, and reinstall. If heavily coated, replace.
Colored coatings heavily deposited on the portion of the plug projecting into the chamber and on the side facing the intake valve.	• Leaking seals if condition is found in only one or two cylinders.	• Check the seals. Replace if necessary. Clean, regap, and reinstall the plugs.
Shiny yellow glaze coating on the insulator.	• Melted by-products of combustion.	• Avoid sudden acceleration with wide-open throttle after long periods of low speed driving. Replace the plugs.
Burned or blistered insulator tips and badly eroded electrodes.	• Overheating.	• Check the cooling system. • Check for sticking heat riser valves. Refer to Section 1. • Lean air-fuel mixture. • Check the heat range of the plugs. May be too hot. • Check ignition timing. May be over-advanced. • Check the torque value of the plugs to ensure good plug-engine seat contact.
Broken or cracked insulator tips.	• Heat shock from sudden rise in tip temperature under severe operating conditions. Improper gapping of plugs.	• Replace the plugs. Gap correctly.

3

ENGINE AND ENGINE OVERHAUL

GENERAL ENGINE SPECIFICATIONS

Year	Engine ID/VIN	Engine Displacement liter (cc)	Fuel System Type	Net Horsepower @ rpm	Net Torque @ rpm (ft. lbs.)	Bore × Stroke (in.)	Com-pression Ratio	Oil Pressure @ rpm
1989	A	2.3 (2300)	EFI	88 @ 4000	132 @ 2600	3.78 × 3.12	9.5:1	40–60 @ 2000 RPM
	E	5.0 HO (5000)	EFI	225 @ 4200	300 @ 3200	4.00 × 3.00	9.0:1	40–60 @ 2000 RPM
1990	A	2.3 (2300)	EFI	88 @ 4000	132 @ 2600	3.78 × 3.12	9.5:1	40–60 @ 2000 RPM
	E	5.0 HO (5000)	EFI	225 @ 4200	300 @ 3200	4.00 × 3.00	9.0:1	40–60 @ 2000 RPM
1991	M	2.3 (2300)	EFI	105 @ 4600	135 @ 2600	3.78 × 3.12	9.5:1	40–60 @ 2000 RPM
	E	5.0 HO (5000)	EFI	225 @ 4200	300 @ 3200	4.00 × 3.00	9.0:1	40–60 @ 2000 RPM
1992	M	2.3 (2300)	EFI	105 @ 4600	135 @ 2600	3.78 × 3.12	9.5:1	40–60 @ 2000 RPM
	E	5.0 HO (5000)	EFI	225 @ 4200	300 @ 3200	4.00 × 3.00	9.0:1	40–60 @ 2000 RPM

NOTE: Horsepower and torque are SAE net figures. They are measured at the rear of the transmission with all accessories installed and operating. Since the figures vary when a given engine is installed in different models, some are representative rather than exact.
EFI—Electronic Fuel Injection
HO—High Output

VALVE SPECIFICATIONS

Year	Engine ID/VIN	Engine Displacement liter (cc)	Seat Angle (deg.)	Face Angle (deg.)	Spring Test Pressure (lbs. @ in.)	Spring Installed Height (in.)	Stem-to-Guide Clearance (in.)		Stem Diameter (in.)	
							Intake	Exhaust	Intake	Exhaust
1989	A	2.3 (2300)	45	44	128–141 @ 1.12	1.52	0.0010–0.0027	0.0015–0.0032	0.3416–0.3423	0.3411–0.3418
	E	5.0 HO (5000)	45	44	①	②	0.0010–0.0027	0.0015–0.0032	0.3416–0.3423	0.3411–0.3418
1990	A	2.3 (2300)	45	44	128–141 @ 1.12	1.52	0.0010–0.0027	0.0015–0.0032	0.3416–0.3423	0.3411–0.3418
	E	5.0 HO (5000)	45	44	①	②	0.0010–0.0027	0.0015–0.0032	0.3416–0.3423	0.3411–0.3418
1991	M	2.3 (2300)	45	44	128–141 @ 1.12	1.52	0.0010–0.0027	0.0015–0.0032	0.3416–0.3423	0.3411–0.3418
	E	5.0 HO (5000)	45	44	①	②	0.0010–0.0027	0.0015–0.0032	0.3416–0.3423	0.3411–0.3418
1992	M	2.3 (2300)	45	44	128–141 @ 1.12	1.52	0.0010–0.0027	0.0015–0.0032	0.3416–0.3423	0.3411–0.3418
	E	5.0 HO (5000)	45	44	①	②	0.0010–0.0027	0.0015–0.0032	0.3416–0.3423	0.3411–0.3418

HO—High Output
① Intake—211-230 @ 1.33
Exhaust—200-226 @ 1.15
② Intake—1.75-1.80
Exhaust—1.58-1.64

CAMSHAFT SPECIFICATIONS

All measurements given in inches.

Year	Engine ID/VIN	Engine Displacement liter (cc)	Journal Diameter					Elevation		Bearing Clearance	Camshaft End Play
			1	2	3	4	5	In.	Ex.		
1989	A	2.3 (2300)	1.7713–1.7720	1.7713–1.7720	1.7713–1.7720	1.7713–1.7720	—	0.400	0.400	0.001–0.003	0.001–0.007
	E	5.0 HO (5000)	2.0805–2.0815	2.0655–2.0665	2.0505–2.0515	2.0355–2.0365	2.0205–2.0215	0.2780	0.2780	0.001–0.003	0.0005–0.0055

CAMSHAFT SPECIFICATIONS

All measurements given in inches.

Year	Engine ID/VIN	Engine Displacement liter (cc)	Journal Diameter 1	2	3	4	5	Elevation In.	Ex.	Bearing Clearance	Camshaft End Play
1990	A	2.3 (2300)	1.7713–1.7720	1.7713–1.7720	1.7713–1.7720	1.7713–1.7720	—	0.400	0.400	0.001–0.003	0.001–0.007
	E	5.0 HO (5000)	2.0805–2.0815	2.0655–2.0665	2.0505–2.0515	2.0355–2.0365	2.0205–2.0215	0.2780	0.2780	0.001–0.003	0.0005–0.0055
1991	M	2.3 (2300)	1.7713–1.7720	1.7713–1.7720	1.7713–1.7720	1.7713–1.7720	—	0.2381	0.2381	0.001–0.003	0.001–0.007
	E	5.0 HO (5000)	2.0805–2.0815	2.0655–2.0665	2.0505–2.0515	2.0355–2.0365	2.0205–2.0215	0.278	0.278	0.001–0.003	0.0005–0.0055
1992	M	2.3 (2300)	1.7713–1.7720	1.7713–1.7720	1.7713–1.7720	1.7713–1.7720	—	0.2381	0.2381	0.001–0.003	0.001–0.007
	E	5.0 HO (5000)	2.0805–2.0815	2.0655–2.0665	2.0505–2.0515	2.0355–2.0365	2.0205–2.0215	0.278	0.278	0.001–0.003	0.0005–0.0055

HO—High Output

CRANKSHAFT AND CONNECTING ROD SPECIFICATIONS

All measurements are given in inches.

Year	Engine ID/VIN	Engine Displacement liter (cc)	Crankshaft Main Brg. Journal Dia.	Main Brg. Oil Clearance	Shaft End-play	Thrust on No.	Connecting Rod Journal Diameter	Oil Clearance	Side Clearance
1989	A	2.3 (2300)	2.3990–2.3982	0.0008–0.0015	0.004–0.008	3	2.0464–2.0472	0.0008–0.0015	0.0035–0.0105
	E	5.0 HO (5000)	2.2482–2.2490	0.0004–0.0015	0.004–0.008	3	2.1228–2.1236	0.0008–0.0015	0.010–0.020
1990	A	2.3 (2300)	2.3990–2.3982	0.0008–0.0015	0.004–0.008	3	2.0465–2.0472	0.0008–0.0015	0.0035–0.0105
	E	5.0 HO (5000)	2.2482–2.2490	0.0004–0.0015	0.004–0.008	3	2.1228–2.1236	0.0008–0.0015	0.010–0.020
1991	M	2.3 (2300)	2.2051–2.2059	0.0008–0.0015	0.003–0.008	3	2.0462–2.0472	0.0008–0.0015	0.0035–0.0105
	E	5.0 HO (5000)	2.2482–2.2490	0.0004–0.0015	0.004–0.008	3	2.1228–2.1236	0.0008–0.0015	0.010–0.020
1992	M	2.3 (2300)	2.2051–2.2059	0.0008–0.0015	0.003–0.008	3	2.0462–2.0472	0.0008–0.0015	0.0035–0.0105
	E	5.0 HO (5000)	2.2482–2.2490	0.0004–0.0015	0.004–0.008	3	2.1228–2.1236	0.0008–0.0015	0.010–0.020

HO—High Output

PISTON AND RING SPECIFICATIONS

All measurements are given in inches.

Year	Engine ID/VIN	Engine Displacement liter (cc)	Piston Clearance	Ring Gap Top Compression	Bottom Compression	Oil Control	Ring Side Clearance Top Compression	Bottom Compression	Oil Control
1989	A	2.3 (2300)	0.0030–0.0038	0.0100–0.0200	0.0100–0.0200	0.0150–0.0550	0.0020–0.0040	0.0020–0.0040	Snug
	E	5.0 HO (5000)	0.0030–0.0038	0.0100–0.0200	0.0100–0.0200	0.0150–0.0550	0.0020–0.0040	0.0020–0.0040	Snug

PISTON AND RING SPECIFICATIONS

All measurements are given in inches.

Year	Engine ID/VIN	Engine Displacement liter (cc)	Piston Clearance	Ring Gap Top Compression	Ring Gap Bottom Compression	Ring Gap Oil Control	Ring Side Clearance Top Compression	Ring Side Clearance Bottom Compression	Ring Side Clearance Oil Control
1990	A	2.3 (2300)	0.0030–0.0038	0.0100–0.0200	0.0100–0.0200	0.0150–0.0550	0.0020–0.0040	0.0020–0.0040	Snug
	E	5.0 HO (5000)	0.0030–0.0038	0.0100–0.0200	0.0100–0.0200	0.0150–0.0550	0.0020–0.0040	0.0020–0.0040	Snug
1991	M	2.3 (2300)	0.0024–0.0033	0.0100–0.0200	0.0100–0.0200	0.0150–0.0490	0.0016–0.0033	0.0016–0.0033	Snug
	E	5.0 HO (5000)	0.0030–0.0038	0.0100–0.0200	0.0100–0.0200	0.0150–0.0550	0.0020–0.0040	0.0020–0.0040	Snug
1992	M	2.3 (2300)	0.0024–0.0033	0.0100–0.0200	0.0100–0.0200	0.0150–0.0490	0.0016–0.0033	0.0016–0.0033	Snug
	E	5.0 HO (5000)	0.0030–0.0038	0.0100–0.0200	0.0100–0.0200	0.0150–0.0550	0.0020–0.0040	0.0020–0.0040	Snug

HO—High Output

TORQUE SPECIFICATIONS

All readings in ft. lbs.

Year	Engine ID/VIN	Engine Displacement liter (cc)	Cylinder Head Bolts	Main Bearing Bolts	Rod Bearing Bolts	Crankshaft Damper Bolts	Flywheel Bolts	Manifold Intake	Manifold Exhaust	Spark Plugs	Lug Nut
1989	A	2.3 (2300)	①	②	③	103–133	56–64	20–29	④	5–10	80–105
	E	5.0 HO (5000)	⑤	60–70	19–24	70–90	75–85	⑥	18–24	5–10	80–105
1990	A	2.3 (2300)	①	②	③	103–133	56–64	20–29	④	5–10	80–105
	E	5.0 HO (5000)	⑤	60–70	19–24	70–90	75–85	⑥	18–24	5–10	80–105
1991	M	2.3 (2300)	①	②	③	114–151	56–64	19–28	④	5–10	80–105
	E	5.0 HO (5000)	⑤	60–70	19–24	70–90	75–85	⑥	18–24	5–10	80–105
1992	M	2.3 (2300)	①	②	③	114–151	56–64	19–28	④	5–10	80–105
	E	5.0 HO (5000)	⑤	60–70	19–24	70–90	75–85	⑥	18–24	5–10	80–105

HO—High Output

① Torque in sequence in 2 steps:
Step 1—50-60 ft. lbs.
Step 2—80-90 ft. lbs.

② Tighten in 2 steps:
Step 1—50-60 ft. lbs.
Step 2—75-85 ft. lbs.

③ Tighten in 2 steps:
Step 1—25-30 ft. lbs.
Step 2—30-36 ft. lbs.

④ Tighten in 2 steps:
Step 1—178-204 inch lbs.
Step 2—20-30 ft. lbs.

⑤ Tighten in 2 steps:
Step 1—55-65 ft. lbs.
Step 2—65-72 ft. lbs.

⑥ Tighten in 2 steps:
Step 1—15-20 ft. lbs.
Step 2—23-25 ft. lbs.

ENGINE ELECTRICAL

Understanding the Engine Electrical System

The engine electrical system can be broken down into three separate and distinct systems:

1. The starting system.
2. The charging system.
3. The ignition system.

BATTERY AND STARTING SYSTEM

Basic Operating Principles

The battery is the first link in the chain of mechanisms which work together to provide cranking of the automobile engine. In most modern vehicles, the battery is a lead/acid electrochemical device consisting of six 2v subsections connected in series so the unit is capable of producing approximately 12v of electrical pressure. Each subsection, or cell, consists of a series of positive and negative plates held a short distance apart in a solution of sulfuric acid and water. The two types of plates are of dissimilar metals. This causes a chemical reaction to be set up, and it is this reaction which produces current flow from the battery when its positive and negative terminals are connected to an electrical appliance such as a lamp or motor. The continued transfer of electrons would eventually convert the sulfuric acid in the electrolyte to water, and make the two plates identical in chemical composition. As electrical energy is removed from the battery, its voltage output tends to drop. Thus, measuring battery voltage and battery electrolyte composition are two ways of checking the ability of the unit to supply power. During the starting of the engine, electrical energy is removed from the battery. However, if the charging circuit is in good condition and the operating conditions are normal, the power removed from the battery will be replaced by the generator (or alternator) which will force electrons back through the battery, reversing the normal flow, and restoring the battery to its original chemical state.

The battery and starting motor are linked by very heavy electrical cables designed to minimize resistance to the flow of current. Generally, the major power supply cable that leaves the battery goes directly to the starter, while other electrical system needs are supplied by a smaller cable. During starter operation, power flows from the battery to the starter and is grounded through the vehicle's frame and the battery's negative ground strap.

The starting motor is a specially designed, direct current electric motor capable of producing a very great amount of power for its size. One thing that allows the motor to produce a great deal of power is its tremendous rotating speed. It drives the engine through a tiny pinion gear (attached to the starter's armature), which drives the very large flywheel ring gear at a greatly reduced speed. Another factor allowing it to produce so much power is that only intermittent operation is required of it. This, little allowance for air circulation is required, and the windings can be built into a very small space.

The starter solenoid is a magnetic device which employs the small current supplied by the starting switch circuit of the ignition switch. This magnetic action moves a plunger which mechanically engages the starter and electrically closes the heavy switch which connects it to the battery. The starting switch circuit consists of the starting switch contained within the ignition switch, a transmission neutral safety switch or clutch pedal switch, and the wiring necessary to connect these in series with the starter solenoid or relay.

A pinion, which is a small gear, is mounted to a one-way drive clutch. This clutch is splined to the starter armature shaft. When the ignition switch is moved to the **start** position, the solenoid plunger slides the pinion toward the flywheel ring gear via a collar and spring. If the teeth on the pinion and flywheel match properly, the pinion will engage the flywheel immediately. If the gear teeth butt one another, the spring will be compressed and will force the gears to mesh as soon as the starter turns far enough to allow them to do so. As the solenoid plunger reaches the end of its travel, it closes the contacts that connect the battery and starter and then the engine is cranked.

As soon as the engine starts, the flywheel ring gear begins turning fast enough to drive the pinion at an extremely high rate of speed. At this point, the one-way clutch begins allowing the pinion to spin faster than the starter shaft so that the starter will not operate at excessive speed. When the ignition switch is released from the starter position, the solenoid is de-energized, and a spring contained within the solenoid assembly pulls the gear out of mesh and interrupts the current flow to the starter.

Some starter employ a separate relay, mounted away from the starter, to switch the motor and solenoid current on and off. The relay thus replaces the solenoid electrical switch, buy does not eliminate the need for a solenoid mounted on the starter used to mechanically engage the starter drive gears. The relay is used to reduce the amount of current the starting switch must carry.

THE CHARGING SYSTEM

Basic Operating Principles

The automobile charging system provides electrical power for operation of the vehicle's ignition and starting systems and all the electrical accessories. The battery services as an electrical surge or storage tank, storing (in chemical form) the energy originally produced by the engine driven generator. The system also provides a means of regulating generator output to protect the battery from being overcharged and to avoid excessive voltage to the accessories.

The storage battery is a chemical device incorporating parallel lead plates in a tank containing a sulfuric acid/water solution. Adjacent plates are slightly dissimilar, and the chemical reaction of the two dissimilar plates produces electrical energy when the battery is connected to a load such as the starter motor. The chemical reaction is reversible, so that when the generator is producing a voltage (electrical pressure) greater than that produced by the battery, electricity is forced into the battery, and the battery is returned to its fully charged state.

The vehicle's generator is driven mechanically, through V-belts, by the engine crankshaft. It consists of two coils of fine wire, one stationary (the stator), and one movable (the rotor). The rotor may also be known as the armature, and consists of fine wire wrapped around an iron core which is mounted on a shaft. The electricity which flows through the two coils of wire (provided initially by the battery in some cases) creates an intense magnetic field around both rotor and stator, and the interaction between the two fields creates voltage, allowing the generator to power the accessories and charge the battery.

There are two types of generators: the earlier is the direct current (DC) type. The current produced by the DC generator is generated in the armature and carried off the spinning armature by stationary brushes contacting the commutator. The commutator is a series of smooth metal contact plates on the end of the armature. The commutator is a series of smooth metal contact plates on the end of the armature. The commutator plates, which are separated from one another by a very short gap, are connected to the armature circuits so that current will flow in one directions only in the wires carrying the generator output. The generator stator consists of two stationary coils of wire which draw some of the output current of the generator to form a powerful magnetic field and create the interaction of fields which generates the voltage. The generator field is wired in series with the regulator.

Newer automobiles use alternating current generators or alternators, because they are more efficient, can be rotated at higher speeds, and have fewer brush problems. In an alternator, the field rotates while all the current produced passes only through the stator winding. The brushes bear against continuous slip rings rather than a commutator. This causes the current produced to periodically reverse the direction of its flow. Diodes (electrical one-way switches) block the flow of current from traveling in the wrong direction. A series of diodes is wired together to permit the alternating flow of the stator to be converted to a pulsating, but unidirectional flow at the alternator output. The alternator's field is wired in series with the voltage regulator.

The regulator consists of several circuits. Each circuit has a core, or magnetic coil of wire, which operates a switch. Each switch is connected to ground through one or more resistors. The coil of wire responds directly to system voltage. When the voltage reaches the required level, the magnetic field created by the winding of wire closes the switch and inserts a resistance into the generator field circuit, thus reducing the output. The contacts of the switch cycle open and close many times each second to precisely control voltage.

While alternators are self-limiting as far as maximum current is concerned, DC generators employ a current regulating circuit which responds directly to the total amount of current flowing through the generator circuit rather than to the output voltage. The current regulator is similar to the voltage regulator except that all system current must flow through the energizing coil on its way to the various accessories.

Alternator

OPERATION

The alternator charging system is a negative (–) ground system which consists of an alternator, a regulator, a charge indicator, a storage battery and wiring connecting the components, and fuse link wire.

The alternator is belt-driven from the engine. Energy is supplied from the alternator/regulator system to the rotating field through two brushes to two slip-rings. The slip-rings are mounted on the rotor shaft and are connected to the field coil. This energy supplied to the rotating field from the battery is called excitation current and is used to initially energize the field to begin the generation of electricity. Once the alternator starts to generate electricity, the excitation current comes from its own output rather than the battery.

The alternator produces power in the form of alternating current. The alternating current is rectified by 6 diodes into direct current. The direct current is used to charge the battery and power the rest of the electrical system.

When the ignition key is turned on, current flows from the battery, through the charging system indicator light on the instrument panel, to the voltage regulator, and to the alternator. Since the alternator is not producing any current, the alternator warning light comes on. When the engine is started, the alternator begins to produce current and turns the alternator light off. As the alternator turns and produces current, the current is divided in two ways: part to the battery to charge the battery and power the electrical components of the vehicle, and part is returned to the alternator to enable it to increase its output. In this situation, the alternator is receiving current from the battery and from itself. A voltage regulator is wired into the current supply to the alternator to prevent it from receiving too much current which would cause it to put out too much current. Conversely, if the voltage regulator does not allow the alternator to receive enough current, the battery will not be fully charged and will eventually go dead.

The battery is connected to the alternator at all times, whether the ignition key is turned on or not. If the battery were shorted to ground, the alternator would also be shorted. This would damage the alternator. To prevent this, a fuse link is installed in the wiring between the battery and the alternator. If the battery is shorted, the fuse link is melted, protecting the alternator.

ALTERNATOR PRECAUTIONS

To prevent damage to the alternator and regulator, the following precautions should be taken when working with the electrical system.

1. Never reverse the battery connections.
2. Booster batteries for starting must be connected properly: positive-to-positive and negative-to-ground.
3. Disconnect the battery cables before using a fast charger; the charger has a tendency to force current through the diodes in the opposite direction for which they were designed. This burns out the diodes.
4. Never use a fast charger as a booster for starting the vehicle.
5. Never disconnect the voltage regulator while the engine is running.
6. Avoid long soldering times when replacing diodes or transistors. Prolonged heat is damaging to AC generators.
7. Do not use test lamps of more than 12 volts (V) for checking diode continuity.
8. Do not short across or ground any of the terminals on the AC generator.
9. The polarity of the battery, generator, and regulator must be matched and considered before making any electrical connections within the system.
10. Never operate the alternator on an open circuit. make sure that all connections within the circuit are clean and tight.
11. Disconnect the battery terminals when performing any service on the electrical system. This will eliminate the possibility of accidental reversal of polarity.
12. Disconnect the battery ground cable if arc welding is to be done on any part of the vehicle.

CHARGING SYSTEM TROUBLESHOOTING

There are many possible ways in which the charging system can malfunction. Often the source of a problem is difficult to diagnose, requiring special equipment and a good deal of experience. This is usually not the case, however, where the charging system fails completely and causes the dash board warning light to come on or the battery to become dead. To troubleshoot a complete system failure only two pieces of equipment are needed: a test light, to determine that current is reaching a certain point; and a current indicator (ammeter), to determine the direction of the current flow and its measurement in amps.

This test works under three assumptions:

1. The battery is known to be good and fully charged.
2. The alternator belt is in good condition and adjusted to the proper tension.
3. All connections in the system are clean and tight.

➡ **In order for the current indicator to give a valid reading, the vehicle must be equipped with battery cables which are of the same gauge size and quality as original equipment battery cables.**

1. Turn off all electrical components on the vehicle. Make sure the doors of the vehicle are closed. If the vehicle is equipped with a clock, disconnect the clock by removing the lead wire from the rear of the clock. Disconnect the positive battery cable from the battery and connect the ground wire on a test light to the disconnected positive battery cable. Touch the probe end of the test light to the positive battery post. The test light should not light. If the test light does light, there is a short or open circuit on the vehicle.
2. Disconnect the voltage regulator wiring harness connector at the voltage regulator. Turn on the ignition key. Connect the wire on a test light to a good ground (engine bolt). Touch the probe end of a test light to the ignition wire connector into the voltage regulator wiring connector. This wire corresponds to the **I** terminal on the regulator. If the test light goes on, the charging system warning light circuit is complete. If the test light does not come on and the warning light on the instrument panel is on, either the resistor wire, which is parallel with the warning light, or the wiring to the voltage regulator, is defective. If the test light does not come on and the warning light is not on, either the bulb is defective or the power supply wire form the battery through the ignition switch to the bulb has an open circuit. Connect the wiring harness to the regulator.
3. Examine the fuse link wire in the wiring harness from the starter relay to the alternator. If the insulation on the wire is cracked or split, the fuse link may be melted. Connect a test light to the fuse link by attaching the ground wire on the test light to an engine bolt and touching the probe end of the light to the bottom of the fuse link wire

where it splices into the alternator output wire. If the bulb in the test light does not light, the fuse link is melted.

4. Start the engine and place a current indicator on the positive battery cable. Turn off all electrical accessories and make sure the doors are closed. If the charging system is working properly, the gauge will show a draw of less than 5 amps. If the system is not working properly, the gauge will show a draw of more than 5 amps. A charge moves the needle toward the battery, a draw moves the needle away from the battery. Turn the engine off.

5. Disconnect the wiring harness from the voltage regulator at the regulator at the regulator connector. Connect a male spade terminal (solderless connector) to each end of a jumper wire. Insert one end of the wire into the wiring harness connector which corresponds to the **A** terminal on the regulator. Insert the other end of the wire into the wiring harness connector which corresponds to the **F** terminal on the regulator. Position the connector with the jumper wire installed so that it cannot contact any metal surface under the hood. Position a current indicator gauge on the positive battery cable. Have an assistant start the engine. Observe the reading on the current indicator. Have your assistant slowly raise the speed of the engine to about 2,000 rpm or until the current indicator needle stops moving, whichever comes first. Do not run the engine for more than a short period of time in this condition. If the wiring harness connector or jumper wire becomes excessively hot during this test, turn off the engine and check for a grounded wire in the regulator wiring harness. If the current indicator shows a charge of about three amps less than the output of the alternator, the alternator is working properly. If the previous tests showed a draw, the voltage regulator is defective. If the gauge does not show the proper charging rate, the alternator is defective.

BELT TENSION ADJUSTMENT

➧ SEE FIG. 1-3

All vehicles are equipped with an automatic belt tensioner. No adjustment is necessary or possible. The belt tensioner is equipped with a belt wear indicator; when 1 percent belt stretch is indicated, the drive belt must be replaced. If the wear indicator is difficult to see on the 5.0L HO engine, locate the tab on the tensioner face plate. The tab should be approximately between the stops.

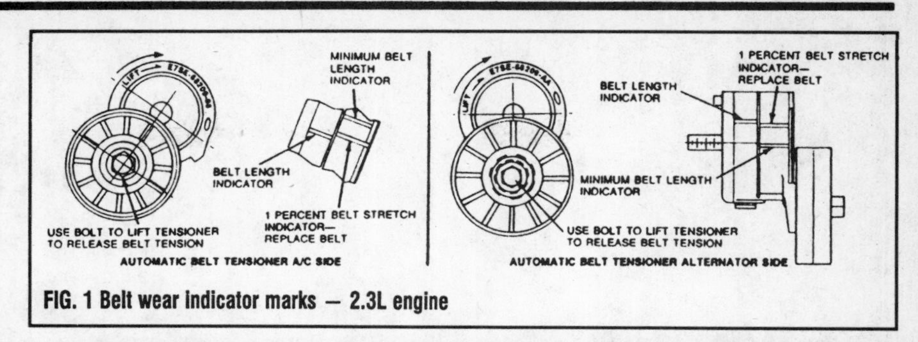

FIG. 1 Belt wear indicator marks — 2.3L engine

FIG. 2 Automatic belt adjuster — 2.3L engine

REMOVAL & INSTALLATION

➧ SEE FIGS. 4-6

1. Disconnect the negative battery cable.
2. Tag and disconnect the wiring connectors from the rear of the alternator. To disconnect push-on type terminals, depress the lock tab and pull straight off.
3. Loosen the alternator pivot bolt and remove the adjusting bolt. Disengage the drive belt from the alternator pulley.
4. Remove the alternator pivot bolt and the alternator.
5. Installation is the reverse of the removal procedure.

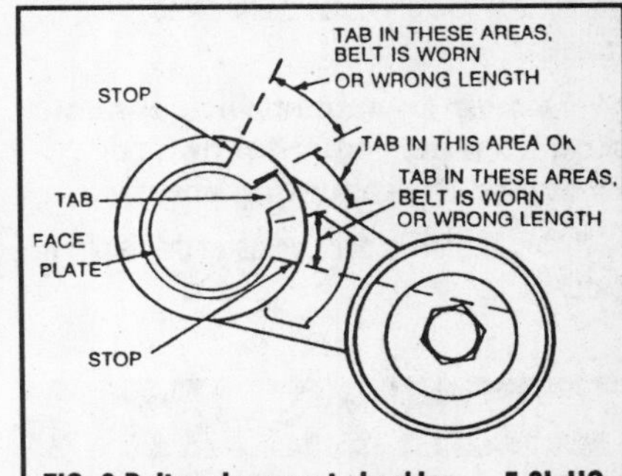

FIG. 3 Belt replacement checking — 5.0L HO engine

FIG. 4 Removing the rear wiring connector form the alternator

FIG. 5 Loosening the alternator pivot bolt

FIG. 6 Removing the alternator attaching bolt

Voltage Regulator

REMOVAL & INSTALLATION

➧ SEE FIG. 7

1. Disconnect the negative battery cable. The regulator is located on the back of the alternator.
2. Remove the regulator mounting screws, unlock the wire connectors and remove the regulator.

➡ **Always disconnect the connector plug from the regulator before removing the mounting screws.**

3. Installation is the reverse of the removal procedure.

Battery

REMOVAL & INSTALLATION

1. Loosen the nuts which secure the cable ends to the battery terminals. Lift the negative battery cables from the terminals first with a twisting motion, then the positive cables. If there is a battery cable puller available, make use of it.
2. Remove the holddown nuts from the battery holddown bracket and remove the bracket and the battery. Lift the battery straight up and out of the vehicle, being sure to keep the battery level to avoid spilling the battery acid.
3. Before installing the battery in the vehicle, make sure that the battery terminals are clean and free from corrosion. Use a battery terminal cleaner on the terminals and on the inside of the battery cable ends. If a cleaner is not available, use coarse grade sandpaper to remove the corrosion. A mixture of baking soda and water poured over the terminals and cable ends will help remove and neutralize any acid buildup.

**** CAUTION**

Take great care to avoid getting any of the baking soda solution inside the battery. If any solution gets inside the battery a violent reaction will take place and/or the battery will be damaged.

4. Before installing the cables onto the terminals, cut a piece of felt cloth, or something similar into a circle about 3 in. (76mm) across. Cut a hole in the middle about the size of the battery terminals at their base. Push the cloth pieces over the terminals so that they lay flat on the top of the battery. Soak the pieces of cloth with oil. This will keep oxidation to a minimum.
5. Place the battery in the vehicle. Install the cables onto the terminals.
6. Tighten the nuts on the cable ends.

➡ See Section 1 for battery maintenance illustrations.

7. Smear a light coating of grease on the cable ends and tops of the terminals. This will further prevent the buildup of oxidation on the terminals and the cable ends.
8. Install and tighten the nuts of the battery holddown bracket.

Starter Motor

REMOVAL & INSTALLATION

➧ SEE FIG. 7-9

1. Disconnect the negative battery cable.
2. Raise the front of the vehicle and install jackstands beneath the frame. Firmly apply the parking brake and place blocks in back of the rear wheels.
3. Tag and disconnect the wiring at the starter.
4. Remove the starter mounting bolts and remove the starter.
5. Reverse the above procedure to install. Observe the following torques:

- Mounting bolts: 12–15 ft. lbs. on starters with 3 mounting bolts and 15–20 ft. lbs. on starters with 2 mounting bolts.

Make sure that the nut securing the heavy cable to the starter is snugged down tightly.

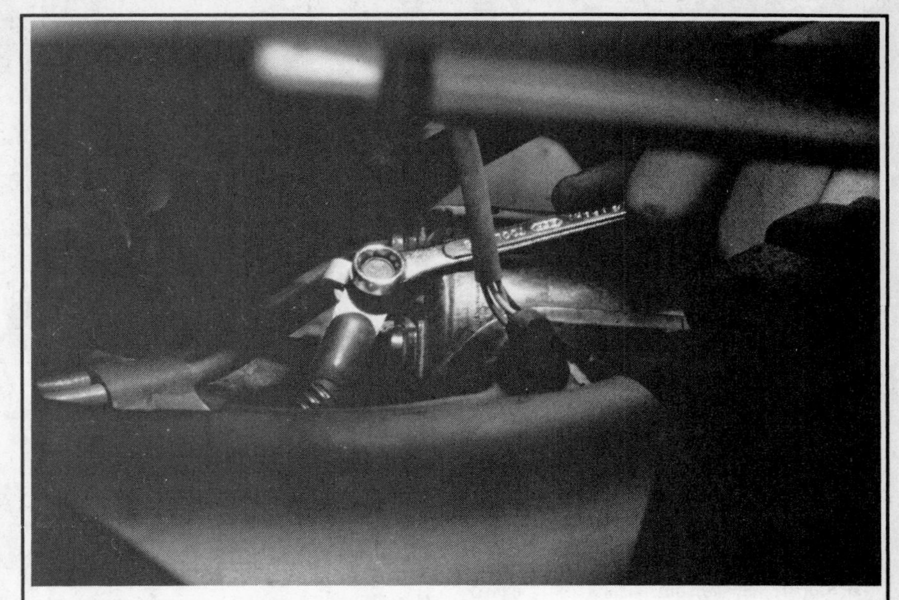

FIG. 7 Disconnecting the battery cable from the starter

FIG. 8 Removing the starter mounting bolts

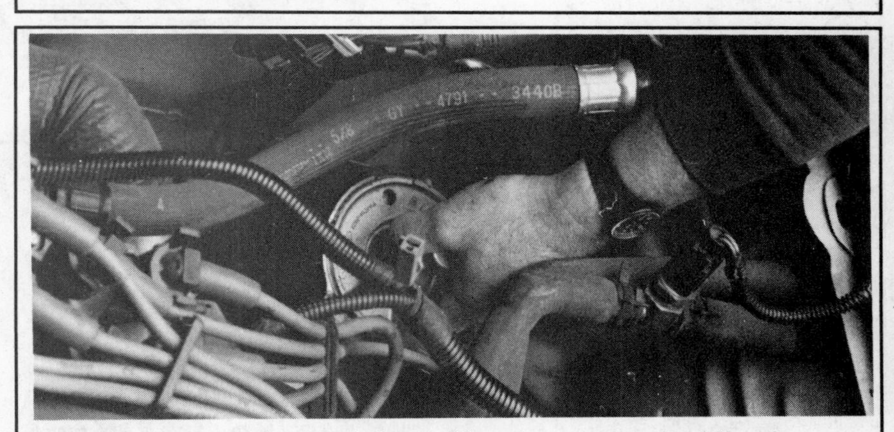

FIG. 9 Removing the starter out-through-the-top of the engine compartment — 2.3L engine

OVERHAUL

➧ SEE FIGS. 10-11

Brush Replacement

1. Remove the starter from the engine as previously outlined.
2. Remove the starter drive plunger lever cover and gasket.
3. Loosen and remove the brush cover band and remove the brushes from their holder.
4. Remove the two through-bolts from the starter frame.
5. Separate the drive end housing, starter frame and brush end plate assemblies.
6. Remove the starter drive plunger lever and pivot pin, and remove the armature.
7. Remove the ground brush retaining screws from the frame and remove the brushes.
8. Cut the insulated brush leads from the field coils, as close to the field connection point as possible.
9. Clean and inspect the starter motor.
10. Replace the brush end plate if the insulator between the field brush holder and the end plate is cracked or broken.
11. Position the new insulated field brushes lead on the field coil connection. Position and crimp the clip provided with the brushes to hold the brush lead to the connection. Solder the lead, clip, and connection together using resin core solder. Use a 300 watt soldering iron.
12. Install the ground brush leads to the frame with the retaining screws.
13. Install the starter drive plunger lever and pivot pin, and install the armature.
14. Assemble the drive end housing, starter frame and brush end plate assemblies.
15. Install the two through-bolts in the starter frame. Torque the through-bolts to 55–75 inch lbs.
16. Install the brushes in their holders and install the brush cover band.
17. Install the starter drive plunger lever cover and gasket.
18. Install the starter on the engine as previously outlined.

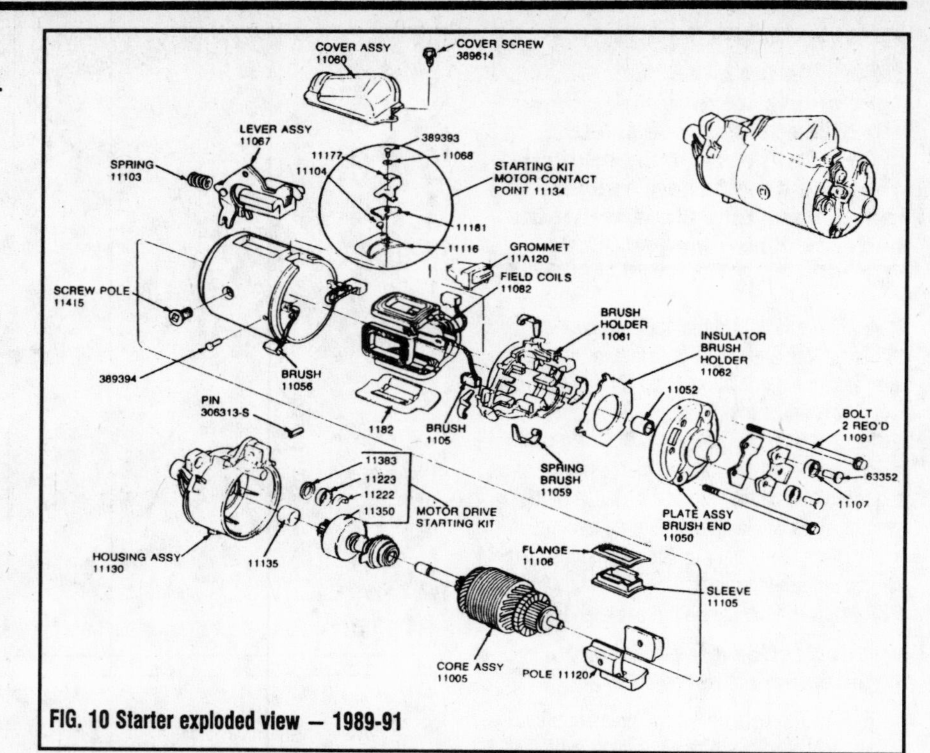

FIG. 10 Starter exploded view — 1989-91

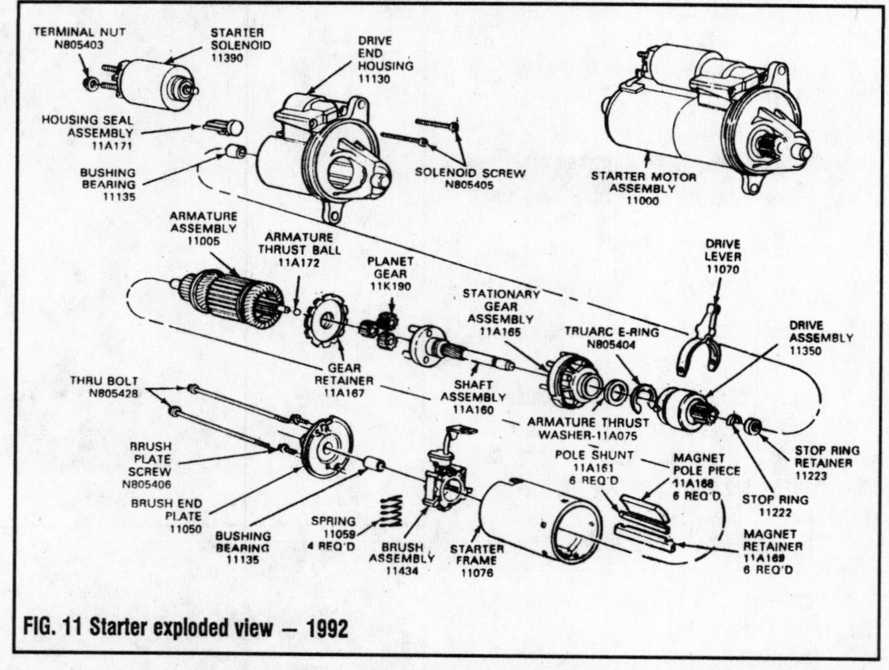

FIG. 11 Starter exploded view — 1992

Drive Replacement

1. Remove the starter as outlined previously.
2. Remove the starter drive plunger lever and gasket and the brush cover band.
3. Remove the two through-bolts from the starter frame.
4. Separate the drive end housing from the starter frame.
5. The starter drive plunger lever return spring may fall out after detaching the drive end housing. If not, remove it.
6. Remove the pivot pin which attaches the starter drive plunger lever to the starter frame and remove the lever.
7. Remove the stop ring retainer and stop ring from the armature shaft.
8. Slide the starter drive off the armature shaft.
9. Examine the wear pattern on the starter drive teeth. There should be evidence of full contact between the starter drive teeth and the flywheel ring gear teeth. If there is evidence of irregular wear, examine the flywheel ring gear for damage and replace if necessary.
10. Apply a thin coat of white grease to the armature shaft before installing the drive gear. Place a small amount of grease in the drive end housing bearing. Slide the starter drive on the armature shaft.
11. Install the stop ring retainer and stop ring on the armature shaft.
12. Install the starter drive plunger lever on the starter frame and install the pin.

13. Assemble the drive end housing on the starter frame.
14. install the two through-bolts in the starter frame. Tighten the starter through bolts to 55–75 inch lbs.
15. Install the starter drive plunger lever and gasket and the brush cover band.
16. Install the starter as outlined previously.

Starter Relay

REMOVAL & INSTALLATION

1989-91

1. Disconnect the positive battery cable from the battery terminal. With dual batteries, disconnect the connecting cable at both ends.
2. Remove the nut securing the positive battery cable to the relay.
3. Remove the positive cable and any other wiring under that cable.
4. Tag and remove the push-on wires from the front of the relay.
5. Remove the nut and disconnect the cable from the starter side of the relay.
6. Remove the relay attaching bolts and remove the relay.
7. Installation is the reverse of removal.

ENGINE MECHANICAL

Engine Overhaul Tips

Most engine overhaul procedures are fairly standard. In addition to specific parts replacement procedures and complete specifications for your individual engine, this section also is a guide to accept rebuilding procedures. Examples of standard rebuilding practice are shown and should be used along with specific details concerning your particular engine.

Competent and accurate machine shop services will ensure maximum performance, reliability and engine life.

In most instances it is more profitable for the do-it-yourself mechanic to remove, clean and inspect the component, buy the necessary parts and deliver these to a shop for actual machine work.

On the other hand, much of the rebuilding work (crankshaft, block, bearings, piston rods, and other components) is well within the scope of the do-it-yourself mechanic.

TOOLS

The tools required for an engine overhaul or parts replacement will depend on the depth of your involvement. With a few exceptions, they will be the tools found in a mechanic's tool kit. More in-depth work will require any or all of the following:

- a dial indicator (reading in thousandths) mounted on a universal base
- micrometers and telescope gauges
- jaw and screw-type pullers
- scraper
- valve spring compressor
- ring groove cleaner
- piston ring expander and compressor

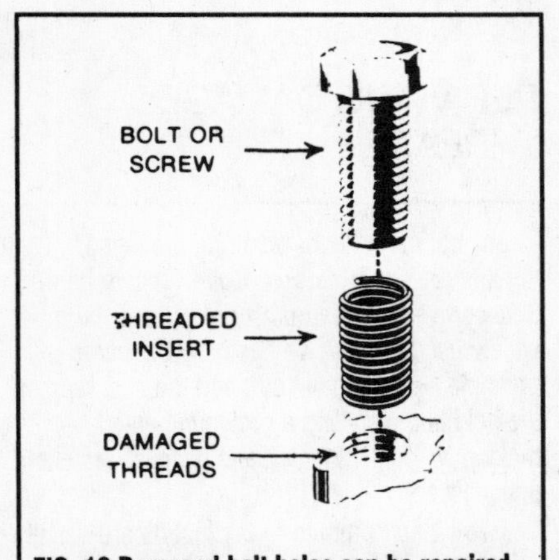

FIG. 12 Damaged bolt holes can be repaired with thread Standard thread repair insert

- ridge reamer
- cylinder hone or glaze breaker
- Plastigage®
- engine stand

The use of most of these tools is illustrated in this section. Many can be rented for a one-time use from a local parts jobber or tool supply house specializing in automotive work.

Occasionally, the use of special tools is called for. See the information on Special Tools and Safety Notice in the front of this book before substituting another tool.

INSPECTION TECHNIQUES

Procedures and specifications are given in this section for inspecting, cleaning and assessing the wear limits of most major components. Other procedures such as Magnaflux® and Zyglo® can be used to locate material flaws and stress cracks. Magnaflux® is a magnetic process applicable only to ferrous materials. The Zyglo® process coats the material with a fluorescent dye penetrant and can be used on any material Check for suspected surface cracks can be more readily made using spot check dye. The dye is sprayed onto the suspected area, wiped off and the area sprayed with a developer. Cracks will show up brightly.

FIG. 13 Spark plug thread insert shown

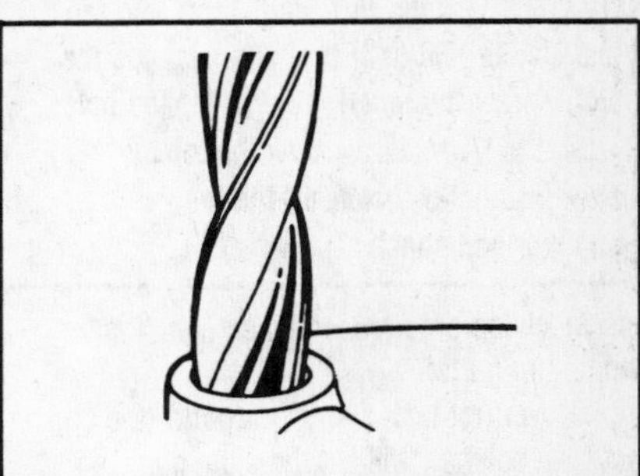
FIG. 14 Drill out the damaged threads with the specified drill. Drill completely through the hole or to the bottom of a blind hole

OVERHAUL TIPS

➧ SEE FIGS. 12-16

Aluminum has become extremely popular for use in engines, due to its low weight. Observe

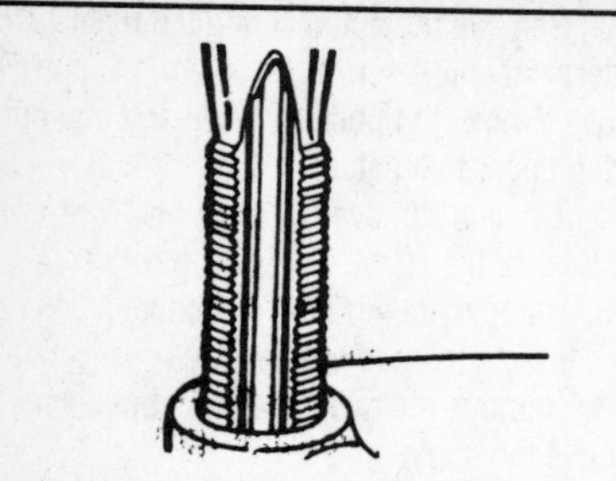

FIG. 15 With the tap supplied, tap the hole to receive the thread insert. Keep the tap well oiled and back it out frequently to avoid clogging the threads

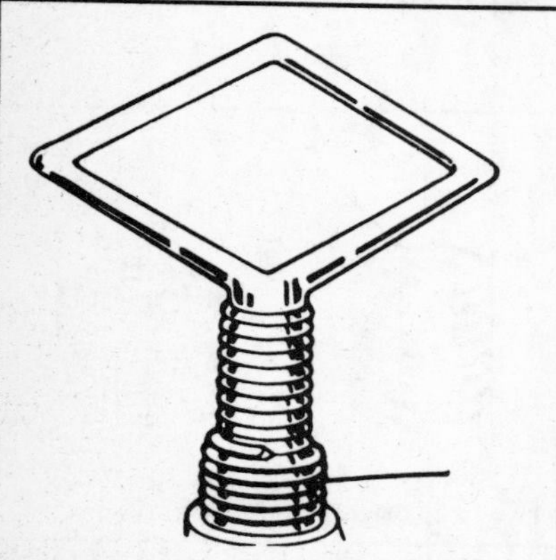

FIG. 16 Screw the threaded insert into the installation tool until the tang engages the slot. Screw the insert into the tapped hole until it is 1/4-1/2 turn below the top surface, after installation break off the tang with a hammer and punch

the following precautions when handling aluminum parts:

• Never hot tank aluminum parts (the caustic hot tank solution will eat the aluminum.

• Remove all aluminum parts (identification tag, etc.) from engine parts prior to the tanking.

• Always coat threads lightly with engine oil or anti-seize compounds before installation, to prevent seizure.

• Never overtorque bolts or spark plugs especially in aluminum threads.

Stripped threads in any component can be repaired using any of several commercial repair kits (Heli-Coil®, Microdot®, Keenserts®, etc.).

When assembling the engine, any parts that will be frictional contact must be prelubed to provide lubrication at initial start-up. Any product specifically formulated for this purpose can be used, but engine oil is not recommended as a prelube.

When semi-permanent (locked, but removable) installation of bolts or nuts is desired, threads should be cleaned and coated with Loctite® or other similar, commercial non-hardening sealant.

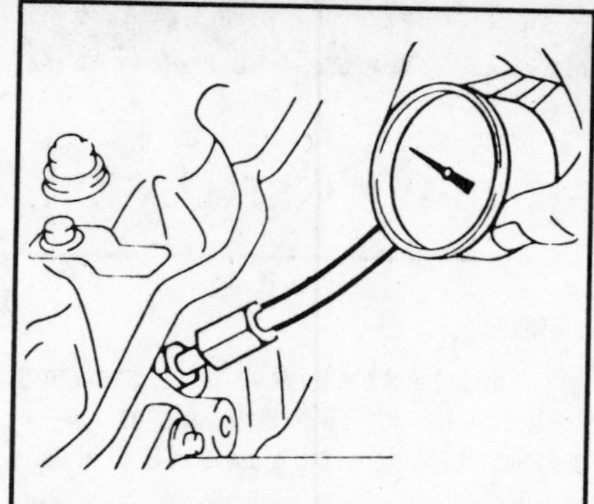

FIG. 17 The screw-in type compression gauge is more accurate

REPAIRING DAMAGED THREADS

Several methods of repairing damaged threads are available. Heli-Coil® (shown here), Keenserts® and Microdot® are among the most widely used. All involve basically the same principle — drilling out stripped threads, tapping the hole and installing a prewound insert — making welding, plugging and oversize fasteners unnecessary.

Two types of thread repair inserts are usually supplied: a standard type for most Inch Coarse, Inch Fine, Metric Course and Metric Fine thread sizes and a spark lug type to fit most spark plug port sizes. Consult the individual manufacturer's catalog to determine exact applications. Typical thread repair kits will contain a selection of prewound threaded inserts, a tap (corresponding to the outside diameter threads of the insert) and an installation tool. Spark plug inserts usually differ because they require a tap equipped with pilot threads and a combined reamer/tap section. Most manufacturers also supply blister-packed thread repair inserts separately in addition to a master kit containing a variety of taps and inserts plus installation tools.

Before effecting a repair to a threaded hole, remove any snapped, broken or damaged bolts or studs. Penetrating oil can be used to free frozen threads. The offending item can be removed with locking pliers or with a screw or stud extractor. After the hole is clear, the thread can be repaired, as shown in the series of accompanying illustrations.

Checking Engine Compression

➧ SEE FIG. 17

A noticeable lack of engine power, excessive oil consumption and/or poor fuel mileage measured over an extended period are all indicators of internal engine war. Worn piston rings, scored or worn cylinder bores, blown head gaskets, sticking or burnt valves and worn valve seats are all possible culprits here. A check of each cylinder's compression will help you locate the problems.

As mentioned earlier, a screw-in type compression gauge is more accurate that the type you simply hold against the spark plug hole, although it takes slightly longer to use. It's worth it to obtain a more accurate reading. Follow the procedures below.

Gasoline Engines

1. Warm up the engine to normal operating temperature.
2. Remove all the spark plugs.
3. Disconnect the high tension lead from the ignition coil.
4. On fully open the throttle either by operating the carburetor throttle linkage by hand or by having an assistant floor the accelerator pedal.
5. Screw the compression gauge into the no.1 spark plug hole until the fitting is snug.

WARNING

Be careful not to crossthread the plug hole. On aluminum cylinder heads use extra care, as the threads in these heads are easily ruined.

6. Ask an assistant to depress the accelerator pedal fully on both carbureted and fuel injected vehicles. Then, while you read the compression gauge, ask the assistant to crank the engine two or three times in short bursts using the ignition switch.
7. Read the compression gauge at the end of each series of cranks, and record the highest of these readings. Repeat this procedure for each of the engine's cylinders. Compare the highest reading to the reading in each cylinder.

A cylinder's compression pressure is usually acceptable if it is not less than 80% of of the highest reading. For example, if the highest reading is 150 psi, the lowest should be no lower than 120 psi.

No cylinder should have a reading below 100 psi.

8. If a cylinder is unusually low, pour a tablespoon of clean engine oil into the cylinder through the spark plug hole and repeat the compression test. If the compression comes up after adding the oil, it appears that the cylinder's piston rings or bore are damaged or worn. If the pressure remains low, the valves may not be seating properly (a valve job is needed), or the head gasket may be blown near that cylinder. If compression in any two adjacent cylinders is low, and if the addition of oil doesn't help the compression, there is leakage past the head gasket. Oil and coolant water in the combustion chamber can result from this problem. There may be evidence of water droplets on the engine dipstick when a head gasket has blown.

Engine

REMOVAL & INSTALLATION

CAUTION

When draining the coolant, keep in mind that cats and dogs are attracted by the ethylene glycol antifreeze, and are quite likely to drink any that is left in an uncovered container or in puddles on the ground. This will prove fatal in sufficient quantity. Always drain the coolant into a sealable container. Coolant should be reused unless it is contaminated or several years old.

2.3L Engine

1. Disconnect the negative battery cable and relieve the fuel system pressure.
2. Drain the cooling system and the crankcase.
3. Mark the position of the hood on the hinges and remove the hood.
4. Remove the air cleaner outlet hose.
5. Remove the radiator upper and lower hoses. Disconnect the electrical connector to the cooling fan and remove the fan and shroud. If equipped with automatic transmission, disconnect the oil cooler lines from the radiator. Remove the radiator.
6. Disconnect the heater hose from the water pump. Tag and disconnect the wires from the alternator and starter. Disconnect the accelerator cable from the throttle body.
7. If equipped with air conditioning, remove the compressor from the mounting bracket and position it out of the way, leaving the refrigerant lines attached.
8. If equipped with power steering, remove the pump and position out of the way, leaving the hoses attached.
9. Disconnect the flexible fuel line at the fuel rail and plug the fuel line.
10. Disconnect the coil primary wire, the water temperature sending unit connector and the injector wiring harness connectors from the main wiring harness.
11. Remove the starter and remove the engine mount bolts.
12. Raise and safely support the vehicle. Remove the flywheel or converter housing upper retaining bolts.
13. Disconnect the muffler inlet pipe at the exhaust manifold. Disconnect the engine right and left mounts at the No. 2 crossmember pedestals. Remove the flywheel or converter housing cover.
14. If equipped with a manual transmission, remove the flywheel housing lower retaining bolts. If equipped with an automatic transmission, disconnect the converter from the flywheel and disconnect the transmission oil cooler lines, if attached to the engine at the pan rail. Remove the converter housing lower retaining bolts.
15. Lower the vehicle. Support the transmission and flywheel or converter housing with a jack.
16. Attach suitable engine lifting equipment to the existing lifting brackets. carefully lift the engine out of the engine compartment and install on a work stand.

To Install:

17. Install the clutch, if removed.
18. Carefully lower the engine into the engine compartment. Make sure the studs on the exhaust manifold or turbocharger are aligned with the holes in the muffler inlet pipe.
19. If equipped with an automatic transmission, start the converter pilot into the crankshaft. If equipped with a manual transmission, start the transmission input shaft into the clutch disc. It may be necessary to adjust the position of the transmission in relation to the engine if the input shaft will not enter the clutch disc.

➡ If the engine hangs up after the shaft enters, turn the crankshaft slowly in a clockwise direction, with the transmission in gear, until the shaft splines mesh with the clutch disc splines.

20. Install the flywheel or converter housing upper retaining bolts. Remove the engine lifting equipment.
21. Remove the jack from the transmission. Raise and safely support the vehicle.
22. Install the flywheel or converter housing lower retaining bolts. If equipped with an automatic transmission, attach the converter to the flywheel and tighten the retaining nuts to 20–34 ft. lbs. (27–46 Nm).
23. Install the flywheel or converter housing dust cover. Install the left and right engine mounts to the No. 2 crossmember pedestal. Tighten the nuts and bolts to 80–106 ft. lbs.
24. Connect the muffler inlet pipe to the manifold or turbocharger. Connect the fuel line to the fuel rail.
25. Install the starter and connect the starter cable.
26. Lower the vehicle. Connect the oil pressure and water temperature sending unit connectors. Connect the coil and alternator wires. Connect the accelerator cable and the heater hoses.
27. If equipped with air conditioning, install the compressor in the mounting bracket. If equipped with power steering, install the pump. Install the drive belt.
28. Install the radiator, cooling fan and shroud. Connect the fan electrical connector. If equipped with automatic transmission, connect the oil cooler lines to the radiator. Install the upper and lower radiator hoses.
29. Install the air cleaner outlet hose.
30. Fill the crankcase with the proper type and quantity of oil. Fill and bleed the cooling system.
31. Connect the negative battery cable, start the engine and bring to normal operating temperature. Check for leaks. Check all fluid levels.
32. Align the hood on the hinges with the marks that were made during removal. Secure with the mounting bolts.

5.0L Engine

1. Disconnect the battery cables. Drain the crankcase and the cooling system.
2. Relieve the fuel system pressure and discharge the air conditioning system.
3. Mark the position of the hood on the hinges and remove the hood. Disconnect the battery ground cables from the cylinder block.
4. Remove the air intake duct and the air cleaner, if engine mounted.
5. Disconnect the upper radiator hose from the thermostat housing and the lower hose from the water pump. If equipped with an automatic transmission, disconnect the oil cooler lines from the radiator.

6. Remove the bolts attaching the radiator fan shroud to the radiator. Remove the radiator. Remove the fan, belt pulley and shroud.

7. Remove the alternator bolts and position the alternator out of the way.

8. Disconnect the oil pressure sending unit wire from the sending unit. Disconnect the flexible fuel line at the fuel tank line. Plug the fuel tank line.

9. Disconnect the accelerator cable from the carburetor or throttle body. Disconnect the TV rod if equipped with an automatic transmission. Disconnect the cruise control cable, if equipped.

10. Disconnect the throttle valve vacuum line from the intake manifold, if equipped. Disconnect the transmission filler tube bracket from the cylinder block.

11. If equipped with air conditioning, disconnect the lines and electrical connectors at the compressor and remove the compressor. Plug the lines and the compressor fittings to prevent the entrance of dirt and moisture.

12. Disconnect the power steering pump bracket from the cylinder head. Remove the drive belt. Position the power steering pump out of the way in a position that will prevent the fluid from leaking.

13. Disconnect the power brake vacuum line from the intake manifold.

14. Disconnect the heater hoses from the heater tubes. Disconnect the electrical connector from the coolant temperature sending unit.

15. Remove the converter housing-to-engine upper bolts.

16. Disconnect the wiring to the solenoid on the left rocker cover. Remove the wire harness from the left rocker arm cover and position the wires out of the way. Disconnect the ground strap from the block.

17. Disconnect the wiring harness at the two 10-pin connectors.

18. Raise and safely support the vehicle. Disconnect the starter cable from the starter and remove the starter.

19. Disconnect the muffler inlet pipes from the exhaust manifolds. Disconnect the engine support insulators from the chassis. Disconnect the downstream thermactor tubing and check valve from the right exhaust manifold stud, if equipped.

20. Disconnect the transmission cooler lines from the retainer and remove the converter housing inspection cover. Disconnect the flywheel from the converter and secure the converter assembly in the housing. Remove the remaining converter housing-to-engine bolts.

21. Lower the vehicle and then support the transmission. Attach engine lifting equipment and hoist the engine.

22. Raise the engine slightly and carefully pull it from the transmission. carefully lift the engine out of the engine compartment. Avoid bending or damaging the rear cover plate or other components. Install the engine on a workstand.

To Install:

23. Attach the engine lifting equipment and remove the engine from the workstand.

24. Lower the engine carefully into the engine compartment. Make sure the exhaust manifolds are properly aligned with the muffler inlet pipes.

25. Start the converter pilot into the crankshaft. Align the paint mark on the flywheel to the paint mark on the torque converter.

26. Install the converter housing upper bolts, making sure the dowels in the cylinder block engage the converter housing.

27. Install the engine support insulator-to-chassis attaching fasteners and remove the engine lifting equipment.

28. Raise and safely support the vehicle. Connect both muffler inlet pipes to the exhaust manifolds. Install the starter and connect the starter cable.

29. Remove the retainer holding the converter in the housing. Attach the converter to the flywheel. Install the converter housing inspection cover and install the remaining converter housing attaching bolts.

30. Remove the support from the transmission and lower the vehicle.

31. Connect the wiring harness at the two 10-pin connectors.

32. Connect the coolant temperature sending unit wire and connect the heater hoses. Connect the wiring to the metal heater tubes, engine coolant temperature, air charge temperature and oxygen sensors.

33. Connect the transmission filler tube bracket. Connect the manual shift rod and the retracting spring. Connect the throttle valve vacuum line, if equipped.

34. Connect the accelerator cable and TV cable. Connect the cruise control cable, if equipped.

35. Remove the plug from the fuel tank line and connect the fuel line and the oil pressure sending unit wire.

36. Install the pulley, water pump belt and fan clutch assembly.

37. Position the alternator bracket and install the alternator bolts. Connect the alternator and ground cables. Adjust the drive belt tension.

38. Install the air conditioning compressor. Unplug and connect the refrigerant lines and connect the electrical connector to the compressor.

39. Install the power steering drive belt and power steering pump bracket. Connect the power brake vacuum line.

40. Install the fan on the water pump pulley. Place the shroud over the fan and install the radiator. Connect the radiator hoses and the transmission oil cooler lines. Position the shroud and install the bolts.

41. Connect the heater hoses to the heater tubes. Fill and bleed the cooling system. Fill the crankcase with the proper type and quantity of engine oil. Adjust the transmission throttle linkage.

42. Connect the negative battery cable. Start the engine and bring to normal operating temperature. Check for leaks. Check all fluid levels.

43. Install the air intake duct assembly. Install the hood, aligning the marks that were made during removal.

44. Leak test, evacuate and charge the air conditioning system according to the proper procedure. Observe all safety precautions.

Engine Mounts

REMOVAL & INSTALLATION

➧ SEE FIG. 18-20

2.3L Engine

FRONT

1. Disconnect the negative battery cable. Raise and safely support the vehicle. Support the engine using a wood block and jack placed under the engine.

2. Remove the through bolts attaching both insulators to the No. 2 crossmember pedestal bracket. On Mustang convertible, remove nuts.

3. Disconnect shift linkage.

4. Raise the engine sufficiently to disengage the insulator from the crossmember pedestal bracket.

5. Remove the bolts attaching the insulator and bracket assembly to the engine. Remove the insulator and bracket assembly.

To Install:

6. Position the insulator and bracket assembly to the engine. Install the attaching bolts. Tighten to 33–45 ft. lbs. (45–61 Nm).

7. Lower the engine into position making sure that the insulators are seated flat on the No. 2 crossmember. Hand start the bolts, lower the engine completely, then tighten the through bolts to 33–45 ft. lbs. (45–61 Nm).

8. On Mustang convertible, tighten the flange nut to 80–106 ft. lbs. (108–144 Nm).

9. Install fuel pump shield attaching screw to left engine support, if equipped.

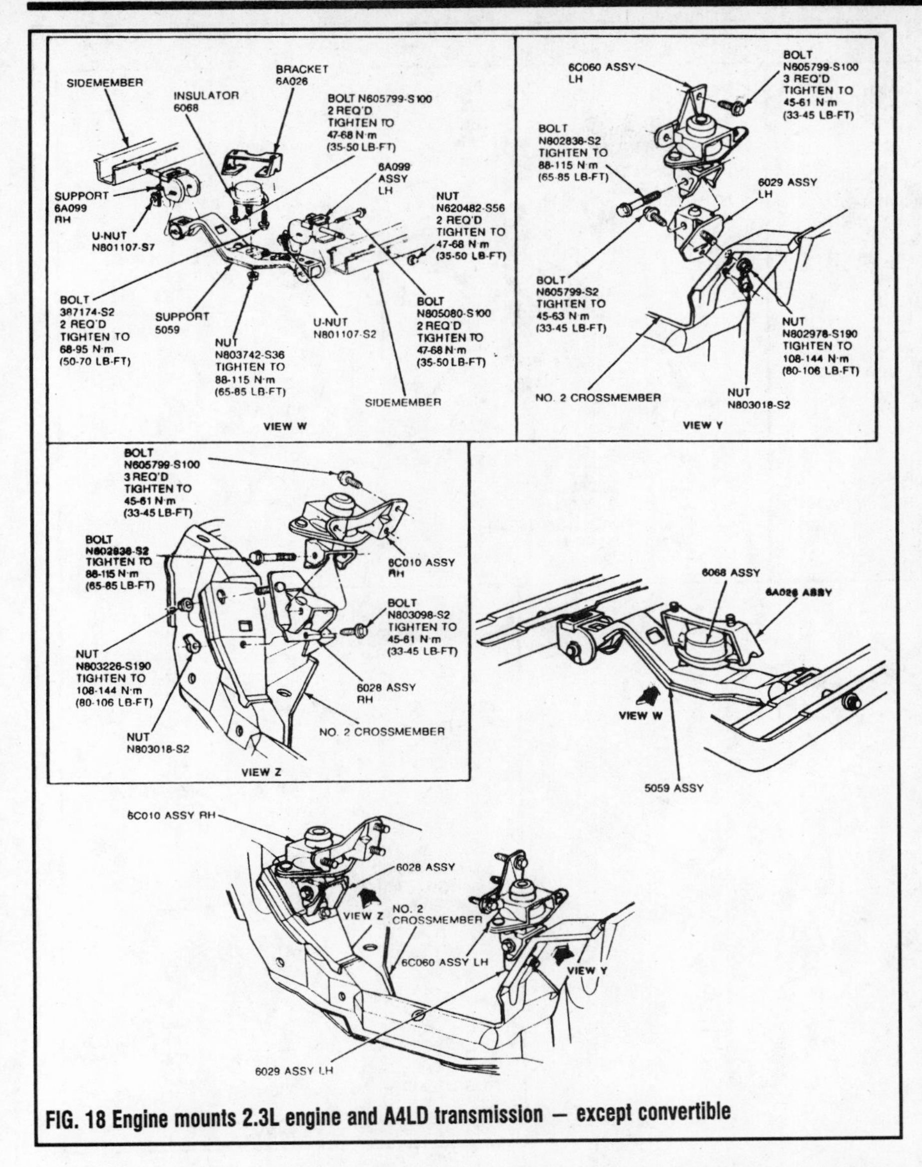

FIG. 18 Engine mounts 2.3L engine and A4LD transmission — except convertible

10. Install shift linkage. Lower the vehicle and connect the negative battery cable.

Rear

1. Disconnect the negative battery cable. Raise and safely support the vehicle.
2. Support the transmission with a jack and a wood block. Remove the nut(s) retaining the rear insulator to the crossmember.
3. Remove the 2 bolts and nuts retaining the crossmember to the body brackets. Remove the crossmember by raising the transmission slightly with the jack.
4. Remove the 2 bolts retaining the rear insulator to the transmission and remove the insulator and retainer. If equipped with automatic transmission, remove the 2 bolts retaining the rear insulator to the intermediate bracket.

To install:

5. Position the rear insulator and retainer on the transmission. Install the 2 retaining bolts and tighten to 50–70 ft. lbs. (68–95 Nm). If equipped with automatic transmission, tighten the 2 bolts to 33–45 ft. lbs. (46–61 Nm).
6. Install the crossmember to the body brackets. Tighten the retaining nuts and bolts to 35–50 ft. lbs. (48–68 Nm).
7. Lower the transmission and install the insulator to crossmember retaining nuts. Tighten to 25–35 ft. lbs. (34–47 Nm). If equipped with automatic transmission, tighten the nut to 65–85 ft. lbs. (85–115 Nm).
8. Lower the vehicle. Connect the negative battery cable.

5.0L Engines

Front

➧ SEE FIGS. 21-22

1. Disconnect the negative battery cable. Remove fan shroud attaching screws.
2. Raise and safely support the vehicle.

Support the engine using a jack and wood block placed under the engine.

3. Remove the nuts or bolts attaching the insulators to the No. 2 crossmember.
4. Disconnect shift linkage.
5. Raise the engine sufficiently with the jack to disengage the insulator from the crossmember. If equipped, remove the transmission brace attached at the left or right engine mount bracket.
6. Remove the engine insulator and bracket assembly to the cylinder block attaching bolts. Remove the engine insulator assembly.

To install:

7. Position the insulator assembly on the engine and install the attaching bolts. Tighten the bolts to 35–60 ft. lbs. (48–81 Nm).
8. Attach the transmission brace to the right or left engine mount, if equipped. Tighten the nut to 35–60 ft. lbs. (48–81 Nm).
9. Lower the engine into position making sure that the insulators are seated flat on the No. 2 crossmember and the insulator studs are at the bottom of the slots.
10. Install and tighten the insulator nuts to 80–105 ft. lbs. (109–142 Nm).
11. Lower the vehicle and install the fan shroud attaching screws. Connect the negative battery cable.

Rear

➧ SEE FIG. 23

1. Disconnect the negative battery cable. Raise and safely support the vehicle.
2. Support the transmission with a jack and wood block. Remove the 2 nuts attaching the insulator to the crossmember.
3. Remove the 2 bolts and nuts attaching the crossmember to the body brackets and remove the crossmember by raising the transmission slightly with the jack.
4. Remove the 2 bolts attaching the rear insulator to the transmission and remove the insulator and retainer.

To install:

5. Position the rear insulator and retainer on the transmission. Install the 2 attaching bolts and tighten to 50–70 ft. lbs. (68–95 Nm).
6. Install the crossmember to the body brackets. Tighten the attaching nuts to 35–50 ft. lbs. (48–68 Nm).
7. Lower the transmission and install the insulator to crossmember attaching nuts. Tighten to 25–35 ft. lbs. (34–48 Nm).
8. Lower the vehicle and connect the negative battery cable.

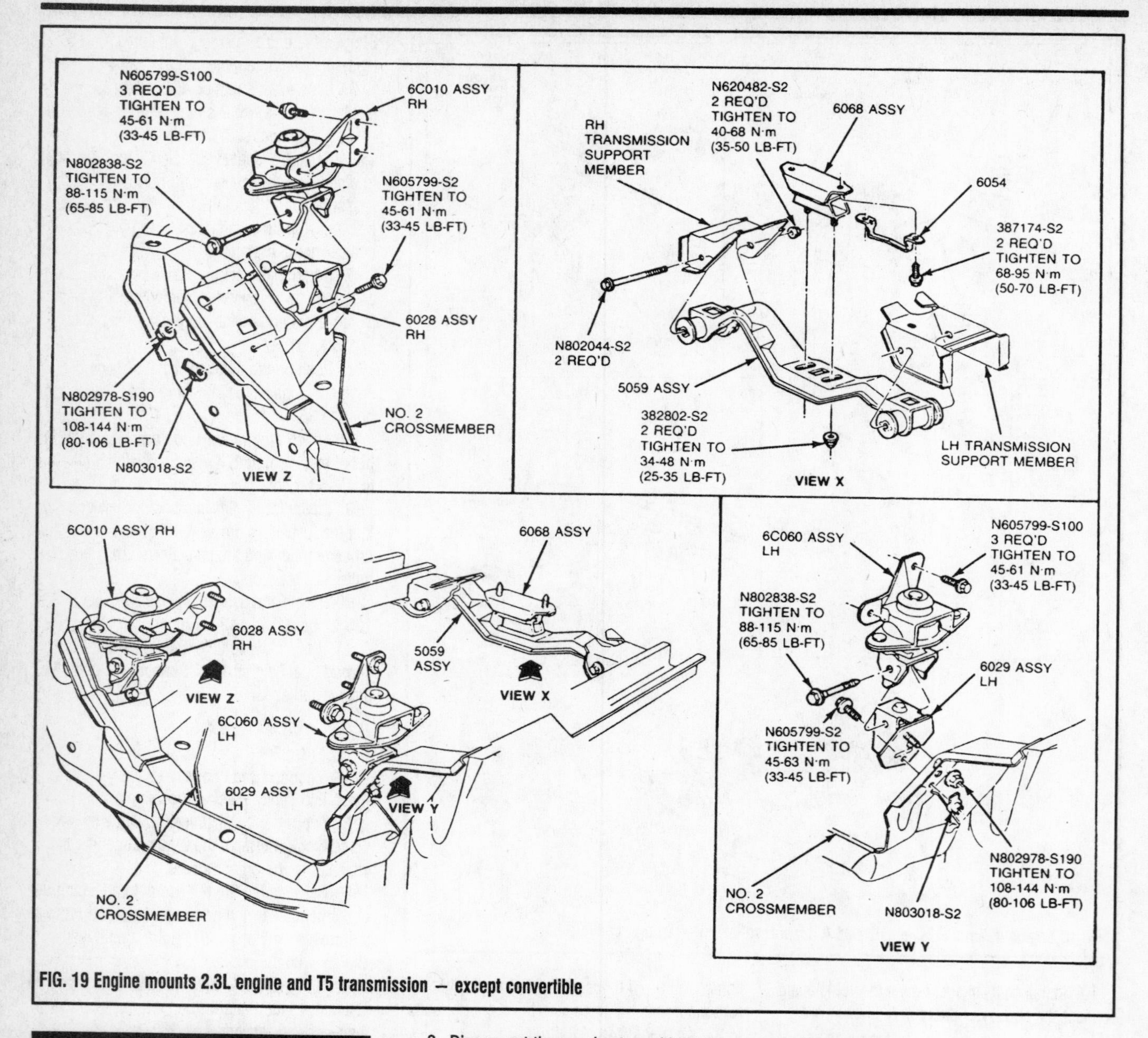

FIG. 19 Engine mounts 2.3L engine and T5 transmission – except convertible

Rocker (Camshaft) Cover(s)

REMOVAL & INSTALLATION

2.3L Engine

➧ SEE FIG. 24

1. Disconnect the inlet hose at the crankcase filler cap.
2. Label for identification and remove all wires and vacuum hoses interfering with the valve cover removal.
3. Disconnect the accelerator cable at the throttle body. Remove the cable retracting spring. Remove the accelerator cable bracket from the upper intake manifold and position the cable and bracket out of the way.
4. Remove the rocker arm cover.
5. Remove and discard the gasket.

To install:

6. Clean the mating surfaces for the cover and head thoroughly.
7. Coat both mating surfaces with gasket sealer and place the new gasket o the head with the locating tabs downward.
8. Place the cover on the head making sure the gasket is evenly seated. Torque the bolts to 62-97 inch lbs.
9. The remainder of installation is the reverse of removal.

5.0L Engine

➧ SEE FIG. 25

1. Disconnect the battery ground cable.
2. Remove the air cleaner and inlet duct.
3. Remove the coil.
4. For the right cover, remove the lifting eye; for the left cover, remove the oil filler pipe attaching bolt.
5. Mark and remove the spark plug wires.
6. Remove any vacuum lines, wires or pipes in the way. Make sure that you tag them for identification.
7. Remove the cover bolts and lift off the cover. It may be necessary to break the cover loose by rapping on it with a rubber mallet. NEVER pry the cover off!

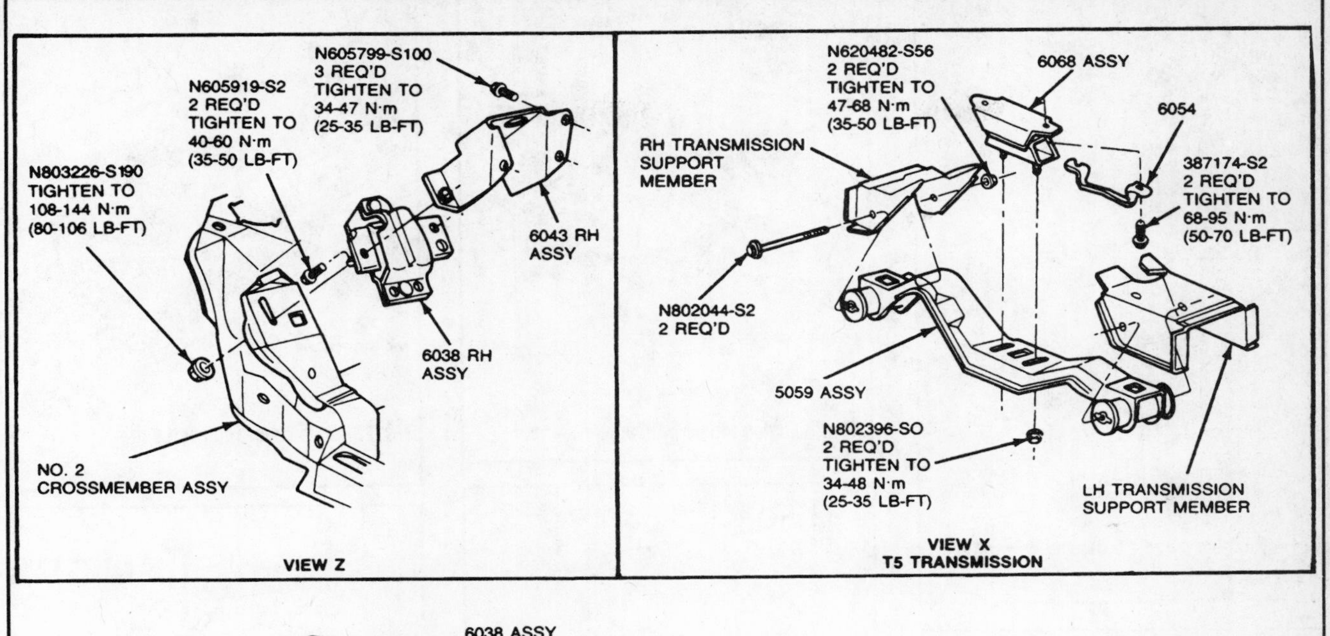

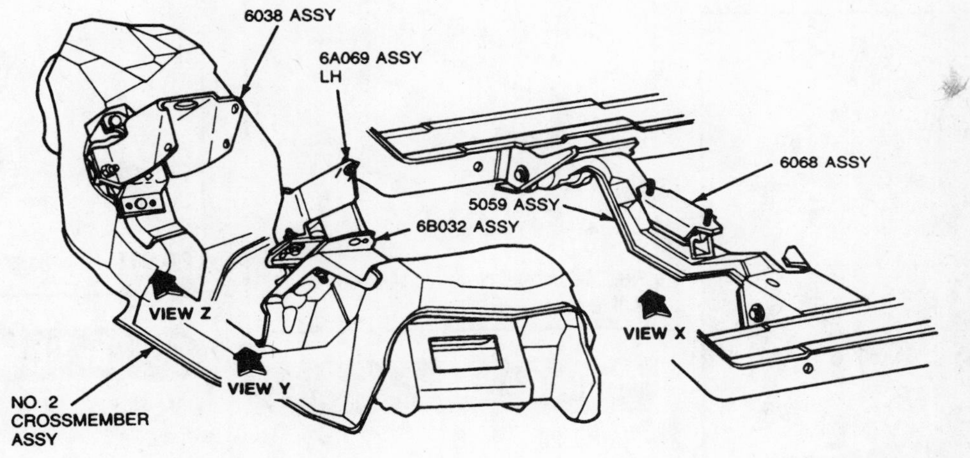

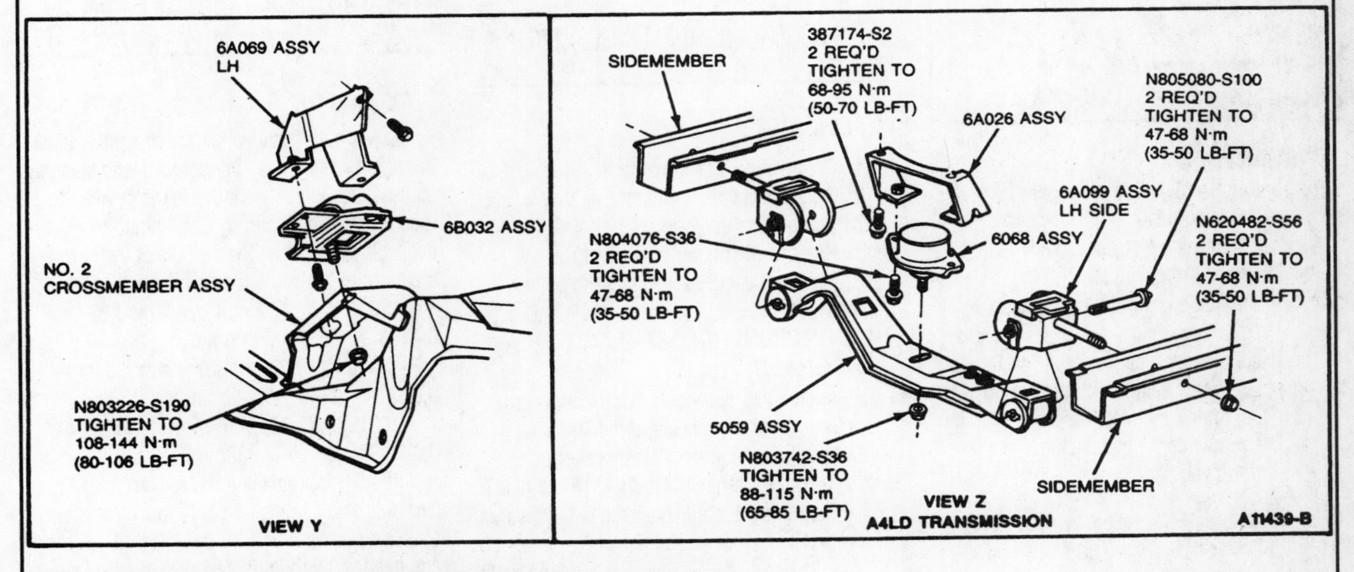

FIG. 20 Engine mounts 2.3L engine and A4LD & T5 transmissions — convertible

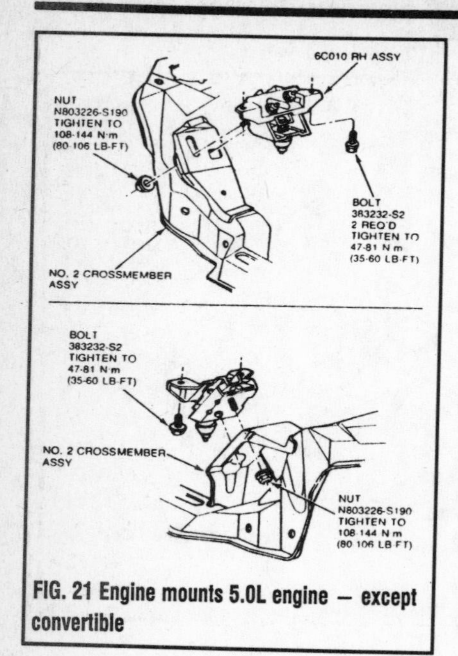

FIG. 21 Engine mounts 5.0L engine — except convertible

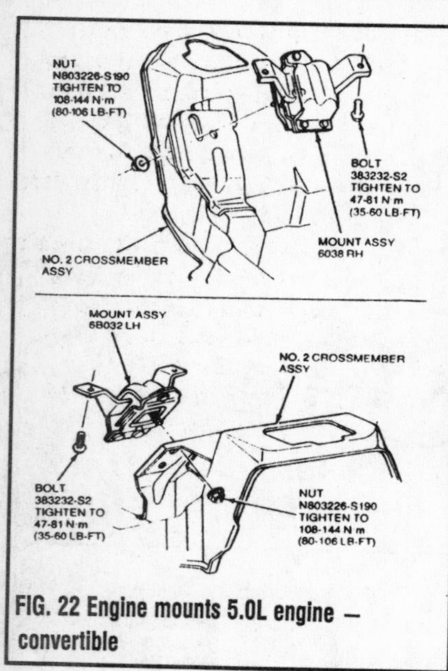

FIG. 22 Engine mounts 5.0L engine — convertible

To Install:

8. Thoroughly clean the mating surfaces of both the cover and head.
9. Coat both mating surfaces with gasket sealer and place the new gasket(s) in the cover(s) with the locating tabs engaging the slots.
10. Place the cover on the head making sure the gasket is evenly seated. Torque the bolts to 10-13 ft. lbs. After 2 minutes, retighten the bolts.
11. The remainder of installation is the reverse of removal.

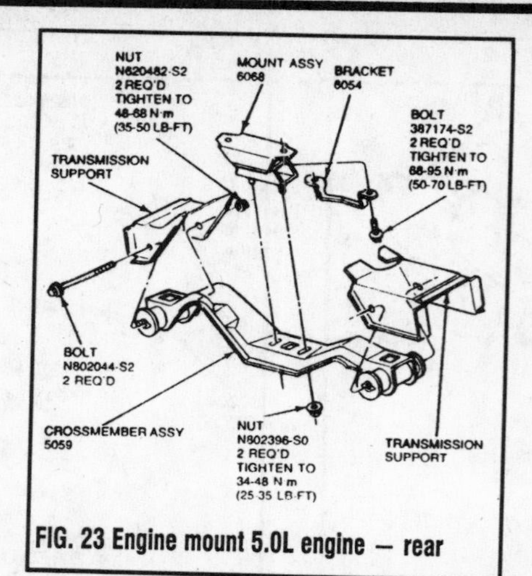

FIG. 23 Engine mount 5.0L engine — rear

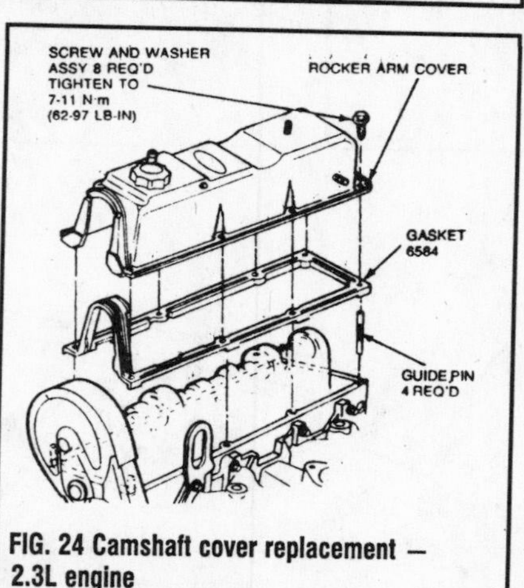

FIG. 24 Camshaft cover replacement — 2.3L engine

Rocker Arms

SEE FIG. 26

REMOVAL & INSTALLATION

5.0L Engine

1. Remove the intake manifold.
2. Disconnect the Thermactor air supply hose at the pump (if so equipped).
3. Remove the rocker arm covers.
4. Loosen the rocker arm fulcrum bolts, fulcrum seats and rocker arms. KEEP ALL PARTS IN ORDER FOR INSTALLATION!

To Install:

5. Apply multi-purpose grease to the valve stem tips, the fulcrum seats and sockets.
6. Install the fulcrum guides, rocker arms, seats and bolts. Torque the bolts to 18–25 ft. lbs.
7. The remainder of installation is the reverse of removal.

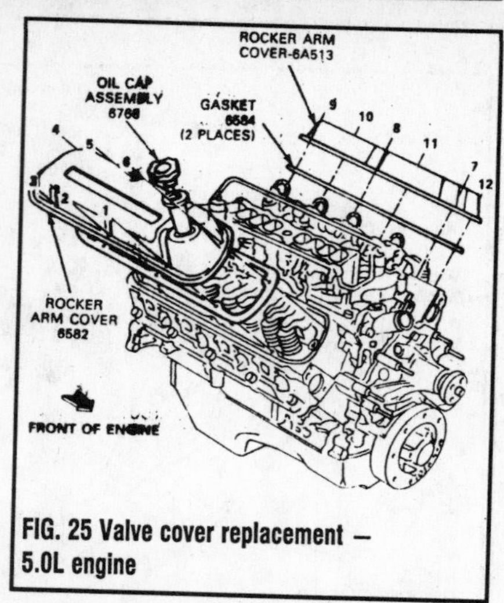

FIG. 25 Valve cover replacement — 5.0L engine

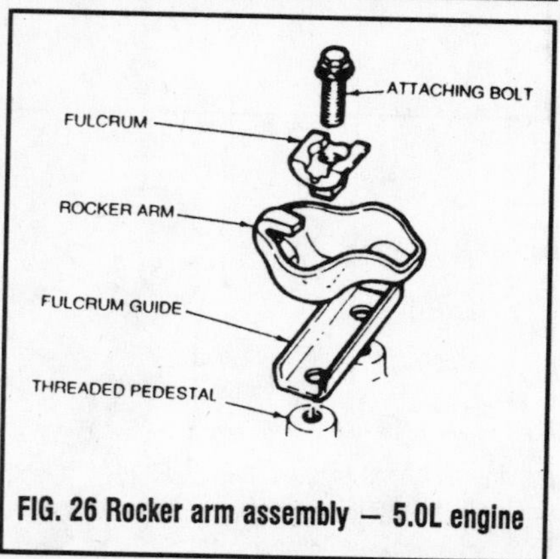

FIG. 26 Rocker arm assembly — 5.0L engine

Rocker Studs

SEE FIGS. 27-28

REMOVAL & INSTALLATION

5.0L Engine

Rocker arm studs which are broken or have damaged threads may be replaced with standard studs. Studs which are loose in the cylinder head must be replaced with oversize studs which are available for service. The amount of oversize and diameter of the studs are as follows:

- 0.006 in. (0.152mm) oversize: 0.3774–0.3781 in. (9.586–9.604mm)
- 0.010 in. (0.254mm) oversize: 0.3814–0.3821 in. (9.688–9.705mm)
- 0.015 in. (0.381mm) oversize: 0.3864–0.3871 in. (9.815–9.832mm)

A tool kit for replacing the rocker studs is available and contains a stud remover and two oversize reamers: one for 0.006 in. (0.152mm) and one for 0.015 in. (0.381mm) oversize studs. For 0.010 in. (0.254mm) oversize studs, use

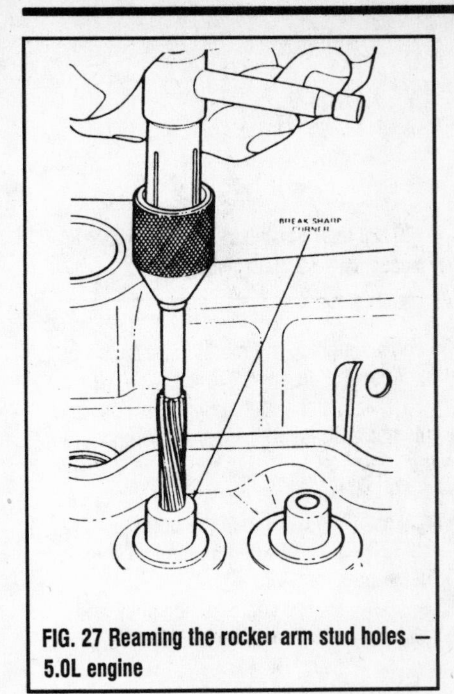

FIG. 27 Reaming the rocker arm stud holes — 5.0L engine

reamer tool T66P–6A527–B. To press the replacement studs into the cylinder head, use the stud replacer tool T69P–6049–D. Use the smaller reamer tool first when boring the hole for oversize studs.

1. Remove the valve rocker cover(s) by moving all hoses aside and unbolting the cover(s). Position the sleeve of the rocker arm stud remover over the stud with the bearing end down. When working on a 5.0L V8, cut the threaded part of the stud off with a hacksaw. Thread the puller into the sleeve and over the stud until it is fully bottomed. Hold the sleeve with a wrench and rotate the puller clockwise to remove the stud.

An alternate method of removing the rocker studs without the special tool is to put spacers over the stud until just enough threads are left showing at the top so a nut can be screwed onto the top of the rocker arm stud and get a full bite. Turn the nut clockwise until the stud is removed, adding spacers under the nut as necessary.

➡ **If the rocker stud was broken off flush with the stud boss, use an easy-out tool to remove the broken off part of the stud from the cylinder head.**

2. If a loose rocker arm stud is being replaced, ream the stud bore for the selected oversize stud.

➡ **Keep all metal particles away from the valves.**

3. Coat the end of the stud with Lubriplate®. Align the stud and installer with the stud bore and top the sliding driver until it bottoms. When the

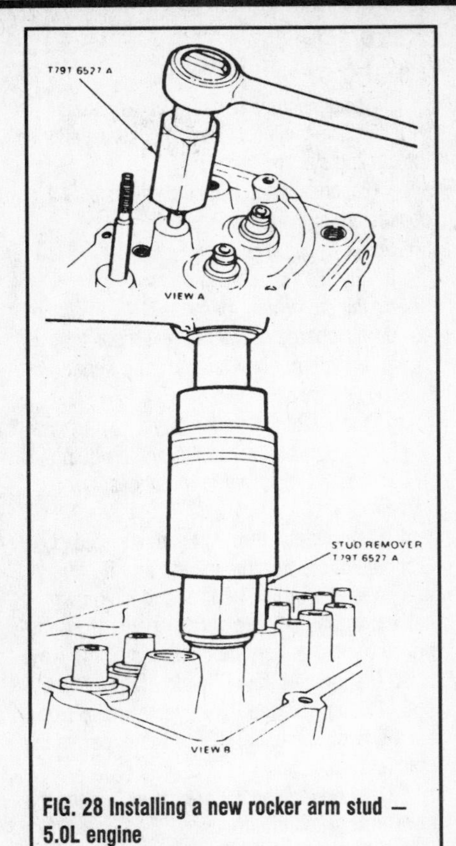

FIG. 28 Installing a new rocker arm stud — 5.0L engine

installer contacts the stud boss, the stud is installed to its correct height.

Thermostat

➧ SEE FIG. 29

REMOVAL & INSTALLATION

CAUTION

When draining the coolant, keep in mind that cats and dogs are attracted by the ethylene glycol antifreeze, and are quite likely to drink any that is left in an uncovered container or in puddles on the ground. This will prove fatal in sufficient quantity. Always drain the coolant into a sealable container. Coolant should be reused unless it is contaminated or several years old.

2.3L Engine

1. Drain the cooling system to a level below the thermostat.

2. Remove the upper radiator hose and disconnect the heater hose at the thermostat housing located on the left front lower side of the engine.

3. Remove the thermostat housing retaining bolts and remove the housing. Remove the thermostat by rotating counterclockwise in the housing until the thermostat becomes free to remove. Do not pry out the thermostat.

4. Remove and discard the gasket.

To install:

5. Clean all gasket mating surfaces and position a new gasket on the cylinder head opening. The gasket must be positioned on the cylinder head, before the thermostat is installed.

6. Install the thermostat into the housing with the bridge section in the housing. Turn the thermostat clockwise to lock it into position on the flats cast into the housing.

➡ **It is important that the rubber thermostat gasket be pressed and the correct thermostat installation alignment be made to provide coolant flow to the heater. Insert and rotate the thermostat to the left or right until it stops in the thermostat housing. Visually check for full width of heater outlet tube opening to be visible within the thermostat port in assembly. This port alignment at assembly is required to provide maximum coolant flow to the heater.**

7. Position the thermostat housing against the gasket on the cylinder head. Install and tighten the retaining bolts to 14–21 ft. lbs. (19–29 Nm).

8. Connect the upper radiator hose and the heater hose to the thermostat housing. Fill the cooling system. Start the engine and bring to normal operating temperature. Check for leaks.

5.0L Engine

1. Drain the cooling system to a level below the thermostat.

2. Disconnect the upper radiator hose and the bypass hose at the thermostat housing.

3. To gain access to the thermostat housing, either mark the location of the distributor, loosen the hold-down clamp and rotate the distributor, or remove the distributor cap and rotor.

4. Remove the thermostat housing retaining bolts and the housing and gasket. Remove the thermostat.

To install:

5. Clean the gasket mating surfaces. Position a new gasket on the intake manifold.

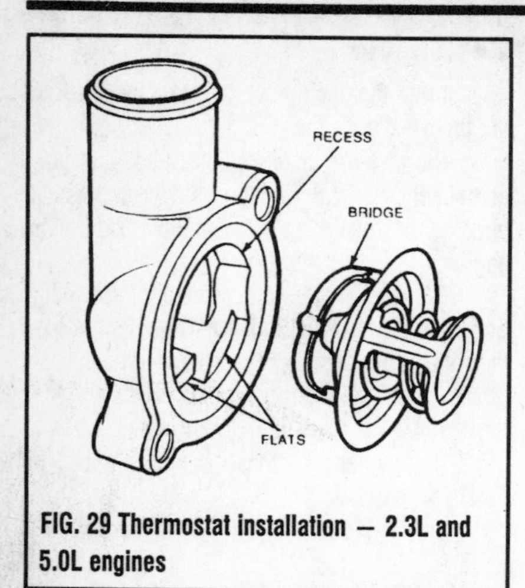

FIG. 29 Thermostat installation — 2.3L and 5.0L engines

6. Install the thermostat in the housing, rotating slightly to lock the thermostat in place on the flats cast into the housing. Install the housing on the manifold and tighten the bolts to 12–18 ft. lbs. (16–24 Nm).

➡ If the thermostat has a bleeder valve, the thermostat should be positioned with the bleeder valve at the 12 o'clock position as viewed from the front of the engine.

7. Install the distributor cap and rotor, or reposition the distributor for correct ignition timing, as necessary. Tighten the hold-down bolt to 18–26 ft. lbs. (24–35 Nm).

8. Connect the bypass hose and the upper radiator hose to the thermostat housing. Fill the cooling system.

9. Start the engine and bring to normal operating temperature. Check for leaks.

Intake Manifold

REMOVAL & INSTALLATION

CAUTION

When draining the coolant, keep in mind that cats and dogs are attracted by the ethylene glycol antifreeze, and are quite likely to drink any that is left in an uncovered container or in puddles on the ground. This will prove fatal in sufficient quantity. Always drain the coolant into a sealable container. Coolant should be reused unless it is contaminated or several years old.

2.3L Engine

➧ SEE FIG. 30

1. Disconnect the negative battery cable.

2. Relieve the fuel system pressure and drain the cooling system.

3. Disconnect and label the electrical connectors at the following:

   a. air bypass valve
   b. throttle positioning sensor
   c. injector wiring harness
   d. air charge temperature sensor
   e. engine coolant temperature sensor
   f. EGR valve, if necessary
   g. fan switch, if necessary
   h. ignition control assembly, if equipped

3. Tag and disconnect the necessary vacuum lines.

4. Remove the throttle linkage shield. Disconnect the throttle linkage and if equipped, the cruise control and kickdown cables. Unbolt the accelerator cable from the bracket and position the cable out of the way.

5. Disconnect the air intake hose.

6. Disconnect the PCV system hose from the fitting on the underside of the upper intake manifold.

7. Disconnect the water bypass hose at the lower intake manifold.

8. Loosen the EGR flange nut and disconnect the EGR tube.

9. Remove the engine oil dipstick bracket retaining bolt.

10. Remove the upper intake manifold retaining bolts and/or studs and remove the upper intake manifold assembly.

11. Disconnect the fuel lines from the fuel supply manifold.

12. Disconnect the electrical connectors from the fuel injectors and move the harness aside.

13. Remove the fuel supply manifold retaining bolts and remove the manifold carefully. Injectors can be removed at this time by exerting a slight twisting/pulling motion.

14. Remove the lower intake manifold retaining bolts and remove the lower intake manifold. The front 2 bolts also secure an engine lift bracket.

To install:

15. Clean all gasket mating surfaces. Clean and oil the manifold bolt threads. Install a new intake manifold gasket.

16. Position the lower intake manifold to the head with the engine lift bracket. Install the manifold retaining bolts finger-tight.

17. On 1989–90 vehicles, tighten the manifold retaining bolts, in sequence, in 2 steps, first to 5–7 ft. lbs. (7–10 Nm) and then to 20–29 ft. lbs. (26–38 Nm). On 1991–92 vehicles, tighten the manifold retaining bolts, in sequence, to 15–22 ft. lbs. (20–30 Nm).

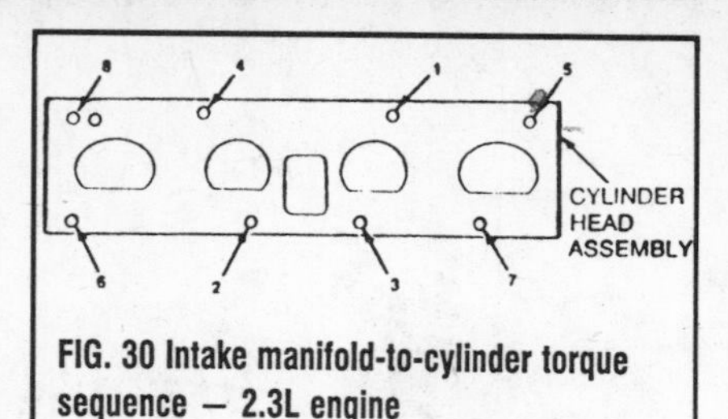

FIG. 30 Intake manifold-to-cylinder torque sequence — 2.3L engine

18. Install the fuel supply manifold and injectors. Connect the electrical connectors to the injectors.

19. Install a new gasket and the upper intake manifold. Tighten the bolts to 15–22 ft. lbs. (20–30 Nm). Connect the fuel lines to the fuel supply manifold.

20. Install the engine oil dipstick and retaining bolt. Connect the EGR tube, water bypass line and PCV hose.

21. Connect the electrical connectors and vacuum lines to their original locations. Connect the throttle linkage.

22. Fill and bleed the cooling system. Connect the negative battery cable, start the engine and check for leaks.

5.0L Engine

➧ SEE FIG. 31-32

1. Disconnect the negative battery cable.

2. Drain the cooling system and relieve the fuel system pressure.

3. Disconnect the accelerator cable and cruise control linkage, if equipped, from the throttle body. Disconnect the TV cable, if equipped. Tag and disconnect the vacuum lines at the intake manifold fitting.

4. Tag and disconnect the spark plug wires from the spark plugs. Remove the wires and bracket assembly from the rocker arm cover attaching stud. Remove the distributor cap and wires assembly.

5. Disconnect the fuel lines and the distributor wiring connector. Mark the position of the rotor on the distributor housing and the distributor housing in the block. Remove the hold-down bolt and remove the distributor.

6. Disconnect the upper radiator hose at the thermostat housing and the water temperature sending unit wire at the sending unit. Disconnect the heater hose from the intake manifold and disconnect the 2 throttle body cooler hoses.

7. Disconnect the water pump bypass hose from the thermostat housing. Tag and disconnect the connectors from the engine coolant temperature, air charge temperature, throttle position and EGR sensors and the idle speed control solenoid. Disconnect the injector

wire connections and the fuel charging assembly wiring.

8. Remove the PCV valve from the grommet at the rear of the lower intake manifold. Disconnect the fuel evaporative purge hose from the plastic connector at the front of the upper intake manifold.

9. Remove the upper intake manifold cover plate and upper intake bolts. Remove the upper intake manifold.

10. If equipped, remove the heater tube assembly from the lower intake manifold studs. If necessary, remove the alternator and air conditioner braces from the intake studs. Disconnect the heater hose from the lower intake manifold.

11. Remove the lower intake manifold retaining bolts and remove the lower intake manifold.

➡ **If it is necessary to pry the intake manifold away from the cylinder heads, be careful to avoid damaging the gasket sealing surfaces.**

To install:

12. Clean all gasket mating surfaces. Apply a 1/8 in. (3mm) bead of silicone sealer to the points where the cylinder block rails meet the cylinder heads.

13. Position new seals on the cylinder block and new gaskets on the cylinder heads with the gaskets interlocked with the seal tabs. Make sure the holes in the gaskets are aligned with the holes in the cylinder heads.

14. Apply a 3/16 in. (5mm) bead of sealer to the outer end of each intake manifold seal for the full width of the seal.

15. Using guide pins to ease installation, carefully lower the intake manifold into position on the cylinder block and cylinder heads.

➡ **After the intake manifold is in place, run a finger around the seal area to make sure the seals are in place. If the seals are not in place, remove the intake manifold and position the seals.**

16. Make sure the holes in the manifold gaskets and the manifold are in alignment. Remove the guide pins. Install the intake manifold attaching bolts and tighten, in sequence, to 23–25 ft. lbs. (31–34 Nm).

17. If required, install the heater tube assembly to the lower intake manifold studs.

18. Install the water pump bypass hose on the thermostat housing. Install the hoses to the heater tubes.

19. Install the distributor, aligning the housing and rotor with the marks that were made during removal. Install the distributor cap. Position the spark plug wires in the harness brackets on the rocker arm cover attaching stud and connect the wires to the spark plugs.

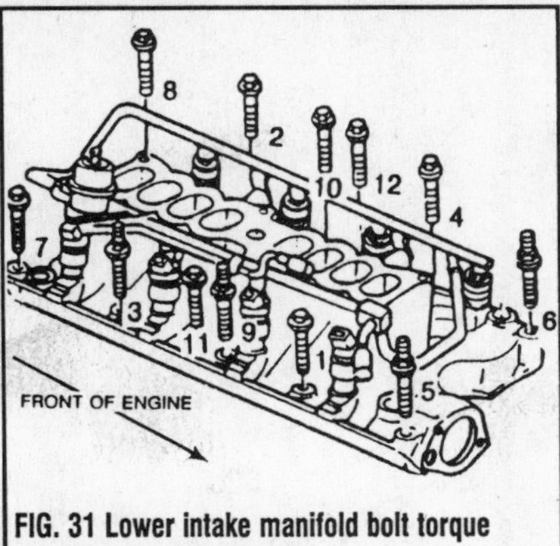

FIG. 31 Lower intake manifold bolt torque sequences — 5.0L engine

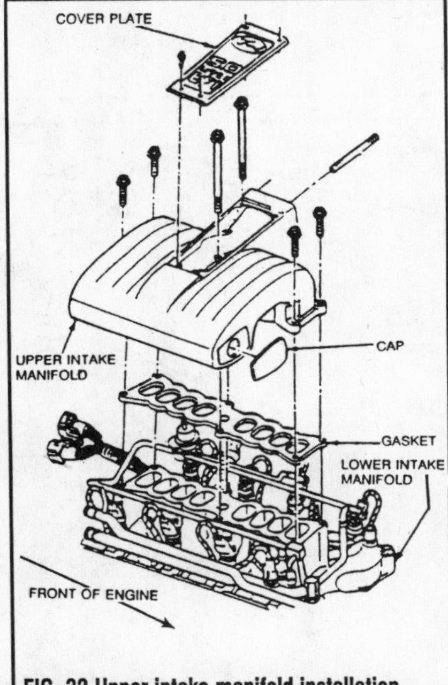

FIG. 32 Upper intake manifold installation — 5.0L engine

20. Install a new gasket and the upper intake manifold. Tighten the bolts to 12–18 ft. lbs. (16–24 Nm). Install the cover plate and connect the crankcase vent tube.

21. Connect the TV cable and cruise control cable, if equipped, to the throttle body. Connect the electrical connectors and vacuum lines.

22. Connect the coolant hoses to the EGR spacer. Fill and bleed the cooling system.

23. Connect the negative battery cable, start the engine and check for leaks. Check the ignition timing.

24. Operate the engine at fast idle. When engine temperatures have stabilized, tighten the intake manifold bolts to 23–25 ft. lbs. (31–34 Nm).

25. Connect the air intake duct and the crankcase vent hose.

Exhaust Manifold

REMOVAL & INSTALLATION

2.3L Engine

➧ SEE FIG. 33

1. Disconnect the negative battery cable.

2. Remove the air cleaner and duct assembly.

3. Remove the EGR tube at the exhaust manifold and loosen the EGR valve.

4. Disconnect and, if necessary, remove the oxygen sensor from the exhaust manifold.

5. Raise and safely support the vehicle. Remove the 2 exhaust pipe bolts and lower the vehicle.

6. Remove the 8 exhaust manifold bolts and remove the exhaust manifold.

7. Installation is the reverse of the removal procedure. Tighten the manifold bolts, in sequence, in 2 steps, first to 15–17 ft. lbs. (20–30 Nm) and then to 20–30 ft. lbs. (27–41 Nm). Tighten the exhaust pipe bolts to 25–34 ft. lbs. (36–46 Nm).

5.0L Engine

1. Disconnect the negative battery cable.

2. Remove the thermactor hardware from the right exhaust manifold. Remove the air cleaner and inlet duct, if necessary.

3. Tag and disconnect the spark plug wires. Remove the spark plugs.

4. Disconnect the engine oil dipstick tube from the exhaust manifold stud.

5. Raise and safely support the vehicle. Disconnect the exhaust pipes from the exhaust manifolds.

6. Remove the engine oil dipstick tube by carefully tapping upward on the tube. Disconnect the oxygen sensor connector.

7. Lower the vehicle. If equipped, remove the nuts attaching the alternator rear brace to the right exhaust manifold and remove the brace.

8. Remove the attaching bolts and washers and remove the exhaust manifolds.

9. Installation is the reverse of the removal procedure. Working from the center to the ends, tighten the exhaust manifold attaching bolts to 18–24 ft. lbs. (24–32 Nm).

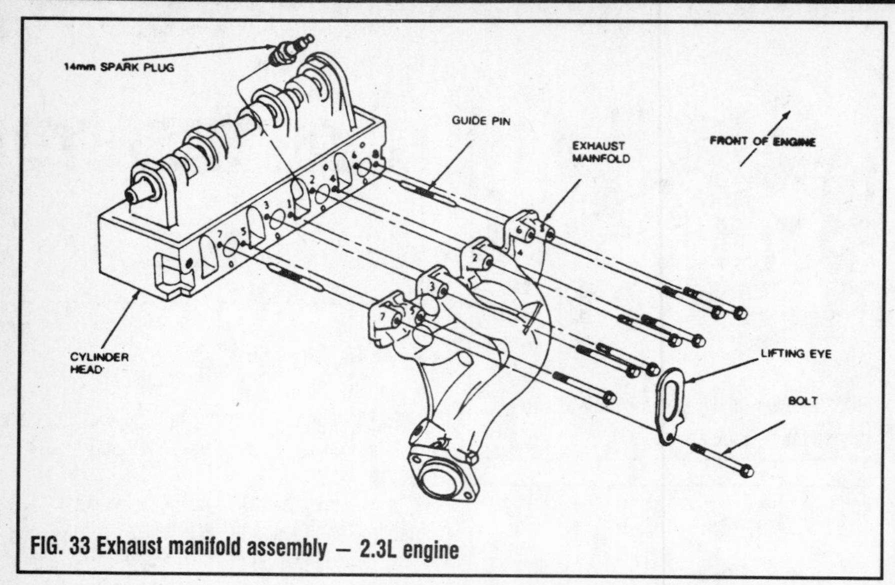

FIG. 33 Exhaust manifold assembly — 2.3L engine

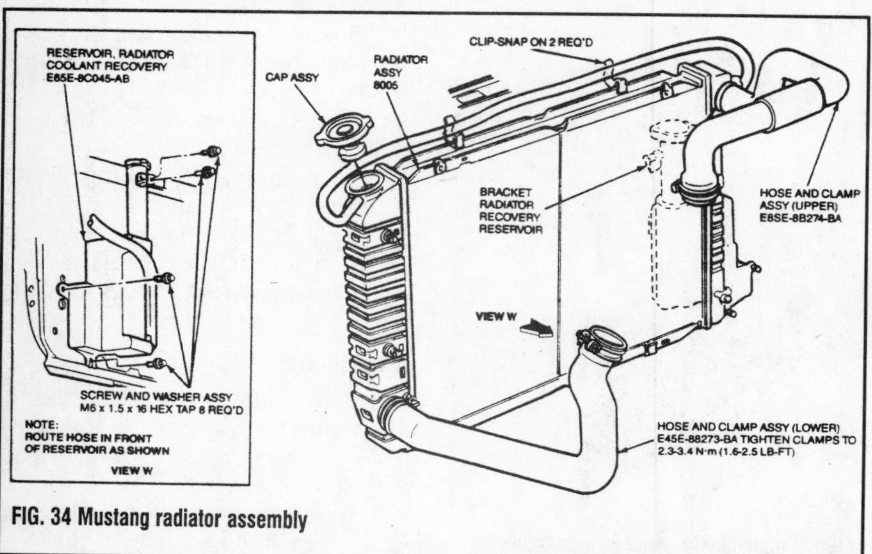

FIG. 34 Mustang radiator assembly

Radiator

➧ SEE FIG. 34

REMOVAL & INSTALLATION

1. Disconnect the negative battery cable. Drain the cooling system.

⁂ CAUTION

When draining the coolant, keep in mind that cats and dogs are attracted by the ethylene glycol antifreeze, and are quite likely to drink any that is left in an uncovered container or in puddles on the ground. This will prove fatal in sufficient quantity. Always drain the coolant into a sealable container. Coolant should be reused unless it is contaminated or several years old.

2. Disconnect the upper, lower and overflow hoses at the radiator.
3. If equipped with an automatic transmission, disconnect the fluid cooler lines at the radiator.
4. On the 2.3L engine, remove the electric cooling fan/shroud assembly. On the 5.0L engine, remove the 2 upper fan shroud retaining bolts at the radiator support, lift the fan shroud sufficiently to disengage the lower retaining clips and lay the shroud back over the fan.
5. Remove the radiator upper support retaining bolts and remove the supports. Lift the radiator from the vehicle.

To install:

6. If a new radiator is to be installed, transfer the petcock from the old radiator to the new one. If equipped with automatic transmission, transfer the fluid cooler line fittings from the old radiator.
7. Position the radiator assembly into the vehicle. Install the upper supports and the retaining bolts. If equipped with automatic transmission, connect the fluid cooler lines.
8. On the 2.3L engine, install the electric cooling fan/shroud assembly. On the 5.oL engine, place the fan shroud into the clips on the lower radiator support and install the 2 upper shroud retaining bolts. Position the shroud to maintain approximately 1 in. (25mm) clearance between the fan blades and the shroud.
9. Connect the radiator hoses. Close the radiator petcock. Fill and bleed the cooling system.
10. Start the engine and bring to operating temperature. Check for coolant and transmission fluid leaks.
11. Check the coolant and transmission fluid levels.

Engine Fan and Fan Clutch

REMOVAL & INSTALLATION

➧ SEE FIG. 35

1. Disconnect the negative battery cable.
2. Remove the fan shroud retaining screws.
3. Loosen the fan belt. Remove the bolt and washers retaining the fan clutch to the water pump hub.
4. Remove the fan clutch and the fan as an assembly

at the same time removing the fan shroud.

5. Remove the retaining bolts and washers and separate the fan from the drive clutch.

To install:

6. Position the fan on the drive clutch. Install the bolts and washers, then tighten to 12–18 ft. lbs.
7. Position the fan drive clutch, the fan assembly and the fan shroud, then bolt the fan and clutch assembly to the water pump hub. Install and tighten the bolts alternately to 15–22 ft. lbs. Adjust the fan belt tension.
8. Adjust the fan shroud clearance for equal clearance around the fan with a minimum

clearance of 0.38 in. (9.6mm). Install and tighten the retaining screws.

Electric Cooling Fan

TESTING

1. Disconnect the electrical connector at the cooling fan motor.
2. Connect a jumper wire between the negative motor lead and ground.
3. Connect another jumper wire between the positive motor lead and the positive terminal of the battery.
4. If the cooling fan motor does not run, it must be replaced.

REMOVAL & INSTALLATION

2.3L Engine

➧ SEE FIG. 36

1. Disconnect the negative battery cable.
2. Remove the fan wiring harness from the routing clip. Disconnect the wiring harness from the fan motor connector by pulling up on the single lock finger to separate the connectors.
3. Remove the 4 mounting bracket attaching screws and remove the fan assembly from the vehicle.
4. Remove the retaining clip from the end of the motor shaft and remove the fan.

➡ A metal burr may be present on the motor after the retaining clip is removed. Deburring of the shaft may be required to remove the fan.

5. Remove the nuts attaching the fan motor to the mounting bracket.
6. Installation is the reverse of the removal procedure.

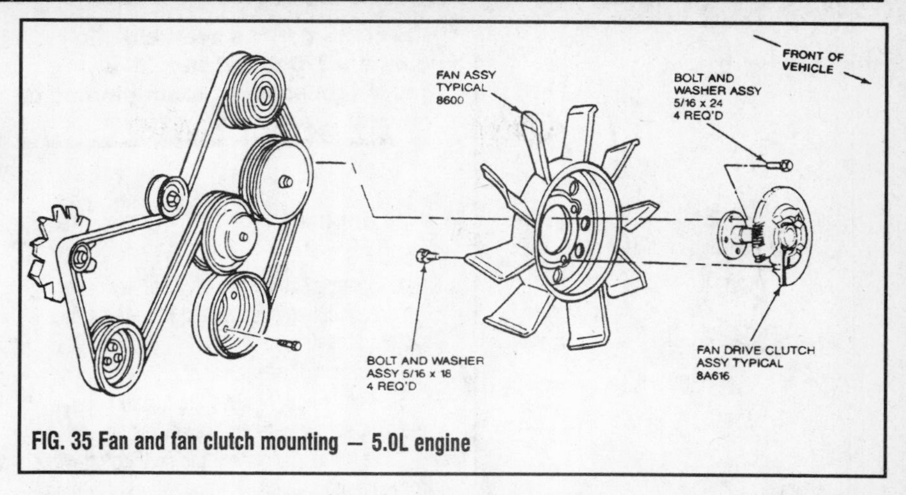

FIG. 35 Fan and fan clutch mounting — 5.0L engine

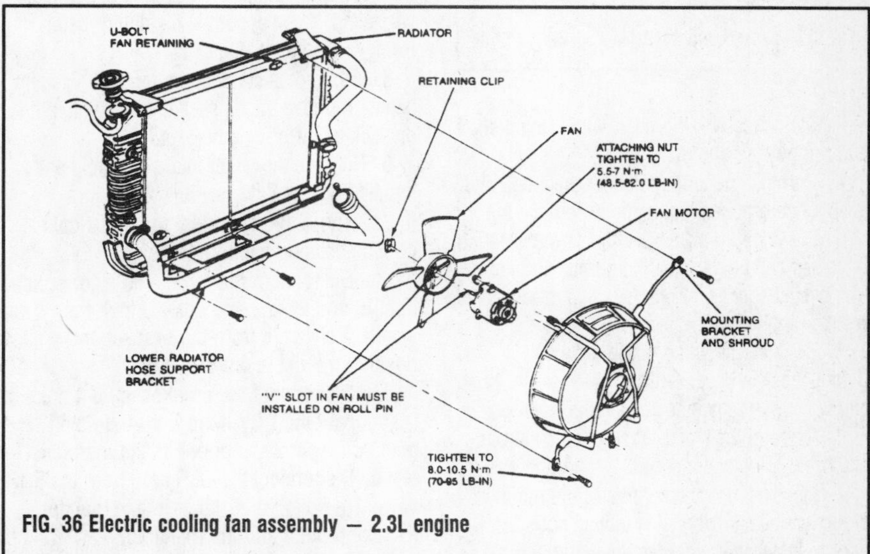

FIG. 36 Electric cooling fan assembly — 2.3L engine

Water Pump

REMOVAL & INSTALLATION

✻✻ CAUTION

When draining the coolant, keep in mind that cats and dogs are attracted by the ethylene glycol antifreeze, and are quite likely to drink any that is left in an uncovered container or in puddles on the ground. This will prove fatal in sufficient quantity. Always drain the coolant into a sealable container. Coolant should be reused unless it is contaminated or several years old.

2.3L Engine

➧ SEE FIG. 37

1. Disconnect the negative battery cable and drain the cooling system.
2. Remove the 4 bolts retaining the pulley to the water pump shaft. Remove the fan and shroud.
3. Remove the air conditioning and power steering belts. Remove the water pump pulley.
4. Remove the heater hose to the water pump and the lower radiator hose.
5. Remove the timing belt outer cover bolt, release the interlocking tabs and remove the cover.
6. Remove the water pump retaining bolts and remove the water pump.
7. Installation is the reverse of the removal procedure. Clean all gasket mating surfaces prior to installation. Apply pipe sealant to the water pump bolts and tighten to 14–21 ft. lbs. (20–30 Nm). Tighten the pulley retaining bolts to 15–22 ft. lbs. (20–30 Nm).
8. Fill and bleed the cooling system. Operate the engine until normal operating temperatures have been reached and check for leaks.

5.0L Engine

1. Disconnect the negative battery cable.
2. Drain the cooling system. Remove the air inlet tube, if equipped.
3. Remove the fan shroud attaching bolts and position the shroud over the fan. Remove the fan

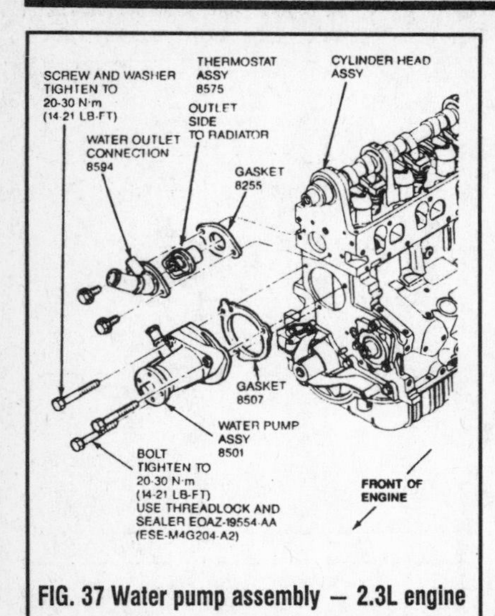

FIG. 37 Water pump assembly — 2.3L engine

and clutch assembly from the water pump shaft and remove the shroud.

4. Remove the air conditioner drive belt and idler pulley bracket, if equipped. Remove the alternator drive belt. Remove the power steering drive belt and power steering pump, if equipped. Remove all accessory brackets that attach to the water pump.

5. Remove the water pump pulley. Disconnect the lower radiator hose, heater hose and water pump bypass hose at the water pump.

6. Remove the water pump attaching bolts and remove the water pump. Discard the gasket.

7. Installation is the reverse of the removal procedure. Clean all gasket mating surfaces prior to installation. Tighten the water pump attaching bolts to 12–18 ft. lbs. (16–24 Nm).

8. Fill and bleed the cooling system. Operate the engine until normal operating temperatures have been reached and check for leaks.

Cylinder Head

REMOVAL & INSTALLATION

CAUTION

When draining the coolant, keep in mind that cats and dogs are attracted by the ethylene glycol antifreeze, and are quite likely to drink any that is left in an uncovered container or in puddles on the ground. This will prove fatal in sufficient quantity. Always drain the coolant into a sealable container. Coolant should be reused unless it is contaminated or several years old.

2.3L Engine

➧ SEE FIG. 38

1. Disconnect the negative battery cable. Drain the cooling system and relieve the fuel system pressure.

2. Remove the air cleaner.

3. If equipped, remove the heater hose retaining screw to the rocker cover.

4. Tag and disconnect the spark plug wires from the spark plugs. Remove the spark plug wires and, if equipped, the distributor cap. Remove the spark plugs.

5. Tag and disconnect the required vacuum hoses. Remove the dipstick and disconnect the dipstick tube from the bracket.

6. Remove the upper intake manifold and throttle body as follows:

a. Tag and disconnect the electrical connectors and vacuum hoses.

b. Disconnect the throttle linkage, cruise control and kickdown cable. Unbolt the accelerator cable from the bracket and position the cable out of the way.

c. Disconnect the crankcase vent hose. Disconnect the PCV hose from the fitting on the underside of the upper intake manifold.

d. Disconnect the EGR tube from the EGR valve. Remove the 4 upper intake manifold mounting bolts and the manifold.

7. Remove the rocker cover retaining bolts and remove the cover. Remove the intake manifold retaining bolts.

8. Remove the alternator belt and swing the alternator aside. Remove the bracket mounting bolts, if necessary.

9. Remove the upper radiator hose. Remove the timing belt cover retaining bolts and remove the cover.

10. Loosen the timing belt idler retaining bolts. Position the idler in the unloaded position and tighten the retaining bolts.

11. Remove the timing belt from the camshaft pulley and the auxiliary pulley.

12. Remove the 8 exhaust manifold retaining bolts. Remove the timing belt idler and 2 bracket bolts. Remove the timing belt idler spring stop from the cylinder head.

13. Disconnect the oil sending unit wire, if necessary.

14. Remove the cylinder head bolts and the cylinder head. Clean all gasket mating surfaces and blow the oil out of the cylinder head bolt block holes.

15. Check the cylinder head for flatness using a straight edge and a feeler gauge. If the head gasket surface is warped greater than 0.006 in. (0.152mm), it must be resurfaced. Do not grind more than 0.010 in. (0.254mm) from the cylinder head.

To install:

16. Position the head gasket on the block. Position the camshaft with the pin approximately 30 degrees to the right of the 6 o' clock position when facing the front of the cylinder head. The camshaft must be positioned this way to protect protruding valves.

17. Position the cylinder head on the block and install the cylinder head bolts. Tighten the bolts, in sequence, in 2 steps, first to 50–60 ft. lbs. (60–81 Nm) and then to 80–90 ft. lbs. (108–122 Nm).

18. Connect the oil sending unit wire, if necessary. Install the timing belt tensioner spring stop to the cylinder head.

19. Position the timing belt tensioner and tensioner spring to the cylinder head and install the retaining bolts. Rotate the tensioner against the spring with belt tensioner tool T74P–6254–A or equivalent, and temporarily tighten.

20. Install the 8 exhaust manifold retaining bolts. Tighten the bolts, in sequence, in 2 steps, first to 178–204 inch lbs. (20–23 Nm) and then to 20–30 ft. lbs. (27–40 Nm).

21. If equipped with a distributor, align the distributor rotor with the No. 1 plug location on the distributor cap. Align the camshaft sprocket with the pointer and align the crankshaft pulley with the pointer on the timing belt cover.

22. Install the timing belt over the sprockets. Loosen the tensioner retaining bolts, rotate the engine by hand 1 complete revolution and check the timing alignment.

23. Tighten the 10mm tensioner bolt to 28–40 ft. lbs. (38–54 Nm) and the 8mm bolt to 14–21 ft. lbs. (19–29 Nm).

24. Install the timing belt cover and tighten the retaining bolts to 6–9 ft. lbs. (8–12 Nm).

25. Install the rocker arm cover and tighten the retaining bolts to 62–97 inch lbs. (7–11 Nm).

26. Install the intake manifold. Tighten the bolts, in sequence, to 19–28 ft. lbs. (26–38 Nm).

27. Install the upper intake manifold and throttle body in the reverse order of removal. Tighten the upper intake-to-lower intake bolts to 15–22 ft. lbs. (20–30 Nm).

28. Position the alternator and install the drive belt. Install the upper radiator hose.

29. Install the dipstick and connect the necessary vacuum hoses. Install the spark

plugs, spark plug wires and distributor cap, if equipped.

30. Position and connect the engine and alternator wiring harnesses. Install the hose from the air cleaner to the throttle body. If equipped, install the retaining heater hose screw to the rocker cover.

31. Fill and bleed the cooling system. Connect the negative battery cable, start the engine and bring to normal operating temperature. Check for leaks. If equipped with distributor ignition, check the ignition timing.

5.0L Engine

➧ SEE FIG. 39

1. Disconnect the negative battery cable.
2. Drain the cooling system and relieve the fuel system pressure.
3. Remove the upper and lower intake manifold and throttle body assembly.
4. If the air conditioning compressor is in the way of a cylinder head that is to be removed, proceed as follows:
   a. Discharge the air conditioning system.
   b. Disconnect and plug the refrigerant lines at the compressor. Cap the openings on the compressor.
   c. Disconnect the electrical connector to the compressor.
   d. Remove the compressor and the necessary mounting brackets.
5. If the left cylinder head is to be removed, disconnect the power steering pump bracket from the cylinder head and remove the drive belt from the pump pulley. Position the pump out of the way in a position that will prevent the oil from draining out.
6. Disconnect the oil level indicator tube bracket from the exhaust manifold stud, if necessary.
7. If the right cylinder head is to be removed, on some vehicles it is necessary to disconnect the alternator mounting bracket from the cylinder head.
8. If equipped, remove the fuel line from the clip at the front of the right cylinder head.
9. Raise and safely support the vehicle. Disconnect the exhaust manifolds from the muffler inlet pipes. Lower the vehicle.
10. Loosen the rocker arm fulcrum bolts so the rocker arms can be rotated to the side. Remove the pushrods in sequence so they may be installed in their original positions.
11. Remove the cylinder head attaching bolts and the cylinder heads. If necessary, remove the exhaust manifolds to gain access to the lower bolts. Remove and discard the head gaskets.
12. Clean all gasket mating surfaces. Check the flatness of the cylinder head using a straight edge and a feeler gauge. The

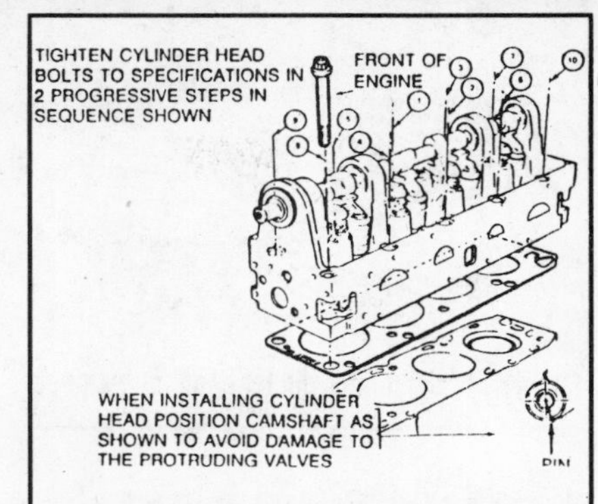

FIG. 38 Cylinder head tightening sequence — 2.3L engine

cylinder head must not be warped any more than 0.003 in. (0.076mm) in any 6 in. (152mm) span; 0.006 in. (0.152mm) overall. Machine as necessary.

To install:

13. Position the new cylinder head gasket over the dowels on the block. Position the cylinder heads on the block and install the attaching bolts.
14. Tighten the bolts, in sequence, in 2 steps, first to 55–65 ft. lbs. (75–88 Nm), then to 65–72 ft. lbs. (88–97 Nm).

➡ When the cylinder head bolts have been tightened following this procedure, it is not necessary to retighten the bolts after extended operation.

15. If removed, install the exhaust manifolds. Tighten the retaining bolts to 18–24 ft. lbs. (24–32 Nm).
16. Clean the pushrods, making sure the oil passages are clean. Check the ends of the pushrods for wear. Visually check the pushrods for straightness or check for runout using a dial indicator. Replace pushrods, as necessary.
17. Apply a suitable grease to the ends of the pushrods and install them in their original positions. Position the rocker arms over the pushrods and the valves.
18. Before tightening each fulcrum bolt, bring the lifter for the fulcrum bolt to be tightened onto the base circle of the camshaft by rotating the engine. When the lifter is on the base circle of the camshaft, tighten the fulcrum bolt to 18–25 ft. lbs. (24–34 Nm).

➡ If all the original valve train parts are reinstalled, a valve clearance check is not necessary. If any valve train components are replaced, a valve clearance check must be performed.

19. Install new rocker arm cover gaskets on

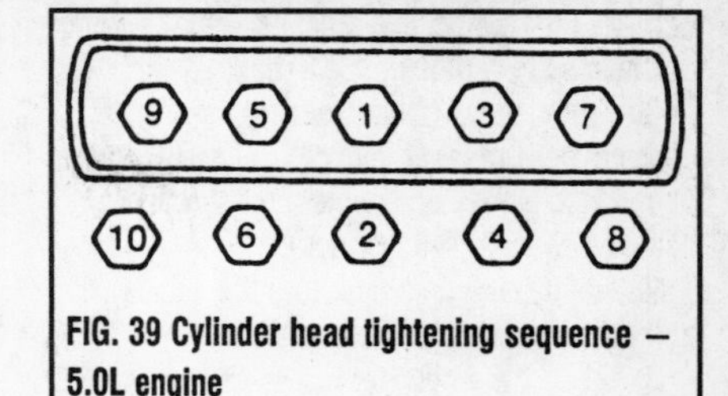

FIG. 39 Cylinder head tightening sequence — 5.0L engine

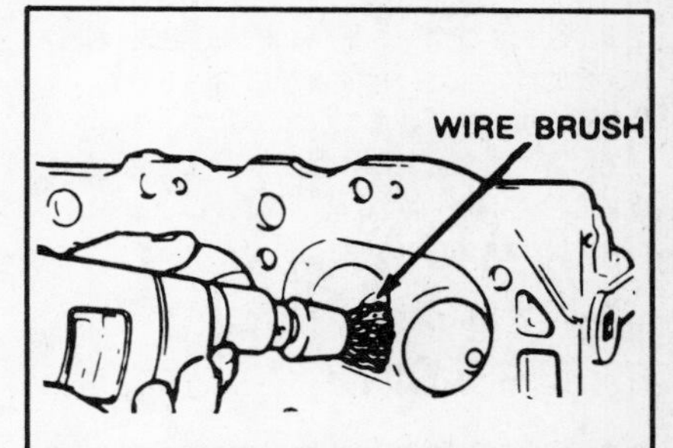

FIG. 40 Remove combustion chamber carbon from the cylinder head with a wire brush and electric drill. Make sure all carbon is removed and not just burnished

the rocker arm covers and install the covers on the cylinder heads.

20. Raise and safely support the vehicle. Connect the exhaust manifolds to the muffler inlet pipes. Lower the vehicle.
21. If necessary, install the air conditioning compressor and brackets. Connect the refrigerant lines and electrical connector to the compressor.
22. If necessary, install the alternator bracket.
23. If the left cylinder head was removed, install the power steering pump.
24. Install the drive belts. Install the thermactor tube at the rear of the cylinder heads.
25. Install the intake manifold. Fill and bleed the cooling system.
26. Connect the negative battery cable, start the engine and bring to normal operating temperature. Check for leaks. Check all fluid levels.
27. If necessary, leak test, evacuate and charge the air conditioning system according to the proper procedure. Observe all safety precautions.

CLEANING AND INSPECTION

➧ SEE FIG. 40

1. With the valves installed to protect the valve seats, remove deposits from the combustion chambers and valve heads with a scraper and a wire brush. Be careful not to

damage the cylinder head gasket surface. After the valves are removed, clean the valve guide bores with a valve guide cleaning tool. Using cleaning solvent to remove dirt, grease and other deposits, clean all bolts holes; be sure the oil passage is clean (V8 engines).

2. Remove all deposits from the valves with a fine wire brush or buffing wheel.

3. Inspect the cylinder heads for cracks or excessively burned areas in the exhaust outlet ports.

4. Check the cylinder head for cracks and inspect the gasket surface for burrs and nicks. Replace the head if it is cracked.

5. On cylinder heads that incorporate valve seat inserts, check the inserts for excessive wear, cracks, or looseness.

RESURFACING

➧ SEE FIG. 41

Cylinder Head Flatness

When the cylinder head is removed, check the flatness of the cylinder head gasket surfaces.

1. Place a straightedge across the gasket surface of the cylinder head. Using feeler gauges, determine the clearance at the center of the straightedge.

2. If warpage exceeds 0.003 in. (0.076mm) in a 6 in. (152mm) span, or 0.006 in. (0.152mm) over the total length, the cylinder head must be resurfaced.

3. If necessary to refinish the cylinder head gasket surface, do not plane or grind off more than 0.254mm (0.010 in.) from the original gasket surface.

➡ When milling the cylinder heads of V8 engines, the intake manifold mounting position is altered, and must be corrected by milling the manifold flange a proportionate amount. Consult an experienced machinist about this.

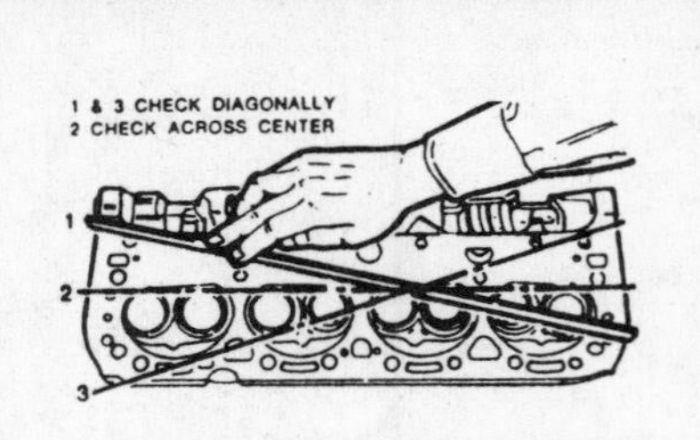

FIG. 41 Checking the cylinder head for warpage

Valves and Springs

REMOVAL & INSTALLATION

➧ SEE FIGS. 42–44

1. Block the head on its side, or install a pair of head-holding brackets made especially for valve removal.

2. Use a socket slightly larger than the valve stem and keepers, place the socket over the valve stem and gently hit the socket with a plastic hammer to break loose any varnish buildup.

3. Remove the valve keepers, retainer, spring shield and valve spring using a valve spring compressor (the locking C-clamp type is the easiest kind to use).

4. Put the parts in a separate container numbered for the cylinder being worked on; do not mix them with other parts removed.

5. Remove and discard the valve stem oil seals. A new seal will be used at assembly time.

6. Remove the valves from the cylinder head and place them, in order, through numbered holes punched in a stiff piece of cardboard or wood valve holding stick.

➡ The exhaust valve stems, on some engines, are equipped with small metal caps. Take care not to lose the caps. Make sure to reinstall them at assembly time. Replace any caps that are worn.

7. Use an electric drill and rotary wire brush to clean the intake and exhaust valve ports, combustion chamber and valve seats. In some cases, the carbon will need to be chipped away. Use a blunt pointed drift for carbon chipping. Be careful around the valve seat areas.

8. Use a wire valve guide cleaning brush and safe solvent to clean the valve guides.

9. Clean the valves with a revolving wires brush. Heavy carbon deposits may be removed with the blunt drift.

➡ When using a wire brush to clean carbon on the valve ports, valves etc., be sure that the deposits are actually removed, rather than burnished.

10. Wash and clean all valve springs, keepers, retaining caps etc., in safe solvent.

11. Clean the head with a brush and some safe solvent and wipe dry.

12. Check the head for cracks. Cracks in the cylinder head usually start around an exhaust valve seat because it is the hottest part of the combustion chamber. If a crack is suspected but cannot be detected visually have the area

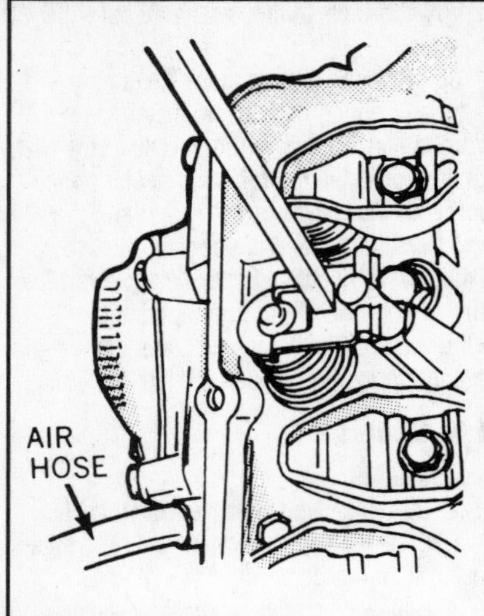

FIG. 42 Compressing engine valve spring. Note the spring compressor position and the air hose. The cylinder must be at TDC

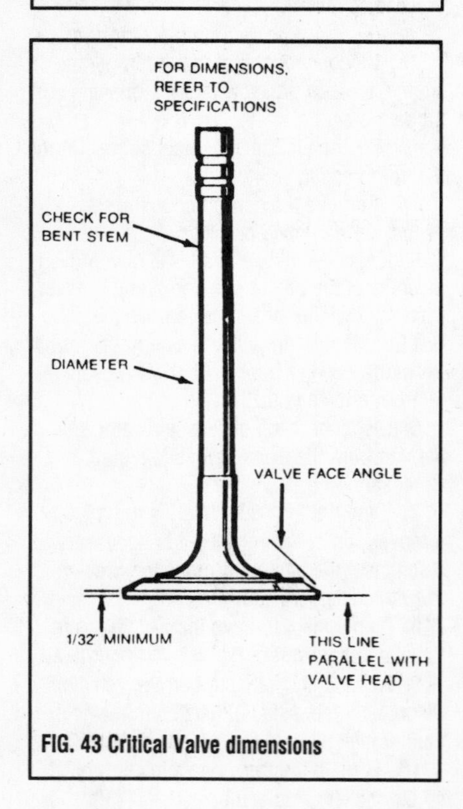

FIG. 43 Critical Valve dimensions

checked with dye penetrant or other method by the machine shop.

13. After all cylinder head parts are reasonably clean, check the valve stem-to-guide clearance. If a dial indicator is not on hand, a visual inspection can give you a fairly good idea if the guide, valve stem or both are worn.

14. Insert the valve into the guide until slight away from the valve seat. Wiggle the valve sideways. A small amount of wobble is normal,

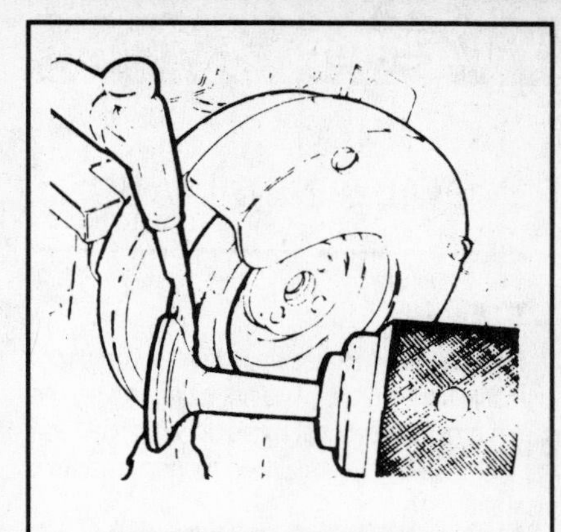

FIG. 44 A well equipped machine shop can handle valve refacing jobs

excessive wobble means a worn guide or valve stem. If a dial indicator is on hand, mount the indicator so that the stem of the valve is at 90° to the valve stem, as close to the valve guide as possible. Move the valve off the seat, and measure the valve guide-to-stem clearance by rocking the stem back and forth to actuate the dial indicator. Measure the valve stem using a micrometer and compare to specifications to determine whether stem or guide wear is causing excessive clearance.

15. The valve guide, if worn, must be repaired before the valve seats can be resurfaced. Ford supplies valves with oversize stems to fit valve guides that are reamed to oversize for repair. The machine shop will be able to handle the guide reaming for you. In some cases, if the guide is not too badly worn, knurling may be all that is required.

16. Reface, or have the valves and valve seats refaced. The valve seats should be a true 45° angle. Remove only enough material to clean up any pits or grooves. Be sure the valve seat is not too wide or narrow. Use a 60° grinding wheel to remove material from the bottom of the seat for raising and a 30° grinding wheel to remove material from the top of the seat to narrow.

17. After the valves are refaced by machine, hand lap them to the valve seat. Clean the grinding compound off and check the position of face-to-seat contact. Contact should be close to the center of the valve face. If contact is close to the top edge of the valve, narrow the seat; if too close to the bottom edge, raise the seat.

18. Valves should be refaced to a true angle of 44°. Remove only enough metal to clean up the valve face or to correct runout. If the edge of a valve head, after machining, is $\frac{1}{32}$ in. (0.8mm) or less replace the valve. The tip of the valve stem should also be dressed on the valve grinding machine, however, do not remove more than 0.010 in. (0.254mm).

19. After all valve and valve seats have been machined, check the remaining valve train parts (springs, retainers, keepers, etc.) for wear. Check the valve springs for straightness and tension.

20. Install the valves in the cylinder head and metal caps.

21. Install new valve stem oil seals.

22. Install the valve keepers, retainer, spring shield and valve spring using a valve spring compressor (the locking C-clamp type is the easiest kind to use).

23. Check the valve spring installed height, shim or replace as necessary.

CHECKING VALVE SPRINGS

➧ SEE FIGS. 45–46

Place the valve spring on a flat surface next to a carpenter's square. Measure the height of the spring, and rotate the spring against the edge of the square to measure distortion. If the spring height varies (by comparison) by more than $\frac{1}{16}$ in. (1.6mm) or if the distortion exceeds $\frac{1}{16}$ in. (1.6mm), replace the spring.

Have the valve springs tested for spring pressure at the installed and compressed (installed height minus valve lift) height using a valve spring tester. Springs should be within one pound, plus or minus each other. Replace springs as necessary.

VALVE SPRING INSTALLED HEIGHT

After installing the valve spring, measure the distance between the spring mounting pad and the lower edge of the spring retainer. Compare the measurement to specifications. If the installed height is incorrect, add shim washers between the spring mounting pad and the spring. Use only washers designed for valve springs, available at most parts houses.

VALVE STEM OIL SEALS

When installing valve stem oil seals, ensure that a small amount of oil is able to pass the seal to lubricate the valve stems and guide walls, otherwise, excessive wear will occur.

VALVE SEATS

➧ SEE FIGS. 47–49

If the valve seat is damaged or burnt and cannot be serviced by refacing, it may be possible to have the seat machined and an insert installed. Consult an automotive machine shop for their advice.

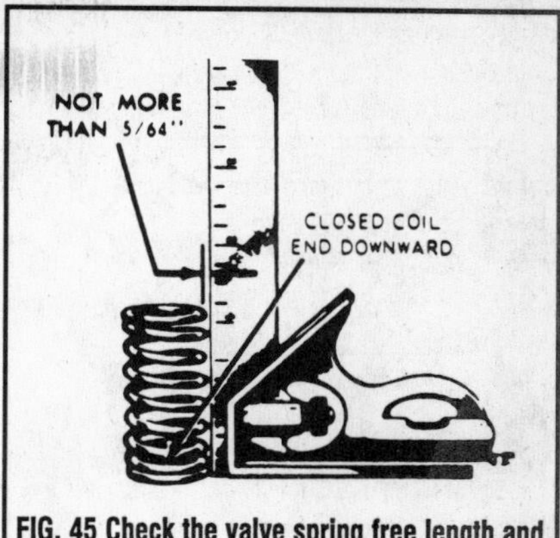

FIG. 45 Check the valve spring free length and squareness

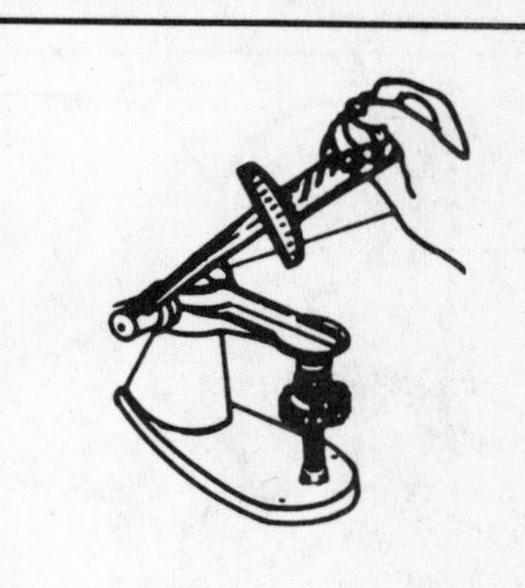

FIG. 46 Have the valve spring test pressure checked at a machine shop. Make sure the readings are within specifications

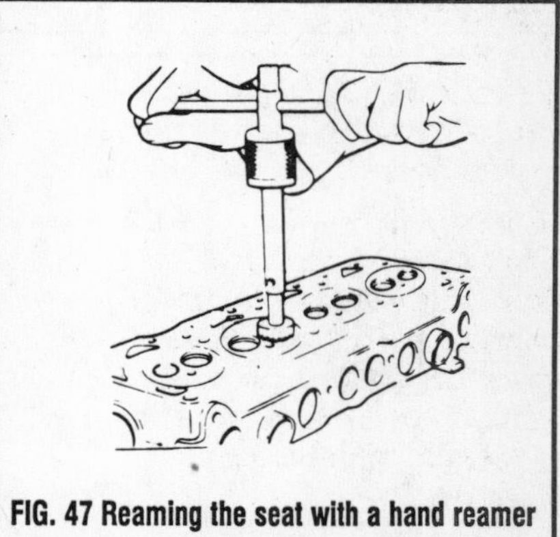

FIG. 47 Reaming the seat with a hand reamer

VALVE GUIDES

➧ SEE FIGS. 50–53

Worn valve guides can, in most cases, be reamed to accept a valve with an oversized

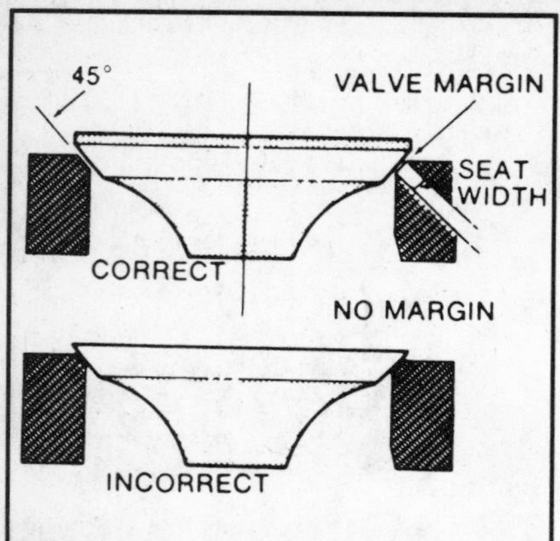

FIG. 48 Valve seat width centering after proper reaming

FIG. 49 Checking valve seat concentricity with a dial gauge

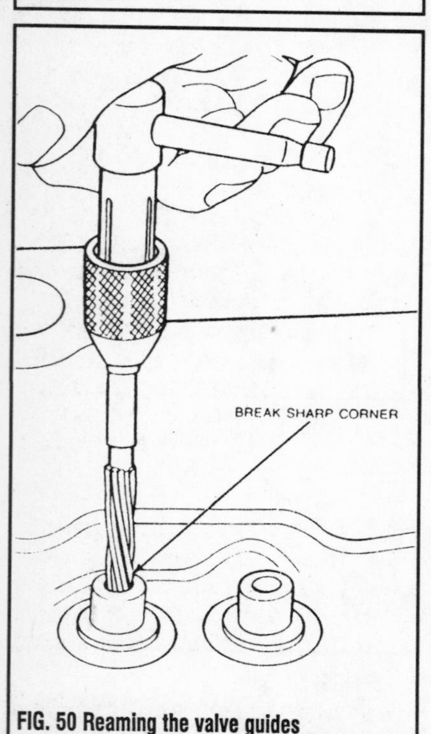

FIG. 50 Reaming the valve guides

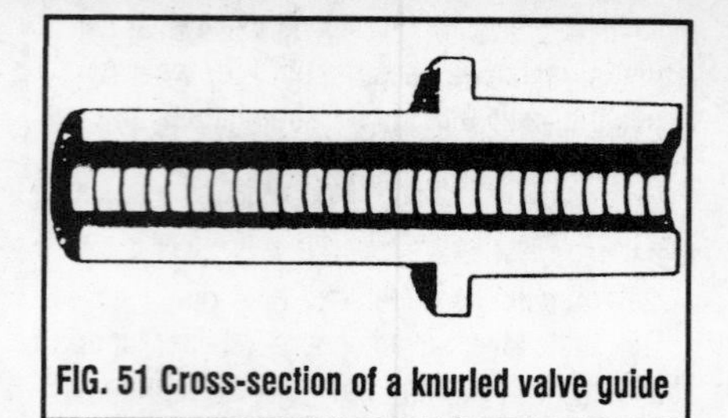

FIG. 51 Cross-section of a knurled valve guide

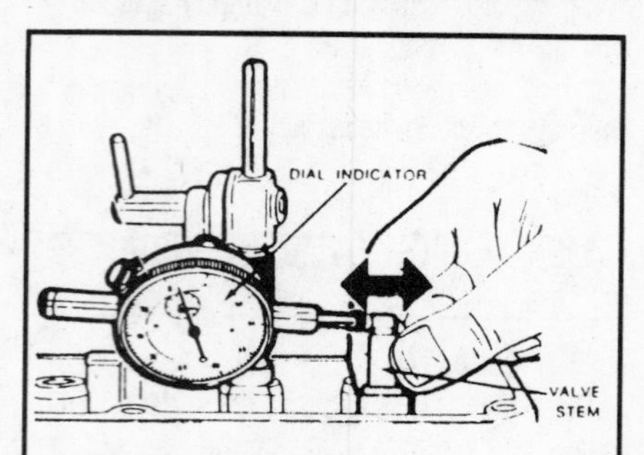

FIG. 52 Measuring valve stem-to-guide clearance. Make sure the indicator is mounted at 90° to the valve stem and as close to the guide as possible

FIG. 53 Lapping the valves by hand. When done, the finish on both valve faces and seats should be smooth and evenly shiny

stem. Valve guides that are not excessively worn or distorted may, in some cases, be knurled rather than reamed. However, if the valve stem is worn reaming for an oversized valve stem is the answer since a new valve would be required.

Knurling is a process in which metal is displaced and raised, thereby reducing clearance. Knurling also produces excellent oil control. The possibility of knurling instead of reaming the valve guides should be discussed with a machinist.

Valve Lifters

REMOVAL & INSTALLATION

2.3L Engine

The 2.3L engine is equipped with hydraulic lash adjusters which, while not being exactly the same as a conventional hydraulic lifter, perform the same function — maintain proper valve train clearance.

1. Disconnect the negative battery cable.
2. If equipped with distributor ignition, tag and disconnect the spark plug wires from the spark plugs. Move the wires out of the way.
3. Remove the hose and the retaining bolts from the rocker arm cover and remove the cover.
4. Rotate the camshaft so the base circle of the cam is facing the cam follower to be removed.
5. Using valve spring compressor tool T88T–6565–BH or equivalent, compress the lash adjuster as required and/or depress the valve spring if necessary and slide the cam follower over the lash adjuster and out.
6. Lift out the hydraulic lash adjuster.

To install:

7. Rotate the camshaft so the base circle of the camshaft is facing the lash adjuster and cam follower to be installed. Place the hydraulic lash adjuster in position in the bore.
8. Using valve spring compressor tool T88T–6565–BH or equivalent, compress the lash adjuster, as necessary, to position the cam follower over the lash adjuster and the valve stem.
9. Before rotating the camshaft to the next position, make sure the lash adjuster just installed is fully compressed and released.
10. Clean the gasket mating surface of the rocker arm cover and cylinder head. Install a new gasket and the rocker arm cover. Install the mounting screws and tighten to 62–97 inch lbs. (7–11 Nm).
11. Install the remaining components in the reverse order of removal. Start the engine and check for oil leaks.

5.0L Engine

1. Disconnect the negative battery cable. Remove the intake manifold and related parts.
2. Remove the crankcase ventilation hoses, PCV valve and elbows from the valve rocker arm covers.
3. Remove the valve rocker arm covers. Loosen the valve rocker arm fulcrum bolts and rotate the rocker arms to the side.
4. Remove the valve pushrods and identify

them so that they can be installed in their original position.

5. If equipped with roller lifters, remove the lifter guide retainer bolts. Remove the retainer and lifter guide plates. Identify the guide plates so they may be reinstalled in their original positions.

6. Using a magnet, remove the lifters and place them in a rack so that they can be installed in their original bores.

➡ **If the lifters are stuck in the bores due to excessive varnish or gum deposits, it may be necessary to use a claw-type tool to aid removal. When using a remover tool, rotate the lifter back and forth to loosen it from gum or varnish that may have formed on the lifter.**

To Install:

7. Lubricate the lifters and install them in their original bores. If new lifters are being installed, check them for free fit in their respective bores.

8. If equipped with roller lifters, install the lifter guide plates in their original positions, then install the guide plate retainer.

9. Install the pushrods in their original positions. Apply grease to the ends prior to installation.

10. Lubricate the rocker arms and fulcrum seats with heavy engine oil. Position the rocker arms over the pushrods and install the fulcrum bolts.

11. Before tightening each fulcrum bolt, rotate the crankshaft until the lifter is on the base circle of the cam. Tighten the fulcrum bolt to 18–25 ft. lbs. (24–34 Nm). Check the valve clearance.

12. Install the rocker arm covers and the intake manifold. Connect the negative battery cable, start the engine and check for leaks.

Oil Pan

REMOVAL & INSTALLATION

2.3L Engine

1989–90

➧ SEE FIG. 54

1. Disconnect the negative battery cable and drain the cooling system.

2. Disconnect the electrical connector to the cooling fan and remove the fan and shroud assembly. Disconnect the radiator hoses at the radiator. If equipped with an automatic transmission, disconnect the oil cooler lines at the radiator.

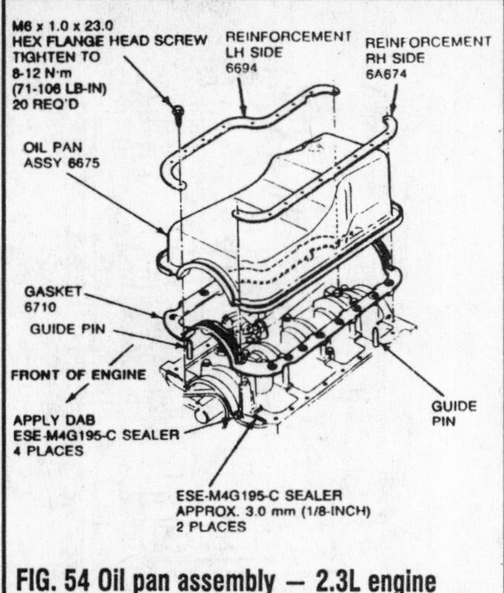

FIG. 54 Oil pan assembly — 2.3L engine 1989–90

3. Raise and safely support the vehicle. Drain the crankcase and disconnect the low oil level sensor, if equipped.

4. Remove the right and left engine support through bolts, except convertible. On convertible, remove the nuts. Using a jack and a block of wood, raise the engine as high as it will go. Place wood blocks between the mounts and the pedestal brackets. Remove the jack.

5. Remove the shake brace. Remove the sway bar retaining bolts and lower the sway bar.

6. Disconnect the cable at the starter and remove the starter. Remove the steering gear retaining bolts and lower the gear.

7. Remove the oil pan retaining bolts and allow the oil pan to drop to the crossmember. Rotate the crankshaft to position No. 4 piston up in the cylinder bore so the oil pan clears the crankshaft throw. Remove the oil pan.

To Install:

8. Clean the oil pan and the gasket mating surfaces. Remove and clean the oil pickup tube and screen assembly. After cleaning, reinstall.

9. Apply silicone sealer to the points where the rear main bearing cap meets the cylinder block, to the corners of the engine front cover and to where the front cover meets the cylinder block. Position the oil pan gasket to the cylinder block.

10. Position the oil pan and pan reinforcements to the cylinder block and install the retaining bolts. Tighten to 71–106 inch lbs. (8–12 Nm).

11. Position the steering gear and install the bolts and nuts. Install the starter and connect the cable.

12. Raise the engine enough to remove the wood blocks. Lower the engine and remove the jack. Install the shake brace.

13. Install the right and left engine through bolts, except convertible. Tighten the bolts to 65–85 ft. lbs. (88–119 Nm). On convertible, install the right and left engine support nuts. Tighten the nuts to 80–106 ft. lbs. (108–144 Nm).

14. Install the sway bar. Connect the low oil level sensor, if equipped. Install a new oil filter.

15. Lower the vehicle. Install the cooling fan and shroud assembly. Connect the cooling fan electrical connector and the radiator hoses. If equipped with an automatic transmission, connect the oil cooler lines.

16. Fill the engine with the proper type and quantity of engine oil. Fill and bleed the cooling system.

17. Connect the negative battery cable, start the engine and check for leaks.

1991–92

➧ SEE FIG. 55

1. Disconnect the negative battery cable. Remove the air cleaner outlet tube at the throttle body.

2. Remove the engine oil dipstick. Remove the engine mount retaining nuts.

3. Remove the oil cooler lines at the radiator, if equipped. Disconnect the electrical connector to the cooling fan and remove the cooling fan and shroud.

4. Raise and safely support the vehicle. Drain the crankcase.

5. Remove the starter cable from the starter and remove the starter.

6. Disconnect the exhaust manifold tube to the inlet pipe bracket. Disconnect the catalytic converter at the inlet pipe.

7. Remove the insulator and retainer assembly at the transmission. Remove the transmission mount retaining nuts to the crossmember. If equipped with an automatic transmission, remove the oil cooler lines from the retainer at the block.

8. Position a jack and a wood block under the engine. Raise the engine approximately 2.5 in. (63.5mm) high and place wood blocks between the mounts and crossmember.

9. Remove the jack and position it under the transmission. Raise the jack slightly.

10. Remove the oil pan retaining bolts and lower the pan to the chassis. Remove the oil pump drive and pickup tube assembly. Remove the oil pan with the oil pump.

To Install:

11. Clean the oil pan and the gasket mating surfaces. Clean the oil pump pickup tube screen.

12. Install the oil pan gasket in the groove in the oil pan. Position the oil pan assembly on the crossmember and install the oil pump drive and pickup tube assembly.

13. Apply silicone sealer to the points where the rear main bearing cap meets the cylinder

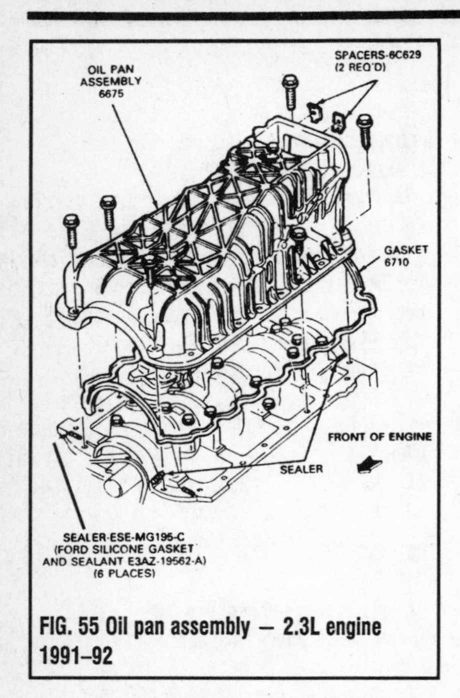

FIG. 55 Oil pan assembly — 2.3L engine 1991–92

block, to the corners of the engine front cover and to where the front cover meets the cylinder block.

14. Install the oil pan assembly. Install the oil pan flange bolts tight enough to compress the oil pan gasket to the point that the 2 transmission holes are aligned with the 2 tapped holes in the oil pan, but loose enough to allow movement of the pan, relative to the block.

15. Install the 2 oil pan/transmission bolts and tighten to 30–39 ft. lbs. (40–50 Nm) to align the oil pan with the transmission, then loosen the bolts 1/2 turn.

16. Tighten all oil pan flange bolts to 90–120 inch lbs. (10–13 Nm). Tighten the 2 oil pan/ transmission bolts to 30–39 ft. lbs. (40–54 Nm).

17. Install a new oil filter. Position the jack and wood block under the engine, raise the engine and remove the wood blocks from under the mounts. Shift the engine/transmission backward to its original position.

18. Install the insulator/bracket assembly to the crossmember and lower the engine on the insulators. Raise the transmission with the jack and install the insulator.

19. Connect the automatic transmission oil cooler line to the engine, if equipped. Install the transmission mount retaining nuts at the crossmember.

20. Connect the rear exhaust pipe just behind the catalytic converter. Install the starter and starter cable and lower the vehicle.

21. Install the cooling fan and shroud on the radiator. Connect the electrical connector to the cooling fan. If equipped, connect the oil cooler lines to the radiator.

22. Install the engine mount retaining nuts.

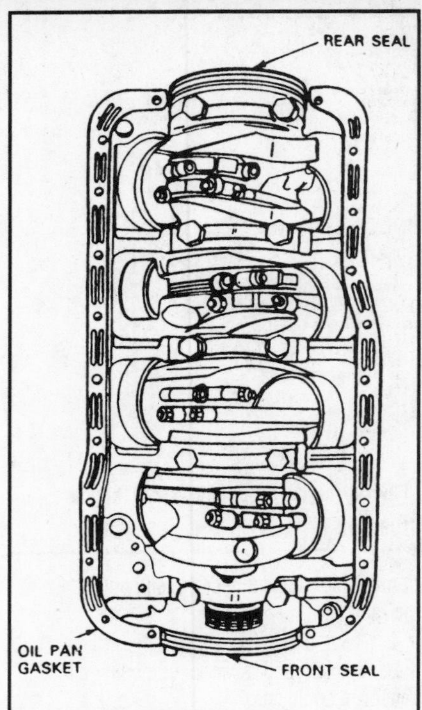

FIG. 56 Oil pan gaskets and seals — 5.0L engine

Install the dipstick and fill the crankcase with the proper type and quantity of engine oil.

23. Connect the negative battery cable, start the engine and check for leaks.

5.0L Engine

➧ SEE FIG. 56

1. Disconnect the negative battery cable and remove the air cleaner tube.

2. Remove the oil level indicator from the left side of the cylinder block. Remove the fan shroud and position the shroud over the fan.

3. Raise and safely support the vehicle. Drain the crankcase and remove the oil level sensor wiring from the oil pan.

4. Disconnect the electrical connectors from the starter and remove the starter. Remove the catalytic converter and muffler inlet pipes.

5. Remove the engine mount-to-No. 2 crossmember attaching bolts or nuts. Support the transmission and remove the No. 3 crossmember and rear insulator support assemblies.

6. Remove the steering gear attaching bolts and position the steering gear forward out of the way.

7. Position a jack and wood block under the oil pan. Raise the engine and install wood blocks between the engine mounts and frame. Lower the engine onto the wood blocks and remove the jack.

8. Remove the oil pan attaching bolts and lower the pan to the crossmember. Remove the oil pump and pickup tube assembly and allow to drop into the pan. Remove the pan.

To install:

9. Clean the oil pan and the gasket mating surfaces. Apply gasket sealer to the gasket mating surfaces and install new oil pan gaskets.

10. With the oil pump and pickup tube assembly positioned in the oil pan, raise the pan onto the crossmember. Install the oil pump and then the pan. Tighten the oil pan bolts to 6–9 ft. lbs. (8–12 Nm).

11. Position the oil pan and the wood block under the oil pan. Raise the engine and remove the wood blocks. Lower the engine and remove the jack. Install the engine mount-to-No. 2 crossmember attaching nuts or bolts.

12. Position the steering gear and install the retaining bolts. Install the starter and connect the electrical connectors. Connect the oil level sensor wire to the oil pan.

13. Install the rear insulator and the No. 3 crossmember. Install the catalytic converter and muffler inlet pipes. Lower the vehicle.

14. Install the fan shroud and install the oil level indicator to the side of the cylinder block. Install the air cleaner assembly.

15. Fill the crankcase with the proper type and quantity of engine oil. Connect the negative battery cable, start the engine and check for leaks.

Oil Pump

REMOVAL & INSTALLATION

1. Disconnect the negative battery cable. Remove the oil pan.

2. Remove the oil pump inlet tube and screen assembly.

3. Remove the oil pump attaching bolts and gasket. Remove the oil pump intermediate shaft.

To install:

4. Prime the oil pump by filling either the inlet or outlet ports with engine oil and rotating the pump shaft to distribute the oil within the pump body.

5. Position the intermediate driveshaft into the distributor socket. With the shaft firmly seated in the distributor socket, the stop on the shaft should touch the roof of the crankcase. Remove the shaft and position the stop, as necessary.

6. Position a new gasket on the pump body, insert the intermediate shaft into the oil pump and install the pump and shaft as an assembly.

➡ **Do not attempt to force the pump into position if it will not seat readily. The driveshaft hex may be misaligned with the distributor shaft. To align, rotate the intermediate shaft into a new position.**

7. Tighten the oil pump attaching screws to 14–21 ft. lbs. (19–29 Nm) on the 2.3L engine and 22–32 ft. lbs. (30–43 Nm) on the 5.0L engine.
8. Clean and install the oil pump inlet tube and screen assembly.
9. Install the oil pan and the remaining components.

CHECKING

1. Check the inside of the pump housing and the inner and outer gears for damage or excessive wear.
2. Check the mating surface of the pump cover for wear. Minor scuff marks are normal, but if the cover, gears or housing surfaces are excessively worn, scored or grooved, replace the pump. Inspect the rotor for nicks, burrs or score marks. Remove minor imperfections with an oil stone.
3. Measure the inner to outer rotor tip clearance. With the rotor assembly removed from the pump and resting on a flat surface, the inner and outer rotor tip clearance must not exceed 0.012 in. (0.30mm) with the feeler gauge inserted 0.5 in. (13mm) minimum.
4. With the rotor assembly installed in the housing, place a straight edge over the rotor assembly and the housing. Measure the rotor endplay between the straight edge and both the inner and outer race. The maximum clearance must not exceed 0.005 in. (0.13mm).
5. Inspect the relief valve spring to see if it is collapsed or worn. Check the relief valve spring tension. Specifications are as follows:
   - 2.3L engine — 12.6–14.5 lbs. @ 1.20 in.
   - 5.0L engine — 10.6–12.2 lbs. @ 1.704 in.
6. If the spring tension is not within specification and/or the spring is worn or damaged, replace the pump. Check the relief valve piston for free operation in the bore.

➡ **Internal oil pump components are not serviced. If any component is out of specification, the entire pump must be replaced.**

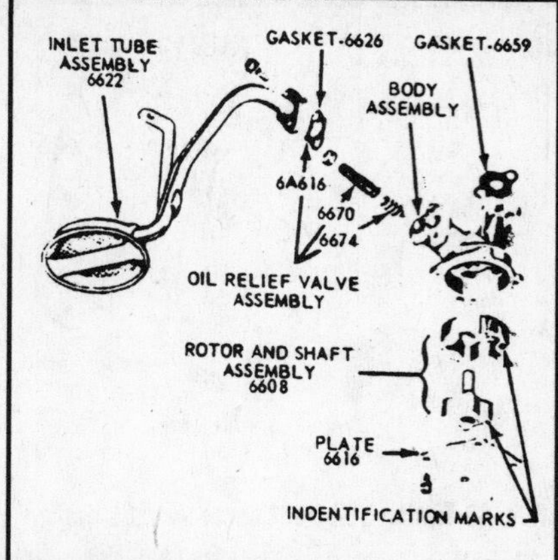

FIG. 57 Explode view of the oil pump — 5.0L oil pump

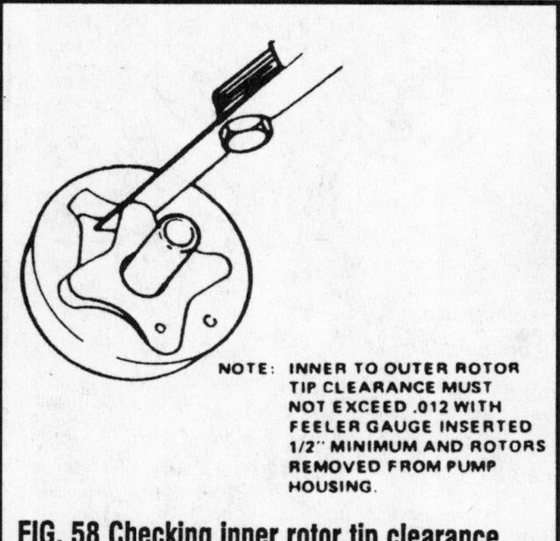

FIG. 58 Checking inner rotor tip clearance

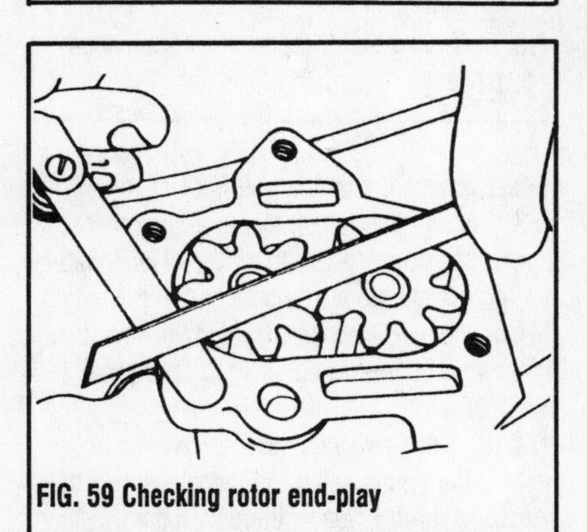

FIG. 59 Checking rotor end-play

OVERHAUL

➧ SEE FIGS. 57–59

1. Wash all parts in solvent and dry them thoroughly with compressed air. Use a brush to clean the inside of the pump housing and the pressure relief valve chamber. Be sure all dirt and metal particles are removed.
2. Check the inside of the pump housing and the outer race and rotor for damage or excessive wear or scoring.
3. Check the mating surface of the pump cover for wear. If the cover mating surface is worn, scored, or grooved, replace the pump.
4. Measure the inner rotor tip clearance.
5. With the rotor assembly installed in the housing, place a straight edge over the rotor assembly and the housing. Measure the clearance (rotor end play) between the straight edge and the rotor and the outer race.
6. Check the drive shaft to housing bearing clearance by measuring the OD of the shaft and the ID of the housing bearing.
7. Components of the oil pump are not serviced. If any part of the pump requires replacement, replace the complete pump assembly.
8. Inspect the relief valve spring to see if it is collapsed or worn.
9. Check the relief valve piston for scores and free operation in the bore.

Timing Belt Front Cover

REMOVAL & INSTALLATION

CAUTION

When draining the coolant, keep in mind that cats and dogs are attracted by the ethylene glycol antifreeze, and are quite likely to drink any that is left in an uncovered container or in puddles on the ground. This will prove fatal in sufficient quantity. Always drain the coolant into a sealable container. Coolant should be reused unless it is contaminated or several years old.

➧ SEE FIGS. 60–61

2.3L Engine

1. Disconnect the negative battery cable and drain the cooling system.
2. Remove the automatic belt tensioner and accessory drive belt. Remove the upper radiator hose.
3. Remove the crankshaft pulley bolt and pulley. Remove the thermostat housing and gasket.
4. Remove the timing belt outer cover

retaining bolt(s). Release the cover interlocking tabs, if equipped, and remove the cover.

To install:

5. Position the timing belt front cover. Snap the interlocking tabs into place, if necessary. Install the timing belt outer cover retaining bolt(s) and tighten to 71–106 inch lbs. (8–12 Nm).

6. Install the thermostat housing and a new gasket. Install the upper radiator hose.

7. Install the crankshaft pulley and retaining bolt. Tighten to 103–133 ft. lbs. (140–180 Nm).

8. Install the water pump pulley and the automatic belt tensioner. Install the accessory drive belt.

9. Connect the negative battery cable, start the engine and check for leaks.

OIL SEAL REPLACEMENT

2.3L Engine

1. Disconnect the negative battery cable.

2. Remove the timing belt front cover and timing belt.

3. Use a suitable puller to remove the crankshaft, camshaft and auxiliary shaft sprockets.

4. Use seal remover tool T74P–6700–B or equivalent, to remove the crankshaft, camshaft and auxiliary shaft seals. Position the tool so the jaws are gripping the thin edge of the seal. Operate the jackscrew on the tool to remove the seal.

To install:

5. Lubricate the lips of the new seals with clean engine oil.

6. Use seal replacer tool T74P–6150–A or equivalent, to install the seals.

7. Install the crankshaft, camshaft and auxiliary shaft sprockets. Tighten the camshaft sprocket retaining bolt to 52–70 ft. lbs. (70–95 Nm) and the auxiliary sprocket retaining bolt to 28–40 ft. lbs. (38–54 Nm).

8. Install the timing belt and timing belt front cover.

9. Connect the negative battery cable, start the engine and check for leaks.

Timing Belt and Tensioner

REMOVAL & INSTALLATION

2.3L Engine

➧ SEE FIGS. 62–63

1. Disconnect the negative battery cable.

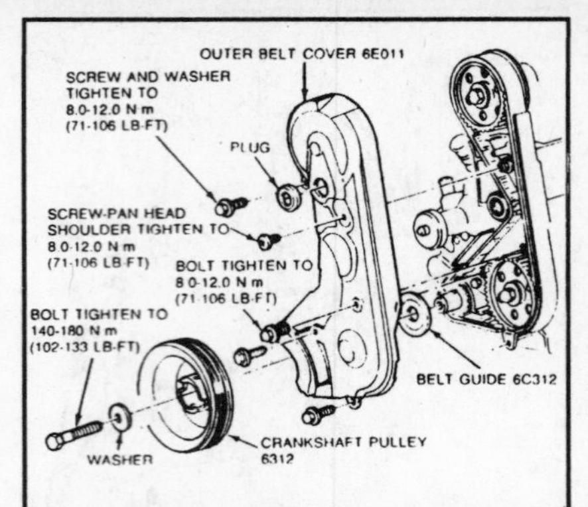

FIG. 60 Timing belt front cover — 2.3L engine 1989-90

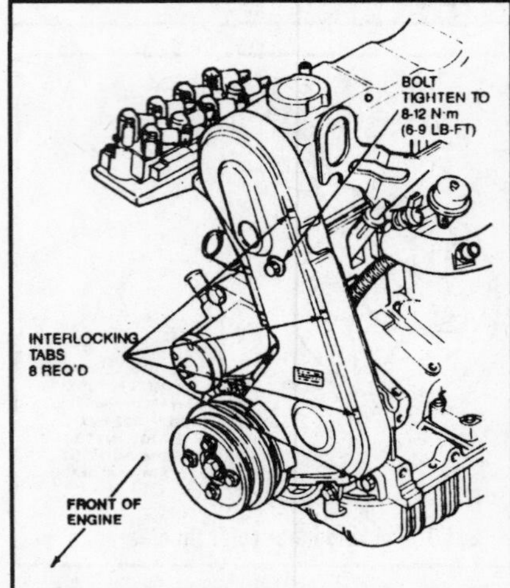

FIG. 61 Timing belt front cover — 2.3L engine 1991-92

2. Remove the timing belt front cover.

3. Loosen the belt tensioner adjustment screw, position belt tensioner tool T74P–6254–A or equivalent, on the tension spring roll pin and release the belt tensioner. Tighten the adjustment screw to hold the tensioner in the released position.

4. On 1991–92 vehicles, remove the bolts holding the timing sensor in place and pull the sensor assembly free of the dowel pin.

5. Remove the crankshaft pulley, hub and belt guide. Remove the timing belt. If the belt is to be reused, mark the direction of rotation so it may be reinstalled in the same direction.

To install:

6. Position the crankshaft sprocket to align with the TDC mark and the camshaft sprocket to align with the camshaft timing pointer. On 1989–90 vehicles, remove the distributor cap and set the rotor to the No. 1 firing position by turning the auxiliary shaft.

7. Install the timing belt over the crankshaft sprocket and then counterclockwise over the auxiliary and camshaft sprockets. Align the belt fore-and-aft on the sprockets.

8. Loosen the tensioner adjustment bolt to allow the tensioner to move against the belt. If the spring does not have enough tension to move the roller against the belt, it may be necessary to manually push the roller against the belt and tighten the bolt.

9. To make sure the belt does not jump time during rotation in Step 10, remove a spark plug from each cylinder.

10. Rotate the crankshaft 2 complete turns in the direction of normal rotation to remove the slack from the belt. Tighten the tensioner adjustment to 29–40 ft. lbs. (40–55 Nm) and pivot bolts to 14–22 ft. lbs. (20–30 Nm). Check the alignment of the timing marks.

11. Install the crankshaft belt guide.

12. On 1989–90 vehicles, install the crankshaft pulley and tighten the retaining bolt to 103–133 ft. lbs. (140–180 Nm). On 1991–92 vehicles, proceed as follows:

a. Install the timing sensor onto the dowel pin and tighten the 2 longer bolts to 14–22 ft. lbs. (20–30 Nm).

b. Rotate the crankshaft 45 degrees counterclockwise and install the crankshaft pulley and hub assembly. Tighten the bolt to 114–151 ft. lbs. (155–205 Nm).

c. Rotate the crankshaft 90 degrees clockwise so the vane of the crankshaft pulley engages with timing sensor positioner tool T89P–6316–A or equivalent. Tighten the 2 shorter sensor bolts to 14–22 ft. lbs. (20–30 Nm).

d. Rotate the crankshaft 90 degrees counterclockwise and remove the sensor positioner tool.

e. Rotate the crankshaft 90 degrees clockwise and measure the outer vane to sensor air gap. The air gap must be 0.018–0.039 in. (0.458–0.996mm).

13. Install the timing belt front cover, spark plugs and remaining components.

14. Connect the negative battery cable, start the engine and check the ignition timing.

Timing Sprockets

REMOVAL & INSTALLATION

2.3L Engine

1. Disconnect the negative battery cable.

2. Remove the timing belt front cover and the timing belt.

3. Remove the camshaft and auxiliary shaft

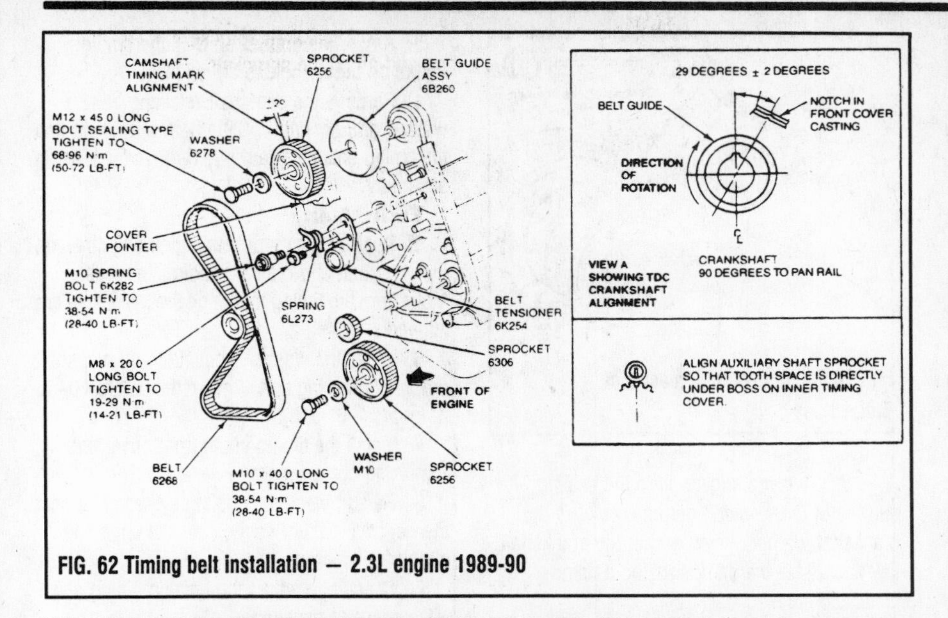

FIG. 62 Timing belt installation — 2.3L engine 1989-90

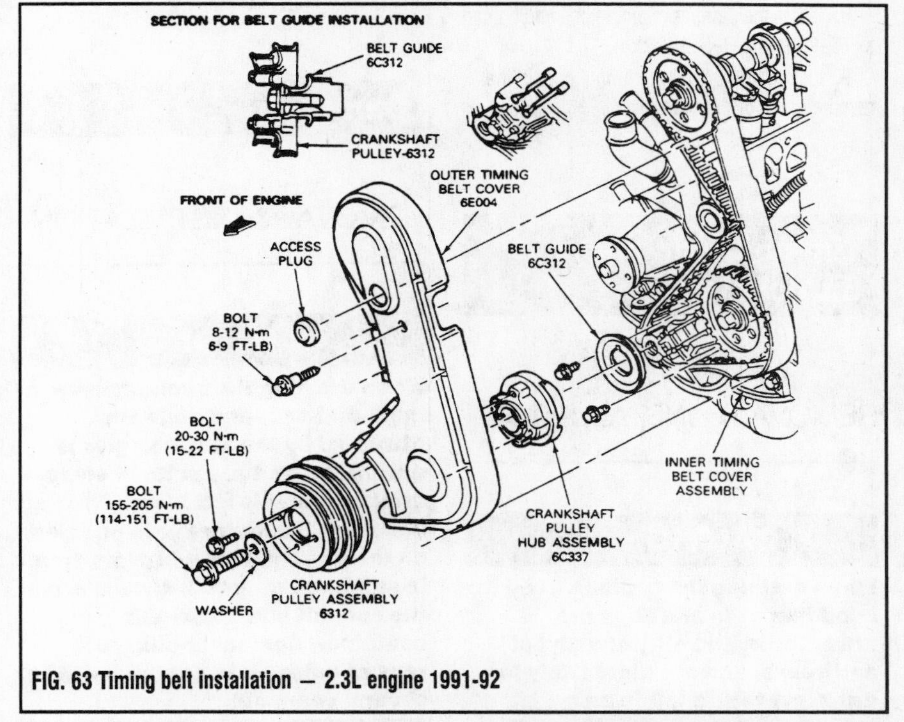

FIG. 63 Timing belt installation — 2.3L engine 1991-92

sprocket retaining bolts. Remove the crankshaft, camshaft and auxiliary shaft sprockets using suitable pullers.

To install:

4. Install the crankshaft, camshaft and auxiliary shaft sprockets. Tighten the camshaft sprocket retaining bolt to 52–70 ft. lbs. (70–95 Nm) and the auxiliary sprocket retaining bolt to 28–40 ft. lbs. (38–54 Nm).
5. Install the timing belt and timing belt front cover.
6. Connect the negative battery cable.

Timing Chain Front Cover

REMOVAL & INSTALLATION

CAUTION

When draining the coolant, keep in mind that cats and dogs are attracted by the ethylene glycol antifreeze, and are quite likely to drink any that is left in an uncovered container or in puddles on the ground. This will prove fatal in sufficient quantity. Always drain the coolant into a sealable container. Coolant should be reused unless it is contaminated or several years old.

5.0L Engine

SEE FIG. 64

1. Disconnect the negative battery cable.
2. Drain the cooling system and the crankcase. Remove the air inlet tube, if equipped.
3. Remove the fan shroud attaching bolts and position the shroud over the fan. Remove the fan and clutch assembly from the water pump shaft and remove the shroud.
4. Remove the air conditioner drive belt and idler pulley bracket, if equipped. Remove the alternator drive belt. Remove the power steering drive belt and power steering pump, if equipped. Remove all accessory brackets that attach to the water pump.
5. Remove the water pump pulley. Disconnect the lower radiator hose, heater hose and water pump bypass hose at the water pump.
6. Remove the crankshaft pulley from the crankshaft vibration damper. Remove the damper attaching bolt and washer and remove the damper using a puller.
7. Remove the fuel line from the clip on the front cover, if equipped.
8. Remove the oil pan-to-front cover attaching bolts. Use a thin blade knife to cut the oil pan gasket flush with the cylinder block face prior to separating the cover from the cylinder block.
9. Remove the cylinder front cover and water pump as an assembly.

Cover the front oil pan opening while the cover assembly is off to prevent foreign material from entering the pan.

To install:

10. If a new front cover is to be installed, remove the water pump from the old front cover and install it on the new front cover.
11. Clean all gasket mating surfaces. Pry the old oil seal from the front cover and install a new 1, using a seal installer.
12. Coat the gasket surface of the oil pan with sealer, cut and position the required sections of a new gasket on the oil pan and apply silicone

sealer at the corners. Apply sealer to a new front cover gasket and install on the block.

13. Position the front cover on the cylinder block. Use care to avoid seal damage or gasket dislocation. It may be necessary to force the cover downward to slightly compress the pan gasket. Use front cover aligner tool T61P–6019–B or equivalent to assist the operation.

14. Coat the threads of the front cover attaching screws with pipe sealant and install. While pushing in on the alignment tool, tighten the oil pan to cover attaching screws to 9–11 ft. lbs. (12–15 Nm).

15. Tighten the front cover to cylinder block attaching bolts to 12–18 ft. lbs. (16–24 Nm). Remove the alignment tool.

16. Apply multi-purpose grease to the sealing surface of the vibration damper. Apply silicone sealer to the keyway of the vibration damper.

17. Line up the vibration damper keyway with the crankshaft key and install the damper using a suitable installation tool. Tighten the retaining bolt to 70–90 ft. lbs. (95–122 Nm). Install the crankshaft pulley.

18. On 5.8L engines, install the fuel pump with a new gasket. Connect the fuel pump outlet line.

19. Install the remaining components in the reverse order of their removal.

20. Fill the crankcase with the proper type and quantity of engine oil. Fill and bleed the cooling system.

21. Connect the negative battery cable, start the engine and check for leaks.

Front Cover Oil Seal

REPLACEMENT

5.0L Engine

1. Disconnect the negative battery cable.
2. Remove the fan shroud and position it back over the fan. Remove the fan/clutch assembly and shroud.
3. Remove the accessory drive belts.
4. Remove the crankshaft pulley from the damper and remove the damper retaining bolt. Remove the damper using a puller.
5. Remove the seal using a seal removal tool.

To Install:

6. Lubricate the seal lip with clean engine oil and install using a seal installer.

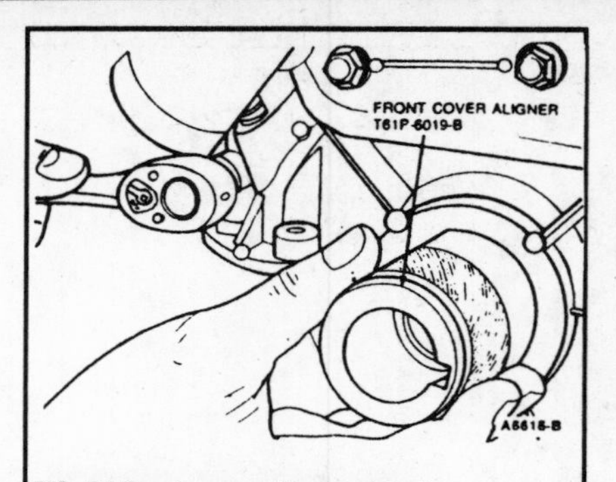

FIG. 64 Front cover alignment tool — 5.0L engine

7. Apply clean engine oil to the sealing surface of the vibration damper. Line up the crankshaft damper keyway with the crankshaft key and install the damper using a damper installation tool.
8. Install the damper retaining bolt and tighten to 70–90 ft. lbs. (95–122 Nm).
9. Install the remaining components in the reverse order of their removal.

Timing Chain and Sprockets

REMOVAL & INSTALLATION

CAUTION

When draining the coolant, keep in mind that cats and dogs are attracted by the ethylene glycol antifreeze, and are quite likely to drink any that is left in an uncovered container or in puddles on the ground. This will prove fatal in sufficient quantity. Always drain the coolant into a sealable container. Coolant should be reused unless it is contaminated or several years old.

5.0L Engine

➧ SEE FIG. 65

1. Disconnect the negative battery cable and drain the cooling system.
2. Remove the timing chain front cover.
3. Rotate the crankshaft until the timing marks on the sprockets are aligned.
4. Remove the camshaft retaining bolt, washer and eccentric. Slide both sprockets and the timing chain forward and remove them as an assembly.

To Install:

5. Position the sprockets and timing chain on the camshaft and crankshaft simultaneously. Make sure the timing marks on the sprockets are aligned.
6. Install the washer, eccentric and camshaft sprocket retaining bolt. Tighten the bolt to 40–45 ft. lbs. (54–61 Nm).
7. Install the timing chain front cover and remaining components.
8. Fill and bleed the cooling system. Connect the negative battery cable, start the engine and check for leaks.
9. Check and adjust the ignition timing and idle speed, as necessary.

Camshaft

REMOVAL & INSTALLATION

CAUTION

When draining the coolant, keep in mind that cats and dogs are attracted by the ethylene glycol antifreeze, and are quite likely to drink any that is left in an uncovered container or in puddles on the ground. This will prove fatal in sufficient quantity. Always drain the coolant into a sealable container. Coolant should be reused unless it is contaminated or several years old.

2.3L Engine

1. Disconnect the negative battery cable and drain the cooling system.
2. Remove the air intake and the throttle body.
3. Disconnect the radiator hoses. Remove the cooling fan and shroud assembly.
4. Tag and disconnect the spark plug wires and position aside.
5. Tag and disconnect the necessary electrical connectors and vacuum lines and position aside.

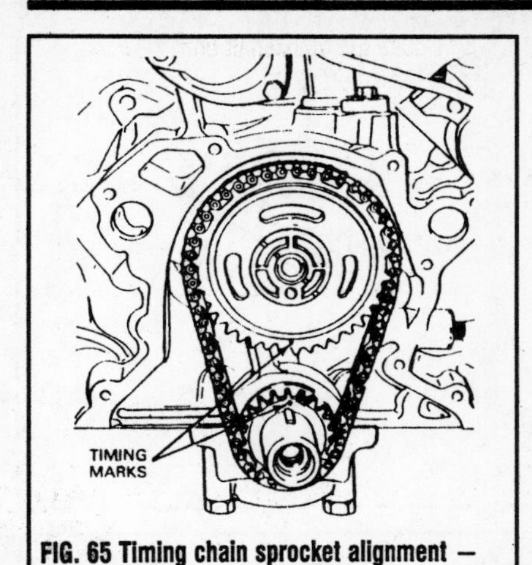

FIG. 65 Timing chain sprocket alignment — 5.0L engine

6. Remove the rocker cover retaining bolts and the rocker cover.
7. Remove the timing belt front cover and the timing belt.
8. Compress the valve springs using valve spring compressor lever T88T–6565–BH or equivalent and remove the cam followers.
9. Remove the camshaft sprocket retaining bolt. Remove the camshaft sprocket using a suitable puller. Remove the camshaft seal using a seal removal tool.
10. Remove the 2 screws and the camshaft rear retainer.
11. Raise and safely support the vehicle. Remove the right and left engine support bolts and nuts.
12. Position a block of wood and a jack under the engine. Raise the engine as high as it will go. Place blocks of wood between the engine mounts and chassis brackets and remove the jack.
13. Lower the vehicle and remove the camshaft.

To install:

14. Make sure the threaded plug is in the rear of the camshaft. If not, remove the plug from the old camshaft and install.
15. Coat the camshaft lobes with multi-purpose grease and lubricate the journals with heavy engine oil before installation. carefully slide the camshaft through the bearings.
16. Install the camshaft rear retainer and tighten the 2 screws to 6–9 ft. lbs. (8–12 Nm). Install a new camshaft seal using a suitable seal installer.
17. Install the camshaft sprocket and tighten the retaining bolt to 52–70 ft. lbs. (70–95 Nm).
18. Install the timing belt and timing belt front cover.
19. Raise and safely support the vehicle. Position a block of wood and a jack and raise the engine. Remove the blocks of wood, lower the engine and remove the jack.
20. Install the engine support bolts and nuts and lower the vehicle.
21. Install the remaining components in the reverse order of removal.
22. Connect the negative battery cable, start the engine and check for leaks. Check the ignition timing, if necessary.

5.0L Engine

1. Disconnect the negative battery cable and drain the cooling system.
2. Relieve the fuel system pressure and discharge the air conditioning system.
3. Remove the radiator. If equipped with air conditioning, remove the condenser.
4. Remove the grille.
5. Remove the intake manifolds and the lifters.
6. Remove the timing chain front cover, the timing chain and spacer.
7. Remove the thrust plate. Remove the camshaft, being careful not to damage the bearing surfaces.

To install:

8. Lubricate the cam lobes and journals with heavy engine oil. Install the camshaft, being careful not to damage the bearing surfaces while sliding into position.
9. Install the thrust plate. Tighten the bolts to 9–12 ft. lbs. (12–16 Nm).
10. Install the timing chain and sprockets. Install the engine front cover.
11. Install the lifters and the intake manifolds.
12. Install the grille. If equipped with air conditioning, install the condenser.
13. Install the radiator. Fill and bleed the cooling system.
14. Connect the negative battery cable. Start the engine and check for leaks.

CHECKING CAMSHAFT

Camshaft Lobe Lift

5.0L engine

Check the lift of each lobe in consecutive order and make a note of the reading.

1. Remove the fresh air inlet tube and the air cleaner. Remove the heater hose and crankcase ventilation hoses. Remove valve rocker arm cover(s).
2. Remove the rocker arm stud nut or fulcrum bolts, fulcrum seat and rocker arm.
3. Make sure the pushrod is in the valve tappet socket. Install a dial indicator D78P–4201–B or equivalent. so that the actuating point of the indicator is in the push rod socket (or the indicator ball socket adaptor tool 6565–AB is on the end of the push rod) and in the same plane as the push rod movement.
4. Disconnect the I terminal and the S terminal at the starter relay. Install an auxiliary starter switch between the battery and S terminals of the start relay. Crank the engine with the ignition switch off. Turn the crankshaft over until the tappet is on the base circle of the camshaft lobe. At this position, the push rod will be in its lowest position.
5. Zero the dial indicator. Continue to rotate the crankshaft slowly until the push rod is in the fully raised position.
6. Compare the total lift recorded on the dial indicator with the specification shown on the Camshaft Specification chart.

To check the accuracy of the original indicator reading, continue to rotate the crankshaft until the indicator reads zero. If the left on any lobe is below specified wear limits listed, the camshaft and the valve tappet operating on the worn lobe(s) must be replaced.

7. Install the dial indicator and auxiliary starter switch.
8. Install the rocker arm, fulcrum seat and stud nut or fulcrum bolts. Check the valve clearance. Adjust if required (refer to procedure in this Section).
9. Install the valve rocker arm cover(s) and the air cleaner.

Camshaft End Play

➡ On all gasoline V8 engines, prying against the aluminum-nylon camshaft sprocket, with the valve train load on the camshaft, can break or damage the sprocket. Therefore, the rocker arm adjusting nuts must be backed off, or the rocker arm and shaft assembly must be loosened sufficiently to free the camshaft. After checking the camshaft end play, check the valve clearance. Adjust if required (refer to procedure in this Section).

1. Push the camshaft toward the rear of the engine. Install a dial indicator (Tool D78P–4201–F, –G or equivalent so that the indicator point is on the camshaft sprocket attaching screw.
2. Zero the dial indicator. Position a prybar between the camshaft gear and the block. Pull the camshaft forward and release it. Compare the dial indicator reading with the specifications.
3. If the end play is excessive, check the spacer for correct installation before it is removed. If the spacer is correctly installed, replace the thrust plate.
4. Remove the dial indicator.

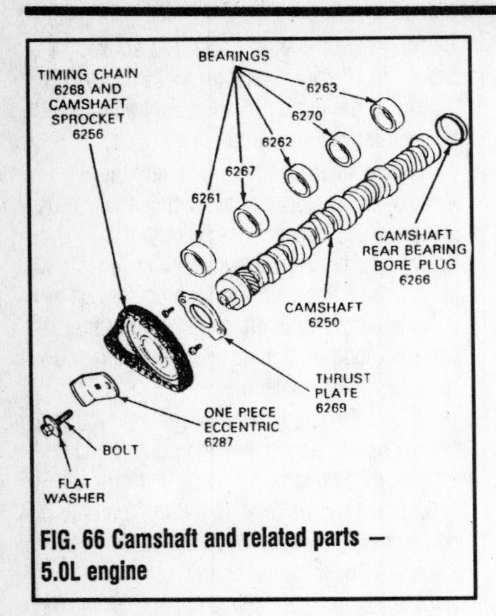

FIG. 66 Camshaft and related parts — 5.0L engine

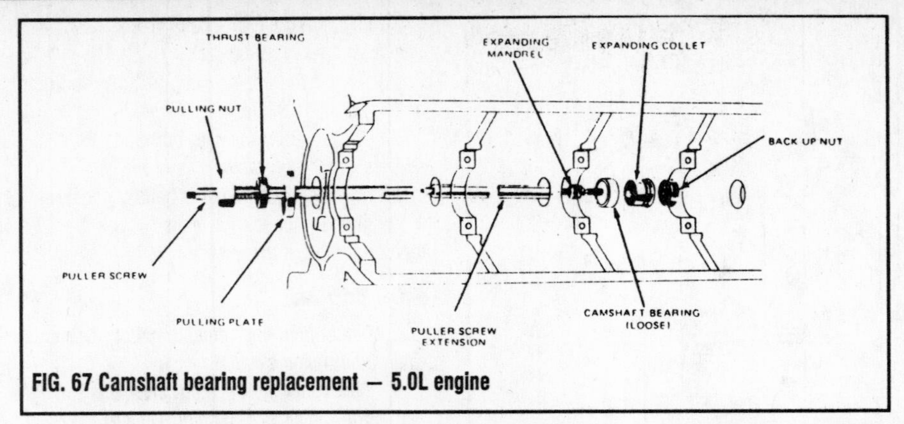

FIG. 67 Camshaft bearing replacement — 5.0L engine

CAMSHAFT BEARING REPLACEMENT

5.0L Engine

➧ SEE FIGS. 66–67

1. Remove the engine following the procedures in this Section and install it on a work stand.
2. Remove the camshaft, flywheel and crankshaft, following the appropriate procedures. Push the pistons to the top of the cylinder.
3. Remove the camshaft rear bearing bore plug. Remove the camshaft bearings with Tool T65L–6250–A or equivalent.
4. Select the proper size expanding collet and back-up nut and assemble on the mandrel. With the expanding collet collapsed, install the collet assembly in the camshaft bearing and tighten the back-up nut on the expanding mandrel until the collet fits the camshaft bearing.
5. Assemble the puller screw and extension (if necessary) and install on the expanding mandrel. Wrap a cloth around the threads of the puller screw to protect the front bearing or journal. Tighten the pulling nut against the thrust bearing and pulling plate to remove the camshaft bearing. Be sure to hold a wrench on the end of the puller screw to prevent it from turning.
6. To remove the front bearing, install the puller from the rear of the cylinder block.
7. Position the new bearings at the bearing bores, and press them in place with tool T65L–6250–A or equivalent. Be sure to center the pulling plate and puller screw to avoid damage to the bearing. Failure to use the correct expanding collet can cause severe bearing damage. Align the oil holes in the bearings with the oil holes in the cylinder block before pressing bearings into place.

➡ Be sure the front bearing is installed 0.005–0.020 in. (0.127–0.508mm) below the front face of the cylinder block.

8. Install the camshaft rear bearing bore plug.
9. Install the camshaft, crankshaft, flywheel and related parts, following the appropriate procedures.
10. Install the engine in the vehicle, following procedures described earlier in this Section.

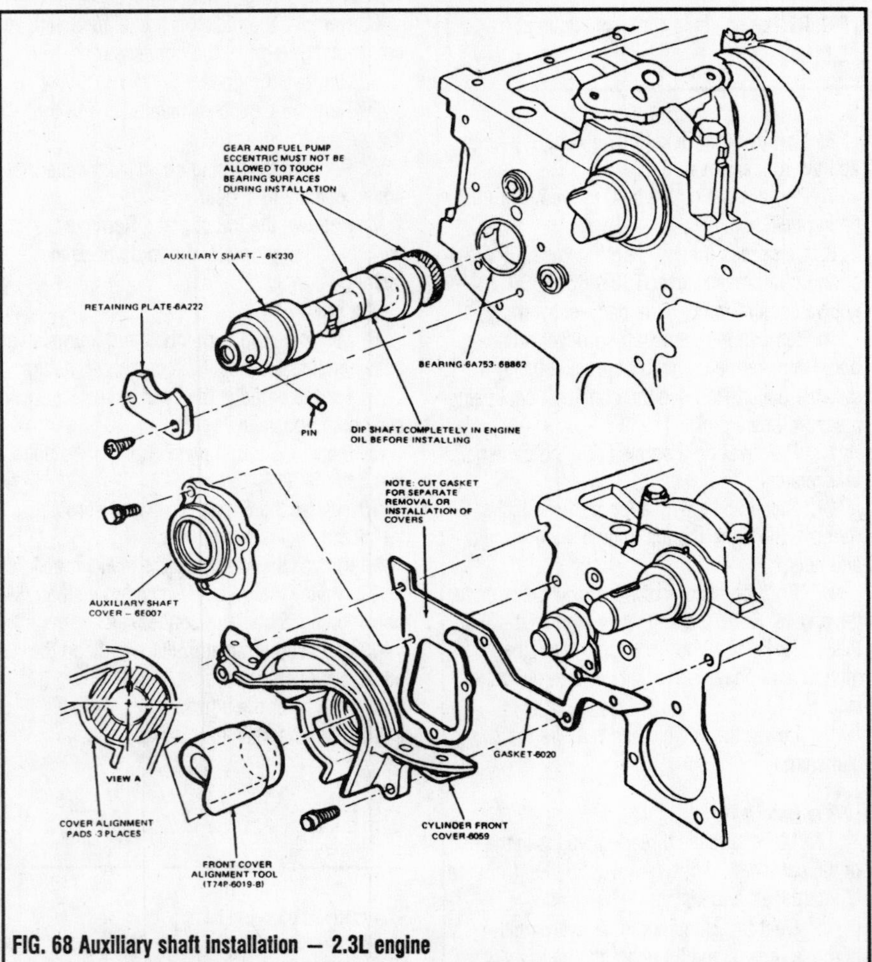

FIG. 68 Auxiliary shaft installation — 2.3L engine

Auxiliary Shaft

REMOVAL & INSTALLATION

2.3L Engine

➧ SEE FIG. 68

1. Disconnect the negative battery cable. Remove the front timing belt cover.
2. Remove the timing belt. Remove the auxiliary shaft sprocket retaining bolt. Remove the sprocket using a puller.
3. On 1989–90 vehicles, remove the distributor.
4. Remove the auxiliary shaft cover and thrust plate.
5. Withdraw the auxiliary shaft from the block being careful not to damage the bearings.

To install:

6. Dip the auxiliary shaft in engine oil before installing. Slide the auxiliary shaft into the cylinder block, being careful not to damage the bearings.
7. Install the thrust plate. Tighten the thrust plate screws to 6–9 ft. lbs. (8–12 Nm).
8. Install a new gasket and auxiliary shaft cover. Tighten the cover screws to 6–9 ft. lbs. (8–12 Nm).

➡ The auxiliary shaft cover and cylinder front cover share a common gasket. Cut off the old gasket around the cylinder cover and use half of the new gasket on the auxiliary shaft cover.

9. Insert the distributor and install the auxiliary shaft sprocket.
10. Align the timing marks and install the timing belt.
11. Install the timing belt cover.
12. Check the ignition timing.

Core (Freeze) Plugs

REPLACEMENT

➧ SEE FIG. 69

Core plugs need replacement only if they are found to be leaking, are excessively rusty, have popped due to freezing or, if the engine is being overhauled.

If the plugs are accessible with the engine in the vehicle, they can be removed as-is. If not, the engine will have to be removed.

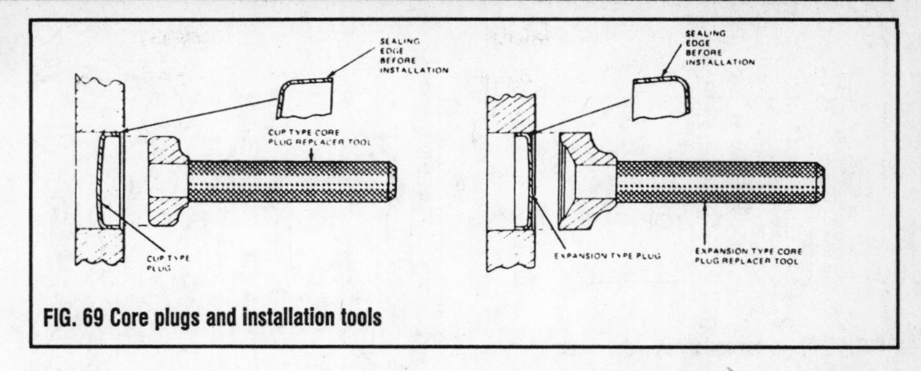

FIG. 69 Core plugs and installation tools

1. If necessary, remove the engine and mount it on a work stand. If the engine is being left in the vehicle, drain the engine coolant and engine oil

✲✲ CAUTION

When draining the coolant, keep in mind that your pets are attracted by the ethylene glycol antifreeze, and are quite likely to drink any that is left in an uncovered container or in puddles on the ground. This will prove fatal in sufficient quantity. Always drain the coolant into a sealable container. Coolant should be reused unless it is contaminated or several years old.

2. Remove anything blocking access to the plug or plugs to be replaced.
3. Drill or center-punch a hole in the plug. For large plugs, drill a 1/2 in. (13mm) hole; for small plugs, drill a 1/4 in. (6mm) hole.
4. For large plugs, using a slide-hammer, thread a machine screw adapter or insert 2-jawed puller adapter into the hole in the plug. Pull the plug from the block; for small plugs, pry the plug out with a pin punch.
5. Thoroughly clean the opening in the block, using steel wool or emery paper to polish the hole rim.
6. Coat the outer diameter of the new plug with sealer and place it in the hole. For cup-type core plugs: These plugs are installed with the flanged end outward. The maximum diameter of this type of plug is located at the outer edge of the flange. carefully and evenly, drive the new plug into place. For expansion-type plugs: These plugs are installed with the flanged end inward. The maximum diameter of this type of plug is located at the base of the flange. It is imperative that the correct type of installation tool is used with this type of plug. Under no circumstances is this type of plug to be driven in using a tool that contacts the crowned portion of the plug. Driving in this plug incorrectly will cause the plug to expand prior to installation. When installed, the trailing (maximum) diameter of the plug MUST be below the chamfered edge of the bore to create an effective seal. If the core plug replacing tool has a depth seating surface, do not seat the tool against a non-machined (casting) surface.
7. Install any removed parts and, if necessary, install the engine in the vehicle.
8. Refill the cooling system and crankcase.
9. Start the engine and check for leaks.

Pistons and Connecting Rods

REMOVAL & INSTALLATION

2.3L Engine

➧ SEE FIG. 70–74

1. Drain the cooling system and the crankcase.

✲✲ CAUTION

The EPA warns that prolonged contact with used engine oil may cause a number of skin disorders, including cancer! You should make every effort to minimize your exposure to used engine oil. Protective gloves should be worn when changing the oil. Wash your hands and any other exposed skin areas as soon as possible after exposure to used engine oil. Soap and water, or waterless hand cleaner should be used.

2. Remove the cylinder head.
3. Remove the oil pan, the oil pump inlet tube and the oil pump.

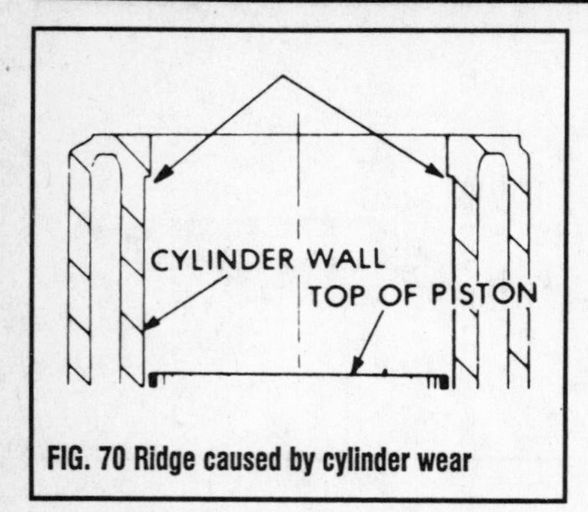

FIG. 70 Ridge caused by cylinder wear

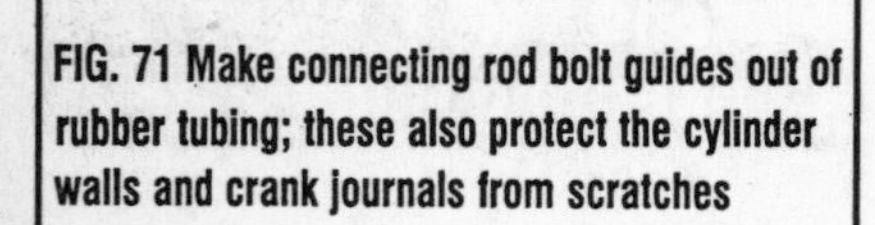
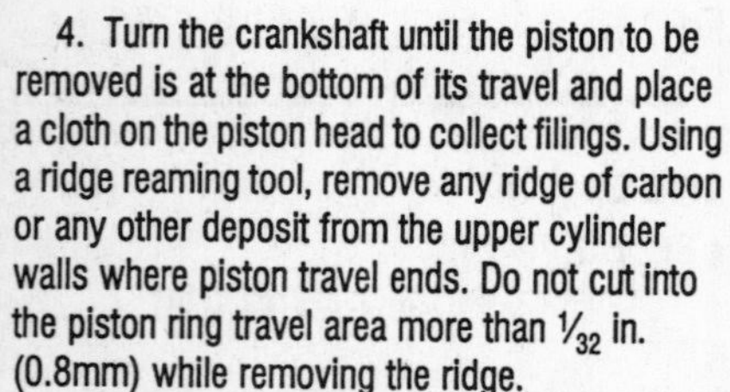

FIG. 71 Make connecting rod bolt guides out of rubber tubing; these also protect the cylinder walls and crank journals from scratches

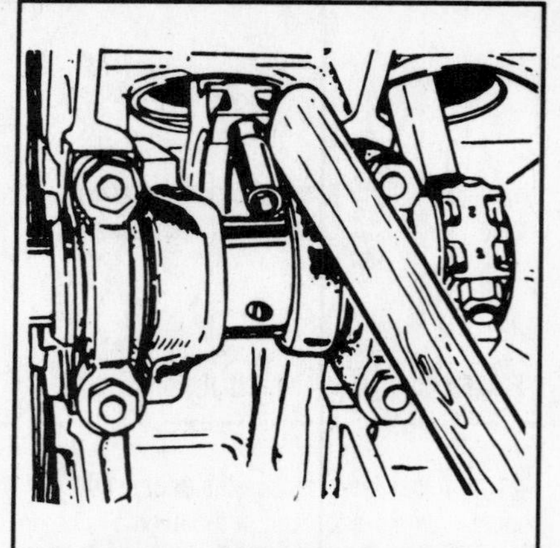

FIG. 72 Push the piston assembly out with a hammer handle

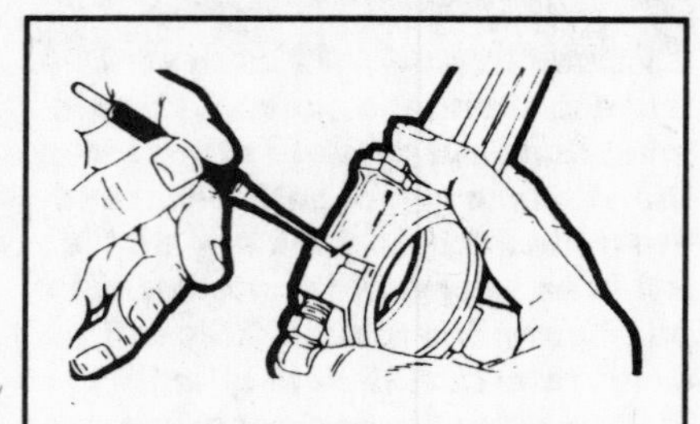

FIG. 73 Match the connecting rods to their caps with a scribe mark for reassembly

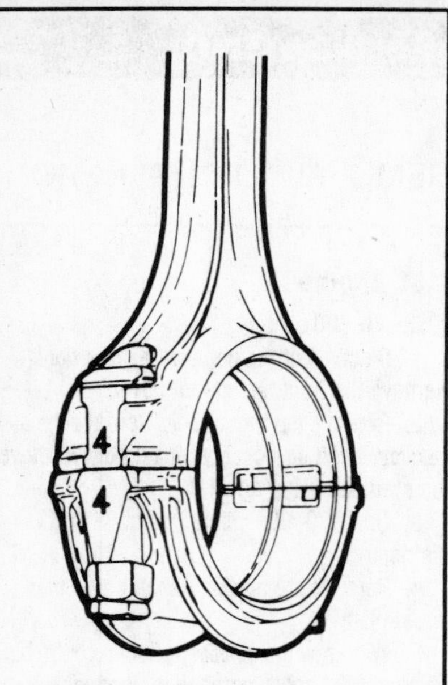

FIG. 74 Number each rod and cap with its cylinder number for correct assembly

4. Turn the crankshaft until the piston to be removed is at the bottom of its travel and place a cloth on the piston head to collect filings. Using a ridge reaming tool, remove any ridge of carbon or any other deposit from the upper cylinder walls where piston travel ends. Do not cut into the piston ring travel area more than $\frac{1}{32}$ in. (0.8mm) while removing the ridge.

5. Mark all of the connecting rod caps so that they can be reinstalled in the original positions from which they are removed and remove the connecting rod bearing cap. Also identify the piston assemblies as they, too, must be reinstalled in the same cylinder from which removed.

6. With the bearing caps removed, the connecting rod bearing bolts are potentially damaging to the cylinder walls during removal. To guard against cylinder wall damage, install 4 in. (101.6mm) or 5 in. (127mm) lengths of $\frac{3}{8}$ in. (9.5mm) rubber tubing onto the connecting rod bolts. These will also protect the crankshaft journal from scratches when the connecting rod is installed, and will serve as a guide for the rod.

7. Squirt some clean engine oil into each cylinder before removing the pistons. Using a wooden hammer handle, push the connecting rod and piston assembly out of the top of the cylinder (pushing from the bottom of the rod). Be careful to avoid damaging both the crank journal and the cylinder wall when removing the rod and piston assembly.

8. Before installing the piston/connecting rod assembly, be sure to clean all gasket mating surfaces, oil the pistons, piston rings and the cylinder walls with light engine oil.

9. Be sure to install the pistons in the cylinders from which they were removed. The connecting rod and bearing caps are numbered from 1 to 4 beginning at the front of the engine. The numbers on the connecting rod and bearing cap must be on the same side when installed in the cylinder bore. If a connecting rod is ever transposed from one engine or cylinder to another, new bearings should be fitted and the connecting rod should be numbered to correspond with the new cylinder number. The notch on the piston head goes toward the front of the engine.

10. Make sure the ring gaps are properly spaced around the circumference of the piston. Make sure rubber hose lengths are fitted to the rod bolts. Fit a piston ring compressor around the piston and slide the piston and connecting rod assembly down into the cylinder bore, pushing it in with the wooden hammer handle. Push the piston down until it is only slightly below the top of the cylinder bore. Guide the connecting rods onto the crankshaft bearing journals carefully, using the rubber hose lengths, to avoid damaging the crankshaft.

11. Check the bearing clearance of all the rod bearings, fitting them to the crankshaft bearing journals.

12. After the bearings have been fitted, apply a light coating of engine oil to the journals and bearings.

13. Turn the crankshaft until the appropriate bearing journal is at the bottom of its stroke, then push the piston assembly all the way down until the connecting rod bearing seats on the crankshaft journal. Be careful not to allow the bearing cap screws to strike the crankshaft bearing journals and damage them.

14. After the piston and connecting rod assemblies have been installed, check the connecting rod side clearance on each crankshaft journal.

15. Prime and install the oil pump and the oil pump intake tube, then install the oil pan.

16. Reassemble the rest of the engine in the reverse order of disassembly.

5.0L Engine

1. Drain the cooling system and the crankcase.

CAUTION

The EPA warns that prolonged contact with used engine oil may cause a number of skin disorders, including cancer! You should make every effort to minimize your exposure to used engine oil. Protective gloves should be worn when changing the oil. Wash your hands and any other exposed skin areas as soon as possible after exposure to used engine oil. Soap and water, or waterless hand cleaner should be used.

2. Remove the intake manifold.
3. Remove the cylinder heads.
4. Remove the oil pan.
5. Remove the oil pump.
6. Turn the crankshaft until the piston to be removed is at the bottom of its travel, then place a cloth on the piston head to collect filings.
7. Remove any ridge of deposits at the end of the piston travel from the upper cylinder bore, using a ridge reaming tool. Do not cut into the piston ring travel area more than $\frac{1}{32}$ in. (0.8mm) when removing the ridge.
8. Make sure that all of the connecting rod bearing caps can be identified, so they will be reinstalled in their original positions.
9. Turn the crankshaft until the connecting rod that is to be removed is at the bottom of its stroke and remove the connecting rod nuts and bearing cap.
10. With the bearing caps removed, the connecting rod bearing bolts are potentially damaging to the cylinder walls during removal. To guard against cylinder wall damage, install 4 or 5 in. (102mm or 127mm) lengths of $\frac{3}{8}$ in. (9.5mm) rubber tubing onto the connecting rod bolts. These will also protect the crankshaft journal from scratches when the connecting rod is installed, and will serve as a guide for the rod.
11. Squirt some clean engine oil into each cylinder before removing the piston assemblies. Using a wooden hammer handle, push the connecting rod and piston assembly out of the top of the cylinder (pushing from the bottom of the rod). Be careful to avoid damaging both the crank journal and the cylinder wall when removing the rod and piston assembly.
12. Remove the bearing inserts from the connecting rod and cap if the bearings are to be replace, and place the cap onto the piston/rod assembly from which it was removed.
13. Install the piston/rod assemblies in the same manner as that for the 6-cylinder engines. See the procedure given for the 2.3L engine.

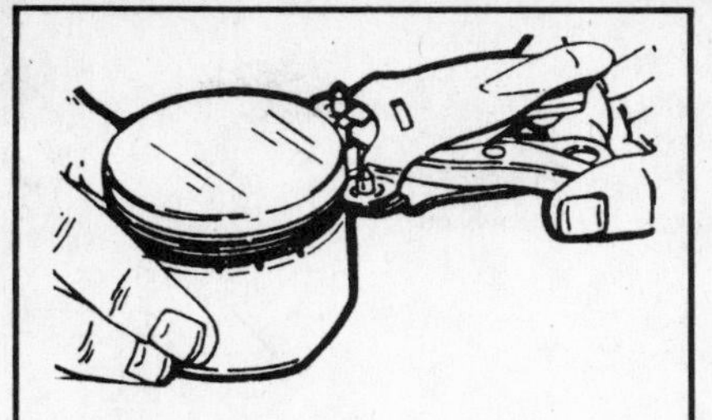

FIG. 75 Remove and install the rings with a ring expander

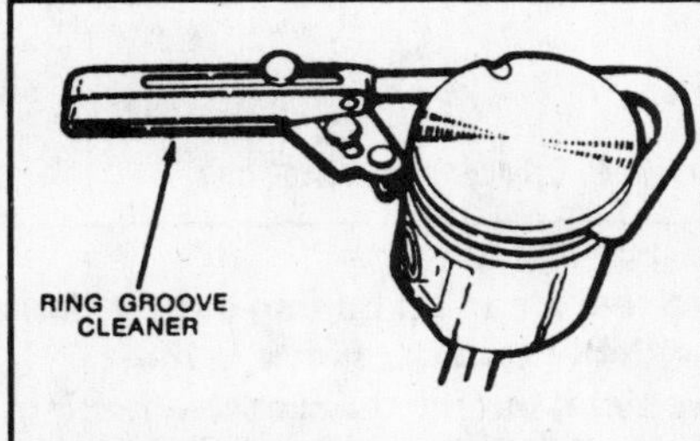

FIG. 76 Clean the ring grooves with this tool or the edge of an old ring

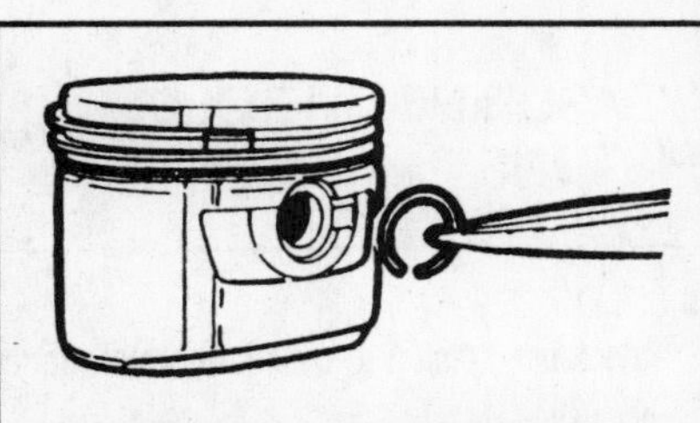

FIG. 77 Use needle-nose or snapring pliers to remove the piston pin clips

➡ The connecting rod and bearing caps are numbered from 1 to 4 in the right bank and from 5 to 8 in the left bank, beginning at the front of the engine. The numbers on the rod and cap must be on the same side when they are installed in the cylinder bore. Also, the largest chamfer at the bearing end of the rod should be positioned toward the crank pin thrust face of the crankshaft and the notch in the head of the piston faces toward the front of the engine.

14. See the appropriate component procedures to assemble the engine.

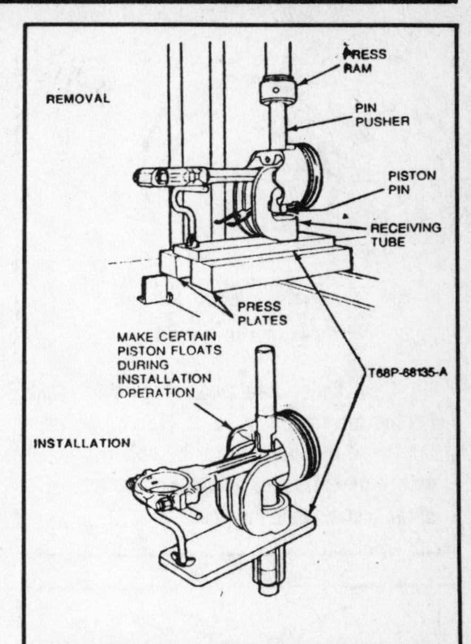

FIG. 78 Have the piston pins pressed in and out with a press

Piston Ring and Wrist Pin

➧ SEE FIG. 75–78

REMOVAL

All of the Ford gasoline engines covered in this guide utilize pressed-in wrist pins, which can only be removed by an arbor press. The piston/connecting rod assemblies should be taken to an engine specialist or qualified machinist for piston removal and installation.

A piston ring expander is necessary for removing the piston rings without damaging them; any other method (screwdriver blades, pliers, etc.) usually results in the rings being bent, scratched or distorted, or the piston itself being damaged. When the rings are removed, clean the ring grooves using an appropriate ring groove cleaning tool, using care not to cut too deeply. Thoroughly clean all carbon and varnish from the piston with solvent.

WARNING

Do not use a wire brush or caustic solvent (acids, etc.) on pistons.

Inspect the pistons for scuffing, scoring, cracks, pitting, or excessive ring groove wear. If these are evident, the piston must be replaced.

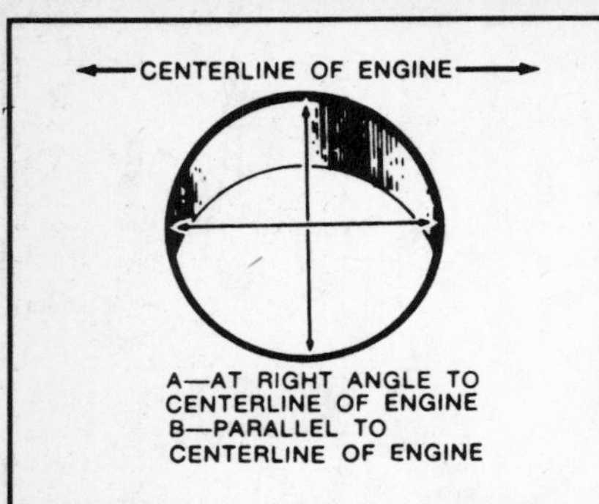

FIG. 79 Cylinder bore measuring points. Take the top measurement 1/2 in. (13mm) below the top of the block deck, the bottom measurement 1/2 in. (13mm) above the top of the piston when the piston is at BDC

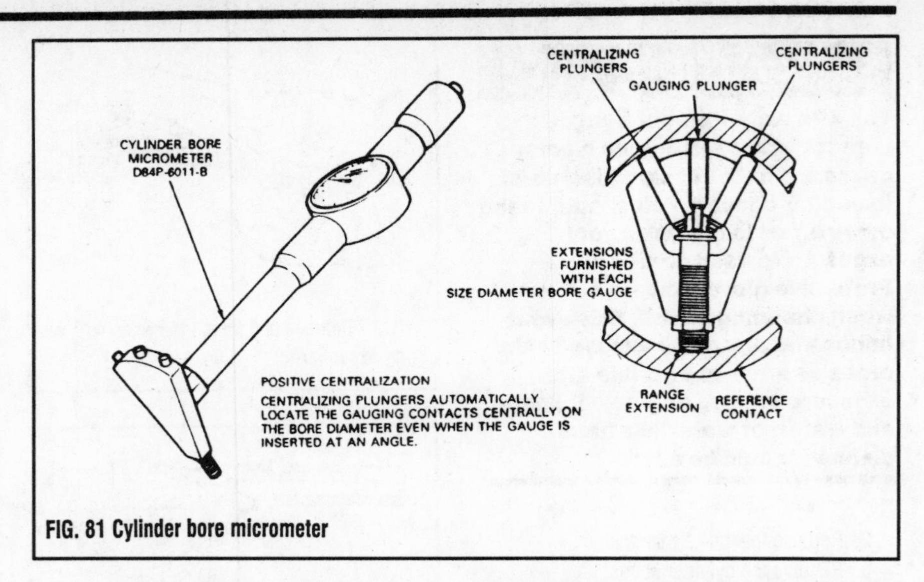

FIG. 81 Cylinder bore micrometer

FIG. 80 Measuring the cylinder bore with a dial gauge

The piston should also be checked in relation to the cylinder diameter. Using a telescoping gauge and micrometer, or a dial gauge, measure the cylinder bore diameter perpendicular (90%) to the piston pin, 2 1/2 in. (64mm) below the cylinder block deck (surface where the block mates with the heads). Then, with the micrometer, measure the piston, perpendicular to its wrist pin on the skirt. the difference between the two measurements is the piston clearance. If the clearance is within specifications or slightly below (after the cylinders have been bored or hones), finish honing is all that is necessary. If the clearance is excessive, try to obtain a slightly larger piston to bring clearance to within specifications. If this is not possible, obtain the first oversize piston and hone (or if necessary, bore) the cylinder to size. Generally, if the cylinder bore is tapered 0.005 in. (0.127mm) or more or is out-of-round 0.003 in. (0.076mm) or more, it is advisable to rebore for the smallest possible oversize piston and rings.

After measuring, mark pistons with a felt tip pen for reference and for assembly.

➡ **Cylinder honing and/or boring should be performed by a reputable, professional mechanic with the proper equipment. In some cases, clean-up honing can be done with the cylinder block in the vehicle, but most excessive honing and all cylinder boring must be done with the block stripped and removed from the vehicle.**

MEASURING THE OLD PISTONS

➧ SEE FIGS. 79–81

Check used piston-to-cylinder bore clearance as follows:

1. Measure the cylinder bore diameter with a telescope gauge.
2. Measure the piston diameter. When measuring the pistons for size or taper, measurements must be made with the piston pin removed.
3. Subtract the piston diameter from the cylinder bore diameter to determine piston-to-bore clearance.
4. Compare the piston-to-bore clearances obtained with those clearances recommended. Determine if the piston-to-bore clearance is in the acceptable range.
5. When measuring taper, the largest reading must be at the bottom of the skirt.

SELECTING NEW PISTONS

1. If the used piston is not acceptable, check the service piston size and determine if a new piston can be selected. (Service pistons are available in standard, high limit and standard oversize.

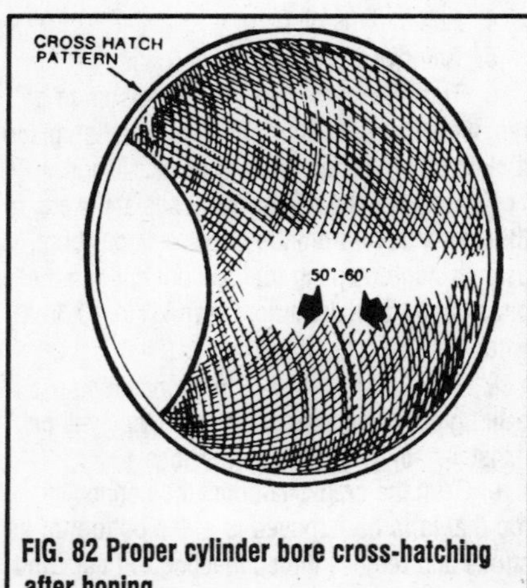

FIG. 82 Proper cylinder bore cross-hatching after honing

2. If the cylinder bore must be reconditioned, measure the new piston diameter, then hone the cylinder bore to obtain the preferred clearance.
3. Select a new piston and mark the piston to identify the cylinder for which it was fitted. (On some vehicles, oversize pistons may be found. These pistons will be 0.254mm [0.010 in.] oversize).

CYLINDER HONING

➧ SEE FIG. 82

1. When cylinders are being honed, follow the manufacturer's recommendations for the use of the hone.
2. Occasionally, during the honing operation, the cylinder bore should be thoroughly cleaned and the selected piston checked for correct fit.
3. When finish-honing a cylinder bore, the hone should be moved up and down at a

sufficient speed to obtain a very fine uniform surface finish in a cross-hatch pattern of approximately 45–65° included angle. The finish marks should be clean but not sharp, free from embedded particles and torn or folded metal.

4. Permanently mark the piston for the cylinder to which it has been fitted and proceed to hone the remaining cylinders.

WARNING

Handle the pistons with care. Do not attempt to force the pistons through the cylinders until the cylinders have been honed to the correct size. Pistons can be distorted through careless handling.

5. Thoroughly clean the bores with hot water and detergent. Scrub well with a stiff bristle brush and rinse thoroughly with hot water. It is extremely essential that a good cleaning operation be performed. If any of the abrasive material is allowed to remain in the cylinder bores, it will rapidly wear the new rings and cylinder bores. The bores should be swabbed several times with light engine oil and a clean cloth and then wiped with a clean dry cloth. CYLINDERS SHOULD NOT BE CLEANED WITH KEROSENE OR GASOLINE! Clean the remainder of the cylinder block to remove the excess material spread during the honing operation.

PISTON RING END GAP

➧ SEE FIGS. 83–84

Piston ring end gap should be checked while the rings are removed from the pistons. Incorrect end gap indicates that the wrong size rings are being used; ring breakage could occur.

Compress the piston rings to be used in a cylinder, one at a time, into that cylinder. Squirt clean oil into the cylinder, so that the rings and the top 2 in. (51mm) of cylinder wall are coated. Using an inverted piston, press the rings approximately 1 in. (25mm) below the deck of the block. Measure the ring end gap with the feeler gauge, and compare to the Ring Gap chart in this Section. carefully pull the ring out of the cylinder and file the ends squarely with a fine file to obtain the proper clearance.

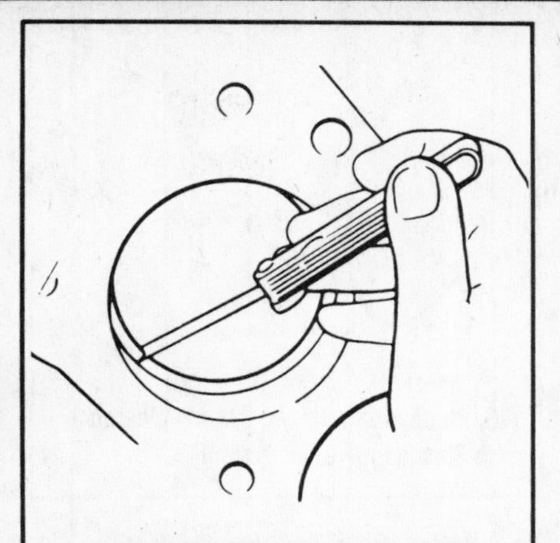

FIG. 83 Check the piston ring end gap with a feeler gauge, with the ring positioned in the cylinder one inch below the deck of the block

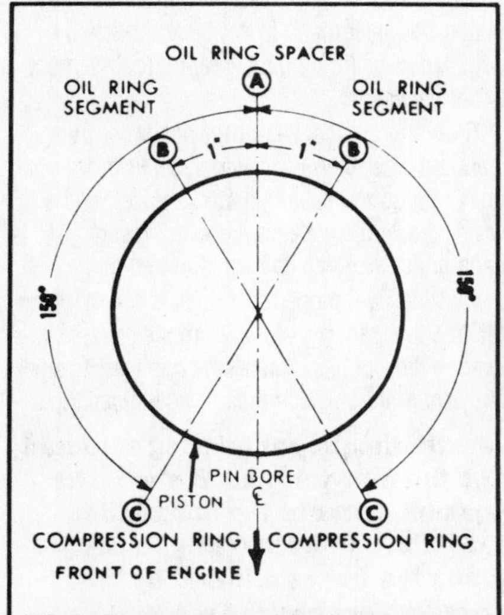

FIG. 84 Proper spacing of the piston ring gaps around the circumference of the piston

PISTON RING SIDE CLEARANCE CHECK AND INSTALLATION

➧ SEE FIG. 85

Check the pistons to see that the ring grooves and oil return holes have been properly cleaned. Slide a piston ring into its groove, and check the side clearance with a feeler gauge. On gasoline engines, make sure you insert the gauge between the ring and its lower land (lower edge of the groove), because any wear that occurs forms a step at the inner portion of the lower land. If the piston grooves have worn to the extend that relatively high steps exist on the lower land, the piston grooves have worn to the extent that relatively high steps exist on the lower

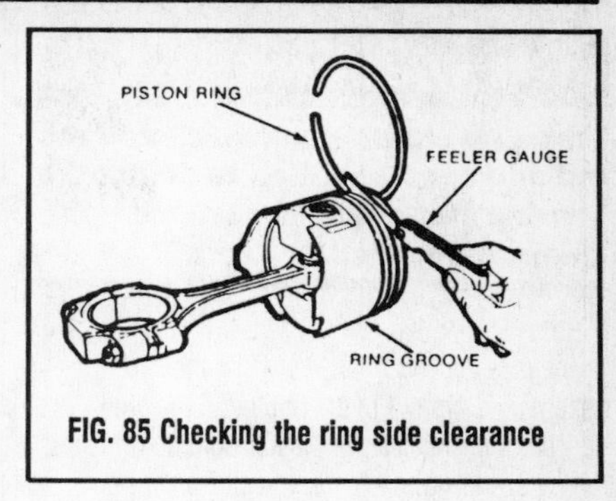

FIG. 85 Checking the ring side clearance

land, the piston should be replaced, because these will interfere with the operation of the new rings and ring clearance will be excessive. Piston rings are not furnished in oversize widths to compensate for ring groove wear.

Install the rings on the piston, lowest ring first, using a piston ring expander. There is a high risk of breaking or distorting the rings, or scratching the piston, if the rings are installed by hand or other means.

Position the rings on the piston as illustrated; spacing of the various piston ring gaps is crucial to proper oil retention and even cylinder wear. When installing new rings, refer to the installation diagram furnished with the new parts.

Connecting Rod Bearings

INSPECTION

Connecting rod bearings for the engines covered in this guide consist of two halves or shells which are interchangeable in the rod and cap. when the shells are placed in position, the ends extend slightly beyond the rod and cap surfaces so that when the rod bolts are torqued the shells will be clamped tightly in place to insure positive seating and to prevent turning. A tang holds the shells in place.

➡ The ends of the bearing shells must never be filed flush with the mating surfaces of the rod and cap.

If a rod bearing becomes noisy or is worn so that its clearance on the crank journal is sloppy, a new bearing of the correct undersize must be selected and installed since there is a provision for adjustment.

WARNING

Under no circumstances should the rod end or cap be filed to adjust the bearing clearance, nor should shims of any kind be used.

Inspect the rod bearings while the rod assemblies are out of the engine. If the shells are scored or show flaking, they should be replaced. If they are in good shape, check for proper clearance on the crank journal (see below). Any scoring or ridges on the crank journal means the crankshaft must be reground and fitted with undersized bearings, or replaced.

CHECKING BEARING CLEARANCE AND REPLACING BEARINGS

➧ SEE FIG. 86

➡ Make sure connecting rods and their caps are kept together, and that the caps are installed in the proper direction.

Replacement bearings are available in standard size, and in undersizes for reground crankshaft. Connecting rod-to-crankshaft bearing clearance is checked using Plastigage® at either the top or bottom of each crank journal. the Plastigage® has a range of 0 to 0.003 in. (0.076mm).

1. Remove the rod cap with the bearing shell. Completely clean the bearing shell and the crank journal, and blow any oil from the oil hole in the crankshaft.

➡ The journal surfaces and bearing shells must be completely free of oil, because Plastigage® is soluble in oil.

2. Place a strip of Plastigage® lengthwise along the bottom center of the lower bearing shell, then install the cap with shell and torque the bolt or nuts to specification. DO NOT TURN the crankshaft with the Plastigage® installed in the bearing.
3. Remove the bearing cap with the shell. The flattened Plastigage® will be found sticking to either the bearing shell or crank journal. Do not remove it yet.
4. Use the printed scale on the Plastigage® envelope to measure the flattened material at its widest point. The number within the scale which most closely corresponds to the width of the Plastigage® indicated bearing clearance in thousandths of an inch or hundreths of a millimeter.

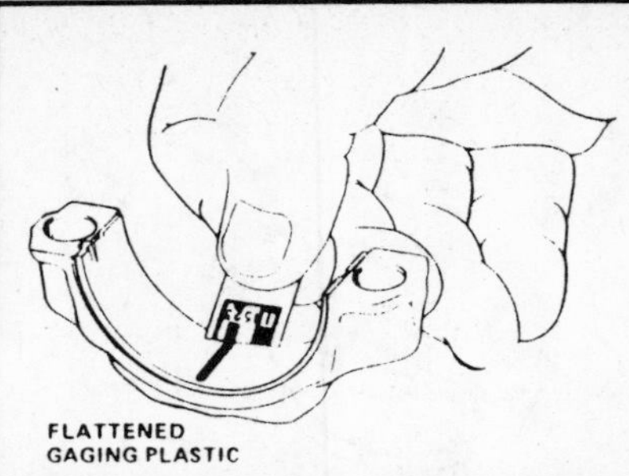

FIG. 86 Checking the rod bearing clearance with Plastigage® or equivalent

5. Check the specifications chart in this Section for the desired clearance. It is advisable to install a new bearing if clearance exceeds 0.003 in. (0.076mm); however, if the bearing is in good condition and is not being checked because of bearing noise, bearing replacement is not necessary.
6. If you are installing new bearings, try a standard size, then each undersize in order until one is found that is within the specified limits when checked for clearance with Plastigage®. Each under size has its size stamped on it.
7. When the proper size shell is found, clean off the Plastigage® material from the shell, oil the bearing thoroughly, reinstall the cap with its shell and torque the rod bolt nuts to specification.

➡ With the proper bearing selected and the nuts torqued, it should be possible to move the connecting rod back and forth freely on the crank journal as allowed by the specified connecting rod end clearance. If the rod cannot be moved, either the rod bearing is too far undersize or the rod is misaligned.

Piston and Connecting Rod

ASSEMBLY AND INSTALLATION

➧ SEE FIG. 87–90

Install the connecting rod to the piston making sure piston installation notches and any marks on the rod are in proper relation to one another. Lubricate the wrist pin with clean engine oil and install the pin into the rod and piston assembly by using an arbor press as required. Install the wrist pin snaprings if equipped, and rotate them in their grooves to make sure they are seated. To install the piston and rod assemblies:

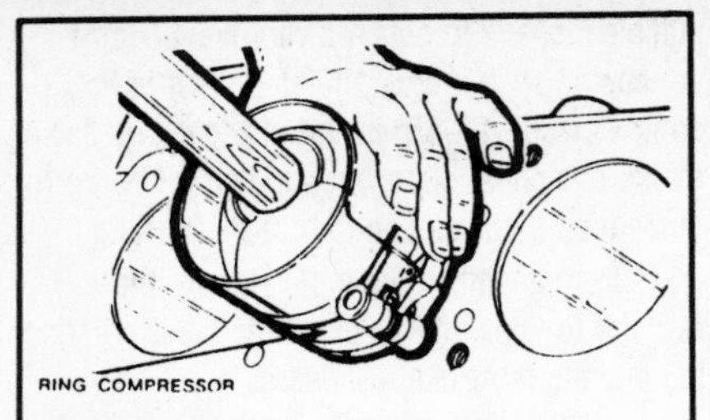

FIG. 87 Tap the piston assembly into the cylinder with a wooden hammer handle. Notches on the piston crown face the front of the engine

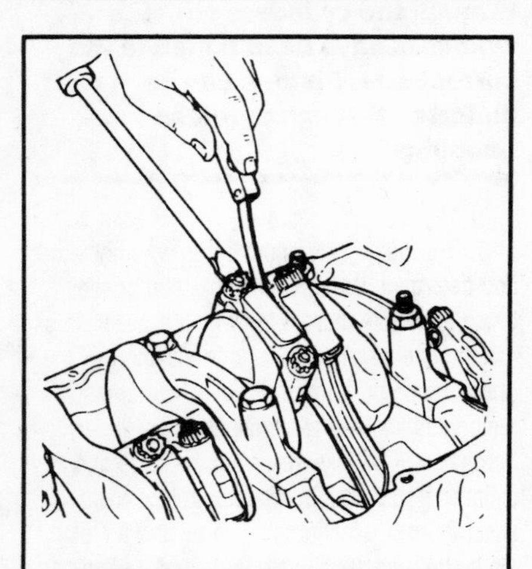

FIG. 88 Checking the connecting rod side clearance with a feeler gauge. Use a small pry bar to spread the connecting rods

1. Make sure the connecting rod big bearings (including end cap) are of the correct size and properly installed.
2. Fit rubber hoses over the connecting rod bolt to protect the crankshaft journals, as in the Piston Removal procedure. Coat the rod bearings with clean oil.
3. Using the proper ring compressor, insert the piston assembly into the cylinder so that the notch in the top of the piston faces the front of the engine (this assumes that the dimple(s) or other markings on the connecting rods are in correct relation to the piston notch(s)).
4. From beneath the engine, coat each crank journal with clean oil. Pull the connecting rod, with the bearing shell in place, into position against the crank journal.
5. Remove the rubber hoses. Install the bearing cap and cap nuts and torque to specification.

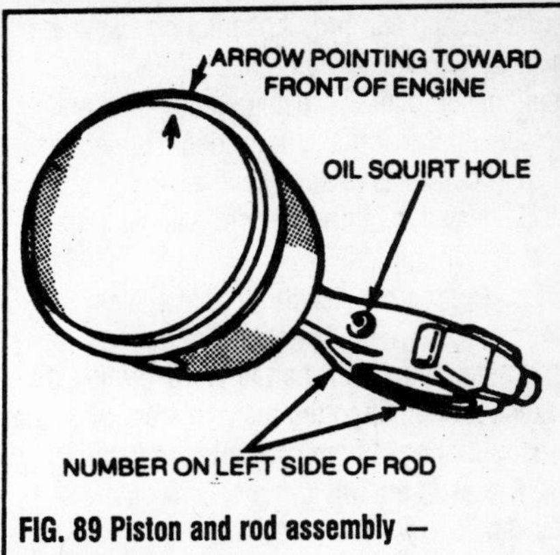

FIG. 89 Piston and rod assembly — 2.3L engine

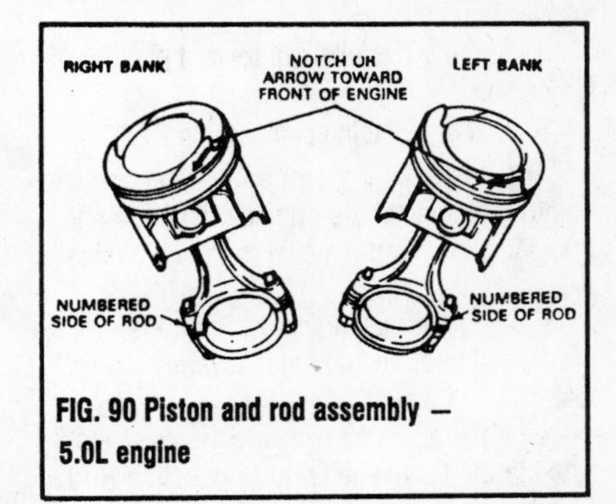

FIG. 90 Piston and rod assembly — 5.0L engine

➡ **When more than one rod and piston assembly is being installed, the connecting rod cap attaching nuts should only be tightened enough to keep each rod in position until all have been installed. This will ease the installation of the remaining piston assemblies.**

6. Check the clearance between the sides of the connecting rods and the crankshaft using a feeler gauge. Spread the rods slightly with a screwdriver to insert the gauge. If clearance is below the minimum tolerance, the rod may be machined to provide adequate clearance. If clearance is excessive, substitute an unworn rod, and recheck. If clearance is still outside specifications, the crankshaft must be welded and reground, or replaced.

7. Replace the oil pump if removed, and the oil pan.

8. Install the cylinder head(s) and intake manifold.

Crankshaft and Main Bearings

REMOVAL & INSTALLATION

Engine Removed

➧ SEE FIGS. 91–98

1. With the engine removed from the vehicle and placed in a work stand, disconnect the spark plug wires from the spark plugs and remove the wires and bracket assembly from the attaching stud on the valve rocker arm cover(s) if so equipped. Disconnect the coil to distributor high tension lead at the coil. Remove the distributor cap and spark plug wires as an assembly. Remove the spark plugs to allow easy rotation of the crankshaft.

2. Remove the fuel pump and oil filter. Slide the water pump by-pass hose clamp (if so equipped) toward the water pump. Remove the alternator and mounting brackets.

3. Remove the crankshaft pulley from the crankshaft vibration damper. Remove the capscrew and washer from the end of the crankshaft. Install a universal puller, Tool T58P–6316–D or equivalent on the crankshaft vibration damper and remove the damper.

4. Remove the cylinder front cover and crankshaft gear, refer to Cylinder Front Cover and Timing Chain in this Section.

5. Invert the engine on the work stand. Remove the clutch pressure plate and disc (manual shift transmission). Remove the flywheel and engine rear cover plate. Remove the oil pan and gasket. Remove the oil pump.

6. Make sure all bearing caps (main and connecting rod) are marked so that they can be installed in their original locations. Turn the crankshaft until the connecting rod from which the cap is being removed is down, and remove the bearing cap. Push the connecting rod and piston assembly up into the cylinder. Repeat this procedure until all the connecting rod bearing caps are removed.

7. Remove the main bearings caps.

8. Carefully lift the crankshaft out of the block so that the thrust bearing surfaces are not damaged. Handle the crankshaft with care to avoid possible fracture to the finished surfaces.

9. Remove the rear journal seal from the block and rear main bearing cap.

10. Remove the main bearing inserts from the block and bearing caps.

11. Remove the connecting rod bearing inserts from the connecting rods and caps.

12. If the crankshaft main bearing journals have been refinished to a definite undersize, install the correct undersize bearings. Be sure the bearing inserts and bearing bores are clean. Foreign material under the inserts will distort the bearing and cause a failure.

13. Place the upper main bearing inserts in position in the bores with the tang fitting in the slot. Be sure the oil holes in the bearing inserts are aligned with the oil holes in the cylinder block.

14. Install the lower main bearing inserts in the bearing caps.

15. Clean the rear journal oil seal groove and the mating surfaces of the block and rear main bearing cap.

16. Dip the lip-type seal halves in clean engine oil. Install the seals in the bearing cap and block with the undercut side of the seal toward the front of the engine.

➡ **This procedure applies only to engines with two piece rear main bearing oil seals. those having one piece seals (2.3L engine) will be installed after the crankshaft is in place.**

17. Carefully lower the crankshaft into place. Be careful not to damage the bearing surfaces.

CHECKING MAIN BEARING CLEARANCES

18. Check the clearance of each main bearing by using the following procedure:

a. Place a piece of Plastigage® or its equivalent, on bearing surface across full width of bearing cap and about 1/4 in. (6mm) off center.

b. Install cap and tighten bolts to specifications. Do not turn crankshaft while Plastigage® is in place.

c. Remove the cap. Using Plastigage® scale, check width of Plastigage® at widest point to get the minimum clearance. Check at narrowest point to get maximum clearance. Difference between readings is taper of journal.

d. If clearance exceeds specified limits, try a 0.001 in. (0.0254mm) or 0.002 in. (0.051mm) undersize bearing in combination with the standard bearing. Bearing clearance must be within specified limits. If standard and 0.002 in. (0.051mm) undersize bearing does not bring clearance within desired limits, refinish crankshaft journal, then install undersize bearings.

➡ **Refer to Rear Main Oil Seal removal and installation, for special instructions in applying RTV sealer to rear main bearing cup.**

19. Install all the bearing caps except the

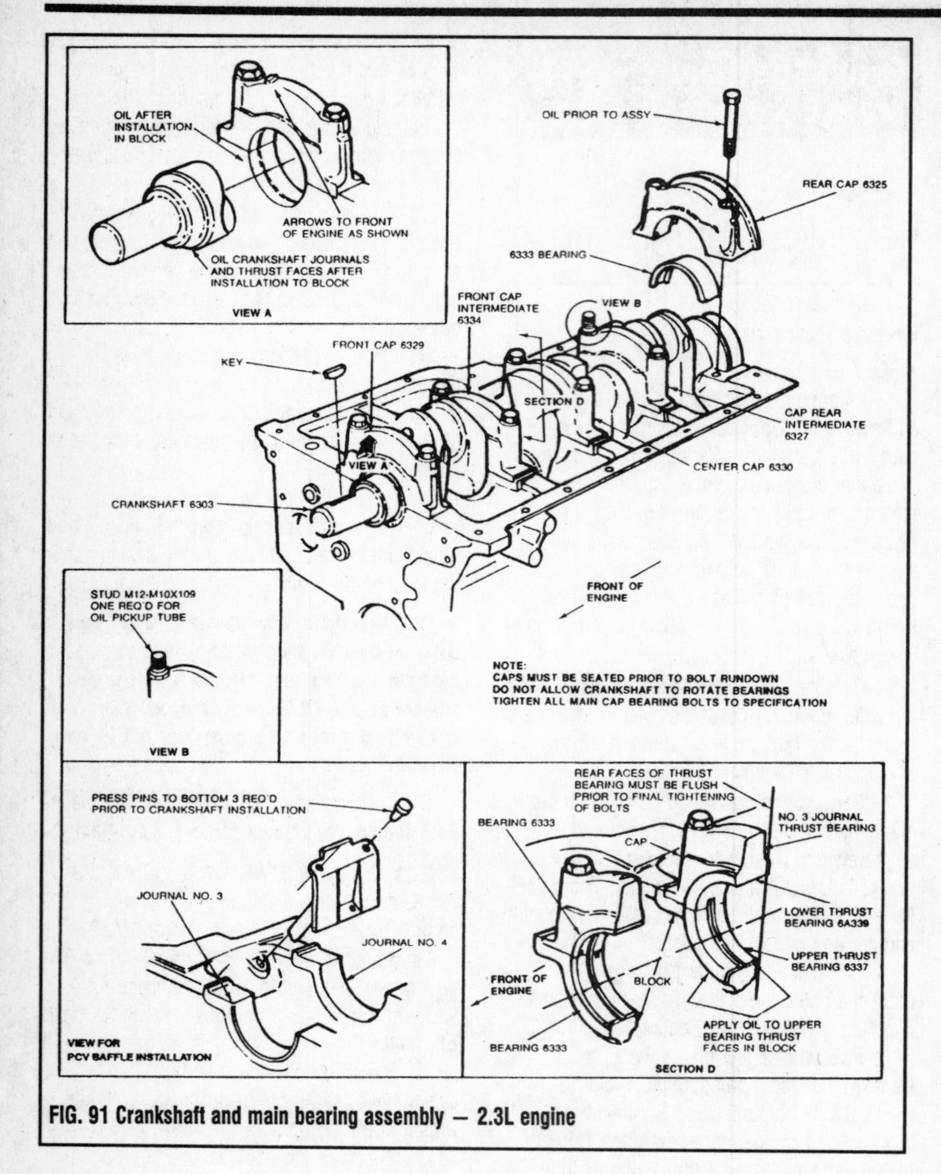

FIG. 91 Crankshaft and main bearing assembly — 2.3L engine

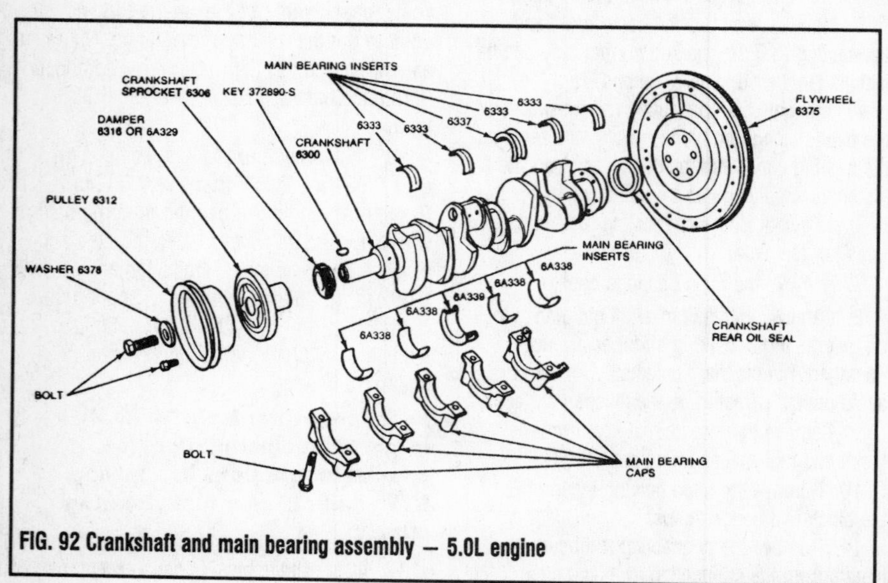

FIG. 92 Crankshaft and main bearing assembly — 5.0L engine

thrust bearing cap (no. 3 bearing on all except the 2.3L which use the no. 5 as the thrust bearing). BE sure the main bearing caps are installed in their original locations. Tighten the bearing cap bolts to specifications.

21. install the thrust bearing cap with the bolts finger tight.

22. Pry the crankshaft forward against the thrust surface of the upper half of the bearing.

23. hold the crankshaft forward and pry the thrust bearing cap to the rear. This will align the thrust surfaces of both halves of the bearing.

24. Retain the forward pressure on the crankshaft. Tighten the cap bolts to specifications.

25. Check the crankshaft end play using the following procedures:

a. Force the crankshaft toward the rear of the engine.

b. Install a dial indicator (tools D78P–4201–F, –G or equivalent) so that the contact point rests against the crankshaft flange and the indicator axis is parallel to the crankshaft axis.

c. Zero the dial indicator. Push the crankshaft forward and note the reading on the dial.

d. If the end play exceeds the wear limit listed in the Crankshaft and Connecting Rod Specifications chart, replace the thrust bearing. If the end play is less than the minimum limit, inspect the thrust bearing faces for scratches, burrs, nicks, or dirt. If the thrust faces are not damaged or dirty, then they probably were not aligned properly. Lubricate and install the new thrust bearing and align the faces following procedures 21 through 24.

26. On engines with one piece rear main bearing oil seal (2.3L engine), coat a new crankshaft rear oil seal with oil and install using Tool T65P–6701–A or equivalent. Inspect the seal to be sure it was not damaged during installation.

27. Install new bearing inserts in the connecting rods and caps. Check the clearance of each bearing, following the procedure (18a through 18d).

28. After the connecting rod bearings have been fitted, apply a light coat of engine oil to the journals and bearings.

29. Turn the crankshaft throw to the bottom of its stroke. Push the piston all the way down until the rod bearing seats on the crankshaft journal.

30. Install the connecting rod cap. Tighten the nuts to specification.

31. After the piston and connecting rod assemblies have been installed, check the side clearance with a feeler gauge between the connecting rods on each connecting rod

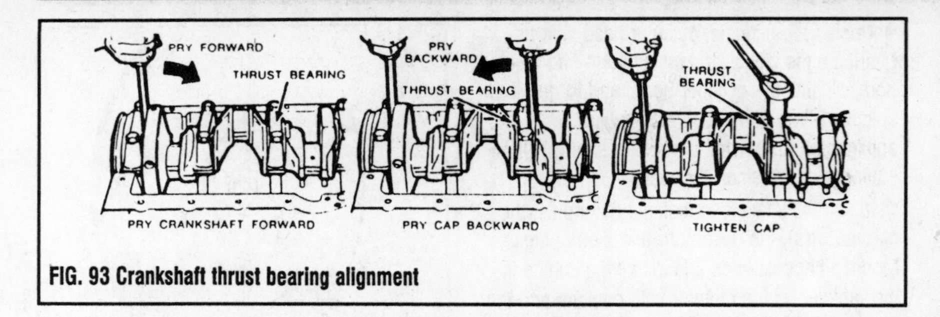

FIG. 93 Crankshaft thrust bearing alignment

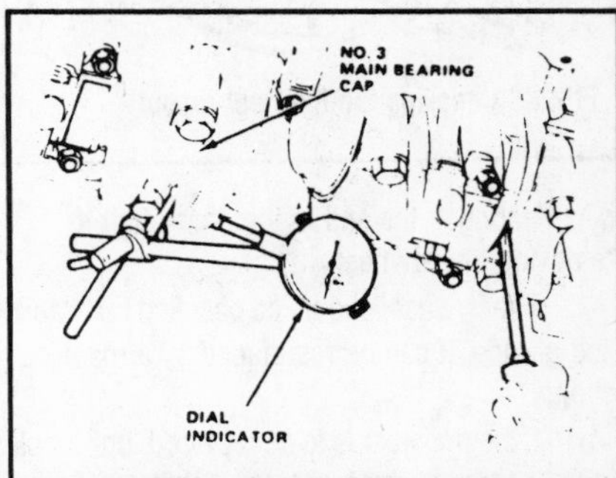

FIG. 94 Checking crankshaft end-play with a dial indicator

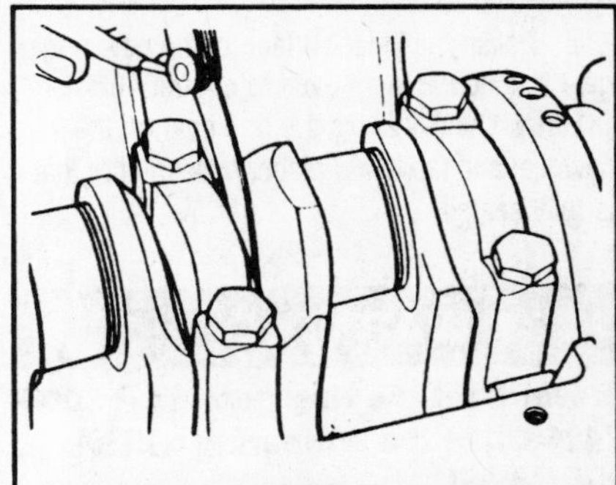

FIG. 95 Checking crankshaft end-play with a feeler gauge

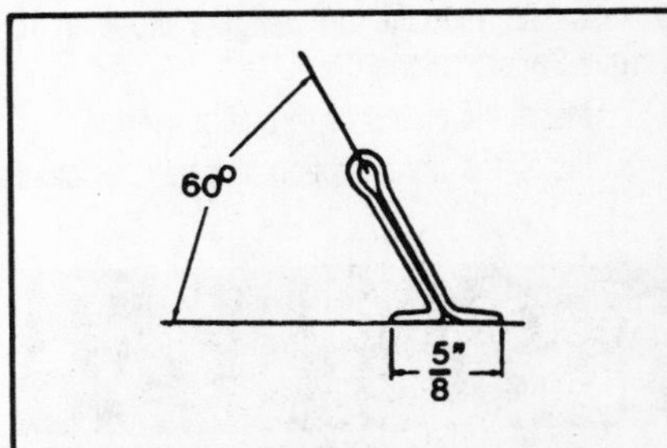

FIG. 96 Make a bearing roll-out pin from a cotter pin

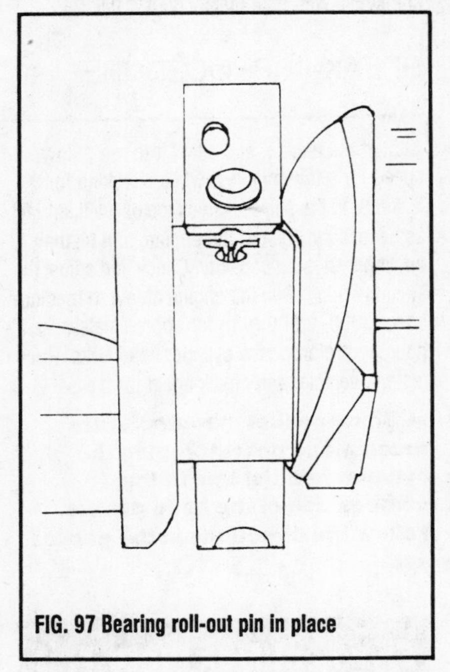

FIG. 97 Bearing roll-out pin in place

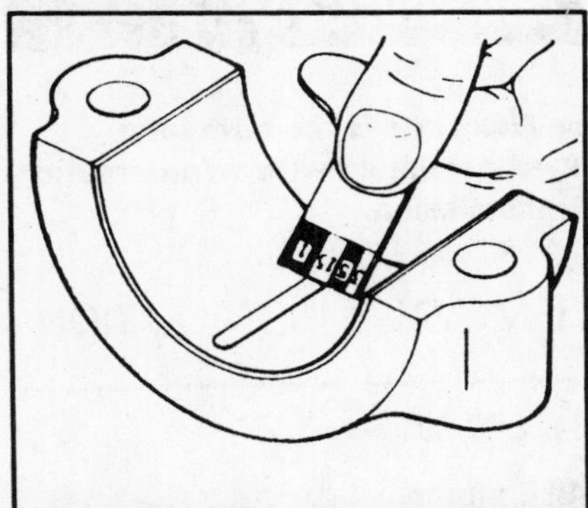

FIG. 98 Checking the main bearing clearance with Plastigage® or equivalent

crankshaft journal. Refer to Crankshaft and Connecting Rod specifications chart in this Section.

32. Install the timing chain and sprockets or gears, cylinder front cover and crankshaft pulley and adapter, following steps under Cylinder Front Cover and Timing Chain Installation in this Section.

Engine in the vehicle

1. With the oil pan, oil pump and spark plugs removed, remove the cap from the main bearing needing replacement and remove the bearing from the cap.

2. Make a bearing roll-out pin, using a bent cotter pin as shown in the illustration. Install the end of the pin in the oil hole in the crankshaft journal.

3. Rotate the crankshaft clockwise as viewed from the front of the engine. This will roll the upper bearing out of the block.

4. Lube the new upper bearing with clean engine oil and insert the plain (un-notched) end between the crankshaft and the indented or notched side of the block. Roll the bearing into place, making sure that the oil holes are aligned. Remove the roll pin from the oil hole.

5. Lube the new lower bearing and install it in the main bearing cap. Install the main bearing cap onto the block, making sure it is positioned in proper direction with the matchmarks in alignment.

6. Torque the main bearing cap to specification.

➡ See Crankshaft Installation for thrust bearing alignment.

CRANKSHAFT CLEANING AND INSPECTION

➡ handle the crankshaft carefully to avoid damage to the finish surfaces.

1. Clean the crankshaft with solvent, and blow out all oil passages with compressed air.

2. Use crocus cloth to remove any sharp edges, burrs or other imperfections which might damage the oil seal during installation or cause premature seal wear.

➡ Do not use crocus cloth to polish the seal surfaces. A finely polished surface may produce poor sealing or cause premature seal wear.

3. Inspect the main and connecting rod journals for cracks, scratches, grooves or scores.

4. Measure the diameter of each journal at least four places to determine out-of-round, taper or undersize condition.

5. On an engine with a manual transmission, check the fit of the clutch pilot bearing in the bore of the crankshaft. A needle roller bearing and adapter assembly is used as a clutch pilot bearing. It is inserted directly into the engine crankshaft. The bearing and adapter assembly cannot be serviced separately. A new bearing must be installed whenever a bearing is removed.

6. Inspect the pilot bearing, when used, for roughness, evidence of overheating or loss of lubricant. Replace if any of these conditions are found.

7. Inspect the rear oil seal surface of the crankshaft for deep grooves, nicks, burrs, porosity, or scratches which could damage the oil seal lip during installation. Remove all nicks and burrs with crocus cloth.

Main Bearings

1. Clean the bearing inserts and caps thoroughly in solvent, and dry them with compressed air.

➡ **Do not scrape varnish or gum deposits from the bearing shells.**

2. Inspect each bearing carefully. Bearings that have a scored, chipped, or worn surface should be replaced.
3. The copper-lead bearing base may be visible through the bearing overlay in small localized areas. This may not mean that the bearing is excessively worn. It is not necessary to replace the bearing if the bearing clearance is within recommended specifications.
4. Check the clearance of bearings that appear to be satisfactory with Plastigage® or its equivalent. Fit the new bearings following the procedure Crankshaft and Main Bearings removal and installation, they should be reground to size for the next undersize bearing.
5. Regrind the journals to give the proper clearance with the next undersize bearing. If the journal will not clean up to maximum undersize bearing available, replace the crankshaft.
6. Always reproduce the same journal shoulder radius that existed originally. Too small a radius will result in fatigue failure of the crankshaft. Too large a radius will result in bearing failure due to radius ride of the bearing.
7. After regrinding the journals, chamfer the oil holes, then polish the journals with a #320 grit polishing cloth and engine oil. Crocus cloth may also be used as a polishing agent.

COMPLETING THE REBUILDING PROCESS

Fill the oil pump with oil, to prevent cavitating (sucking air) on initial engine start up. Install the oil pump and the pickup tube on the engine. Coat the oil pan gasket as necessary, and install the gasket and the oil pan. Mount the flywheel and the crankshaft vibration damper or pulley on the crankshaft.

➡ **Always use new bolts when installing the flywheel. Inspect the clutch shaft pilot bushing in the crankshaft. If the bushing is excessively worn, remove it with an expanding puller and a slide hammer, and tap a new bushing into place.**

Position the engine, cylinder head side up. Lubricate the lifters, and install them into their bores. Install the cylinder head, and torque it as specified. Insert the pushrods (where applicable), and install the rocker shaft(s) (if so equipped) or position the rocker.

Install the intake and exhaust manifolds, the carburetor(s), the distributor and spark plugs. Mount all accessories and install the engine in the vehicle. Fill the radiator with coolant, and the crankcase with high quality engine oil.

BREAK-IN PROCEDURE

Start the engine, and allow it to run at low speed for a few minutes, while checking for leaks. Stop the engine, check the oil level, and fill as necessary. Restart the engine, and fill the cooling system to capacity. Check and adjust the ignition timing. Run the engine at low to medium speed (800–2,500 rpm) for approximately 1/2 hour, and retorque the cylinder head bolts. Road test the vehicle, and check again for leaks.

➡ **Some gasket manufacturers recommend not retorquing the cylinder head(s) due to the composition of the head gasket. Follow the directions in the gasket set.**

Flywheel/Flex Plate and Ring Gear

➡ **Flex plate is the term for a flywheel mated with an automatic transmission.**

REMOVAL & INSTALLATION

➧ SEE FIG. 99

All Engines

➡ **The ring gear is replaceable only on engines mated with a manual transmission. Engines with automatic transmissions have ring gears which are welded to the flex plate.**

1. Remove the transmission.
2. Remove the clutch, if equipped, or torque converter from the flywheel. The flywheel bolts should be loosened a little at a time in a cross pattern to avoid warping the flywheel. On vehicles with manual transmissions, replace the pilot bearing in the end of the crankshaft if removing the flywheel.

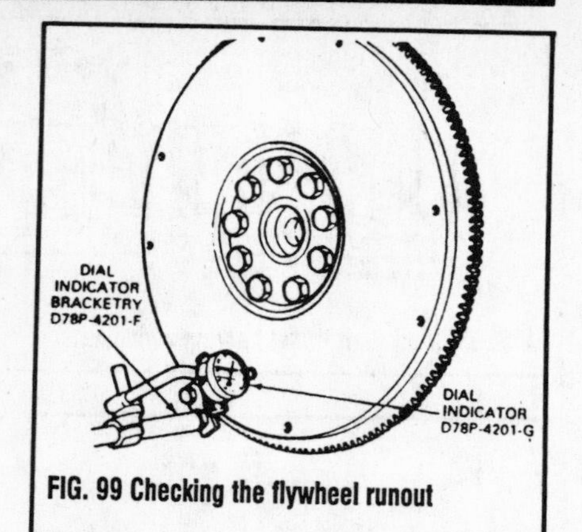

FIG. 99 Checking the flywheel runout

3. The flywheel should be checked for cracks and glazing. It can be resurfaced by a machine shop.
4. If the ring gear is to be replaced, drill a hole in the gear between two teeth, being careful not to contact the flywheel surface. Using a cold chisel at this point, crack the ring gear and remove it.
5. Polish the inner surface of the new ring gear and heat it in an oven to about 600°F (316°C). Quickly place the ring gear on the flywheel and tap it into place, making sure that it is fully seated.

⁂ WARNING

Never heat the ring gear past 800°F (426°C), or the tempering will be destroyed.

6. Position the flywheel on the end of the crankshaft. Torque the bolts a little at a time, in a cross pattern, to the torque figure shown in the Torque Specifications Chart.
7. Install the clutch or torque converter.
8. Install the transmission and transfer case.

Rear Main Oil Seal

REPLACEMENT ONE PIECE SEAL

2.3L Engine

➧ SEE FIG. 100

1. Remove the transmission, clutch assembly or converter and flywheel.

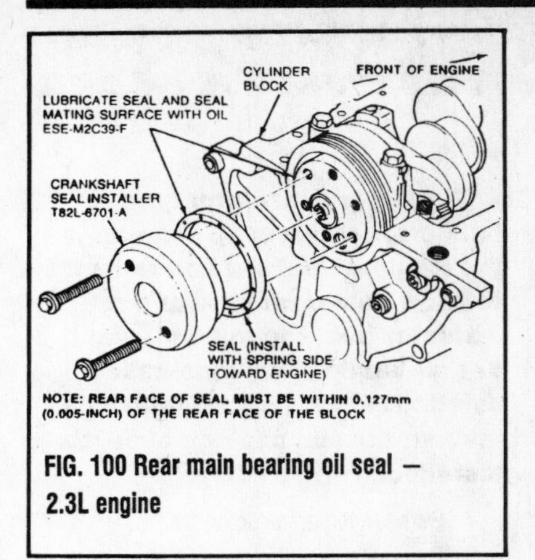

FIG. 100 Rear main bearing oil seal — 2.3L engine

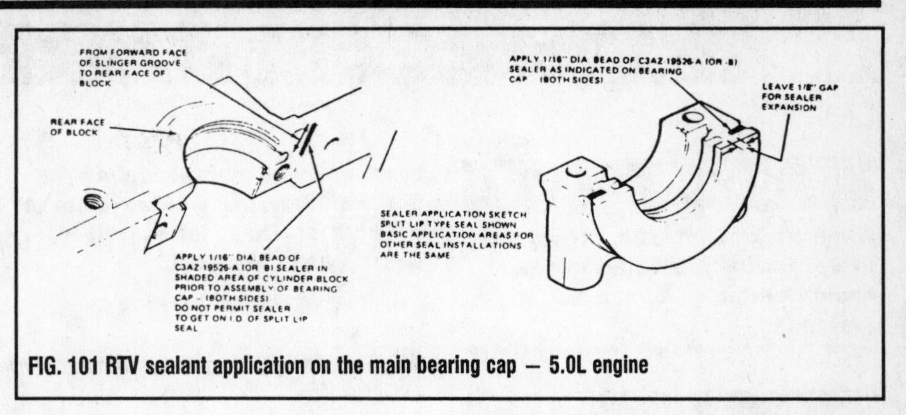

FIG. 101 RTV sealant application on the main bearing cap — 5.0L engine

2. Lower the oil pan if necessary for working room.

3. Use an awl to punch two small holes on opposite sides of the seal just above the split between the main bearing cap and engine block. Install a sheet metal screw in each hole. Use two small pry bars and pry evenly on both screws using two small blocks of wood as a fulcrum point for the pry bars. Use caution throughout to avoid scratching or damage to the oil seal mounting surfaces.

4. When the seal has been removed, clean the mounting recess.

5. Coat the seal and block mounting surfaces with oil. Apply white lube to the contact surface of the seal and crankshaft. Start the seal into the mounting recess and install with seal mounting tool Ford number T82L–6701–A or equivalent.

6. Install the remaining components in the reverse.

7. Reinstall in reverse order of removal.

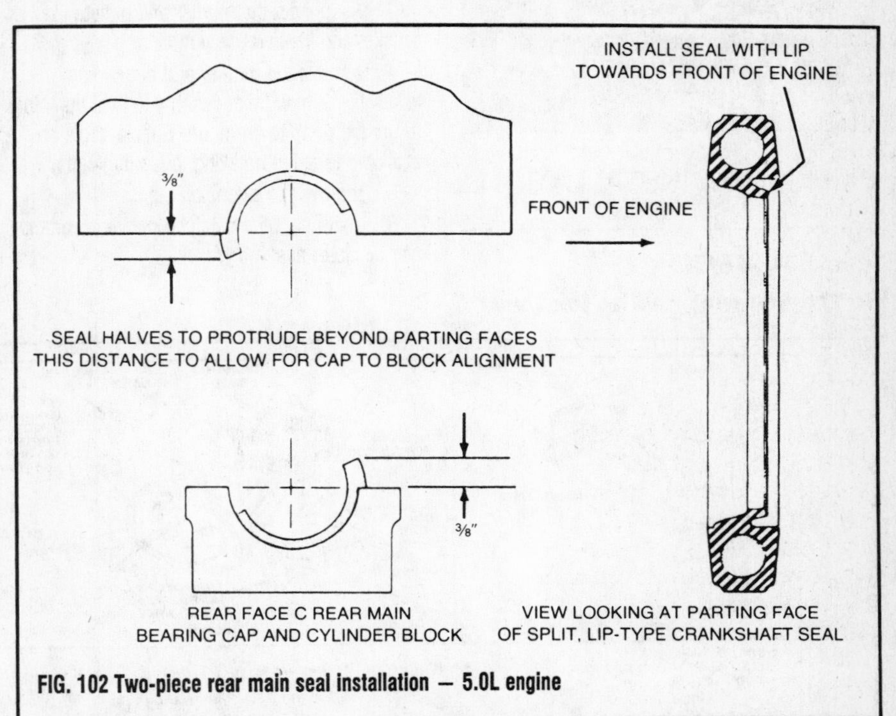

FIG. 102 Two-piece rear main seal installation — 5.0L engine

REPLACEMENT — TWO PIECE SEAL

5.0L Engine

➧ SEE FIGS. 101–102

1. Remove the oil pan and the oil pump (if required).

2. Loosen all the main bearing cap bolts, thereby lowering the crankshaft slightly but not to exceed $\frac{1}{32}$ in. (0.8mm).

3. Remove the rear main bearing cap, and remove the oil seal from the bearing cap and cylinder block. On the block half of the seal use a seal removal tool, or install a small metal screw in one end of the seal, and pull on the screw to remove the seal. Exercise caution to prevent scratching or damaging the crankshaft seal surfaces.

4. Remove the oil seal retaining pin from the bearing cap if so equipped. The pin is not used with the split-lip seal.

5. Carefully clean the seal groove in the cap and block with a brush and solvent such as lacquer thinner, spot remover, or equivalent, or trichlorethylene. Also, clean the area thoroughly, so that no solvent touches the seal.

6. Dip the split lip-type seal halves in clean engine oil.

7. Carefully install the upper seal (cylinder block) into its groove with undercut side of the seal toward the FRONT of the engine, by rotating it on the seal journal of the crankshaft until approximately 3/8 in. (9.5mm) protrudes below the parting surface. Be sure no rubber has been shaved from the outside diameter of the seal by the bottom edge of the groove. Do not allow oil to get on the sealer area.

8. Tighten the remaining bearing cap bolts to the specifications listed in the Torque chart at the beginning of this Section.

9. Install the lower seal in the rear main bearing cap under undercut side of seal toward the FRONT of the engine, allow the seal to protrude approximately 3/8 in. (9.5mm) above the parting surface to mate with the upper seal when the cap is installed.

10. Apply an even $\frac{1}{16}$ in. (1.6mm) bead of RTV silicone rubber sealer, to the areas shown, following the procedure given in the illustration.

➡ **This sealer sets up in 15 minutes.**

11. Install the rear main bearing cap. Tighten the cap bolts to specifications.

12. Install the oil pump and oil pan. Fill the crankcase with the proper amount and type of oil.

13. Operate the engine and check for oil leaks.

EXHAUST SYSTEM

CAUTION

When working on exhaust systems, ALWAYS wear protective goggles! Avoid working on a hot exhaust system!

Muffler, Catalytic Converter, Inlet and Outlet Pipes

REMOVAL & INSTALLATION

SEE FIGS. 103–104

The following applies to exhaust systems using clamped joints. Some models, use welded joints at the muffler. These joints will, of course, have to be cut.

1. Raise and support the vehicle on jackstands.
2. Remove the U-clamps securing the muffler and outlet pipe.
3. Disconnect the muffler and outlet pipe bracket and insulator assemblies.
4. Remove the muffler and outlet pipe assembly. It may be necessary to heat the joints to get the parts to come off. Special tools are available to aid in breaking loose the joints.
5. Remove the extension pipe.
6. Disconnect the catalytic converter bracket and insulator assembly.

For rod and insulator type hangers, apply a soap solution to the insulator surface and rod ends to allow easier removal of the insulator from the rod end. Don't use oil-based or silicone-based solutions since they will allow the insulator to slip back off once it's installed.

7. Remove the catalytic converter.
8. Remove the inlet pipe assembly.
9. Install the components making sure that all the components in the system are properly aligned before tightening any fasteners. Make sure all tabs are indexed and all parts are clear of surrounding body panels. See the accompanying illustrations for proper clearances and alignment.

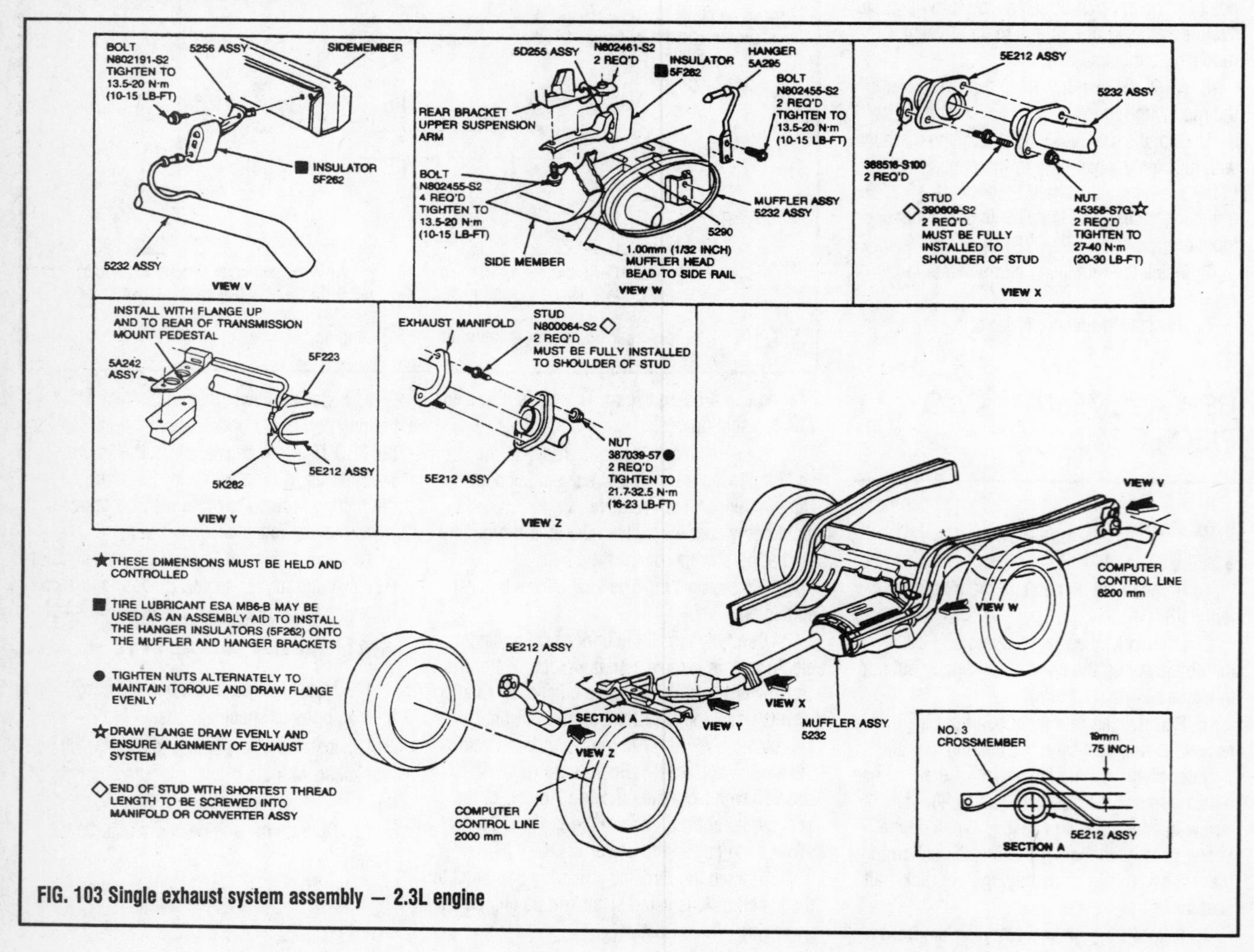

FIG. 103 Single exhaust system assembly — 2.3L engine

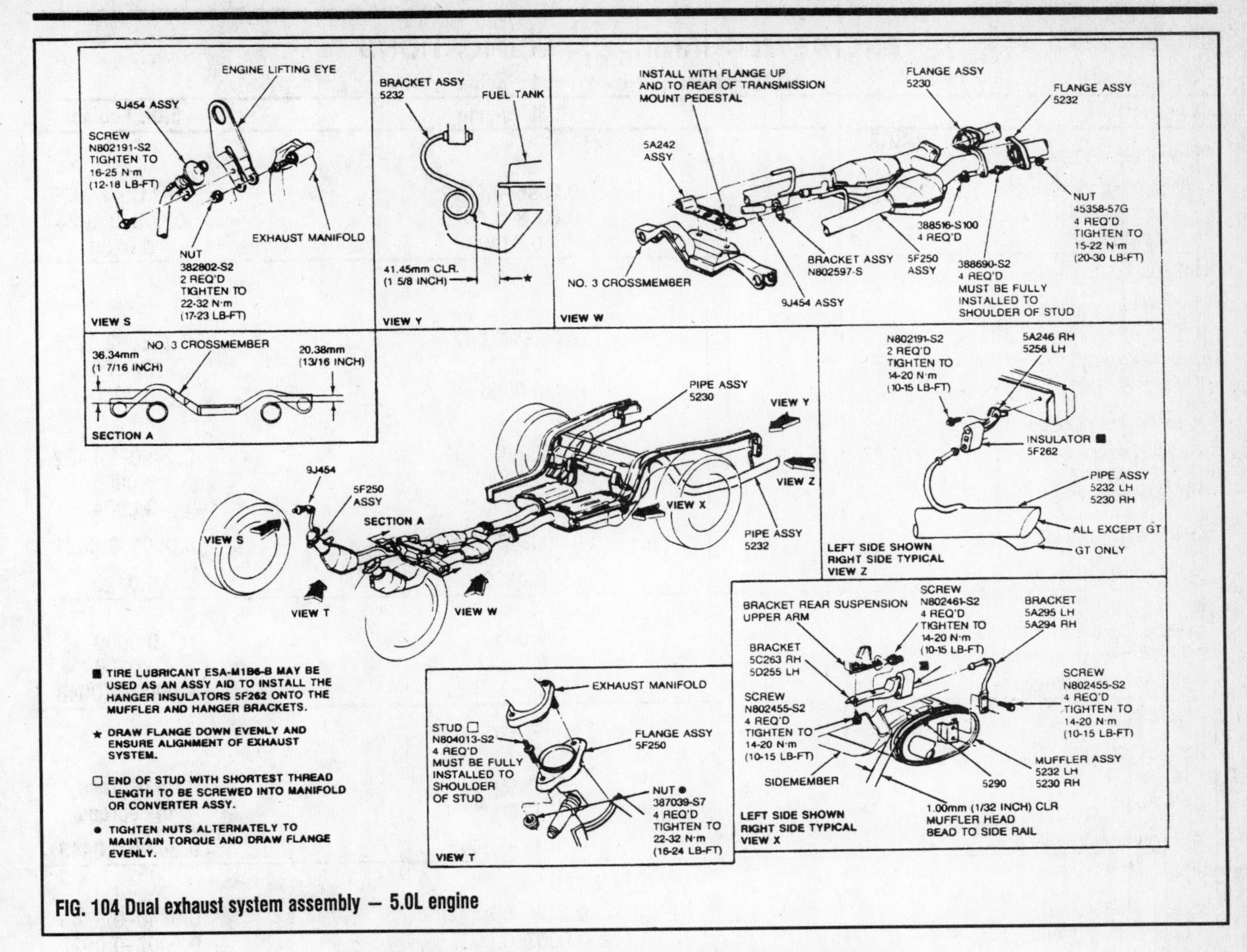

FIG. 104 Dual exhaust system assembly — 5.0L engine

ENGINE MECHANICAL SPECIFICATIONS

All specifications are given in inches

Component	2.3L Engine	5.0L Engine
Camshaft		
Maximum end-play:	0.0090	0.0090
Bearing diameter		
No. 1:	—	2.0825–2.0835
No. 2:	—	2.0675–2.0685
No. 3:	—	2.0525–2.0535
No. 4:	—	2.0375–2.0385
No. 5:	—	2.0225–2.0235
Journal diameter		
No. 1:	1.7713–1.7720	2.0805–2.0815
No. 2:	1.7713–1.7720	2.0655–2.0665
No. 3:	1.7713–1.7720	2.0505–2.0515
No. 4:	1.7713–1.7720	2.0355–2.0365
No. 5:		2.0205–2.0215
Bearing clearance:	0.0060	0.0060
Lobe lift		
Intake:	0.2381	0.2780
Exhaust:	0.2381	0.2780

ENGINE MECHANICAL SPECIFICATIONS

All specifications are given in inches

Component	2.3L Engine	5.0L Engine
Connecting rod		
Piston pin bore diameter:	0.9096–0.9012	0.9112–0.9096
Bearing oil clearance:	0.0008–0.0026	0.008–0.0024
Side clearance:	0.0140	0.0230
Crankshaft		
Connecting rod journal		
Diameter:	2.0462–2.0472	2.1288–2.1236
Out-of-round (max):	0.0006	0.0006
Taper (max):	0.0006	0.0006
Main bearing journal		
Diameter:	2.2059–2.2051	2.2490–2.2482
Out-of-round (max):	0.0006	0.0006
Taper (max):	0.0006	0.0004
Main bearing oil clearance:	0.0008–0.0026	0.0004–0.0021
Crankshaft end-play:	0.0120	0.0120
Cylinder block		
Cylinder bore out-of-round limit:	0.0050	0.0050
Cylinder bore maximum taper:	0.0100	0.0100
Cylinder bore diameter:	3.7795–3.7825	4.0000–4.0048
Cylinder head		
Valve seat angle		
Intake:	45 degrees	45 degrees
Exhaust:	45 degrees	45 degrees
Valve guide bore diameter:	0.3433–0.3443	0.3433–0.3443
Hydraulic lifters		
Body diameter:	0.8422–0.8427	0.8740–0.8745
Clearance in block:	0.0007–0.0027	0.0007–0.0027
Collapsed tappet gap:	0.0350–0.0550	0.098–0.1980
Oil pump		
Relief valve spring tension (lbs. at spec. length):	12.6–14.5 @ 1.20	10.6–12.2 @ 1.704
Driveshaft-to-housing bearing clearance:	0.0015–0.0030	0.0015–0.0030
Relief valve-to-bore clearance:	0.0015–0.0030	0.0015–0.0030
Rotor assembly end clearance (max):	0.0040	0.0040
Outer race-to-housing clearance:	0.001–0.0130	0.001–0.0130
Pistons		
Ring end gap		
Compression (top):	0.0100–0.0200	0.0100–0.0200
Compression (bottom):	0.0100–0.0200	0.0100–0.0200
Oil (steel rail):	0.0150–0.0490	0.0150–0.0550
Ring side clearance		
Compression (top):	0.0016–0.0033	0.002–0.004
Compression (bottom):	0.0016–0.0033	0.002–0.004
Ring width		
Compression (top):	0.0580–0.0590	0.0577–0.0587
Compression (bottom):	0.0580–0.0590	0.0577–0.0587
Piston-to-bore clearance:	0.0024–0.0034	0.0030–0.0038

ENGINE MECHANICAL SPECIFICATIONS

All specifications are given in inches

Component	2.3L Engine	5.0L Engine
Ring groove width		
Compression (top):	0.0600–0.0610	0.0600–0.0610
Compression (bottom):	0.0600–0.0610	0.0600–0.0610
Oil:	0.1596–0.1589	0.1587–0.1597
Piston pin length:	3.0100–3.0400	3.0100–3.0400
Piston pin diameter:	0.9118–0.9124	0.9119–0.9124
Piston-to-pin clearance:	0.0003–0.0005	0.0002–0.0004
Valves		
Face angle		
Intake:	44 degrees	44 degrees
Exhaust:	44 degrees	44 degrees
Head diameter		
Intake:	1.7300–1.7470	1.7700–1.7940
Exhaust:	1.4900–1.5100	1.4530–1.4680
Spring test pressure (lbs. @ spec. length)		
Intake:	64–74 @ 1.52	74–84 @ 1.78
Exhaust:	128–142 @ 1.12	77–85 @ 1.60
Spring installed height		
Intake:	1.4900–1.5500	1.7500–1.8000
Exhaust:	1.4900–1.5500	1.5800–1.6400
Spring free length		
Intake:	1.8770	2.0200
Exhaust:	1.8770	1.7900
Stem-to-guide clearance		
Intake:	0.0055	0.0055
Exhaust:	0.0055	0.0055
Stem diameter		
Intake:	0.3416–0.3423	0.3416–0.3423
Exhaust:	0.3411–0.3418	0.3411–0.3418

ENGINE TORQUE SPECIFICATIONS

Component	English	Metric
Auxiliary shaft gear bolt		
2.3L engine	28–44 ft. lbs.	35–54 Nm
Auxiliary shaft thrust plate bolt		
2.3L engine	72–108 inch lbs.	8–12 Nm
Camshaft sprocket bolt		
2.3L engine	50–71 ft. lbs.	68–96 Nm
5.0L engine	40–45 ft. lbs.	54–61 Nm
Camshaft thrust plate		
2.3L engine	72–96 lbs.	8–12 Nm
5.0L engine	9–12 ft. lbs.	12–16 Nm
Connecting rod bearing cap nuts		
2.3L engine		
step 1:	25–30 ft. lbs.	34–41 Nm
step 2:	30–36 ft. lbs.	41–48 Nm
5.0L engine	19–24 ft. lbs.	25–33 Nm

ENGINE TORQUE SPECIFICATIONS

Component	English	Metric
Crankshaft pulley bolt		
2.3L engine		
1988–90	103–133 ft. lbs.	137–180 Nm
1991–92	114–151 ft. lbs.	155–205 Nm
5.0L engine	70–90 ft. lbs.	95–122 Nm
Cylinder Head bolt		
2.3L engine		
step 1:	50–60 ft. lbs.	55–81 Nm
step 2:	80–90 ft. lbs.	108–122 Nm
5.0L engine		
step 1:	55–65 ft. lbs.	75–88 Nm
step 2:	65–72 ft. lbs.	88–98 Nm
EGR valve		
2.3L engine	14–21 ft. lbs.	19–29 Nm
5.0L engine	15–22 ft. lbs.	20–30 Nm
EGR valve-to-spacer plate		
5.0L engine	12–18 ft. lbs.	16–24 Nm
EGR supply tube		
2.3L engine	9–12 ft. lbs.	12–16 Nm
Engine mounts		
2.3L engine		
rear mount-to-trans bolts	35–45 ft. lbs.	41–61 Nm
front engine mount bolts	33–45 ft. lbs.	45–61 Nm
mount-to-body nuts	80–106 ft. lbs.	108–144 Nm
bracket-to-engine	30–42 ft. lbs.	40–60 Nm
5.0L engine		
rear mount-to-trans bolts	47–64 ft. lbs.	35–50 Nm
front engine mount bolts	15–22 ft. lbs.	20–30 Nm
mount-to-body bolts	35–50 ft. lbs.	47–67 Nm
bracket-to-engine	30–42 ft. lbs.	40–60 Nm
Exhaust flange-to-manifold nuts		
2.3L engine	35–48 ft. lbs.	46–65 Nm
5.0L engine	20–27 ft. lbs.	27–34 Nm
Exhaust Manifold		
2.3L engine		
step 1:	15–17 ft. lbs.	20–23 Nm
step 2:	20–30 ft. lbs.	27–41 Nm
5.0L engine	18–24 ft. lbs.	25–33 Nm
Exhaust manifold-to-heat shield nuts	15–22 ft. lbs.	20–30 Nm
Flywheel/flex plate-to-crankshaft bolts		
2.3L engine	56–64 ft. lbs.	76–87 Nm
5.0L engine	75–85 ft. lbs.	102–116 Nm
Flywheel-to-converter bolts	20–34 ft. lbs.	27–46 Nm
Fuel rail-to-intake manifold bolts		
2.3L engine	14–21 ft. lbs.	19–29 Nm
5.0L engine	70–105 ft. lbs.	8–12 Nm
Intake Manifold		
2.3L engine		
1989–90 engine	20–29 ft. lbs.	27–41 Nm
1991–92 engine	19–28 ft. lbs.	26–40 Nm
5.0L HO engine		
step 1:	15–20 ft. lbs.	20–30 Nm
step 2:	23–25 ft. lbs.	31–34 Nm

ENGINE TORQUE SPECIFICATIONS

Component	English	Metric
Intake Plenum-to-lower manifold		
2.3L engine	15–22 ft. lbs.	20–30 Nm
5.0L engine	12–18 ft. lbs.	16–24 Nm
Main bearing cap bolts		
2.3L engine		
step 1:	50–60 ft. lbs.	68–81 Nm
step 2:	75–85 ft. lbs.	102–116 Nm
5.0L engine	60–70 ft. lbs.	81–95 Nm
Oil filter adapter	15–22 ft. lbs.	20–30 Nm
Oil Pan-to-block		
2.3L engine	10–13 ft. lbs.	13–18 Nm
5.0L engine	9–11 ft. lbs.	12–15 Nm
Oil pan drain plug	15–25 ft. lbs.	20–34 Nm
Oil level low sensor	20–30 ft. lbs.	27–41 Nm
Oil pressure sending unit	12–16 ft. lbs.	16–22 Nm
Oil pump attaching bolts		
2.3L engine	14–21 ft. lbs.	19–29 Nm
5.0L engine	22–32 ft. lbs.	30–43 Nm
Oil pump cover		
2.3L engine	90–130 inch lbs.	10–15 Nm
Rocker arm bolt		
5.0L engine	18–25 ft. lbs.	24–34 Nm
Rocker (Valve) cover		
2.3L engine	60–96 inch lbs.	7–11 Nm
5.0L engine	36–60 inch lbs.	4–7 Nm
Spark plug	60–144 ft. lbs.	7–16 Nm
Starter-to-block bolts	15–20 ft. lbs.	20–27 Nm
Thermostat housing		
2.3L engine	14–21 ft. lbs.	19–29 Nm
5.0L engine	15–22 ft. lbs.	20–30 Nm
Throttle body nuts	15–22 ft. lbs.	20–30 Nm
Timing belt tensioner bolt	14–19 ft. lbs.	19–29 Nm
Timing cover		
2.3L engine	72–108 inch lbs.	8–12 Nm
5.0L engine	12–18 ft. lbs.	16–24 Nm
Water pump		
2.3L engine	14–21 ft. lbs.	19–29 Nm
5.0L engine	12–18 ft. lbs.	16–24 Nm
Water pump pulley hub bolts		
2.3L engine	12–18 ft. lbs.	16–24 Nm
5.0L engine	36–47 ft. lbs.	48–64 Nm
Water pump tensioner bolt		
2.3L engine	35–48 ft. lbs.	47–65 Nm

Troubleshooting Basic Charging System Problems

Problem	Cause	Solution
Noisy alternator	• Loose mountings • Loose drive pulley • Worn bearings • Brush noise • Internal circuits shorted (High pitched whine)	• Tighten mounting bolts • Tighten pulley • Replace alternator • Replace alternator • Replace alternator
Squeal when starting engine or accelerating	• Glazed or loose belt	• Replace or adjust belt
Indicator light remains on or ammeter indicates discharge (engine running)	• Broken fan belt • Broken or disconnected wires • Internal alternator problems • Defective voltage regulator	• Install belt • Repair or connect wiring • Replace alternator • Replace voltage regulator
Car light bulbs continually burn out—battery needs water continually	• Alternator/regulator overcharging	• Replace voltage regulator/alternator
Car lights flare on acceleration	• Battery low • Internal alternator/regulator problems	• Charge or replace battery • Replace alternator/regulator
Low voltage output (alternator light flickers continually or ammeter needle wanders)	• Loose or worn belt • Dirty or corroded connections • Internal alternator/regulator problems	• Replace or adjust belt • Clean or replace connections • Replace alternator or regulator

Troubleshooting Basic Starting System Problems

Problem	Cause	Solution
Starter motor rotates engine slowly	• Battery charge low or battery defective • Defective circuit between battery and starter motor • Low load current • High load current	• Charge or replace battery • Clean and tighten, or replace cables • Bench-test starter motor. Inspect for worn brushes and weak brush springs. • Bench-test starter motor. Check engine for friction, drag or coolant in cylinders. Check ring gear-to-pinion gear clearance.
Starter motor will not rotate engine	• Battery charge low or battery defective • Faulty solenoid • Damage drive pinion gear or ring gear • Starter motor engagement weak • Starter motor rotates slowly with high load current • Engine seized	• Charge or replace battery • Check solenoid ground. Repair or replace as necessary. • Replace damaged gear(s) • Bench-test starter motor • Inspect drive yoke pull-down and point gap, check for worn end bushings, check ring gear clearance • Repair engine

Troubleshooting Basic Starting System Problems

Problem	Cause	Solution
Starter motor drive will not engage (solenoid known to be good)	• Defective contact point assembly • Inadequate contact point assembly ground • Defective hold-in coil	• Repair or replace contact point assembly • Repair connection at ground screw • Replace field winding assembly
Starter motor drive will not disengage	• Starter motor loose on flywheel housing • Worn drive end busing • Damaged ring gear teeth • Drive yoke return spring broken or missing	• Tighten mounting bolts • Replace bushing • Replace ring gear or driveplate • Replace spring
Starter motor drive disengages prematurely	• Weak drive assembly thrust spring • Hold-in coil defective	• Replace drive mechanism • Replace field winding assembly
Low load current	• Worn brushes • Weak brush springs	• Replace brushes • Replace springs

Troubleshooting Engine Mechanical Problems

Problem	Cause	Solution
External oil leaks	• Fuel pump gasket broken or improperly seated • Cylinder head cover RTV sealant broken or improperly seated • Oil filler cap leaking or missing	• Replace gasket • Replace sealant; inspect cylinder head cover sealant flange and cylinder head sealant surface for distortion and cracks • Replace cap
External oil leaks	• Oil filter gasket broken or improperly seated • Oil pan side gasket broken, improperly seated or opening in RTV sealant • Oil pan front oil seal broken or improperly seated • Oil pan rear oil seal broken or improperly seated • Timing case cover oil seal broken or improperly seated • Excess oil pressure because of restricted PCV valve • Oil pan drain plug loose or has stripped threads • Rear oil gallery plug loose • Rear camshaft plug loose or improperly seated • Distributor base gasket damaged	• Replace oil filter • Replace gasket or repair opening in sealant; inspect oil pan gasket flange for distortion • Replace seal; inspect timing case cover and oil pan seal flange for distortion • Replace seal; inspect oil pan rear oil seal flange; inspect rear main bearing cap for cracks, plugged oil return channels, or distortion in seal groove • Replace seal • Replace PCV valve • Repair as necessary and tighten • Use appropriate sealant on gallery plug and tighten • Seat camshaft plug or replace and seal, as necessary • Replace gasket

Troubleshooting Engine Mechanical Problems (cont.)

Problem	Cause	Solution
Excessive oil consumption	• Oil level too high	• Drain oil to specified level
	• Oil with wrong viscosity being used	• Replace with specified oil
	• PCV valve stuck closed	• Replace PCV valve
	• Valve stem oil deflectors (or seals) are damaged, missing, or incorrect type	• Replace valve stem oil deflectors
	• Valve stems or valve guides worn	• Measure stem-to-guide clearance and repair as necessary
	• Poorly fitted or missing valve cover baffles	• Replace valve cover
	• Piston rings broken or missing	• Replace broken or missing rings
	• Scuffed piston	• Replace piston
	• Incorrect piston ring gap	• Measure ring gap, repair as necessary
	• Piston rings sticking or excessively loose in grooves	• Measure ring side clearance, repair as necessary
	• Compression rings installed upside down	• Repair as necessary
	• Cylinder walls worn, scored, or glazed	• Repair as necessary
	• Piston ring gaps not properly staggered	• Repair as necessary
	• Excessive main or connecting rod bearing clearance	• Measure bearing clearance, repair as necessary
No oil pressure	• Low oil level	• Add oil to correct level
	• Oil pressure gauge, warning lamp or sending unit inaccurate	• Replace oil pressure gauge or warning lamp
	• Oil pump malfunction	• Replace oil pump
	• Oil pressure relief valve sticking	• Remove and inspect oil pressure relief valve assembly
	• Oil passages on pressure side of pump obstructed	• Inspect oil passages for obstruction
	• Oil pickup screen or tube obstructed	• Inspect oil pickup for obstruction
	• Loose oil inlet tube	• Tighten or seal inlet tube
Low oil pressure	• Low oil level	• Add oil to correct level
	• Inaccurate gauge, warning lamp or sending unit	• Replace oil pressure gauge or warning lamp
	• Oil excessively thin because of dilution, poor quality, or improper grade	• Drain and refill crankcase with recommended oil
	• Excessive oil temperature	• Correct cause of overheating engine
	• Oil pressure relief spring weak or sticking	• Remove and inspect oil pressure relief valve assembly
	• Oil inlet tube and screen assembly has restriction or air leak	• Remove and inspect oil inlet tube and screen assembly. (Fill inlet tube with lacquer thinner to locate leaks.)
	• Excessive oil pump clearance	• Measure clearances
	• Excessive main, rod, or camshaft bearing clearance	• Measure bearing clearances, repair as necessary

Troubleshooting Engine Mechanical Problems (cont.)

Problem	Cause	Solution
High oil pressure	• Improper oil viscosity	• Drain and refill crankcase with correct viscosity oil
	• Oil pressure gauge or sending unit inaccurate	• Replace oil pressure gauge
	• Oil pressure relief valve sticking closed	• Remove and inspect oil pressure relief valve assembly
Main bearing noise	• Insufficient oil supply	• Inspect for low oil level and low oil pressure
	• Main bearing clearance excessive	• Measure main bearing clearance, repair as necessary
	• Bearing insert missing	• Replace missing insert
	• Crankshaft end play excessive	• Measure end play, repair as necessary
	• Improperly tightened main bearing cap bolts	• Tighten bolts with specified torque
	• Loose flywheel or drive plate	• Tighten flywheel or drive plate attaching bolts
	• Loose or damaged vibration damper	• Repair as necessary
Connecting rod bearing noise	• Insufficient oil supply	• Inspect for low oil level and low oil pressure
	• Carbon build-up on piston	• Remove carbon from piston crown
	• Bearing clearance excessive or bearing missing	• Measure clearance, repair as necessary
	• Crankshaft connecting rod journal out-of-round	• Measure journal dimensions, repair or replace as necessary
	• Misaligned connecting rod or cap	• Repair as necessary
	• Connecting rod bolts tightened improperly	• Tighten bolts with specified torque
Piston noise	• Piston-to-cylinder wall clearance excessive (scuffed piston)	• Measure clearance and examine piston
	• Cylinder walls excessively tapered or out-of-round	• Measure cylinder wall dimensions, rebore cylinder
	• Piston ring broken	• Replace all rings on piston
	• Loose or seized piston pin	• Measure piston-to-pin clearance, repair as necessary
	• Connecting rods misaligned	• Measure rod alignment, straighten or replace
	• Piston ring side clearance excessively loose or tight	• Measure ring side clearance, repair as necessary
	• Carbon build-up on piston is excessive	• Remove carbon from piston
Valve actuating component noise	• Insufficient oil supply	• Check for: (a) Low oil level (b) Low oil pressure (c) Plugged push rods (d) Wrong hydraulic tappets (e) Restricted oil gallery (f) Excessive tappet to bore clearance
	• Push rods worn or bent	• Replace worn or bent push rods
	• Rocker arms or pivots worn	• Replace worn rocker arms or pivots
	• Foreign objects or chips in hydraulic tappets	• Clean tappets

Troubleshooting Engine Mechanical Problems (cont.)

Problem	Cause	Solution
Valve actuating component noise	• Excessive tappet leak-down	• Replace valve tappet
	• Tappet face worn	• Replace tappet; inspect corresponding cam lobe for wear
	• Broken or cocked valve springs	• Properly seat cocked springs; replace broken springs
	• Stem-to-guide clearance excessive	• Measure stem-to-guide clearance, repair as required
	• Valve bent	• Replace valve
	• Loose rocker arms	• Tighten bolts with specified torque
	• Valve seat runout excessive	• Regrind valve seat/valves
	• Missing valve lock	• Install valve lock
	• Push rod rubbing or contacting cylinder head	• Remove cylinder head and remove obstruction in head
	• Excessive engine oil (four-cylinder engine)	• Correct oil level

4

EMISSION CONTROLS

AIR POLLUTION

The earth's atmosphere, at or near sea level, consists of 78% nitrogen, 21% oxygen and 1% other gases, approximately. If it were possible to remain in this state, 100% clean air would result. However, many varied causes allow other gases and particulates to mix with the clean air, causing the air to become unclean or polluted.

Certain of these pollutants are visible while others are invisible, with each having the capability of causing distress to the eyes, ears, throat, skin and respiratory system. Should these pollutants be concentrated in a specific area and under the right conditions, death could result due to the displacement or chemical change of the oxygen content in the air. These pollutants can cause much damage to the environment and to the many man made objects that are exposed to the elements.

To better understand the causes of air pollution, the pollutants can be categorized into 3 separate types, natural, industrial and automotive.

Natural Pollutants

Natural pollution has been present on earth before man appeared and is still a factor to be considered when discussing air pollution, although it causes only a small percentage of the present overall pollution problem existing in our country. It is the direct result of decaying organic matter, wind born smoke and particulates from such natural events as plains and forest fires (ignited by heat or lightning), volcanic ash, sand and dust which can spread over a large area of the countryside.

Such a phenomenon of natural pollution has been recent volcanic eruptions, with the resulting plume of smoke, steam and volcanic ash blotting out the sun's rays as it spreads and rises higher into the atmosphere, where the upper air currents catch and carry the smoke and ash, while condensing the steam back into water vapor. As the water vapor, smoke and ash traveled on their journey, the smoke dissipates into the atmosphere while the ash and moisture settle back to earth in a trail hundred of miles long. In many cases, lives are lost and millions of dollars of property damage result, and ironically, man can only stand by and watch it happen.

Industrial Pollution

Industrial pollution is caused primarily by industrial processes, the burning of coal, oil and natural gas, which in turn produces smoke and fumes. Because the burning fuels contain much sulfur, the principal ingredients of smoke and fumes are sulfur dioxide (SO_2) and particulate matter. This type of pollutant occurs most severely during still, damp and cool weather, such as at night. Even in its less severe form, this pollutant is not confined to just cities. Because of air movements, the pollutants move for miles over the surrounding countryside, leaving in its path a barren and unhealthy environment for all living things.

Working with Federal, State and Local mandated rules, regulations and by carefully monitoring the emissions, industries have greatly reduced the amount of pollutant emitted from their industrial sources, striving to obtain an acceptable level. Because of the mandated industrial emission clean up, many land areas and streams in and around the cities that were formerly barren of vegetation and life, have now begun to move back in the direction of nature's intended balance.

Automotive Pollutants

The third major source of air pollution is the automotive emissions. The emissions from the internal combustion engine were not an appreciable problem years ago because of the small number of registered vehicles and the nation's small highway system. However, during the early 1950's, the trend of the American people was to move from the cities to the surrounding suburbs. This caused an immediate problem in the transportation areas because the majority of the suburbs were not afforded mass transit conveniences. This lack of transportation created an attractive market for the automobile manufacturers, which resulted in a dramatic increase in the number of vehicles produced and sold, along with a marked increase in highway construction between cities and the suburbs. Multi-vehicle families emerged with much emphasis placed on the individual vehicle per family member. As the increase in vehicle ownership and usage occurred, so did the pollutant levels in and around the cities, as the suburbanites drove daily to their businesses and employment in the city and its fringe area, returning at the end of the day to their homes in the suburbs.

It was noted that a fog and smoke type haze was being formed and at times, remained in suspension over the cities and did not quickly dissipate. At first this "smog", derived from the words "smoke" and "fog", was thought to result from industrial pollution but it was determined that the automobile emissions were largely to blame. It was discovered that as normal automobile emissions were exposed to sunlight for a period of time, complex chemical reactions would take place.

It was found the smog was a photo chemical layer and was developed when certain oxides of nitrogen (NOx) and unburned hydrocarbons (HC) from the automobile emissions were exposed to sunlight and was more severe when the smog would remain stagnant over an area in which a warm layer of air would settle over the top of a cooler air mass at ground level, trapping and holding the automobile emissions, instead of the emissions being dispersed and diluted through normal air flows. This type of air stagnation was given the name "Temperature Inversion".

Temperature Inversion

In normal weather situations, the surface air is warmed by the heat radiating from the earth's surface and the sun's rays and will rise upward, into the atmosphere, to be cooled through a convection type heat expands with the cooler upper air. As the warm air rises, the surface pollutants are carried upward and dissipated into the atmosphere.

When a temperature inversion occurs, we find the higher air is no longer cooler but warmer than the surface air, causing the cooler surface air to become trapped and unable to move. This warm air blanket can extend from above ground level to a few hundred or even a few thousand feet into the air. As the surface air is trapped, so are the pollutants, causing a severe smog condition. Should this stagnant air mass extend to a few thousand feet high, enough air movement with the inversion takes place to allow the smog layer to rise above ground level but the pollutants still cannot dissipate. This inversion can remain for days over an area, with only the smog level rising or lowering from ground level to a few hundred feet high. Meanwhile, the pollutant levels increases, causing eye irritation, respirator problems, reduced visibility, plant damage and in some cases, cancer type diseases.

This inversion phenomenon was first noted in the Los Angeles, California area. The city lies in a basin type of terrain and during certain weather conditions, a cold air mass is held in the basin while a warmer air mass covers it like a lid.

Because this type of condition was first documented as prevalent in the Los Angeles area, this type of smog was named Los Angeles Smog, although it occurs in other areas where a large concentration of automobiles are used and the air remains stagnant for any length of time.

Internal Combustion Engine Pollutants

Consider the internal combustion engine as a machine in which raw materials must be placed so a finished product comes out. As in any machine operation, a certain amount of wasted material is formed. When we relate this to the internal combustion engine, we find that by putting in air and fuel, we obtain power from this mixture during the combustion process to drive the vehicle. The by-product or waste of this power is, in part, heat and exhaust gases with which we must concern ourselves.

HEAT TRANSFER

The heat from the combustion process can rise to over 4000°F (2204°C). The dissipation of this heat is controlled by a ram air effect, the use of cooling fans to cause air flow and having a liquid coolant solution surrounding the combustion area and transferring the heat of combustion through the cylinder walls and into the coolant. The coolant is then directed to a thin-finned, multi-tubed radiator, from which the excess heat is transferred to the outside air by 1 or all of the 3 heat transfer methods, conduction, convection or radiation.

The cooling of the combustion area is an important part in the control of exhaust emissions. To understand the behavior of the combustion and transfer of its heat, consider the air/fuel charge. It is ignited and the flame front burns progressively across the combustion chamber until the burning charge reaches the cylinder walls. Some of the fuel in contact with the walls is not hot enough to burn, thereby snuffing out or Quenching the combustion process. This leaves unburned fuel in the combustion chamber. This unburned fuel is then forced out of the cylinder along with the exhaust gases and into the exhaust system.

Many attempts have been made to minimize the amount of unburned fuel in the combustion chambers due to the snuffing out or "Quenching", by increasing the coolant temperature and lessening the contact area of the coolant around the combustion area. Design limitations within the combustion chambers prevent the complete burning of the air/fuel charge, so a certain amount of the unburned fuel is still expelled into the exhaust system, regardless of modifications to the engine.

EXHAUST EMISSIONS

Composition Of The Exhaust Gases

The exhaust gases emitted into the atmosphere are a combination of burned and unburned fuel. To understand the exhaust emission and its composition review some basic chemistry.

When the air/fuel mixture is introduced into the engine, we are mixing air, composed of nitrogen (78%), oxygen (21%) and other gases (1%) with the fuel, which is 100% hydrocarbons (HC), in a semi-controlled ratio. As the combustion process is accomplished, power is produced to move the vehicle while the heat of combustion is transferred to the cooling system. The exhaust gases are then composed of nitrogen, a diatomic gas (N_2), the same as was introduced in the engine, carbon dioxide (CO2), the same gas that is used in beverage carbonation and water vapor (H_2O). The nitrogen (N_2), for the most part passes through the engine unchanged, while the oxygen (O_2) reacts (burns) with the hydrocarbons (HC) and produces the carbon dioxide (CO_2) and the water vapors (H_2O). If this chemical process would be the only process to take place, the exhaust emissions would be harmless. However, during the combustion process, other pollutants are formed and are considered dangerous. These pollutants are carbon monoxide (CO), hydrocarbons (HC), oxides of nitrogen (NOx) oxides of sulfur (SOx) and engine particulates.

Lead (Pb), is considered 1 of the particulates and is present in the exhaust gases whenever leaded fuels are used. Lead (Pb) does not dissipate easily. Levels can be high along roadways when it is emitted from vehicles and can pose a health threat. Since the increased usage of unleaded gasoline and the phasing out of leaded gasoline for fuel, this pollutant is gradually diminishing. While not considered a major threat lead is still considered a dangerous pollutant.

HYDROCARBONS

Hydrocarbons (HC) are essentially unburned fuel that have not been successfully burned during the combustion process or have escaped into the atmosphere through fuel evaporation. The main sources of incomplete combustion are rich air/fuel mixtures, low engine temperatures and improper spark timing. The main sources of hydrocarbon emission through fuel evaporation come from the vehicle's fuel tank and carburetor bowl.

To reduce combustion hydrocarbon emission, engine modifications were made to minimize dead space and surface area in the combustion chamber. In addition the air/fuel mixture was made more lean through improved carburetion, fuel injection and by the addition of external controls to aid in further combustion of the hydrocarbons outside the engine. Two such methods were the addition of an air injection system, to inject fresh air into the exhaust manifolds and the installation of a catalytic converter, a unit that is able to burn traces of hydrocarbons without affecting the internal combustion process or fuel economy.

To control hydrocarbon emissions through fuel evaporation, modifications were made to the fuel tank and carburetor bowl to allow storage of the fuel vapors during periods of engine shut-down, and at specific times during engine operation, to purge and burn these same vapors by blending them with the air/fuel mixture.

CARBON MONOXIDE

Carbon monoxide is formed when not enough oxygen is present during the combustion process to convert carbon (C) to carbon dioxide (CO_2). An increase in the carbon monoxide (CO) emission is normally accompanied by an increase in the hydrocarbon (HC) emission because of the lack of oxygen to completely burn all of the fuel mixture.

Carbon monoxide (CO) also increases the rate at which the photo chemical smog is formed by speeding up the conversion of nitric oxide (NO) to nitrogen dioxide (NO_2). To accomplish this, carbon monoxide (CO) combines with oxygen (O_2) and nitrogen dioxide (NO_2) to produce carbon dioxide (CO_2) and nitrogen dioxide (NO_2). ($CO + O_2 + NO = CO_2 + NO_2$).

The dangers of carbon monoxide, which is an odorless, colorless toxic gas are many. When carbon monoxide is inhaled into the lungs and passed into the blood stream, oxygen is replaced by the carbon monoxide in the red blood cells, causing a reduction in the amount of oxygen being supplied to the many parts of the body. This lack of oxygen causes headaches, lack of coordination, reduced mental alertness and should the carbon monoxide concentration be high enough, death could result.

NITROGEN

Normally, nitrogen is an inert gas. When heated to approximately 2500°F (1371°C) through the combustion process, this gas becomes active and causes an increase in the nitric oxide (NOx) emission.

Oxides of nitrogen (NOx) are composed of approximately 97–98% nitric oxide (NO2). Nitric oxide is a colorless gas but when it is passed into the atmosphere, it combines with oxygen and forms nitrogen dioxide (NO2). The nitrogen dioxide then combines with chemically active hydrocarbons (HC) and when in the presence of sunlight, causes the formation of photo chemical smog.

OZONE

To further complicate matters, some of the nitrogen dioxide (NO_2) is broken apart by the sunlight to form nitric oxide and oxygen. (NO_2 + sunlight = $NO + O$). This single atom of oxygen then combines with diatomic (meaning 2 atoms) oxygen (O_2) to form ozone (O_3). Ozone is 1 of the smells associated with smog. It has a pungent and offensive odor, irritates the eyes and lung tissues, affects the growth of plant life and causes rapid deterioration of rubber products. Ozone can be formed by sunlight as well as electrical discharge into the air.

The most common discharge area on the automobile engine is the secondary ignition electrical system, especially when inferior quality spark plug cables are used. As the surge of high voltage is routed through the secondary cable, the circuit builds up an electrical field around the wire, acting upon the oxygen in the surrounding air to form the ozone. The faint glow along the cable with the engine running that may be visible on a dark night, is called the "corona discharge." It is the result of the electrical field passing from a high along the cable, to a low in the surrounding air, which forms the ozone gas. The combination of corona and ozone has been a major cause of cable deterioration. Recently, different types and better quality insulating materials have lengthened the life of the electrical cables.

Although ozone at ground level can be harmful, ozone is beneficial to the earth's inhabitants. By having a concentrated ozone layer called the 'ozonosphere', between 10 and 20 miles (16–32km) up in the atmosphere much of the ultra violet radiation from the sun's rays are absorbed and screened. If this ozone layer were not present, much of the earth's surface would be burned, dried and unfit for human life.

There is much discussion concerning the ozone layer and its density. A feeling exists that this protective layer of ozone is slowly diminishing and corrective action must be directed to this problem. Much experimenting is presently being conducted to determine if a problem exists and if so, the short and long term effects of the problem and how it can be remedied.

OXIDES OF SULFUR

Oxides of sulfur (SOx) were initially ignored in the exhaust system emissions, since the sulfur content of gasoline as a fuel is less than $\frac{1}{10}$ of 1%. Because of this small amount, it was felt that it contributed very little to the overall pollution problem. However, because of the difficulty in solving the sulfur emissions in industrial pollutions and the introduction of catalytic converter to the automobile exhaust systems, a change was mandated. The automobile exhaust system, when equipped with a catalytic converter, changes the sulfur dioxide (SO_2) into the sulfur trioxide (SO_3).

When this combines with water vapors (H_2O), a sulfuric acid mist (H_2SO_4) is formed and is a very difficult pollutant to handle and is extremely corrosive. This sulfuric acid mist that is formed, is the same mist that rises from the vents of an automobile storage battery when an active chemical reaction takes place within the battery cells.

When a large concentration of vehicles equipped with catalytic converters are operating in an area, this acid mist will rise and be distributed over a large ground area causing land, plant, crop, paints and building damage.

PARTICULATE MATTER

A certain amount of particulate matter is present in the burning of any fuel, with carbon constituting the largest percentage of the particulates. In gasoline, the remaining percentage of particulates is the burned remains of the various other compounds used in its manufacture. When a gasoline engine is in good internal condition, the particulate emissions are low but as the engine wears internally, the particulate emissions increase. By visually inspecting the tail pipe emissions, a determination can be made as to where an engine defect may exist. An engine with light gray smoke emitting from the tail pipe normally indicates an increase in the oil consumption through burning due to internal engine wear. Black smoke would indicate a defective fuel delivery system, causing the engine to operate in a rich mode. Regardless of the color of the smoke, the internal part of the engine or the fuel delivery system should be repaired to a "like new" condition to prevent excess particulate emissions.

Diesel and turbine engines emit a darkened plume of smoke from the exhaust system because of the type of fuel used. Emission control regulations are mandated for this type of emission and more stringent measures are being used to prevent excess emission of the particulate matter. Electronic components are being introduced to control the injection of the fuel at precisely the proper time of piston travel, to achieve the optimum in fuel ignition and fuel usage. Other particulate after-burning components are being tested to achieve a cleaner particular emission.

Good grades of engine lubricating oils should be used, meeting the manufacturers specification. "Cut-rate" oils can contribute to the particulate emission problem because of their low "flash" or ignition temperature point. Such oils burn prematurely during the combustion process causing emissions of particulate matter.

The cooling system is an important factor in the reduction of particulate matter. With the cooling system operating at a temperature specified by the manufacturer, the optimum of combustion will occur. The cooling system must be maintained in the same manner as the engine oiling system, as each system is required to perform properly in order for the engine to operate efficiently for a long time.

Other Automobile Emission Sources

Before emission controls were mandated on the internal combustion engines, other sources of engine pollutants were discovered, along with the exhaust emission. It was determined the engine combustion exhaust produced 60% of the total emission pollutants, fuel evaporation from the fuel tank and carburetor vents produced 20%, with the another 20% being produced through the crankcase as a by-product of the combustion process.

CRANKCASE EMISSIONS

Crankcase emissions are made up of water, acids, unburned fuel, oil fumes and particulates. The emissions are classified as hydrocarbons (HC) and are formed by the small amount of unburned, compressed air/fuel mixture entering the crankcase from the combustion area during the compression and power strokes, between the cylinder walls and piston rings. The head of the compression and combustion help to form the remaining crankcase emissions.

Since the first engines, crankcase emissions were allowed to go into the air through a road draft tube, mounted on the lower side of the engine block. Fresh air came in through an open oil filler cap or breather. The air passed through the crankcase mixing with blow-by gases. The motion of the vehicle and the air blowing past the open end of the road draft tube caused a low pressure area at the end of the tube. Crankcase emissions were simply drawn out of the road draft tube into the air.

To control the crankcase emission, the road draft tube was deleted. A hose and/or tubing was routed from the crankcase to the intake manifold so the blow-by emission could be burned with the air/fuel mixture. However, it was found that intake manifold vacuum, used to draw the crankcase emissions into the manifold, would vary in strength at the wrong time and not allow the proper emission flow. A regulating type valve was needed to control the flow of air through the crankcase.

Testing, showed the removal of the blow-by gases from the crankcase as quickly as possible, was most important to the longevity of the engine. Should large accumulations of blow-by gases remain and condense, dilution of the engine oil would occur to form water, soots, resins, acids and lead salts, resulting in the formation of sludge and varnishes. This condensation of the blow-by gases occur more frequently on vehicles used in numerous starting and stopping conditions, excessive idling and when the engine is not allowed to attain normal operating temperature through short runs. The crankcase purge control or PCV system will be described in detail later in this section.

FUEL EVAPORATIVE EMISSIONS

Gasoline fuel is a major source of pollution, before and after it is burned in the automobile engine. From the time the fuel is refined, stored, pumped and transported, again stored until it is pumped into the fuel tank of the vehicle, the gasoline gives off unburned hydrocarbons (HC) into the atmosphere. Through redesigning of the storage areas and venting systems, the pollution factor has been diminished but not eliminated, from the refinery standpoint. However, the automobile still remained the primary source of vaporized, unburned hydrocarbon (HC) emissions.

Fuel pumped form an underground storage tank is cool but when exposed to a warner ambient temperature, will expand. Before controls were mandated, an owner would fill the fuel tank with fuel from an underground storage tank and park the vehicle for some time in warm area, such as a parking lot. As the fuel would warm, it would expand and should no provisions or area be provided for the expansion, the fuel would spill out the filler neck and onto the ground, causing hydrocarbon (HC) pollution and creating a severe fire hazard. To correct this condition, the vehicle manufacturers added overflow plumbing and/or gasoline tanks with built in expansion areas or domes.

However, this did not control the fuel vapor emission from the fuel tank and the carburetor bowl. It was determined that most of the fuel evaporation occurred when the vehicle was stationary and the engine not operating. Most vehicles carry 5–25 gallons (19–95 liters) of gasoline. Should a large concentration of vehicles be parked in one area, such as a large parking lot, excessive fuel vapor emissions would take place, increasing as the temperature increases.

To prevent the vapor emission from escaping into the atmosphere, the fuel system is designed to trap the fuel vapors while the vehicle is stationary, by sealing the fuel system from the atmosphere. A storage system is used to collect and hold the fuel vapors from the carburetor and the fuel tank when the engine is not operating. When the engine is started, the storage system is then purged of the fuel vapors, which are drawn into the engine and burned with the air/fuel mixture.

The components of the fuel evaporative system will be described in detail later in this section.

TERMS USED IN THIS SECTION

ACT: Air Charge Temperature
CO: Carbon Monoxide
CO_2: Carbon Dioxide
DIS: Distributorless Ignition System
DPFE: Differential Pressure Feedback EGR
ECU: Electronic Control Unit
EEC: Electronic Engine Control
EEGR: Electronic EGR
EFI: Electronic Fuel Injection
EGR: Exhaust Gas Recirculation
EVP: EGR Valve Position
EVR: EGR Vacuum Regulator
H_2: Hydrogen
H_2O: Water Vapor
HC: Hydrocarbons
HEGO: Heated Exhaust Gas Oxygen
MAF: Mass Air Flow
MAP: Manifold Absolute Pressure
N_2: Nitrogen
NOx: Nitrogen Oxides
PCV: Positive Crankcase

CUSTOMER MAINTENANCE SCHEDULE A

Follow this Schedule if your driving habits MAINLY include one or more of the following conditions:

- Short trips of less than 16 km (10 miles) when outside temperatures remain below freezing.
- Operating during HOT WEATHER
  — Driving in stop-and-go "rush hour" traffic.
- Towing a trailer or using a car-top carrier.
- Operating in severe dust conditions.
- Extensive idling, such as police, taxi or door-to-door delivery service.

SERVICE INTERVAL Perform at the months or distances shown, whichever comes first. Miles × 1000	3	6	9	12	15	18	21	24	27	30	33	36	39	42	45	48	51	54	57	60
Kilometers × 1000	4.8	9.6	14.4	19.2	24	28.8	33.6	38.4	43.2	48	52.8	57.6	62.4	67.2	72	76.8	81.6	86.4	91.2	96
EMISSION CONTROL SERVICE																				
Replace Engine Oil and Oil Filter Every 3 Months OR	X	X	X	X	X	X	X	X	X	X	X	X	X	X	X	X	X	X	X	X
Replace Spark Plugs										X										X
Inspect Accessory Drive Belt(s)										X										X
Replace PCV Valve and Crankcase Emission Filter — 5.0L					(X)					(X)					(X)					X
Replace Air Cleaner Filter①										X										X
Replace Engine Coolant, EVERY 36 Months OR										X										X
Check Engine Coolant Protection, Hoses and Clamps										ANNUALLY										
GENERAL MAINTENANCE																				
Inspect Exhaust Heat Shields										X										X
Change Automatic Transmission Fluid②										X										X
Lubricate Tie Rods										X										X
Inspect Disc Brake Pads and Rotors②										X										X
Inspect Brake Linings and Drums (Rear)③										X										X
Inspect and Repack Front Wheel Bearings										X										X
Rotate Tires		X					X					X					X			

① If operating in severe dust, more frequent intervals may be required. Consult your dealer.
② Change automatic transmission fluid if your driving habits frequently include one or more of the following conditions:
- Operation during hot weather (above 32°C (90°F)) carrying heavy loads and in hilly terrain.
- Towing a trailer or using a car top carrrier.
- Police, taxi or door to door delivery service.

③ If your driving includes continuous stop-and-go driving or driving in mountainous areas, more frequent intervals may be required.
X All items designated by an X must be performed in all states.
(X) This item not required to be performed, however, Ford recommends that you also perform maintenance on items designated by an (X) in order to achieve best vehicle operation. Failure to perform this recommended maintenance will not invalidate the vehicle emissions warranty or manufacturer recall liability.

FIG. 1 Emission maintenance schedule A — Light Service

CUSTOMER MAINTENANCE SCHEDULE B

Follow Maintenance Schedule B if, generally, you drive your vehicle on a daily basis for more than 16 Km (10 miles) and NONE OF THE UNIQUE DRIVING CONDITIONS SHOWN IN SCHEDULE A APPLY TO YOUR DRIVING HABITS.

SERVICE INTERVALS Perform at the months or distances shown, whichever comes first. Miles x 1000	7.5	15	22.5	30	37.5	45	52.5	60
Kilometers x 1000	12	24	36	48	60	72	84	96
EMISSIONS CONTROL SERVICE								
Replace Engine Oil and Filter Every 6 Months OR 7,500 Miles Whichever Occurs First	X	X	X	X	X	X	X	X
Replace Spark Plugs				X				X
Replace Crankcase Emission Filter①				X				X
Inspect Accessory Drive Belt(s)				X				X
Replace Air Cleaner Filter①				X				X
Replace PCV Valve and Crankcase Emission Filter — 5.0L		(X)		(X)		(X)		X
Replace Engine Coolant Every 36 Months OR				X				X
Check Engine Coolant Protection, Hoses and Clamps				ANNUALLY				
GENERAL MAINTENANCE								
Check Exhaust Heat Shields				X				X
Lube Tie Rods		X③		X		X③		X
Inspect Disc Brake Pads and Rotors②				X				X
Inspect Brake Linings and Drums (Rear)②				X②				X②
Inspect and Repack Front Wheel Bearings				X				X
Rotate Tires	X		X		X		X	

① If operating in severe dust, more frequent intervals may be required. Consult your dealer.
② If your driving includes continuous stop-and-go driving or driving in mountainous areas, more frequent intervals may be required.
③ All vehicles.
X All items designated by an X must be performed in all states.
(X) This item not required to be performed, however, Ford recommends that you also perform maintenance on items designated by an (X) in order to achieve best vehicle operation. Failure to perform this recommended maintenance will not invalidate the vehicle emissions warranty or manufacturer recall liability.

FIG. 2 Emission maintenance schedule B — Medium/Heavy Service

Ventilation
Pd: Palladium
PFE: Pressure Feedback Electronic
Pt: Platinum
RH: Rhodium
SEFI: Sequential Electronic Fuel Injection
STI: Self Test Input
STO: Self-Test Output
TFI: Thick Film Integrated
TIV: Thermactor Idle Vacuum
TP: Throttle Position
WOT: Wide Open Throttle

POSITIVE CRANKCASE VENTILATION SYSTEM

PCV Valve System

GENERAL INFORMATION

➧ SEE FIG. 3–4

The PCV valve system vents crankcase gases into the engine air intake where they are burned with the fuel and air mixture. The PCV valve system keeps pollutants from being released into the atmosphere, and also helps to keep the engine oil clean, by ridding the crankcase of moisture and corrosive fumes. The PCV valve system consists of the PCV valve, it's mounting grommet, the nipple in the air intake and the connecting hoses. On some engine applications, the PCV valve system is connected with the evaporative emission system.

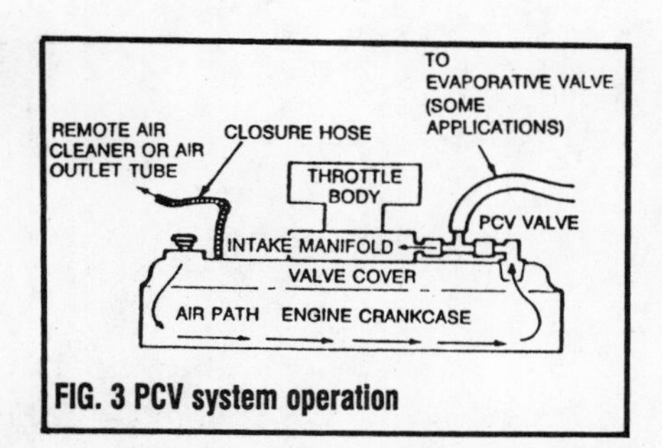

FIG. 3 PCV system operation

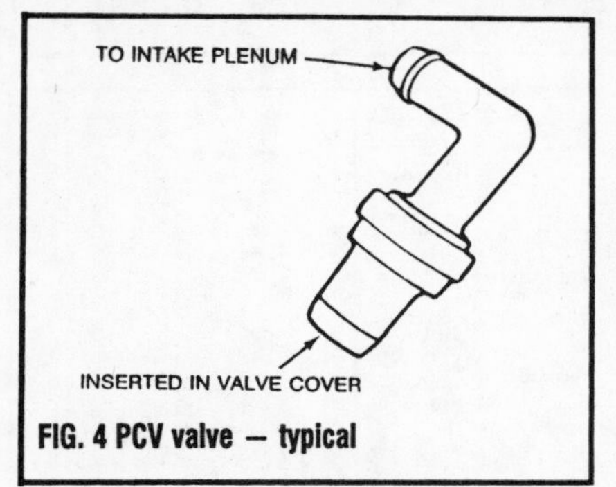

FIG. 4 PCV valve — typical

DESCRIPTION

The PCV valve controls the amount of vapors pulled into the intake manifold from the crankcase and acts as a check valve by preventing air flow from entering the crankcase in the opposite direction. The PCV valve also prevents combustion backfiring from entering the crankcase in order to prevent detonation of the accumulated crankcase gases.

TESTING

1. Remove the PCV valve from the valve cover grommet. On the 5.0L engine, the PCV valve is located on top of the block by the firewall.
2. Shake the PCV valve. If the valve rattles when shaken, reinstall and proceed to Step 3. If the valve does not rattle, it is sticking and must be replaced.
3. Start the engine and bring it to normal operating temperature.
4. Disconnect the hose from the remote air cleaner or air outlet tube, the tube connecting the mass air flow meter and throttle body.
5. Place a stiff piece of paper over the hose end and wait 1 minute.
6. If vacuum holds the paper in place, the system is okay; reconnect the hose.
7. If the paper is not held in place, check for loose hose connections, vacuum leaks or blockage. Correct as necessary.

REMOVAL & INSTALLATION

1. Disconnect the vacuum hose from the PCV valve.
2. Remove the PCV valve from it's mounting grommet.
3. Installation is the reverse of the removal procedure.

FUEL EVAPORATIVE EMISSION CONTROL SYSTEM

General Information

➧ SEE FIG. 5–6

Fuel vapors trapped in the sealed fuel tank are vented through the orifice vapor valve assembly in the top of the tank. The vapors leave the valve assembly through a single vapor line and continue to the carbon canister for storage until they are purged to the engine for burning.

Purging the carbon canister removes the fuel vapor stored in the carbon canister. The fuel vapor is purged via a purge port system, where the vapors flow from the carbon canister to the throttle body or via an EEC-IV controlled system, where the flow of vapors from the canister to the engine is controlled by a purge solenoid or vacuum controlled purge valve. Purging occurs when the engine is at operating temperature and, in most systems, off idle.

The evaporative emission control system consists of the carbon canister, fuel tank vapor orifice and rollover valve assembly, purge control valve and/or purge solenoid valve and the pressure/vacuum relief valve.

Carbon Canister

➧ SEE FIG. 7

The fuel vapors from the fuel tank are stored in the carbon canister until the vehicle is operated, at which time, the vapors will purge

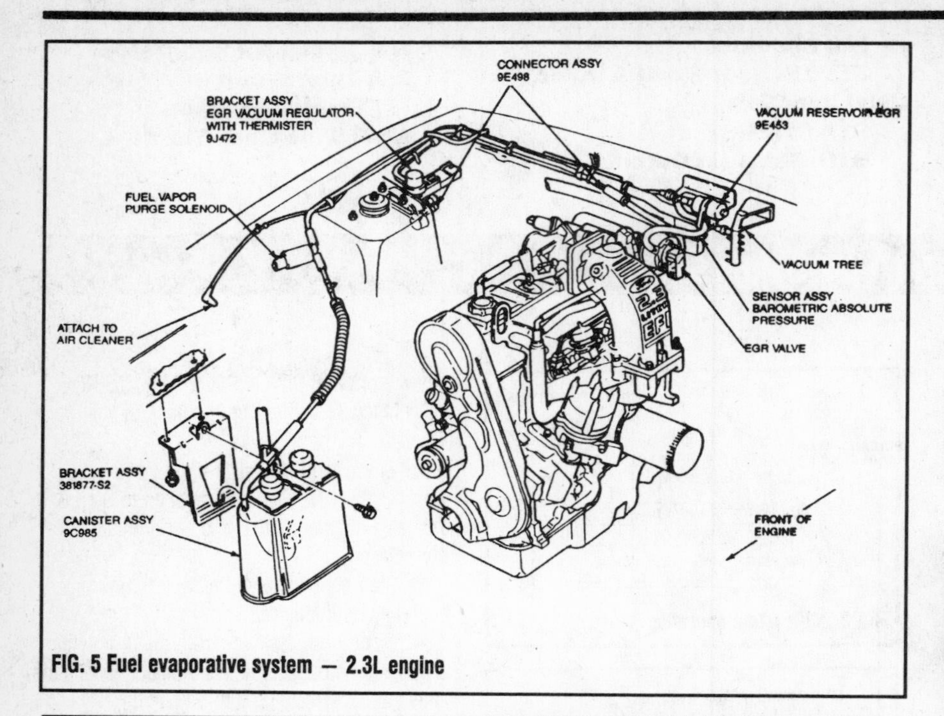

FIG. 5 Fuel evaporative system — 2.3L engine

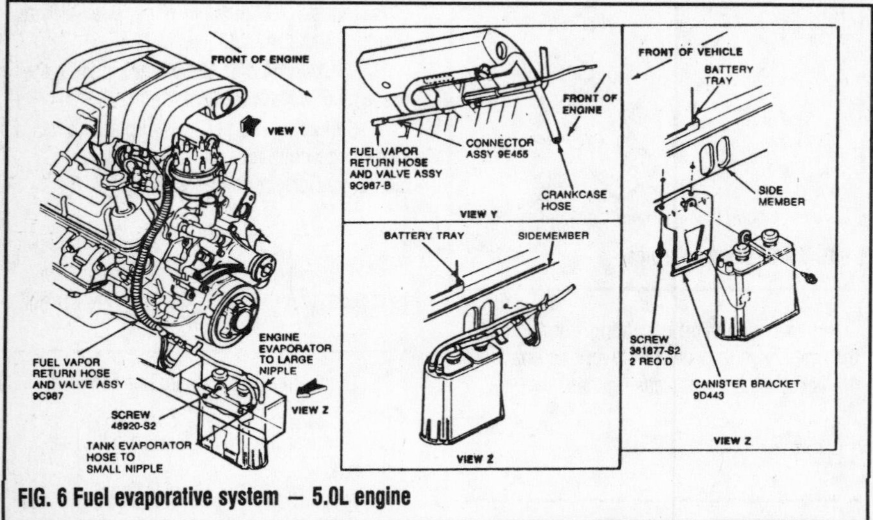

FIG. 6 Fuel evaporative system — 5.0L engine

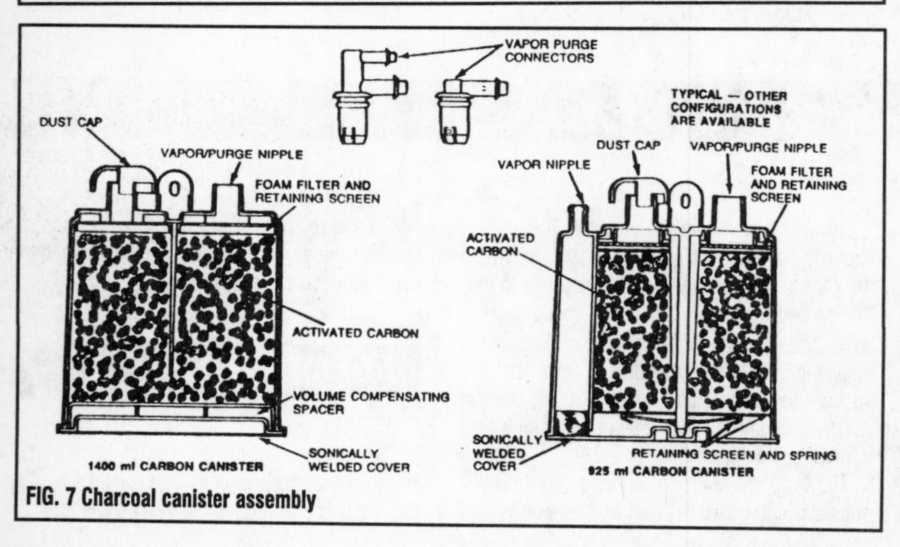

FIG. 7 Charcoal canister assembly

from the canister into the engine for consumption. The carbon canister contains activated carbon, which absorbs the fuel vapor. The canister is located in the engine compartment or along the frame rail.

REMOVAL & INSTALLATION

1. Disconnect the vapor hoses from the carbon canister.
2. Remove the carbon canister attaching screws and remove the carbon canister.
3. Installation is the reverse of the removal procedure.

Fuel Tank Vapor Orifice And Rollover Valve Assembly

➧ SEE FIG. 8

Fuel vapor in the fuel tank is vented to the carbon canister through the vapor valve assembly. The valve is mounted in a rubber grommet at a central location in the upper surface of the fuel tank. A vapor space between the fuel level and the tank upper surface is combined with a small orifice and float shut-off valve in the vapor valve assembly to prevent liquid fuel from passing to the carbon canister. The vapor space also allows for thermal expansion of the fuel.

REMOVAL & INSTALLATION

1. Disconnect the negative battery cable.
2. Relieve the fuel system pressure as follows:
   a. Remove the fuel filler cap.
   b. Attach fuel pressure gauge T80L–9974–B or equivalent, to the fuel diagnostic valve on the fuel supply manifold.
   c. Relieve the fuel pressure into a suitable container.
   d. Remove the fuel pressure gauge.
3. Drain the fuel from the fuel tank as completely as possible by siphoning or pumping the fuel out through the fuel filler pipe.

➡ There are reservoirs inside the fuel tank to maintain fuel near the fuel pickup during cornering and under low fuel operating conditions. These reservoirs could block siphon tubes or hoses from

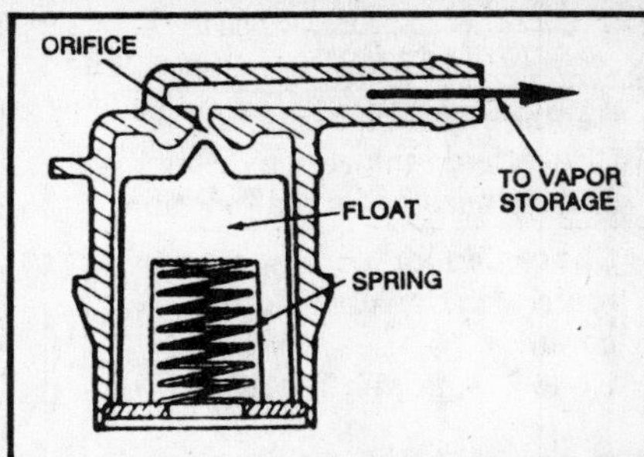

FIG. 8 Fuel tank vapor orifice and rollover valve assembly

reaching the bottom of the fuel tank. A little perseverance combined with different hose orientations will overcome this situation.

4. Raise and safely support the vehicle.
5. Disconnect the fuel hoses and tubes.
6. On vehicles equipped with a metal retainer that fastens the filler pipe to the fuel tank, remove the screw attaching the retainer to the fuel tank flange.
7. Position a suitable jack to support the fuel tank and remove the bolts from the fuel tank straps.
8. Lower the tank just enough to gain access to the vapor orifice. If necessary, disconnect the electrical connectors to the fuel pump and sending unit.
9. Remove the vapor orifice valve.
10. Installation is the reverse of the removal procedure.

Purge Control Valve

➧ SEE FIG. 9

The purge control valve is in-line with the carbon canister and controls the flow of fuel vapors out of the canister.

When the engine is stopped, vapors from the fuel tank flow into the canister. On controlled systems, the vacuum signal is strong enough during normal cruise to actuate the purge control valve and vapors are drawn from the canister to the engine vacuum connection. At the same time, vapors from the fuel tank are routed directly into the engine. On some systems where purging does not affect idle quality, the purge control valve is connected to engine manifold vacuum and opens any time there is enough manifold vacuum.

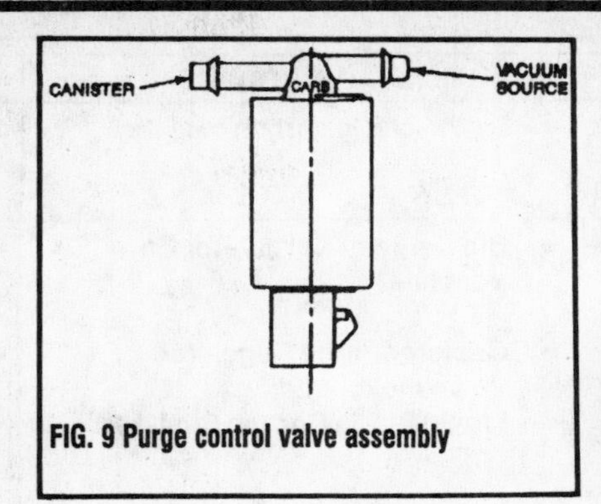

FIG. 9 Purge control valve assembly

REMOVAL & INSTALLATION

1. Disconnect the vapor hoses from the purge control valve.
2. Remove the purge control valve from the vehicle.
3. Installation is the reverse of the removal procedure.

Purge Solenoid Valve

The purge solenoid valve is in-line with the carbon canister and controls the flow of fuel vapors out of the canister. It is normally closed. When the engine is shut off, the vapors from the fuel tank flow into the canister. After the engine is started, the solenoid is engaged and opens, purging the vapors into the engine. With the valve open, vapors from the fuel tank are routed directly into the engine.

REMOVAL & INSTALLATION

1. Disconnect the vapor hoses from the purge solenoid valve.
2. Disconnect the electrical connector from the purge solenoid valve.
3. Remove the purge solenoid valve from the vehicle.
4. Installation is the reverse of the removal procedure.

Pressure/Vacuum Relief Fuel Cap

The fuel cap contains an integral pressure and vacuum relief valve. The vacuum valve acts to allow air into the fuel tank to replace the fuel as it is used, while preventing vapors from escaping the tank through the atmosphere. The vacuum relief valve opens after a vacuum of –0.5 psi. The pressure valve acts as a backup pressure relief valve in the event the normal venting system is overcome by excessive generation of internal pressure or restriction of the normal venting system. The pressure relief range is 1.6–2.1 psi. Fill cap damage or contamination that stops the pressure vacuum valve from working may result in deformation of the fuel tank.

REMOVAL & INSTALLATION

1. Unscrew the fuel filler cap. The cap has a pre-vent feature that allows the tank to vent for the first 3/4 turn before unthreading.
2. Remove the screw retaining the fuel cap tether and remove the fuel cap.
3. Installation is the reverse of the removal procedure. When installing the cap, continue to turn clockwise until the ratchet mechanism gives off 3 or more loud clicks.

Evaporative Emission Symptom Diagnosis Chart

CONDITION	POSSIBLE SOURCE	ACTION
• Surge at Steady Speed	• Liquid fuel in carbon canister.	• Replace carbon canister. Check fuel tank vent system and carburetor for malfunction.
• Gas Smell	• Blockage of Carburetor Bowl Vent line. • Canister Purge Regulator Valve malfunction. • Liquid fuel in Carbon Canister. • Fuel Tank Vent System blocked. • Hole or cut in Carburetor Bowl Vent Line or Fuel Tank Vent Line.	• Check line for blockage and route with downhill slope to canister. • Go to Diagnostic Test EE1. • Replace canister. Check fuel tank vent system and carburetor for malfunction. • Check fuel tank vent system. • Visually inspect and replace damaged line.

NOTE: **System component fault in order of occurrence.**

DIAGNOSTIC TEST NUMBER	SOURCE COMPONENT	DIAGNOSTIC ACTION
EE1	Canister Purge Regulator Valve	•With the Canister Purge Regulator Valve de-energized, apply 17 kPa (5 in-Hg) to "vacuum source" port. Valve should not pass air; if it does, replace valve. While applying 9 to 14 volts DC to valve, the valve will open and pass air. If it does not, replace valve (Figure 5). VACUUM SOURCE / CARB / CANISTER *Figure 5 Canister Purge Regulator Valve*

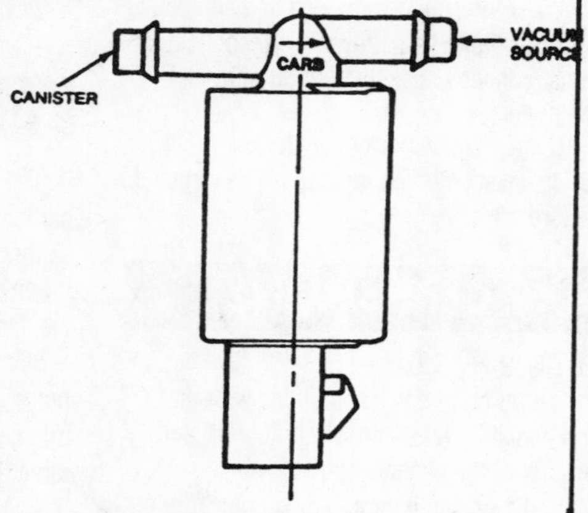

FIG. 10 Evaporative Emission Symptom Diagnosis Chart

EXHAUST EMISSION CONTROL SYSTEM

General Information

The exhaust emission control system begins at the air intake and ends at the tailpipe. Ford rear wheel drive vehicles are equipped with the following systems or components to manage exhaust emission control: thermostatic air inlet system, thermactor air injection system, exhaust gas recirculation system and exhaust catalyst. All vehicles do not share all systems or all components.

Thermostatic Air Inlet System

GENERAL INFORMATION

➧ SEE FIG. 11

The thermostatic air inlet system is used on the 2.3L engine. The thermostatic air inlet system regulates the air inlet temperature by drawing air in from a cool air source as well as heated air from a heat shroud which is mounted on the exhaust manifold. The system consists of the following components: duct and valve assembly, heat shroud, bimetal sensor and the necessary vacuum lines and air ducts.

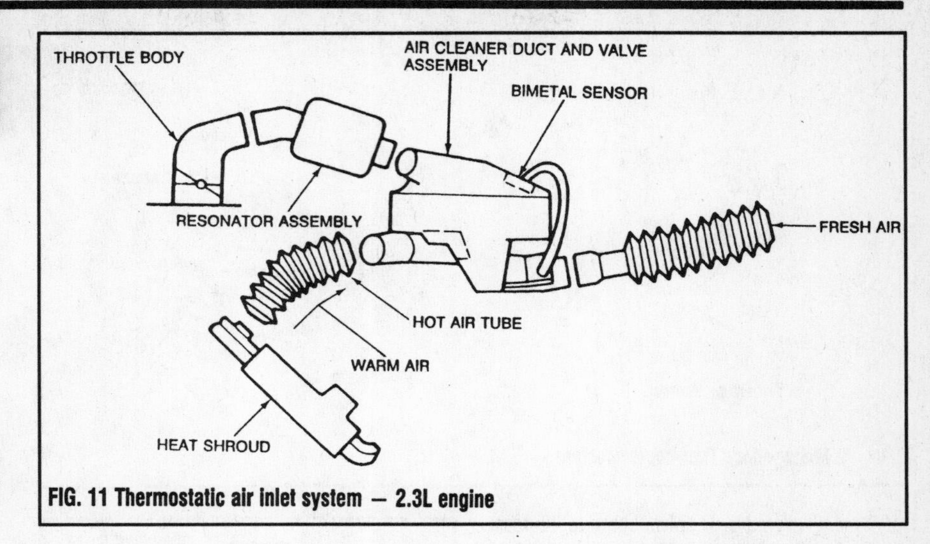

FIG. 11 Thermostatic air inlet system — 2.3L engine

Duct and Valve Assembly

The duct and valve assembly which regulates the air flow from the cool and heated air sources is attached to the air cleaner. The flow is regulated by means of a door that is operated by a vacuum motor. The operation of the motor is controlled by the bimetal sensor.

TESTING

1. If the duct door is in the closed to fresh air position, remove the hose from the air cleaner vacuum motor.
2. The door should go to the open to fresh air position. If it sticks or binds, service or replace, as required.
3. If the door is in the open to fresh air position, check the door by applying 8 in. Hg or greater of vacuum to the vacuum motor.
4. The door should move freely to the closed to fresh air position. If it binds or sticks, service or replace, as required.

➡ **Make sure the vacuum motor is functional before changing the duct and valve assembly.**

REMOVAL & INSTALLATION

1. Disconnect the vacuum hose from the vacuum motor.
2. Separate the vacuum motor from the vacuum operated door and remove the vacuum motor.
3. Installation is the reverse of the removal procedure.

Bimetal Sensor

The core of the bimetal sensor is made of 2 different types of metals bonded together, each having different temperature expansion rates. At a given increase in temperature, the shape of the sensor core changes, bleeding off vacuum available at the vacuum motor. This permits the vacuum motor to open the duct door to allow fresh air in while shutting off full heat. The bimetal sensor is calibrated according to the needs of each particular application.

TESTING

1. Bring the temperature of the bimetal sensor below 75°F (24°C) and apply 16 in. Hg of vacuum with a vacuum pump at the vacuum source port of the sensor.
2. The duct door should stay closed. If not, replace the bimetal sensor.
3. The sensor will bleed off vacuum to allow the duct door to open and let in fresh air at or above the following temperatures:
   a. Brown — 75°F (24°C)
   b. Pink, black or red — 90°F (32.2°C)
   c. Blue, yellow or green — 105°F (40.6°C)

➡ **Do not cool the bimetal sensor while the engine is running.**

REMOVAL & INSTALLATION

1. Remove the air cleaner housing lid to gain access to the sensor.
2. Disconnect the vacuum hoses from the sensor. It may be necessary to move the air cleaner housing to accomplish this.
3. Remove the sensor from the air cleaner housing.
4. Installation is the reverse of the removal procedure.

Thermactor Air Injection System

GENERAL INFORMATION

➧ SEE FIG. 12

The thermactor air injection system reduces the hydrocarbon and carbon monoxide content of the exhaust gases by continuing the combustion of unburned gases after they leave the combustion chamber. This is done by injecting fresh air into the hot exhaust stream leaving the exhaust ports or into the catalyst. At this point, the fresh air mixes with hot exhaust gases to promote further oxidation of both the hydrocarbons and carbon monoxide, thereby reducing their concentration and converting some of them into harmless carbon dioxide and water.

All Ford rear wheel drive vehicles equipped with the 5.0L engine are equipped with a managed air thermactor system. This system is utilized in electronic control systems to divert thermactor air either upstream to the exhaust

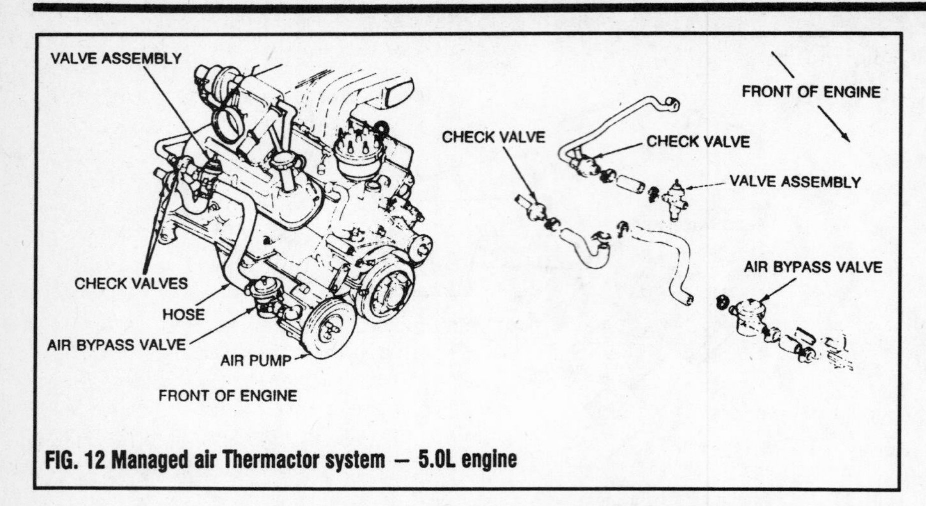

FIG. 12 Managed air Thermactor system — 5.0L engine

manifold check valve or downstream to the rear section check valve and dual bed catalyst. The system will also dump thermactor air to atmosphere during some operating modes.

The Thermactor air injection system consists of the air supply pump, air bypass valve, check valves, air supply control valve, combination air bypass/air control valve, solenoid vacuum valve, thermactor idle vacuum valve and air pump resonator. Components will vary according to application.

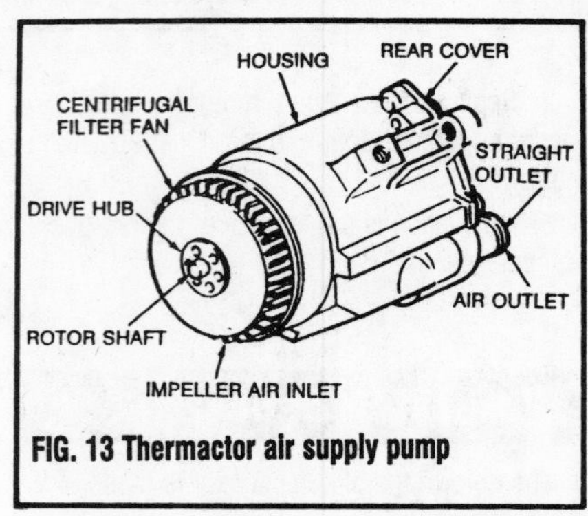

FIG. 13 Thermactor air supply pump

Air Supply Pump

➧ SEE FIG. 13

The air supply pump is a belt-driven, positive displacement, vane-type pump that provides air for the thermactor system. It is available in 19 cu. in. (311cc) and 22 cu. in. (360cc) sizes, either of which may be driven with different pulley ratios for different applications. Pumps receive air from a remote silencer filter on the rear side of the engine air cleaner attached to the pumps' air inlet nipple or through an impeller-type centrifugal filter fan.

TESTING

1. Check belt tension and adjust if needed.

➡ Do not pry on the pump to adjust the belt. The aluminum housing is likely to collapse.

2. Disconnect the air supply hose from the bypass control valve.
3. The pump is operating properly if airflow is felt at the pump outlet and the flow increases as engine speed increases.
4. If the pump is not operating as described in Step 3 and the system is equipped with a silencer/filter, check the silencer/filter for possible obstruction before replacing the pump.

REMOVAL & INSTALLATION

1. Disconnect the negative battery cable.
2. Remove the drive belt from the air pump pulley.
3. Disconnect the air hose(s) from the air pump.
4. Remove the mounting bolts and if necessary the mounting brackets.
5. Remove the air pump from the vehicle.
6. Installation is the reverse of the removal procedure.

Air Bypass Valve

➧ SEE FIG. 14

The air bypass valve supplies air to the exhaust system with medium and high applied vacuum signals when the engine is at normal operating temperature. With low or no vacuum applied, the pumped air is dumped through the silencer ports of the valve or through the dump port.

TESTING

1. Turn the ignition key **OFF**.
2. Remove the control vacuum line from the bypass valve.
3. Start the engine and bring to normal operating temperature.
4. Check for vacuum at the vacuum line. If there is no vacuum, check the solenoid vacuum valve assembly. If vacuum is present, inspect the air bypass valve.
5. Turn the engine **OFF** and disconnect the air hose at the bypass valve outlet.
6. Inspect the outlet for damage from the hot exhaust gas.
7. If the valve is damaged, replace it. If the valve is not damaged, check the bypass valve diaphragm.
8. Connect a vacuum pump to the bypass valve and apply 10 in. Hg of vacuum.
9. If the valve holds vacuum, leave the vacuum applied and proceed to Step 10. If the valve does not hold vacuum, it must be replaced.
10. Start the engine and increase the engine speed to 1500 rpm.
11. Check for air flow at the valve outlet, either audibly or by feel. If there is air flow, proceed to Step 12. If there is no air flow, replace the air bypass valve.
12. Release the vacuum applied by the vacuum pump and check that the air flow switches from the valve outlet to the dump port or silencer ports, either audibly or by feel.
13. If the air flow does not switch, replace the air by pass valve. If the air flow switches, the air bypass valve is okay, check the air supply control valve.

REMOVAL & INSTALLATION

1. Disconnect the negative battery cable.
2. Disconnect the air inlet and outlet hoses and the vacuum hose from the bypass valve.
3. Remove the bypass valve from the vehicle.
4. Installation is the reverse of the removal procedure.

Check Valve

➧ SEE FIG. 15

The air check valve is a 1-way valve that allows thermactor air to pass into the exhaust system while preventing exhaust gases from passing in the opposite direction.

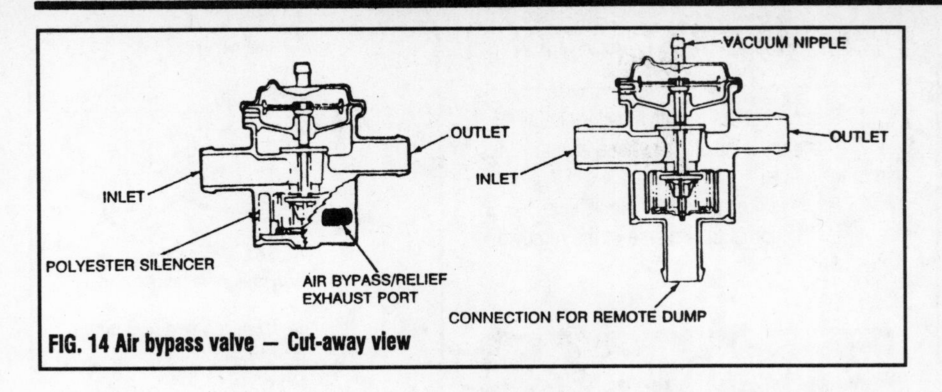

FIG. 14 Air bypass valve — Cut-away view

TESTING

1. Turn the ignition key **OFF**.
2. Visually inspect the thermactor system hoses, tubes, control valve(s) and check valve(s) for leaks or external signs of damage, from the back flow of hot exhaust gases.
3. If the hoses and valves are okay, proceed to Step 4. If they are not, service or replace the damaged parts, including the check valve.
4. Remove the hose from the check valve inlet and visually check the inside of the hose for damage from hot exhaust gas.
5. If the hose is clean and undamaged, proceed to Step 6. If not, replace the hose and check valve.
6. Start the engine and listen for escaping exhaust gas from the check valve. Feel for the gas only if the engine temperature is at an acceptable level.
7. If any exhaust gas is escaping, replace the check valve.

REMOVAL & INSTALLATION

1. Disconnect the negative battery cable.
2. Disconnect the input hose from the check valve.
3. Remove the check valve from the connecting tube.
4. Installation is the reverse of the removal procedure.

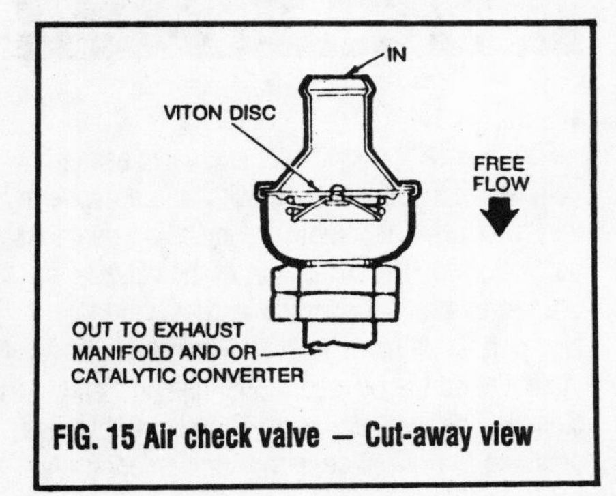

FIG. 15 Air check valve — Cut-away view

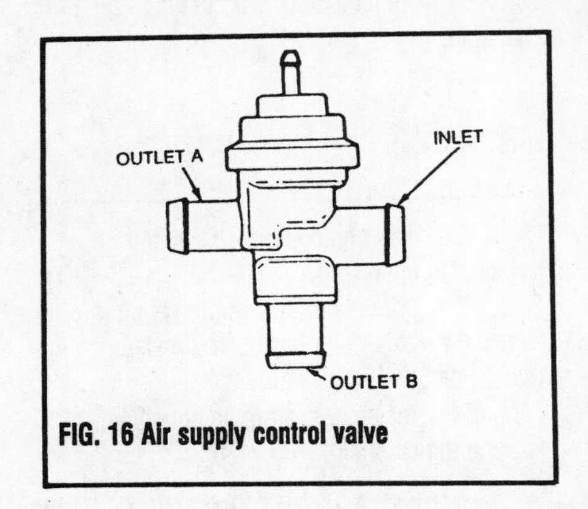

FIG. 16 Air supply control valve

Air Supply Control Valve

➧ SEE FIG. 16

The air supply control valve directs air pump output to the exhaust manifold or downstream to the catalyst system depending upon the engine control strategy. It may also be used to dump air to the air cleaner or dump silencer.

TESTING

1. Turn the ignition key **OFF**.
2. Remove the hoses from the air control valve outlets and inspect the outlets for damage from hot exhaust gases.
3. If the air supply control valve is damaged, it must be replaced, then check the air check valve. If the air supply control valve is not damaged, proceed to Step 4.
4. Remove the vacuum line from the air supply control valve. Start the engine and bring to normal operating temperature, then shut the engine off.
5. Restart the engine and immediately check for vacuum at the hose. If vacuum was present at the start, proceed to Step 6. If vacuum was not present at the start, check the solenoid vacuum valve.
6. Start the engine and let it run. Check for the vacuum to change from high to low.
7. If the vacuum dropped to 0 within a few minutes after the engine started, proceed to Step 8. If not, check the solenoid vacuum valve.
8. Connect a vacuum pump to the air supply control valve and apply 10 in. Hg of vacuum.
9. If the valve holds vacuum, proceed to Step 10. If it does not hold vacuum, replace the air supply control valve.
10. Start the engine and bring to normal operating temperature. Make sure that air is being supplied to the air supply control valve.
11. If air is present, proceed to Step 12. If air is not present, check air pump operation.
12. Leave the engine running and apply 10 in. Hg of vacuum to the air supply control valve. Increase engine speed to 1500 rpm.
13. If air flow comes out of outlet A, proceed to Step 14. If not, replace the air supply control valve.
14. Leave the engine running. Vent the vacuum pump until there is 0 vacuum.
15. If the air flow switches from outlet A to outlet B, the air supply control valve is okay. If the air flow does not switch, replace the air supply control valve.

REMOVAL & INSTALLATION

1. Disconnect the negative battery cable.
2. Disconnect the air hoses and the vacuum line from the air control valve.
3. Remove the air control valve from the vehicle.
4. Installation is the reverse of the removal procedure.

Combination Air Bypass/Air Control Valve

➧ SEE FIG. 17

The combination air control/bypass valve combines the secondary air bypass and air control functions. The valve is located in the air supply line between the air pump and the upstream/downstream air supply check valves.

The air bypass portion controls the flow of thermactor air to the exhaust system or allows thermactor air to be bypassed to atmosphere. When air is not being bypassed, the air control portion of the valve switches the air injection point to either an upstream or downstream location.

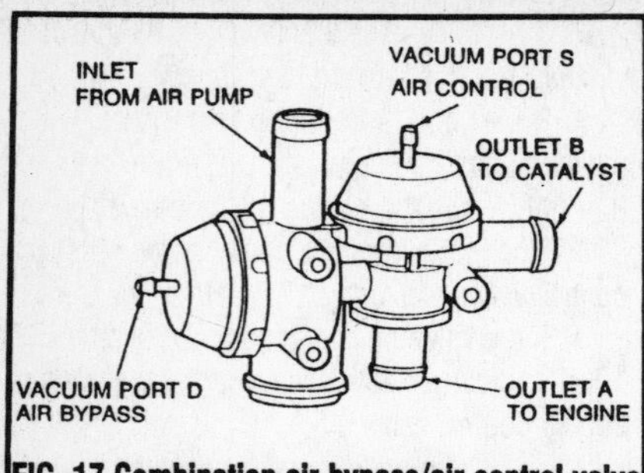

FIG. 17 Combination air bypass/air control valve

TESTING

1. Turn the ignition key **OFF**.
2. Remove the hoses from the combination air control valve outlets A and B and inspect the outlets for damage from hot exhaust gases.
3. If the valve appears damaged, replace it, then check the air check valve. If the valve is not damaged, proceed to Step 4.
4. Leave the hoses disconnected from the valve. Disconnect and plug the vacuum line to port D.
5. Start the engine and run at 1500 rpm. If air flow is present at the valve, proceed to Step 6. If it is not, check the air pump. If the air pump is okay, replace the combination air control valve.
6. Leave the engine running. Disconnect both vacuum lines to ports D and S.
7. Measure the manifold vacuum at both ports. If the proper vacuum is present, proceed to Step 8. If not, check the solenoid vacuum valve.
8. Turn the ignition key **OFF**. Reconnect the vacuum line to port D, but leave the vacuum line to port S disconnected and plugged.
9. Start the engine and run it at 1500 rpm. If air flow is present at outlet B, but not at outlet A, proceed to Step 10. If not, replace the combination air control valve and reconnect all hoses.
10. Turn the ignition key **OFF** and leave the vacuum line to port S disconnected and unplugged.
11. Apply 8–10 in. Hg of vacuum to port S on the combination valve. Start the engine and run at 1500 rpm. If air flow is present at outlet A, the combination valve is okay. If not, replace the combination air control valve.

➡ **If the combination valve is a bleed type, this will affect the amount of air flow.**

REMOVAL & INSTALLATION

1. Disconnect the negative battery cable.
2. Disconnect the air hoses and vacuum lines from the valve.
3. Remove the valve from the vehicle.
4. Installation is the reverse of the removal procedure.

Solenoid Vacuum Valve Assembly

➧ SEE FIG. 18

The normally closed solenoid valve assembly consists of 2 vacuum ports with an atmospheric vent. The valve assembly can be with or without control bleed. The outlet port of the valve is opened to atmospheric vent and closed to the inlet port when de-energized. When energized, the outlet port is opened to the inlet port and closed to atmospheric vent. The control bleed is provided to prevent contamination entering the solenoid valve assembly from the intake manifold.

TESTING

1. The ports should flow air when the solenoid is energized.
2. Check the resistance at the solenoid terminals with an ohmmeter. The resistance should be 51–108 ohms.
3. If the resistance is not as specified, replace the solenoid.

➡ **The valve can be expected to have a very small leakage rate when energized or de-energized. This leakage is not measurable in the field and is not detrimental to valve function.**

REMOVAL & INSTALLATION

1. Disconnect the negative battery cable.
2. Disconnect the electrical connector and the vacuum lines from the solenoid valve.
3. Remove the mounting bolts and remove the solenoid valve.
4. Installation is the reverse of the removal procedure.

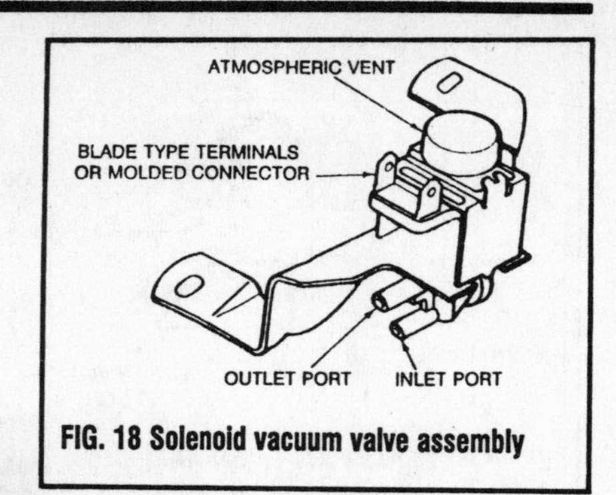

FIG. 18 Solenoid vacuum valve assembly

Air Pump Resonator

➧ SEE FIG. 19

The air pump resonator reduces air dump noise during cold start and some cruise modes.

TESTING

1. Visually inspect the resonator for holes.
2. Remove the hoses and check for blocked or restricted ports.
3. Replace the resonator if it has holes or the ports are blocked or restricted.
4. Reconnect the hoses and install and tighten the clamps.

REMOVAL & INSTALLATION

1. Disconnect the negative battery cable.
2. Loosen the clamps and remove the hoses.
3. Remove the mounting bolts and remove the resonator.
4. Installation is the reverse of the removal procedure.

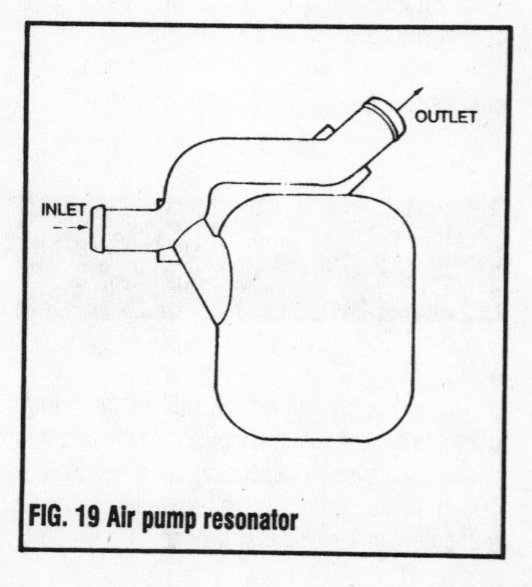

FIG. 19 Air pump resonator

CONDITION	POSSIBLE SOURCE	ACTION
• Backfire (Exhaust)	• Air bypass valve malfunction.	• Perform bypass valve diagnosis.
	• Air control valve malfunction.	• Perform air control valve diagnosis.
	• Combination air bypass / control valve malfunction.	• Perform combination valve diagnosis.
	• Thermactor solenoid valve malfunction.	• Perform solenoid diagnosis.
	• Exhaust manifolds or pipes loose.	• Inspect and tighten nuts or bolts to specification.
• Surge at Steady Speed	• Air control valve malfunction.	• Perform control valve diagnosis.
	• Combination air bypass / control valve malfunction.	• Perform combination valve diagnosis.
	• Thermactor solenoid malfunction.	• Perform solenoid diagnosis.
• Engine Noise - (Hiss)	• Thermactor hose leaks or disconnects.	• Visual inspection of hoses and connections.
• Engine Noise - (Rap or Roar)	• Thermactor hose or valves leak exhaust.	• Visual inspection of hoses and valves. Perform air check valve diagnosis.
• Poor Fuel Economy	• Air control valve malfunction.	• Perform air control valve diagnosis.
	• Combination air bypass / control valve malfunction.	• Perform combination valve diagnosis.
	• Thermactor solenoid valve malfunction.	• Perform solenoid diagnosis.
	• Disconnected vacuum or electrical connections for thermactor components.	• Visual inspection.

FIG. 20 Exhaust Emission Control System Diagnosis Chart

Thermactor System Noise Test

CAUTION

Do not use a pry bar to move the air pump for belt adjustment.

NOTE: **The Thermactor system is not completely noiseless. Under normal conditions, noise rises in pitch as engine speed increases. To determine if noise is the fault of the air injection system, disconnect the belt drive (only after verifying that belt tension is correct), and operate the engine. If the noise disappears, proceed with the following diagnosis.**

Diagnosis

CONDITION	POSSIBLE SOURCE	ACTION
• Excessive Belt Noise	• Loose belt.	• Tighten to specification using Tool T75L-9480-A or equivalent to hold belt tension and Belt Tension Gauge T63L-8620-A or equivalent. **CAUTION: Do not use a pry bar to move air pump.**
	• Seized pump.	• Replace pump.
	• Loose pulley.	• Replace pulley and / or pump if damaged. Tighten bolts to 13.6-17.0 N·m (120-150 lb-in).
	• Loose or broken mounting brackets or bolts.	• Replace parts as required and tighten bolts to specification.
• Excessive Mechanical Noise, Chirps, Squeaks, Clicks or Ticks	• Overtightened mounting bolt.	• Tighten to 34 N·m (25 lb-ft).
	• Overtightened drive belt.	• Same as loose belt.
	• Excessive flash on the air pump adjusting arm boss.	• Remove flash from the boss.
	• Distorted adjusting arm.	• Replace adjusting arm.
	• Pump or pulley mounting fasteners loose.	• Tighten fasteners to specifications.

FIG. 21 Exhaust Emission Control System Diagnosis Chart

CONDITION	POSSIBLE SOURCE	ACTION
• Excessive Thermactor System Noise (Putt-Putt, Whirring or Hissing)	• Leak in hose.	• Locate source of leak using soap solution and replace hoses as necessary.
	• Loose, pinched or kinked hose.	• Reassemble, straighten or replace hose and clamps as required.
	• Hose touching other engine parts.	• Adjust hose to prevent contact with other engine parts.
	• Bypass valve inoperative.	• Test the valve.
	• Check valve inoperative.	• Test the valve.
	• Restricted or bent pump outlet fitting.	• Inspect fitting and remove any flash blocking the air passage way. Replace bent fittings.

CONDITION	POSSIBLE SOURCE	ACTION
	• Air dumping through bypass valve (at idle only).	• On many vehicles, the Thermactor system has been designed to dump air at idle to prevent overheating the catalyst. This condition is normal. Determine that the noise persists at higher speeds before proceeding.
	• Air dumping through bypass valve (decel and cruise).	• On many vehicles, the thermactor air is dumped in the air cleaner or in remote silencer. Make sure hoses are connected and not cracked.
	• Air pump resonator leaking or blocked.	• Check resonator for hole or restricted inlet / outlet tubes.
• Excessive Pump Noise - (Chirps, Squeaks and Ticks)	• Worn or damaged pump.	• Check the Thermactor system for wear or damage and make necessary corrections.
• Engine Noise - (Rap or Roar)	• Hose disconnected.	• Audible and visual inspection to assure all hoses are connected.
• State Emissions Test Failure	• Restricted hose.	• Inspect hoses for crimped and / or kinked hoses.
	• Plugged pulse air silencer.	• Remove inlet hose and inspect silencer inlet for dirt and foreign material. Clean or replace silencer as appropriate.
	• Pulse air valve malfunction, leaking or restricted.	• Perform pulse air check valve diagnosis.
	• Pulse air control valve malfunction.	• Perform pulse air control valve diagnosis.

FIG. 22 Exhaust Emission Control System Diagnosis Chart

EXHAUST GAS RECIRCULATION SYSTEM

General Information

➧ SEE FIG. 23

The Exhaust Gas Recirculation (EGR) system is designed to reintroduce exhaust gas into the combustion cycle, thereby lowering combustion temperatures and reducing the formation of nitrous oxide. There are 2 different EGR systems used on Ford rear wheel drive vehicles.

All 2.3L and 5.0L engines use the Electronic EGR valve (EEGR) system. In the EEGR system, EGR flow is controlled according to computer demands by means of an EGR Valve Position (EVP) sensor attached to the valve. The valve is operated by a vacuum signal from the electronic vacuum regulator which actuates the valve diaphragm. As supply vacuum overcomes the spring load, the diaphragm is actuated. This lifts the pintle off of it's seat allowing exhaust gas to recirculate. The amount of flow is proportional to the pintle position. The EVP sensor mounted on the valve sends an electrical signal of it's position to the ECU.

Electronic EGR Valve

➧ SEE FIGS. 24–25

The electronic EGR valve is vacuum operated, lifting the pintle off of it's seat to allow exhaust gas to recirculate when the vacuum signal is strong enough. The EVP sensor which is mounted on top of the electronic EGR valve. The electronic EGR valve assembly is not serviceable. The EVP sensor and the EGR valve must be serviced separately.

REMOVAL & INSTALLATION

1. Disconnect the negative battery cable.
2. Disconnect the vacuum line from the EGR valve and the connector from the EVP sensor.

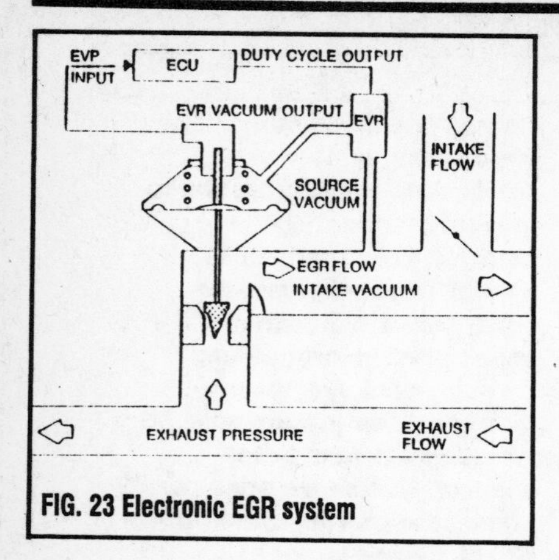

FIG. 23 Electronic EGR system

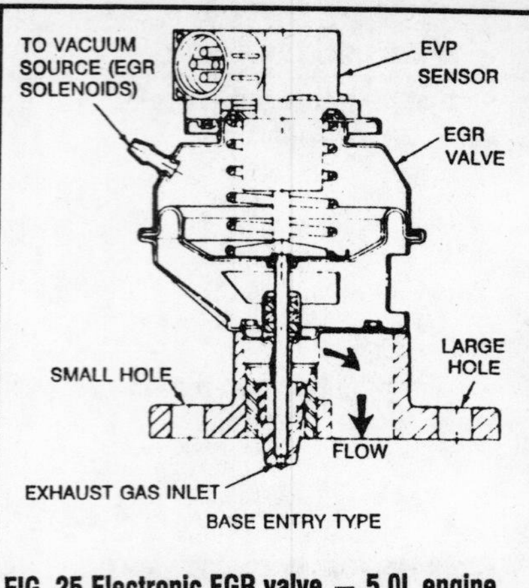

FIG. 25 Electronic EGR valve — 5.0L engine

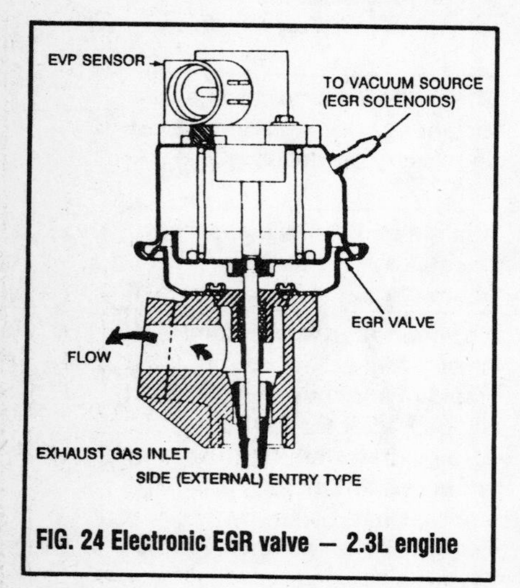

FIG. 24 Electronic EGR valve — 2.3L engine

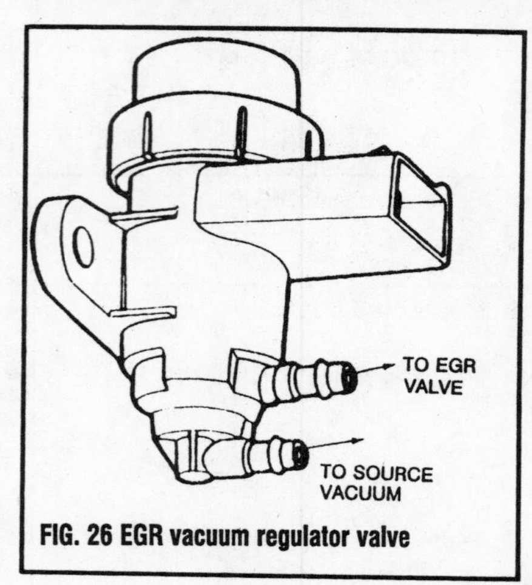

FIG. 26 EGR vacuum regulator valve

3. Remove the mounting bolts and remove the EGR valve.
4. Remove the EVP sensor from the EGR valve.
5. Installation is the reverse of the removal procedure. Be sure to remove all old gasket material before installation. Use a new gasket during installation.

EGR Vacuum Regulator (EVR)

➧ SEE FIG. 26

The EVR is an electromagnetic device which controls vacuum output to the EGR valve. The EVR replaces the EGR solenoid vacuum vent valve assembly. An electric current in the coil induces a magnetic field in the armature which pulls on a disk, closing the vent to atmosphere. The ECU outputs a duty cycle to the EVR which regulates the vacuum level to the EGR valve. As the duty cycle is increased, an increased vacuum signal goes to the EGR valve. The vacuum source is manifold vacuum.

On the 2.3L engine, a current control thermistor device is also used to compensate for extreme temperature operation. The EVR and thermistor are serviced as an assembly.

REMOVAL & INSTALLATION

1. Disconnect the negative battery cable.
2. Disconnect the electrical connector and the vacuum lines from the regulator.
3. Remove the regulator mounting bolts and remove the regulator.
4. Installation is the reverse of the removal procedure.

Exhaust Gas Recirculation System Diagnosis Charts

CONDITION	POSSIBLE SOURCE	ACTION
• Rough Idle Cold	• EGR valve malfunction.	• Run EEC-IV Quick Test,
	• EGR flange gasket leaking.	• Replace flange gasket and tighten valve attaching nuts or bolts to specification.
	• EGR valve attaching nuts or bolts loose or missing.	• Replace flange gasket and tighten valve attaching nuts or bolts to specification.
	• Vacuum leak at EVP sensor.	• Replace O-ring seal and tighten EVP sensor attaching nuts to specification.
	• EVR solenoid malfunction.	• Run EEC-IV Quick Test
	• EGR valve contamination.	• Clean EGR valve.

CONDITION	POSSIBLE SOURCE	ACTION
• Rough Idle Hot	• EGR valve malfunction. • EGR flange gasket leaking. • EGR valve attaching nuts or bolts loose or missing. • Vacuum leak at EVP sensor. • EVR solenoid malfunction. • EGR valve contamination.	• Run EEC-IV Quick Test • Replace flange gasket and tighten valve attaching nuts or bolts to specification. • Replace flange gasket and tighten valve attaching nuts or bolts to specification. • Replace O-ring seal and tighten EVP sensor attaching nuts to specification. • Run EEC-IV Quick Test • Clean EGR valve.
• Rough Running, Surge, Hesitation, Poor Part Throttle Performance—Hot	• EGR valve malfunction / erratic operation. • EGR valve contamination. • EVR solenoid malfunction. • Pressure / Vacuum signal hose(s) leak (PFE / DPFE).	• Perform EGR valve diagnosis. • Clean EGR valve and if necessary, replace EGR valve. • Run EEC-IV Quick Test • Replace hose(s).
• Engine Stalls On Deceleration—Hot	• EGR valve malfunction. • EVR solenoid malfunction. • EGR valve contamination.	• Perform EGR valve diagnosis. • Run EEC-IV Quick Test • Clean EGR valve and if necessary, replace EGR valve.
• Engine Spark Knock or Ping	• EGR valve malfunction. • EGR valve attaching nuts or bolts loose or missing. • Blocked or restricted passages in valve or spacer (Electronic EGR).	• Perform EGR valve diagnosis. • Replace flange gasket and tighten valve attaching nuts or bolts to specification. • Clean passages in EGR spacer and EGR valve.

FIG. 27 Exhaust Gas Recirculation System Diagnosis Chart

CONDITION	POSSIBLE SOURCE	ACTION
• Engine Stalls At Idle—Cold	• EGR valve malfunction. • EGR flange gasket leaking. • EGR valve attaching nuts or bolts loose or missing. • EVR solenoid malfunction. • EGR valve contamination.	• Perform EGR valve diagnosis. • Replace flange gasket and tighten valve attaching nuts or bolts to specification. • Replace flange gasket and tighten valve attaching nuts or bolts to specification. • Run EEC-IV Quick Test • Clean EGR valve.
• Engine Stalls At Idle—Hot	• EGR valve malfunction. • EGR flange gasket leaking. • EGR valve attaching nuts or bolts loose or missing.	• Perform EGR valve diagnosis. • Replace flange gasket and tighten valve attaching nuts or bolts to specification. • Replace flange gasket and tighten valve attaching nuts or bolts to specification.

CONDITION	POSSIBLE SOURCE	ACTION
• Engine Stalls At Idle—Cold	• EGR valve contamination.	• Clean EGR valve and if necessary, replace EGR valve.
	• Vacuum leak at EVP sensor.	• Replace O-ring seal and tighten EVP sensor attaching nuts to specification.
	• EVR solenoid malfunction.	• Run EEC-IV Quick Test
• Engine Starts But Will Not Run—Engine Hard To Start Or Will Not Start	• EGR valve malfunction.	• Perform EGR valve diagnosis.
	• EGR flange gasket leaking.	• Replace flange gasket and tighten valve attaching nuts or bolts to specification.
	• EGR valve attaching nuts or bolts loose or missing.	• Replace flange gasket and tighten valve attaching nuts or bolts to specification.
	• EVR solenoid malfunction.	• Run EEC-IV Quick Test
	• EGR valve contamination.	• Clean EGR valve.

FIG. 28 Exhaust Gas Recirculation System Diagnosis Chart

TEST STEP		RESULT	▶ ACTION TO TAKE
EEGR1	CHECK SYSTEM INTEGRITY		
	• Check vacuum hoses and connections for looseness, pinching, leakage, splitting, blockage, and proper routing. • Inspect EGR valve for loose attaching bolts or damaged flange gasket. • **Does system appear to be in good condition and vacuum hoses properly routed?**	Yes ▶ No ▶	GO to EEGR2. SERVICE EGR system as required. RE-EVALUATE symptom.
EEGR2	CHECK EGR VACUUM AT IDLE		
	• Run engine until normal operating temperature is reached. • With engine running at idle, disconnect EGR vacuum supply at the EGR valve and check for a vacuum signal. NOTE: **The EVR solenoid has a constant internal leak. You may notice a small vacuum signal. This signal should be less than 3.4 kPa (1.0 in-Hg) at idle.** • **Is EGR vacuum signal less than 3.4 kPa (1.0 in-Hg) at idle?**	Yes ▶ No ▶	GO to EEGR3. RECONNECT EGR vacuum hose. INSPECT EVR solenoid for leakage

FIG. 29 Exhaust Gas Recirculation System Diagnosis Chart — Electronic EGR Valve

CATALYTIC CONVERTERS

General Information

➧ SEE FIG. 30

Engine exhaust consists mainly of Nitrogen (N_2), however, it also contains Carbon Monoxide (CO), Carbon Dioxide (CO_2), Water Vapor (H_2O), Oxygen (O_2), Nitrogen Oxides (NOx) and Hydrogen, as well as various, unburned Hydrocarbons (HC). Three of these exhaust components, CO, NOx and HC, are major air pollutants, so their emission to the atmosphere has to be controlled.

The catalytic converter, mounted in the engine exhaust stream, plays a major role in the emission control system. The converter works as a gas reactor and it's catalytic function is to speed up the heat producing chemical reaction between the exhaust gas components in order to reduce the air pollutants in the engine exhaust. The catalyst material, contained inside the converter, is made of a ceramic substrate that is coated with a high surface area alumina and impregnated with catalytically active, precious metals.

Catalytic Converter

All Ford Mustang's use a 3-way catalyst and some also use this in conjunction with a conventional oxidation catalyst. The conventional oxidation catalyst, containing Platinum (Pt) and Palladium (Pd), is effective for catalyzing the oxidation reactions of HC and CO. The 3-way catalyst, containing Platinum (Pt) and Rhodium (RH) or Palladium (Pd) and Rhodium (RH), is not only effective for catalyzing the oxidation reactions of HC and CO, but it also catalyzes the reduction of NOx.

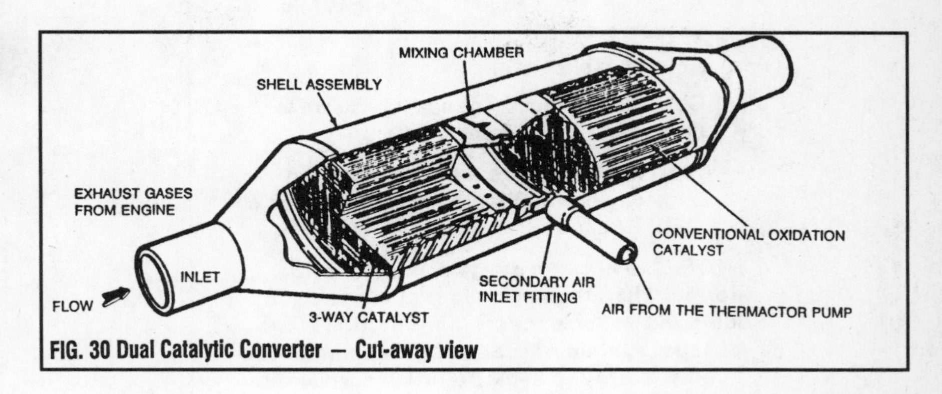

FIG. 30 Dual Catalytic Converter — Cut-away view

The catalytic converter assembly consists of a structured shell containing a ceramic, honeycomb construction. In order to maintain the converter's exhaust oxygen content at a high level to obtain the maximum oxidation for producing the heated chemical reaction, the oxidation catalyst sometimes requires the use of a secondary air source. This is provided by the thermactor air injection system.

The catalytic converter is protected by several devices that block out the air supply from the thermactor air injection system when the engine is laboring under one or more of the following conditions:

- Cold engine operation with rich choke mixture.
- Abnormally high engine coolant temperatures above 225°F (107°C), which may result from a condition such as an extended, hot idle on a hot day.
- Wide-open throttle.
- Engine deceleration.
- Extended idle operation.

REMOVAL & INSTALLATION

✻✻ CAUTION

Catalytic converters operate at extremely high temperatures. Do not attempt to remove the converter until it has been allowed to cool, or bodily injury may result.

1. Raise and safely support the vehicle.
2. Disconnect the secondary air supply tube from the fitting on the converter, if necessary.
3. Remove the retaining clamps or mounting bolts, as necessary and remove the converter.
4. Installation is the reverse of the removal procedure.

Catalytic Converter Diagnosis Charts

	TEST STEP	RESULT	▶	ACTION TO TAKE
B1	VISUAL INSPECTION			
	• **Visually inspect exhaust system.** • **Is exhaust system visually OK?**	Yes	▶	GO to B2.
		No	▶	REPLACE any collapsed exhaust components. If problem is not corrected, GO to B2.

TEST STEP	RESULT ▶	ACTION TO TAKE
B2 VACUUM TEST		
• Attach vacuum gauge to intake manifold vacuum source. • Hook up tachometer. • Observe the vacuum gauge needle while performing the following: — Start engine and gradually increase the engine rpm to 2000 with the transmission in NEUTRAL. NOTE: **The vacuum gauge reading may be normal when the engine is first started and idled. However, excessive restriction in the exhaust system will cause the vacuum gauge needle to drop to a low point even while the engine is idled.** • Decrease engine speed to base idle rpm. • **Did manifold vacuum reach above 16 inches of mercury with the engine rpm at 2000?**	Yes ▶ No ▶	**No restriction in the exhaust system.** GO to B3.

FIG. 31 Catalytic Converter Diagnosis Chart

TEST STEP	RESULT ▶	ACTION TO TAKE
B3 VACUUM TEST—RATE OF VACUUM GAUGE NEEDLE RETURN MOVEMENT		
• Vacuum gauge attached to intake manifold vacuum source. • Tachometer installed. • Increase the engine speed gradually from base idle rpm to 2000 rpm with the transmission in NEUTRAL. • Observe the rate of speed of the vacuum gauge needle as it falls and rises, while maintaining the increased engine rpm. NOTES: — **On a non-restricted system, the vacuum gauge needle will drop to zero and then quickly return to the normal setting without delay.** — **On a restricted system, as the engine rpm is increased to 2000, the vacuum gauge needle will slowly drop to zero. As the increased rpm is maintained, the needle will slowly rise to normal.** — **The rate of speed at which the vacuum gauge needle returns to the normal setting is much slower on a restricted system than on a non-restricted system.** • Decrease engine speed to base idle rpm. • **Is rate of speed that the vacuum gauge needle returns to the normal setting much slower than that of a non-restricted system?**	Yes ▶ No ▶	GO to B4. **No restriction in the exhaust system.**

TEST STEP		RESULT	►	ACTION TO TAKE
B4	VACUUM TEST—EXHAUST DISCONNECTED			
	• Turn engine off. • Disconnect exhaust system at exhaust manifold(s). • Repeat vacuum test. • **Is manifold vacuum above 16 inches of mercury?**	Yes No	► ►	GO to B5. GO to B6.

FIG. 32 Catalytic Converter Diagnosis Chart

TEST STEP		RESULT	►	ACTION TO TAKE
B5	VACUUM TEST—CATALYTIC CONVERTER(S) ON/MUFFLER(S) OFF			
	• Turn engine off. • Reconnect exhaust system at exhaust manifold(s). • Disconnect muffler(s). • Repeat vacuum test. • **Is the manifold vacuum above 16 inches of mercury?**	Yes No	► ►	REPLACE muffler(s). REPLACE catalytic converter and inspect muffler to be sure converter debris has not entered muffler.
B6	EXHAUST MANIFOLD RESTRICTED			
	• Remove the exhaust manifold(s). Inspect the ports for casting flash by dropping a length of chain into each port. NOTE: **Do not use a wire or lamp to check ports. The restriction may be large enough for them to pass through but small enough to cause excessive back pressure at high engine rpm.** • **Is a restriction present?**	Yes	►	REMOVE casting flash. If flash cannot be removed, REPLACE exhaust manifold(s).

FIG. 33 Catalytic Converter Diagnosis Chart

SELF DIAGNOSTIC SYSTEMS

General Description

Ford Mustang vehicles employ the 4th generation Electronic Engine Control system, commonly designated EEC-IV, to manage fuel, ignition and emissions on vehicle engines.

ENGINE CONTROL SYSTEM

The Engine Control Assembly (ECA) is given responsibility for the operation of the emission control devices, cooling fans, ignition and advance and in some cases, automatic transmission functions. Because the EEC-IV oversees both the ignition timing and the fuel injector operation, a precise air/fuel ratio will be maintained under all operating conditions. The ECA is a microprocessor or small computer which receives electrical inputs from several sensors, switches and relays on and around the engine.

Based on combinations of these inputs, the ECA controls outputs to various devices concerned with engine operation and emissions. The engine control assembly relies on the signals to form a correct picture of current vehicle operation. If any of the input signals is incorrect, the ECA reacts to what ever picture is painted for it. For example, if the coolant temperature sensor is inaccurate and reads too low, the ECA may see a picture of the engine never warming up. Consequently, the engine settings will be maintained as if the engine were cold. Because so many inputs can affect one output, correct diagnostic procedures are essential on these systems.

One part of the ECA is devoted to monitoring both input and output functions within the system. This ability forms the core of the self-diagnostic system. If a problem is detected within a circuit, the controller will recognize the fault, assign it an identification code, and store the code in a memory section. Depending on the year and model, the fault code(s) may be represented by two or three digit numbers. The stored code(s) may be retrieved during diagnosis.

While the EEC-IV system is capable of recognizing many internal faults, certain faults will not be recognized. Because the computer system sees only electrical signals, it cannot sense or react to mechanical or vacuum faults affecting engine operation. Some of these faults may affect another component which will set a code. For example, the ECA monitors the output signal to the fuel injectors, but cannot detect a partially clogged injector. As long as the output driver responds correctly, the computer will read the system as functioning correctly. However, the improper flow of fuel may result in a lean mixture. This would, in turn, be detected by the oxygen sensor and noticed as a constantly lean signal by the ECA. Once the signal falls outside the pre-programmed limits, the engine control assembly would notice the fault and set an identification code.

Additionally, the EEC-IV system employs adaptive fuel logic. This process is used to compensate for normal wear and variability within the fuel system. Once the engine enters steady-state operation, the engine control assembly watches the oxygen sensor signal for a bias or tendency to run slightly rich or lean. If such a bias is detected, the adaptive logic corrects the fuel delivery to bring the air/fuel mixture towards a centered or 14.7:1 ratio. This compensating shift is stored in a non-volatile memory which is retained by battery power even with the ignition switched off. The correction factor is then available the next time the vehicle is operated.

➡ If the battery cable(s) is disconnected for longer than 5 minutes, the adaptive fuel factor will be lost. After repair it will be necessary to drive the car at least 10 miles to allow the processor to relearn the correct factors. The driving period should include steady-throttle open road driving if possible. During the drive, the vehicle may exhibit driveability symptoms not noticed before. These symptoms should clear as the ECA computes the correction factor. The ECA will also store Code 19 indicating loss of power to the controller.

Failure Mode Effects Management (FMEM)

The engine controller assembly contains back-up programs which allow the engine to operate if a sensor signal is lost. If a sensor input is seen to be out of range — either high or low — the FMEM program is used. The processor substitutes a fixed value for the missing sensor signal. The engine will continue to operate, although performance and driveability may be noticeably reduced. This function of the controller is sometimes referred to as the limp-in or fail-safe mode. If the missing sensor signal is restored, the FMEM system immediately returns the system to normal operation. The dashboard warning lamp will be lit when FMEM is in effect.

Hardware Limited Operation Strategy (HLOS)

This mode is only used if the fault is too extreme for the FMEM circuit to handle. In this mode, the processor has ceased all computation and control; the entire system is run on fixed values. The vehicle may be operated but performance and driveabilty will be greatly reduced. The fixed or default settings provide minimal calibration, allowing the vehicle to be carefully driven in for service. The dashboard warning lamp will be lit when HLOS is engaged. Codes cannot be read while the system is operating in this mode.

DASHBOARD WARNING LAMP (MIL)

The CHECK ENGINE or SERVICE ENGINE SOON dashboard warning lamp is referred to as the Malfunction Indicator Lamp (MIL). The lamp is connected to the engine control assembly and will alert the driver to certain malfunctions within the EEC-IV system. When the lamp is lit, the ECA has detected a fault and stored an identity code in memory. The engine control system will usually enter either FMEM or HLOS mode and driveability will be impaired.

The light will stay on as long as the fault causing it is present. Should the fault self-correct, the MIL will extinguish but the stored code will remain in memory.

Under normal operating conditions, the MIL should light briefly when the ignition key is turned **ON**. As soon as the ECA receives a signal that the engine is cranking, the lamp will be extinguished. The dash warning lamp should remain out during the entire operating cycle.

Tools and Equipment

➧ SEE FIGS. 34–38

Although stored codes may be read through the flashing of the CHECK ENGINE or SERVICE ENGINE SOON lamp, the use of hand-held scan tools such as Ford's Self-Test Automatic Readout (STAR) tester or the second generation SUPER STAR II tester or their equivalent is highly recommended. There are many manufacturers of these tools; the purchaser must be certain that the tool is proper for the intended use.

Both the STAR and SUPER STAR testers are designed to communicate directly with the EEC-IV system and interpret the electrical signals. The SUPER STAR tester may be used to read either 2 or 3 digit codes; the original STAR tester will not read the 3 digit codes used on many 1990–92 vehicles.

The scan tool allows any stored faults to be read from the engine controller memory. Use of the scan tool provides additional data during troubleshooting but does not eliminate the use of the charts. The scan tool makes collecting information easier; the data must be correctly interpreted by an operator familiar with the system.

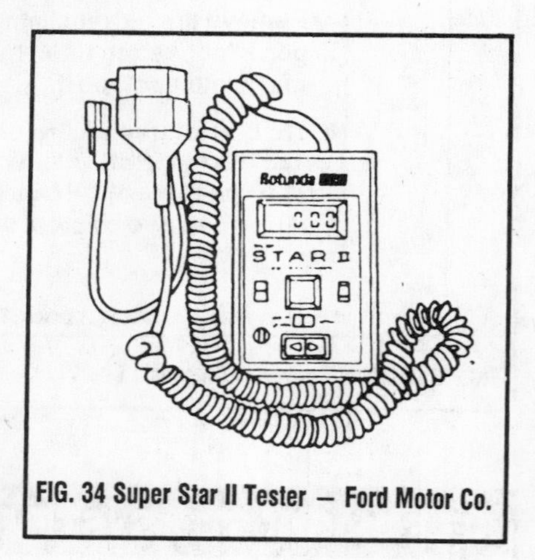

FIG. 34 Super Star II Tester — Ford Motor Co.

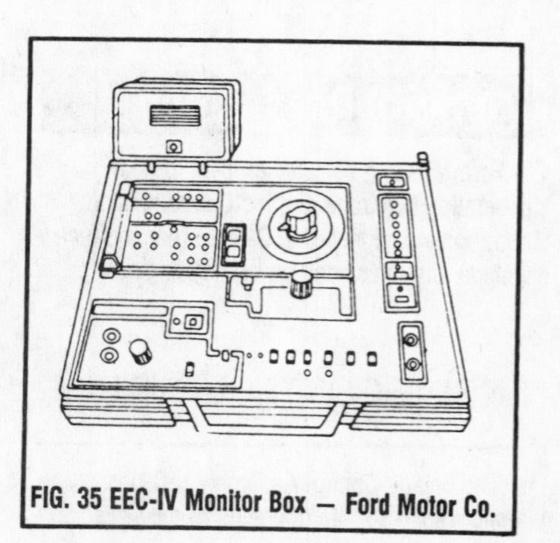
FIG. 35 EEC-IV Monitor Box — Ford Motor Co.

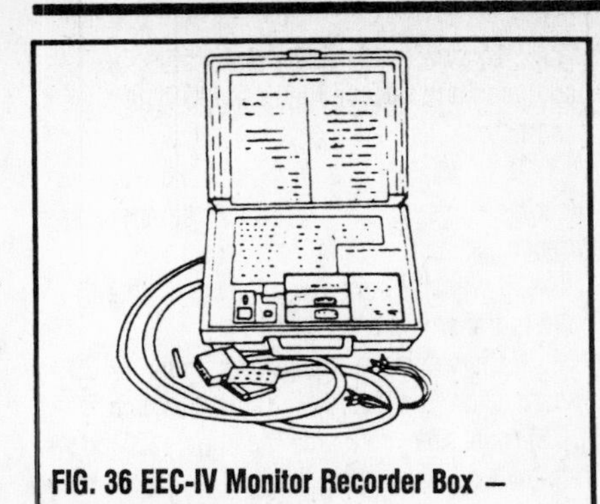

FIG. 36 EEC-IV Monitor Recorder Box — Ford Motor Co.

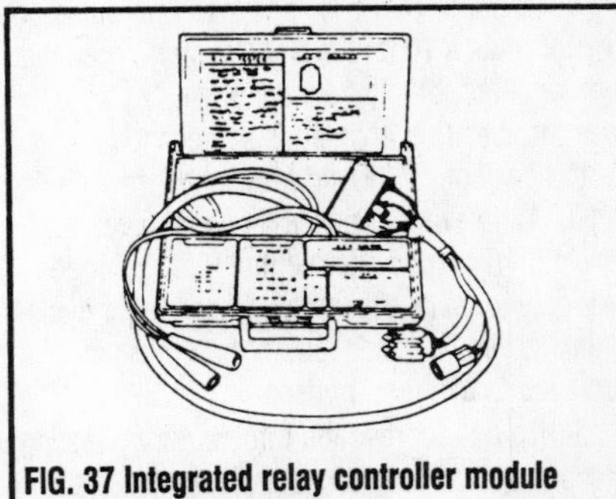

FIG. 37 Integrated relay controller module tester — Ford Motor Co.

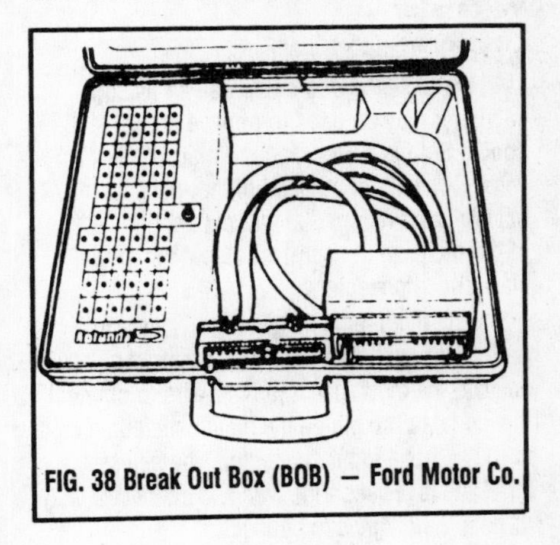

FIG. 38 Break Out Box (BOB) — Ford Motor Co.

ELECTRICAL TOOLS

The most commonly required electrical diagnostic tool is the Digital Multimeter, allowing voltage, ohmmage (resistance) and amperage to be read by one instrument. Many of the diagnostic charts require the use of a volt or ohmmeter during diagnosis.

The multimeter must be a high impedance unit, with 10 megohms of impedance in the voltmeter. This type of meter will not place an additional load on the circuit it is testing; this is extremely important in low voltage circuits. The multimeter must be of high quality in all respects. It should be handled carefully and protected from impact or damage. Replace the batteries frequently in the unit.

Additionally, an analog (needle type) voltmeter may be used to read stored fault codes if the STAR tester is not available. The codes are transmitted as visible needle sweeps on the face of the instrument.

Almost all diagnostic procedures will require the use of the Breakout Box, a device which connects into the EEC-IV harness and provides testing ports for the 60 wires in the harness. Direct testing of the harness connectors at the terminals or by backprobing is not recommended; damage to the wiring and terminals is almost certain to occur.

Other necessary tools include a quality tachometer with inductive (clip-on) pickup, a fuel pressure gauge with system adapters and a vacuum gauge with an auxiliary source of vacuum.

Diagnosis and Testing

Diagnosis of a driveability problem requires attention to detail and following the diagnostic procedures in the correct order. Resist the temptation to begin extensive testing before completing the preliminary diagnostic steps. The preliminary or visual inspection must be completed in detail before diagnosis begins. In many cases this will shorten diagnostic time and often cure the problem without electronic testing.

VISUAL INSPECTION

This is possibly the most critical step of diagnosis. A detailed examination of all connectors, wiring and vacuum hoses can often lead to a repair without further diagnosis. Performance of this step relies on the skill of the technician performing it; a careful inspector will check the undersides of hoses as well as the integrity of hard-to-reach hoses blocked by the air cleaner or other components. Wiring should be checked carefully for any sign of strain , burning, crimping or terminal pull-out from a connector.

Checking connectors at components or in harnesses is required; usually, pushing them together will reveal a loose fit. Pay particular attention to ground circuits, making sure they are not loose or corroded. Remember to inspect connectors and hose fittings at components not mounted on the engine, such as the evaporative canister or relays mounted on the fender aprons. Any component or wiring in the vicinity of a fluid leak or spillage should be given extra attention during inspection.

Additionally, inspect maintenance items such as belt condition and tension, battery charge and condition and the radiator cap carefully. Any of these very simple items may affect the system enough to set a fault.

READING FAULTS OR FAULT CODES

➧ SEE FIG. 39

The EEC-IV system may be interrogated for stored codes using the Quick Test procedures. These tests will reveal faults immediately present during the test as well as any intermittent codes set within the previous 80 warm up cycles. If a code was set before a problem self-corrected (such as a momentarily loose connector), the code will be erased if the problem does not reoccur within 80 warm-up cycles.

The Quick Test procedure is divided into 2 sections, Key On Engine Off (KOEO) and Key On Engine Running (KOER). These 2 procedures must be performed correctly if the system is to run the internal self-checks and provide accurate fault codes. Codes will be output and displayed as numbers on the hand scan tool, i.e. 23. If the codes are being read through the dashboard warning lamp, the codes will be displayed as groups of flashes separated by pauses. Code 23 would be shown as two flashes, a pause and three more flashes. A longer pause will occur between codes. If the codes are being read on an analog voltmeter, the needle sweeps indicate the code digits in the same manner as the lamp flashes.

In all cases, the codes 11 or 111 are used to indicate PASS during testing. Note that the PASS code may appear, followed by other stored codes. These are codes from the continuous memory and may indicate intermittent faults, even though the system does not presently contain the fault. The PASS designation only indicates the system passes all internal tests at the moment.

Once the Quick Test has been performed and all fault codes recorded, refer to the code charts. The charts direct the use of specific pinpoint tests for the appropriate circuit and will allow complete circuit testing.

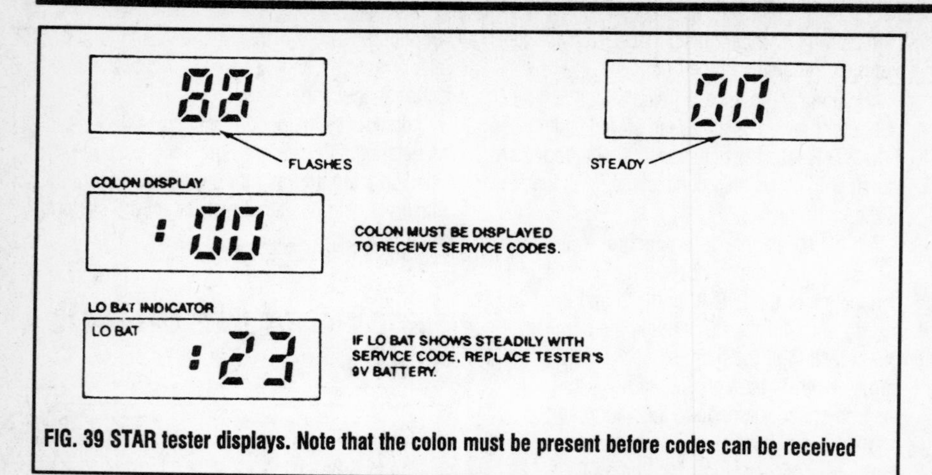

FIG. 39 STAR tester displays. Note that the colon must be present before codes can be received

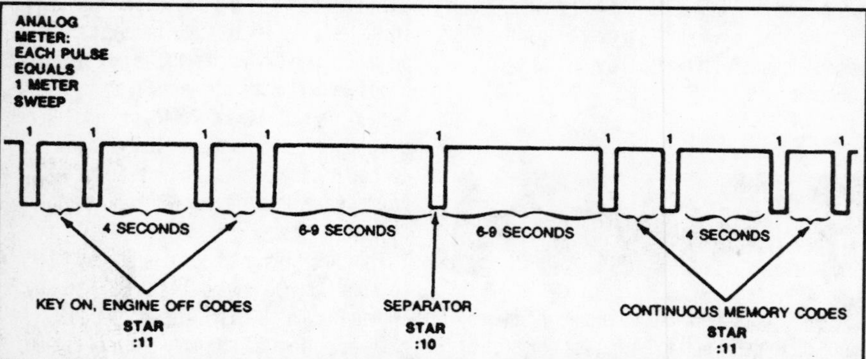

FIG. 40 Code transmission during KOEO test. Note that the continuous memory codes are transmitted after a pause and a separator pause

CAUTION

To prevent injury and/or property damage, always block the drive wheels, firmly apply the parking brake, place the transmission in Park or Neutral and turn all electrical loads off before performing the Quick Test procedures.

Key On Engine Off Test

➧ SEE FIG. 40

1. Connect the scan tool to the self-test connectors. Make certain the test button is unlatched or up.
2. Start the engine and run it until normal operating temperature is reached.
3. Turn the engine **OFF** for 10 seconds.
4. Activate the test button on the STAR tester.
5. Turn the ignition switch **ON** but do not start the engine. For vehicles with 4.9L engines, depress the clutch during the entire test. For vehicles with the 7.3L diesel engine, hold the accelerator to the floor during the test.
6. The KOEO codes will be transmitted. Six to nine seconds after the last KOEO code, a single separator pulse will be transmitted. Six to nine seconds after this pulse, the codes from the Continuous Memory will be transmitted.
7. Record all service codes displayed. Do not depress the throttle on gasoline engines during the test.

Key On Engine Running Test

➧ SEE FIG. 41

1. Make certain the self-test button is released or de-activated on the STAR tester.
2. Start the engine and run it at 2000 rpm for two minutes. This action warms up the oxygen sensor.
3. Turn the ignition switch **OFF** for 10 seconds.
4. Activate or latch the self-test button on the scan tool.
5. Start the engine. The engine identification code will be transmitted. This is a single digit number representing 1/2 the number of cylinders in a gasoline engine. On the STAR tester, this number may appear with a zero, i.e., 20 = 2. The code is used to confirm that the correct processor is installed and that the self-test has begun.
6. If the vehicle is equipped with a Brake On/Off (BOO) switch, the brake pedal must be depressed and released after the ID code is transmitted.
7. If the vehicle is equipped with a Power Steering Pressure Switch (PSPS), the steering wheel must be turned at least 1/2 turn and released within 2 seconds after the engine ID code is transmitted.
8. If the vehicle is equipped with the E4OD transmission, the Overdrive Cancel Switch (OCS) must be cycled after the engine ID code is transmitted.
9. Certain Ford vehicles will display a Dynamic Response code 6–20 seconds after the engine ID code. This will appear as one pulse on a meter or as a 10 on the STAR tester. When this code appears, briefly take the engine to wide open throttle. This allows the system to test the throttle position, MAF and MAP sensors.
10. All relevant codes will be displayed and should be recorded. Remember that the codes refer only to faults present during this test cycle. Codes stored in Continuous Memory are not displayed in this test mode.
11. Do not depress the throttle during testing unless a dynamic response code is displayed.

Reading Codes With Analog Voltmeter

➧ SEE FIGS. 42–43

In the absence of a scan tool, an analog voltmeter may be used to retrieve stored fault codes. Set the meter range to read DC 0–15 volts. Connect the + lead of the meter to the battery positive terminal and connect the – lead of the meter to the self-test output pin of the diagnostic connector.

Follow the directions given previously for performing the KOEO and KOER tests. To activate the tests, use a jumper wire to connect the signal return pin on the diagnostic connector to the self-test input connector. The self-test input line is the separate wire and connector with or near the diagnostic connector.

The codes will be transmitted as groups of needle sweeps. This method may be used to read either 2 or 3 digit codes. The Continuous Memory codes are separated from the KOEO codes by 6 seconds, a single sweep and another 6 second delay.

Reading Codes With MIL

➧ SEE FIG. 44–45

The Malfunction Indicator Lamp on the dashboard may also be used to retrieve the stored codes. This method displays only the stored codes and does not allow any system investigation. It should only be used in field conditions where a quick check of stored codes is needed.

Follow the directions given previously for

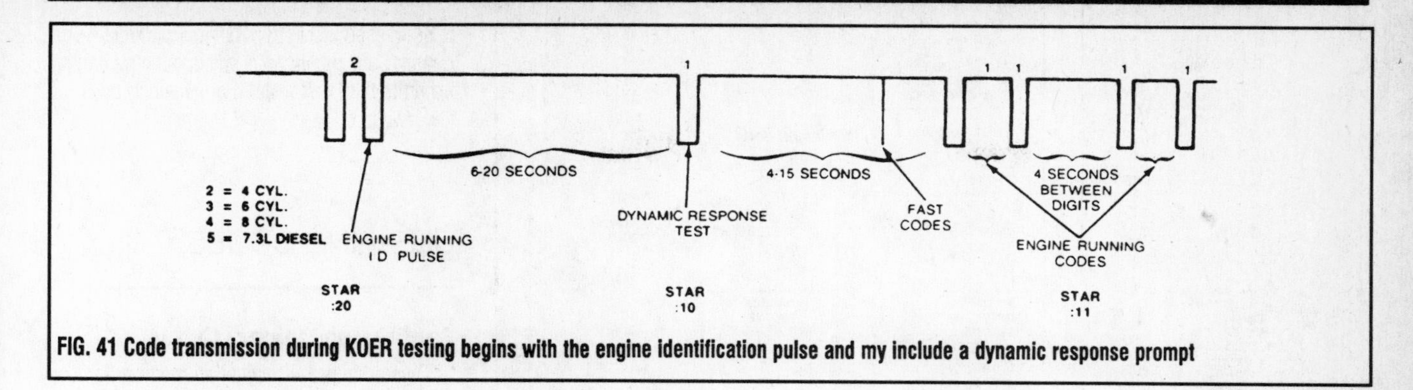

FIG. 41 Code transmission during KOER testing begins with the engine identification pulse and my include a dynamic response prompt

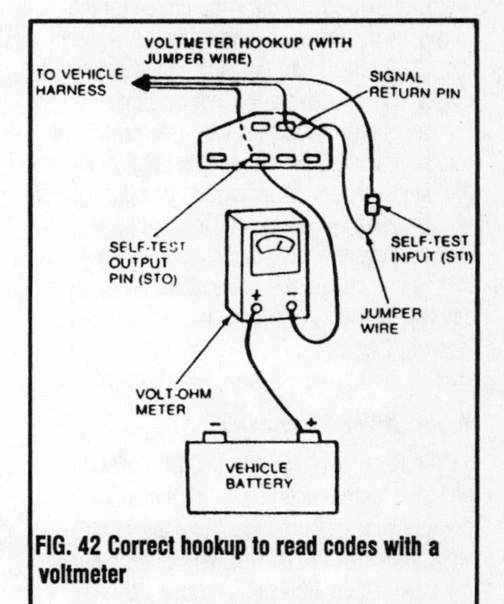

FIG. 42 Correct hookup to read codes with a voltmeter

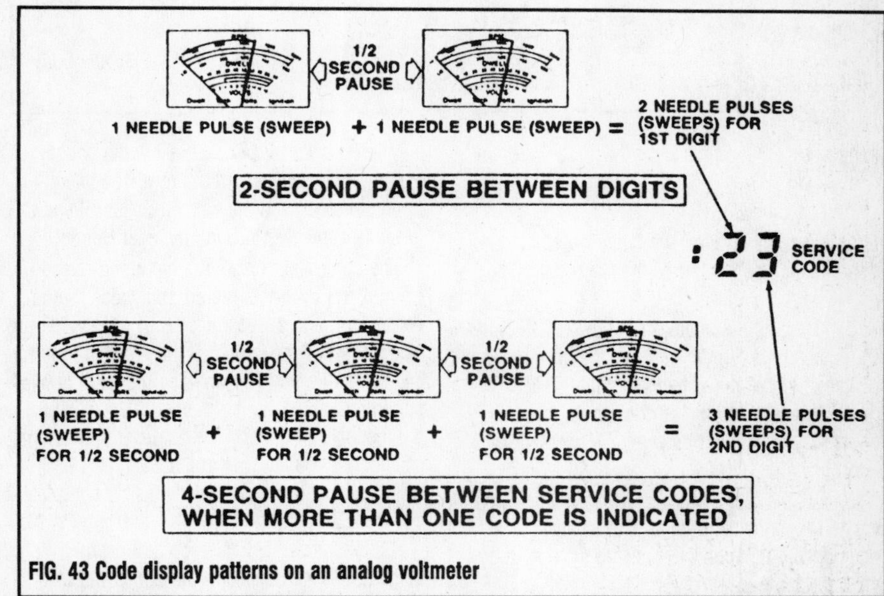

FIG. 43 Code display patterns on an analog voltmeter

performing the KOEO and KOER tests. To activate the tests, use a jumper wire to connect the signal return pin on the diagnostic connector to the self-test input connector. The self-test input line is the separate wire and connector with or near the diagnostic connector.

Codes are transmitted by place value with a pause between the digits; Code 32 would be sent as 3 flashes, a pause and 2 flashes. A slightly longer pause divides codes from each other. Be ready to count and record codes; the only way to repeat a code is to re-cycle the system. This method may be used to read either 2 or 3 digit codes. The Continuous Memory codes are separated from the KOEO codes by 6 seconds, a single flash and another 6 second delay.

Other Test Modes

CONTINUOUS MONITOR OR WIGGLE TEST MODE

Once entered, this mode allows the technician to attempt to recreate intermittent faults by wiggling or tapping components, wiring or connectors. The test may be performed during either KOEO or KOER procedures. The test

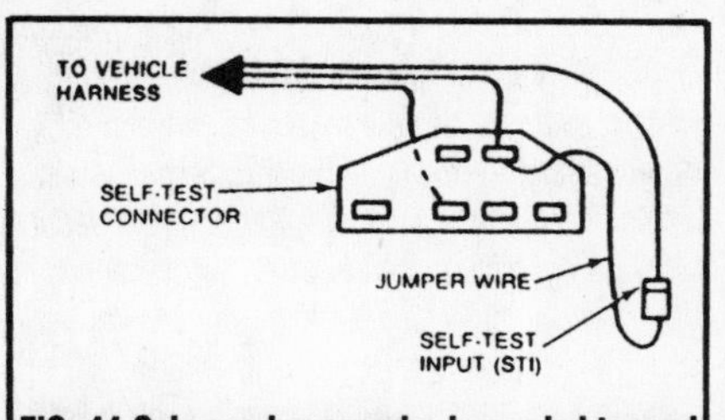

FIG. 44 Only one jumper wire is needed to read codes through the MIL or the message center

requires the use of either an analog voltmeter or a hand scan tool.

To enter the continuous monitor mode during KOEO testing, turn the ignition switch **ON**. Activate the test, wait 10 seconds, then deactivate and reactivate the test; the system will enter the continuous monitor mode. Tap, move or wiggle the harness, component or connector suspected of causing the problem; if a fault is detected, the code will store in the memory. When the fault occurs, the dash warning lamp will illuminate, the STAR tester will light a red indicator (and possibly beep) and the analog meter needle will sweep once.

To enter this mode in the KOER test:

1. Start the engine and run it at 2000 rpm for two minutes. This action warms up the oxygen sensor.
2. Turn the ignition switch **OFF** for 10 seconds.
3. Start the engine.
4. Activate the test, wait 10 seconds, then deactivate and reactivate the test; the system will enter the continuous monitor mode.
5. Tap, move or wiggle the harness, component or connector suspected of causing the problem; if a fault is detected, the code will store in the memory.
6. When the fault occurs, the dash warning lamp will illuminate, the STAR tester will light a red indicator (and possibly beep) and the analog meter needle will sweep once.

OUTPUT STATE CHECK

This testing mode allows the operator to energize and de-energize most of the outputs controlled by the EEC-IV system. Many of the outputs may be checked at the component by listening for a click or feeling the item move or

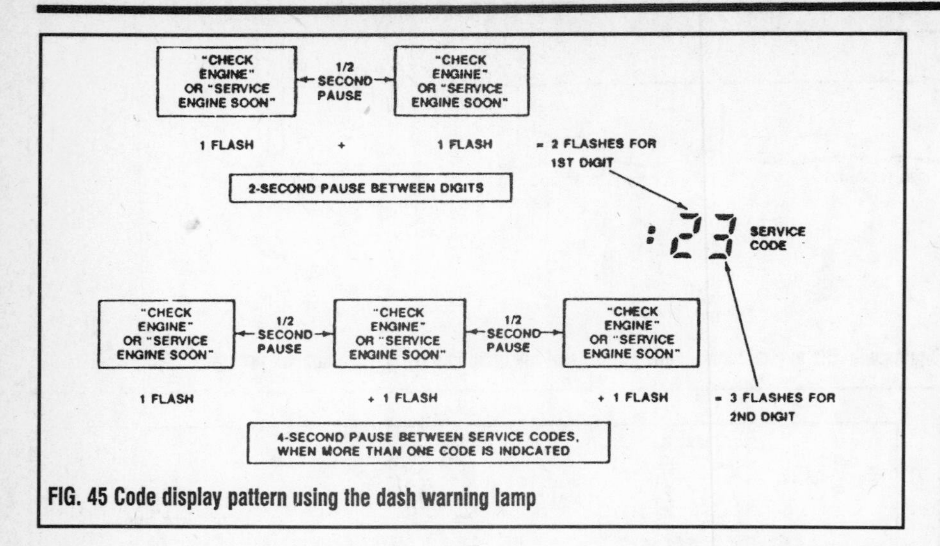

FIG. 45 Code display pattern using the dash warning lamp

engage by a hand placed on the case. To enter this check:

1. Enter the KOEO test mode.
2. When all codes have been transmitted, depress the accelerator all the way to the floor and release it.
3. The output actuators are now all ON. Depressing the throttle pedal to the floor again switches the all the actuator outputs OFF.
4. This test may be performed as often as necessary, switching between ON and OFF by depressing the throttle.
5. Exit the test by turning the ignition switch **OFF**, disconnecting the jumper at the diagnostic connector or releasing the test button on the scan tool.

CYLINDER BALANCE TEST — SEFI ENGINES ONLY

The EEC-IV system allows a cylinder balance test to be performed on engines equipped with the Sequential Electronic Fuel Injection system. Cylinder balance testing identifies a weak or non-contributing cylinder.

Enter the cylinder balance test by depressing and releasing the throttle pedal within 2 minutes of the last code output in the KOER test. The idle speed will become fixed and engine rpm is recorded for later reference. The engine control assembly will shut off the fuel to the highest numbered cylinder (4 or 8), allow the engine to stabilize and then record the rpm. The injector is turned back on and the next one shut off and the process continues through cylinder No. 1.

The controller selects the highest rpm drop from all the cylinders tested, multiplies it by a percentage and arrives at an rpm drop value for all cylinders. For example, if the greatest drop for any cylinder was 150 rpm, the processor applies a multiple of 65% and arrives at 98 rpm. The processor then checks the recorded rpm drops, checking that each was at least 98 rpm. If all cylinders meet the criteria, the test is complete and the ECA outputs Code 90 indicating PASS.

If one cylinder did not drop at least this amount, then the cylinder number is output instead of the 90 code. The cylinder number will be followed by a zero, so 30 indicates cylinder No. 3 did not meet the minimum rpm drop.

The test may be repeated a second time by depressing and releasing the throttle pedal within 2 minutes of the last code output. For the second test, the controller uses a lower percentage (and thus a lower rpm) to determine the minimum acceptable rpm drop. Again, either Code 90 or the number of the weak cylinder will be output.

Performing a third test causes the ECA to select an even lower percentage and rpm drop. If a cylinder is shown as weak in the third test, it should be considered non-contributing. The tests may be repeated as often as needed if the throttle is depressed within two minutes of the last code output. Subsequent tests will use the percentage from the third test instead of selecting even lower values.

CLEARING CODES

Continuous Memory Codes

These codes are retained in memory for 80 warm-up cycles. To clear the codes for the purposes of testing or confirming repair, perform the KOEO test. When the fault codes begin to be displayed, de-activate the test by either disconnecting the jumper wire (meter, MIL or message center) or releasing the test button on the hand scanner. Stopping the test during code transmission will erase the Continuous Memory. Do not disconnect the negative battery cable to clear these codes; the Keep Alive memory will be cleared and a new code, 19, will be stored for loss of ECA power.

Keep Alive Memory

The Keep Alive Memory (KAM) contains the adaptive factors used by the processor to compensate for component tolerances and wear. It should not be routinely cleared during diagnosis. If an emissions related part is replaced during repair, the KAM must be cleared. Failure to clear the KAM may cause severe driveability problems since the correction factor for the old component will be applied to the new component.

To clear the Keep Alive Memory, disconnect the negative battery cable for at least 5 minutes. After the memory is cleared and the battery reconnected, the vehicle must be driven at least 10 miles so that the processor may relearn the needed correction factors. The distance to be driven depends on the engine and vehicle, but all drives should include steady-throttle cruise on open roads. Certain driveability problems may be noted during the drive because the adaptive factors are not yet functioning.

1991–92 FORD EEC-IV VEHICLES
3-DIGIT QUICK TEST CODES—PASSENGER CARS

Service Codes — ENGINE (Liters) / FUEL SYSTEM	Quick Test Mode	2.3L OHC EFI	5.0L ① SEFI
111—System Pass	O/R/C	✓	✓
112—ACT sensor circuit grounded or reads 254°F	O/C	✓	✓
112—ACT sensor circuit grounded	O/R		
113—ACT sensor circuit open	O/R		
113—ACT sensor circuit open or reads −40°F	O/C	✓	✓
114—ACT outside test limits during KOEO or KOER tests	O/R	✓	✓
116—ECT outside test limits during KOEO or KOER tests	O/R	✓	✓
117—ECT sensor circuit grounded	O/C	✓	✓
118—ECT sensor circuit above maximum voltage or reads −40°F	O/C		✓
118—ECT sensor circuit open	O/C	✓	
121—Closed throttle voltage higher or lower than expected	O/R/C	✓	✓
122—TP sensor circuit below minimum voltage	O/C	✓	✓
123—TP sensor above maximum voltage	O/C	✓	✓
124—TP sensor voltage higher than expected, in range	C	✓	✓
125—TP sensor voltage lower than expected, in range	C	✓	✓
126—BP or MAP sensor higher or lower than expected	O/R/C	✓	
129—Insufficient MAF change during Dynamic Response test	R	✓	✓
136—HEGO shows system always lean (front)	R		
136—HEGO shows system always lean (left)	R		✓
137—HEGO shows system always rich (front)	R		
137—HEGO shows system always rich (left)	R		✓
139—No HEGO switching (front)	C		
139—No HEGO switching (left)	C		✓
144—No HEGO switching (right)	C		✓
144—No HEGO switching	C		
144—No HEGO switching detected	C	✓	
157—MAF sensor circuit below minimum voltage	C	✓	✓
158—MAF sensor circuit above maximum voltage	O/C	✓	✓
158—MAF sensor circuit above maximum voltage	O/R/C		
159—MAF higher or lower than expected during KOEO and KOER test	O/R	✓	✓
167—Insufficient TP change during Dynamic Response test	R	✓	✓
171—Fuel system at adaptive limit, HEGO unable to switch	C	✓	
171—Fuel system at adaptive limit, HEGO unable to switch (right)	C		
171—No HEGO switching; system at adaptive limit (rear)	C		
172—HEGO shows system always lean (rear)	R/C		
172—No HEGO switching seen; indicates lean	R/C	✓	
172—No HEGO switching seen; indicates lean (right)	R/C	✓	✓
173—HEGO shows system always rich (rear)	R/C		✓
173—No HEGO switching seen; indicates rich	R/C	✓	✓
173—No HEGO switching seen; indicates rich (right)	R/C		
174—HEGO switching time is slow (right)	C		✓
175—No HEGO switching; system at adaptive limit (front)	C		
175—No HEGO switching; system at adaptive limit (left)	C		✓
176—HEGO shows system always lean (front)	C		
176—HEGO shows system always lean (left)	C	✓	
177—HEGO shows system always lean (front)	C		
177—HEGO shows system always lean (left)	C		✓
178—HEGO switching time is slow (left)	C		✓
179—Fuel at lean adaptive limit at part throttle; system rich	C	✓	
179—System at lean adaptive limit at part throttle; system rich (rear)	C		
179—System at lean adaptive limit at part throttle; system rich (right)	C		✓
181—Fuel at rich adaptive limit at part throttle; system rich	C	✓	
181—System at rich adaptive limit at part throttle; system rich (rear)	C		
181—System at rich adaptive limit at part throttle; system rich (right)	C		✓
182—Fuel at lean adaptive limit at idle; system rich	C	✓	
182—System at lean adaptive limit at idle; system rich (rear)	C		
182—System at lean adaptive limit at idle; system rich (right)	C		✓
183—Fuel at rich adaptive limit at idle; system lean	C	✓	
183—System at rich adaptive limit at idle; system lean (rear)	C		
183—System at rich adaptive limit at idle; system lean (right)	C		✓
184—MAF higher than expected	C	✓	✓
185—MAF lower than expected	C	✓	✓
186—Injector pulse width higher than expected	C	✓	✓
187—Injector pulse width lower than expected	C	✓	✓
188—System at lean adaptive limit at part throttle; system rich (front)	C		
188—System at lean adaptive limit at part throttle; system rich (left)	C		✓
189—System at rich adaptive limit at part throttle; system rich (front)	C		
189—System at rich adaptive limit at part throttle; system rich (left)	C		✓
191—System at lean adaptive limit at idle; system rich (front)	C		
191—System at lean adaptive limit at idle; system rich (left)	C		✓
192—System at rich adaptive limit at idle; system rich (front)	C		
192—System at rich adaptive limit at idle; system rich (left)	C		✓
211—PIP circuit fault	C	✓	✓
212—Loss of IDM input to ECA or SPOUT circuit grounded	C	✓	✓
213—SPOUT circuit open	R	✓	✓
214—Cylinder identification circuit failure	C	✓	

1991-92 FORD EEC-IV VEHICLES
3-DIGIT QUICK TEST CODES—PASSENGER CARS

Service Codes / ENGINE (Liters) / FUEL SYSTEM	Quick Test Mode	2.3L OHC EFI	5.0L ① SEFI
215—EEC processor detected Coil 1 primary circuit failure	C	✓	
216—EEC processor detected Coil 2 primary circuit fauilre	C	✓	
218—Loss of IDM signal, left side	C	✓	
219—Spark timing defaulted to 10°BTDC or SPOUT circ. open	C		
222—Loss of IDM signal, right side		✓	
223—Loss of dual plug inhibit control	C		
224—Erratic IDM input to rpocessor	C		
225—Knock not sensed during Dynamic Response test	R		
311—Thermactor air system inoperative (right)	R		✓
313—Thermactor air not bypassed during self-test	R		✓
314—Thermactor air system inoperative (left)	R		✓
326—PFE or DPFE circiut voltage lower than expected	R/C		
327—EVP or DPFE circuit below minimum voltage	O/R/C	✓	✓
328—EGR closed voltage lower than expected	O/R/C	✓	✓
332—Insufficient EGR flow detected	R/C	✓	✓
334—EGR closed voltage higher than expected	O/R/C	✓	✓
335—PFE or DPFE sensor voltage out of self-test range	O		
336—PFE sensor voltage higher than expected	R/C		
337—EVP or DPFE circuit above maximum voltage	O/R/C	✓	✓
341—Octane adjust service pin in use	O	✓	
411—Cannot control rpm during KOER low rpm check	R	✓	✓
412—Cannot control rpm during KOER high rpm check	R	✓	✓
452—Insufficient input from vehicle speed sensor	C	✓	✓
511—EEC processor ROM test failed	O	✓	✓
512—EEC processor Keep Alive Memory test failed	O		
512—EEC processor Keep Alive Memory test failed	C	✓	✓
513—Failure in EEC processor internal voltage	O		✓
519—Power steering pressure switch circuit open	O	✓	
521—Power steering pressure switch did not change state	R	✓	
522—Vehicle not in Park or Neutral during KOEO test	O	✓	✓
525—Vehicle in gear or A/C on during self-test	O		
528—Clutch switch circuit failure	C	✓	
536—Brake On/Off circuit failure/not actuated during KOER test	R/C	✓	
538—Insufficient rpm change during KOER Dynamic Response test	R	✓	✓
539—A/C on or Defroster on during KOEO test	O	✓	✓
542—Fuel pump secondary circuit failure: ECA to ground	O/C	✓	✓
543—Fuel pump secondary circuit failure: Batt to ECA	O/C	✓	✓
552—Air management 1 circuit failure	O		✓
556—Fuel pump primary circuit failure	O/C	✓	✓
558—EGR vacuum regulator circuit failure	O	✓	✓
563—High speed electro-drive fan circuit failure	O		
564—Electro-drive fan circiut failure	O	✓	
565—Canister purge circuit failure	O	✓	✓
566—3-4 shift solenoid circuit failure	O	✓	
621—Shift solenoid 1 circuit failure	O		
622—Shift solenoid 2 circuit failure	O		
624—EPC solenoid or driver circuit faiulre	O/C		
625—EPC driver open in ECA	O		
628—Lock-up solenoid failure: excessive clutch slippage	C		
629—Converter clutch control circuit failure	O	✓	
629—Lock-up solenoid failure	O		
634—MLP sensor voltage out of self-test range	C		
636—TOT sensor voltage out of self-test range	O/R		
637—TOT sensor circuit above maximum voltage	O/C		
638—TOT sensor circuit below maximum voltage	O/C		
639—Insufficient input from turbine speed sensor	R/C		
641—Shift solenoid 3 circuit failure	O		
645—Incorrect gear ratio obtained for 1st gear	C		
646—Incorrect gear ratio obtained for 2nd gear	C		
647—Incorrect gear ratio obtained for 3rd gear	C		
648—Incorrect gear ratio obtained for 4th gear	C		
649—EPC range failure	C		
651—EPC circuit failure	C		
998—Hard fault present	R	✓	✓

Codes not listed: Do not apply to vehicle being tested
No codes: Cannot perform self-test or cannot transmit codes
O—Key off, engine off test
R—Key on engine running test
C—Continuous codes
① Thunderbird and Cougar

1989–92 FORD EEC-IV VEHICLES
2-DIGIT QUICK TEST CODES—PASSENGER CARS

Service Codes — ENGINE (Liters) / FUEL SYSTEM	Quick Test Mode	2.3L (1) OHC EFI	5.0L SEFI (INC. MA)
11—System Pass	O/R/C	✓	✓
12—Rpm unable to achieve upper test limit	R	✓	✓
13—D.C. motor movement not detected	O		
13—Rpm unable to achieve lower test limit	R	✓	✓
13—D.C. motor did not follow dashpot	C		
14—PIP circuit failure	C	✓	✓
15—ECA read only memory test failed	O	✓	✓
15—ECA keep alive memory test failed	C	✓	✓
16—Idle rpm high with ISC off	R		
16—Idle too low to perform EGO test	R	✓	✓
17—Idle rpm low with ISC off	R		
18—SPOUT circuit open or spark angle word failure	R		✓
18—IDM circuit failure or SPOUT circuit grounded	C	✓	✓
19—Failure in ECA internal voltage	O	✓	✓
19—CID circuit failure	C		
19—Rpm dropped too low in ISC off test	R		
19—Rpm for EGR test not achieved	R		
21—ECT out of self-test range	O/R	✓	✓
22—BP sensor out of self-test range	O/C		✓
22—BP or map out of self-test range	O/R/C	✓	
23—TP out of self-test range	O/R	✓	✓
23—TP out of self-test range	O/R/C		
24—ACT sensor out of self-test range	O/R	✓	✓
25—Knock not sensed during dynamic test	R	✓	✓
26—VAF/MAF out of self-test range	O/R		✓
28—VAT out of self-test range	O/R		
29—Insufficient input from vehicle speed sensor	C		✓
31—PFE, EVP or EVR circuit below minimum voltage	O/R/C	✓	✓
32—EPT circuit voltage low (PFE)	R/C		
32—EVP voltage below closed limit	O/R/C		✓
32—EGR not controlling	R	✓	
33—EGR valve opening not detected	R/C		✓
33—EGR not closing fully	R	✓	
34—Defective PFE sensor or voltage out of range	O		
34—EPT sensor voltage high (PFE)	R/C		
34—EVP voltage above closed limit	O/R/C		✓
34—EGR opening not detected	R	✓	
35—PFE or EVP circuit above maximum voltage	O/R/C		✓
35—Rpm too low to perform EGR test	R	✓	
38—Idle tracking switch circuit open	C		
39—AXOD lock up failed	C		
41—HEGO sensor circuit indicates system lean	R	✓	✓ (3)
41—No HEGO switching detected	R	✓	✓ (3)
42—HEGO sensor circuit indicates system rich	R	✓	✓ (3)
42—No HEGO switching detected—reads rich	C		
43—HEGO lean at wide open throttle	C		
44—Thermactor air system inoperative—right side	R		✓
45—Thermactor air upstream during self-test	R		✓
45—Coil 1 primary circuit failure	C		
46—Thermactor air not bypassed during self-test	R		✓
46—Coil 2 primary circuit failure	C		
47—Measured airflow low at base idle	R		
48—Coil 3 primary circuit failure	C		
48—Measured airflow high at base idle	R		
49—SPOUT signal defaulted to 10°BTDC or SPOUT open	C		
51—ECT/ACT reads −40°F or circuit open	O/C	✓	✓
52—Power steering pressure switch circuit open	O	✓	
52—Power steering pressure switch always open or closed	R		
53—TP circuit above maximum voltage	O/C	✓	✓
54—ACT sensor circuit open	O/C	✓	✓
55—Keypower circuit open	R		
56—VAF or MAF circuit above maximum voltage	O/C		✓
56—MAF circuit above maximum voltage	O/R/C		
57—Octane adjust service pin in use	O		
57—AXOD neutral pressure switch circuit failed open	C		
58—Idle tracking switch circuit open	O		
58—Idle tracking switch closed/circuit grounded	R		
58—VAT reads −40°F or circuit open	O/C		
59—Idle adjust service pin in use	O		
59—AXOD 4/3 pressure switch circuit failed open	C		
59—Low speed fuel pump circuit open—Battery to ECA	O/C		
59—AXOD 4/3 pressure switch failed closed	O		

1989–92 FORD EEC-IV VEHICLES
2-DIGIT QUICK TEST CODES—PASSENGER CARS

Service Codes — ENGINE (Liters) / FUEL SYSTEM	Quick Test Mode	2.3L ① OHC EFI	5.0L SEFI (INC. MA)
61—ECT reads 254°F or circuit grounded	O/C	✓	✓
62—AXOD 4/3 or 3/2 pressure switch circuit grounded	O		
63—TP circuit below minimum voltage	O/C	✓	✓
64—ACT sensor input below test minimum or grounded	O/C	✓	✓
65—Never went to closed loop fuel control	C		
66—MAF sensor input below minimum voltage	C		✓
66—VAF sensor below minimum voltage	O/C		
66—MAF circuit below minimum voltage	R/C		
67—Neutral/drive switch open or A/C on	O	✓	✓
67—Clutch switch circuit failure	C		
67—Neutral/drive switch open or A/C on	O/R		
68—Idle tracking switch closed or circuit grounded	O		
68—Idle tracking switch circuit open	R		
68—AXOD transmission temperature switch failed open	O/R/C		
68—VAT reads 254°F or circuit grounded	O/C		
69—AXOD 3/2 pressure switch circuit failed closed	O		
69—AXOD 3/4 pressure switch circuit failed open	C		
70—ECA DATA communications link circuit failure	C		
71—Software re-initialization detected	C		
71—Idle tracking switch shorted to ground	C		
71—Cluster control assembly circuit failed	C		
72—Insufficient MAF/MAP change during dynamic test	R	✓	
72—Power interrupt or re-initialization detected	C		
72—Message Center control assembly circuit failed	C		
73—Insufficient throttle position change	O		
73—Insufficient TP change during dynamic test	R	✓	
74—Brake on/off switch failure or not actuated	R	✓	✓
75—Brake on/off switch circuit closed or ECA input open	R	✓	✓
76—Insufficient VAF change during dynamic test	R		
77—No WOT seen in self-test or operator error	R	✓	✓
79—A/C or defrost on during self-test	O		✓
81—IAS circuit failure	O		
81—Air management 2 circuit failure	O		✓
82—Air management 1 circuit failure	O		
82—Supercharger bypass circuit failure	O		
83—High-speed electro drive fan circuit failure	O		
83—Low speed primary fuel pump circuit failure	O/C		
84—EGR vacuum solenoid circuit failure	O	✓	✓
84—EGR vacuum regulator circuit failure	O/R		
85—Canister purge circuit failure	O/R		
85—Canister purge solenoid circuit failure	O		✓
85—Adaptive fuel lean limit reached	C		
86—3-4 shift solenoid circuit failure	O		
86—Adaptive fuel rich limit reached	C		
87—Fuel pump primary circuit failure	O/C		✓
87—Fuel pump primary circuit failure	O/C/R		
87—Fuel pump primary circuit failure	O	✓	
88—Electro drive fan circuit failure	O		
89—Converter clutch override circuit failure	O	✓	
89—Lock-up solenoid circuit failure	O		
91—HEGO sensor indicates system lean	R		✓④
91—No HEGO switching detected	C		✓④
92—HEGO sensor indicates system rich	R		✓④
93—TP sensor input low at maximum motor travel	O		
94—Thermactor air system inoperative—left side	R		✓
95—Fuel pump secondary circuit failure—ECA to ground	O/C		✓
96—Fuel pump secondary circuit failure—Battery to ECA	O/C		✓
96—High speed fuel pump circuit open	O/C		
98—Hard fault present	R	✓	✓
99—EEC has not learned to control idle: ignore codes 12 & 13	R		

No Codes: Cannot begin self-test or cannot transmit codes
Codes Not Listed: Do not apply to vehicle being tested
O—Key on, engine off test
R—Key on, engine running test
C—Continuous memory
① 1991–92: 3 digit codes
② Front HEGO
③ Right HEGO
④ Left HEGO
⑤ Rear HEGO

CALIBRATION: 8–25F–R00 — 2.3L HSC-EFI

FORD MOTOR COMPANY
VEHICLE EMISSION CONTROL INFORMATION

THIS VEHICLE IS EQUIPPED WITH ELECTRONIC FUEL INJECTION. IDLE MIXTURE, COLD ENGINE IDLE SPEED AND COLD ENGINE FUEL ENRICHMENT NOT ADJUSTABLE.

SET PARKING BRAKE AND BLOCK WHEELS. DISCONNECT AUTOMATIC PARKING BRAKE RELEASE (IF SO EQUIPPED). MAKE ALL ADJUSTMENTS WITH ENGINE AT NORMAL OPERATING TEMPERATURE, TRANSMISSION IN NEUTRAL AND ACCESSORIES OFF.

IGNITION TIMING–
(1) TURN OFF ENGINE
(2) DISCONNECT THE IN-LINE SPOUT CONNECTOR (⊏⊐).
(3) RE-START PREVIOUSLY WARMED-UP ENGINE.
(4) ADJUST IGNITION TIMING TO 15° BTDC OR D.
(5) TURN OFF ENGINE AND RESTORE ELECTRICAL CONNECTION.

THIS ENGINE IS EQUIPPED WITH AUTOMATIC IDLE SPEED CONTROL. IDLE RPM IS NOT ADJUSTABLE. IF NOT WITHIN SPECIFIED RPM RANGE, SEE SHOP MANUAL:
MANUAL TRANS. IN NEUTRAL: 820-880 RPM
AUTO. TRANS. IN DRIVE: 690-750 RPM

USE SAE 5W-30 OIL-API CATEGORY SG, SG/CC OR SG/CD.

THIS VEHICLE CONFORMS TO U.S. EPA AND CALIFORNIA REGULATIONS APPLICABLE TO 1989 MODEL YEAR NEW MOTOR VEHICLES INTRODUCED INTO COMMERCE SOLELY FOR SALE IN CALIFORNIA.

VACUUM HOSE ROUTING

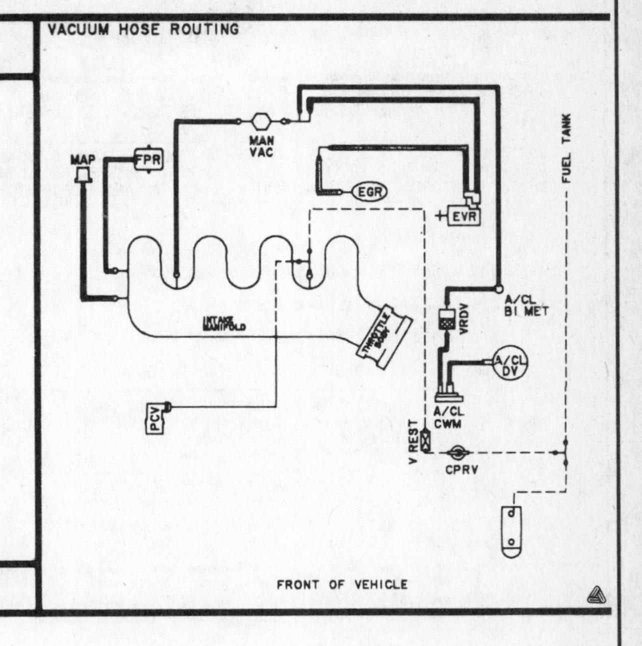

E9AE-9C485- CATALYST SPARK PLUG: AWSF-42C GAP- .052-.056
2.3L-9HM
KFM2.3V5HEH6 - EGS/EGR/AIV/FI/TWC

FRONT OF VEHICLE

Vacuum Diagram 2.3L engine — 1989

CALIBRATION: 9–25F–R10 — 2.3L HSC-EFI

FORD MOTOR COMPANY
VEHICLE EMISSION CONTROL INFORMATION

THIS VEHICLE IS EQUIPPED WITH ELECTRONIC FUEL INJECTION. IDLE MIXTURE, COLD ENGINE IDLE SPEED AND COLD ENGINE FUEL ENRICHMENT NOT ADJUSTABLE.

SET PARKING BRAKE AND BLOCK WHEELS. DISCONNECT AUTOMATIC PARKING BRAKE RELEASE (IF SO EQUIPPED). MAKE ALL ADJUSTMENTS WITH ENGINE AT NORMAL OPERATING TEMPERATURE, TRANSMISSION IN NEUTRAL AND ACCESSORIES OFF.

IGNITION TIMING–
(1) TURN OFF ENGINE
(2) DISCONNECT THE IN-LINE SPOUT CONNECTOR (⊏⊐).
(3) RE-START PREVIOUSLY WARMED-UP ENGINE.
(4) ADJUST IGNITION TIMING TO 15° BTDC OR D.
(5) TURN OFF ENGINE AND RESTORE ELECTRICAL CONNECTION.

THIS ENGINE IS EQUIPPED WITH AUTOMATIC IDLE SPEED CONTROL. IDLE RPM IS NOT ADJUSTABLE. IF NOT WITHIN SPECIFIED RPM RANGE, SEE SHOP MANUAL:
MANUAL TRANS. IN NEUTRAL: 820-880 RPM
AUTO. TRANS. IN DRIVE: 690-750 RPM

USE SAE 5W-30 OIL - API CATEGORY SG, SG/CC OR SG/CD.

THIS VEHICLE CONFORMS TO U.S. EPA REGULATIONS APPLICABLE TO 1989 MODEL YEAR NEW MOTOR VEHICLES.

VACUUM HOSE ROUTING

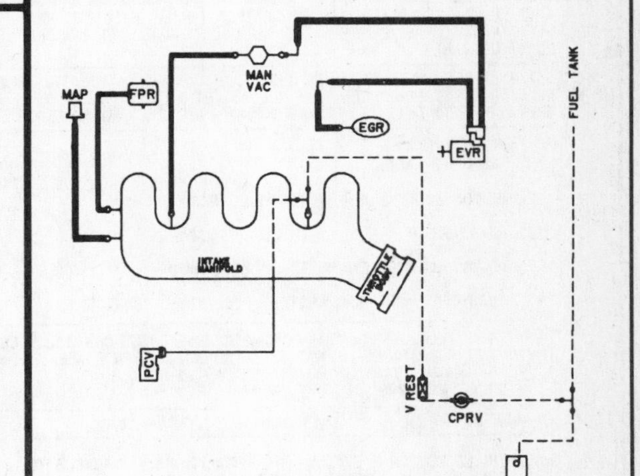

E9AF-9C485- CATALYST SPARK PLUG: AWSF-42C GAP- .052-.056
2.3L-9HM
KFM2.5V5HEF4 - EGS/EGR/AIV/FI/TWC

FRONT OF VEHICLE

Vacuum Diagram 2.3L engine — 1989

CALIBRATION: 8–26E–R00 — 2.3L HSC-EFI

FORD MOTOR COMPANY
VEHICLE EMISSION CONTROL INFORMATION

THIS VEHICLE IS EQUIPPED WITH ELECTRONIC FUEL INJECTION. IDLE MIXTURE, COLD ENGINE IDLE SPEED AND COLD ENGINE FUEL ENRICHMENT NOT ADJUSTABLE.

SET PARKING BRAKE AND BLOCK WHEELS. DISCONNECT AUTOMATIC PARKING BRAKE RELEASE (IF SO EQUIPPED). MAKE ALL ADJUSTMENTS WITH ENGINE AT NORMAL OPERATING TEMPERATURE, TRANSMISSION IN NEUTRAL AND ACCESSORIES OFF.

IGNITION TIMING-

(1) TURN OFF ENGINE

(2) DISCONNECT THE IN-LINE SPOUT CONNECTOR (=□◁).

(3) RE-START PREVIOUSLY WARMED-UP ENGINE.

(4) ADJUST IGNITION TIMING TO 15° BTDC OR D.

(5) TURN OFF ENGINE AND RESTORE ELECTRICAL CONNECTION.

THIS ENGINE IS EQUIPPED WITH AUTOMATIC IDLE SPEED CONTROL. IDLE RPM IS NOT ADJUSTABLE. IF NOT WITHIN SPECIFIED RPM RANGE, SEE SHOP MANUAL:
MANUAL TRANS. IN NEUTRAL: 820-880 RPM
AUTO. TRANS. IN DRIVE: 690-750 RPM

USE SAE 5W-30 OIL - API CATEGORY SG, SG/CC OR SG/CD.

THIS VEHICLE CONFORMS TO U.S. EPA REGULATIONS APPLICABLE TO 1989 MODEL YEAR NEW MOTOR VEHICLES.

E9AE-9C485- CATALYST SPARK PLUG: AWSF-42C GAP- .052-.056
2.3L-9HM
KFM2.3V5HEF4 - EGS/EGR/AIV/FI/TWC

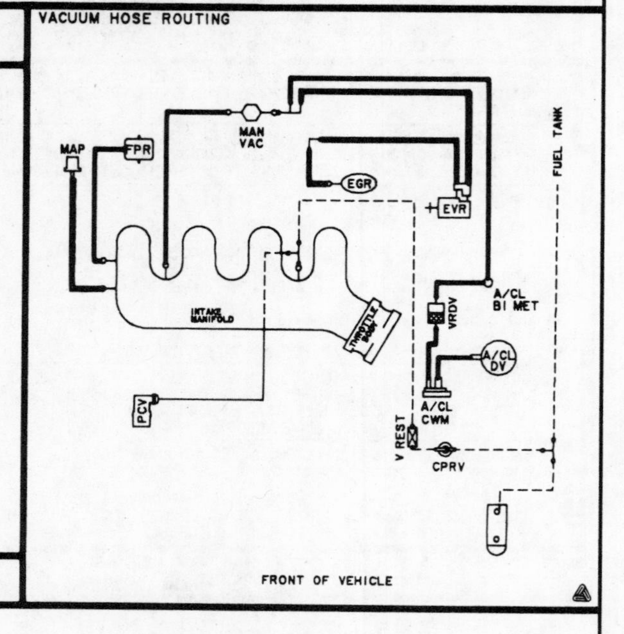

Vacuum Diagram 2.3L engine — 1989

CALIBRATION: 9–26E–R10 — 2.3L HSC-EFI

FORD MOTOR COMPANY
VEHICLE EMISSION CONTROL INFORMATION

THIS VEHICLE IS EQUIPPED WITH ELECTRONIC FUEL INJECTION. IDLE MIXTURE, COLD ENGINE IDLE SPEED AND COLD ENGINE FUEL ENRICHMENT NOT ADJUSTABLE.

SET PARKING BRAKE AND BLOCK WHEELS. DISCONNECT AUTOMATIC PARKING BRAKE RELEASE (IF SO EQUIPPED). MAKE ALL ADJUSTMENTS WITH ENGINE AT NORMAL OPERATING TEMPERATURE, TRANSMISSION IN NEUTRAL AND ACCESSORIES OFF.

IGNITION TIMING-

(1) TURN OFF ENGINE

(2) DISCONNECT THE IN-LINE SPOUT CONNECTOR (=□◁).

(3) RE-START PREVIOUSLY WARMED-UP ENGINE.

(4) ADJUST IGNITION TIMING TO 15° BTDC OR D.

(5) TURN OFF ENGINE AND RESTORE ELECTRICAL CONNECTION.

THIS ENGINE IS EQUIPPED WITH AUTOMATIC IDLE SPEED CONTROL. IDLE RPM IS NOT ADJUSTABLE. IF NOT WITHIN SPECIFIED RPM RANGE, SEE SHOP MANUAL:
MANUAL TRANS. IN NEUTRAL: 820-880 RPM
AUTO. TRANS. IN DRIVE: 690-750 RPM

USE SAE 5W-30 OIL - API CATEGORY SG, SG/CC OR SG/CD.

THIS VEHICLE CONFORMS TO U.S. EPA REGULATIONS APPLICABLE TO 1989 MODEL YEAR NEW MOTOR VEHICLES.

E9AE-9C485- CATALYST SPARK PLUG: AWSF-42C GAP- .052-.056
2.3L-9HM
KFM2.5V5HEF4 - EGS/EGR/AIV/FI/TWC

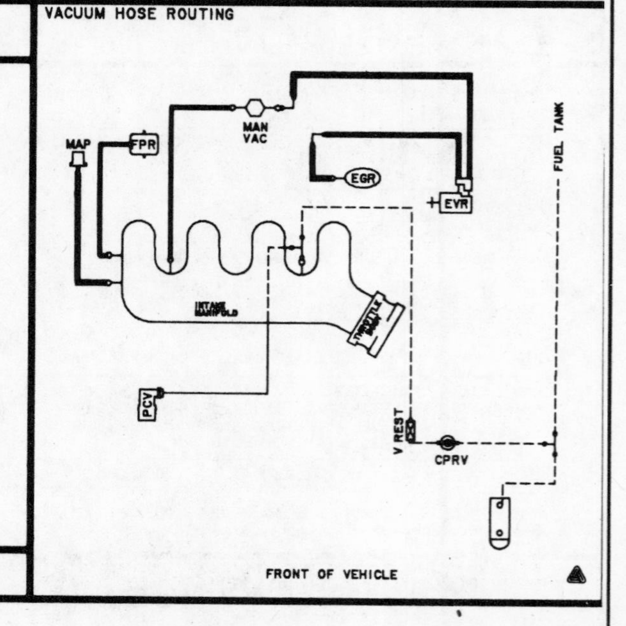

Vacuum Diagram 2.3L engine — 1989

CALIBRATION: 9–21A–R00 — 5.0L EFI

FORD MOTOR COMPANY
VEHICLE EMISSION CONTROL INFORMATION

THIS VEHICLE IS EQUIPPED WITH ELECTRONIC FUEL INJECTION. IDLE MIXTURE, COLD ENGINE IDLE SPEED AND COLD ENGINE FUEL ENRICHMENT ARE NOT ADJUSTABLE.

SET PARKING BRAKE AND BLOCK WHEELS. DISCONNECT AUTOMATIC PARKING BRAKE RELEASE, IF SO EQUIPPED. MAKE ALL ADJUSTMENTS WITH ENGINE AT NORMAL OPERATING TEMPERATURE AND ACCESSORIES OFF.

IGNITION TIMING- ADJUST WITH MAN. TRANS. IN NEUTRAL AND AUTO. TRANS. IN "D".

(1) TURN OFF ENGINE.
(2) DISCONNECT THE IN-LINE SPOUT CONNECTOR (=◻◻).
(3) RE-START PREVIOUSLY WARMED-UP ENGINE.
(4) ADJUST IGNITION TIMING TO 10° BTDC.
(5) TURN OFF ENGINE AND RESTORE ELECTRICAL CONNECTION.

THIS ENGINE IS EQUIPPED WITH AUTOMATIC IDLE SPEED CONTROL. IDLE RPM IS NOT ADJUSTABLE. IF NOT WITHIN SPECIFIED RPM RANGE, SEE SHOP MANUAL.
AUTO. TRANS. IN DRIVE: 575-725 RPM
MANUAL TRANS. IN NEUTRAL: 625-775 RPM

FIRING ORDER-1-3-7-2-6-5-4-8

THIS VEHICLE CONFORMS TO U.S. EPA REGULATIONS APPLICABLE TO 1989 MODEL YEAR NEW MOTOR VEHICLES.

E9AE-9C485- CATALYST SPARK PLUG: ASF-42C GAP- .052-.056
5.0L-9HM
KFM5.0V5HBF4 - FI/EGR/EGS/AIP/TWC

VACUUM HOSE ROUTING

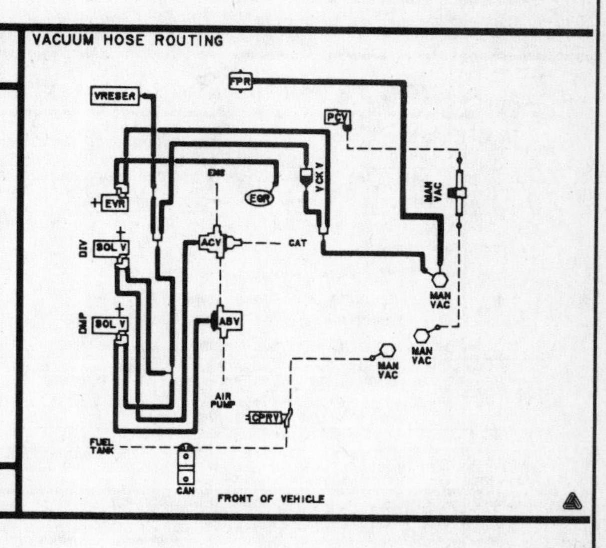

Vacuum Diagram 5.0L engine — 1989

CALIBRATION: 9–21C–R00 — 5.0L EFI

FORD MOTOR COMPANY
VEHICLE EMISSION CONTROL INFORMATION

THIS VEHICLE IS EQUIPPED WITH ELECTRONIC FUEL INJECTION. IDLE MIXTURE, COLD ENGINE IDLE SPEED AND COLD ENGINE FUEL ENRICHMENT ARE NOT ADJUSTABLE.

SET PARKING BRAKE AND BLOCK WHEELS. DISCONNECT AUTOMATIC PARKING BRAKE RELEASE, IF SO EQUIPPED. MAKE ALL ADJUSTMENTS WITH ENGINE AT NORMAL OPERATING TEMPERATURE AND ACCESSORIES OFF.

IGNITION TIMING- ADJUST WITH MAN. TRANS. IN NEUTRAL AND AUTO. TRANS. IN "D".

(1) TURN OFF ENGINE.
(2) DISCONNECT THE IN-LINE SPOUT CONNECTOR (=◻◻).
(3) RE-START PREVIOUSLY WARMED-UP ENGINE.
(4) ADJUST IGNITION TIMING TO 10° BTDC.
(5) TURN OFF ENGINE AND RESTORE ELECTRICAL CONNECTION.

THIS ENGINE IS EQUIPPED WITH AUTOMATIC IDLE SPEED CONTROL. IDLE RPM IS NOT ADJUSTABLE. IF NOT WITHIN SPECIFIED RPM RANGE, SEE SHOP MANUAL.
AUTO. TRANS. IN DRIVE: 575-725 RPM
MANUAL TRANS. IN NEUTRAL: 625-775 RPM

FIRING ORDER-1-3-7-2-6-5-4-8

THIS VEHICLE CONFORMS TO U.S. EPA REGULATIONS APPLICABLE TO 1989 MODEL YEAR NEW MOTOR VEHICLES.

E9AE-9C485- CATALYST SPARK PLUG: ASF-42C GAP- .052-.056
5.0L-9HM
KFM5.0V5HBF4 - FI/EGR/EGS/AIP/TWC

VACUUM HOSE ROUTING

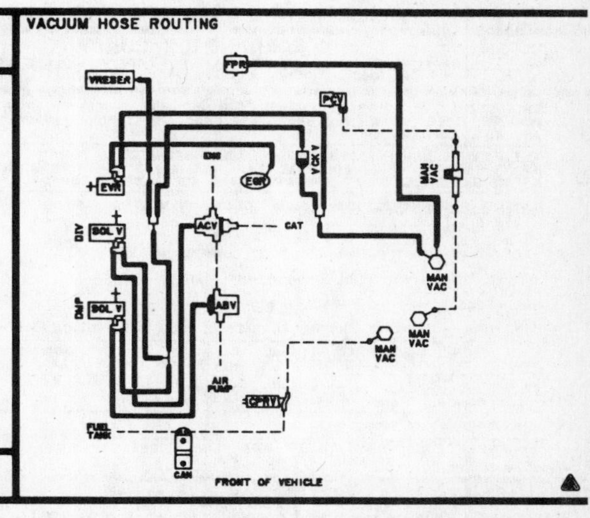

Vacuum Diagram 5.0L engine — 1989

CALIBRATION: 8–21P–R10 — 5.0L EFI

FORD MOTOR COMPANY
VEHICLE EMISSION CONTROL INFORMATION

THIS VEHICLE IS EQUIPPED WITH ELECTRONIC FUEL INJECTION. IDLE MIXTURE, COLD ENGINE IDLE SPEED AND COLD ENGINE FUEL ENRICHMENT ARE NOT ADJUSTABLE.

SET PARKING BRAKE AND BLOCK WHEELS. DISCONNECT AUTOMATIC PARKING BRAKE RELEASE, IF SO EQUIPPED. MAKE ALL ADJUSTMENTS WITH ENGINE AT NORMAL OPERATING TEMPERATURE AND ACCESSORIES OFF.

IGNITION TIMING- ADJUST WITH MAN. TRANS. IN NEUTRAL AND AUTO. TRANS. IN "D".

(1) TURN OFF ENGINE.
(2) DISCONNECT THE IN-LINE SPOUT CONNECTOR (=□◁).
(3) RE-START PREVIOUSLY WARMED-UP ENGINE.
(4) ADJUST IGNITION TIMING TO 10° BTDC.
(5) TURN OFF ENGINE AND RESTORE ELECTRICAL CONNECTION.

THIS ENGINE IS EQUIPPED WITH AUTOMATIC IDLE SPEED CONTROL. IDLE RPM IS NOT ADJUSTABLE. IF NOT WITHIN SPECIFIED RPM RANGE, SEE SHOP MANUAL:
AUTO. TRANS. IN DRIVE: 575-725 RPM
MANUAL TRANS. IN NEUTRAL: 625-775 RPM

FIRING ORDER-1-3-7-2-6-5-4-8

THIS VEHICLE CONFORMS TO U.S. EPA AND CALIFORNIA REGULATIONS APPLICABLE TO 1989 MODEL YEAR NEW MOTOR VEHICLES INTRODUCED INTO COMMERCE SOLELY FOR SALE IN CALIFORNIA.

E9AE-9C485- CATALYST SPARK PLUG: ASF-42C GAP- .052-.056
5.0L-9HM
KFM5.0V5HBC1 - FI/EGR/EGS/AIP/TWC

VACUUM HOSE ROUTING

FPR
VRESER
PCV
ENG
EGR
VCK V
EVR
MAN VAC
DIV
SOL V
ACV
CAT
DMP
SOL V
ABV
MAN VAC
MAN VAC
MAN VAC
AIR PUMP
CPRV
FUEL TANK
CAN
FRONT OF VEHICLE

Vacuum Diagram 5.0L engine — 1989

CALIBRATION: 9–22A–R00 — 5.0L EFI

FORD MOTOR COMPANY
VEHICLE EMISSION CONTROL INFORMATION

THIS VEHICLE IS EQUIPPED WITH ELECTRONIC FUEL INJECTION. IDLE MIXTURE, COLD ENGINE IDLE SPEED AND COLD ENGINE FUEL ENRICHMENT ARE NOT ADJUSTABLE.

SET PARKING BRAKE AND BLOCK WHEELS. DISCONNECT AUTOMATIC PARKING BRAKE RELEASE, IF SO EQUIPPED. MAKE ALL ADJUSTMENTS WITH ENGINE AT NORMAL OPERATING TEMPERATURE AND ACCESSORIES OFF.

IGNITION TIMING- ADJUST WITH MAN. TRANS. IN NEUTRAL AND AUTO. TRANS. IN "D".

(1) TURN OFF ENGINE.
(2) DISCONNECT THE IN-LINE SPOUT CONNECTOR (=□◁).
(3) RE-START PREVIOUSLY WARMED-UP ENGINE.
(4) ADJUST IGNITION TIMING TO 10° BTDC.
(5) TURN OFF ENGINE AND RESTORE ELECTRICAL CONNECTION.

THIS ENGINE IS EQUIPPED WITH AUTOMATIC IDLE SPEED CONTROL. IDLE RPM IS NOT ADJUSTABLE. IF NOT WITHIN SPECIFIED RPM RANGE, SEE SHOP MANUAL.
AUTO. TRANS. IN DRIVE: 575-725 RPM
MANUAL TRANS. IN NEUTRAL: 625-775 RPM

FIRING ORDER-1-3-7-2-6-5-4-8

THIS VEHICLE CONFORMS TO U.S. EPA REGULATIONS APPLICABLE TO 1989 MODEL YEAR NEW MOTOR VEHICLES.

E9AE-9C485- CATALYST SPARK PLUG: ASF-42C GAP- .052-.056
5.0L-9HM
KFM5.0V5HBF4 - FI/EGR/EGS/AIP/TWC

VACUUM HOSE ROUTING

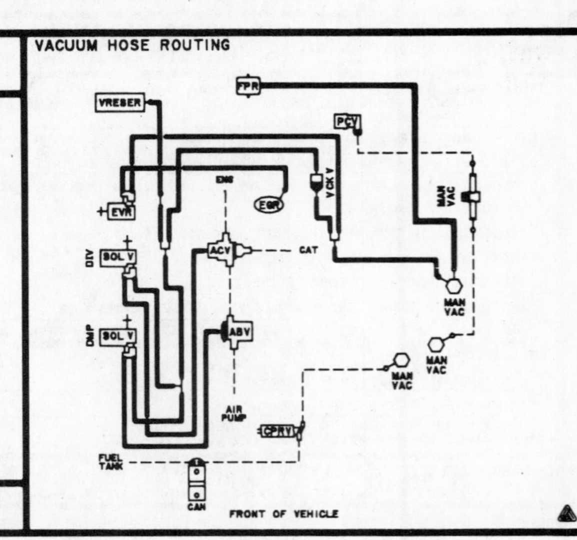

Vacuum Diagram 5.0L engine — 1989

CALIBRATION: 8–22B–R00 — 5.0L EFI

FORD MOTOR COMPANY
VEHICLE EMISSION CONTROL INFORMATION

THIS VEHICLE IS EQUIPPED WITH ELECTRONIC FUEL INJECTION. IDLE MIXTURE, COLD ENGINE IDLE SPEED AND COLD ENGINE FUEL ENRICHMENT ARE NOT ADJUSTABLE.

SET PARKING BRAKE AND BLOCK WHEELS. DISCONNECT AUTOMATIC PARKING BRAKE RELEASE, IF SO EQUIPPED. MAKE ALL ADJUSTMENTS WITH ENGINE AT NORMAL OPERATING TEMPERATURE AND ACCESSORIES OFF.

IGNITION TIMING- ADJUST WITH AUTO. TRANS. IN "D".

(1) TURN OFF ENGINE.

(2) DISCONNECT THE IN-LINE SPOUT CONNECTOR (=◻◅).

(3) RE-START PREVIOUSLY WARMED-UP ENGINE.

(4) ADJUST IGNITION TIMING TO 10° BTDC.

(5) TURN OFF ENGINE AND RESTORE ELECTRICAL CONNECTION.

THIS ENGINE IS EQUIPPED WITH AUTOMATIC IDLE SPEED CONTROL. IDLE RPM IS NOT ADJUSTABLE. IF NOT WITHIN SPECIFIED RPM RANGE, SEE SHOP MANUAL.
AUTO. TRANS. IN DRIVE: 525-650 RPM

FIRING ORDER-1-5-4-2-6-3-7-8

THIS VEHICLE CONFORMS TO U.S. EPA REGULATIONS APPLICABLE TO 1989 MODEL YEAR NEW MOTOR VEHICLES.

E9AE-9C485- **CATALYST** SPARK PLUG: AWSF-44C GAP- .048-.052
5.0L-9HM
KFM5.0V5HBF4 - FI/EGR/EGS/AIP/TWC

VACUUM HOSE ROUTING

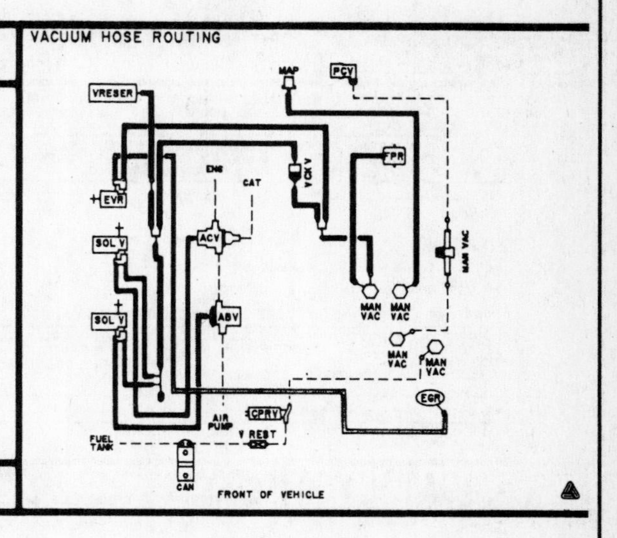

Vacuum Diagram 5.0L engine — 1989

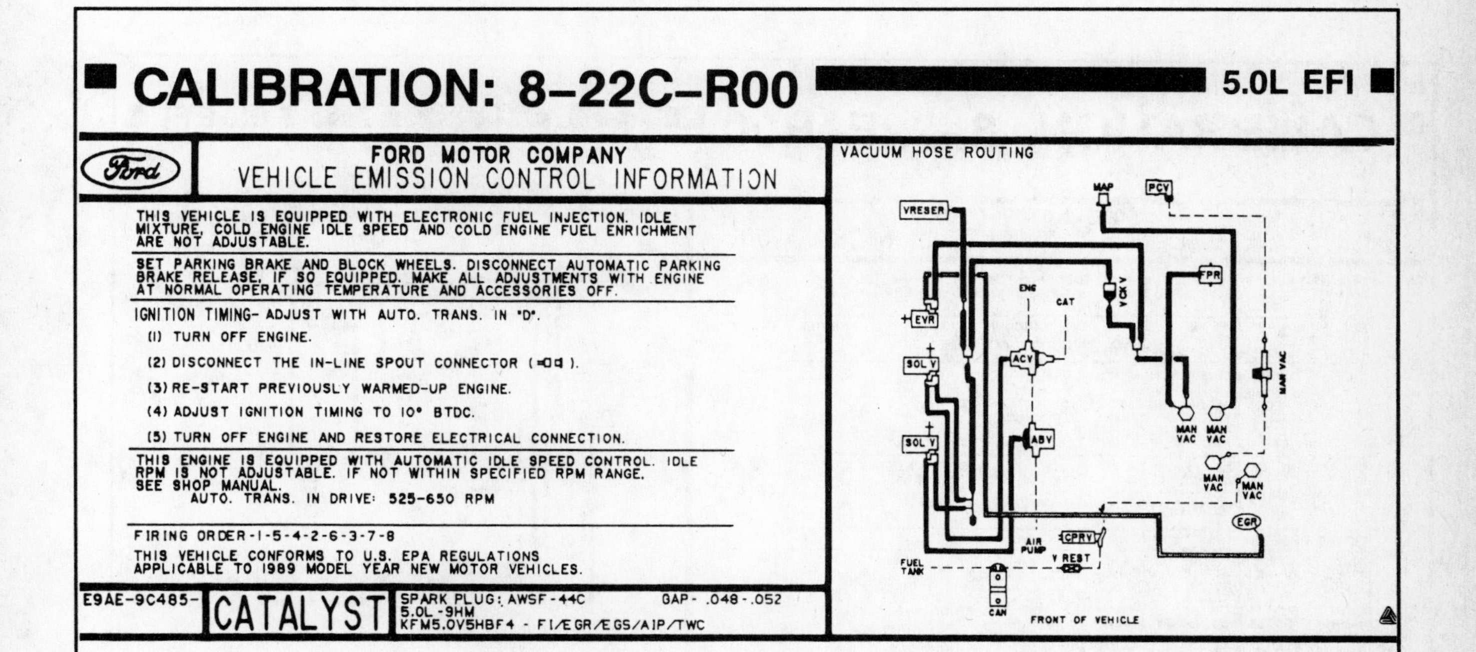

CALIBRATION: 8–22C–R00 — 5.0L EFI

FORD MOTOR COMPANY
VEHICLE EMISSION CONTROL INFORMATION

THIS VEHICLE IS EQUIPPED WITH ELECTRONIC FUEL INJECTION. IDLE MIXTURE, COLD ENGINE IDLE SPEED AND COLD ENGINE FUEL ENRICHMENT ARE NOT ADJUSTABLE.

SET PARKING BRAKE AND BLOCK WHEELS. DISCONNECT AUTOMATIC PARKING BRAKE RELEASE, IF SO EQUIPPED. MAKE ALL ADJUSTMENTS WITH ENGINE AT NORMAL OPERATING TEMPERATURE AND ACCESSORIES OFF.

IGNITION TIMING- ADJUST WITH AUTO. TRANS. IN "D".

(1) TURN OFF ENGINE.

(2) DISCONNECT THE IN-LINE SPOUT CONNECTOR (=◻◅).

(3) RE-START PREVIOUSLY WARMED-UP ENGINE.

(4) ADJUST IGNITION TIMING TO 10° BTDC.

(5) TURN OFF ENGINE AND RESTORE ELECTRICAL CONNECTION.

THIS ENGINE IS EQUIPPED WITH AUTOMATIC IDLE SPEED CONTROL. IDLE RPM IS NOT ADJUSTABLE. IF NOT WITHIN SPECIFIED RPM RANGE, SEE SHOP MANUAL.
AUTO. TRANS. IN DRIVE: 525-650 RPM

FIRING ORDER-1-5-4-2-6-3-7-8

THIS VEHICLE CONFORMS TO U.S. EPA REGULATIONS APPLICABLE TO 1989 MODEL YEAR NEW MOTOR VEHICLES.

E9AE-9C485- **CATALYST** SPARK PLUG: AWSF-44C GAP- .048-.052
5.0L-9HM
KFM5.0V5HBF4 - FI/EGR/EGS/AIP/TWC

VACUUM HOSE ROUTING

Vacuum Diagram 5.0L engine — 1989

CALIBRATION: 8–22D–R00 5.0L EFI

FORD MOTOR COMPANY
VEHICLE EMISSION CONTROL INFORMATION

THIS VEHICLE IS EQUIPPED WITH ELECTRONIC FUEL INJECTION. IDLE MIXTURE, COLD ENGINE IDLE SPEED AND COLD ENGINE FUEL ENRICHMENT ARE NOT ADJUSTABLE.

SET PARKING BRAKE AND BLOCK WHEELS. DISCONNECT AUTOMATIC PARKING BRAKE RELEASE, IF SO EQUIPPED. MAKE ALL ADJUSTMENTS WITH ENGINE AT NORMAL OPERATING TEMPERATURE AND ACCESSORIES OFF.

IGNITION TIMING- ADJUST WITH AUTO. TRANS. IN "D".

(1) TURN OFF ENGINE.
(2) DISCONNECT THE IN-LINE SPOUT CONNECTOR (=▯◁).
(3) RE-START PREVIOUSLY WARMED-UP ENGINE.
(4) ADJUST IGNITION TIMING TO 10° BTDC.
(5) TURN OFF ENGINE AND RESTORE ELECTRICAL CONNECTION.

THIS ENGINE IS EQUIPPED WITH AUTOMATIC IDLE SPEED CONTROL. IDLE RPM IS NOT ADJUSTABLE. IF NOT WITHIN SPECIFIED RPM RANGE, SEE SHOP MANUAL.
AUTO. TRANS. IN DRIVE: 550-675 RPM

FIRING ORDER-1-3-7-2-6-5-4-8

THIS VEHICLE CONFORMS TO U.S. EPA REGULATIONS APPLICABLE TO 1989 MODEL YEAR NEW MOTOR VEHICLES.

E9AE-9C485- CATALYST SPARK PLUG: ASF-42C GAP- .052-.056
5.0L-9HM
KFM5.0V5HBF4 - FI/EGR/EGS/AIP/TWC

VACUUM HOSE ROUTING

FPR MAP VRESER PCV VCK V ENG EGR EVR MAN VAC DIV SOL V ACV CAT MAN VAC MAN VAC DMP SOL V ABV MAN VAC MAN VAC AIR PUMP CPRV FUEL TANK CAN FRONT OF VEHICLE

Vacuum Diagram 5.0L engine — 1989

CALIBRATION: 8–22E–R00 5.0L EFI

FORD MOTOR COMPANY
VEHICLE EMISSION CONTROL INFORMATION

THIS VEHICLE IS EQUIPPED WITH ELECTRONIC FUEL INJECTION. IDLE MIXTURE, COLD ENGINE IDLE SPEED AND COLD ENGINE FUEL ENRICHMENT NOT ADJUSTABLE.

SET PARKING BRAKE AND BLOCK WHEELS. DISCONNECT AUTOMATIC PARKING BRAKE RELEASE, IF SO EQUIPPED. MAKE ALL ADJUSTMENTS WITH ENGINE AT NORMAL OPERATING TEMPERATURE AND ACCESSORIES OFF.

IGNITION TIMING- ADJUST WITH AUTO TRANS. IN "D".

(1) TURN OFF ENGINE.
(2) DISCONNECT THE IN-LINE SPOUT CONNECTOR (=▯◁).
(3) RE-START PREVIOUSLY WARMED-UP ENGINE.
(4) ADJUST IGNITION TIMING TO 10° BTDC.
(5) TURN OFF ENGINE AND RESTORE ELECTRICAL CONNECTION.

THIS ENGINE IS EQUIPPED WITH AUTOMATIC IDLE SPEED CONTROL. IDLE RPM IS NOT ADJUSTABLE. IF NOT WITHIN SPECIFIED RPM RANGE, SEE SHOP MANUAL.
AUTO. TRANS. IN DRIVE: 525-650 RPM

FIRING ORDER-1-5-4-2-6-3-7-8

THIS VEHICLE CONFORMS TO U.S. EPA REGULATIONS APPLICABLE TO 1989 MODEL YEAR NEW MOTOR VEHICLES. COMPLIANCE DEMONSTRATED AND DESIGNED FOR PRINCIPAL USE BELOW 4000 FEET. THIS VEHICLE IS EXEMPT FROM MEETING EMISSION STANDARDS ABOVE 4000 FEET BECAUSE OF POSSIBLY UNSUITABLE PERFORMANCE, AND THE EMISSIONS PERFORMANCE WARRANTY DOES NOT APPLY ABOVE 4000 FEET.

E9AE-9C485- CATALYST SPARK PLUG: AWSF-44C GAP- .048-.052
5.0L-9HM
KFM5.0V5HBF4 - FI/EGR/EGS/AIP/TWC

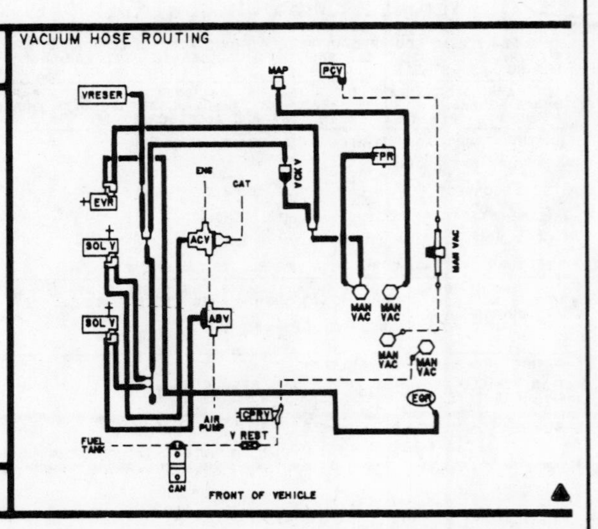

Vacuum Diagram 5.0L engine — 1989

■ CALIBRATION: 8–22G–R00 5.0L EFI ■

FORD MOTOR COMPANY
VEHICLE EMISSION CONTROL INFORMATION

THIS VEHICLE IS EQUIPPED WITH ELECTRONIC FUEL INJECTION. IDLE MIXTURE, COLD ENGINE IDLE SPEED AND COLD ENGINE FUEL ENRICHMENT ARE NOT ADJUSTABLE.

SET PARKING BRAKE AND BLOCK WHEELS. DISCONNECT AUTOMATIC PARKING BRAKE RELEASE, IF SO EQUIPPED. MAKE ALL ADJUSTMENTS WITH ENGINE AT NORMAL OPERATING TEMPERATURE AND ACCESSORIES OFF.

IGNITION TIMING- ADJUST WITH AUTO. TRANS. IN "D".

(1) TURN OFF ENGINE.

(2) DISCONNECT THE IN-LINE SPOUT CONNECTOR (=Dd).

(3) RE-START PREVIOUSLY WARMED-UP ENGINE.

(4) ADJUST IGNITION TIMING TO 10° BTDC.

(5) TURN OFF ENGINE AND RESTORE ELECTRICAL CONNECTION.

THIS ENGINE IS EQUIPPED WITH AUTOMATIC IDLE SPEED CONTROL. IDLE RPM IS NOT ADJUSTABLE. IF NOT WITHIN SPECIFIED RPM RANGE, SEE SHOP MANUAL.
AUTO. TRANS. IN DRIVE: 525-650 RPM

FIRING ORDER-1-5-4-2-6-3-7-8

THIS VEHICLE CONFORMS TO U.S. EPA REGULATIONS APPLICABLE TO 1989 MODEL YEAR NEW MOTOR VEHICLES.

E9AE-9C485- CATALYST SPARK PLUG: AWSF-44C GAP- .048-.052
5.0L-9HM
KFM5.0V5HBF4 - FI/EGR/EGS/AIP/TWC

VACUUM HOSE ROUTING

MAP PCV VRESER FPR ENG CAT VCKV EVR SOL V ACV ABV MAN VAC EGR AIR PUMP CPRV V REST FUEL TANK CAN FRONT OF VEHICLE

Vacuum Diagram 5.0L engine — 1989

■ CALIBRATION: 8–22I–R00 5.0L EFI ■

FORD MOTOR COMPANY
VEHICLE EMISSION CONTROL INFORMATION

THIS VEHICLE IS EQUIPPED WITH ELECTRONIC FUEL INJECTION. IDLE MIXTURE, COLD ENGINE IDLE SPEED AND COLD ENGINE FUEL ENRICHMENT ARE NOT ADJUSTABLE.

SET PARKING BRAKE AND BLOCK WHEELS. DISCONNECT AUTOMATIC PARKING BRAKE RELEASE, IF SO EQUIPPED. MAKE ALL ADJUSTMENTS WITH ENGINE AT NORMAL OPERATING TEMPERATURE AND ACCESSORIES OFF.

IGNITION TIMING- ADJUST WITH AUTO. TRANS. IN "D".

(1) TURN OFF ENGINE.

(2) DISCONNECT THE IN-LINE SPOUT CONNECTOR (=Dd).

(3) RE-START PREVIOUSLY WARMED-UP ENGINE.

(4) ADJUST IGNITION TIMING TO 10° BTDC.

(5) TURN OFF ENGINE AND RESTORE ELECTRICAL CONNECTION.

THIS ENGINE IS EQUIPPED WITH AUTOMATIC IDLE SPEED CONTROL. IDLE RPM IS NOT ADJUSTABLE. IF NOT WITHIN SPECIFIED RPM RANGE, SEE SHOP MANUAL.
AUTO. TRANS. IN DRIVE: 525-650 RPM

FIRING ORDER-1-5-4-2-6-3-7-8

THIS VEHICLE CONFORMS TO U.S. EPA REGULATIONS APPLICABLE TO 1989 MODEL YEAR NEW MOTOR VEHICLES.

E9AE-9C485- CATALYST SPARK PLUG: AWSF-44C GAP- .048-.052
5.0L-9HM
KFM5.0V5HBF4 - FI/EGR/EGS/AIP/TWC

VACUUM HOSE ROUTING

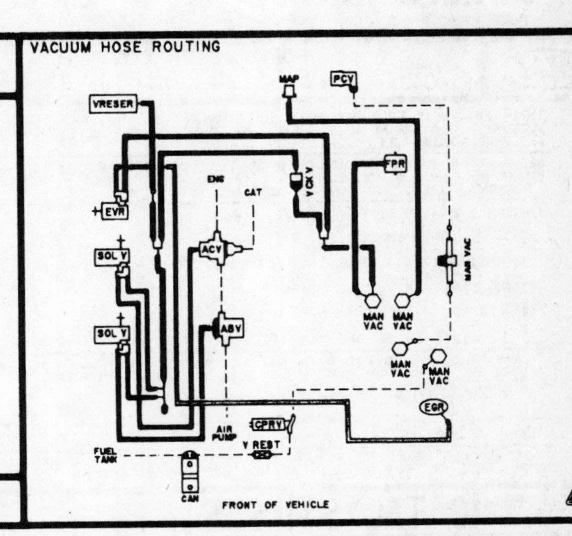

Vacuum Diagram 5.0L engine — 1989

CALIBRATION: 8–22J–R00 — 5.0L EFI

FORD MOTOR COMPANY
VEHICLE EMISSION CONTROL INFORMATION

THIS VEHICLE IS EQUIPPED WITH ELECTRONIC FUEL INJECTION. IDLE MIXTURE, COLD ENGINE IDLE SPEED AND COLD ENGINE FUEL ENRICHMENT ARE NOT ADJUSTABLE.

SET PARKING BRAKE AND BLOCK WHEELS. DISCONNECT AUTOMATIC PARKING BRAKE RELEASE, IF SO EQUIPPED. MAKE ALL ADJUSTMENTS WITH ENGINE AT NORMAL OPERATING TEMPERATURE AND ACCESSORIES OFF.

IGNITION TIMING- ADJUST WITH AUTO. TRANS. IN "D".

(1) TURN OFF ENGINE.
(2) DISCONNECT THE IN-LINE SPOUT CONNECTOR (=0d).
(3) RE-START PREVIOUSLY WARMED-UP ENGINE.
(4) ADJUST IGNITION TIMING TO 10° BTDC.
(5) TURN OFF ENGINE AND RESTORE ELECTRICAL CONNECTION.

THIS ENGINE IS EQUIPPED WITH AUTOMATIC IDLE SPEED CONTROL. IDLE RPM IS NOT ADJUSTABLE. IF NOT WITHIN SPECIFIED RPM RANGE, SEE SHOP MANUAL.
AUTO. TRANS. IN DRIVE: 525-650 RPM

FIRING ORDER-1-5-4-2-6-3-7-8

THIS VEHICLE CONFORMS TO U.S. EPA REGULATIONS APPLICABLE TO 1989 MODEL YEAR NEW MOTOR VEHICLES.

E9AE-9C485- CATALYST SPARK PLUG: AWSF-44C GAP- .048-.052
5.0L-9HM
KFM5.0V5HBF4 - FI/EGR/EGS/AIP/TWC

VACUUM HOSE ROUTING

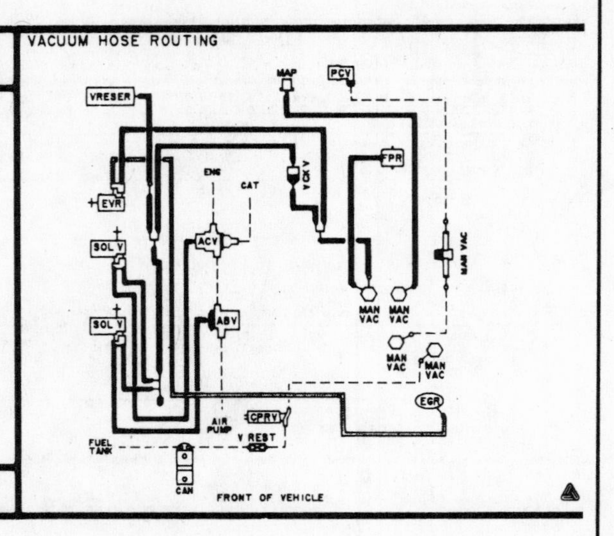

Vacuum Diagram 5.0L engine — 1989

CALIBRATION: 8–22L–R00 — 5.0L EFI

FORD MOTOR COMPANY
VEHICLE EMISSION CONTROL INFORMATION

THIS VEHICLE IS EQUIPPED WITH ELECTRONIC FUEL INJECTION. IDLE MIXTURE, COLD ENGINE IDLE SPEED AND COLD ENGINE FUEL ENRICHMENT ARE NOT ADJUSTABLE.

SET PARKING BRAKE AND BLOCK WHEELS. DISCONNECT AUTOMATIC PARKING BRAKE RELEASE, IF SO EQUIPPED. MAKE ALL ADJUSTMENTS WITH ENGINE AT NORMAL OPERATING TEMPERATURE AND ACCESSORIES OFF.

IGNITION TIMING- ADJUST WITH AUTO. TRANS. IN "D".

(1) TURN OFF ENGINE.
(2) DISCONNECT THE IN-LINE SPOUT CONNECTOR (=0d).
(3) RE-START PREVIOUSLY WARMED-UP ENGINE.
(4) ADJUST IGNITION TIMING TO 10° BTDC.
(5) TURN OFF ENGINE AND RESTORE ELECTRICAL CONNECTION.

THIS ENGINE IS EQUIPPED WITH AUTOMATIC IDLE SPEED CONTROL. IDLE RPM IS NOT ADJUSTABLE. IF NOT WITHIN SPECIFIED RPM RANGE, SEE SHOP MANUAL.
AUTO. TRANS. IN DRIVE: 550-675 RPM

FIRING ORDER-1-3-7-2-6-5-4-8

THIS VEHICLE CONFORMS TO U.S. EPA REGULATIONS APPLICABLE TO 1989 MODEL YEAR NEW MOTOR VEHICLES.

E9AE-9C485- CATALYST SPARK PLUG: ASF-42C GAP- .052-.056
5.0L-9HM
KFM5.0V5HBF4 - FI/EGR/EGS/AIP/TWC

VACUUM HOSE ROUTING

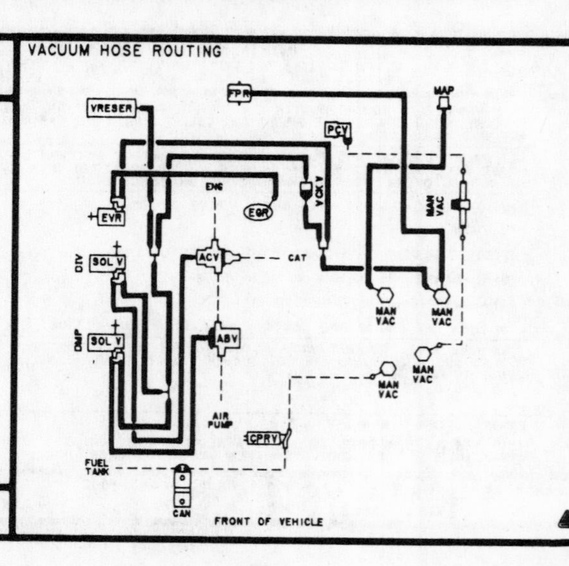

Vacuum Diagram 5.0L engine — 1989

CALIBRATION: 8–22M–R00 — 5.0L EFI

FORD MOTOR COMPANY
VEHICLE EMISSION CONTROL INFORMATION

THIS VEHICLE IS EQUIPPED WITH ELECTRONIC FUEL INJECTION. IDLE MIXTURE, COLD ENGINE IDLE SPEED AND COLD ENGINE FUEL ENRICHMENT NOT ADJUSTABLE.

SET PARKING BRAKE AND BLOCK WHEELS. DISCONNECT AUTOMATIC PARKING BRAKE RELEASE, IF SO EQUIPPED. MAKE ALL ADJUSTMENTS WITH ENGINE AT NORMAL OPERATING TEMPERATURE AND ACCESSORIES OFF.

IGNITION TIMING- ADJUST WITH AUTO. TRANS. IN DRIVE.

(1) TURN OFF ENGINE.
(2) DISCONNECT THE IN-LINE SPOUT CONNECTOR (-Dd).
(3) RE-START PREVIOUSLY WARMED-UP ENGINE.
(4) ADJUST IGNITION TIMING TO 10° BTDC.
(5) TURN OFF ENGINE AND RESTORE ELECTRICAL CONNECTION.

THIS ENGINE IS EQUIPPED WITH AUTOMATIC IDLE SPEED CONTROL. IDLE RPM IS NOT ADJUSTABLE. IF NOT WITHIN SPECIFIED RPM RANGE, SEE SHOP MANUAL.
AUTO. TRANS. IN DRIVE: 525-650 RPM

FIRING ORDER-1-5-4-2-6-3-7-8
THIS VEHICLE CONFORMS TO U.S. EPA AND CALIFORNIA REGULATIONS APPLICABLE TO 1989 MODEL YEAR NEW MOTOR VEHICLES INTRODUCED INTO COMMERCE SOLELY FOR SALE IN CALIFORNIA.

E9AE-9C485- CATALYST SPARK PLUG: AWSF-44C GAP- .048-.052
5.0L-9HM
KFM5.0V5HBC1 - FI/EGR/EGS/AIP/TWC

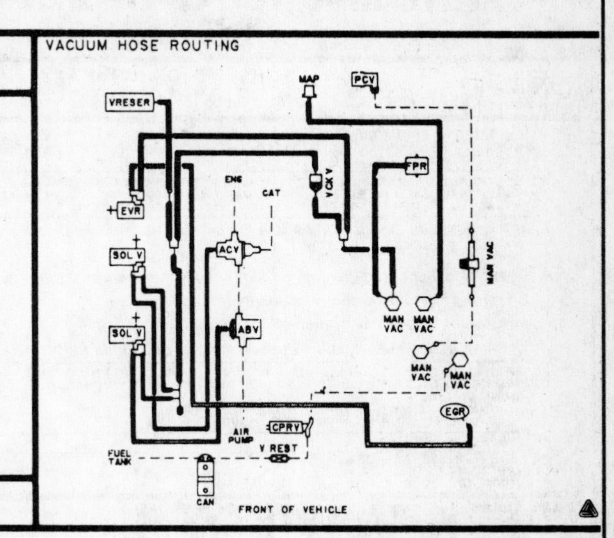

Vacuum Diagram 5.0L engine — 1989

CALIBRATION: 8–22N–R00 — 5.0L EFI

FORD MOTOR COMPANY
VEHICLE EMISSION CONTROL INFORMATION

THIS VEHICLE IS EQUIPPED WITH ELECTRONIC FUEL INJECTION. IDLE MIXTURE, COLD ENGINE IDLE SPEED AND COLD ENGINE FUEL ENRICHMENT NOT ADJUSTABLE.

SET PARKING BRAKE AND BLOCK WHEELS. DISCONNECT AUTOMATIC PARKING BRAKE RELEASE, IF SO EQUIPPED. MAKE ALL ADJUSTMENTS WITH ENGINE AT NORMAL OPERATING TEMPERATURE AND ACCESSORIES OFF.

IGNITION TIMING- ADJUST WITH AUTO. TRANS. IN DRIVE.

(1) TURN OFF ENGINE.
(2) DISCONNECT THE IN-LINE SPOUT CONNECTOR (-Dd).
(3) RE-START PREVIOUSLY WARMED-UP ENGINE.
(4) ADJUST IGNITION TIMING TO 10° BTDC.
(5) TURN OFF ENGINE AND RESTORE ELECTRICAL CONNECTION.

THIS ENGINE IS EQUIPPED WITH AUTOMATIC IDLE SPEED CONTROL. IDLE RPM IS NOT ADJUSTABLE. IF NOT WITHIN SPECIFIED RPM RANGE, SEE SHOP MANUAL.
AUTO. TRANS. IN DRIVE: 525-650 RPM

FIRING ORDER-1-5-4-2-6-3-7-8
THIS VEHICLE CONFORMS TO U.S. EPA AND CALIFORNIA REGULATIONS APPLICABLE TO 1989 MODEL YEAR NEW MOTOR VEHICLES INTRODUCED INTO COMMERCE SOLELY FOR SALE IN CALIFORNIA.

E9AE-9C485- CATALYST SPARK PLUG: AWSF-44C GAP- .048-.052
5.0L-9HM
KFM5.0V5HBC1 - FI/EGR/EGS/AIP/TWC

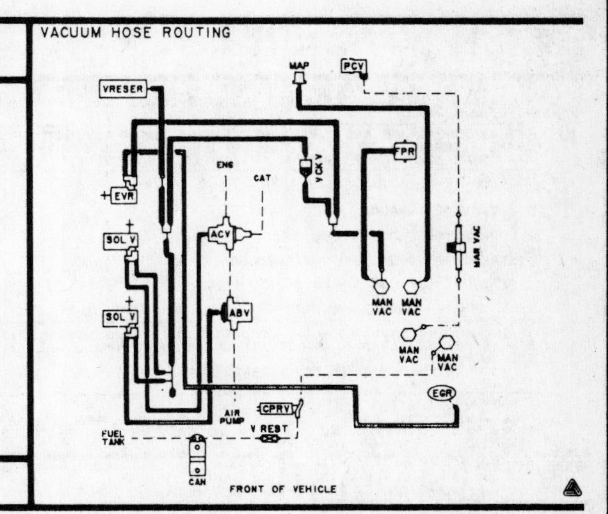

Vacuum Diagram 5.0L engine — 1989

CALIBRATION: 8–22P–R12 — 5.0L EFI

FORD MOTOR COMPANY
VEHICLE EMISSION CONTROL INFORMATION

THIS VEHICLE IS EQUIPPED WITH ELECTRONIC FUEL INJECTION. IDLE MIXTURE, COLD ENGINE IDLE SPEED AND COLD ENGINE FUEL ENRICHMENT ARE NOT ADJUSTABLE.

SET PARKING BRAKE AND BLOCK WHEELS. DISCONNECT AUTOMATIC PARKING BRAKE RELEASE, IF SO EQUIPPED. MAKE ALL ADJUSTMENTS WITH ENGINE AT NORMAL OPERATING TEMPERATURE AND ACCESSORIES OFF.

IGNITION TIMING- ADJUST WITH MAN. TRANS. IN NEUTRAL AND AUTO. TRANS. IN "D".

(1) TURN OFF ENGINE.

(2) DISCONNECT THE IN-LINE SPOUT CONNECTOR (=Dd).

(3) RE-START PREVIOUSLY WARMED-UP ENGINE.

(4) ADJUST IGNITION TIMING TO 10° BTDC.

(5) TURN OFF ENGINE AND RESTORE ELECTRICAL CONNECTION.

THIS ENGINE IS EQUIPPED WITH AUTOMATIC IDLE SPEED CONTROL. IDLE RPM IS NOT ADJUSTABLE. IF NOT WITHIN SPECIFIED RPM RANGE, SEE SHOP MANUAL:
AUTO. TRANS. IN DRIVE: 575-725 RPM
MANUAL TRANS. IN NEUTRAL: 625-775 RPM

FIRING ORDER-1-3-7-2-6-5-4-8

THIS VEHICLE CONFORMS TO U.S. EPA AND CALIFORNIA REGULATIONS APPLICABLE TO 1989 MODEL YEAR NEW MOTOR VEHICLES INTRODUCED INTO COMMERCE SOLELY FOR SALE IN CALIFORNIA.

E9AE-9C485- CATALYST SPARK PLUG: ASF-42C GAP- .052-.056
5.0L-9HM
KFM5.0V5HBC1 - FI/EGR/EGS/AIP/TWC

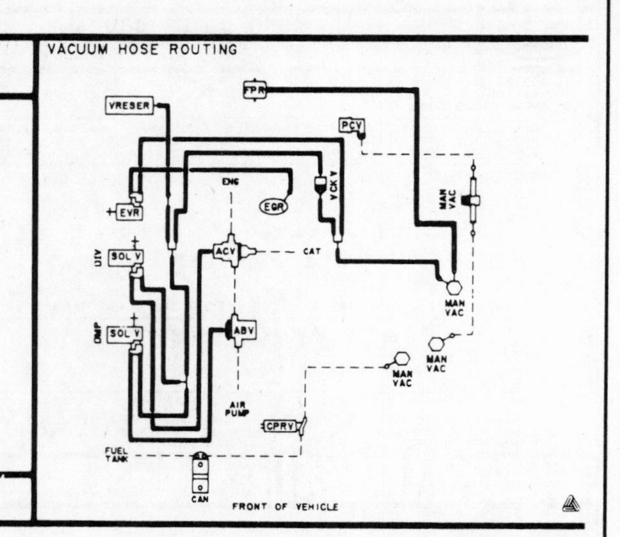

Vacuum Diagram 5.0L engine — 1989

CALIBRATION: 8–22Q–R00 — 5.0L EFI

FORD MOTOR COMPANY
VEHICLE EMISSION CONTROL INFORMATION

THIS VEHICLE IS EQUIPPED WITH ELECTRONIC FUEL INJECTION. IDLE MIXTURE, COLD ENGINE IDLE SPEED AND COLD ENGINE FUEL ENRICHMENT NOT ADJUSTABLE.

SET PARKING BRAKE AND BLOCK WHEELS. DISCONNECT AUTOMATIC PARKING BRAKE RELEASE, IF SO EQUIPPED. MAKE ALL ADJUSTMENTS WITH ENGINE AT NORMAL OPERATING TEMPERATURE AND ACCESSORIES OFF.

IGNITION TIMING- ADJUST WITH AUTO. TRANS. IN DRIVE.

(1) TURN OFF ENGINE.

(2) DISCONNECT THE IN-LINE SPOUT CONNECTOR (=Dd).

(3) RE-START PREVIOUSLY WARMED-UP ENGINE.

(4) ADJUST IGNITION TIMING TO 10° BTDC.

(5) TURN OFF ENGINE AND RESTORE ELECTRICAL CONNECTION.

THIS ENGINE IS EQUIPPED WITH AUTOMATIC IDLE SPEED CONTROL. IDLE RPM IS NOT ADJUSTABLE. IF NOT WITHIN SPECIFIED RPM RANGE, SEE SHOP MANUAL.
AUTO. TRANS. IN DRIVE: 525-650 RPM

FIRING ORDER-1-5-4-2-6-3-7-8

THIS VEHICLE CONFORMS TO U.S. EPA AND CALIFORNIA REGULATIONS APPLICABLE TO 1989 MODEL YEAR NEW MOTOR VEHICLES INTRODUCED INTO COMMERCE SOLELY FOR SALE IN CALIFORNIA.

E9AE-9C485- CATALYST SPARK PLUG: AWSF-44C GAP- .048-.052
5.0L-9HM
KFM5.0V5HBC1 - FI/EGR/EGS/AIP/TWC

VACUUM HOSE ROUTING

Vacuum Diagram 5.0L engine — 1989

CALIBRATION: 8–22R–R00 — 5.0L EFI

FORD MOTOR COMPANY
VEHICLE EMISSION CONTROL INFORMATION

THIS VEHICLE IS EQUIPPED WITH ELECTRONIC FUEL INJECTION. IDLE MIXTURE, COLD ENGINE IDLE SPEED AND COLD ENGINE FUEL ENRICHMENT ARE NOT ADJUSTABLE.

SET PARKING BRAKE AND BLOCK WHEELS. DISCONNECT AUTOMATIC PARKING BRAKE RELEASE, IF SO EQUIPPED. MAKE ALL ADJUSTMENTS WITH ENGINE AT NORMAL OPERATING TEMPERATURE AND ACCESSORIES OFF.

IGNITION TIMING- ADJUST WITH AUTO. TRANS. IN "D".

(1) TURN OFF ENGINE.

(2) DISCONNECT THE IN-LINE SPOUT CONNECTOR (=0d).

(3) RE-START PREVIOUSLY WARMED-UP ENGINE.

(4) ADJUST IGNITION TIMING TO 10° BTDC.

(5) TURN OFF ENGINE AND RESTORE ELECTRICAL CONNECTION.

THIS ENGINE IS EQUIPPED WITH AUTOMATIC IDLE SPEED CONTROL. IDLE RPM IS NOT ADJUSTABLE. IF NOT WITHIN SPECIFIED RPM RANGE, SEE SHOP MANUAL:
AUTO. TRANS. IN DRIVE: 550-675 RPM

FIRING ORDER-1-3-7-2-6-5-4-8

THIS VEHICLE CONFORMS TO U.S. EPA AND CALIFORNIA REGULATIONS APPLICABLE TO 1989 MODEL YEAR NEW MOTOR VEHICLES INTRODUCED INTO COMMERCE SOLELY FOR SALE IN CALIFORNIA.

E9AE-9C485- CATALYST

SPARK PLUG: ASF-42C GAP- .052-.056
5.0L-9HM
KFM5.0V5HBC1 - FI/EGR/EGS/AIP/TWC

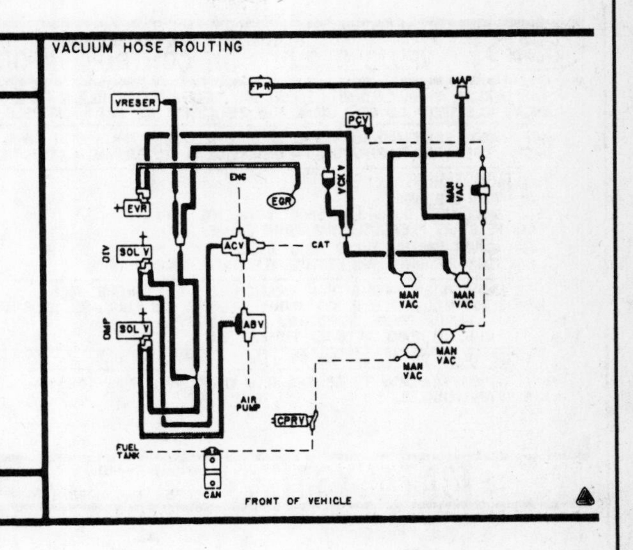

Vacuum Diagram 5.0L engine — 1989

CALIBRATION: 8–22S–R00 — 5.0L EFI

FORD MOTOR COMPANY
VEHICLE EMISSION CONTROL INFORMATION

THIS VEHICLE IS EQUIPPED WITH ELECTRONIC FUEL INJECTION. IDLE MIXTURE, COLD ENGINE IDLE SPEED AND COLD ENGINE FUEL ENRICHMENT NOT ADJUSTABLE.

SET PARKING BRAKE AND BLOCK WHEELS. DISCONNECT AUTOMATIC PARKING BRAKE RELEASE, IF SO EQUIPPED. MAKE ALL ADJUSTMENTS WITH ENGINE AT NORMAL OPERATING TEMPERATURE AND ACCESSORIES OFF.

IGNITION TIMING- ADJUST WITH AUTO. TRANS. IN DRIVE.

(1) TURN OFF ENGINE.

(2) DISCONNECT THE IN-LINE SPOUT CONNECTOR (=0d).

(3) RE-START PREVIOUSLY WARMED-UP ENGINE.

(4) ADJUST IGNITION TIMING TO 10° BTDC.

(5) TURN OFF ENGINE AND RESTORE ELECTRICAL CONNECTION.

THIS ENGINE IS EQUIPPED WITH AUTOMATIC IDLE SPEED CONTROL. IDLE RPM IS NOT ADJUSTABLE. IF NOT WITHIN SPECIFIED RPM RANGE, SEE SHOP MANUAL.
AUTO. TRANS. IN DRIVE: 525-650 RPM

FIRING ORDER-1-5-4-2-6-3-7-8

THIS VEHICLE CONFORMS TO U.S. EPA AND CALIFORNIA REGULATIONS APPLICABLE TO 1989 MODEL YEAR NEW MOTOR VEHICLES INTRODUCED INTO COMMERCE SOLELY FOR SALE IN CALIFORNIA.

E9AE-9C485- CATALYST

SPARK PLUG: AWSF-44C GAP- .048-.052
5.0L-9HM
KFM5.0V5HBC1 - FI/EGR/EGS/AIP/TWC

VACUUM HOSE ROUTING

Vacuum Diagram 5.0L engine — 1989

■ CALIBRATION: 8-05A-R10 — 2.3L OHC-EFI

FORD MOTOR COMPANY
VEHICLE EMISSION CONTROL INFORMATION

THIS VEHICLE IS EQUIPPED WITH ELECTRONIC FUEL INJECTION. IDLE MIXTURE, COLD ENGINE IDLE SPEED AND COLD ENGINE FUEL ENRICHMENT ARE NOT ADJUSTABLE.

SET PARKING BRAKE AND BLOCK WHEELS. MAKE ALL ADJUSTMENTS WITH ENGINE AT NORMAL OPERATING TEMPERATURE, TRANSMISSION IN NEUTRAL AND ACCESSORIES OFF.

IGNITION TIMING-
(1) TURN OFF ENGINE.
(2) DISCONNECT THE IN-LINE SPOUT CONNECTOR (=▯◁).
(3) RE-START PREVIOUSLY WARMED-UP ENGINE.
(4) ADJUST IGNITION TIMING TO 10° BTDC.
(5) TURN OFF ENGINE AND RESTORE ELECTRICAL CONNECTION.

THIS ENGINE IS EQUIPPED WITH AUTOMATIC IDLE SPEED CONTROL. IDLE RPM IS NOT ADJUSTABLE. IF NOT WITHIN SPECIFIED RPM RANGE, SEE SHOP MANUAL:
MANUAL TRANS. IN NEUTRAL: 770-830 RPM
AUTO. TRANS. IN DRIVE: 770-830 RPM

USE SAE 5W-30 OIL API SERVICE SG - ENERGY CONSERVING II.

THIS VEHICLE CONFORMS TO U.S. EPA REGULATIONS APPLICABLE TO 1990 MODEL YEAR NEW MOTOR VEHICLES.

CATALYST
SPARK PLUG: AWSF-44C GAP- .042-.046
2.3L-9HM
LFM2.3V5FYF5 - TWC/HO2S/EGR/MPI

VACUUM HOSE ROUTING

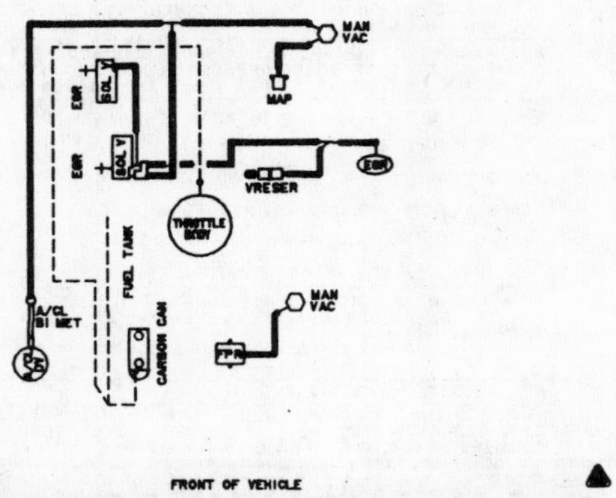

Vacuum Diagram 2.3L engine — 1990

■ CALIBRATION: 0-05S-R00 — 2.3L OHC-EFI

FORD MOTOR COMPANY
VEHICLE EMISSION CONTROL INFORMATION

THIS VEHICLE IS EQUIPPED WITH ELECTRONIC FUEL INJECTION. IDLE MIXTURE, COLD ENGINE IDLE SPEED AND COLD ENGINE FUEL ENRICHMENT ARE NOT ADJUSTABLE.

SET PARKING BRAKE AND BLOCK WHEELS. MAKE ALL ADJUSTMENTS WITH ENGINE AT NORMAL OPERATING TEMPERATURE, TRANSMISSION IN NEUTRAL AND ACCESSORIES OFF.

IGNITION TIMING-
(1) TURN OFF ENGINE
(2) DISCONNECT THE IN-LINE SPOUT CONNECTOR (=▯◁).
(3) RE-START PREVIOUSLY WARMED-UP ENGINE.
(4) ADJUST IGNITION TIMING TO 10° BTDC.
(5) TURN OFF ENGINE AND RESTORE ELECTRICAL CONNECTION.

THIS ENGINE IS EQUIPPED WITH AUTOMATIC IDLE SPEED CONTROL. IDLE RPM IS NOT ADJUSTABLE. IF NOT WITHIN SPECIFIED RPM RANGE, SEE SHOP MANUAL:
MAN. TRANS. IN NEUTRAL: 830-890 RPM
AUTO. TRANS. IN DRIVE: 790-850 RPM

USE SAE 5W-30 OIL API SERVICE SG - ENERGY CONSERVING II.

THIS VEHICLE CONFORMS TO U.S. EPA AND CALIFORNIA REGULATIONS APPLICABLE TO 1990 MODEL YEAR NEW MOTOR VEHICLES INTRODUCED INTO COMMERCE SOLELY FOR SALE IN CALIFORNIA. OBD EXEMPT.

CATALYST
SPARK PLUG: AWSF-44C GAP- .042-.046
2.3L-9HM
LFM2.3V5FYC2 - TWC/HO2S/EGR/MPI

VACUUM HOSE ROUTING

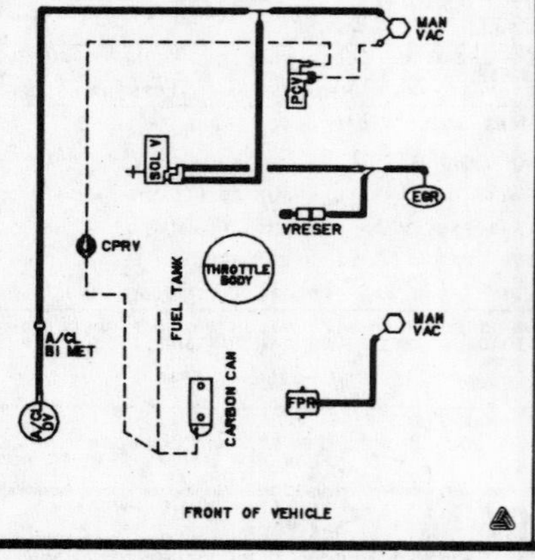

Vacuum Diagram 2.3L engine — 1990

CALIBRATION: 0-06S-R00 — 2.3L OHC-EFI

FORD MOTOR COMPANY
VEHICLE EMISSION CONTROL INFORMATION

THIS VEHICLE IS EQUIPPED WITH ELECTRONIC FUEL INJECTION. IDLE MIXTURE, COLD ENGINE IDLE SPEED AND COLD ENGINE FUEL ENRICHMENT ARE NOT ADJUSTABLE.

SET PARKING BRAKE AND BLOCK WHEELS. MAKE ALL ADJUSTMENTS WITH ENGINE AT NORMAL OPERATING TEMPERATURE, TRANSMISSION IN NEUTRAL AND ACCESSORIES OFF.

IGNITION TIMING-
(1) TURN OFF ENGINE
(2) DISCONNECT THE IN-LINE SPOUT CONNECTOR (=□ᑫ).
(3) RE-START PREVIOUSLY WARMED-UP ENGINE.
(4) ADJUST IGNITION TIMING TO 10° BTDC.
(5) TURN OFF ENGINE AND RESTORE ELECTRICAL CONNECTION.

THIS ENGINE IS EQUIPPED WITH AUTOMATIC IDLE SPEED CONTROL. IDLE RPM IS NOT ADJUSTABLE. IF NOT WITHIN SPECIFIED RPM RANGE, SEE SHOP MANUAL:
MAN. TRANS. IN NEUTRAL: 830-890 RPM
AUTO. TRANS. IN DRIVE: 790-850 RPM

USE SAE 5W-30 OIL API SERVICE SG - ENERGY CONSERVING II.

THIS VEHICLE CONFORMS TO U.S. EPA AND CALIFORNIA REGULATIONS APPLICABLE TO 1990 MODEL YEAR NEW MOTOR VEHICLES INTRODUCED INTO COMMERCE SOLELY FOR SALE IN CALIFORNIA. OBD EXEMPT.

CATALYST
SPARK PLUG: AWSF-44C GAP- .042-.046
2.3L -9HM
LFM2.3V5FYC2 - TWC/HO2S/EGR/MPI

VACUUM HOSE ROUTING
MAN VAC
PCV
SOL V
EGR
VRESER
CPRV
THROTTLE BODY
FUEL TANK
MAN VAC
A/CL BI MET
CARBON CAN
A/CL DV
FPR
FRONT OF VEHICLE
23V5FYCC

Vacuum Diagram 2.3L engine — 1990

CALIBRATION: 8-06A-R10 — 2.3L OHC-EFI

FORD MOTOR COMPANY
VEHICLE EMISSION CONTROL INFORMATION

THIS VEHICLE IS EQUIPPED WITH ELECTRONIC FUEL INJECTION. IDLE MIXTURE, COLD ENGINE IDLE SPEED AND COLD ENGINE FUEL ENRICHMENT ARE NOT ADJUSTABLE.

SET PARKING BRAKE AND BLOCK WHEELS. MAKE ALL ADJUSTMENTS WITH ENGINE AT NORMAL OPERATING TEMPERATURE, TRANSMISSION IN NEUTRAL AND ACCESSORIES OFF.

IGNITION TIMING-
(1) TURN OFF ENGINE.
(2) DISCONNECT THE IN-LINE SPOUT CONNECTOR (=□ᑫ).
(3) RE-START PREVIOUSLY WARMED-UP ENGINE.
(4) ADJUST IGNITION TIMING TO 10° BTDC.
(5) TURN OFF ENGINE AND RESTORE ELECTRICAL CONNECTION.

THIS ENGINE IS EQUIPPED WITH AUTOMATIC IDLE SPEED CONTROL. IDLE RPM IS NOT ADJUSTABLE. IF NOT WITHIN SPECIFIED RPM RANGE, SEE SHOP MANUAL:
MANUAL TRANS. IN NEUTRAL: 770-830 RPM
AUTO. TRANS. IN DRIVE: 770-830 RPM

USE SAE 5W-30 OIL API SERVICE SG - ENERGY CONSERVING II.

THIS VEHICLE CONFORMS TO U.S. EPA REGULATIONS APPLICABLE TO 1990 MODEL YEAR NEW MOTOR VEHICLES.

CATALYST
SPARK PLUG: AWSF-44C GAP- .042-.046
2.3L -9HM
LFM2.3V5EYF5 - TWC/HO2S/EGR/MPI

VACUUM HOSE ROUTING

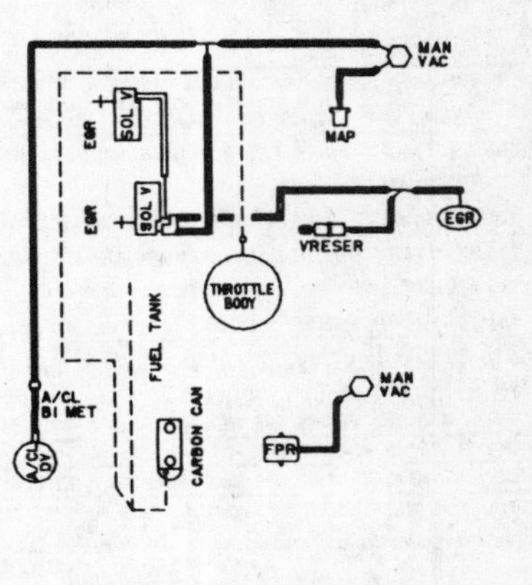

FRONT OF VEHICLE

Vacuum Diagram 2.3L engine — 1990

CALIBRATION: 9-21C-R00 — 5.0L HO + EFI

FORD MOTOR COMPANY
VEHICLE EMISSION CONTROL INFORMATION

THIS VEHICLE IS EQUIPPED WITH ELECTRONIC FUEL INJECTION. IDLE MIXTURE, COLD ENGINE IDLE SPEED AND COLD ENGINE FUEL ENRICHMENT ARE NOT ADJUSTABLE.

SET PARKING BRAKE AND BLOCK WHEELS. DISCONNECT AUTOMATIC PARKING BRAKE RELEASE, IF SO EQUIPPED. MAKE ALL ADJUSTMENTS WITH ENGINE AT NORMAL OPERATING TEMPERATURE AND ACCESSORIES OFF.

IGNITION TIMING- ADJUST WITH TRANSMISSION IN NEUTRAL.

(1) TURN OFF ENGINE.
(2) DISCONNECT THE IN-LINE SPOUT CONNECTOR (=⊐⊏).
(3) RE-START PREVIOUSLY WARMED-UP ENGINE.
(4) ADJUST IGNITION TIMING TO 10° BTDC.
(5) TURN OFF ENGINE AND RESTORE ELECTRICAL CONNECTION.

THIS ENGINE IS EQUIPPED WITH AUTOMATIC IDLE SPEED CONTROL. IDLE RPM IS NOT ADJUSTABLE. IF NOT WITHIN SPECIFIED RPM RANGE, SEE SHOP MANUAL.
AUTO. TRANS. IN DRIVE: 575-725 RPM
MANUAL TRANS. IN NEUTRAL: 625-775 RPM

FIRING ORDER-1-3-7-2-6-5-4-8

USE SAE 10W-30 OIL API SERVICE SG - ENERGY CONSERVING II.

THIS VEHICLE CONFORMS TO U.S. EPA REGULATIONS APPLICABLE TO 1990 MODEL YEAR NEW MOTOR VEHICLES.

CATALYST
SPARK PLUG: ASF-42C GAP- .052-.056
5.0L-9HM
LFM5.0V5HBG6 - TWC+OC/AIR/HO2S/EGR/SMPI

VACUUM HOSE ROUTING

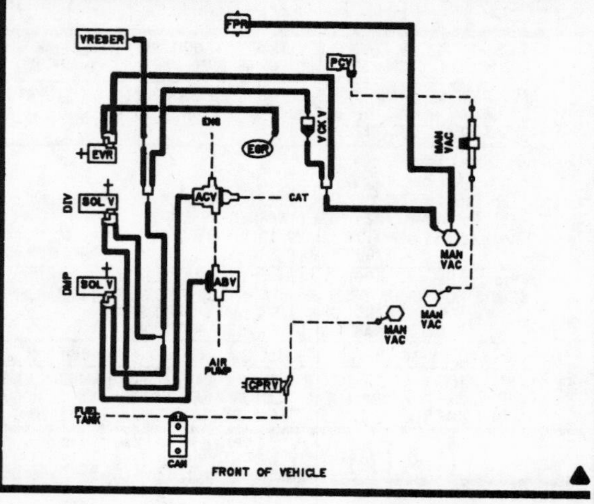

FPR, VRESER, PCV, EVR, ENG, EGR, VCKV, MAN VAC, DIV, SOL V, ACV, CAT, DMP, ABV, AIR PUMP, CPRV, FUEL TANK, CAN, FRONT OF VEHICLE

Vacuum Diagram 5.0L engine — 1990

CALIBRATION: 9-22A-R00 — 5.0L HO + EFI

FORD MOTOR COMPANY
VEHICLE EMISSION CONTROL INFORMATION

THIS VEHICLE IS EQUIPPED WITH ELECTRONIC FUEL INJECTION. IDLE MIXTURE, COLD ENGINE IDLE SPEED AND COLD ENGINE FUEL ENRICHMENT ARE NOT ADJUSTABLE.

SET PARKING BRAKE AND BLOCK WHEELS. DISCONNECT AUTOMATIC PARKING BRAKE RELEASE, IF SO EQUIPPED. MAKE ALL ADJUSTMENTS WITH ENGINE AT NORMAL OPERATING TEMPERATURE AND ACCESSORIES OFF.

IGNITION TIMING- ADJUST WITH TRANSMISSION IN NEUTRAL.

(1) TURN OFF ENGINE.
(2) DISCONNECT THE IN-LINE SPOUT CONNECTOR (=⊐⊏).
(3) RE-START PREVIOUSLY WARMED-UP ENGINE.
(4) ADJUST IGNITION TIMING TO 10° BTDC.
(5) TURN OFF ENGINE AND RESTORE ELECTRICAL CONNECTION.

THIS ENGINE IS EQUIPPED WITH AUTOMATIC IDLE SPEED CONTROL. IDLE RPM IS NOT ADJUSTABLE. IF NOT WITHIN SPECIFIED RPM RANGE, SEE SHOP MANUAL.
AUTO. TRANS. IN DRIVE: 575-725 RPM
MANUAL TRANS. IN NEUTRAL: 625-775 RPM

FIRING ORDER-1-3-7-2-6-5-4-8

USE SAE 10W-30 OIL API SERVICE SG - ENERGY CONSERVING II.

THIS VEHICLE CONFORMS TO U.S. EPA REGULATIONS APPLICABLE TO 1990 MODEL YEAR NEW MOTOR VEHICLES.

CATALYST
SPARK PLUG: ASF-42C GAP- .052-.056
5.0L-9HM
LFM5.0V5HBF5 - TWC+OC/AIR/HO2S/EGR/SMPI

VACUUM HOSE ROUTING

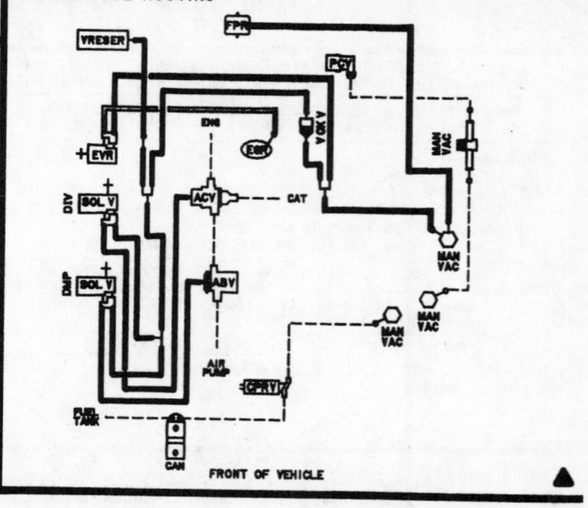

FPR, VRESER, PCV, EVR, ENG, EGR, VCKV, MAN VAC, DIV, SOL V, ACV, CAT, DMP, ABV, AIR PUMP, CPRV, FUEL TANK, CAN, FRONT OF VEHICLE

Vacuum Diagram 5.0L engine — 1990

CALIBRATION: 8-22D-R00 — 5.0L HO + EFI

FORD MOTOR COMPANY
VEHICLE EMISSION CONTROL INFORMATION

THIS VEHICLE IS EQUIPPED WITH ELECTRONIC FUEL INJECTION. IDLE MIXTURE, COLD ENGINE IDLE SPEED AND COLD ENGINE FUEL ENRICHMENT ARE NOT ADJUSTABLE.

SET PARKING BRAKE AND BLOCK WHEELS. DISCONNECT AUTOMATIC PARKING BRAKE RELEASE, IF SO EQUIPPED. MAKE ALL ADJUSTMENTS WITH ENGINE AT NORMAL OPERATING TEMPERATURE AND ACCESSORIES OFF.

IGNITION TIMING- ADJUST WITH TRANSMISSION IN NEUTRAL.

(1) TURN OFF ENGINE.
(2) DISCONNECT THE IN-LINE SPOUT CONNECTOR (=□d).
(3) RE-START PREVIOUSLY WARMED-UP ENGINE.
(4) ADJUST IGNITION TIMING TO 10° BTDC.
(5) TURN OFF ENGINE AND RESTORE ELECTRICAL CONNECTION.

THIS ENGINE IS EQUIPPED WITH AUTOMATIC IDLE SPEED CONTROL. IDLE RPM IS NOT ADJUSTABLE. IF NOT WITHIN SPECIFIED RPM RANGE, SEE SHOP MANUAL.
AUTO. TRANS. IN DRIVE: 550-675 RPM

FIRING ORDER-1-3-7-2-6-5-4-8

USE SAE 10W-30 OIL API SERVICE SG - ENERGY CONSERVING II.

THIS VEHICLE CONFORMS TO U.S. EPA REGULATIONS APPLICABLE TO 1990 MODEL YEAR NEW MOTOR VEHICLES.

CATALYST — SPARK PLUG: ASF-42C GAP- .052-.056
5.0L-9HM
LFM5.0V5HBG6 - TWC+OC/AIR/HO2S/EGR/SMPI

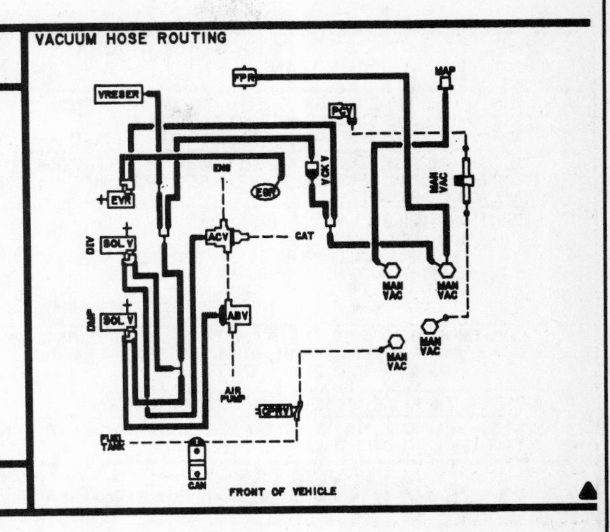

Vacuum Diagram 5.0L engine — 1990

CALIBRATION: 8-22L-R00 — 5.0L HO + EFI

FORD MOTOR COMPANY
VEHICLE EMISSION CONTROL INFORMATION

THIS VEHICLE IS EQUIPPED WITH ELECTRONIC FUEL INJECTION. IDLE MIXTURE, COLD ENGINE IDLE SPEED AND COLD ENGINE FUEL ENRICHMENT ARE NOT ADJUSTABLE.

SET PARKING BRAKE AND BLOCK WHEELS. DISCONNECT AUTOMATIC PARKING BRAKE RELEASE, IF SO EQUIPPED. MAKE ALL ADJUSTMENTS WITH ENGINE AT NORMAL OPERATING TEMPERATURE AND ACCESSORIES OFF.

IGNITION TIMING- ADJUST WITH TRANSMISSION IN NEUTRAL.

(1) TURN OFF ENGINE.
(2) DISCONNECT THE IN-LINE SPOUT CONNECTOR (=□d).
(3) RE-START PREVIOUSLY WARMED-UP ENGINE.
(4) ADJUST IGNITION TIMING TO 10° BTDC.
(5) TURN OFF ENGINE AND RESTORE ELECTRICAL CONNECTION.

THIS ENGINE IS EQUIPPED WITH AUTOMATIC IDLE SPEED CONTROL. IDLE RPM IS NOT ADJUSTABLE. IF NOT WITHIN SPECIFIED RPM RANGE, SEE SHOP MANUAL.
AUTO. TRANS. IN DRIVE: 550-675 RPM

FIRING ORDER-1-3-7-2-6-5-4-8

USE SAE 10W-30 OIL API SERVICE SG - ENERGY CONSERVING II.

THIS VEHICLE CONFORMS TO U.S. EPA REGULATIONS APPLICABLE TO 1990 MODEL YEAR NEW MOTOR VEHICLES.

CATALYST — SPARK PLUG: ASF-42C GAP- .052-.056
5.0L-9HM
LFM5.0V5HBF5 - TWC+OC/AIR/HO2S/EGR/SMPI

VACUUM HOSE ROUTING

Vacuum Diagram 5.0L engine — 1990

CALIBRATION: 0-50T-R10 — 2.3L EFI

FORD MOTOR COMPANY
IMPORTANT VEHICLE INFORMATION

THIS VEHICLE IS EQUIPPED WITH EEC IV, EFI AND DIS SYSTEMS. IDLE SPEEDS, IDLE MIXTURES AND IGNITION TIMING ARE NOT ADJUSTABLE. SEE SHOP MANUAL FOR ADDITIONAL INFORMATION.

SET PARKING BRAKE AND BLOCK WHEELS. MAKE ALL VERIFICATIONS WITH ENGINE AT NORMAL OPERATING TEMPERATURE, TRANSMISSION IN NEUTRAL AND ACCESSORIES OFF.

IGNITION TIMING IS NOT ADJUSTABLE
(1) TURN OFF ENGINE.
(2) DISCONNECT SMALL IN-LINE SPOUT CONNECTOR (=▯◁).
(3) RE-START PREVIOUSLY WARMED-UP ENGINE.
(4) VERIFY THAT THE IGNITION TIMING IS 10° BTDC. IF NOT SEE SHOP MANUAL.
(5) TURN OFF ENGINE AND RESTORE ELECTRICAL CONNECTION.

THIS ENGINE IS EQUIPPED WITH AUTOMATIC IDLE SPEED CONTROL. IDLE RPM IS NOT ADJUSTABLE. IF NOT WITHIN SPECIFIED RPM RANGE, SEE SHOP MANUAL:
AUTO. TRANS.IN DRIVE: 575-725 RPM

USE SAE 5W-30 OIL API SERVICE SG - ENERGY CONSERVING II.

THIS VEHICLE CONFORMS TO U.S. EPA REGULATIONS APPLICABLE TO 1991 MODEL YEAR NEW LIGHT-DUTY TRUCKS.

CATALYST | SPARK PLUG: AWSF-32C GAP- .042-.046
2.3L-9HM
MFM2.3T5FMG0 - TWC/HO2S/MPI

VACUUM HOSE ROUTING

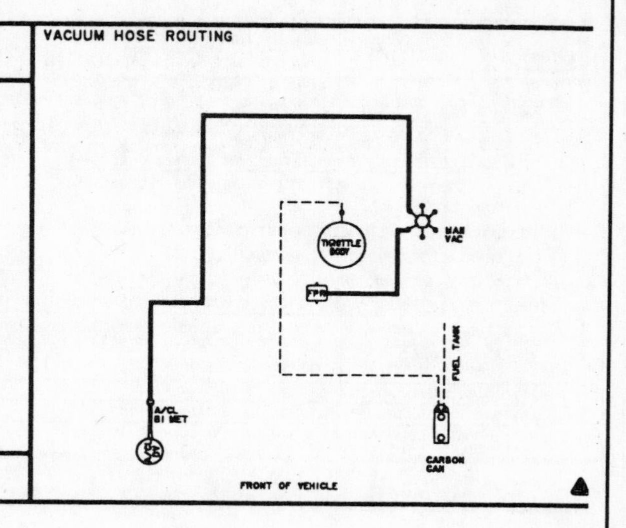

Vacuum Diagram 2.3L engine — 1991

CALIBRATION: 1-22F-R00 — 5.0L SEFI-HO

FORD MOTOR COMPANY
VEHICLE EMISSION CONTROL INFORMATION

THIS VEHICLE IS EQUIPPED WITH ELECTRONIC FUEL INJECTION. IDLE MIXTURE, COLD ENGINE IDLE SPEED AND COLD ENGINE FUEL ENRICHMENT ARE NOT ADJUSTABLE.

SET PARKING BRAKE AND BLOCK WHEELS. DISCONNECT AUTOMATIC PARKING BRAKE RELEASE, IF SO EQUIPPED. MAKE ALL ADJUSTMENTS WITH ENGINE AT NORMAL OPERATING TEMPERATURE AND ACCESSORIES OFF.

IGNITION TIMING- ADJUST WITH TRANSMISSION IN NEUTRAL.
(1) TURN OFF ENGINE.
(2) DISCONNECT THE IN-LINE SPOUT CONNECTOR (=▯◁).
(3) RE-START PREVIOUSLY WARMED-UP ENGINE.
(4) ADJUST IGNITION TIMING TO 10° BTDC.
(5) TURN OFF ENGINE AND RESTORE ELECTRICAL CONNECTION.

THIS ENGINE IS EQUIPPED WITH AUTOMATIC IDLE SPEED CONTROL. IDLE RPM IS NOT ADJUSTABLE. IF NOT WITHIN SPECIFIED RPM RANGE, SEE SHOP MANUAL.
AUTO. TRANS. IN DRIVE: 610 RPM

FIRING ORDER-1-3-7-2-6-5-4-8

USE SAE 10W-30 OIL API SERVICE SG - ENERGY CONSERVING II.

THIS VEHICLE CONFORMS TO U.S. EPA REGULATIONS APPLICABLE TO 1991 MODEL YEAR NEW MOTOR VEHICLES.

CATALYST | SPARK PLUG: ASF-42C GAP- .052-.056
5.0L-9HM
MFM5.0V5FXFX - TWC/AIR/HO2S/EGR/SMPI

VACUUM HOSE ROUTING

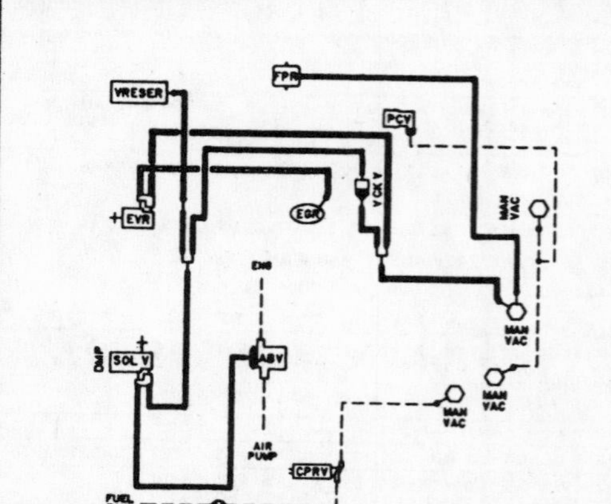

Vacuum Diagram 5.0L engine — 1991

CALIBRATION: 1-22T-R00 — 5.0L SEFI-HO

FORD MOTOR COMPANY
VEHICLE EMISSION CONTROL INFORMATION

THIS VEHICLE IS EQUIPPED WITH ELECTRONIC FUEL INJECTION. IDLE MIXTURE, COLD ENGINE IDLE SPEED AND COLD ENGINE FUEL ENRICHMENT ARE NOT ADJUSTABLE.

SET PARKING BRAKE AND BLOCK WHEELS. DISCONNECT AUTOMATIC PARKING BRAKE RELEASE, IF SO EQUIPPED. MAKE ALL ADJUSTMENTS WITH ENGINE AT NORMAL OPERATING TEMPERATURE AND ACCESSORIES OFF.

IGNITION TIMING- ADJUST WITH TRANSMISSION IN NEUTRAL.

(1) TURN OFF ENGINE.
(2) DISCONNECT THE IN-LINE SPOUT CONNECTOR (=▯◁).
(3) RE-START PREVIOUSLY WARMED-UP ENGINE.
(4) ADJUST IGNITION TIMING TO 10° BTDC.
(5) TURN OFF ENGINE AND RESTORE ELECTRICAL CONNECTION.

THIS ENGINE IS EQUIPPED WITH AUTOMATIC IDLE SPEED CONTROL. IDLE RPM IS NOT ADJUSTABLE. IF NOT WITHIN SPECIFIED RPM RANGE, SEE SHOP MANUAL:
AUTO. TRANS. IN DRIVE: 610 RPM

FIRING ORDER-1-3-7-2-6-5-4-8

USE SAE 10W-30 OIL API SERVICE SG - ENERGY CONSERVING II.

THIS VEHICLE CONFORMS TO U.S. EPA AND CALIFORNIA REGULATIONS APPLICABLE TO 1991 MODEL YEAR NEW MOTOR VEHICLES INTRODUCED INTO COMMERCE SOLELY FOR SALE IN CALIFORNIA.

CATALYST | SPARK PLUG: ASF-42C GAP- .052-.056
5.0L 9HM
MFM5.0V5FXC7 - TWC/AIR/HO2S/EGR/SMPI

VACUUM HOSE ROUTING

FPR
VRESER
PCV
EGR
VCKV
EVR
MAN VAC
EMS
DMP
SOL V
ABV
MAN VAC
AIR PUMP
CPRV
FUEL TANK
CAN
FRONT OF VEHICLE

50V5FXCD

Vacuum Diagram 5.0L engine — 1991

CALIBRATION: 1-05A-R05 — 2.3L EFI

Ford Motor Company
VEHICLE EMISSION CONTROL INFORMATION

This vehicle is equipped with EEC-IV, EFI and DIS systems. Engine idle speed, idle mixture, and ignition timing are not adjustable. See Engine/Emissions Diagnosis Shop Manual for additional information.

To check engine timing set parking brake and block wheels. Engine must be at normal operating temperature, transmission in neutral, and accessories off.

(1) Turn off engine.
(2) Disconnect the in-line Spout Connector (=▯◁).
(3) Re-start previously warmed-up engine.
(4) Verify that the ignition timing is 10° BTDC. If not see shop manual.
(5) Turn engine off and restore electrical connection.

Use SAE 5W-30 Oil API Service SG - Energy Conserving II.

This vehicle conforms to U.S. EPA regulations applicable to 1992 model year new motor vehicles.

F2AE-9C485-
H D B | Catalyst | Spark Plug: AWSF-32C Gap: .042-.046
2.3L-9HM
NFM2.3V5FYF7-TWC/HO2S/EGR/MPI

VACUUM HOSE ROUTING

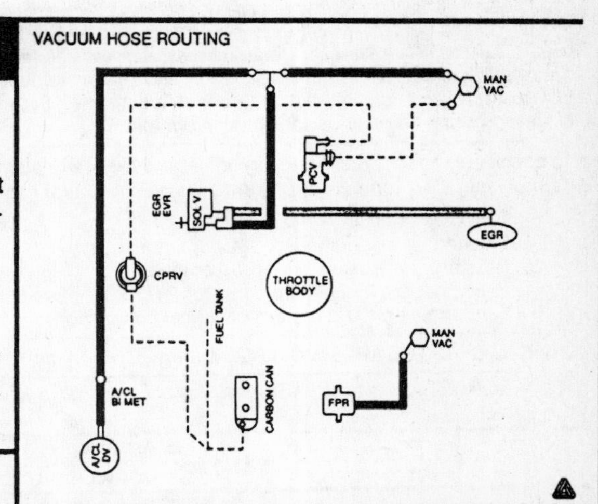

Vacuum Diagram 2.3L engine — 1992

CALIBRATION: 1-05S-R05 2.3L EFI

Ford Motor Company
VEHICLE EMISSION CONTROL INFORMATION

This vehicle is equipped with EEC-IV, EFI and DIS systems. Engine idle speed, idle mixture, and ignition timing are not adjustable. See Engine/Emissions Diagnosis Shop Manual for additional information.

To check engine timing set parking brake and block wheels. Engine must be at normal operating temperature, transmission in neutral, and accessories off.

(1) Turn off engine.
(2) Disconnect the in-line Spout Connector (=▯◁).
(3) Re-start previously warmed-up engine.
(4) Verify that the ignition timing is 10° BTDC. If not see shop manual.
(5) Turn engine off and restore electrical connection.

Use SAE 5W-30 Oil API Service SG - Energy Conserving II.

This vehicle conforms to U.S. EPA and California regulations applicable to 1992 model year new motor vehicles introduced into commerce solely for sale in California.

F2AE-9C485- HDD	Catalyst	Spark Plug: AWSF-32C Gap: .042-.046 2.3L-9HM NFM2.3V5FYC4-TWC/HO2S/EGR/MPI

VACUUM HOSE ROUTING

MAN VAC, PCV, EGR EVR, SOL V, EGR, CPRV, THROTTLE BODY, FUEL TANK, MAN VAC, A/CL BI MET, CARBON CAN, FPR, A/CL DV

23V5FYCC

Vacuum Diagram 2.3L engine — 1992

CALIBRATION: 1-06A-R05 2.3L EFI

Ford Motor Company
VEHICLE EMISSION CONTROL INFORMATION

This vehicle is equipped with EEC-IV, EFI and DIS systems. Engine idle speed, idle mixture, and ignition timing are not adjustable. See Engine/Emissions Diagnosis Shop Manual for additional information.

To check engine timing set parking brake and block wheels. Engine must be at normal operating temperature, transmission in neutral, and accessories off.

(1) Turn off engine.
(2) Disconnect the in-line Spout Connector (=▯◁).
(3) Re-start previously warmed-up engine.
(4) Verify that the ignition timing is 10° BTDC. If not see shop manual.
(5) Turn engine off and restore electrical connection.

Use SAE 5W-30 Oil API Service SG - Energy Conserving II.

This vehicle conforms to U.S. EPA regulations applicable to 1992 model year new motor vehicles.

F2AE-9C485- HDB	Catalyst	Spark Plug: AWSF-32C Gap: .042-.046 2.3L-9HM NFM2.3V5FYF7-TWC/HO2S/EGR/MPI

VACUUM HOSE ROUTING

MAN VAC, PCV, EGR EVR, SOL V, EGR, CPRV, THROTTLE BODY, FUEL TANK, MAN VAC, A/CL BI MET, CARBON CAN, FPR, A/CL DV

Vacuum Diagram 2.3L engine — 1992

CALIBRATION: 1-06S-R05 — 2.3L EFI

Ford Motor Company
VEHICLE EMISSION CONTROL INFORMATION

This vehicle is equipped with EEC-IV, EFI and DIS systems. Engine idle speed, idle mixture, and ignition timing are not adjustable. See Engine/Emissions Diagnosis Shop Manual for additional information.

To check engine timing set parking brake and block wheels. Engine must be at normal operating temperature, transmission in neutral, and accessories off.

(1) Turn off engine.
(2) Disconnect the in-line Spout Connector (=▯◁).
(3) Re-start previously warmed-up engine.
(4) Verify that the ignition timing is 10° BTDC. If not see shop manual.
(5) Turn engine off and restore electrical connection.

Use SAE 5W-30 Oil API Service SG - Energy Conserving II.

This vehicle conforms to U.S. EPA and California regulations applicable to 1992 model year new motor vehicles introduced into commerce solely for sale in California.

F2AE-9C485- HDD | Catalyst | Spark Plug: AWSF-32C Gap: .042-.046 2.3L-9HM NFM2.3V5FYC4-TWC/HO2S/EGR/MPI

VACUUM HOSE ROUTING

MAN VAC
PCV
EGR EVR
SOL V
EGR
CPRV
THROTTLE BODY
FUEL TANK
MAN VAC
CARBON CAN
A/CL BI MET
FPR
A/CL DV

23V5FYCC

Vacuum Diagram 2.3L engine — 1992

CALIBRATION: 2-49S-R00 — 2.3L EFI

Ford Motor Company
IMPORTANT VEHICLE INFORMATION

This vehicle is equipped with EEC-IV, EFI and DIS systems. Engine idle speed, idle mixture, and ignition timing are not adjustable. See Engine/Emissions Diagnosis Shop Manual for additional information.

To check engine timing set parking brake and block wheels. Engine must be at normal operating temperature, transmission in neutral, and accessories off.

(1) Turn off engine.
(2) Disconnect the in-line Spout Connector (=▯◁).
(3) Re-start previously warmed-up engine.
(4) Verify that the ignition timing is 10° BTDC. If not see shop manual.
(5) Turn engine off and restore electrical connection.

Use SAE 5W-30 Oil API Service SG - Energy Conserving II.

This vehicle conforms to U.S. EPA regulations applicable to 1992 model year new light-duty trucks.

F2AE-9C485- HFL | Catalyst | Spark Plug: AWSF-32C Gap: .042-.046 2.3L-9HM NFM2.3T5FMG1-TWC/HO2S/EGR/MPI

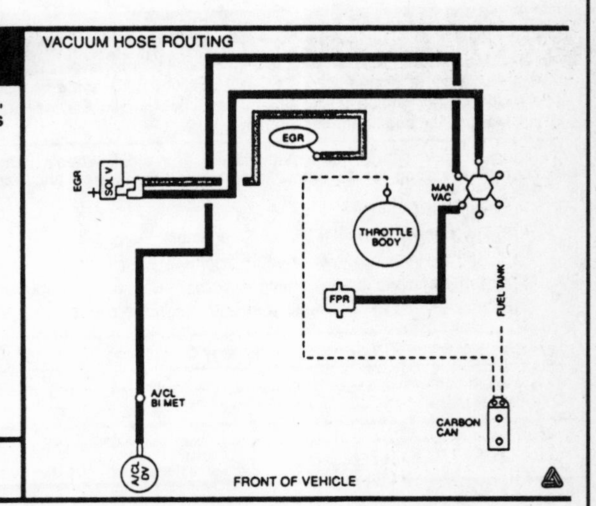

Vacuum Diagram 2.3L engine — 1992

CALIBRATION: 2-49T-R00 — 2.3L EFI

Ford Motor Company
IMPORTANT VEHICLE INFORMATION

This vehicle is equipped with EEC-IV, EFI and DIS systems. Engine idle speed, idle mixture, and ignition timing are not adjustable. See Engine/Emissions Diagnosis Shop Manual for additional information.

To check engine timing set parking brake and block wheels. Engine must be at normal operating temperature, transmission in neutral, and accessories off.

(1) Turn off engine.
(2) Disconnect the in-line Spout Connector (=▯◁).
(3) Re-start previously warmed-up engine.
(4) Verify that the ignition timing is 10° BTDC. If not see shop manual.
(5) Turn engine off and restore electrical connection.

Use SAE 5W-30 Oil API Service SG - Energy Conserving II.

This vehicle conforms to U.S. EPA and California regulations applicable to 1992 model year new light-duty trucks introduced into commerce solely for sale in California.

F2AE-9C485- HFT	Catalyst	Spark Plug: AWSF-32C Gap: .042-.046 2.3L-9HM NFM2.3T5FML8-TWC/HO2S/EGR/MPI

VACUUM HOSE ROUTING

EGR
SOL V
EGR
MAN VAC
THROTTLE BODY
FPR
FUEL TANK
A/CL BI MET
CARBON CAN
A/CL DV
FRONT OF VEHICLE
23T5FMLC

Vacuum Diagram 2.3L engine — 1992

CALIBRATION: 2-50S-R00 — 2.3L EFI

Ford Motor Company
IMPORTANT VEHICLE INFORMATION

This vehicle is equipped with EEC-IV, EFI and DIS systems. Engine idle speed, idle mixture, and ignition timing are not adjustable. See Engine/Emissions Diagnosis Shop Manual for additional information.

To check engine timing set parking brake and block wheels. Engine must be at normal operating temperature, transmission in neutral, and accessories off.

(1) Turn off engine.
(2) Disconnect the in-line Spout Connector (=▯◁).
(3) Re-start previously warmed-up engine.
(4) Verify that the ignition timing is 10° BTDC. If not see shop manual.
(5) Turn engine off and restore electrical connection.

Use SAE 5W-30 Oil API Service SG - Energy Conserving II.

This vehicle conforms to U.S. EPA regulations applicable to 1992 model year new light-duty trucks.

F2AE-9C485- HFP	Catalyst	Spark Plug: AWSF-32C Gap:.042-.046 2.3L-9HM NFM2.3T5FMG1-TWC/HO2S/MPI

VACUUM HOSE ROUTING

MAN VAC
THROTTLE BODY
FPR
FUEL TANK
A/CL BI MET
A/CL DV
FRONT OF VEHICLE
CARBON CAN

Vacuum Diagram 2.3L engine — 1992

CALIBRATION: 1-22F-R00 — 5.0L SEFI-HO

Ford Motor Company
VEHICLE EMISSION CONTROL INFORMATION

This vehicle is equipped with EEC-IV, EFI systems. Engine idle speed, idle mixture, and ignition timing are not adjustable. See Engine/Emissions Diagnosis Shop Manual for additional information.

To check engine timing set parking brake and block wheels. Engine must be at normal operating temperature, transmission in neutral, and accessories off.

(1) Turn off engine.
(2) Disconnect the in-line Spout Connector (=▯d).
(3) Re-start previously warmed-up engine.
(4) Verify that the ignition timing is 10° BTDC. If not see shop manual.
(5) Turn engine off and restore electrical connection.

Firing Order-1-3-7-2-6-5-4-8

Use SAE 10W-30 Oil API Service SG - Energy Conserving II.

This vehicle conforms to U.S. EPA regulations applicable to 1992 model year new motor vehicles.

F2AE-9C485- **HAT**	Catalyst	Spark Plug: ASF-42C Gap: .052-.056 5.0L-9HM NFM5.0V5FXF0-TWC/AIR/HO2S/EGR/SMPI

VACUUM HOSE ROUTING
FPR
VRESER
PCV
MAN VAC
V CK V
EVR
EGR
ENG
ABV
DMP
SOL V
AIR PUMP
CPRV
FUEL TANK
CAN
FRONT OF VEHICLE

Vacuum Diagram 5.0L engine — 1992

CALIBRATION: 1-22T-R00 — 5.0L SEFI-HO

Ford Motor Company
VEHICLE EMISSION CONTROL INFORMATION

This vehicle is equipped with EEC-IV, EFI systems. Engine idle speed, idle mixture, and ignition timing are not adjustable. See Engine/Emissions Diagnosis Shop Manual for additional information.

To check engine timing set parking brake and block wheels. Engine must be at normal operating temperature, transmission in neutral, and accessories off.

(1) Turn off engine.
(2) Disconnect the in-line Spout Connector (=▯d).
(3) Re-start previously warmed-up engine.
(4) Verify that the ignition timing is 10° BTDC. If not see shop manual.
(5) Turn engine off and restore electrical connection.

Use SAE 10W-30 Oil API Service SG - Energy Conserving II.

This vehicle conforms to U.S. EPA and California regulations applicable to 1992 model year new motor vehicles introduced into commerce solely for sale in California.

F2AE-9C485- **HAY**	Catalyst	Spark Plug: ASF-42C Gap: .052-.056 5.0L-9HM NFM5.0V5FXC7-TWC/AIR/HO2S/EGR/SMPI

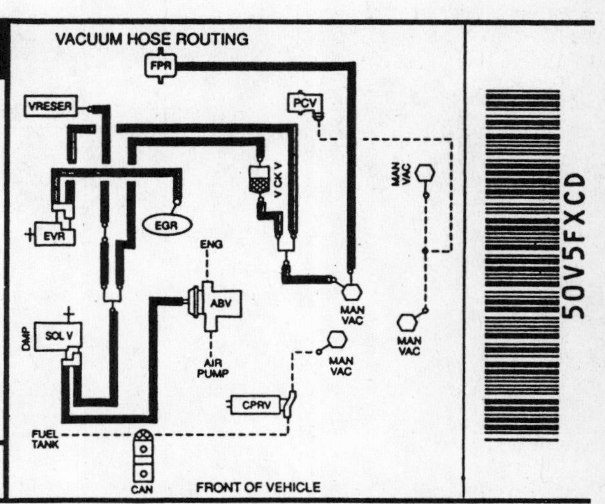

Vacuum Diagram 5.0L engine — 1992

■ CALIBRATION: 2-22G-R00 — 5.0L SEFI-HO ■

Ford Motor Company
VEHICLE EMISSION CONTROL INFORMATION

This vehicle is equipped with EEC-IV, EFI systems. Engine idle speed, idle mixture, and ignition timing are not adjustable. See Engine/Emissions Diagnosis Shop Manual for additional information.

To check engine timing set parking brake and block wheels. Engine must be at normal operating temperature, transmission in neutral, and accessories off.

(1) Turn off engine.
(2) Disconnect the in-line Spout Connector (=▯d).
(3) Re-start previously warmed-up engine.
(4) Verify that the ignition timing is 10° BTDC. If not see shop manual.
(5) Turn engine off and restore electrical connection.

Firing Order-1-3-7-2-6-5-4-8

Use SAE 10W-30 Oil API Service SG - Energy Conserving II.

This vehicle conforms to U.S. EPA regulations applicable to 1992 model year new motor vehicles.

F2AE-9C485- HAT	Catalyst	Spark Plug: ASF-42C Gap: .052-.056 5.0L-9HM NFM5.0V5FXF0-TWC/AIR/HO2S/EGR/SMPI

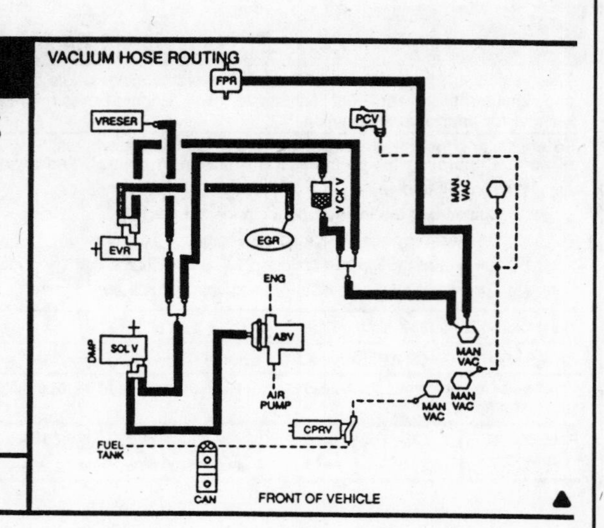

Vacuum Diagram 5.0L engine — 1992

5

FUEL SYSTEM

ELECTRONIC FUEL INJECTION SYSTEM APPLICATION CHART

Model	Body VIN ①	Year	Engine Liter	Engine VIN ②	Ignition Type	Fuel Injection Type
Mustang	4	1989–90	2.3 OHC	A	TFI	EFI
		1991–92	2.3 OHC	M	DIS	EFI-MA
		1989–92	5.0 HO	E	TFI	SEFI-MA

TFI—Thick Film Integrated Ignition System
DIS—Distributorless Ignition System
EFI—Electronic Fuel Injection System
EFI-MA—Electronic Fuel Injection System with Mass Air Flow Metering
SEFI-MA—Sequential Electronic Fuel Injection System with Mass Air Flow Metering
① Sixth digit of VIN number
② Eighth digit of VIN number

MULTI-POINT (EFI) AND SEQUENTIAL (SEFI) FUEL INJECTION SYSTEMS

Description of Systems

➧ SEE FIGS. 1–2

The Multi-Point (EFI) and Sequential (SEFI) Fuel Injection sub systems include a high pressure inline electric fuel pump, a low-pressure tank-mounted fuel pump, fuel charging manifold, pressure regulator, fuel filter and both solid and flexible fuel lines. The fuel charging manifold includes 4 or 8 electronically controlled fuel injectors, each mounted directly above an intake port in the lower intake manifold. On the 4 cylinder EFI system, all injectors are energized simultaneously and spray once every crankshaft revolution, delivering a predetermined quantity of fuel into the intake air stream. On the V8 EFI engines, the injectors are energized in 2 banks of 4, once each crankshaft revolution. On the SEFI system, each injector fires once every crankshaft revolution, in sequence with the engine firing order.

The fuel pressure regulator maintains a constant pressure drop across the injector nozzles. The regulator is referenced to intake manifold vacuum and is connected parallel to the fuel injectors and positioned on the far end of the fuel rail. Any excess fuel supplied by the pump passes through the regulator and is returned to the fuel tank via a return line.

➡ The pressure regulator reduces fuel pressure to 39–40 psi under normal operating conditions. At idle or high manifold vacuum condition, fuel pressure is reduced to approximately 30 psi.

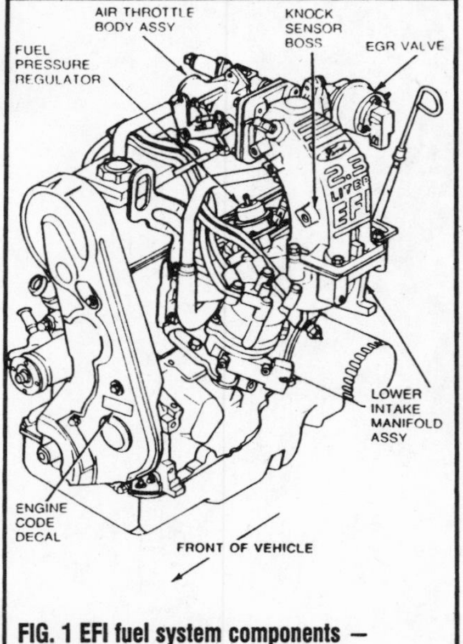

FIG. 1 EFI fuel system components — 2.3L engine

The fuel pressure regulator is a diaphragm operated relief valve in which the inside of the diaphragm senses fuel pressure and the other side senses manifold vacuum. Normal fuel pressure is established by a spring preload applied to the diaphragm. Control of the fuel system is maintained through the EEC power relay and the EEC-IV control unit, although electrical power is routed through the fuel pump relay and an inertia switch. The fuel pump relay is normally located on a bracket somewhere above the Electronic Control Assembly (ECA) and the inertia switch is located in the trunk. The in-line fuel pump is usually mounted on a bracket at the fuel tank, or on a frame rail. Tank-mounted pumps can be either high or low-pressure, depending on the model.

The inertia switch opens the power circuit to the fuel pump in the event of a collision. Once tripped, the switch must be reset manually by pushing the reset button on the assembly. Check that the inertia switch is reset before diagnosing power supply problems to the fuel pump circuit.

Fuel Injectors

➧ SEE FIG. 3

The fuel injectors used with the EFI and SEFI system are electromechanical (solenoid) type designed to meter and atomize fuel delivered to the intake ports of the engine. The injectors are mounted in the lower intake manifold and positioned so that their spray nozzles direct the fuel charge in front of the intake valves. The injector body consists of a solenoid actuated pintle and needle valve assembly. The control unit sends an electrical impulse that activates the solenoid, causing the pintle to move inward off the seat and allow the fuel to flow. The amount of fuel delivered is controlled by the length of time the injector is energized (pulse width), since the fuel flow orifice is fixed and the fuel pressure drop across the injector tip is constant. Correct atomization is achieved by contouring the pintle at the point where the fuel enters the pintle chamber.

➡ Exercise care when handling fuel injectors during service. Be careful not to lose the pintle cap and replace O-rings to assure a tight seal. Never apply direct battery voltage to test a fuel injector.

The injectors receive high pressure fuel from the fuel manifold (fuel rail) assembly. The

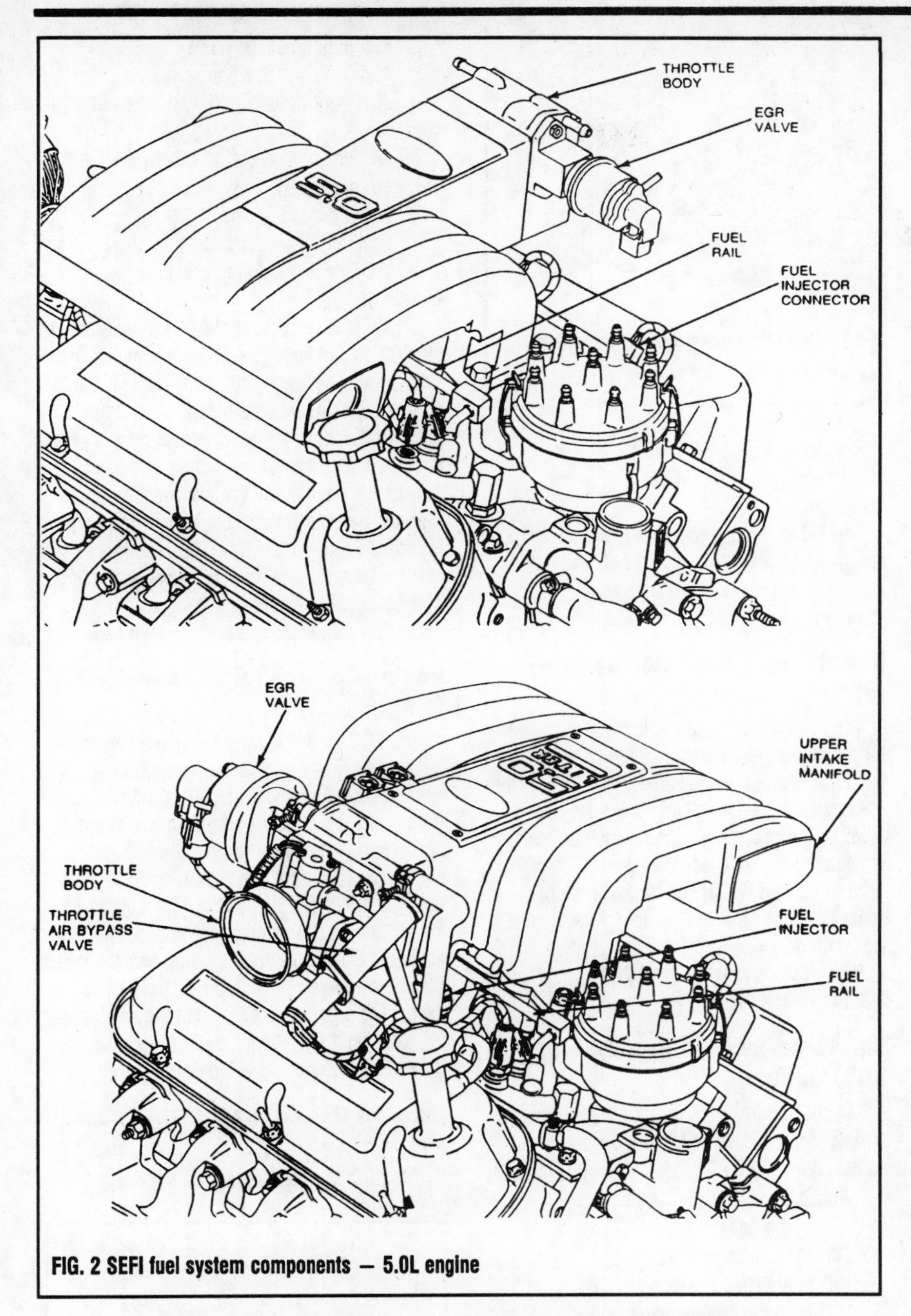

FIG. 2 SEFI fuel system components — 5.0L engine

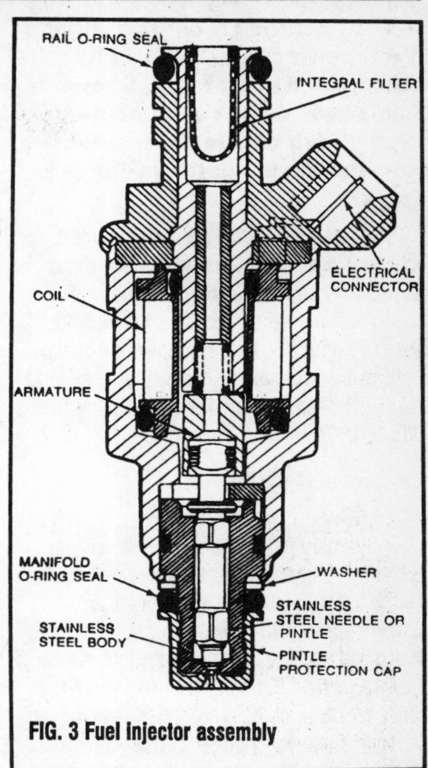

FIG. 3 Fuel injector assembly

complete assembly includes a single, preformed tube with 4 or 8 injector connectors, mounting flange for the pressure regulator, mounting attachments to locate the manifold and provide the fuel injector retainers and a Schrader® quick-disconnect fitting used to perform fuel pressure tests.

The fuel manifold is normally removed with fuel injectors and pressure regulator attached. Fuel injector electrical connectors are plastic and have locking tabs that must be released when disconnecting when disconnecting the wiring harness.

Throttle Air Bypass Valve

The throttle air bypass valve is an electro-mechanical (solenoid) device whose operation is controlled by the EEC-IV control unit. A variable air metering valve controls both cold and warm idle air flow in response to commands from the control unit. The valve operates by bypassing a regulated amount of air around the throttle plate; the higher the voltage signal from the control unit, the more air is bypassed through the valve. In this manner, additional air can be added to the fuel mixture without moving the throttle plate. At curb idle, the valve provides smooth idle for various engine coolant temperatures, compensates for air conditioning load and compensates for transaxle load and no-load conditions. The valve also provides fast idle for start-up, replacing the fast idle cam, throttle kicker and anti-dieseling solenoid common to previous models.

There are no curb idle or fast idle adjustments. As in curb idle operation, the fast idle speed is proportional to engine coolant temperature. Fast idle kick-down will occur when the throttle is kicked. A time-out feature in the ECA will also automatically kick-down fast idle to curb idle after a time period of approximately 15–25 seconds; after coolant has reached approximately 71°C (160°F). The signal duty cycle from the ECA to the valve will be at 100% (maximum current) during the crank to provide maximum air flow to allow no touch starting at any time (engine cold or hot).

Electronic Engine Control

The electronic engine control sub system consists of the ECA and various sensors and actuators. The ECA reads inputs from engine sensors, then outputs a voltage signal to various components (actuators) to control engine functions. The period of time that the injectors are energized ("ON" time or "pulse width") determines the amount of fuel delivered to each cylinder. The longer the pulse width, the richer the fuel mixture.

➡ The operating reference voltage (Vref) between the ECA and its sensors and actuators is 5 volts. This allows these components to work during the crank operation even though the battery voltage drops.

In order for the ECA to properly control engine operation, it must first receive current status reports on various operating conditions. The control unit constantly monitors crankshaft position, throttle plate position, engine coolant temperature, exhaust gas oxygen level, air intake volume and temperature, air conditioning (On/Off), spark knock and barometric pressure.

Universal Distributor

➧ SEE FIG. 4

The primary function of the TFI IV ignition system universal distributor is to direct the high secondary voltage to the spark plugs. In addition, the universal distributor supplies crankshaft position and frequency information to the ECA using a Profile Ignition Pick-up (PIP) sensor in place of the magnetic pick-up or the crankshaft position sensor used on other models. This distributor does not have any mechanical or vacuum advance. The universal distributor assembly is adjustable for resetting base timing, if required, by disconnecting the SPOUT connector.

➡ The PIP replaces the crankshaft position sensor found on other EEC-IV models.

The PIP sensor has an armature with 4 windows and 4 metal tabs that rotates past the stator assembly (Hall effect switch). When a metal tab enters the stator assembly, a positive signal (approximately 10 volts) is sent to the ECA, indicating the 10 degrees BTDC crankshaft position. The ECA calculates the precise time to energize the spark output signal to the TFI module. When the TFI module receives the spark output signal, it shuts off the coil primary current and the collapsing field energizes the secondary output.

➡ Misadjustment of the base timing affects the spark advance in the same manner as a conventional solid-state ignition system.

Thick Film Ignition (TFI IV) Module

The TFI IV ignition module has 6 connector pins at the engine wiring harness that supply the following signals:

- Ignition switch in **RUN** position
- Engine cranking
- Tachometer
- PIP (crankshaft position to ECA)

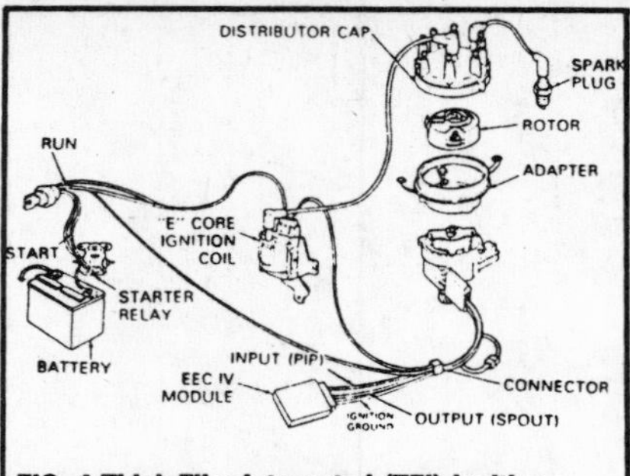

FIG. 4 Thick Film Integrated (TFI) ignition system — 5.0L engine shown

- Spark advance (from ECA)
- Internal ground from the ECA to the distributor

The TFI IV module supplies the spark to the distributor through the ignition coil and calculates the duration. It receives its control signal from the ECA (spark output).

Distributorless Ignition System (DIS)

The DIS ignition systems eliminates the need for a distributor by using multiple coils, which fire 2 spark plugs at the same time. This system uses either a dual function crank sensor or a crank sensor and cam sensor. The cam sensor provides the cylinder identification (CID) signal, used to choose which coil to fire. The crank sensor provides a Profile Ignition Pick-up (PIP) signal for spark timing. The dual function crank sensor provides both CID and PIP signals to the DIS module.

Distributorless Ignition System (DIS) Module

The DIS module controls coil firing from the ECA commands similar to the way the TFI IV module does. The DIS module functions include:

- Selection of Coil(s)
- Drives the coils
- Drives the tachometer
- PIP (crankshaft position to ECA)
- CID (cylinder identification)
- LOS (limited operation strategy—base timing)

Throttle Position (TP) Sensor

➧ SEE FIG. 5

The TP sensor is mounted on the throttle body. This sensor provides the ECA with a signal that indicates the opening angle of the throttle plate. The sensor output signal uses the 5 volt reference voltage (Vref) previously described. From this input, the ECA controls:

1. Operating modes, which are wide open throttle (WOT), part throttle (PT) and closed throttle (CT).
2. Fuel enrichment at WOT.
3. Additional spark advance at WOT.
4. EGR cut off during WOT, deceleration and idle.
5. Air conditioning cut off during WOT (30 seconds maximum).
6. Cold start kick-down.
7. Fuel cut off during deceleration.
8. WOT dechoke during crank mode (starting).

On the EEC-IV system, the TP sensor signal to the ECA only changes the spark timing during the WOT mode. As the throttle plate rotates, the TP sensor varies its voltage output. As the throttle plate moves from a closed throttle position to a WOT Position, the voltage output of the TP sensor will change from a low voltage (approximately 1.0 volt) to a high voltage (approximately 4.75 volts). The TP sensor used is not adjustable and must be replaced if it is out of specification. The EEC-IV programming compensates for differences between sensors.

Engine Coolant Temperature (ECT) Sensor

The ECT sensor is located either in the heater supply tube at the rear of the engine, or in the lower intake manifold. The ECT is a thermistor (changes resistance as temperature changes). The sensor detects the temperature of engine coolant and provides a corresponding signal to the ECA. From this signal, the ECA will modify the air/fuel ratio (mixture), idle speed, spark advance, EGR and Canister purge control. When the engine coolant is cold, the ECT signal causes the ECA to provide enrichment to the air/fuel ratio for good cold drive away as engine coolant warms up, the voltage will drop.

Exhaust Gas Oxygen (EGO) Sensor

➧ SEE FIG. 6

The EGO or HEGO sensor on the EEC-IV

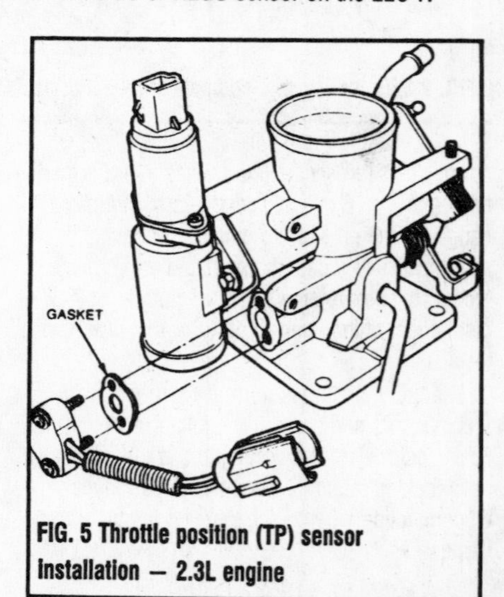

FIG. 5 Throttle position (TP) sensor installation — 2.3L engine

system is a little different from others used and is mounted in its own mounting boss, located between the 2 downstream tubes in the header near the exhaust system. The EGO sensor works between 0–1 volt output, depending on the presence (lean) or absence (rich) of oxygen in the exhaust gas. A voltage reading greater than 0.6 volts indicates a rich air/fuel ratio, while a reading of less than 0.4 volts indicates a lean air/fuel ratio.

➡ Never apply voltage to the EGO sensor because it could destroy the sensor's calibration. This includes the use of an ohmmeter. Before connecting and using a voltmeter, make sure it has a high-input impedance (at least 10 megohms) and is set on the proper resistance range. Any attempt to use a powered voltmeter to measure the EGO voltage output directly will damage or destroy the sensor.

Operation of the sensor is the same as previous models. A difference that should be noted is that the rubber protective cap used on top of the sensor on the earlier models has been replaced with a metal cap. In addition, later model sensors incorporate a heating element (HEGO), to bring the sensor up to operating temperature more quickly and keep it there during extended idle periods to prevent the sensor from cooling off and placing the system into open loop operation.

Vane Meter

The vane meter is actually 2 sensors in 1 assembly—a Vane Air Flow (VAF) sensor and Vane Air Temperature (VAT) sensor. This meter measures air flow to the engine and the temperature of the air stream. The vane meter is located either behind or under the air cleaner.

Air flow through the body moves a vane mounted on a pivot pin. The more air flowing through the meter, the further the vane rotates about the pivot pin. The air vane pivot pin is connected to a variable resistor (potentiometer) on top of the assembly. The vane meter uses the 5 volt reference voltage. The output of the potentiometer to the ECA varies between 0 and Vref (5 volts), depending on the volume of air flowing through the sensor. A higher volume of air will produce a higher voltage output.

The volume of air measured through the meter has to be converted into an air mass value. The mass (weight) of a specific volume of air varies with pressure and temperature. To compensate for these variables, a temperature sensor in front of the vane measures incoming air temperature. The ECA uses the air temperature and a programmed pressure value to convert the VAF

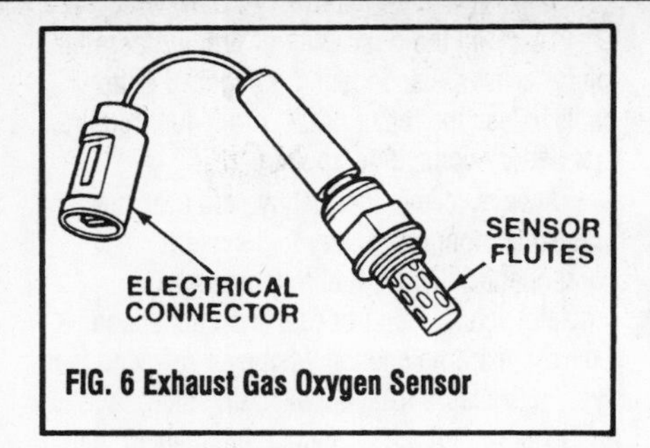

FIG. 6 Exhaust Gas Oxygen Sensor

signal into a mass air flow value. This value is used to calculate the fuel flow necessary for the optimum air/fuel ratio. The VAT also affects spark timing as a function of air temperature.

Air Conditioning Clutch Compressor (ACC) Signal

Anytime battery voltage is applied to the A/C clutch, the same signal is also applied to the ECA. The ECA then maintains the engine idle speed with the throttle air bypass valve control solenoid (fuel injection) to compensate for the added load created by the A/C clutch operation. Shutting down the A/C clutch will have a reverse effect. The ECA will maintain the engine idle speed at 850–950 rpm.

Knock Sensor (KS)

The knock sensor is used to detect detonation. In situations of excessive knock the ECA receives a signal from this sensor and retards the spark accordingly. It is mounted in the lower intake manifold at the rear of the engine.

Barometric (BAP) Sensor

➧ SEE FIG. 7

The barometric sensor is used to compensate for altitude variations. From this signal, the ECA modifies the air/fuel ratio, spark timing, idle speed, and EGR flow. The barometric sensor is a design that produces a frequency based on atmospheric pressure (altitude). The barometric sensor is mounted on the right-hand fender apron.

Manifold Absolute Pressure (MAP) Sensor

The MAP sensor measures manifold vacuum and outputs a variable frequency. This gives the ECA information on engine load. It replaces the BAP sensor by providing the ECA updated barometric pressure readings during key **ON** engine **OFF** and wide open throttle. The MAP sensor output is used by the ECA to control spark advance, EGR flow and air/fuel ratio.

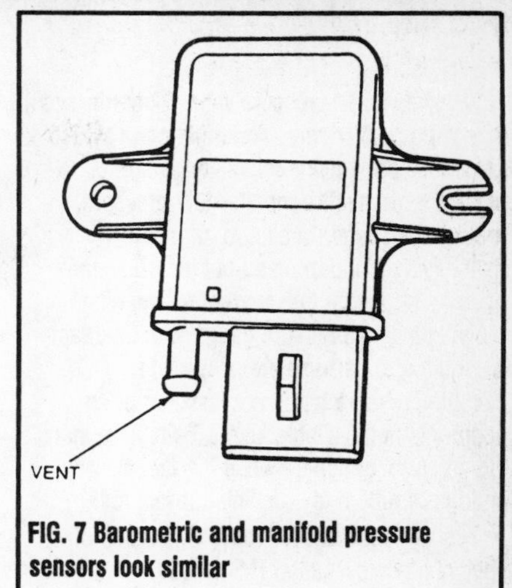

FIG. 7 Barometric and manifold pressure sensors look similar

EGR Shut-Off Solenoid

The electrical signal to the EGR shut-off solenoid is controlled by the ECA. The signal is either **ON** or **OFF**. It is **OFF** during cold start, closed throttle or WOT. It is **ON** at all other times.

➡ The canister purge valve is controlled by vacuum from the EGR solenoid. The purge valve is a standard-type valve and operates the same as in previous systems.

The solenoid is the same as the EGR control solenoid used on previous EEC systems. It is usually mounted on the LH side of the dash panel in the engine compartment, or on the RH shock tower in the engine compartment. The solenoid is normally closed, and the control vacuum from the solenoid is applied to the EGR valve.

Diagnosis and Testing

➡ For Self-Diagnosis System and Accessing Trouble Code Memory, see Section 4 "Electronic Engine Controls".

ELECTRIC FUEL DELIVERY SYSTEMS

✻✻ CAUTION

Fuel pressure must be relieved before attempting to disconnect any fuel lines.

Pressure Tests

SEE FIG. 8

The diagnostic pressure valve (Schrader type) is located on the fuel rail on multi-point systems. This valve provides a convenient point for service personnel to monitor fuel pressure, release the system pressure prior to maintenance, and to bleed out air which may become trapped in the system during filter replacement. A pressure gauge with an adapter is required to perform pressure tests.

If the pressure tap is not installed or an adapter is not available, use a T-fitting to install the pressure gauge between the fuel filter line and the throttle body fuel inlet or fuel rail.

Testing fuel pressure requires the use of a special pressure gauge (T80L-9974-A or equivalent) that attaches to the diagnostic pressure tap fitting. Depressurize the fuel system before disconnecting any lines.

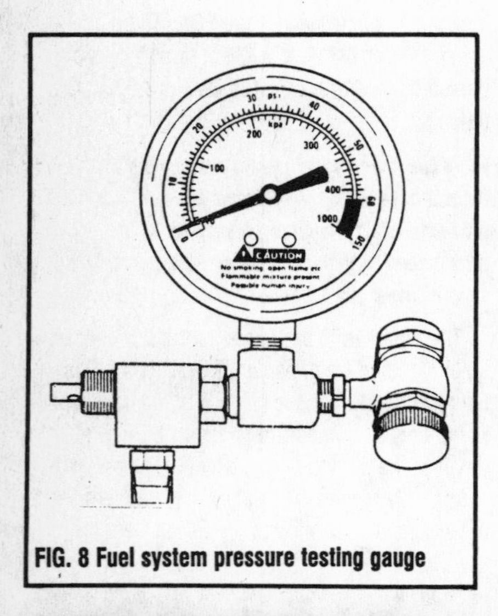

FIG. 8 Fuel system pressure testing gauge

Component Replacement

FUEL SYSTEM SERVICE PRECAUTIONS

Safety is the most important factor when performing not only fuel system maintenance but any type of maintenance. Failure to conduct maintenance and repairs in a safe manner may result in serious personal injury or death. Maintenance and testing of the vehicle's fuel system components can be accomplished safely and effectively by adhering to the following rules and guidelines.

- To avoid the possibility of fire and personal injury, always disconnect the negative battery cable unless the repair or test procedure requires that battery voltage be applied.
- Always relieve the fuel system pressure prior to disconnecting any fuel system component (injector, fuel rail, pressure regulator, etc.), fitting or fuel line connection. Exercise extreme caution whenever relieving fuel system pressure to avoid exposing skin, face and eyes to fuel spray. Please be advised that fuel under pressure may penetrate the skin or any part of the body that it contacts.
- Always place a shop towel or cloth around the fitting or connection prior to loosening to absorb any excess fuel due to spillage. Ensure that all fuel spillage (should it occur) is quickly removed from engine surfaces. Ensure that all fuel soaked cloths or towels are deposited into a suitable waste container.
- Always keep a dry chemical (Class B) fire extinguisher near the work area.
- Do not allow fuel spray or fuel vapors to come into contact with a spark or open flame.
- Always use a backup wrench when loosening and tightening fuel line connection fittings. This will prevent unnecessary stress and torsion to fuel line piping. Always follow the proper torque specifications.
- Always replace worn fuel fitting O-rings with new. Do not substitute fuel hose or equivalent where fuel pipe is installed.

RELIEVING FUEL SYSTEM PRESSURE

All EFI and SEFI fuel injected engines are equipped with a pressure relief valve located on the fuel supply manifold. Remove the fuel tank cap and attach fuel pressure gauge T80L-9974-A, or equivalent, to the valve to release the fuel pressure. If a suitable pressure gauge is not available, disconnect the vacuum hose from the fuel pressure regulator and attach a hand vacuum pump. Apply about 25 in. Hg (84 kPa) of vacuum to the regulator to vent the fuel system pressure into the fuel tank through the fuel return hose. Note that this procedure will remove the fuel pressure from the lines, but not the fuel. Take precautions to avoid the risk of fire and use clean rags to soak up any spilled fuel when the lines are disconnected.

QUICK CONNECT FUEL LINE FITTINGS

Removal and Installation

SEE FIGS. 9-11

Quick Connect (push) type fuel line fittings must be disconnected using proper procedures or the fitting may be damaged. There are 2 types of retainers are used on the push connect fittings. Line sizes of $^3/_8$ and $^5/_{16}$ in. use a hairpin clip retainer. The $^1/_4$ in. line connectors use a duck bill clip retainer. In addition, some engines use spring lock connections secured by a garter spring which requires a special tool (T81P-19623-G or equivalent) for removal.

HAIRPIN CLIP

1. Clean all dirt and/or grease from the fitting. Spread the 2 clip legs about $^1/_8$ in. (3mm) each to disengage from the fitting and pull the clip outward from the fitting. Use finger pressure only, do not use any tools.
2. Grasp the fitting and hose assembly and pull away from the steel line. Twist the fitting and hose assembly slightly while pulling, if necessary, when a sticking condition exists.
3. Inspect the hairpin clip for damage, replace the clip if necessary. Reinstall the clip in position on the fitting.
4. Inspect the fitting and inside of the connector to insure freedom of dirt or obstruction. Install fitting into the connector and push together. A click will be heard when the hairpin snaps into proper connection. Pull on the line to insure full engagement.

DUCK BILL CLIP

1. A special tool is available from Ford and other tool manufacturers for removing the retaining clips (tool No. T82L-9500-AH or equivalent). If the tool is not on hand see Step 2. Align the slot on the push connector disconnect tool with either tab on the retaining clip. Pull the line from the connector.
2. If the special clip tool is not available, use a pair of narrow 6 in. (152mm) channel lock pliers with a jaw width of 0.2 in. (5mm) or less. Align the jaws of the pliers with the openings of the fitting case and compress the part of the retaining clip that engages the case. Compressing the retaining clip will release the fitting which may be pulled from the connector. Both sides of the clip must be compressed at the same time to disengage.
3. Inspect the retaining clip, fitting end and

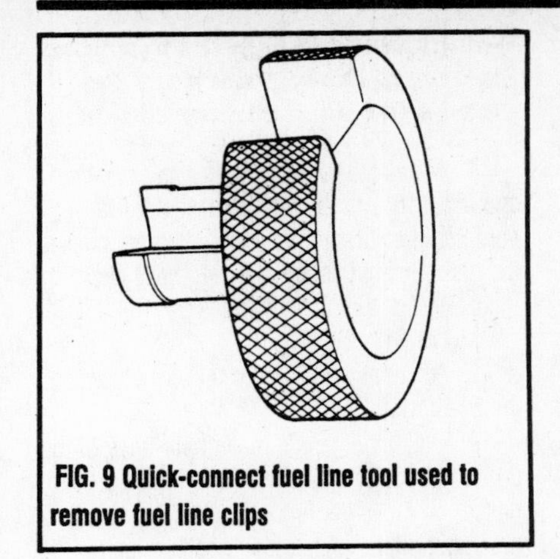

FIG. 9 Quick-connect fuel line tool used to remove fuel line clips

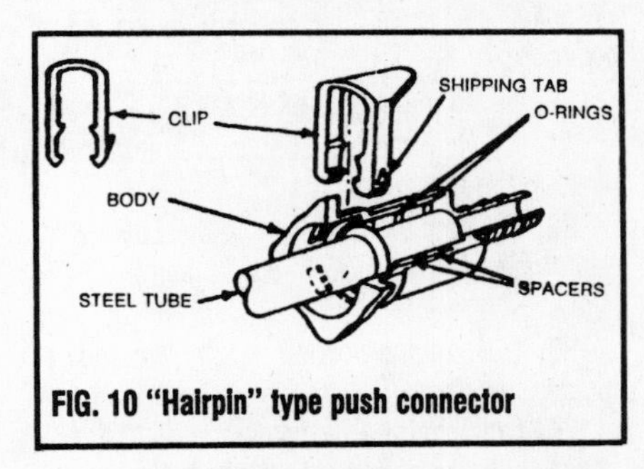

FIG. 10 "Hairpin" type push connector

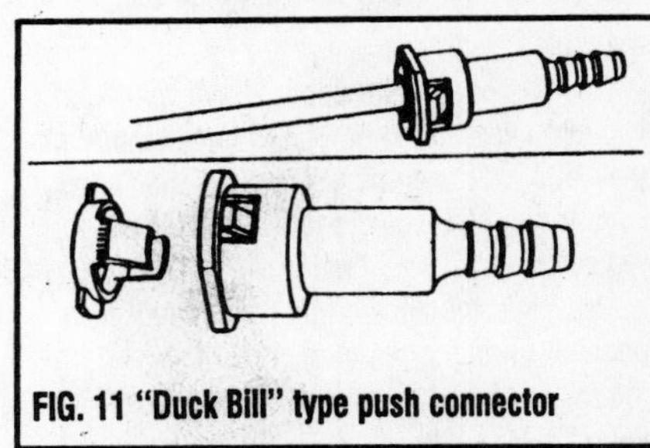

FIG. 11 "Duck Bill" type push connector

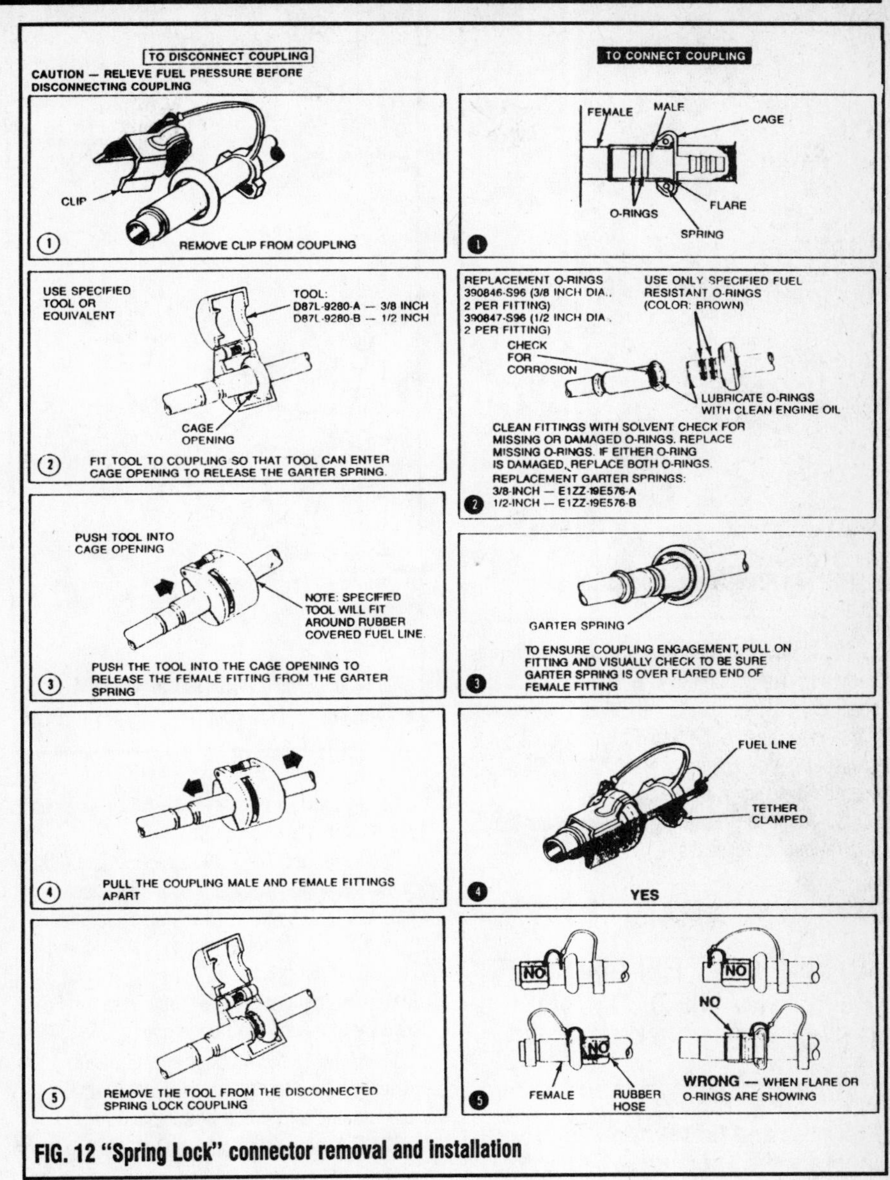

FIG. 12 "Spring Lock" connector removal and installation

connector. Replace the clip if any damage is apparent.

4. Push the line into the steel connector until a click is heard, indicting the clip is in place. Pull on the line to check engagement.

SPRING LOCK COUPLING

The spring lock coupling is a fuel line coupling held together by a garter spring inside a circular cage. When the coupling is connected together, the flared end of the female fitting slips behind the garter spring inside the cage of the male fitting. The garter spring and cage then prevent the flared end of the female fitting from pulling out of the cage. As an additional locking feature, most vehicles have a horseshoe shaped retaining clip that improves the retaining reliability of the spring lock coupling.

FUEL PUMP

Removal and Installation

➧ SEE FIG. 13

1. It is necessary to remove the fuel tank.
2. Depressurize the fuel system.
3. Remove fuel form the fuel tank by pumping out through the filter tube.
4. Disconnect the supply and return line fittings and the vent line.
5. Disconnect and remove the fuel filter tube.
6. Disconnect the electrical connections to both the fuel sender and the fuel pump wiring harness.
7. Remove the fuel tank support straps and remove the fuel tank.
8. Remove any dirt that has accumulated around the fuel pump attaching flange, to prevent it from entering the tank during service procedures.
9. Turn the fuel pump locking ring counter-clockwise with the necessary tool and remove the locking ring.
10. Remove the fuel pump and bracket assembly.
11. Remove the seal gasket and discard.

To install:

12. Put a light coating of heavy grease on a new seal ring to hold it in place during assembly. Install it in fuel tank ring groove.
13. Install the tank in the vehicle.
14. Install the electrical connector.
15. Install the fuel line fittings and tighten to 30–40 ft. lbs (40–54 Nm).
16. Install a minimum of 10 gallons of fuel and inspect for leaks.

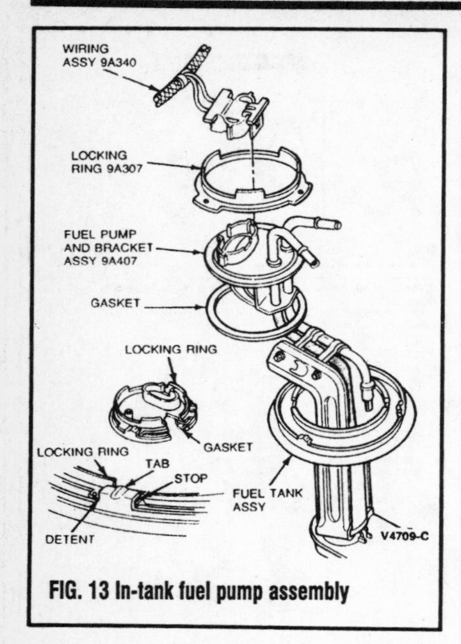

FIG. 13 In-tank fuel pump assembly

17. Install pressure gauge on valve on throttle body and turn ignition to **ON** position for 3 seconds. Turn ignition key **OFF** and back on for 3 seconds repeatedly, 5 to 10 times, until pressure gauge shows at least 30 psi. Re-inspect for leaks at fittings.

18. Remove pressure gauge. Start engine and re-inspect for leaks.

MULTI-POINT EFI AND SEFI INTAKE MANIFOLD AND THROTTLE BODY ASSEMBLY

Removal and Installation

2.3L ENGINE

➧ SEE FIG. 14

1. Drain the cooling system.
2. Make certain ignition switch is **OFF** and disconnect the negative battery cable.
3. Remove fuel cap and relieve the fuel system pressure.
4. Disconnect the electrical connectors to the throttle position sensor, knock sensor, air charge temperature sensor, coolant temperature sensor and the injector wiring harness.
5. Tag and disconnect the vacuum lines at the upper intake manifold vacuum tree, the EGR valve vacuum line and the fuel pressure regulator vacuum line.
6. Remove the throttle linkage shield and disconnect the throttle linkage and speed control cable, if equipped. Unbolt the accelerator cable from the bracket and position the cable out of the way.

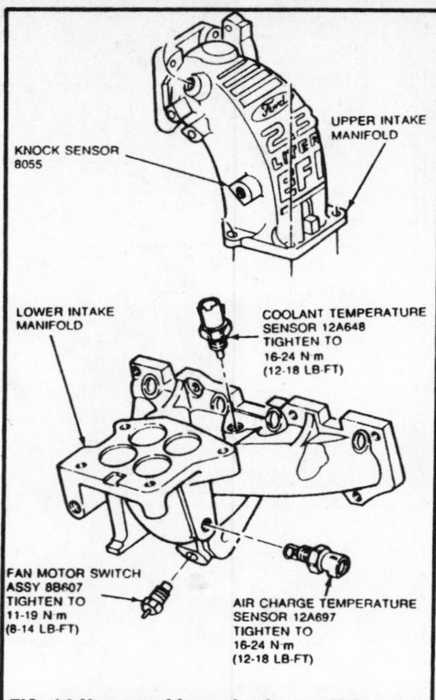

FIG. 14 Upper and lower intake manifold assemblies — 2.3L engine

7. Disconnect the air intake hose, air bypass hose and crankcase vent hose.
8. Disconnect the PCV hose from the fitting on the underside of the upper intake manifold.
9. Loosen the hose clamp on the coolant bypass line at the lower intake manifold and disconnect the hose.
10. Disconnect the EGR tube from the EGR valve by removing the flange nut.
11. Remove the 4 upper intake manifold retaining nuts, then remove the upper intake manifold and air throttle body assembly.
12. Disconnect the push connect fitting at the fuel supply manifold and fuel return lines, then disconnect the fuel return line from the fuel supply manifold.
13. Remove the engine oil dipstick bracket retaining bolt, as needed.
14. Disconnect the electrical connectors from all 4 fuel injectors and move the harness aside.
15. Remove the 2 fuel supply manifold retaining bolt, then carefully remove the fuel supply manifold with the injectors attached. The injectors may be removed from the fuel supply manifold at this time by exerting a slight twisting/pulling motion.

To Install:

16. Lubricate new injector O-rings with a light grade engine oil and install 2 on each injector. If the injectors were not removed from the fuel supply manifold, only 1 O-ring will be necessary.

➡ Do not use silicone grease on the O-rings as it will clog the injectors. Make sure the injector caps are clean and free of contamination.

17. Install the fuel injector supply manifold and injectors into the intake manifold, making sure the injectors are fully seated, then secure the fuel manifold assembly with the 2 retaining bolts. Tighten the retaining bolts to 15–22 ft. lbs. (20–30 Nm).
18. Reconnect the electrical connectors to the injectors.
19. Clean the gasket mating surfaces of the upper and lower intake manifold. Place a new gasket on the lower intake manifold, then place the upper intake manifold in position. Install the 4 retaining bolts and tighten them, in sequence, to 15–22 ft. lbs. (20–30 Nm).
20. Install the engine oil dipstick.
21. Connect the fuel supply and return fuel lines to the fuel supply manifold.
22. Connect the EGR tube to the EGR valve and tighten the fitting to 6–8.5 ft. lbs. (8–11.5 Nm).
23. Connect the coolant bypass line and tighten the clamp.
24. Connect the PCV system hose to the fitting on the underside of the upper intake manifold.
25. Reconnect the upper intake manifold vacuum lines, being careful to install them in their original locations. Reconnect the vacuum lines to the EGR valve and fuel pressure regulator.
26. Hold the accelerator cable bracket in position on the upper intake manifold and install the retaining bolt. Tighten the bolt to 10–15 ft. lbs. (13–20 Nm).
27. Install the accelerator cable to the bracket.
28. If the air intake throttle body was removed from the upper intake manifold, position a new gasket on the mounting flange and install the throttle body.
29. Connect the accelerator cable and speed control cable. Install the throttle linkage shield.
30. Reconnect the electrical connectors to the throttle position sensor, knock sensor, air charge temperature sensor, coolant temperature sensor and injector wiring harness.
31. Connect the air intake hose, air bypass hose and crankcase vent hose.
32. Connect the negative battery cable.
33. Refill the cooling system.
34. Build up fuel pressure by turning the ignition switch **ON** and **OFF** at least 6 times, leaving the ignition **ON** for at least 5 seconds each time. Check for fuel leaks.
35. Replace fuel cap and start the engine and allow it to reach normal operating temperature, then check for coolant leaks.

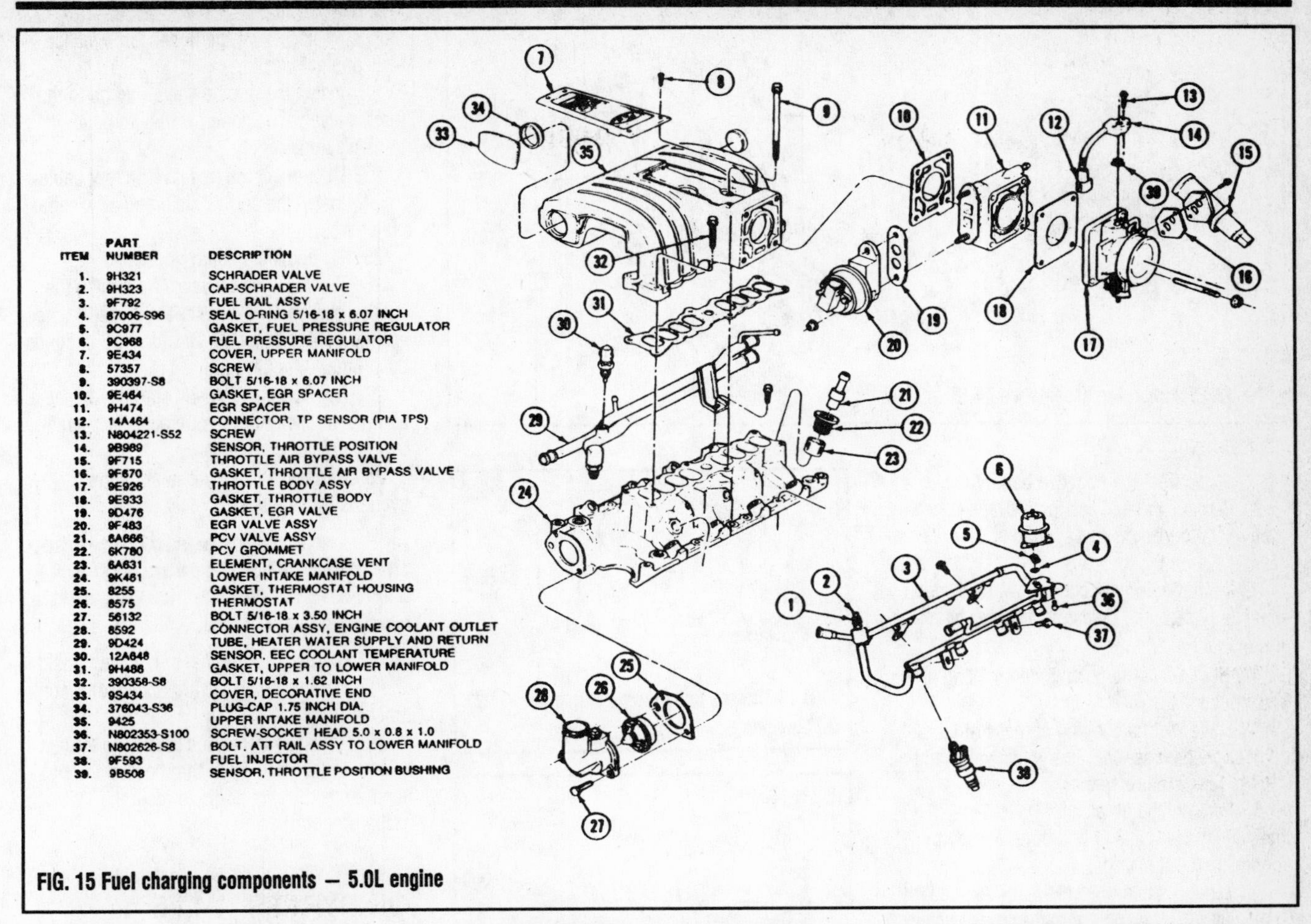

ITEM	PART NUMBER	DESCRIPTION
1.	9H321	SCHRADER VALVE
2.	9H323	CAP-SCHRADER VALVE
3.	9F792	FUEL RAIL ASSY
4.	87006-S96	SEAL O-RING 5/16-18 x 6.07 INCH
5.	9C977	GASKET, FUEL PRESSURE REGULATOR
6.	9C968	FUEL PRESSURE REGULATOR
7.	9E434	COVER, UPPER MANIFOLD
8.	57357	SCREW
9.	390397-S8	BOLT 5/16-18 x 6.07 INCH
10.	9E464	GASKET, EGR SPACER
11.	9H474	EGR SPACER
12.	14A464	CONNECTOR, TP SENSOR (PIA TPS)
13.	N804221-S52	SCREW
14.	9B989	SENSOR, THROTTLE POSITION
15.	9F715	THROTTLE AIR BYPASS VALVE
16.	9F670	GASKET, THROTTLE AIR BYPASS VALVE
17.	9E926	THROTTLE BODY ASSY
18.	9E933	GASKET, THROTTLE BODY
19.	9D476	GASKET, EGR VALVE
20.	9F483	EGR VALVE ASSY
21.	6A666	PCV VALVE ASSY
22.	6K780	PCV GROMMET
23.	6A631	ELEMENT, CRANKCASE VENT
24.	9K461	LOWER INTAKE MANIFOLD
25.	8255	GASKET, THERMOSTAT HOUSING
26.	8575	THERMOSTAT
27.	56132	BOLT 5/16-18 x 3.50 INCH
28.	8592	CONNECTOR ASSY, ENGINE COOLANT OUTLET
29.	9D424	TUBE, HEATER WATER SUPPLY AND RETURN
30.	12A648	SENSOR, EEC COOLANT TEMPERATURE
31.	9H486	GASKET, UPPER TO LOWER MANIFOLD
32.	390358-S8	BOLT 5/16-18 x 1.62 INCH
33.	9S434	COVER, DECORATIVE END
34.	376043-S36	PLUG-CAP 1.75 INCH DIA.
35.	9425	UPPER INTAKE MANIFOLD
36.	N802353-S100	SCREW-SOCKET HEAD 5.0 x 0.8 x 1.0
37.	N802626-S8	BOLT, ATT RAIL ASSY TO LOWER MANIFOLD
38.	9F593	FUEL INJECTOR
39.	9B508	SENSOR, THROTTLE POSITION BUSHING

FIG. 15 Fuel charging components — 5.0L engine

36. Perform EEC-IV self-test to check systems functions.

5.0L SEFI ENGINE

➧ SEE FIGS. 15–16

1. Make certain ignition switch is **OFF** and disconnect the negative battery cable.
2. Remove fuel cap, relieve the fuel system pressure and disconnect the fuel lines.
3. Disconnect electrical connectors at air bypass valve, throttle position sensor and EGR position sensor.
4. Disconnect throttle linkage at throttle ball and transmission linkage from throttle body. Remove 2 bolts securing bracket to intake manifold and position bracket cables out of way.
5. Disconnect upper intake manifold vacuum fitting connections by disconnecting all vacuum lines to vacuum tree, vacuum lines to EGR valve, and vacuum line to fuel pressure regulator.
6. Disconnect PCV system by disconnecting hose from fitting on rear of upper manifold. Remove canister purge lines from fittings on throttle body.
7. Disconnect the water heater lines from the throttle body.
8. Disconnect the EGR tube from the EGR valve by removing the flange nut.
9. Remove the bolt from the upper intake support bracket to the upper.
10. Remove the 6 upper intake manifold retaining bolts and remove upper intake and throttle body as an assembly from lower intake manifold.

To Install:

11. Clean and inspect the mounting faces of the lower and upper intake manifolds.
12. Position new gasket on lower intake mounting face. The use of alignment studs may be helpful.
13. Install upper intake manifold and throttle body assembly to lower manifold. Ensure gasket remains in place if alignment studs are not used.
14. Install upper intake manifold retaining bolts. Tighten to 12–18 ft. lbs. (16–24 Nm).
15. Install the upper intake support bracket to upper manifold attaching bolt.
16. Install the EGR tube.
17. Install canister purge lines to fittings on throttle body.
18. Connect the water heater lines to the throttle body.
19. Connect PCV hose to rear of upper manifold.
20. Connect vacuum lines to vacuum tree, EGR valve, and fuel pressure regulator.
21. Position throttle linkage bracket with cables to upper intake manifold. Install 2 attaching bolts and tighten to specification. Connect throttle cable and AOD transmission cable to throttle body, if equipped.
22. Connect electrical connectors at air bypass valve, TP sensor, and EGR position sensor.

➡ If lower intake manifold was removed, fill and bleed cooling system.

MULTI-POINT (EFI) AND SEFI THROTTLE BODY ASSEMBLY

Removal and Installation

➧ SEE FIGS. 17–18

➡ The following is a general procedure for throttle body removal. All components mentioned may not be used on all engines. Disconnection of fuel supply line may not be necessary.

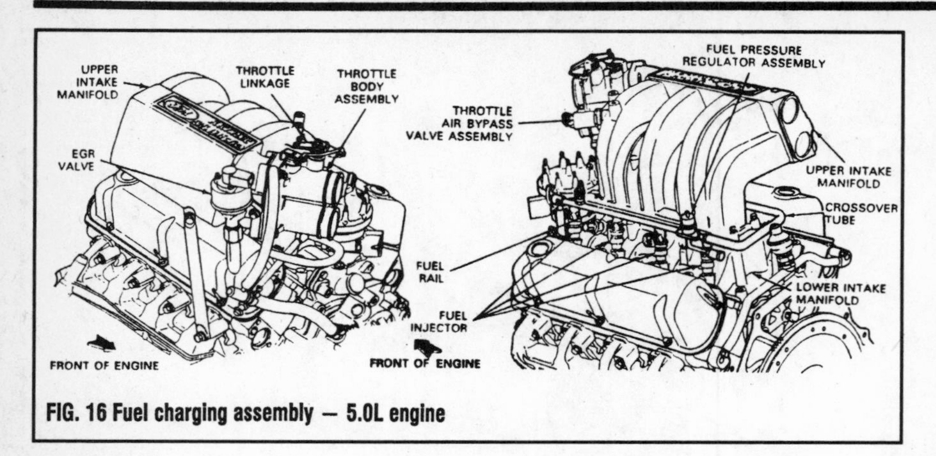

FIG. 16 Fuel charging assembly — 5.0L engine

1. Disconnect the negative battery cable.
2. Remove the fuel cap to vent tank pressure, then depressurize the fuel system at the pressure relief valve on the fuel rail.
3. If throttle body is coolant warmed, drain cooling system and disconnect heater hoses at throttle body.
4. Disconnect the push connect fitting at the fuel supply line.
5. Disconnect the wiring harness at the throttle position sensor, air bypass valve and air charge temperature sensor.
6. Remove the air cleaner outlet tube between the air cleaner and throttle body by loosening the 2 clamps.
7. Remove the snow shield, if equipped, by removing the retaining nut on top of the shield and the bolts on the side.
8. Tag and disconnect the vacuum hoses at the vacuum fittings on the intake manifold.
9. Disconnect and remove the accelerator and speed control cables (if equipped) from the accelerator mounting bracket and throttle lever.
10. Remove the transmission Throttle Valve (TV) linkage from the throttle lever on automatic transmission models.
11. Remove the retaining bolts and lift the air intake/throttle body assembly off the intake manifold and remove the assembly from the engine.
12. Remove and discard the gasket from the intake manifold assembly.

To Install:

13. Clean and inspect the mounting faces of the air intake/throttle body assembly and the intake manifold. Both surfaces must be clean and flat.
14. Clean and oil the manifold stud threads.
15. Install a new gasket on the lower intake manifold.
16. Using the guide pins as locators (if equipped), install the air intake/throttle body assembly to the intake manifold.
17. Install the studs and/or retaining bolts

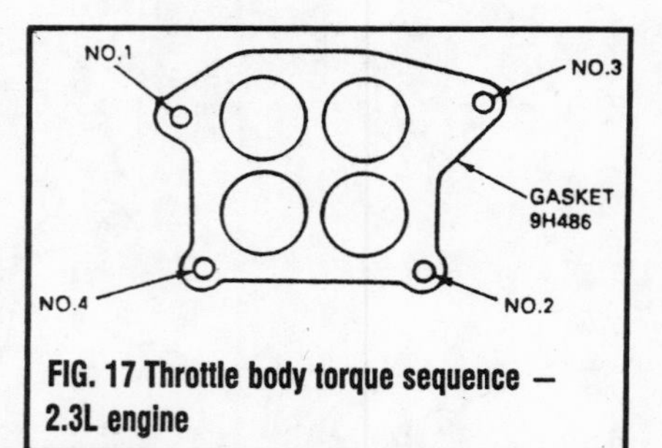

FIG. 17 Throttle body torque sequence — 2.3L engine

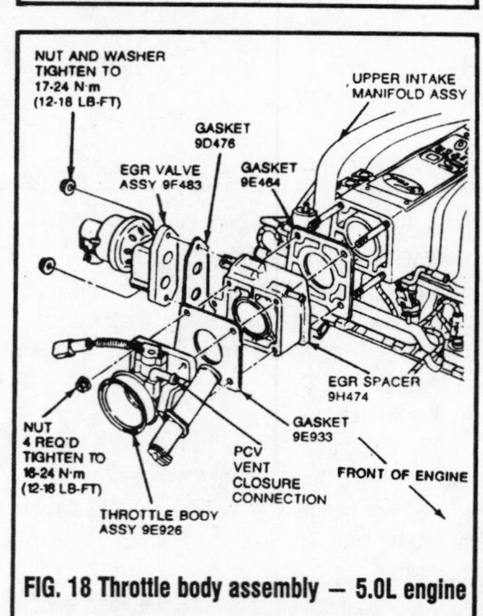

FIG. 18 Throttle body assembly — 5.0L engine

finger tight, then tighten them to specifications in a criss-cross pattern. Torque the mounting bolts as follows:

- 2.3L — 15–22 ft. lbs. (20–30 Nm)
- 5.0L — 12–18 ft. lbs. (17–24 Nm)

18. Connect the fuel supply and return lines to the fuel rail.
19. Connect the wiring harness to the throttle position sensor, air charge temperature sensor and air bypass valve.
20. Install the accelerator cable and speed control cable, if equipped.
21. Install the vacuum hoses to the vacuum fittings, making sure the hoses are installed in their original locations.
22. Install the throttle valve linkage to the throttle lever, if equipped with automatic transmission.
23. Reconnect the negative battery cable.
24. Install the snow shield and air cleaner outlet tube.
25. Fill cooling system, if drained.
26. Build up fuel pressure by turning the ignition switch **ON** and **OFF** at least 6 times, leaving the ignition **ON** for at least 5 seconds each time. Check for fuel leaks.
27. Replace fuel cap and start the engine and allow it to reach normal operating temperature, then check for coolant leaks.
28. Perform EEC-IV self-test to check systems functions.

➡ Resetting automatic transaxle throttle valve linkage must be performed if the air intake throttle body was removed.

29. Hold linkage ratchet in released position and push cable fitting toward accelerator control bracket.
30. Release ratchet and rotate the throttle linkage to the wide-open throttle position by hand, at the throttle body, to reset the TV cable to the proper position.

AIR BYPASS VALVE

Removal and Installation

➧ SEE FIG. 19

1. Disconnect the electrical connector at the air bypass valve.
2. Remove the air cleaner cover, if necessary to gain access to the bypass valve.
3. Separate the air bypass valve and gasket from the air cleaner, throttle body, or intake manifold by removing the mounting bolts.
4. Install the air bypass valve and gasket to the air cleaner cover and tighten the retaining bolts to 6–8.5 ft. lbs. (8–11.5 Nm).
5. Install the air cleaner cover, if removed.
6. Reconnect the air bypass valve electrical connector.

FUEL MANIFOLD AND INJECTORS

Removal and Installation

2.3L ENGINE

For injector and fuel manifold removal, follow

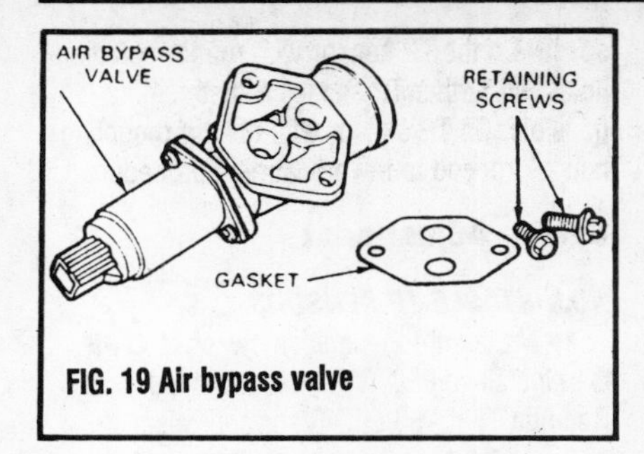

FIG. 19 Air bypass valve

the procedures under "Multi-point EFI And SEFI Intake Manifold and Throttle Body Assembly".

5.0L ENGINE

➧ SEE FIG. 20

1. Remove the upper intake manifold. Be sure to depressurize the fuel system before disconnecting any fuel lines.
2. Disconnect the fuel supply and return line retaining clips.
3. Disconnect the fuel chassis inlet and outlet fuel hoses from the fuel supply manifold.
4. Remove the 4 fuel supply manifold retaining bolts.
5. Carefully disengage the fuel rail assembly from the fuel injectors by lifting and gently rocking the rail.
6. Remove the fuel injectors from the intake manifold by lifting while gently rocking from side to side.
7. Place all removed components on a clean surface to prevent contamination by dirt or grease.

➡ Never use silicone grease, it will clog the injector. Injectors and fuel rail must be handles with extreme care to prevent damage to sealing areas and sensitive fuel metering orifices.

8. Examine the injector O-rings for deterioration or damage, replace as needed.
9. Make sure the injector caps are clean and free from contamination or damage.

To install:

10. Lubricate all O-rings with clean engine oil, then install the injectors in the fuel rail using a light twisting/pushing motion.
11. Carefully install the fuel rail assembly and injectors into the lower intake manifold. Make certain to correctly position insulators. Push down on the fuel rail to make sure the O-rings are seated.
12. Hold the fuel rail assembly in place and install the retaining bolts finger tight. Then, tighten the retaining bolts to 15–22 ft. lbs. (20–30 Nm).
13. Connect the fuel supply and return lines.
14. Connect the fuel injector wiring harness at the injectors.

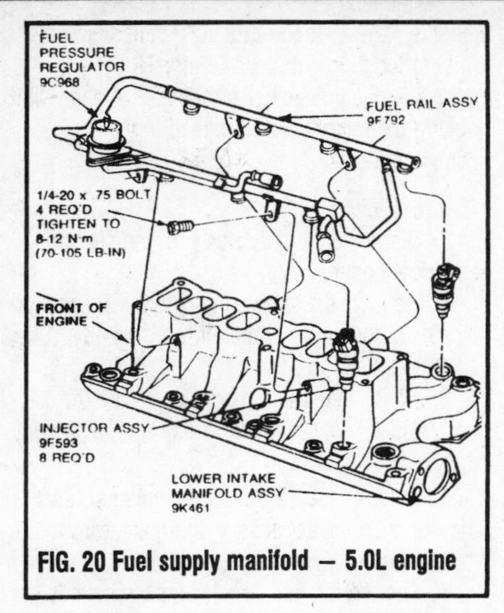

FIG. 20 Fuel supply manifold — 5.0L engine

15. Connect the vacuum line to the fuel pressure regulator, if removed.
16. Install the air intake and throttle body assembly.
17. Run the engine and check for fuel leaks.

FUEL PRESSURE REGULATOR

Removal and Installation

➧ SEE FIG. 21

➡ Procedures vary slightly for screw in type regulators.

1. Depressurize the fuel system, remove shielding as needed.
2. Remove the vacuum line at the pressure regulator.
3. Remove the 3 Allen® retaining screws from the regulator housing.
4. Remove the pressure regulator assembly, gasket and O-ring. Discard the gasket and check the O-ring for signs of cracks or deterioration.
5. Clean the gasket mating surfaces. If scraping is necessary, be careful not to damage the fuel pressure regulator or supply line gasket mating surfaces.
6. Lubricate the pressure regulator O-ring with with light engine oil. Do not use silicone grease; it will clog the injectors.
7. Install the O-ring and a new gasket on the pressure regulator.
8. Install the pressure regulator on the fuel manifold and tighten the retaining screws to 27–40 inch lbs. (3–4 Nm)

➡ On the 2.3L install the fuel manifold shield, and torque bolts to 15–22 ft. lbs. (20–30 Nm).

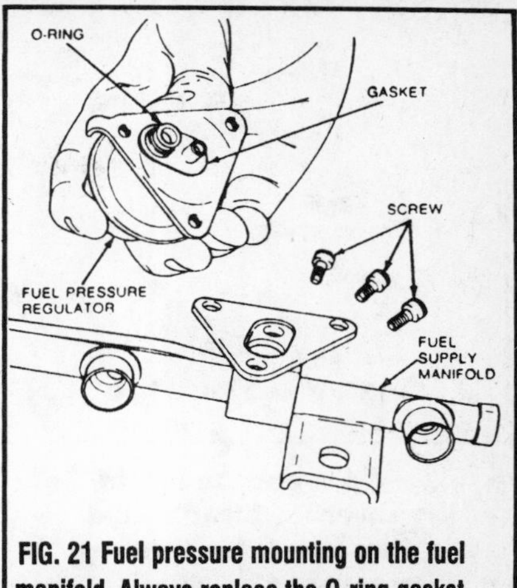

FIG. 21 Fuel pressure mounting on the fuel manifold. Always replace the O-ring gasket

9. Install the vacuum line at the pressure regulator.
10. Build up fuel pressure by turning the ignition switch **ON** and **OFF** at least 6 times, leaving the ignition **ON** for at least 5 seconds each time.
11. Check for system for fuel leaks.

THROTTLE POSITION (TP) SENSOR

Removal and Installation

➧ SEE FIG. 22

2.3L ENGINE

1. Disconnect the throttle position sensor electrical connector.
2. Remove the screw retaining the TP sensor electrical connector to the air throttle body, if equipped.
3. Scribe alignment marks on the air throttle body and TP sensor to indicate proper alignment during installation.
4. Remove the TP sensor retaining screws, then remove the TP sensor and gasket from the throttle body.

To install:

5. Position the TP sensor so that the wiring harness is parallel to the venturi bores. Place the TP sensor and gasket on the throttle body, making sure the rotary tangs on the sensor are aligned with the throttle shaft blade. Slide the rotary tangs into position over the throttle shaft blade, then rotate the throttle position sensor clockwise only to its installed position (align the scribe marks made earlier).

.99 . OHMS

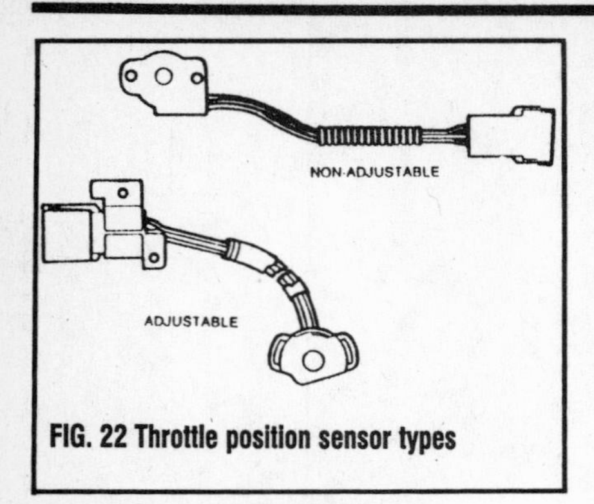

FIG. 22 Throttle position sensor types

➡ Slide the rotary tangs into position over the throttle shaft blade, then rotate the throttle position sensor clockwise only its installed position. Failure to install the TP sensor in this manner may result in excessive idle speeds.

6. Once the scribe marks are aligned, install the TP sensor retaining screws and tighten them to 14–16 inch lbs. (1.6–1.8 Nm)
7. Position the electrical connector over the locating dimple, if equipped, then secure to the throttle body with the retaining screw.
8. Inspect the connector terminals for corrosion, damage and bent pins and repair as necessary. Replace any cracked or burned connector terminals. Reconnect the TP sensor electrical connector, start the engine and check the idle speed. On the rotary TP sensor with round mounting holes, no further TP sensor adjustment is possible. On rotary TP sensor with oblong mounting holes, proceed to the adjustment procedure.

5.0L ENGINE

1. Disconnect the throttle position sensor electrical connector.
2. Remove the screw retaining the TP sensor electrical connector to the air throttle body, if equipped.
3. Scribe alignment marks on the air throttle body and TP sensor to indicate proper alignment during installation.
4. Remove the TP sensor retaining screws, then remove the TP sensor and gasket from the throttle body.
5. The throttle position sensor bushing, if equipped, must be reused.

To install:

6. Install the bushing with the larger diameter facing outward.
7. Install the TP sensor on the throttle shaft and rotate counterclockwise 10–20 degrees to align the screw holes.
8. Install the pair of retaining screws and torque to 11–16 inch lbs. (1.2–1.8 Nm).
9. Cycle throttle lever to wide open throttle. It should return without interference.
10. Reconnect the TP sensor electrical connector, start the engine and check the idle speed. On the TP sensor with round mounting holes, no further TP sensor adjustment is possible. On TP sensor with oblong mounting holes, proceed to the adjustment procedure.

Sensor Adjustment

ADJUSTABLE TP SENSORS

1. Make sure the ignition switch is **OFF**, then install an EEC-IV Breakout Box such as Rotunda T83L–50 EEC-IV, or equivalent.
2. Attach a digital volt/ohmmeter (10 megohms impedance) positive lead (+) to Pin 47 and the negative lead (–) to Pin 46. Set the DVOM on the 20 volt scale.
3. Turn the ignition key to the **ON** position (do not start the engine).
4. Loosen the mounting screws slightly and rotate the TP sensor until the DVOM reads the minimum specified voltage. Maximum voltage on all engines is 4.84 volts.

Specifications:

- 2.3L — 0.34 volts
- 5.0L — 0.20 volts

5. Tighten the TP sensor mounting screws to specification.
6. While watching the DVOM, move the throttle to the wide open position, then back to idle. For proper operation, the DVOM should move from the minimum voltage to the maximum specified voltage and back to the minimum voltage during throttle movement.

1989–90 FORD EEC-IV IDLE SPEED SETTING PROCEDURE

ADJUSTMENT PROCEDURE FOR PASSENGER CAR	2.3L EFI OHC	5.0L SEFI
1. Disconnect Idle Speed Control-Air Bypass Solenoid.	X	
2. Start engine and run at: rpm/sec	1500/30	
3. Place automatic transmission in manual transmission in Neutral	Park	Park
4. Engine off, back out throttle plate stop screw clear off the throttle lever pad.		X
5. With a .010 in. feeler gauge between the throttle plate stop screw and the throttle lever pad turn the screw in until contact is made then turn it an additional		① ②
6. Check/adjust idle rpm: • Loosen the throttle camplate roller bolt • Turn the throttle plate stop screw to (rpm) • Tighten the throttle camplate roller bolt	A 650±25 M 600±25	
7. Shut engine off and disconnect battery for 5 minutes minimum.		X
8. Engine off reconnect idle speed control-air bypass solenoid verify the throttle is not stuck in the bore and linkage not preventing throttle from closing.	X	
9. Start engine and stabilize for 2 minutes then goose engine and let it return to idle, lightly depress and release the accelerator. Let engine idle	X	X
	If idle problem still exists, Go To Section 2 for other possible causes.	

1989–90 FORD EEC-IV IDLE SPEED SETTING PROCEDURE

ADJUSTMENT PROCEDURE FOR PASSENGER CAR	2.3L EFI OHC	5.0L SEFI
10. On Automatic Overdrive Transmission (AOD) applications or Automatic Transaxle (AXOD) application check TV adjustment.		X

NOTE For Step 6: After a time, idle speed may change due to strategy parameter.

① 1½ turns 5.0L HI Output Engine
② 1⅞ turns 5.0L base Engine

1991–92 FORD EEC-IV IDLE SPEED SETTING PROCEDURE

(USE ONLY STEPS MARKED "X" FOR THE ENGINE BEING SERVICED)	2.3L EFI OHC	5.0L HO
1. Engine off, disconnect the negative (−) terminal of the battery for five minutes, then reconnect it.	X	X
2. Start engine and stabilize for two minutes, then goose engine and let it return to idle, lightly depress and release the accelerator and let engine idle. Does engine idle properly? **NOTE: If electric fan comes on wait until it turns itself off.**	X	
3. If engine does not idle properly proceed with this procedure.	X	X①
4. Shut engine off and install a .025″ feeler gauge between plate stop screw and throttle lever.		X
5. Disconnect the idle speed control air bypass solenoid.	X	
6. Start engine and run at rpm/time (no touch start).	1500/30 sec	Idle
7. Place automatic transmission in PARK, manual transmission in NEUTRAL	Park	Park
8. Check idle rpm to the range using a tachometer.	A 650 ± 150 M 600 ± 150	675 ± 50
A. **If rpm is too low**	X	X
a. Do not clean the throttle body, check for the plate orifice plug. If there is no plug, turn screw clockwise to the desired rpm. Rpm ± 25, if there is plug from previous service, remove plug and then adjust screw in either direction as required. Screw must be in contact with the lever pad after adjustment.	X	X②
B. **If rpm is too high**	X	X
a. Turn engine off.	X	X
b. Disconnect air cleaner hose.	X	X
c. Block off the orifice in the throttle plate temporarily with tape. If the orifice already has a plug from previous service, go to Step f.	X	X
d. Restart the engine and check idle speed using a tachometer (mass air applications will require air cleaner hose to be reattached before rpm check. If engine stalled, crack open the plate with stop screw. Do not over adjust.	X	X
e. If rpm continues to be fast, perform test in Step 15. If TPS output code is within range, remove tape, go to ?? other causes. If out of range, adjust screw for proper TPS code. Lever pad must be in contact the screw. If rpm is still fast terminate this procedure and go to other possible causes.	X	
f. If rpm drops to value in Step 12 or below engine or engine stalls turn the engine off, disconnect air cleaner hose, remove the tape.	X	X
g. Install the plug with proper color code depending on orifice size.	X	X
h. Reconnect the air cleaner hose—start the engine. Check idle rpm using a tachometer. Turn the plate stop screw clockwise (do not turn it counter-clockwise as this may cause the throttle plate to stick at idle) to the nominal rpm ± 25 rpm shown in Step 12.	X	X
9. Remove the feeler gauge between plate stop screw and throttle lever.		X②
10. Shut engine off and disconnect battery for 5 minutes minimum.		X

1991–92 FORD EEC-IV IDLE SPEED SETTING PROCEDURE

(USE ONLY STEPS MARKED "X" FOR THE ENGINE BEING SERVICED)	2.3L EFI OHC	5.0L HO
11. Run KOEO Self-Test for proper TPS output code.	X	X
12. Engine off, reconnect idle speed control-air bypass solenoid, verify the throttle is not stuck in the bore and linkage is not preventing throttle from closing.	X	
13. Start engine and stabilize for two minutes then goose engine and let it return to idle, lightly depress and release the accelerator let engine idle.	X If idle problem still persists, go to ?? other possible causes.	X
14. On Automatic Overdrive Transmission (AOD) applications check TV pressure adjustment.		X

① In some cases, even if idle speed is OK, proceed with this procedure if customer symptom persists (idle speed control duty cycle may be out of range).
② Key Off/Restart/Recheck idle for two minutes (eliminates possibility of entering part throttle mode).

FUEL TANK

REMOVAL & INSTALLATION

➧ SEE FIG. 23

1. Disconnect the negative battery cable and relieve the fuel system pressure.
2. Siphon or pump as much fuel as possible out through the fuel filler pipe.

➡ Fuel injected vehicles have reservoirs inside the fuel tank to maintain fuel near the fuel pickup during cornering and under low fuel operating conditions. These reservoirs could block siphon tubes or hoses from reaching the bottom of the fuel tank. Repeated attempts using different hose orientations can overcome this obstacle.

3. Raise and safely support the vehicle.
4. Disconnect the fuel fill and vent hoses connecting the filler pipe to the tank. Disconnect 1 end of the vapor crossover hose at the rear over the driveshaft.
5. On vehicles equipped with a metal retainer which fastens the filler pipe to the fuel tank, remove the screw attaching the retainer to the fuel tank flange.
6. Disconnect the fuel lines and the electrical connector to the fuel tank sending unit. On some vehicles, these are inaccessible on top of the tank. In these case they must be disconnected with tank partially removed.
7. Place a safety support under the fuel tank and remove the bolts from the fuel tank straps. Allow the straps to swing out of the way. Be careful not to deform the fuel tank.
8. Partially remove the tank and disconnect the fuel lines and electrical connector from the sending unit, if required.
9. Remove the tank from the vehicle.

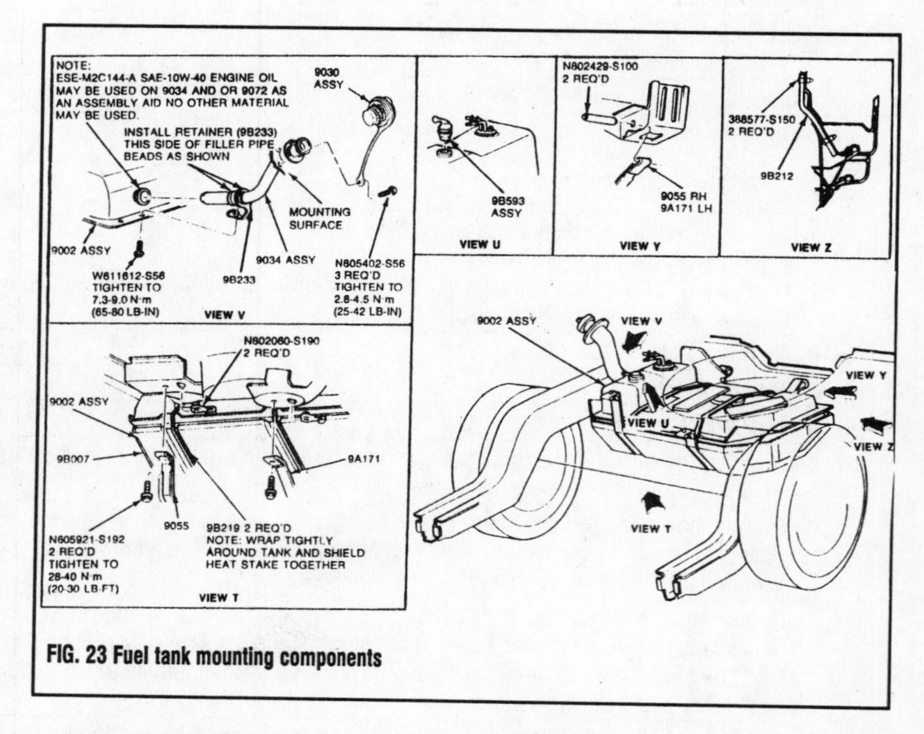

FIG. 23 Fuel tank mounting components

To install:

10. Raise the fuel tank into position in the vehicle. Connect the fuel lines and sending unit electrical connector if it is necessary to connect them before the tank is in the final installed position.
11. Lubricate the fuel filler pipe with water base tire mounting lubricant and install the tank

onto the filler pipe, then bring the tank into final position. Be careful not to deform the tank.

12. Bring the fuel tank straps around the tank and start the retaining nut or bolt. Align the tank with the straps. If equipped, make sure the fuel tank shields are installed with the straps and are positioned correctly on the tank.

13. Check the hoses and wiring mounted on the tank top to make sure they are correctly routed and will not be pinched between the tank and body.

14. Tighten the fuel tank strap retaining nuts or bolts to 20–30 ft. lbs. (28–40 Nm).

15. If not already connected, connect the fuel hoses and lines which were disconnected. Make sure the fuel supply, fuel return, if present, and vapor vent connections are made correctly. If not already connected, connect the sending unit electrical connector.

16. Lower the vehicle. Replace the fuel that was drained from the tank. Check all connections for leaks.

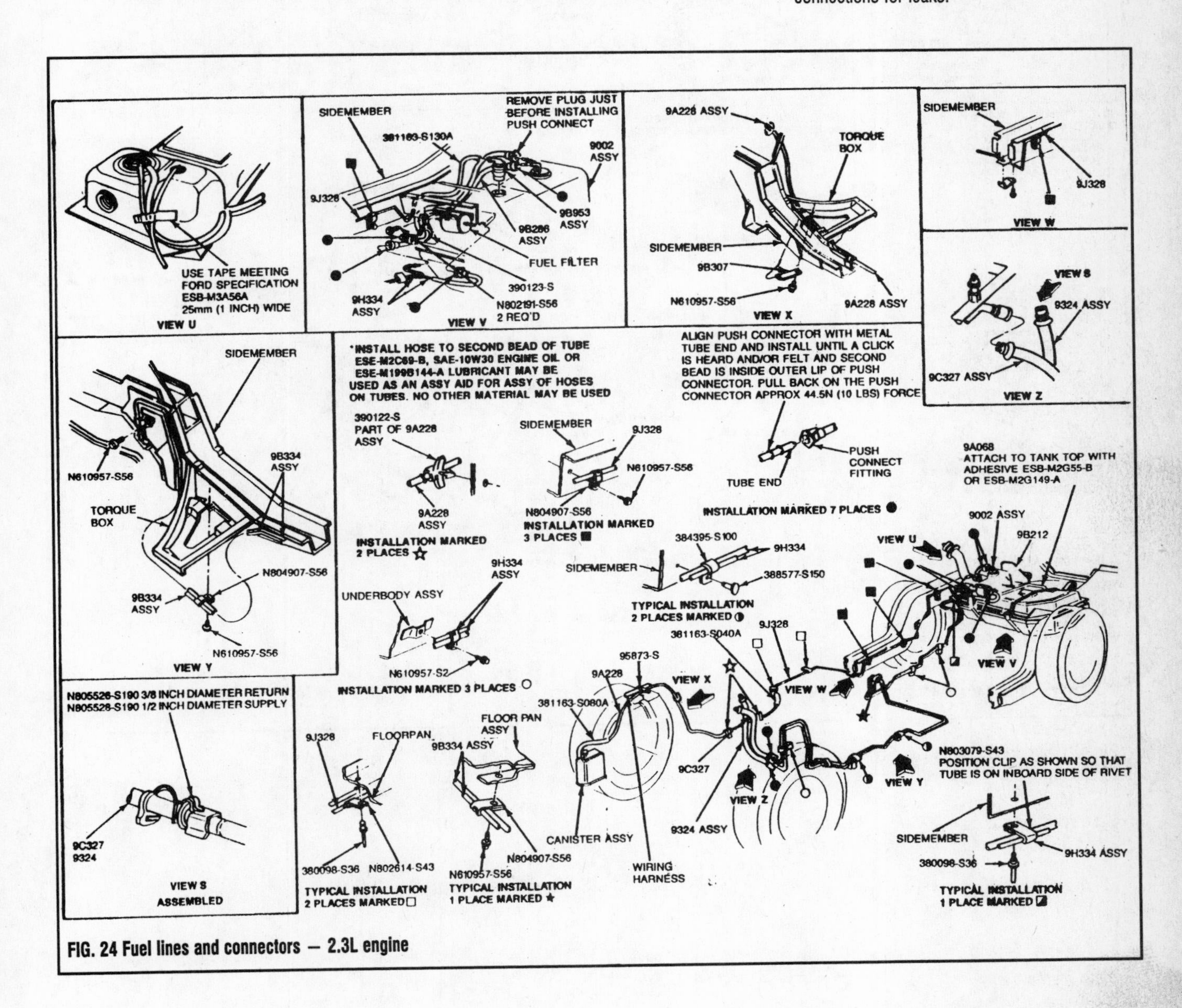

FIG. 24 Fuel lines and connectors — 2.3L engine

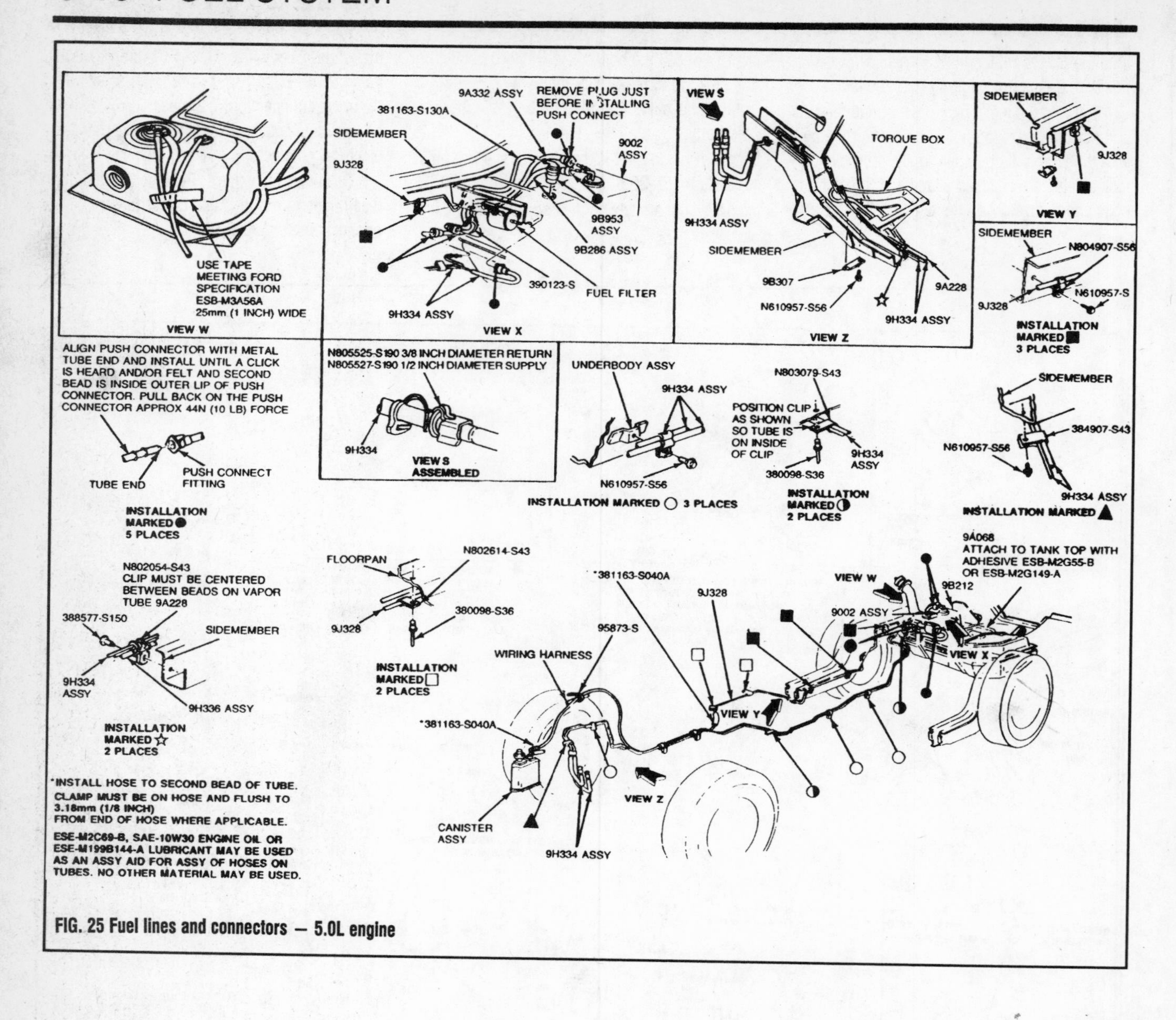

FIG. 25 Fuel lines and connectors — 5.0L engine

6

CHASSIS ELECTRICAL

UNDERSTANDING AND TROUBLESHOOTING ELECTRICAL SYSTEMS

At the rate with which both import and domestic manufacturers are incorporating electronic control systems into their production lines, it won't be long before every new vehicle is equipped with one or more on-board computer. These electronic components (with no moving parts) should theoretically last the life of the vehicle, provided nothing external happens to damage the circuits or memory chips.

While it is true that electronic components should never wear out, in the real world malfunctions do occur. It is also true that any computer-based system is extremely sensitive to electrical voltages and cannot tolerate careless or haphazard testing or service procedures. An inexperienced individual can literally do major damage looking for a minor problem by using the wrong kind of test equipment or connecting test leads or connectors with the ignition switch ON. When selecting test equipment, make sure the manufacturers instructions state that the tester is compatible with whatever type of electronic control system is being serviced. Read all instructions carefully and double check all test points before installing probes or making any test connections.

The following section outlines basic diagnosis techniques for dealing with computerized automotive control systems. Along with a general explanation of the various types of test equipment available to aid in servicing modern electronic automotive systems, basic repair techniques for wiring harnesses and connectors is given. Read the basic information before attempting any repairs or testing on any computerized system, to provide the background of information necessary to avoid the most common and obvious mistakes that can cost both time and money. Although the replacement and testing procedures are simple in themselves, the systems are not, and unless one has a thorough understanding of all components and their function within a particular computerized control system, the logical test sequence these systems demand cannot be followed. Minor malfunctions can make a big difference, so it is important to know how each component affects the operation of the overall electronic system to find the ultimate cause of a problem without replacing good components unnecessarily. It is not enough to use the correct test equipment; the test equipment must be used correctly.

Safety Precautions

** CAUTION

Whenever working on or around any computer based microprocessor control system, always observe these general precautions to prevent the possibility of personal injury or damage to electronic components.

- Never install or remove battery cables with the key ON or the engine running. Jumper cables should be connected with the key OFF to avoid power surges that can damage electronic control units. Engines equipped with computer controlled systems should avoid both giving and getting jump starts due to the possibility of serious damage to components from arcing in the engine compartment when connections are made with the ignition ON.
- Always remove the battery cables before charging the battery. Never use a high output charger on an installed battery or attempt to use any type of "hot shot" (24 volt) starting aid.
- Exercise care when inserting test probes into connectors to insure good connections without damaging the connector or spreading the pins. Always probe connectors from the rear (wire) side, NOT the pin side, to avoid accidental shorting of terminals during test procedures.
- Never remove or attach wiring harness connectors with the ignition switch ON, especially to an electronic control unit.
- Do not drop any components during service procedures and never apply 12 volts directly to any component (like a solenoid or relay) unless instructed specifically to do so. Some component electrical windings are designed to safely handle only 4 or 5 volts and can be destroyed in seconds if 12 volts are applied directly to the connector.
- Remove the electronic control unit if the vehicle is to be placed in an environment where temperatures exceed approximately 176°F (80°C), such as a paint spray booth or when arc or gas welding near the control unit location in the car.

ORGANIZED TROUBLESHOOTING

When diagnosing a specific problem, organized troubleshooting is a must. The complexity of a modern automobile demands that you approach any problem in a logical, organized manner. There are certain troubleshooting techniques that are standard:

1. Establish when the problem occurs. Does the problem appear only under certain conditions? Were there any noises, odors, or other unusual symptoms?
2. Isolate the problem area. To do this, make some simple tests and observations; then eliminate the systems that are working properly. Check for obvious problems such as broken wires, dirty connections or split or disconnected vacuum hoses. Always check the obvious before assuming something complicated is the cause.
3. Test for problems systematically to determine the cause once the problem area is isolated. Are all the components functioning properly? Is there power going to electrical switches and motors? Is there vacuum at vacuum switches and/or actuators? Is there a mechanical problem such as bent linkage or loose mounting screws? Doing careful, systematic checks will often turn up most causes on the first inspection without wasting time checking components that have little or no relationship to the problem.
4. Test all repairs after the work is done to make sure that the problem is fixed. Some causes can be traced to more than one component, so a careful verification of repair work is important to pick up additional malfunctions that may cause a problem to reappear or a different problem to arise. A blown fuse, for example, is a simple problem that may require more than another fuse to repair. If you don't look for a problem that caused a fuse to blow, for example, a shorted wire may go undetected.

Experience has shown that most problems tend to be the result of a fairly simple and obvious cause, such as loose or corroded connectors or air leaks in the intake system; making careful inspection of components during testing essential to quick and accurate troubleshooting. Special, hand held computerized testers designed specifically for diagnosing the EEC-IV system are available from a variety of aftermarket sources, as well as from the vehicle manufacturer, but care should be taken that any test equipment being used is designed to diagnose that particular computer controlled system accurately without damaging the control unit (ECU) or components being tested.

➡ **Pinpointing the exact cause of trouble in an electrical system can sometimes only be accomplished by the use of special test equipment. The following describes commonly used test equipment and explains how to put it to best use in diagnosis. In addition to the information covered below, the manufacturer's instructions booklet provided with the tester should be read and clearly understood before attempting any test procedures.**

TEST EQUIPMENT

Jumper Wires

Jumper wires are simple, yet extremely valuable, pieces of test equipment. Jumper wires are merely wires that are used to bypass sections of a circuit. The simplest type of jumper wire is merely a length of multistrand wire with an alligator clip at each end. Jumper wires are usually fabricated from lengths of standard automotive wire and whatever type of connector (alligator clip, spade connector or pin connector) that is required for the particular vehicle being tested. The well equipped tool box will have several different styles of jumper wires in several different lengths. Some jumper wires are made with three or more terminals coming from a common splice for special purpose testing. In cramped, hard-to-reach areas it is advisable to have insulated boots over the jumper wire terminals in order to prevent accidental grounding, sparks, and possible fire, especially when testing fuel system components.

Jumper wires are used primarily to locate open electrical circuits, on either the ground (-) side of the circuit or on the hot (+) side. If an electrical component fails to operate, connect the jumper wire between the component and a good ground. If the component operates only with the jumper installed, the ground circuit is open. If the ground circuit is good, but the component does not operate, the circuit between the power feed and component is open. You can sometimes connect the jumper wire directly from the battery to the hot terminal of the component, but first make sure the component uses 12 volts in operation. Some electrical components, such as fuel injectors, are designed to operate on about 4 volts and running 12 volts directly to the injector terminals can burn out the wiring. By inserting an inline fuseholder between a set of test leads, a fused jumper wire can be used for bypassing open circuits. Use a 5 amp fuse to provide protection against voltage spikes. When in doubt, use a voltmeter to check the voltage input to the component and measure how much voltage is being applied normally. By moving the jumper wire successively back from the lamp toward the power source, you can isolate the area of the circuit where the open is located. When the component stops functioning, or the power is cut off, the open is in the segment of wire between the jumper and the point previously tested.

✲✲ CAUTION

Never use jumpers made from wire that is of lighter gauge than used in the circuit under test. If the jumper wire is of too small gauge, it may overheat and possibly melt. Never use jumpers to bypass high resistance loads (such as motors) in a circuit. Bypassing resistances, in effect, creates a short circuit which may, in turn, cause damage and fire. Never use a jumper for anything other than temporary bypassing of components in a circuit.

12 Volt Test Light

The 12 volt test light is used to check circuits and components while electrical current is flowing through them. It is used for voltage and ground tests. Twelve volt test lights come in different styles but all have three main parts; a ground clip, a probe, and a light. The most commonly used 12 volt test lights have pick-type probes. To use a 12 volt test light, connect the ground clip to a good ground and probe wherever necessary with the pick. The pick should be sharp so that it can penetrate wire insulation to make contact with the wire, without making a large hole in the insulation. The wrap-around light is handy in hard to reach areas or where it is difficult to support a wire to push a probe pick into it. To use the wrap around light, hook the wire to probed with the hook and pull the trigger. A small pick will be forced through the wire insulation into the wire core.

✲✲ CAUTION

Do not use a test light to probe electronic ignition spark plug or coil wires. Never use a pick-type test light to probe wiring on computer controlled systems unless specifically instructed to do so. Any wire insulation that is pierced by the test light probe should be taped and sealed with silicone after testing.

Like the jumper wire, the 12 volt test light is used to isolate opens in circuits. But, whereas the jumper wire is used to bypass the open to operate the load, the 12 volt test light is used to locate the presence of voltage in a circuit. If the test light glows, you know that there is power up to that point; if the 12 volt test light does not glow when its probe is inserted into the wire or connector, you know that there is an open circuit (no power). Move the test light in successive steps back toward the power source until the light in the handle does glow. When it does glow, the open is between the probe and point previously probed.

➡ **The test light does not detect that 12 volts (or any particular amount of voltage) is present; it only detects that some voltage is present. It is advisable before using the test light to touch its terminals across the battery posts to make sure the light is operating properly.**

Self-Powered Test Light

The self-powered test light usually contains a 1.5 volt penlight battery. One type of self-powered test light is similar in design to the 12 volt test light. This type has both the battery and the light in the handle and pick-type probe tip. The second type has the light toward the open tip, so that the light illuminates the contact point. The self-powered test light is dual purpose piece of test equipment. It can be used to test for either open or short circuits when power is isolated from the circuit (continuity test). A powered test light should not be used on any computer controlled system or component unless specifically instructed to do so. Many engine sensors can be destroyed by even this small amount of voltage applied directly to the terminals.

Open Circuit Testing

To use the self-powered test light to check for open circuits, first isolate the circuit from the vehicle's 12 volt power source by disconnecting the battery or wiring harness connector. Connect the test light ground clip to a good ground and probe sections of the circuit sequentially with the test light. (start from either end of the circuit). If the light is out, the open is between the probe and the circuit ground. If the light is on, the open is between the probe and end of the circuit toward the power source.

Short Circuit Testing

By isolating the circuit both from power and from ground, and using a self-powered test light, you can check for shorts to ground in the circuit. Isolate the circuit from power and ground. Connect the test light ground clip to a good ground and probe any easy-to-reach test point in the circuit. If the light comes on, there is a short somewhere in the circuit. To isolate the short, probe a test point at either end of the isolated circuit (the light should be on). Leave the test light probe connected and open connectors, switches, remove parts, etc., sequentially, until the light goes out. When the light goes out, the short is between the last circuit component opened and the previous circuit opened.

➡ The 1.5 volt battery in the test light does not provide much current. A weak battery may not provide enough power to illuminate the test light even when a complete circuit is made (especially if there are high resistances in the circuit). Always make sure that the test battery is strong. To check the battery, briefly touch the ground clip to the probe; if the light glows brightly the battery is strong enough for testing. Never use a self-powered test light to perform checks for opens or shorts when power is applied to the electrical system under test. The 12 volt vehicle power will quickly burn out the 1.5 volt light bulb in the test light.

Voltmeter

A voltmeter is used to measure voltage at any point in a circuit, or to measure the voltage drop across any part of a circuit. It can also be used to check continuity in a wire or circuit by indicating current flow from one end to the other. Voltmeters usually have various scales on the meter dial and a selector switch to allow the selection of different voltages. The voltmeter has a positive and a negative lead. To avoid damage to the meter, always connect the negative lead to the negative (-) side of circuit (to ground or nearest the ground side of the circuit) and connect the positive lead to the positive (+) side of the circuit (to the power source or the nearest power source). Note that the negative voltmeter lead will always be black and that the positive voltmeter will always be some color other than black (usually red). Depending on how the voltmeter is connected into the circuit, it has several uses.

A voltmeter can be connected either in parallel or in series with a circuit and it has a very high resistance to current flow. When connected in parallel, only a small amount of current will flow through the voltmeter current path; the rest will flow through the normal circuit current path and the circuit will work normally. When the voltmeter is connected in series with a circuit, only a small amount of current can flow through the circuit. The circuit will not work properly, but the voltmeter reading will show if the circuit is complete or not.

Available Voltage Measurement

Set the voltmeter selector switch to the 20V position and connect the meter negative lead to the negative post of the battery. Connect the positive meter lead to the positive post of the battery and turn the ignition switch ON to provide a load. Read the voltage on the meter or digital display. A well charged battery should register over 12 volts. If the meter reads below 11.5 volts, the battery power may be insufficient to operate the electrical system properly. This test determines voltage available from the battery and should be the first step in any electrical trouble diagnosis procedure. Many electrical problems, especially on computer controlled systems, can be caused by a low state of charge in the battery. Excessive corrosion at the battery cable terminals can cause a poor contact that will prevent proper charging and full battery current flow.

Normal battery voltage is 12 volts when fully charged. When the battery is supplying current to one or more circuits it is said to be "under load". When everything is off the electrical system is under a "no-load" condition. A fully charged battery may show about 12.5 volts at no load; will drop to 12 volts under medium load; and will drop even lower under heavy load. If the battery is partially discharged the voltage decrease under heavy load may be excessive, even though the battery shows 12 volts or more at no load. When allowed to discharge further, the battery's available voltage under load will decrease more severely. For this reason, it is important that the battery be fully charged during all testing procedures to avoid errors in diagnosis and incorrect test results.

Voltage Drop

When current flows through a resistance, the voltage beyond the resistance is reduced (the larger the current, the greater the reduction in voltage). When no current is flowing, there is no voltage drop because there is no current flow. All points in the circuit which are connected to the power source are at the same voltage as the power source. The total voltage drop always equals the total source voltage. In a long circuit with many connectors, a series of small, unwanted voltage drops due to corrosion at the connectors can add up to a total loss of voltage which impairs the operation of the normal loads in the circuit.

INDIRECT COMPUTATION OF VOLTAGE DROPS

1. Set the voltmeter selector switch to the 20 volt position.
2. Connect the meter negative lead to a good ground.
3. Probe all resistances in the circuit with the positive meter lead.
4. Operate the circuit in all modes and observe the voltage readings.

DIRECT MEASUREMENT OF VOLTAGE DROPS

1. Set the voltmeter switch to the 20 volt position.
2. Connect the voltmeter negative lead to the ground side of the resistance load to be measured.
3. Connect the positive lead to the positive side of the resistance or load to be measured.
4. Read the voltage drop directly on the 20 volt scale.

Too high a voltage indicates too high a resistance. If, for example, a blower motor runs too slowly, you can determine if there is too high a resistance in the resistor pack. By taking voltage drop readings in all parts of the circuit, you can isolate the problem. Too low a voltage drop indicates too low a resistance. If, for example, a blower motor runs too fast in the MED and/or LOW position, the problem can be isolated in the resistor pack by taking voltage drop readings in all parts of the circuit to locate a possibly shorted resistor. The maximum allowable voltage drop under load is critical, especially if there is more than one high resistance problem in a circuit because all voltage drops are cumulative. A small drop is normal due to the resistance of the conductors.

HIGH RESISTANCE TESTING

1. Set the voltmeter selector switch to the 4 volt position.
2. Connect the voltmeter positive lead to the positive post of the battery.
3. Turn on the headlights and heater blower to provide a load.
4. Probe various points in the circuit with the negative voltmeter lead.

5. Read the voltage drop on the 4 volt scale. Some average maximum allowable voltage drops are:

FUSE PANEL—7 volts
IGNITION SWITCH—5 volts
HEADLIGHT SWITCH—7 volts
IGNITION COIL (+)—5 volts
ANY OTHER LOAD—1.3 volts

➡ **Voltage drops are all measured while a load is operating; without current flow, there will be no voltage drop.**

Ohmmeter

The ohmmeter is designed to read resistance (ohms) in a circuit or component. Although there are several different styles of ohmmeters, all will usually have a selector switch which permits the measurement of different ranges of resistance (usually the selector switch allows the multiplication of the meter reading by 10, 100, 1000, and 10,000). A calibration knob allows the meter to be set at zero for accurate measurement. Since all ohmmeters are powered by an internal battery (usually 9 volts), the ohmmeter can be used as a self-powered test light. When the ohmmeter is connected, current from the ohmmeter flows through the circuit or component being tested. Since the ohmmeter's internal resistance and voltage are known values, the amount of current flow through the meter depends on the resistance of the circuit or component being tested.

The ohmmeter can be used to perform continuity test for opens or shorts (either by observation of the meter needle or as a self-powered test light), and to read actual resistance in a circuit. It should be noted that the ohmmeter is used to check the resistance of a component or wire while there is no voltage applied to the circuit. Current flow from an outside voltage source (such as the vehicle battery) can damage the ohmmeter, so the circuit or component should be isolated from the vehicle electrical system before any testing is done. Since the ohmmeter uses its own voltage source, either lead can be connected to any test point.

➡ **When checking diodes or other solid state components, the ohmmeter leads can only be connected one way in order to measure current flow in a single direction. Make sure the positive (+) and negative (-) terminal connections are as described in the test procedures to verify the one-way diode operation.**

In using the meter for making continuity checks, do not be concerned with the actual resistance readings. Zero resistance, or any resistance readings, indicate continuity in the circuit. Infinite resistance indicates an open in the circuit. A high resistance reading where there should be none indicates a problem in the circuit. Checks for short circuits are made in the same manner as checks for open circuits except that the circuit must be isolated from both power and normal ground. Infinite resistance indicates no continuity to ground, while zero resistance indicates a dead short to ground.

RESISTANCE MEASUREMENT

The batteries in an ohmmeter will weaken with age and temperature, so the ohmmeter must be calibrated or "zeroed" before taking measurements. To zero the meter, place the selector switch in its lowest range and touch the two ohmmeter leads together. Turn the calibration knob until the meter needle is exactly on zero.

➡ **All analog (needle) type ohmmeters must be zeroed before use, but some digital ohmmeter models are automatically calibrated when the switch is turned on. Self-calibrating digital ohmmeters do not have an adjusting knob, but its a good idea to check for a zero readout before use by touching the leads together. All computer controlled systems require the use of a digital ohmmeter with at least 10 megohms impedance for testing. Before any test procedures are attempted, make sure the ohmmeter used is compatible with the electrical system or damage to the on-board computer could result.**

To measure resistance, first isolate the circuit from the vehicle power source by disconnecting the battery cables or the harness connector. Make sure the key is OFF when disconnecting any components or the battery. Where necessary, also isolate at least one side of the circuit to be checked to avoid reading parallel resistances. Parallel circuit resistances will always give a lower reading than the actual resistance of either of the branches. When measuring the resistance of parallel circuits, the total resistance will always be lower than the smallest resistance in the circuit. Connect the meter leads to both sides of the circuit (wire or component) and read the actual measured ohms on the meter scale. Make sure the selector switch is set to the proper ohm scale for the circuit being tested to avoid misreading the ohmmeter test value.

⁂ CAUTION

Never use an ohmmeter with power applied to the circuit. Like the self-powered test light, the ohmmeter is designed to operate on its own power supply. The normal 12 volt automotive electrical system current could damage the meter.

Ammeters

An ammeter measures the amount of current flowing through a circuit in units called amperes or amps. Amperes are units of electron flow which indicate how fast the electrons are flowing through the circuit. Since Ohms Law dictates that current flow in a circuit is equal to the circuit voltage divided by the total circuit resistance, increasing voltage also increases the current level (amps). Likewise, any decrease in resistance will increase the amount of amps in a circuit. At normal operating voltage, most circuits have a characteristic amount of amperes, called "current draw" which can be measured using an ammeter. By referring to a specified current draw rating, measuring the amperes, and comparing the two values, one can determine what is happening within the circuit to aid in diagnosis. An open circuit, for example, will not allow any current to flow so the ammeter reading will be zero. More current flows through a heavily loaded circuit or when the charging system is operating.

An ammeter is always connected in series with the circuit being tested. All of the current that normally flows through the circuit must also flow through the ammeter; if there is any other path for the current to follow, the ammeter reading will not be accurate. The ammeter itself has very little resistance to current flow and therefore will not affect the circuit, but it will measure current draw only when the circuit is closed and electricity is flowing. Excessive current draw can blow fuses and drain the battery, while a reduced current draw can cause motors to run slowly, lights to dim and other components to not operate properly. The ammeter can help diagnose these conditions by locating the cause of the high or low reading.

Multimeters

Different combinations of test meters can be built into a single unit designed for specific tests. Some of the more common combination test devices are known as Volt/Amp testers, Tach/Dwell meters, or Digital Multimeters. The Volt/Amp tester is used for charging system, starting system or battery tests and consists of a voltmeter, an ammeter and a variable resistance carbon pile. The voltmeter will usually have at least two ranges for use with 6, 12 and 24 volt systems. The ammeter also has more than one range for testing various levels of battery loads and starter current draw and the carbon pile can be adjusted to offer different amounts of resistance. The Volt/Amp tester has heavy leads to carry large amounts of current and many later models have an inductive ammeter pickup that clamps around the wire to simplify test connections. On some models, the ammeter also has a zero-center scale to allow testing of charging and starting systems without switching leads or polarity. A digital multimeter is a voltmeter, ammeter and ohmmeter combined in an instrument which gives a digital readout. These are often used when testing solid state circuits because of their high input impedance (usually 10 megohms or more).

The tach/dwell meter combines a tachometer and a dwell (cam angle) meter and is a specialized kind of voltmeter. The tachometer scale is marked to show engine speed in rpm and the dwell scale is marked to show degrees of distributor shaft rotation. In most electronic ignition systems, dwell is determined by the control unit, but the dwell meter can also be used to check the duty cycle (operation) of some electronic engine control systems. Some tach/ dwell meters are powered by an internal battery, while others take their power from the car battery in use. The battery powered testers usually require calibration much like an ohmmeter before testing.

Special Test Equipment

A variety of diagnostic tools are available to help troubleshoot and repair computerized engine control systems. The most sophisticated of these devices are the console type engine analyzers that usually occupy a garage service bay, but there are several types of aftermarket electronic testers available that will allow quick circuit tests of the engine control system by plugging directly into a special connector located in the engine compartment or under the dashboard. Several tool and equipment manufacturers offer simple, hand held testers that measure various circuit voltage levels on command to check all system components for proper operation. Although these testers usually cost about $300-500, consider that the average computer control unit (or ECM) can cost just as much and the money saved by not replacing perfectly good sensors or components in an attempt to correct a problem could justify the purchase price of a special diagnostic tester the first time it's used.

These computerized testers can allow quick and easy test measurements while the engine is operating or while the car is being driven. In addition, the on-board computer memory can be read to access any stored trouble codes; in effect allowing the computer to tell you where it hurts and aid trouble diagnosis by pinpointing exactly which circuit or component is malfunctioning. In the same manner, repairs can be tested to make sure the problem has been corrected. The biggest advantage these special testers have is their relatively easy hookups that minimize or eliminate the chances of making the wrong connections and getting false voltage readings or damaging the computer accidentally.

➡ It should be remembered that these testers check voltage levels in circuits; they don't detect mechanical problems or failed components if the circuit voltage falls within the preprogrammed limits stored in the tester PROM unit. Also, most of the hand held testers are designed to work only on one or two systems made by a specific manufacturer.

A variety of aftermarket testers are available to help diagnose different computerized control systems. Owatonna Tool Company (OTC), for example, markets a device called the OTC Monitor which plugs directly into the assembly line diagnostic link (ALDL). The OTC tester makes diagnosis a simple matter of pressing the correct buttons and, by changing the internal PROM or inserting a different diagnosis cartridge, it will work on any model from full size to subcompact, over a wide range of years. An adapter is supplied with the tester to allow connection to all types of ALDL links, regardless of the number of pin terminals used. By inserting an updated PROM into the OTC tester, it can be easily updated to diagnose any new modifications of computerized control systems.

Wiring Harnesses

The average automobile contains about 1/2 mile of wiring, with hundreds of individual connections. To protect the many wires from damage and to keep them from becoming a confusing tangle, they are organized into bundles, enclosed in plastic or taped together and called wire harnesses. Different wiring harnesses serve different parts of the vehicle. Individual wires are color coded to help trace them through a harness where sections are hidden from view.

A loose or corroded connection or a replacement wire that is too small for the circuit will add extra resistance and an additional voltage drop to the circuit. A ten percent voltage drop can result in slow or erratic motor operation, for example, even though the circuit is complete. Automotive wiring or circuit conductors can be in any one of three forms:

1. Single strand wire
2. Multistrand wire
3. Printed circuitry

Single strand wire has a solid metal core and is usually used inside such components as alternators, motors, relays and other devices. Multistrand wire has a core made of many small strands of wire twisted together into a single conductor. Most of the wiring in an automotive electrical system is made up of multistrand wire, either as a single conductor or grouped together in a harness. All wiring is color coded on the insulator, either as a solid color or as a colored wire with an identification stripe. A printed circuit is a thin film of copper or other conductor that is printed on an insulator backing. Occasionally, a printed circuit is sandwiched between two sheets of plastic for more protection and flexibility. A complete printed circuit, consisting of conductors, insulating material and connectors for lamps or other components is called a printed circuit board. Printed circuitry is used in place of individual wires or harnesses in places where space is limited, such as behind instrument panels.

Wire Gauge

Since computer controlled automotive electrical systems are very sensitive to changes in resistance, the selection of properly sized wires is critical when systems are repaired. The wire gauge number is an expression of the cross section area of the conductor. The most common system for expressing wire size is the American Wire Gauge (AWG) system.

Wire cross section area is measured in circular mils. A mil is 1/1000" (0.001"); a circular mil is the area of a circle one mil in diameter. For example, a conductor 1/4" in diameter is 0.250 in. or 250 mils. The circular mil cross section area of the wire is 250 squared (250ó)or 62,500 circular mils. Imported car models usually use metric wire gauge designations, which is simply the cross section area of the conductor in square millimeters (mmó).

Gauge numbers are assigned to conductors of various cross section areas. As gauge number increases, area decreases and the conductor becomes smaller. A 5 gauge conductor is smaller than a 1 gauge conductor and a 10 gauge is smaller than a 5 gauge. As the cross section area of a conductor decreases, resistance increases and so does the gauge number. A conductor with a higher gauge number will carry less current than a conductor with a lower gauge number.

➡ **Gauge wire size refers to the size of the conductor, not the size of the complete wire. It is possible to have two wires of the same gauge with different diameters because one may have thicker insulation than the other.**

12 volt automotive electrical systems generally use 10, 12, 14, 16 and 18 gauge wire. Main power distribution circuits and larger accessories usually use 10 and 12 gauge wire. Battery cables are usually 4 or 6 gauge, although 1 and 2 gauge wires are occasionally used. Wire length must also be considered when making repairs to a circuit. As conductor length increases, so does resistance. An 18 gauge wire, for example, can carry a 10 amp load for 10 feet without excessive voltage drop; however if a 15 foot wire is required for the same 10 amp load, it must be a 16 gauge wire.

An electrical schematic shows the electrical current paths when a circuit is operating properly. It is essential to understand how a circuit works before trying to figure out why it doesn't. Schematics break the entire electrical system down into individual circuits and show only one particular circuit. In a schematic, no attempt is made to represent wiring and components as they physically appear on the vehicle; switches and other components are shown as simply as possible. Face views of harness connectors show the cavity or terminal locations in all multi-pin connectors to help locate test points.

If you need to backprobe a connector while it is on the component, the order of the terminals must be mentally reversed. The wire color code can help in this situation, as well as a keyway, lock tab or other reference mark.

WIRING REPAIR

Soldering is a quick, efficient method of joining metals permanently. Everyone who has the occasion to make wiring repairs should know how to solder. Electrical connections that are soldered are far less likely to come apart and will conduct electricity much better than connections that are only "pig-tailed" together. The most popular (and preferred) method of soldering is with an electrical soldering gun. Soldering irons are available in many sizes and wattage ratings. Irons with higher wattage ratings deliver higher temperatures and recover lost heat faster. A small soldering iron rated for no more than 50 watts is recommended, especially on electrical systems where excess heat can damage the components being soldered.

There are three ingredients necessary for successful soldering; proper flux, good solder and sufficient heat. A soldering flux is necessary to clean the metal of tarnish, prepare it for soldering and to enable the solder to spread into tiny crevices. When soldering, always use a resin flux or resin core solder which is non-corrosive and will not attract moisture once the job is finished. Other types of flux (acid core) will leave a residue that will attract moisture and cause the wires to corrode. Tin is a unique metal with a low melting point. In a molten state, it dissolves and alloys easily with many metals. Solder is made by mixing tin with lead. The most common proportions are 40/60, 50/50 and 60/40, with the percentage of tin listed first. Low priced solders usually contain less tin, making them very difficult for a beginner to use because more heat is required to melt the solder. A common solder is 40/60 which is well suited for all-around general use, but 60/40 melts easier, has more tin for a better joint and is preferred for electrical work.

Soldering Techniques

Successful soldering requires that the metals to be joined be heated to a temperature that will melt the solder—usually 360-460°F (182-238°C). Contrary to popular belief, the purpose of the soldering iron is not to melt the solder itself, but to heat the parts being soldered to a temperature high enough to melt the solder when it is touched to the work. Melting flux-cored solder on the soldering iron will usually destroy the effectiveness of the flux.

➡ **Soldering tips are made of copper for good heat conductivity, but must be "tinned" regularly for quick transference of heat to the project and to prevent the solder from sticking to the iron. To "tin" the iron, simply heat it and touch the flux-cored solder to the tip; the solder will flow over the hot tip. Wipe the excess off with a clean rag, but be careful as the iron will be hot.**

After some use, the tip may become pitted. If so, simply dress the tip smooth with a smooth file and "tin" the tip again. An old saying holds that "metals well cleaned are half soldered." Flux-cored solder will remove oxides but rust, bits of insulation and oil or grease must be removed with a wire brush or emery cloth. For maximum strength in soldered parts, the joint must start off clean and tight. Weak joints will result in gaps too wide for the solder to bridge.

If a separate soldering flux is used, it should be brushed or swabbed on only those areas that are to be soldered. Most solders contain a core of flux and separate fluxing is unnecessary. Hold the work to be soldered firmly. It is best to solder on a wooden board, because a metal vise will only rob the piece to be soldered of heat and make it difficult to melt the solder. Hold the soldering tip with the broadest face against the work to be soldered. Apply solder under the tip close to the work, using enough solder to give a heavy film between the iron and the piece being soldered, while moving slowly and making sure the solder melts properly. Keep the work level or the solder will run to the lowest part and favor the thicker parts, because these require more heat to melt the solder. If the soldering tip overheats (the solder coating on the face of the tip burns up), it should be retinned. Once the soldering is completed, let the soldered joint stand until cool. Tape and seal all soldered wire splices after the repair has cooled.

Wire Harness and Connectors

The on-board computer (ECM) wire harness electrically connects the control unit to the various solenoids, switches and sensors used by the control system. Most connectors in the engine compartment or otherwise exposed to the elements are protected against moisture and dirt which could create oxidation and deposits on the terminals. This protection is important because of the very low voltage and current levels used by the computer and sensors. All connectors have a lock which secures the male and female terminals together, with a secondary lock holding the seal and terminal into the connector. Both terminal locks must be released when disconnecting ECM connectors.

These special connectors are weather-proof and all repairs require the use of a special terminal and the tool required to service it. This tool is used to remove the pin and sleeve terminals. If removal is attempted with an ordinary pick, there is a good chance that the terminal will be bent or deformed. Unlike standard blade type terminals, these terminals cannot be straightened once they are bent. Make certain that the connectors are properly seated and all of the sealing rings in place when connecting leads. On some models, a hinge-type flap provides a backup or secondary locking feature for the terminals. Most secondary locks are used to improve the connector reliability by retaining the terminals if the small terminal lock tangs are not positioned properly.

Molded-on connectors require complete replacement of the connection. This means splicing a new connector assembly into the harness. All splices in on-board computer systems should be soldered to insure proper contact. Use care when probing the connections or replacing terminals in them as it is possible to short between opposite terminals. If this happens to the wrong terminal pair, it is possible to damage certain components. Always use jumper wires between connectors for circuit checking and never probe through weatherproof seals.

Open circuits are often difficult to locate by sight because corrosion or terminal misalignment are hidden by the connectors. Merely wiggling a connector on a sensor or in the wiring harness may correct the open circuit condition. This should always be considered when an open circuit or a failed sensor is indicated. Intermittent problems may also be caused by oxidized or loose connections. When using a circuit tester for diagnosis, always probe connections from the wire side. Be careful not to damage sealed connectors with test probes.

All wiring harnesses should be replaced with identical parts, using the same gauge wire and connectors. When signal wires are spliced into a harness, use wire with high temperature insulation only. With the low voltage and current levels found in the system, it is important that the best possible connection at all wire splices be made by soldering the splices together. It is seldom necessary to replace a complete harness. If replacement is necessary, pay close attention to insure proper harness routing. Secure the harness with suitable plastic wire clamps to prevent vibrations from causing the harness to wear in spots or contact any hot components.

➡ Weatherproof connectors cannot be replaced with standard connectors. Instructions are provided with replacement connector and terminal packages. Some wire harnesses have mounting indicators (usually pieces of colored tape) to mark where the harness is to be secured.

In making wiring repairs, it's important that you always replace damaged wires with wires that are the same gauge as the wire being replaced. The heavier the wire, the smaller the gauge number. Wires are color-coded to aid in identification and whenever possible the same color coded wire should be used for replacement. A wire stripping and crimping tool is necessary to install solderless terminal connectors. Test all crimps by pulling on the wires; it should not be possible to pull the wires out of a good crimp.

Wires which are open, exposed or otherwise damaged are repaired by simple splicing. Where possible, if the wiring harness is accessible and the damaged place in the wire can be located, it is best to open the harness and check for all possible damage. In an inaccessible harness, the wire must be bypassed with a new insert, usually taped to the outside of the old harness.

When replacing fusible links, be sure to use fusible link wire, NOT ordinary automotive wire. Make sure the fusible segment is of the same gauge and construction as the one being replaced and double the stripped end when crimping the terminal connector for a good contact. The melted (open) fusible link segment of the wiring harness should be cut off as close to the harness as possible, then a new segment spliced in as described. In the case of a damaged fusible link that feeds two harness wires, the harness connections should be replaced with two fusible link wires so that each circuit will have its own separate protection.

➡ Most of the problems caused in the wiring harness are due to bad ground connections. Always check all vehicle ground connections for corrosion or looseness before performing any power feed checks to eliminate the chance of a bad ground affecting the circuit.

Repairing Hard Shell Connectors

Unlike molded connectors, the terminal contacts in hard shell connectors can be replaced. Weatherproof hard-shell connectors with the leads molded into the shell have non-replaceable terminal ends. Replacement usually involves the use of a special terminal removal tool that depress the locking tangs (barbs) on the connector terminal and allow the connector to be removed from the rear of the shell. The connector shell should be replaced if it shows any evidence of burning, melting, cracks, or breaks. Replace individual terminals that are burnt, corroded, distorted or loose.

➡ The insulation crimp must be tight to prevent the insulation from sliding back on the wire when the wire is pulled. The insulation must be visibly compressed under the crimp tabs, and the ends of the crimp should be turned in for a firm grip on the insulation.

The wire crimp must be made with all wire strands inside the crimp. The terminal must be fully compressed on the wire strands with the ends of the crimp tabs turned in to make a firm grip on the wire. Check all connections with an ohmmeter to insure a good contact. There should be no measurable resistance between the wire and the terminal when connected.

Mechanical Test Equipment

Vacuum Gauge

Most gauges are graduated in inches of mercury (in.Hg), although a device called a manometer reads vacuum in inches of water (in. H_2O). The normal vacuum reading usually varies between 18 and 22 in.Hg at sea level. To test engine vacuum, the vacuum gauge must be connected to a source of manifold vacuum. Many engines have a plug in the intake manifold which can be removed and replaced with an adapter fitting. Connect the vacuum gauge to the fitting with a suitable rubber hose or, if no manifold plug is available, connect the vacuum gauge to any device using manifold vacuum, such as EGR valves, etc. The vacuum gauge can be used to determine if enough vacuum is reaching a component to allow its actuation.

Hand Vacuum Pump

Small, hand-held vacuum pumps come in a variety of designs. Most have a built-in vacuum gauge and allow the component to be tested without removing it from the vehicle. Operate the pump lever or plunger to apply the correct amount of vacuum required for the test specified in the diagnosis routines. The level of vacuum in inches of Mercury (in.Hg) is indicated on the pump gauge. For some testing, an additional vacuum gauge may be necessary.

Intake manifold vacuum is used to operate various systems and devices on late model vehicles. To correctly diagnose and solve problems in vacuum control systems, a vacuum source is necessary for testing. In some cases, vacuum can be taken from the intake manifold when the engine is running, but vacuum is normally provided by a hand vacuum pump. These hand vacuum pumps have a built-in vacuum gauge that allow testing while the device is still attached to the component. For some tests, an additional vacuum gauge may be necessary.

HEATER

Blower Motor

REMOVAL & INSTALLATION

➧ SEE FIG. 1

1. Disconnect the negative battery cable. Loosen glove compartment assembly by squeezing the sides together to disengage the retainer tabs.
2. Let the glove compartment and door hang down in front of instrument panel and remove blower motor cooling hose.
3. Disconnect electrical wiring harness. Remove 4 screws attaching motor to housing. Pull motor and wheel out of housing.
4. Installation is the reverse of the removal procedure.

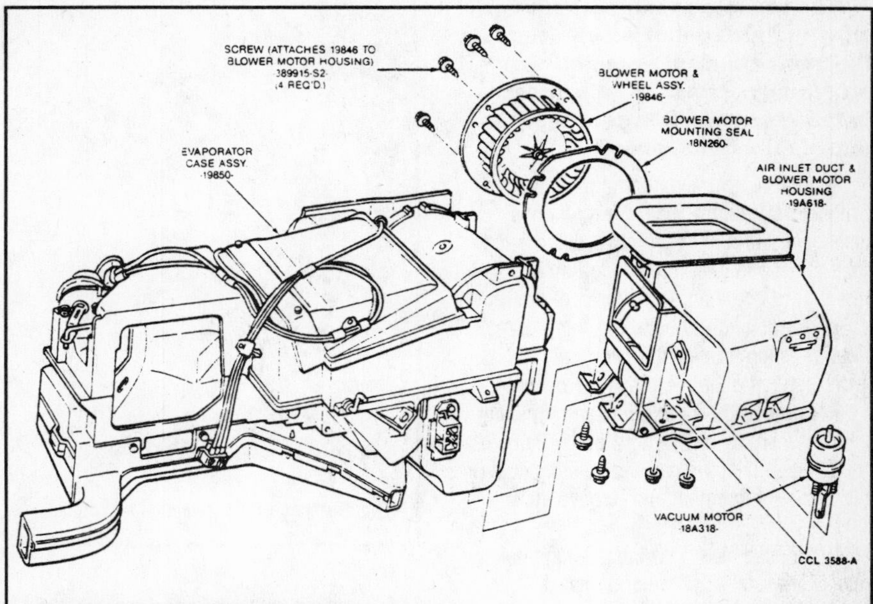

FIG. 1 Blower motor, and air inlet duct assembly

Blower Motor Resistor

REMOVAL & INSTALLATION

➧ SEE FIG. 2

1. Disconnect the negative battery cable.
2. Disconnect the blower motor resistor wiring harness.
3. Remove the 2 screws that attach the resistor board to the evaporator case and remove the resistor assembly.
4. Installation is the reverse of the removal procedure.

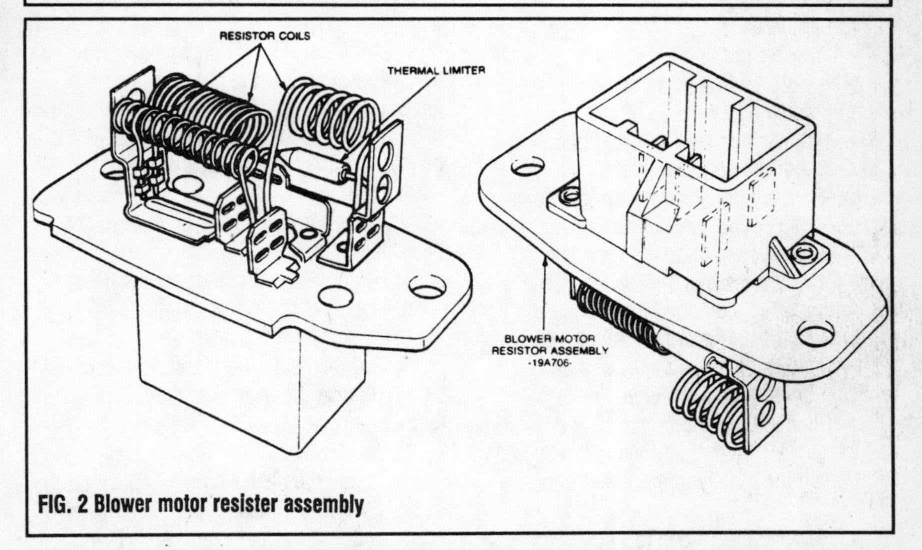

FIG. 2 Blower motor resister assembly

Heater Core

REMOVAL & INSTALLATION

CAUTION

When draining the coolant, keep in mind that cats and dogs are attracted by the ethylene glycol antifreeze, and are quite likely to drink any that is left in an uncovered container or in puddles on the ground. This will prove fatal in sufficient quantity. Always drain the coolant into a sealable container. Coolant should be reused unless it is contaminated or several years old.

Without Air Conditioning

➧ SEE FIG. 3

1. Disconnect the negative battery cable.
2. Remove the floor console and instrument panel as follows:
   a. Remove the 4 screws that attach the console top panel assembly to the console assembly.
   b. Lift the console top panel assembly off the console assembly and disconnect the 2 electrical connectors.
   c. Remove the 2 screws attaching the rear end of the console assembly to the console panel support assembly.
   d. Remove 4 screws attaching the console assembly to the console panel front support.

e. Remove the 4 screws attaching the console assembly to the instrument panel.

f. Lift the console assembly off of the transmission tunnel.

➡ The console assembly includes a snap-in finish panel which conceals the heater control assembly attaching screws. To gain access to these screws, it is necessary to remove the floor console

g. Disconnect all underhood wiring connectors from the main wiring harness. Disengage the rubber grommet seal from the dash panel and push the wiring harness and connectors into the passenger compartment.

h. Remove the 3 bolts attaching the steering column opening cover and reinforcement panel. Remove the cover.

i. Remove the steering column opening reinforcement by removing 2 bolts. Remove the 2 bolts retaining the lower steering column opening reinforcement and remove the reinforcement.

j. Remove the 6 steering column retaining nuts. 2 are retaining the hood release mechanism and 4 retain the column to the lower brake pedal support. Lower the steering column to the floor.

k. Remove the steering column upper and lower shrouds and disconnect the wiring from the multi-function switch.

l. Remove the brake pedal support nut and snap out the defroster grille.

m. Remove the screws from the speaker covers. Snap out the speaker covers. Remove the front screws retaining the right and left scuff plates at the cowl trim panel. Remove the right and left side cowl trim panels.

n. Disconnect the wiring at the right and left cowl sides. Remove the cowl side retaining bolts, 1 on each side.

o. Open the glove compartment door and flex the glove compartment bin tabs inward. Drop down the glove compartment door assembly.

p. Remove the 5 cowl top screw attachments. Gently pull the instrument panel away from the cowl. Disconnect the speedometer cable and wire connectors.

3. Drain the coolant from the cooling system and remove the hoses from the heater core. Plug the hoses and the core.

4. Remove the screw attaching the air inlet duct and blower housing assembly support bracket to the cowl top panel.

5. Disconnect the black vacuum supply hose from the in-line vacuum check valve in the engine compartment.

6. Disconnect the blower motor wire harness from the resistor and motor head.

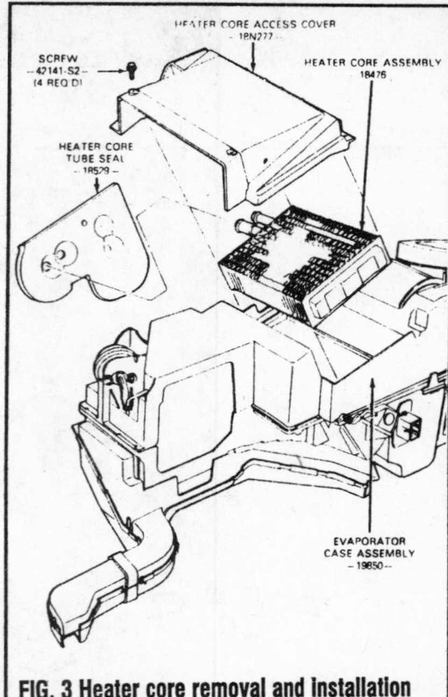

FIG. 3 Heater core removal and installation

7. Working under the hood, remove the 2 nuts retaining the heater assembly to the dash panel.

8. In the passenger compartment, remove the screw attaching the heater assembly support bracket to the cowl top panel. Remove the 1 screw retaining the bracket below the heater assembly to the dash panel.

9. Carefully pull the heater assembly away from the dash panel and remove from the vehicle.

10. Remove the 4 heater core access cover attaching screws and remove the access cover from the case.

11. Lift the heater core and seal from the case. Remove the seal from the heater core tubes.

To install:

12. Install the heater core tube seal on the heater core tubes. Inspect the heater core sealer in the heater case and replace, if necessary.

13. Install the heater core in the case with the seals on the outside of the case. Position the heater core access cover on the case and install the 4 attaching screws.

14. Position the heater assembly in the vehicle. Install the screw attaching the heater assembly support bracket to the cowl top panel.

15. Check the heater assembly drain tube to ensure it is through the dash panel and is not pinched or kinked.

16. Working under the hood, install the 2 nuts retaining the heater assembly to the dash panel. Install the air inlet duct and blower housing support bracket attaching screw. Install 1 screw to the retainer bracket below the heater assembly to the dash panel.

17. Connect the blower motor ground wire to ground and the harness to the resistor and blower motor lead.

18. Connect the black vacuum supply hose to the vacuum check valve in the engine compartment.

19. Install the instrument panel and floor console by reversing the removal procedure.

20. Connect the heater hoses to the heater core and fill the cooling system. Check the system for proper operation.

With Air Conditioning

➧ SEE FIG. 3

1. Disconnect the negative battery cable and drain the cooling system.

2. Discharge the refrigerant from the air conditioning system according to the proper procedure.

3. Remove the evaporator case assembly.

➡ Whenever an evaporator case is replaced, it will be necessary to replace the suction accumulator/drier.

4. Remove the 4 heater core access cover attaching screws and remove the cover from the case.

5. Lift the heater core and seal from the case. Remove the seal from the heater core tubes.

To install:

6. Install the heater core tube seal on the heater core tubes.

7. Inspect the heater core sealer in the evaporator case. Replace with suitable caulking cord, if necessary.

8. Install the heater core in the case with the seals on the outside of the case. Position the heater core access cover on the case and install the 4 attaching screws.

9. Install the evaporator case.

10. Fill the cooling system. Leak test, evacuate and charge the refrigerant system according to the proper procedure. Observe all safety precautions.

11. Connect the negative battery cable and check the system for proper operation.

Control Panel

REMOVAL & INSTALLATION

➧ SEE FIG. 4

1. Disconnect the negative battery cable.

2. Remove the snap-in trim molding in the floor console to expose the 4 control assembly attaching screws. Remove the 4 screws attaching the control assembly to the instrument panel.

3. Roll the control out of the opening in the console. Disconnect the fan switch connectors and temperature control cable. Disconnect the vacuum hose connector and electrical connector from the back of the function selector knob. Disconnect the connector for the control assembly illumination bulbs.

4. Remove the control assembly.

To install:

5. Connect the temperature cable to the geared arm on the temperature control.

6. Install the electrical connector at the following locations: blower switch, control assembly illumination bulbs and function selector switch.

7. Install the vacuum harness connector for the function selector knob.

8. Roll the control assembly into position against the instrument panel and install the 4 attaching screws.

9. Snap the console trim molding into position, connect the negative battery cable and check the system for proper operation.

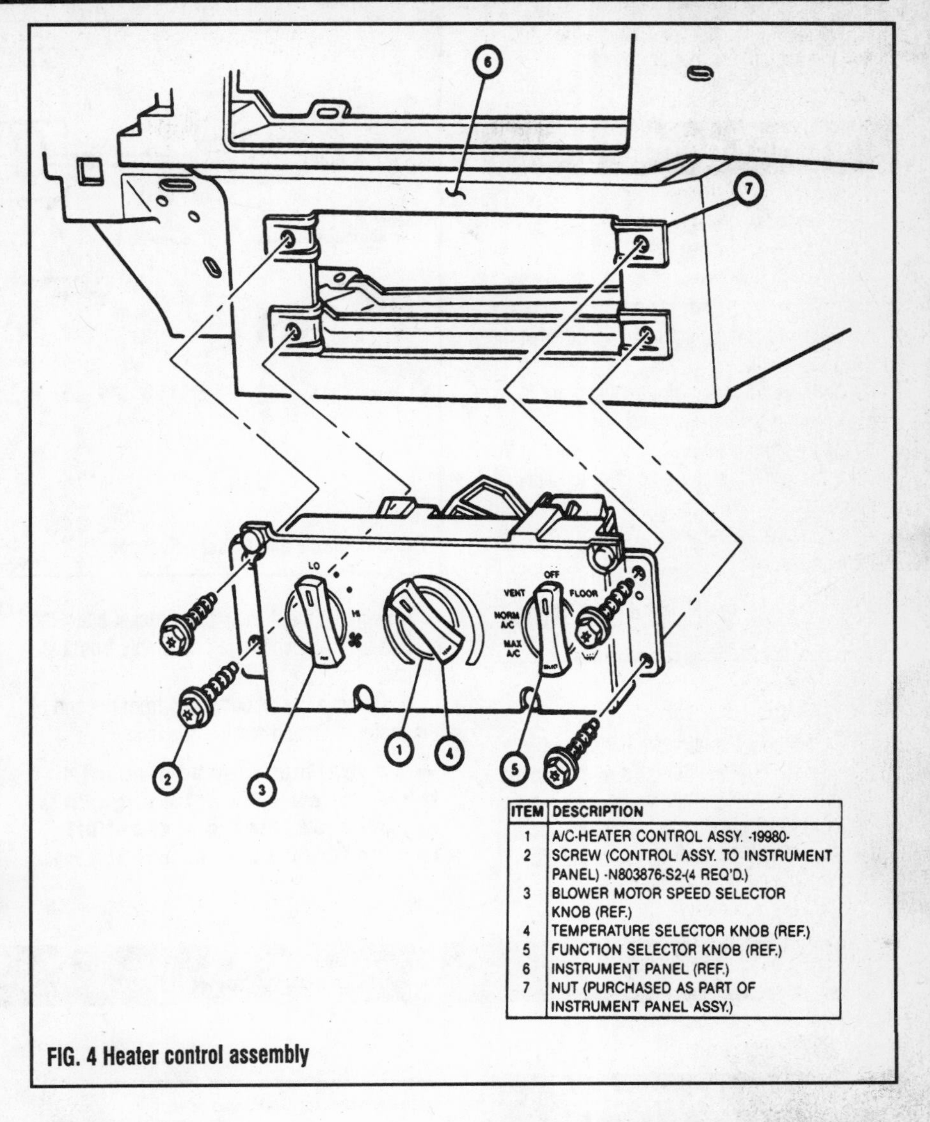

ITEM	DESCRIPTION
1	A/C-HEATER CONTROL ASSY. -19980-
2	SCREW (CONTROL ASSY. TO INSTRUMENT PANEL) -N803876-S2-(4 REQ'D.)
3	BLOWER MOTOR SPEED SELECTOR KNOB (REF.)
4	TEMPERATURE SELECTOR KNOB (REF.)
5	FUNCTION SELECTOR KNOB (REF.)
6	INSTRUMENT PANEL (REF.)
7	NUT (PURCHASED AS PART OF INSTRUMENT PANEL ASSY.)

FIG. 4 Heater control assembly

Control Cables

ADJUSTMENT

➧ SEE FIG. 5

The temperature control cable is self-adjusting with the movement of the temperature selector knob to it's fully clockwise position in the red band on the face of the control assembly. To prevent kinking of the control wire, a preset adjustment should be made before attempting to perform the self-adjustment operation. The preset adjustment may be performed either with the cable installed in the vehicle or before cable installation.

Cable Preset And Self-Adjustment

BEFORE INSTALLATION

1. Insert the end of a suitable tool in the end loop of the temperature control cable, at the temperature door crank arm end.

2. Slide the self-adjusting clip down the control wire, away from the end loop, approximately 1 in. (25mm).

3. Install the cable assembly.

4. Turn the temperature control knob fully clockwise to position the self-adjusting clip.

5. Check for proper control operation.

AFTER INSTALLATION

1. Turn the temperature selector knob to the **COOL** position.

2. Hold the temperature door crank arm firmly in position, insert a suitable tool into the wire end loop and pull the cable wire through the self-adjusting clip until there is a space of approximately 1 in. (25mm) between the clip and the wire end loop.

3. Turn the temperature control knob fully clockwise to position the self-adjusting clip.

4. Check for proper control operation.

REMOVAL & INSTALLATION

1. Disconnect the negative battery cable.

2. Remove the control assembly from the instrument panel.

3. Disengage the temperature control cable from the cable actuator on the control assembly. Disconnect the temperature cable from the plenum temperature blend door crank arm and cable mounting bracket.

4. Note the cable routing and remove the cable from the vehicle.

To install:

5. Check to ensure the self-adjusting clip is at least 1 in. (25.4mm) from the end loop of the control cable.

6. Route the cable behind the instrument panel and connect the control cable to the mounting bracket on the plenum. Install the self-adjusting clip on the temperature blend door crank arm.

7. Engage the cable end with the cable actuator on the control assembly. Install the control assembly in the instrument panel.

8. Turn the temperature control knob all the way to the right, to the **WARM** position, to position the self-adjusting clip on the control cable. Check the temperature control knob for proper operation.

9. Connect the negative battery cable and check the system for proper operation.

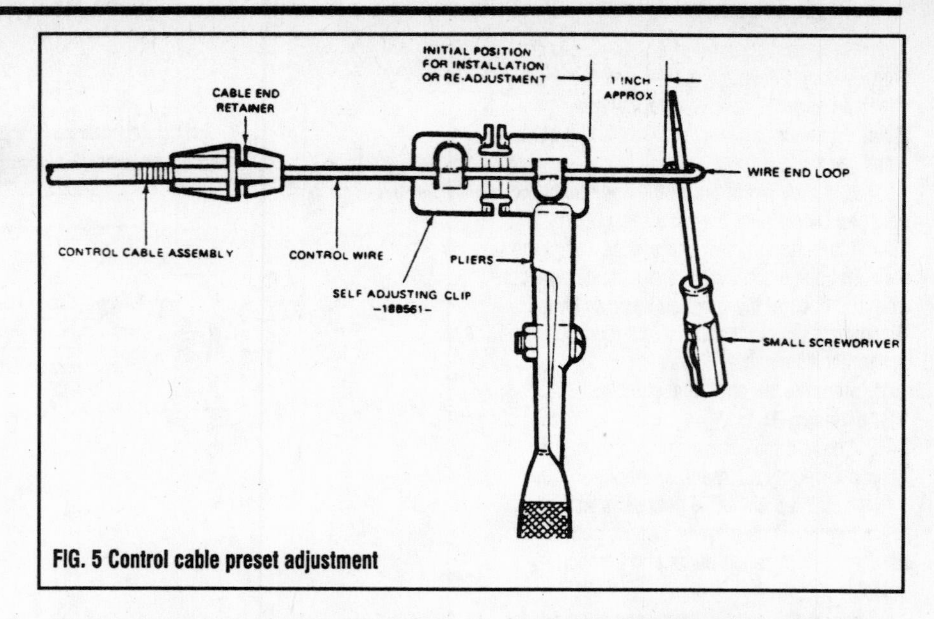

FIG. 5 Control cable preset adjustment

Electric Cooling Fan

TESTING

1. Disconnect the electrical connector at the cooling fan motor.
2. Connect a jumper wire between the negative motor lead and ground.
3. Connect another jumper wire between the positive motor lead and the positive terminal of the battery.
4. If the cooling fan motor does not run, it must be replaced.

REMOVAL & INSTALLATION

2.3L Engine

1. Disconnect the negative battery cable.
2. Remove the fan wiring harness from the routing clip. Disconnect the wiring harness from the fan motor connector by pulling up on the single lock finger to separate the connectors.
3. Remove the 4 mounting bracket attaching screws and remove the fan assembly from the vehicle.
4. Remove the retaining clip from the end of the motor shaft and remove the fan.

➡ A metal burr may be present on the motor after the retaining clip is removed. De Burring of the shaft may be required to remove the fan.

5. Remove the nuts attaching the fan motor to the mounting bracket.
6. Installation is the reverse of the removal procedure. Tighten the fan motor-to-mounting bracket attaching nuts to 48.5–62.0 inch lbs. (5.5–7.0 Nm) and the mounting bracket attaching screws to 70–95 inch lbs. (8.0–10.5 Nm).

AIR CONDITIONING

General Information

➧ SEE FIG. 6

The air conditioning system used on Ford Mustang vehicles is the fixed orifice tube-cycling clutch type. The system components consist of the compressor, magnetic clutch, condenser, evaporator, suction accumulator/drier and the necessary connecting refrigerant lines. The fixed orifice tube assembly is the restriction between the high and low pressure liquid refrigerant and meters the flow of liquid refrigerant into the evaporator core. Evaporator temperature is controlled by sensing the pressure within the evaporator with the clutch cycling pressure switch. The pressure switch controls compressor operation as necessary to maintain the evaporator pressure within specified limits.

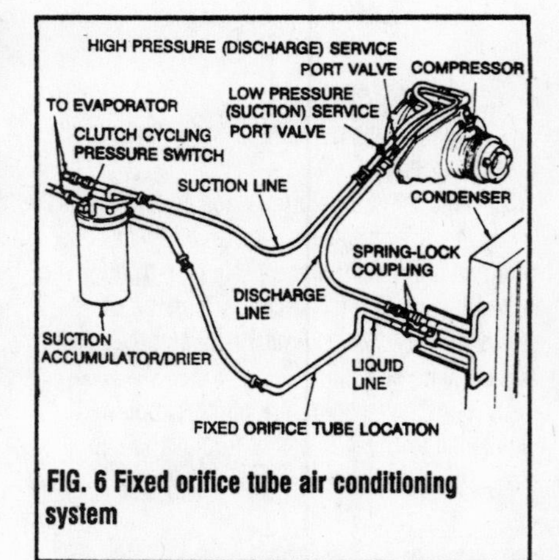

FIG. 6 Fixed orifice tube air conditioning system

Service Valve Location

➧ SEE FIGS. 7–8

The air conditioning system has a high pressure (discharge) and a low pressure (suction) gauge port valve. These are Schrader valves which provide access to both the high and low pressure sides of the system for service hoses and a manifold gauge set so system pressures can be read. The high pressure gauge port valve is located between the compressor and the condenser, in the high pressure vapor (discharge) line. The low pressure gauge port valve is located between the suction accumulator/drier and the compressor, in the low pressure vapor (suction) line.

High side adapter set D81L–19703–A or tool YT–354, 355 or equivalent, is required to connect a manifold gauge set or charging station to the high pressure gauge port valve. Service tee fitting D87P–19703–A, which may be mounted on the clutch cycling pressure switch fitting, is available for use in the low pressure side, to be used in place of the low pressure gauge port valve.

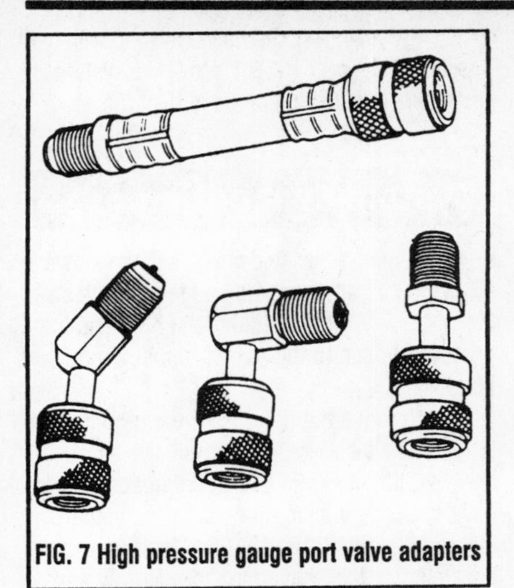

FIG. 7 High pressure gauge port valve adapters

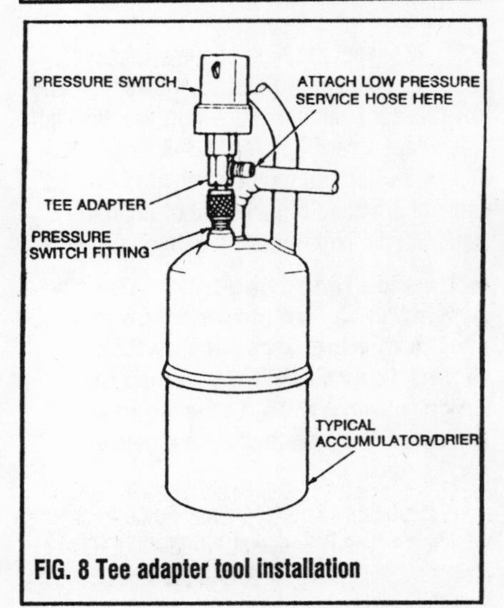

FIG. 8 Tee adapter tool installation

System Discharging

The use of refrigerant recovery systems and recycling stations makes possible the recovery and reuse of refrigerant after contaminants and moisture have been removed. The reuse of refrigerant will minimize the discharge of ozone depleting chlorofluorocarbons into the atmosphere. If a recovery system or recycling station is used, the following general procedures should be observed, in addition to the operating instructions provided by the equipment manufacturer.

1. Connect the refrigerant recycling station hose(s) to the vehicle air conditioning service ports and the recovery station inlet fitting.

➡ Hoses should have shut off devices or check valves within 12 in. (305mm) of the hose end to minimize the introduction of air into the recycling station and to minimize the amount of refrigerant released when the hose(s) is disconnected.

2. Turn the power to the recycling station **ON** to start the recovery process. Allow the recycling station to pump the refrigerant from the system until the station pressure goes into a vacuum. On some stations the pump will be shut off automatically by a low pressure switch in the electrical system. On other units it may be necessary to manually turn **OFF** the pump.
3. Once the recycling station has evacuated the vehicle air conditioning system, close the station inlet valve, if equipped. Then switch **OFF** the electrical power.
4. Allow the vehicle air conditioning system to remain closed for about 2 minutes. Observe the system vacuum level as shown on the gauge. If the pressure does not rise, disconnect the recycling station hose(s).
5. If the system pressure rises, repeat Steps 2, 3 and 4 until the vacuum level remains stable for 2 minutes.

System Flushing

A refrigerant system can become badly contaminated for a number of reasons:

- The compressor may have failed due to damage or wear.
- The compressor may have been run for some time with a severe leak or an opening in the system.
- The system may have been damaged by a collision and left open for some time.
- The system may not have been cleaned properly after a previous failure.
- The system may have been operated for a time with water or moisture in it.

A badly contaminated system contains water, carbon and other decomposition products. When this condition exists, the system must be flushed with a special flushing agent using equipment designed specially for this purpose.

FLUSHING AGENTS

To be suitable as a flushing agent, a refrigerant must remain in liquid state during the flushing operation in order to wash the inside surfaces of the system components. Refrigerant vapor will not remove contaminant particles. They must be flushed with a liquid.

Some refrigerants are better suited to flushing than others. Neither Refrigerant-12 (R-12) nor Refrigerant-114 (R-114) is suitable for flushing a system because of low vaporization (boiling) points: –21.6°F (–29.8°C) for R-12 and 38.4°F (3.5°C) for R-114. Both these refrigerants would be difficult to use and would not do a sufficient job because of the tendency to vaporize rather than remain in a liquid state, especially in high ambient temperatures.

Refrigerant-11 (R-11) and Refrigerant-113 (R-113) are much better suited for use with special flushing equipment. Both have rather high vaporization points: 74.7°F (23.7°C) for R-11 and 117.6°F (47.5°C) for R-113. Both refrigerants also have low closed container pressures. This reduces the danger of an accidental system discharge due to a ruptured hose or fitting. R-113 will do the best job and is recommended as a flushing refrigerant. Both R-11 and R-113 require a propellant or pump type flushing equipment due to their low closed container pressures. R-12 can be used as a propellant with either flushing refrigerant. R-11 is available in pressurized containers. Although not recommended for regular use, it may become necessary to use R-11 if special flushing equipment is not available. R-11 is more toxic than other refrigerants and should be handled with extra care.

SPECIAL FLUSHING EQUIPMENT

Special refrigerant system flushing equipment is available from a number of air conditioning equipment and usually comes in kit form. A flushing kit, such as model 015–00205 or equivalent, consists of a cylinder for the flushing agent, a nozzle to introduce the flushing agent into the system and a connecting hose.

A second type of equipment, which must be connected into the system, allows for the continuous circulation of the flushing agent through the system. Contaminants are trapped by an external filter/drier. If this equipment is used, follow the manufacturer's instructions and observe all safety precautions.

SYSTEM CLEANING AND FLUSHING

➡ Use extreme care and adhere to all safety precautions related to the use of refrigerants when flushing a system.

When it is necessary to flush a refrigerant system, the accumulator/drier must be removed and replaced, because it is impossible to clean. Remove the fixed orifice tube and replace it. If a new tube is not available, carefully wash the contaminated tube in flushing refrigerant or mineral spirits and blow it dry. If the tube does not show signs of damage or deterioration, it may be reused. Install new O-rings.

Any moisture in the evaporator core will be removed during leak testing and system evacuation following the cleaning job. Perform the cleaning procedure carefully as follows:

1. Check the hose connections at the flushing cylinder outlet and flushing nozzle, to make sure they are secure.
2. Make sure the flushing cylinder is filled with approximately 1 pint of R-113 and the valve assembly on top of the cylinder is tightened securely.
3. Connect a can of R-12 to the Schrader valve at the top of the charging cylinder. A refrigerant hose and a special safety-type refrigerant dispensing valve, such as YT–280 refrigerant dispensing valve or equivalent, and a valve retainer are required for connecting the small can to the cylinder. Make sure all connections are secure.
4. Connect a gauge manifold and discharge the system. Disconnect the gauge manifold.
5. Remove and discard the accumulator/drier. Install a new accumulator/drier and connect it to the evaporator. Do not connect it to the suction line from the compressor. Make sure the protective cap is in place on the suction line connection.
6. Replace the fixed orifice tube. Install the protective cap on the evaporator inlet tube as soon as the new orifice tube is in place. The liquid line will be connected later.
7. Remove the compressor from the vehicle for cleaning and service or replacement, whichever is required. If the compressor is cleaned and serviced, add the specified amount of refrigerant oil prior to installing in the vehicle. Place protective caps on the compressor inlet and outlet connections and install it on the mounting brackets in the vehicle. If the compressor is replaced, adjust the oil level. Install the shipping caps on the compressor connections and install the new compressor on the mounting brackets in the vehicle.
8. Back-flush the condenser and liquid line as follows:
   a. Remove the discharge hose from the condenser and clamp a piece of 1/2 in. i.d. heater hose to the condenser inlet line. Make sure the hose is long enough so the free end can be inserted into a suitable waste container to catch the flushing refrigerant.
   b. Move the flushing equipment into position and open the valve on the can of R-12, fully counterclockwise.
   c. Back-flush the condenser and liquid line by introducing the flushing refrigerant into the supported end of the liquid line with the flushing nozzle. Hold the nozzle firmly against the open end of the liquid line.
   d. After the liquid line and condenser have been flushed, lay the charging cylinder on it's side so the R-12 will not force more flushing refrigerant into the liquid line. Press the nozzle firmly to the liquid line and admit R-12 to force all flushing refrigerant from the liquid line and condenser.
   e. Remove the heater hose and clamp from the condenser inlet connection.
   f. Stand the flushing cylinder upright and flush the compressor discharge hose. Secure it so the flushing refrigerant goes into the waste container.
   g. Close the dispensing valve of the R-12 can, full clockwise. If there is any flushing refrigerant in the cylinder, it may be left there until the next flushing job. Put the flushing kit and R-12 can in a suitable storage location.
9. Connect all refrigerant lines. All connections should be cleaned and new O-rings should be used. Lubricate new O-rings with clean refrigerant oil.
10. Connect a charging station or manifold gauge set and charge the system with 1 lb. of R-12. Do not evacuate the system until after it has been leak tested.
11. Leak test all connections and components with flame-type leak detector 023–00006, electronic leak detector 055–00014 or 055–00015 or equivalent. If no leaks are found, proceed to Step 12. If leaks are found, service as necessary, check the system and then go to Step 12.

CAUTION

Fumes from flame-type leak detectors are noxious, avoid inhaling them or personal injury may result.

➡ **Good ventilation is necessary in the area where air conditioning leak testing is to be done. If the surrounding air is contaminated with refrigerant gas, the leak detector will indicate this gas all the time. Odors from other chemicals such as anti-freeze, diesel fuel, disc brake cleaner or other cleaning solvents can cause the same problem. A fan, even in a well ventilated area, is very helpful in removing small traces of air contamination that might affect the leak detector.**

12. Evacuate and charge the system with the specified amount of R-12. Operate the system to make sure it is cooling properly.

System Evacuating

1. Connect a manifold gauge set as follows:
   a. Turn both manifold gauge set valves all the way to the right, to close the high and low pressure hoses to the center manifold and hose.
   b. Remove the caps from the high and low pressure service gauge port valves.
   c. If the manifold gauge set hoses do not have valve depressing pins in them, install fitting adapters T71P–19703–S and R or equivalent, which have pins, on the low and high pressure hoses.
   d. Connect the high and low pressure hoses or adapters, to the respective high and low pressure service gauge port valves. High side adapter set D81L–19703–A or tool YT–354 or 355 or equivalent is required to connect a manifold gauge set or charging station to the high pressure gauge port valve.

➡ **Service tee fitting D87P–19703–A, which may be mounted on the clutch cycling pressure switch fitting, is available for use in the low pressure side, to be used in place of the low pressure gauge port valve.**

2. Leak test all connections and components with flame-type leak detector 023–00006, electronic leak detector 055–00014 or 055–00015 or equivalent.

CAUTION

Fumes from flame-type leak detectors are noxious, avoid inhaling them or personal injury may result.

➡ **Good ventilation is necessary in the area where air conditioning leak testing is to be done. If the surrounding air is contaminated with refrigerant gas, the leak detector will indicate this gas all the time. Odors from other chemicals such as anti-freeze, diesel fuel, disc brake cleaner or other cleaning solvents can cause the same problem. A fan, even in a well ventilated area, is very helpful in removing small traces of air contamination that might affect the leak detector.**

3. Properly discharge the refrigerant system according the to proper procedure.
4. Make sure both manifold gauge valves are turned all the way to the right. Make sure the center hose connection at the manifold gauge is tight.
5. Connect the manifold gauge set center hose to a vacuum pump.
6. Open the manifold gauge set valves and start the vacuum pump.
7. Evacuate the system with the vacuum pump until the low pressure gauge reads at least 25 in. Hg or as close to 30 in. Hg as possible. Continue to operate the vacuum pump for 15 minutes. If a part of the system has been replaced, continue to operate the vacuum pump for another 20–30 minutes.
8. When evacuation of the system is complete, close the manifold gauge set valves and turn the vacuum pump **OFF**.
9. Observe the low pressure gauge for 5 minutes to ensure that system vacuum is held. If vacuum is held, charge the system. If vacuum is not held for 5 minutes, leak test the system, service the leaks and evacuate the system again.

System Charging

1. Connect a manifold gauge set according to the proper procedure. Properly discharge and evacuate the system.
2. With the manifold gauge set valves closed to the center hose, disconnect the vacuum pump from the manifold gauge set.
3. Connect the center hose of the manifold gauge set to a refrigerant drum or a small can refrigerant dispensing valve tool YT–280, 1034 or equivalent. If a charging station is used, follow the instructions of the manufacturer. If a small can dispensing valve is used, install the small can(s) on the dispensing valve.

➡ **Use only a safety type dispensing valve.**

4. Loosen the center hose at the manifold gauge set and open the refrigerant drum valve or small can dispensing valve. Allow the refrigerant to escape to purge air and moisture from the center hose. Then, tighten the center hose connection at the manifold gauge set.
5. Disconnect the wire harness snap lock connector from the clutch cycling pressure switch and install a jumper wire across the 2 terminals of the connector.
6. Open the manifold gauge set low side valve to allow refrigerant to enter the system. Keep the refrigerant can in an upright position.

✲✲ CAUTION

Do not open the manifold gauge set high pressure (discharge) gauge valve when charging with a small container. Opening the valve can cause the small refrigerant container to explode, which can result in personal injury.

7. When no more refrigerant is being drawn into the system, start the engine and set the control assembly for MAX cold and HI blower to draw the remaining refrigerant into the system. If equipped, press the air conditioning switch. Continue to add refrigerant to the system until the specified weight of R-12 is in the system. Then close the manifold gauge set low pressure valve and the refrigerant supply valve.
8. Remove the jumper wire from the clutch cycling pressure switch snap lock connector. Connect the connector to the pressure switch.
9. Operate the system until pressures stabilize to verify normal operation and system pressures.
10. In high ambient temperatures, it may be necessary to operate a high volume fan positioned to blow air through the radiator and condenser to aid in cooling the engine and prevent excessive refrigerant system pressures.
11. When charging is completed and system operating pressures are normal, disconnect the manifold gauge set from the vehicle. Install the protective caps on the service gauge port valves.

Compressor

REMOVAL & INSTALLATION

2.3L Engine

➧ SEE FIG. 9

➡ **The suction accumulator/drier and the fixed orifice tube should also be replaced whenever the compressor is replaced.**

1. Disconnect the negative battery cable.
2. Discharge the refrigerant from the air conditioning system according to the proper procedure.
3. Disconnect the compressor clutch wires at the field coil connector on the compressor.
4. Disconnect the discharge and suction hoses from the compressor manifolds. Cap the refrigerant lines and compressor manifolds to prevent the entrance of dirt and moisture.
5. Remove the screw and washer assembly from the adjusting bracket.
6. Rotate the compressor outboard and remove the drive belts.
7. Remove the compressor mounting bolt attaching the bracket to the compressor lower mounting lug.
8. Remove the compressor and adjusting bracket intact.
9. Remove the clutch field coil and adjusting bracket from the compressor, if the compressor is to be replaced.

To install:

➡ **A new service replacement HR-980 compressor contains 8 oz. (237ml) of refrigerant oil. Prior to installing the replacement compressor, drain 4 oz. (118ml) of refrigerant oil from the compressor. This will maintain the system total oil charge. A new service replacement 10P15C compressor contains 8 oz. (237ml) of refrigerant oil. Prior to installing the replacement compressor, drain the refrigerant oil from the removed compressor into a calibrated container. Then, drain the refrigerant oil from the new compressor into a clean calibrated container. If the amount of oil drained from the removed compressor was between 3–5 oz. (90–148ml), pour the same amount of clean refrigerant oil into the new compressor. If the amount of oil that was removed from the old compressor is greater than 5 oz. (148ml), pour 5 oz. (148ml) of clean refrigerant oil into the new compressor. If the amount of refrigerant oil that was removed from the old compressor is less than 3 oz. (90ml), pour 3 oz. (90ml) of clean refrigerant oil into the new compressor. This will maintain the total system oil charge within specification.**

10. Install the clutch field coil and adjusting bracket if it was removed or the compressor was replaced.
11. Install the compressor mounting brackets and drive belt in the reverse order of their removal.
12. Connect the refrigerant lines to the compressor and connect the clutch wires to the field coil connector.
13. Leak test, evacuate and charge the air conditioning system. Observe all safety precautions.

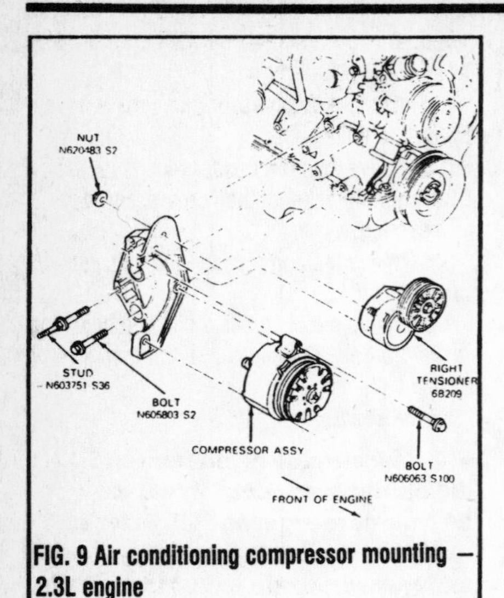

FIG. 9 Air conditioning compressor mounting — 2.3L engine

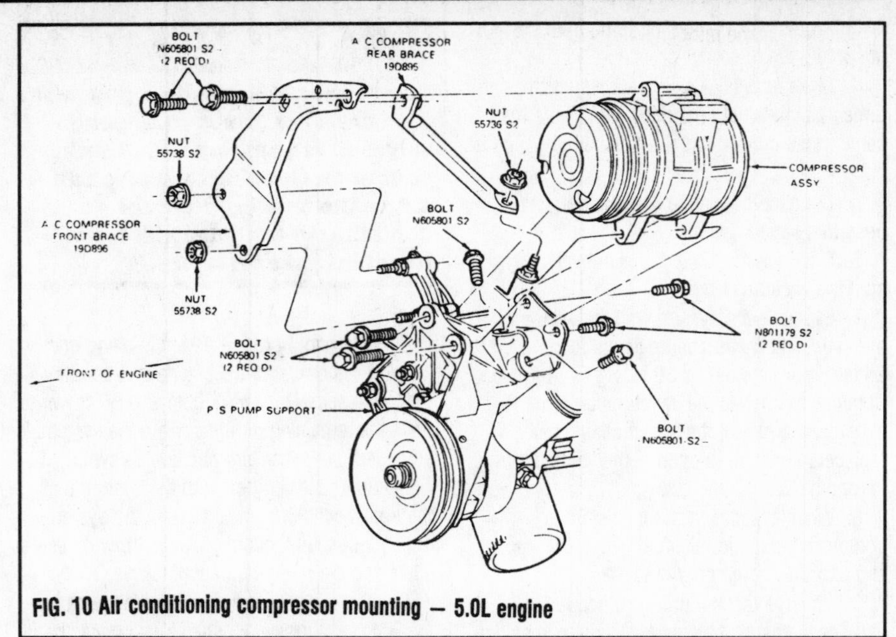

FIG. 10 Air conditioning compressor mounting — 5.0L engine

14. Check the operation of the air conditioning system.

5.0L Engine

➧ SEE FIG. 10

➡ The suction accumulator/drier and the fixed orifice tube should also be replaced whenever the compressor is replaced.

1. Disconnect the negative battery cable.
2. Discharge the refrigerant from the air conditioning system according to the proper procedure.
3. Disconnect the compressor clutch wires at the field coil connector on the compressor.
4. Disconnect the discharge and suction hoses from the compressor manifolds. Cap the refrigerant lines and compressor manifolds to prevent the entrance of dirt and moisture.
5. Remove the 2 screws attaching the brace to the rear mounting bracket.
6. Remove the 2 screws attaching the bracket to the compressor lower front mounting lugs.
7. Remove 1 screw attaching the bracket to the compressor upper mounting lug.
8. Remove the compressor/clutch assembly and rear support from the vehicle as a unit.
9. Remove the 2 screws attaching the compressor to the rear support and remove the rear support.
10. If the compressor is to be replaced, remove the clutch and field coil assembly from the compressor.

To install:

➡ A new service replacement 6P148 compressor contains 10 oz. (300ml) of refrigerant oil. Prior to installing the replacement compressor, drain the refrigerant oil from the removed compressor into a calibrated container. Then, drain the refrigerant oil from the new compressor into a clean calibrated container. If the amount of oil drained from the removed compressor was between 3–5 oz. (90–148ml), pour the same amount of clean refrigerant oil into the new compressor. If the amount of oil that was removed from the old compressor is greater than 5 oz. (148ml), pour 5 oz. (148ml) of clean refrigerant oil into the new compressor. If the amount of refrigerant oil that was removed from the old compressor is less than 3 oz. (90ml), pour 3 oz. (90ml) of clean refrigerant oil into the new compressor. This will maintain the total system oil charge within specification.

11. Install the compressor and brackets in the reverse order of their removal.
12. Using new O-rings lubricated with clean refrigerant oil, connect the suction and discharge lines to the compressor manifolds. Tighten the suction hose fitting to 21–27 ft. lbs. (28–36Nm) and the discharge hose fitting to 15–20 ft. lbs. (20–27Nm).
13. Connect the clutch wires to the field coil connector and install the drive belt.
14. Leak test, evacuate and charge the air conditioning system. Observe all safety precautions.
15. Check the system for proper operation.

Condenser

REMOVAL & INSTALLATION

CAUTION

When draining the coolant, keep in mind that cats and dogs are attracted by the ethylene glycol antifreeze, and are quite likely to drink any that is left in an uncovered container or in puddles on the ground. This will prove fatal in sufficient quantity. Always drain the coolant into a sealable container. Coolant should be reused unless it is contaminated or several years old.

➧ SEE FIG. 11

➡ Whenever the condenser is replaced, it will be necessary to replace the suction accumulator/drier.

1. Disconnect the battery cables and remove the battery and heat shield.
2. Discharge the refrigerant from the air conditioning system according to the proper procedure.
3. Place a clean container under the radiator draincock and drain the coolant from the radiator. Save the coolant for reuse.

4. Disconnect the refrigerant lines at the right side of the radiator according to the spring-lock coupling disconnect procedure.

5. Remove the 2 fan shroud attaching screws. Disengage the fan shroud and position it rearward.

6. Disconnect the upper hose from the radiator, remove the 2 radiator retaining clamps and tilt the radiator rearward.

7. Remove the 2 screws attaching the top of the condenser to the radiator support and lift the condenser from the vehicle. Cap the refrigerant lines to prevent entry of dirt and excessive moisture.

To install:

➡ **When replacing the condenser in the refrigerant system, 1 oz. (29.5ml) of clean refrigerant oil should be added to the new replacement condenser to maintain the total system oil charge.**

8. If the condenser is to be replaced, transfer the rubber isolators from the bottom of the old condenser to the new condenser.

9. Position the condenser assembly to the vehicle making sure the lower isolators are properly seated. Then install the upper 2 condenser attaching screws.

10. Position the radiator to the radiator support and install the 2 retaining clamps.

11. Connect the upper hose to the radiator and fill the radiator with the previously removed coolant.

12. Position the fan shroud to the radiator and install the 2 attaching screws.

13. Connect the refrigerant lines to the condenser according to the spring-lock coupling connect procedure.

14. Install the battery and heat shield, then connect the battery cables.

15. Leak test, evacuate and charge the refrigerant system according to the proper procedure. Observe all safety precautions.

16. Check the operation of the air conditioning system.

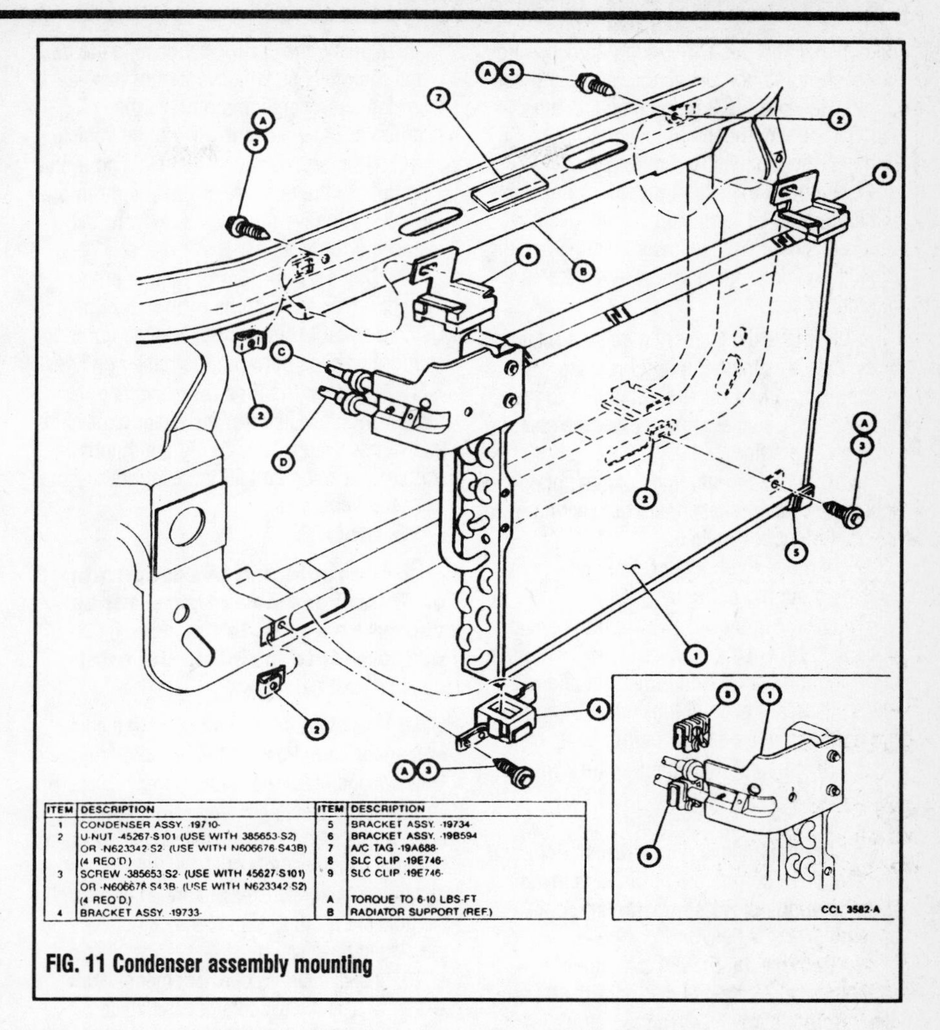

ITEM	DESCRIPTION	ITEM	DESCRIPTION
1	CONDENSER ASSY. -19710-	5	BRACKET ASSY. -19734-
2	U-NUT -45267-S101 (USE WITH 385653-S2) OR -N623342-S2- (USE WITH N606676-S43B) (4 REQ'D.)	6	BRACKET ASSY. -19B594
		7	A/C TAG -19A688-
		8	SLC CLIP -19E746-
3	SCREW -385653-S2- (USE WITH 45627-S101) OR -N606676-S43B- (USE WITH N623342-S2) (4 REQ'D.)	9	SLC CLIP -19E746-
		A	TORQUE TO 6-10 LBS-FT
4	BRACKET ASSY. -19733-	B	RADIATOR SUPPORT (REF.)

FIG. 11 Condenser assembly mounting

Evaporator

REMOVAL & INSTALLATION

CAUTION

When draining the coolant, keep in mind that cats and dogs are attracted by the ethylene glycol antifreeze, and are quite likely to drink any that is left in an uncovered container or in puddles on the ground. This will prove fatal in sufficient quantity. Always drain the coolant into a sealable container. Coolant should be reused unless it is contaminated or several years old.

➧ SEE FIGS. 12–14

➡ **Whenever an evaporator core is replaced, it will be necessary to replace the suction accumulator/drier.**

1. Disconnect the negative battery cable and drain the cooling system.

2. Discharge the refrigerant from the air conditioning system according to the proper procedure.

3. Remove the instrument panel according to the following procedure:

a. Remove the 4 screws that attach the floor console top panel assembly to the floor console assembly.

b. Lift the console top panel assembly off the console assembly and disconnect the 2 electrical connectors.

c. Remove the 2 screws attaching the rear end of the console assembly to the console panel support assembly.

d. Remove 4 screws attaching the console assembly to the console panel front support.

e. Remove the 4 screws attaching the console assembly to the instrument panel.

f. Lift the console assembly off of the transmission tunnel.

➡ **The console assembly includes a snap-in finish panel which conceals the heater and air conditioning control assembly attaching screws. To gain access to these screws, it is necessary to remove the floor console.**

g. Disconnect all underhood wiring connectors from the main wiring harness. Disengage the rubber grommet seal from the dash panel and push the wiring harness and connectors into the passenger compartment.

h. Remove the 3 bolts attaching the steering column opening cover and reinforcement panel. Remove the cover.

i. Remove the steering column opening reinforcement by removing 2 bolts. Remove the 2 bolts retaining the lower steering column opening reinforcement and remove the reinforcement.

j. Remove the 6 steering column retaining nuts. 2 are retaining the hood release mechanism and 4 retain the column to the lower brake pedal support. Lower the steering column to the floor.

k. Remove the steering column upper and lower shrouds and disconnect the wiring from the multi-function switch.

l. Remove the brake pedal support nut and snap out the defroster grille.

m. Remove the screws from the speaker covers. Snap out the speaker covers. Remove the front screws retaining the right and left scuff plates at the cowl trim panel. Remove the right and left side cowl trim panels.

n. Disconnect the wiring at the right and left cowl sides. Remove the cowl side retaining bolts, 1 on each side.

o. Open the glove compartment door and flex the glove compartment bin tabs inward. Drop down the glove compartment door assembly.

p. Remove the 5 cowl top screw attachments. Gently pull the instrument panel away from the cowl. Disconnect the speedometer cable and wire connectors.

4. Disconnect the liquid line and the accumulator/drier inlet tube from the evaporator core at the dash panel. Cap the refrigerant lines and evaporator core tube to prevent the entrance of dirt and excessive moisture.

5. Disconnect the heater hoses from the heater core tubes and plug the hoses and tubes.

6. Remove the screw attaching the air inlet duct and blower housing assembly support brace to the cowl top panel.

7. Disconnect the black vacuum supply hose from the in-line vacuum check valve in the engine compartment. Disconnect the blower motor wires from the wire harness and disconnect the wire harness from the blower motor resistor.

8. Working under the hood, remove the 2 nuts retaining the evaporator case to the dash panel. Inside the passenger compartment, remove the 2 screws attaching the evaporator case support brackets to the cowl top panel.

9. Remove the 1 screw retaining the bracket below the evaporator case to the dash panel. Carefully pull the evaporator case away from the dash panel and remove the evaporator case assembly from the vehicle.

10. Remove the 4 screws retaining the air inlet duct to the evaporator case and remove the duct. Remove the foam seal from the evaporator core tubes.

11. Drill a $\frac{3}{16}$ in. (5mm) hole in both upright tabs on top of the evaporator case. Using a suitable cutting tool, cut the top of the evaporator case between the raised outlines.

12. Remove the 2 screws retaining the blower motor resistor to the evaporator case and remove the resistor. Fold the cutout flap from the opening and lift the evaporator core from the case.

To install:

➡ When replacing the evaporator, 3 oz. (90ml) of clean refrigerant oil should be added to the new evaporator to maintain the total system oil charge.

13. Transfer 2 foam core seals to the new evaporator core. Position the evaporator core in the case and close the cutout cover.

14. Install a spring nut on each of the 2 upright tabs. Make sure the hole in the spring nut is aligned with the hole drilled in the tab. Install and tighten a screw in each spring nut through the hole in the tab to secure the cutout cover in the closed position.

15. Install caulking cord or other suitable sealer, to seal the evaporator case against leakage along the cut line. Using new caulking cord, assemble the air inlet duct to the evaporator case.

16. Install the blower motor resistor and the foam seal over the evaporator core and heater core tubes.

17. Position the evaporator case assembly in the vehicle. Install the screws attaching the evaporator case support brackets to the cowl top panel. Check the evaporator case drain tube to ensure it is through the dash panel and is not pinched or kinked.

18. Install 1 screw retaining the bracket below the evaporator case to the dash panel. Working under the hood, install the 2 nuts retaining the evaporator case to the dash panel. Tighten the 4 nuts and 2 screws in the engine compartment. Tighten the 2 screws in the passenger compartment and the 2 support bracket attaching screws.

19. Connect the blower motor wire harness to the resistor and blower motor. Connect the black vacuum supply hose to the vacuum check valve in the engine compartment.

20. Using new O-rings lubricated with clean refrigerant oil, connect the liquid line and suction accumulator inlet to the evaporator core tubes. Tighten each connection using a backup wrench to prevent component damage.

21. Install the instrument panel by reversing the removal procedure.

22. Connect the heater hoses to the heater core and fill the cooling system.

23. Connect the negative battery cable. Leak test, evacuate and charge the refrigerant system

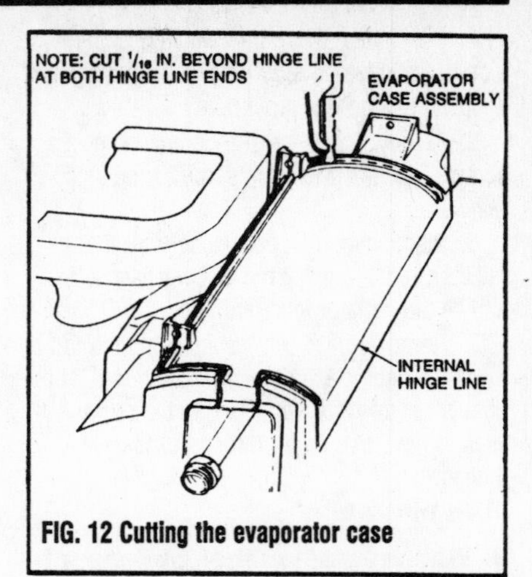

FIG. 12 Cutting the evaporator case

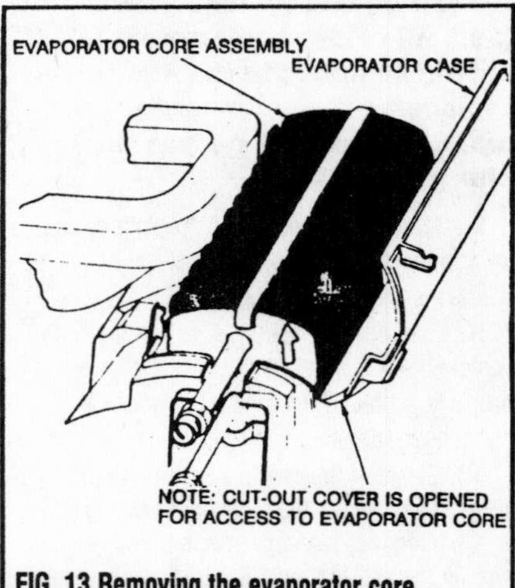

FIG. 13 Removing the evaporator core

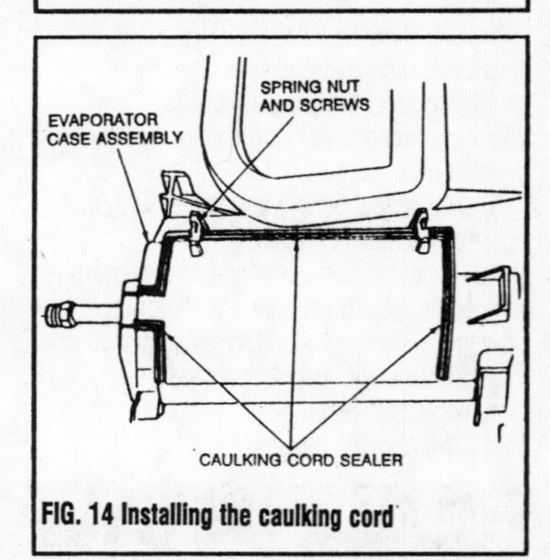

FIG. 14 Installing the caulking cord

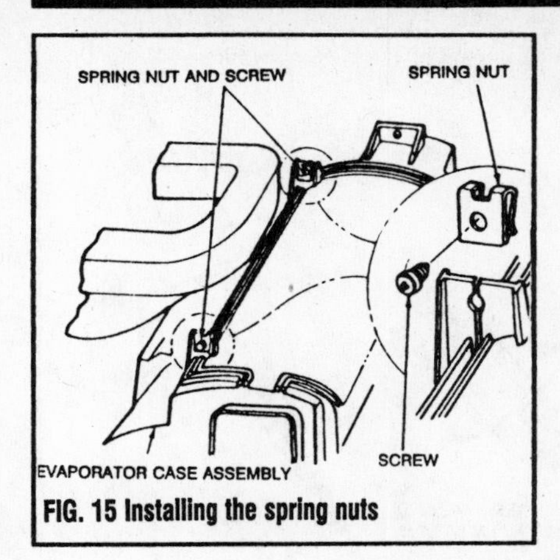

FIG. 15 Installing the spring nuts

according to the proper procedure. Observe all safety precautions.

24. Check the system for proper operation.

Control Panel

REMOVAL & INSTALLATION

➧ SEE FIG. 4

1. Disconnect the negative battery cable.
2. Remove the snap-in trim molding in the floor console to expose the 4 control assembly attaching screws. Remove the 4 screws attaching the control assembly to the instrument panel.
3. Roll the control out of the opening in the console. Disconnect the fan switch connectors and temperature control cable. Disconnect the vacuum hose connector and electrical connector from the back of the function selector knob. Disconnect the connector for the control assembly illumination bulbs.
4. Remove the control assembly.

To install:

5. Connect the temperature cable to the geared arm on the temperature control.
6. Install the electrical connector at the following locations: blower switch, control assembly illumination bulbs and function selector switch.
7. Install the vacuum harness connector for the function selector knob.
8. Roll the control assembly into position against the instrument panel and install the 4 attaching screws.
9. Snap the console trim molding into position, connect the negative battery cable and check the system for proper operation.

Accumulator/Drier

REMOVAL & INSTALLATION

Any time a major component of the air conditioning system is replaced, it is necessary to replace the suction accumulator/drier. A major component would be the condenser, compressor, evaporator or a refrigerant hose/line. A fixed orifice tube or O-ring is not considered a major component but the orifice tube should be replaced whenever the compressor is replaced for lack of performance.

The accumulator/drier should also be replaced, if 1 of the following conditions exist:

- The accumulator/drier is perforated.
- The refrigerant system has been opened to the atmosphere for a period of time longer than required to make a minor repair.
- There is evidence of moisture in the system such as internal corrosion of metal refrigerant lines or the refrigerant oil is thick and dark.

➡ The compressor oil from vehicles equipped with an FX-15 compressor may have a dark color while maintaining a normal oil viscosity. This is normal for this compressor because carbon from the compressor piston rings may discolor the oil.

When replacing the suction accumulator/drier, the following procedure must be used to ensure that the total oil charge in the system is correct after the new accumulator/drier is installed.

1. Drain the oil from the removed accumulator/drier into a suitable measuring container. It may be necessary to drill one or two 1/2 in. (13mm) holes in the bottom of the old accumulator/drier to ensure that all the oil has drained out.
2. Add the same amount of clean new refrigerant oil plus 2 oz. to the new accumulator/drier. Use only the proper type of oil for the vehicle being serviced.

REPLACEMENT

➧ SEE FIG. 16

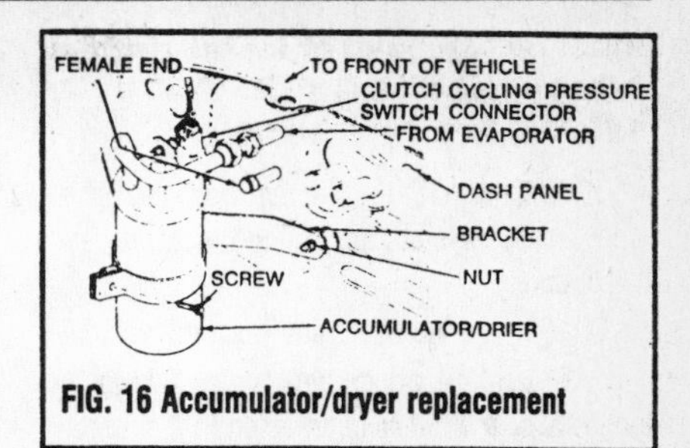

FIG. 16 Accumulator/dryer replacement

1. Disconnect the negative battery cable.
2. Discharge the refrigerant from the air conditioning system according to the proper procedure.
3. On 2.3L engine equipped vehicles, remove the speed control servo, if equipped.
4. Disconnect the suction hose at the compressor. Cap the hose and compressor to prevent the entrance of dirt and moisture.
5. Disconnect the accumulator/drier inlet tube from the evaporator core outlet. Use 2 wrenches to prevent component damage.
6. Disconnect the wire harness connector from the pressure switch on top of the accumulator/drier.
7. Remove the screw holding the accumulator/drier in the accumulator bracket and remove the accumulator/drier.

To install:

8. Position the accumulator/drier to the vehicle and route the suction hose to the compressor.
9. Using a new O-ring lubricated with clean refrigerant oil, connect the accumulator/drier inlet tube to the evaporator core outlet. Tighten the connection using a back-up wrench to prevent component damage.
10. Install the screw in the accumulator/drier bracket
11. Using a new O-ring lubricated with clean refrigerant oil, connect the suction hose to the compressor.
12. On 2.3L engine equipped vehicles, install the speed control servo, if equipped.
13. Leak test, evacuate and charge the air conditioning system. Observe all safety precautions.
14. Check the system for proper operation.

Fixed Orifice Tube

REMOVAL & INSTALLATION

➧ SEE FIG. 17

➡ The fixed orifice tube should be replaced whenever a compressor is replaced. If high pressure reads extremely high and low pressure (suction) is almost a vacuum, the fixed orifice is plugged and must be replaced.
If replacement of the fixed orifice tube is necessary, the liquid line

must be replaced or install orifice tube replacement kit E5VY-190695-A or equivalent.

1. Discharge the refrigerant from the air conditioning system according the to proper procedure.
2. Remove the liquid line from the vehicle.
3. Locate the orifice tube by the 3 indented notches or a circular depression in the metal portion of the liquid line.
4. Note the angular position of the ends of the liquid line so it can be reassembled in the correct position.
5. Cut a 2½ in. (63.5mm) section from the tube at the orifice tube location. Do not cut closer than 1 in. (25.4mm) from the start of a bend in the tube.
6. Remove the orifice tube from the housing with pliers. The orifice tube removal tool cannot be used.
7. Flush the 2 pieces of liquid line to remove any contaminants.
8. Lubricate the O-rings with clean refrigerant oil and assemble the orifice tube kit, with the orifice tube installed, to the liquid line. Make sure the flow direction arrow is pointing toward the evaporator end of the liquid line and the taper of each compression ring is toward the compression nut.

➡ The inlet tube will be positioned against the orifice tube tabs when correctly assembled.

9. While holding the hex of the tube in a suitable vise, tighten each compression nut to 65–70 ft. lbs. (88–94 Nm) with a crow foot wrench.
10. Install the liquid line on the vehicle using new O-rings lubricated with clean refrigerant oil.
11. Leak test, evacuate and charge the system. Observe all safety precautions.
12. Check the system for proper operation.

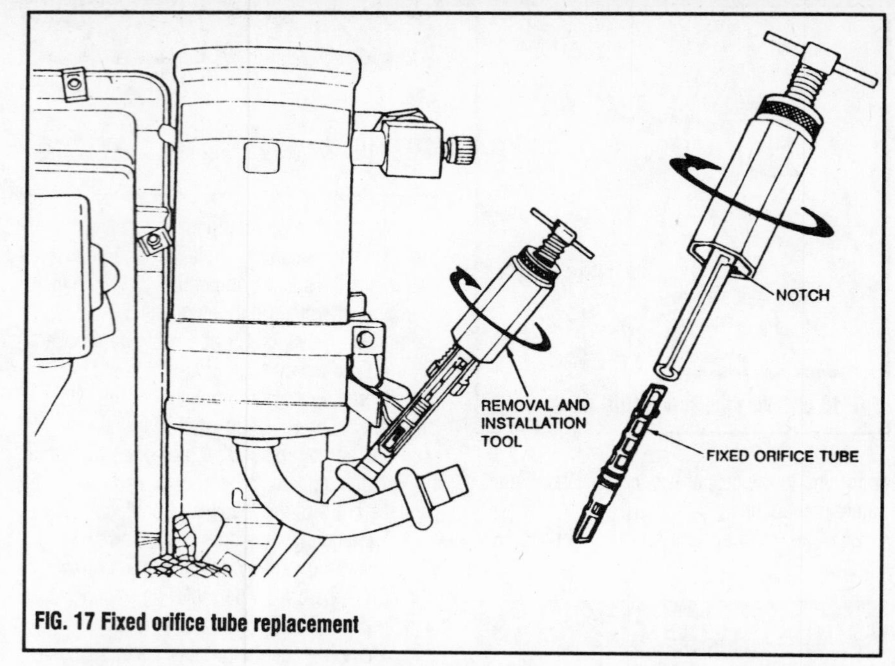

FIG. 17 Fixed orifice tube replacement

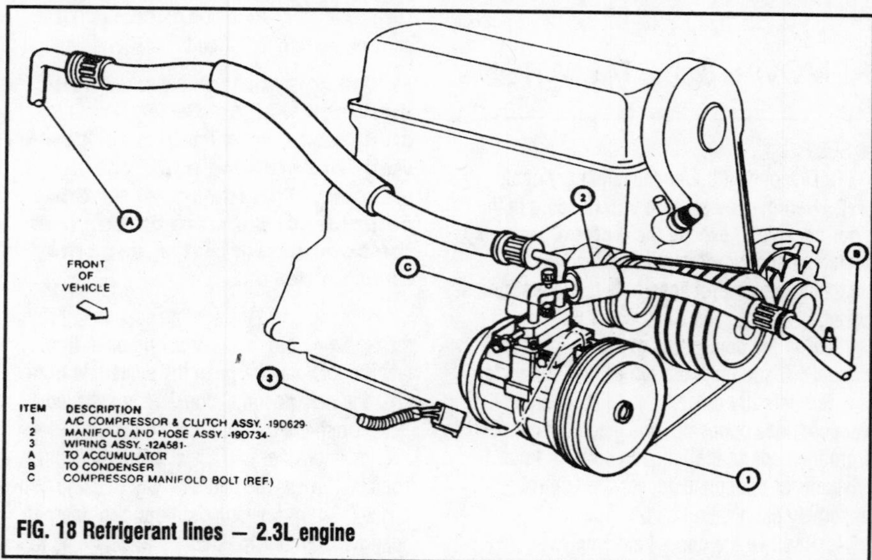

FIG. 18 Refrigerant lines — 2.3L engine

Refrigerant Lines

REMOVAL & INSTALLATION

➧ SEE FIGS. 18–19

➡ Whenever a refrigerant line is replaced, it will be necessary to replace the suction accumulator/drier.

1. Disconnect the negative battery cable. Discharge the refrigerant from the air conditioning system.
2. Disconnect the refrigerant line using a wrench on each side of the fitting or the spring-lock coupling disconnect procedure, whichever is necessary. Remove the refrigerant line.
3. Route the new refrigerant line with the protective caps installed.
4. Connect the refrigerant line into the system using new O-rings lubricated with clean refrigerant oil. Tighten the connection using 2 wrenches or the spring-lock coupling connect procedure, whichever is necessary.
5. Connect the negative battery cable. Leak test, evacuate and charge the refrigerant system according to the proper procedure. Observe all safety precautions.

Vacuum Motors

OPERATION

The vacuum motors operate the doors which in turn direct the airflow through the system. A vacuum selector valve, controlled by the function control lever, distributes the vacuum to the various door vacuum motors.

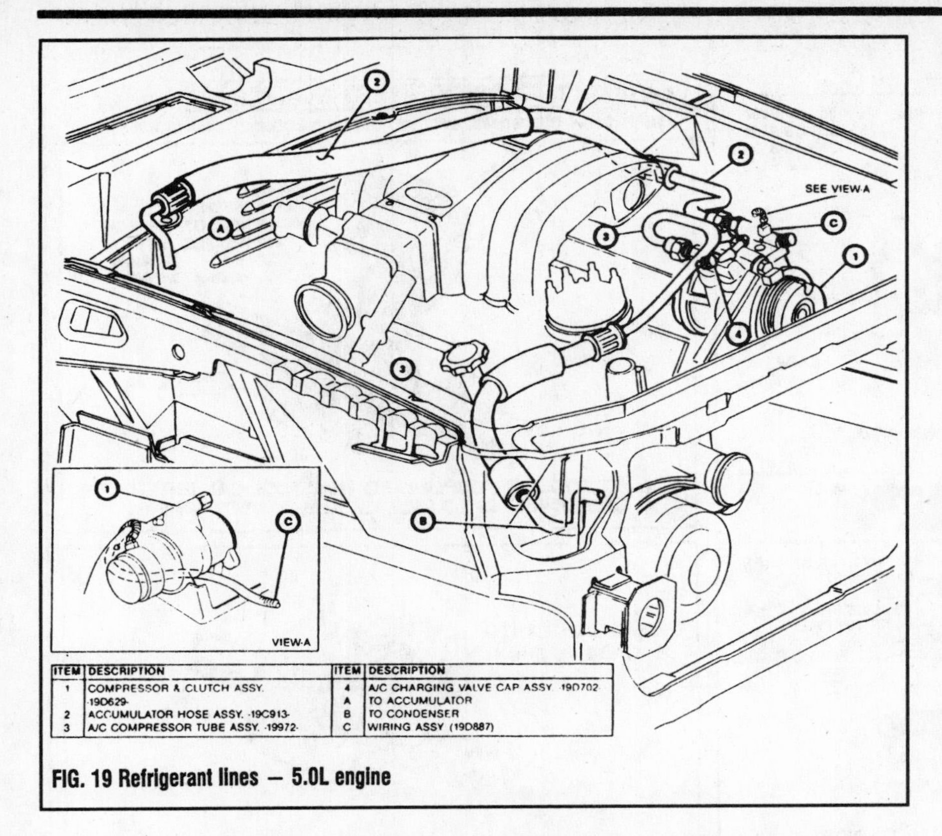

ITEM	DESCRIPTION	ITEM	DESCRIPTION
1	COMPRESSOR & CLUTCH ASSY. -19D629-	4	A/C CHARGING VALVE CAP ASSY. -19D702-
2	ACCUMULATOR HOSE ASSY. -19C913-	A	TO ACCUMULATOR
3	A/C COMPRESSOR TUBE ASSY. -19972-	B	TO CONDENSER
		C	WIRING ASSY (19D887)

FIG. 19 Refrigerant lines — 5.0L engine

REMOVAL & INSTALLATION

Panel/Defrost Door Vacuum Motor

1. Disconnect the negative battery cable.
2. Remove the instrument panel according to the following procedure:
   a. Remove the 4 screws that attach the floor console top panel assembly to the floor console assembly.
   b. Lift the console top panel assembly off the console assembly and disconnect the 2 electrical connectors.
   c. Remove the 2 screws attaching the rear end of the console assembly to the console panel support assembly.
   d. Remove 4 screws attaching the console assembly to the console panel front support.
   e. Remove the 4 screws attaching the console assembly to the instrument panel.
   f. Lift the console assembly off of the transmission tunnel.

➡ The console assembly includes a snap-in finish panel which conceals the heater and air conditioning control assembly attaching screws. To gain access to these screws, it is necessary to remove the floor console.

   g. Disconnect all underhood wiring connectors from the main wiring harness. Disengage the rubber grommet seal from the dash panel and push the wiring harness and connectors into the passenger compartment.
   h. Remove the 3 bolts attaching the steering column opening cover and reinforcement panel. Remove the cover.
   i. Remove the steering column opening reinforcement by removing 2 bolts. Remove the 2 bolts retaining the lower steering column opening reinforcement and remove the reinforcement.
   j. Remove the 6 steering column retaining nuts; 2 retain the hood release mechanism and 4 retain the column to the lower brake pedal support. Lower the steering column to the floor.
   k. Remove the steering column upper and lower shrouds and disconnect the wiring from the multi-function switch.
   l. Remove the brake pedal support nut and snap out the defroster grille.
   m. Remove the screws from the speaker covers. Snap out the speaker covers. Remove the front screws retaining the right and left scuff plates at the cowl trim panel. Remove the right and left side cowl trim panels.
   n. Disconnect the wiring at the right and left cowl sides. Remove the cowl side retaining bolts, 1 on each side.
   o. Open the glove compartment door and flex the glove compartment bin tabs inward. Drop down the glove compartment door assembly.
   p. Remove the 5 cowl top screw attachments. Gently pull the instrument panel away from the cowl. Disconnect the speedometer cable and wire connectors.
3. Remove the spring nut retaining the panel/defrost door vacuum motor arm to the door shaft.
4. Remove the 2 nuts retaining the vacuum motor to the mounting bracket. Remove the vacuum motor from the mounting bracket and disconnect the vacuum hose.

To install:

5. Position the vacuum motor to the mounting bracket and door shaft. Install 2 nuts to attach the panel/defrost vacuum motor to the mounting bracket.
6. Connect the vacuum hose to the panel/defrost vacuum motor.
7. Install the instrument panel by reversing the removal procedure.
8. Connect the negative battery cable.

Floor/Defrost Door Vacuum Motor

1. Disconnect the negative battery cable and drain the cooling system.
2. Discharge the refrigerant from the air conditioning system according the to proper procedure.
3. Remove the instrument panel according to the following procedure:
   a. Remove the 4 screws that attach the floor console top panel assembly to the floor console assembly.
   b. Lift the console top panel assembly off the console assembly and disconnect the 2 electrical connectors.
   c. Remove the 2 screws attaching the rear end of the console assembly to the console panel support assembly.
   d. Remove 4 screws attaching the console assembly to the console panel front support.
   e. Remove the 4 screws attaching the console assembly to the instrument panel.
   f. Lift the console assembly off of the transmission tunnel.

➡ The console assembly includes a snap-in finish panel which conceals the heater and air conditioning control assembly attaching screws. To gain access to these screws, it is necessary to remove the floor console.

   g. Disconnect all underhood wiring connectors from the main wiring harness. Disengage the rubber grommet seal from the dash panel and push the wiring harness and connectors into the passenger compartment.

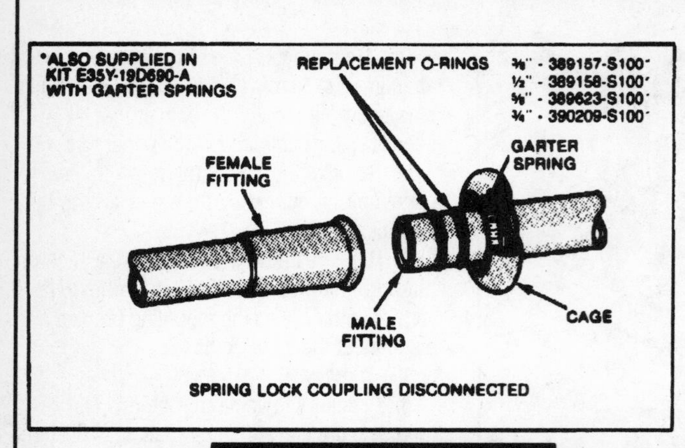

SPRING LOCK COUPLING DISCONNECTED

TO CONNECT COUPLING

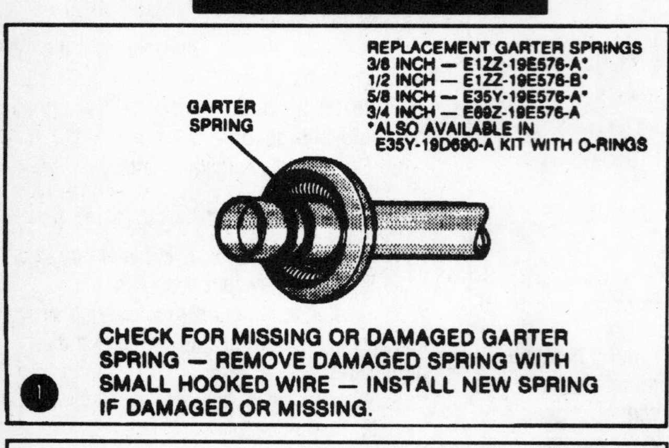

1. CHECK FOR MISSING OR DAMAGED GARTER SPRING — REMOVE DAMAGED SPRING WITH SMALL HOOKED WIRE — INSTALL NEW SPRING IF DAMAGED OR MISSING.

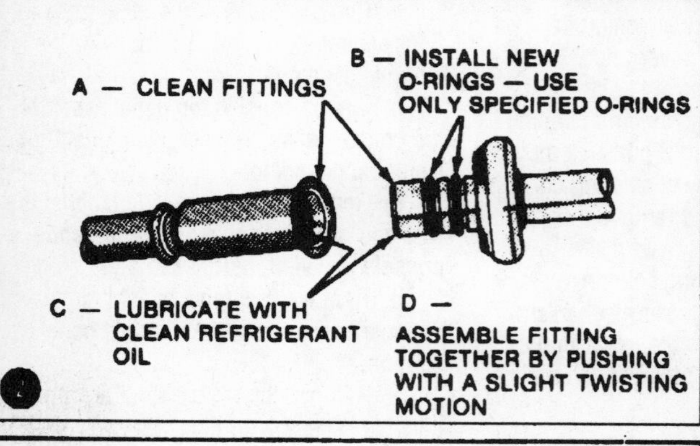

2.

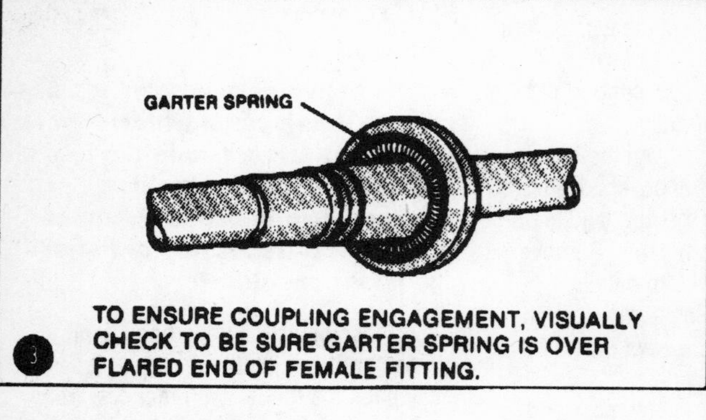

3. TO ENSURE COUPLING ENGAGEMENT, VISUALLY CHECK TO BE SURE GARTER SPRING IS OVER FLARED END OF FEMALE FITTING.

TO DISCONNECT COUPLING

CAUTION — DISCHARGE SYSTEM BEFORE DISCONNECTING COUPLING

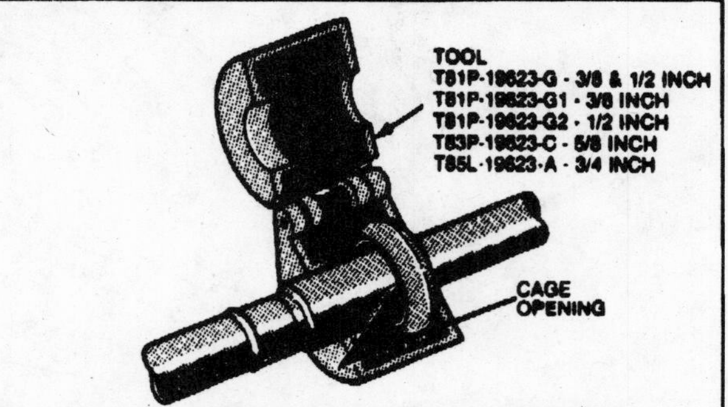

1. FIT TOOL TO COUPLING SO THAT TOOL CAN ENTER CAGE OPENING TO RELEASE THE GARTER SPRING.

2. PUSH THE TOOL INTO THE CAGE OPENING TO RELEASE THE FEMALE FITTING FROM THE GARTER SPRING.

3. PULL THE COUPLING MALE AND FEMALE FITTINGS APART.

4. REMOVE THE TOOL FROM THE DISCONNECTED SPRING LOCK COUPLING.

FIG. 20 Spring-Lock Coupling Procedures

h. Remove the 3 bolts attaching the steering column opening cover and reinforcement panel. Remove the cover.

i. Remove the steering column opening reinforcement by removing 2 bolts. Remove the 2 bolts retaining the lower steering column opening reinforcement and remove the reinforcement.

j. Remove the 6 steering column retaining nuts; 2 retain the hood release mechanism and 4 retain the column to the lower brake pedal support. Lower the steering column to the floor.

k. Remove the steering column upper and lower shrouds and disconnect the wiring from the multi-function switch.

l. Remove the brake pedal support nut and snap out the defroster grille.

m. Remove the screws from the speaker covers. Snap out the speaker covers. Remove the front screws retaining the right and left scuff plates at the cowl trim panel. Remove the right and left side cowl trim panels.

n. Disconnect the wiring at the right and left cowl sides. Remove the cowl side retaining bolts, 1 on each side.

o. Open the glove compartment door and flex the glove compartment bin tabs inward. Drop down the glove compartment door assembly.

p. Remove the 5 cowl top screw attachments. Gently pull the instrument panel away from the cowl. Disconnect the speedometer cable and wire connectors.

4. Disconnect the liquid line and the accumulator/drier inlet tube from the evaporator core at the dash panel. Cap the refrigerant lines and evaporator core tube to prevent the entrance of dirt and excessive moisture.

5. Disconnect the heater hoses from the heater core tubes and plug the hoses and tubes.

6. Remove the screw attaching the air inlet duct and blower housing assembly support brace to the cowl top panel.

7. Disconnect the black vacuum supply hose from the in-line vacuum check valve in the engine compartment. Disconnect the blower motor wires from the wire harness and disconnect the wire harness from the blower motor resistor.

8. Working under the hood, remove the 2 nuts retaining the evaporator case to the dash panel. Inside the passenger compartment, remove the 2 screws attaching the evaporator case support brackets to the cowl top panel.

9. Remove the 1 screw retaining the bracket below the evaporator case to the dash panel. Carefully pull the evaporator case away from the dash panel and remove the evaporator case assembly from the vehicle.

10. Remove the 2 nuts that attach the vacuum motor to the case and disconnect the vacuum hose from the motor.

11. Remove the spring nut that attaches the motor crank arm to the shaft and remove the motor.

To install:

12. Position the motor and install the spring nut that attaches the motor crank arm to the shaft.

13. Connect the vacuum hose to the motor and install the 2 nuts that attach the vacuum motor to the case.

14. Position the evaporator case assembly in the vehicle. Install the screws attaching the evaporator case support brackets to the cowl top panel. Check the evaporator case drain tube to ensure it is through the dash panel and is not pinched or kinked.

15. Install 1 screw retaining the bracket below the evaporator case to the dash panel. Working under the hood, install the 2 nuts retaining the evaporator case to the dash panel. Tighten the 4 nuts and 2 screws in the engine compartment. Tighten the 2 screws in the passenger compartment and the 2 support bracket attaching screws.

16. Connect the blower motor wire harness to the resistor and blower motor. Connect the black vacuum supply hose to the vacuum check valve in the engine compartment.

17. Using new O-rings lubricated with clean refrigerant oil, connect the liquid line and suction accumulator inlet to the evaporator core tubes. Tighten each connection using a backup wrench to prevent component damage.

18. Install the instrument panel by reversing the removal procedure.

19. Connect the heater hoses to the heater core and fill the cooling system.

20. Connect the negative battery cable. Leak test, evacuate and charge the refrigerant system according to the proper procedure. Observe all safety precautions.

21. Check the system for proper operation.

Outside/Recirculating Door Vacuum Motor

1. Remove the glove compartment. Disconnect the vacuum hose from the vacuum motor.

2. Remove the motor arm retainer from the outside/recirculating door shaft.

3. Remove the 2 nuts retaining the vacuum motor to the mounting bracket and remove the motor.

4. Installation is the reverse of the removal procedure.

CRUISE CONTROL

Control Switches

REMOVAL & INSTALLATION

➡ Please refer to "Steering Wheel" Removal and Installation in Section 8 for control switches which are located in the steering wheel. Extreme caution must be observed with vehicles equipped with air bags!

Speed Sensor

REMOVAL & INSTALLATION

➧ SEE FIG. 21

1. Raise the vehicle and support safely on jackstands.

2. Remove the bolt retaining the speed sensor mounting clip to the transmission.

3. Remove the sensor and the drive gear from the transmission.

4. Disconnect the electrical connector and the speedometer cable from the speed sensor.

5. Remove the speedometer cable by pulling it out of the sensor.

➡ DO NOT remove the spring retainer clip with the speedometer cable in the sensor.

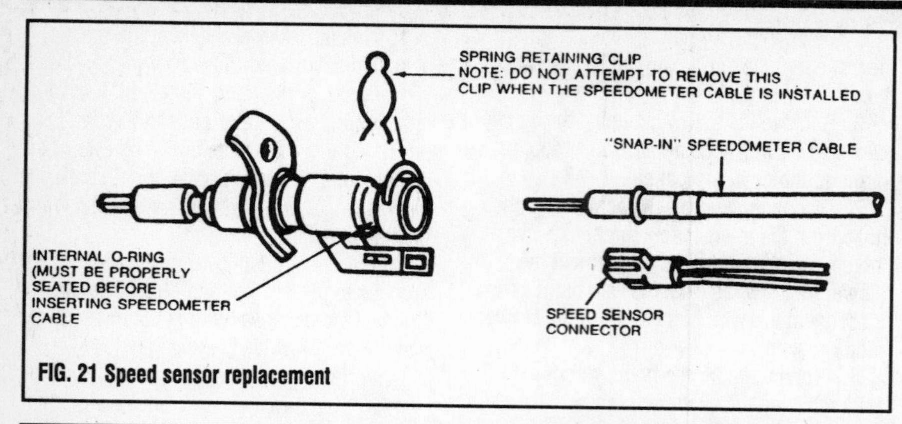

FIG. 21 Speed sensor replacement

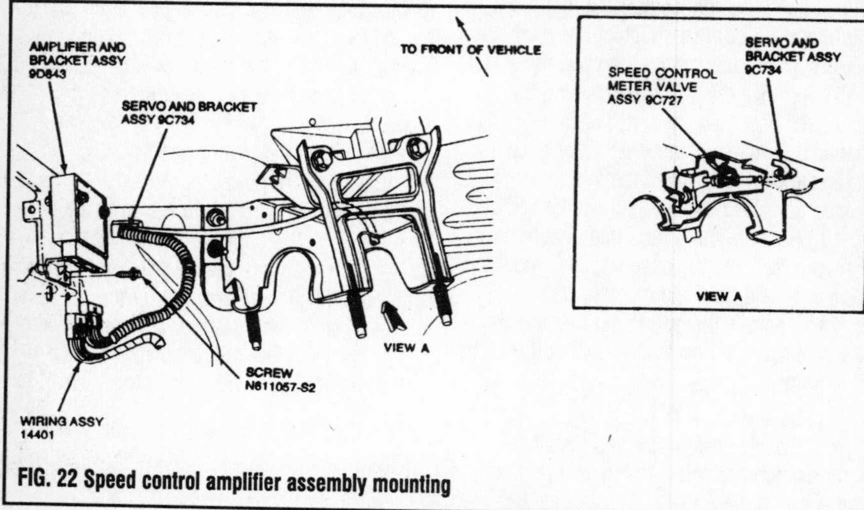

FIG. 22 Speed control amplifier assembly mounting

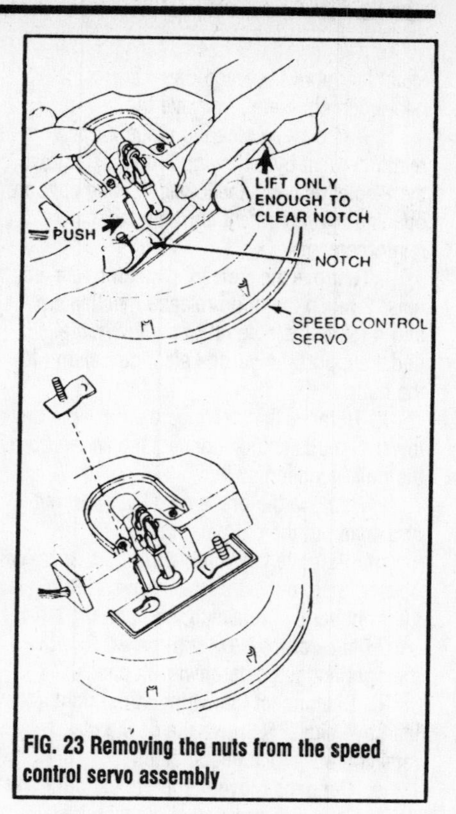

FIG. 23 Removing the nuts from the speed control servo assembly

6. Remove the drive gear retainer. Remove the drive gear from the sensor.

To Install:

7. Position the drive gear to the speed sensor. Install the gear retainer.
8. Connect the electrical connector.
9. Check that the O-ring is properly seated in the sensor housing. Snap the speedometer cable onto the sensor housing.
10. Insert the sensor assembly into the transmission. Install the retaining bolt. Lower the vehicle.

Amplifier

REMOVAL & INSTALLATION

➧ SEE FIG. 22

The amplifier is located inside the passenger compartment, on the left hand side cowl panel around the parking brake.

1. Disconnect the negative battery cable.
2. Remove the screws retaining the amplifier assembly to the mounting bracket.
3. Disconnect the two electrical connectors at the amplifier.

To Install:

4. Connect the two electrical connectors at the amplifier.
5. Position the amplifier assembly on the mounting bracket and install the retaining screws.
6. Connect the negative battery cable.

Servo

REMOVAL & INSTALLATION

➧ SEE FIG. 23–24

1. Disconnect the negative battery cable.
2. Disconnect the servo wiring at the amplifier, and disconnect the white stripe vacuum hose from the dump valve in the passenger compartment.
3. Disconnect the speed control actuator cable from the accelerator cable.
4. Remove the grommet and wiring from the passenger compartment.
5. Raise the vehicle and support safely on jackstands.
6. Remove the left front tire.
7. Remove the inner fender splash shield.
8. Remove the brown stripe vacuum hose from the servo assembly.
9. Remove the two screws from the servo bracket to the A-pillar.
10. Remove the two nuts from the actuator cable cover at the servo. Remove the cable and the cover. Remove the rubber boot.
11. Remove the two nuts retaining the servo to the mounting bracket.
12. If the servo is being replaced, remove the two bolt assemblies from the front of the servo.

To Install:

13. Install the two bolts to the front of the servo.
14. Install the two nuts retaining the servo to the mounting bracket. Tighten to 45–65 inch lbs. (5–7 Nm).
15. Install the rubber boot.
16. Attach the actuator cable to the servo plunger. Install the cable cover to the servo with two nuts. Tighten to 45–65 inch lbs. (5–7 Nm).
17. Install the servo and bracket to the A-pillar with two screws.
18. Insert the servo connector and dump valve hose through the grommet hole in the passenger compartment. Fully seat the wire harness and the hose assembly into the hole.
19. Attach the brown strip vacuum hose to

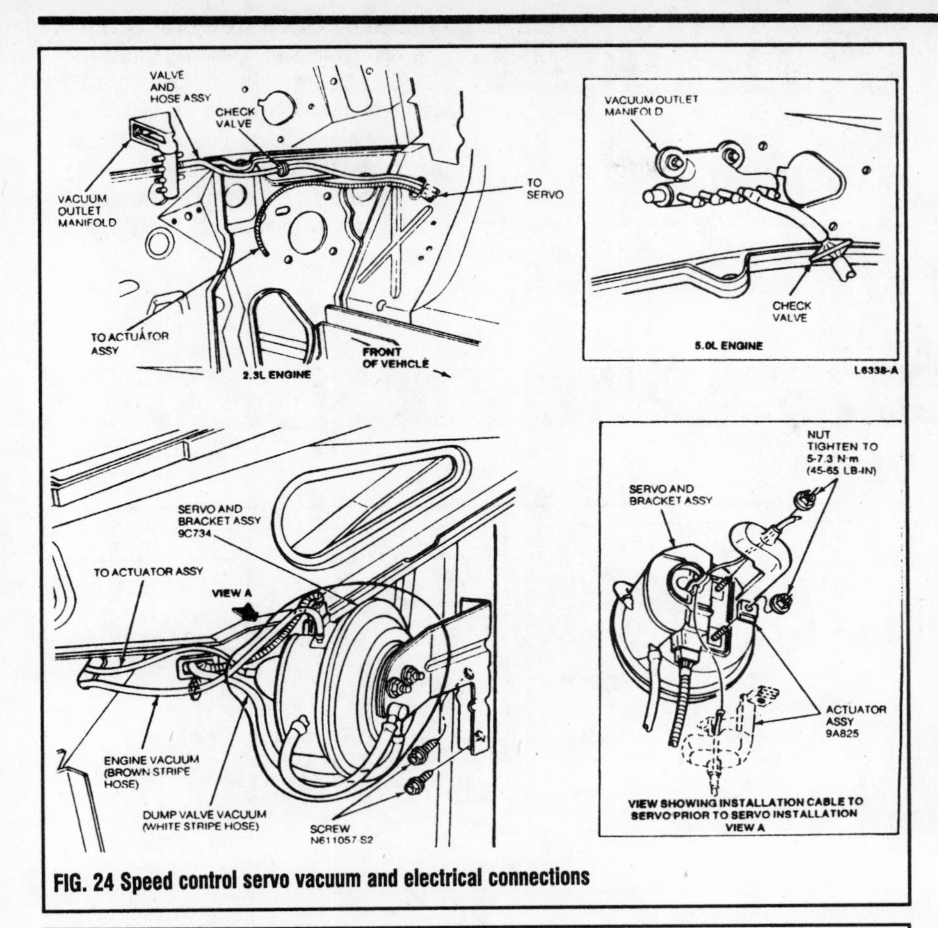

FIG. 24 Speed control servo vacuum and electrical connections

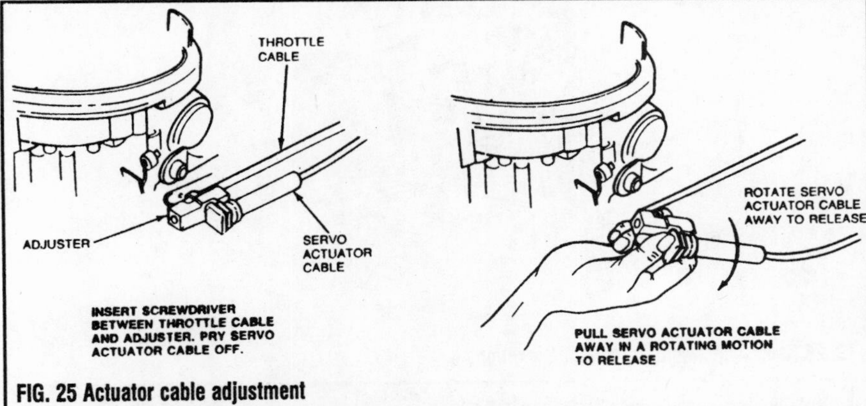

FIG. 25 Actuator cable adjustment

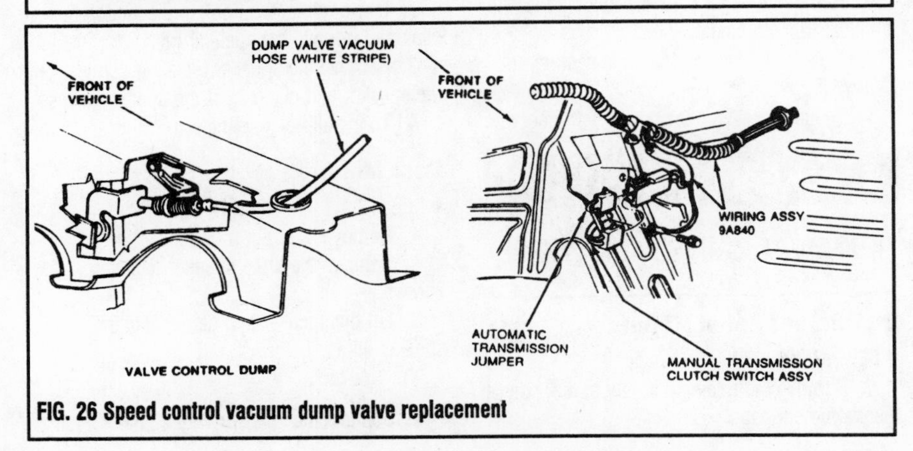

FIG. 26 Speed control vacuum dump valve replacement

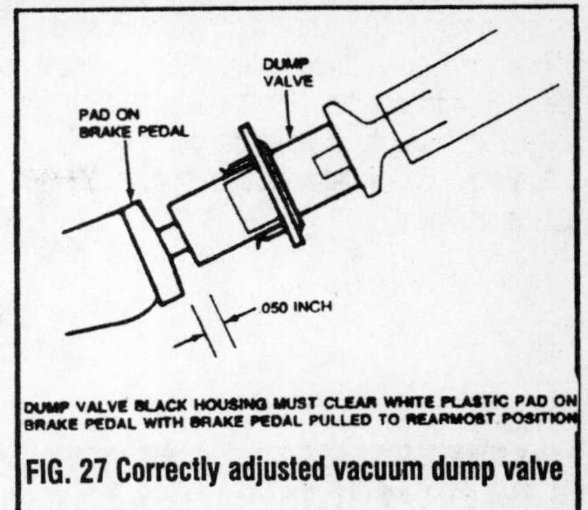

FIG. 27 Correctly adjusted vacuum dump valve

the servo. Adjust the servo boot to protect the servo.

20. Install the inner fender splash shield.
21. Install the tire and tighten the lug bolts to 85–105 ft. lbs. (115–142 Nm).
22. Lower the vehicle.
23. Connect the servo wiring at the amplifier in the passenger compartment.
24. Attach the servo vacuum hose with the white strip to the dump valve in the passenger compartment.
25. Connect the speed control actuator cable to the speedometer cable. Connect the negative battery cable.

ACTUATOR CABLE LINKAGE ADJUSTMENT

➧ SEE FIG. 25

1. Remove the cable retaining clip.
2. Push the cable through the adjuster until a slight tension is felt.
3. Insert the cable retaining clip and snap into place.

Vacuum Dump Valve

REMOVAL & INSTALLATION

➧ SEE FIG. 26

1. Remove the vacuum hose from the valve.
2. Remove the valve from the bracket.

To Install:

3. Install the valve to the bracket.
4. Connect the vacuum hose.
5. Adjust the valve.

ADJUSTMENT

➧ SEE FIG. 27

The vacuum dump valve is movable in its

mounting bracket. It should be adjusted so that it is closed, no vacuum leaks, when the brake pedal is in its normal release position, not depressed, and open when the pedal is depressed. Use a hand vacuum pump to make this adjustment.

Clutch Switch

ADJUSTMENT

1. Prop the clutch pedal in the full-up position, pawl fully released from the sector.
2. Loosen the switch retaining screw.
3. Slide the switch forward toward the clutch pedal until the switch plunger cap is 0.030 in. (0.76mm) from contacting the switch housing. Then, tighten the retaining screw.
4. Remove the prop from the clutch pedal and test drive for clutch switch cancellation of cruise control.

ENTERTAINMENT SYSTEMS

Radio/Tape Player

REMOVAL & INSTALLATION

➧ SEE FIG. 28

➡ These vehicles have "DIN" mount radios, a special tool T87P-19061-A or equivalent is required to release the clips and remove the radio from the vehicle.

1. Disconnect the negative battery cable.
2. Install two radio removing tool T87P-19061-A or equivalent into the radio face plate. Push the tools in approximately one inch (25mm) to release the retaining clips.

➡ DO NOT use excessive force when inserting the radio removal tools, as this will damage the retaining clips.

3. Apply a light spreading force on the tools and pull the radio out of the dash.
4. Disconnect the wiring and antenna connectors.

To Install:

5. Connect the wiring and antenna connectors to the radio.
6. Slide the radio into the instrument panel ensuring that the rear radio bracket is engaged on the upper support rail.
7. Connect the negative battery cable. Check the radio for proper operation.

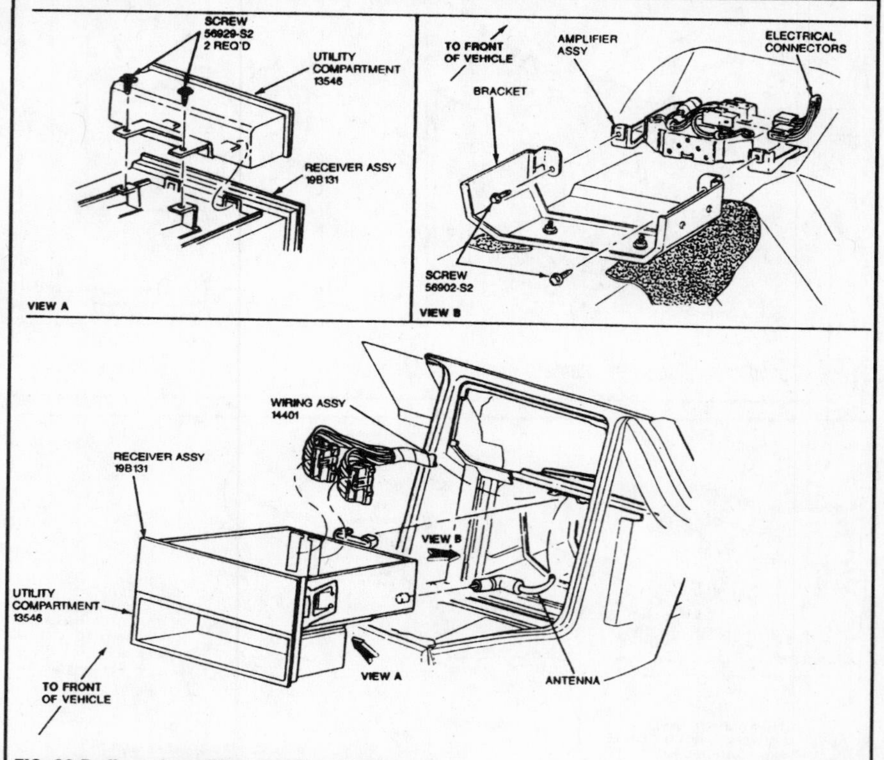

FIG. 28 Radio and amplifier mounting locations

Amplifier

REMOVAL & INSTALLATION

➧ SEE FIG. 28

1. Disconnect the negative battery cable.
2. Remove the radio as outlined previously.
3. Remove the two screws retaining the amplifier to the tunnel bracket.
4. Disconnect the electrical connectors and remove the amplifier.
5. To install, reverse the removal procedures.

Speakers

REMOVAL & INSTALLATION

Instrument Panel Mounted

➧ SEE FIGS. 29–30

1. Remove the retaining screw at the side of the instrument panel.
2. Use a fabricated hook to disengage the clips and remove the speaker grills.
3. Remove the speaker retaining screws. Lift the speaker and disconnect the speaker wires.
4. To install, reverse the removal procedure.

Door Mounted

➧ SEE FIG. 31

1. Remove the door trim panel.
2. Remove the three speaker retaining screws.
3. Lift the speaker and disconnect the speaker wires.

To Install:

4. Connect the speaker wires, position the

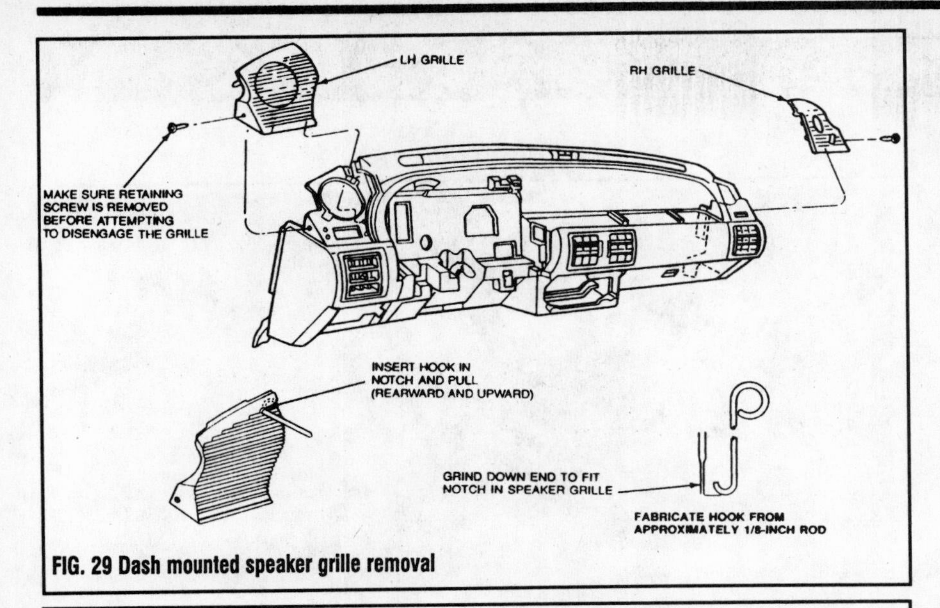

FIG. 29 Dash mounted speaker grille removal

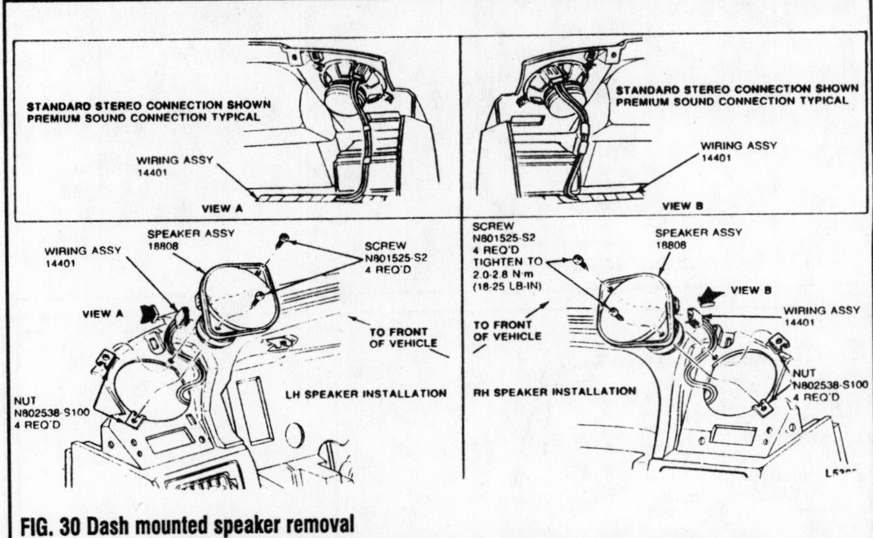

FIG. 30 Dash mounted speaker removal

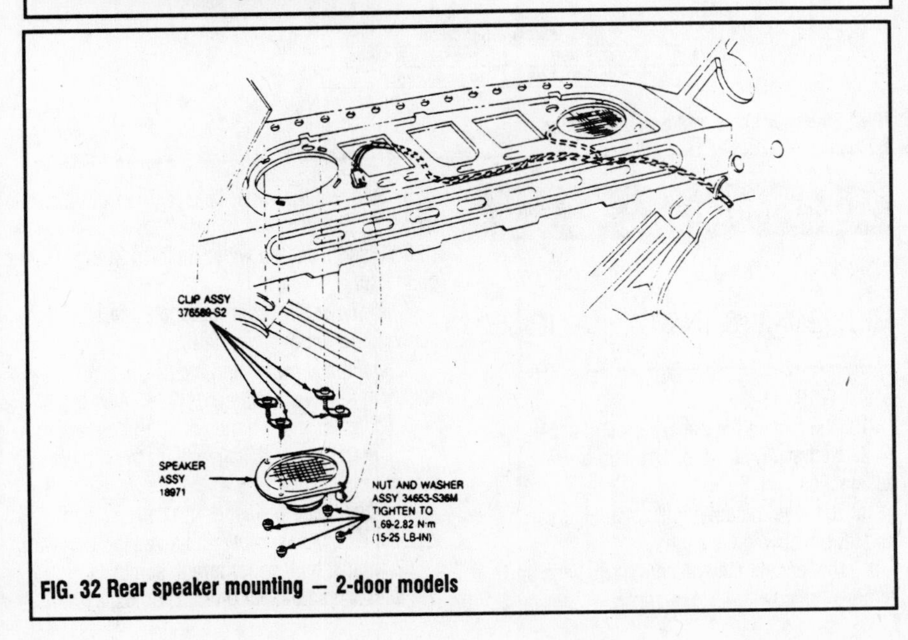

FIG. 32 Rear speaker mounting — 2-door models

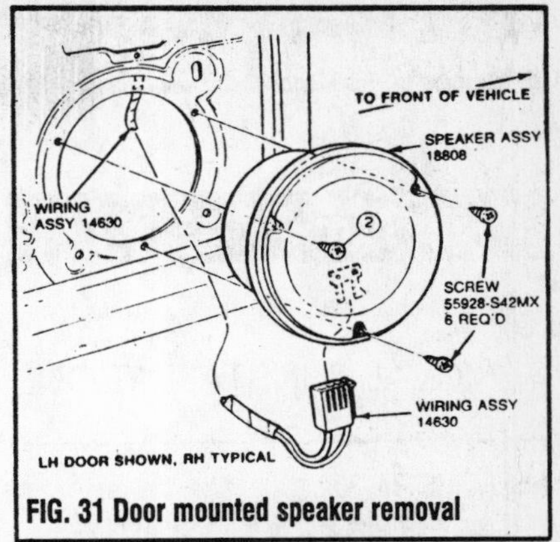

FIG. 31 Door mounted speaker removal

VIEW SHOWING RH INSTALLATION OF PREMIUM SOUND SPEAKER AND GRILLE ASSY LH SYMETRICALLY OPPOSITE CONVENTIONAL SPEAKER AND GRILLE TYPICAL.

TO FRONT OF VEHICLE

TRIM PANEL

WIRING ASSY 14405

SPEAKER AND GRILLE ASSY 18932

WIRING ASSY 14405

SCREW 55962-SW 4 REQ'D

FIG. 33 Rear speaker mounting — 3-door models

speaker to the door side panel and install the retaining screws.

5. Push the lock clip into the hole in the door inner panel and check the speaker operation.

6. Install the door trim panel.

Rear Mounted

2-DOOR

➧ SEE FIG. 32

1. From within the luggage compartment, disconnect the speaker wiring.

2. Remove the speaker cover, speaker retaining nuts and the speaker from the underside of the package shelf.

3. To install, reverse the removal procedure.

3-DOOR

➧ SEE FIG. 33

1. Remove the speaker and grille assembly retaining screws from the quarter trim panel.

2. Lift the speaker and the grille assembly from the trim panel. Disconnect the speaker wires.

3. remove the nuts retaining the speaker to the grille assembly and remove the speaker.

4. To install, reverse the removal procedures.

WINDSHIELD WIPERS AND WASHERS

Wiper Arm

REMOVAL & INSTALLATION

➧ SEE FIGS. 34–35

Raise the blade end of the arm off of the windshield and move the slide latch away from the pivot shaft. This will unlock the wiper arm from the pivot shaft and hold the blade end of the arm off of the glass at the same time. The wiper arm can now be pulled off of the pivot shaft without the aid of any tools.

When installing the wiper arm, the arm must be positioned properly. There is a measurement which can be made to determine the proper blade positioning. With the wiper motor in the PARK position, install the arm so that the distance between the blade-to-arm saddle and the lower windshield molding is as shown in FIG. 35.

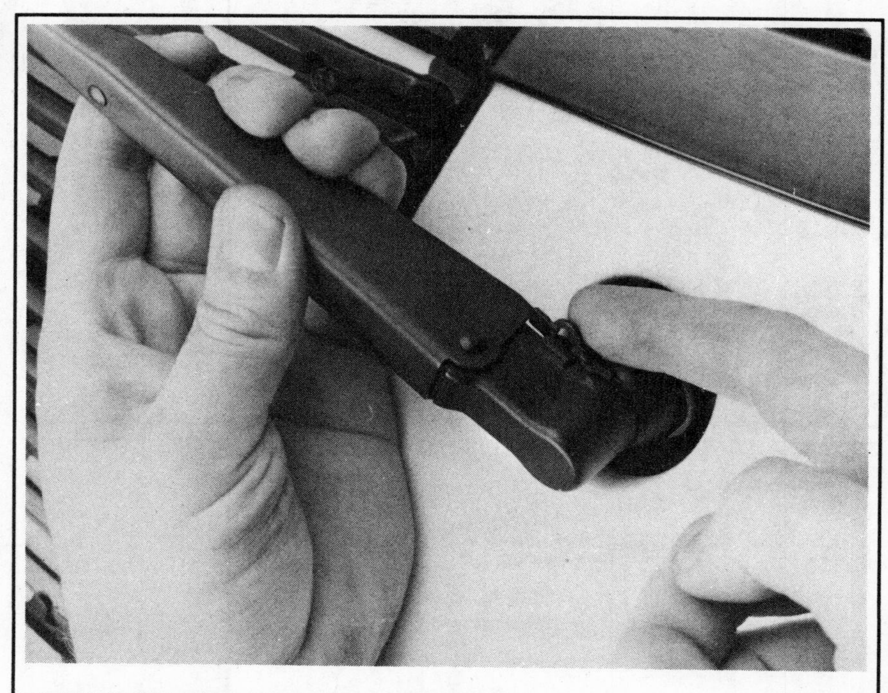

FIG. 34 Slide the latch mechanism for wiper arm removal

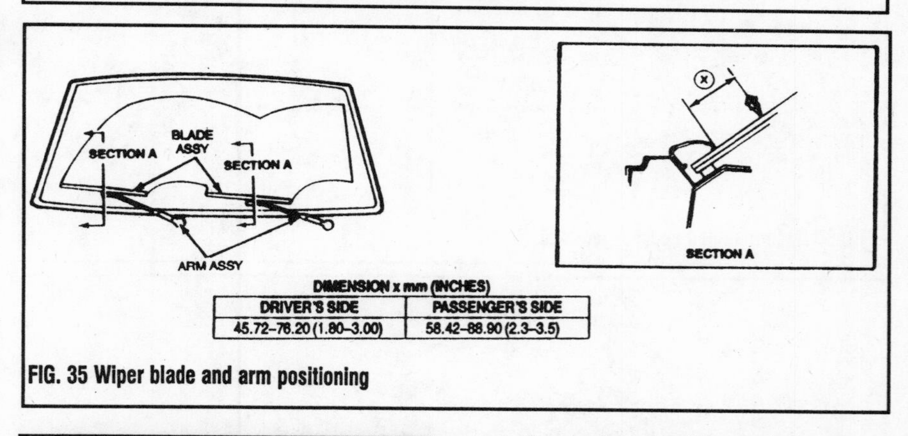

DIMENSION x mm (INCHES)

DRIVER'S SIDE	PASSENGER'S SIDE
45.72–76.20 (1.80–3.00)	58.42–88.90 (2.3–3.5)

FIG. 35 Wiper blade and arm positioning

Wiper Blade

REMOVAL & INSTALLATION

➧ SEE FIG. 36

1. Cycle arm and blade assembly to a position on the windshield where removal of blade assembly can be performed without difficulty. Turn ignition key off at desired position.
2. With the blade assembly resting on windshield, grasp either end of the wiper blade frame and pull away from windshield, then pull blade assembly from pin.

➡ Rubber element extends past frame. To prevent damage to the blade element, be sure to grasp blade frame and not the end of the blade element.

3. To install, push blade assembly onto pin until fully seated. Be sure blade is securely attached to the wiper arm.

Wiper Motor

REMOVAL & INSTALLATION

➧ SEE FIGS. 37–39

1. Disconnect the battery ground cable.
2. Remove both wiper arm and blade assemblies.
3. Remove the cowl grille attaching screws and lift the cowl grille slightly.
4. Disconnect the washer nozzle hose and remove the cowl grille assembly.
5. Remove the wiper linkage clip from the motor output arm.
6. Disconnect the wiper motor's wiring connector.
7. Remove the wiper motor's three attaching screws and remove the motor.
8. Install the motor and attach the three attaching screws. Tighten to 60–85 inch lbs.
9. Connect wiper motor's wiring connector.
10. Install wiper linkage clip to the motors output arm.
11. Connect the washer nozzle hose and install the cowl assembly and attaching screws.
12. Install both wiper arm assemblies.
13. Connect battery ground cable.

Wiper Linkage

REMOVAL & INSTALLATION

1. Disconnect the battery ground cable.
2. Remove both wiper arm assemblies.
3. Remove the cowl grille attaching screws and lift the cowl grille slightly.
4. Disconnect the washer nozzle hose and remove the cowl grille assembly.
5. Remove the wiper linkage clip from the motor output arm and pull the linkage from the output arm.
6. Remove the pivot body to cowl screws and remove the linkage and pivot shaft assembly (three screws on each side). The left and right pivots and linkage are independent and can be serviced separately.
7. Attach the linkage and pivot shaft assembly to cowl with attaching screws.
8. Replace the linkage to the output arm and attach the linkage clip.
9. Connect the washer nozzle hose and cowl grills assembly.
10. Attach cowl grille attaching screws.
11. Replace both wiper arm assemblies.
12. Connect battery ground cable.

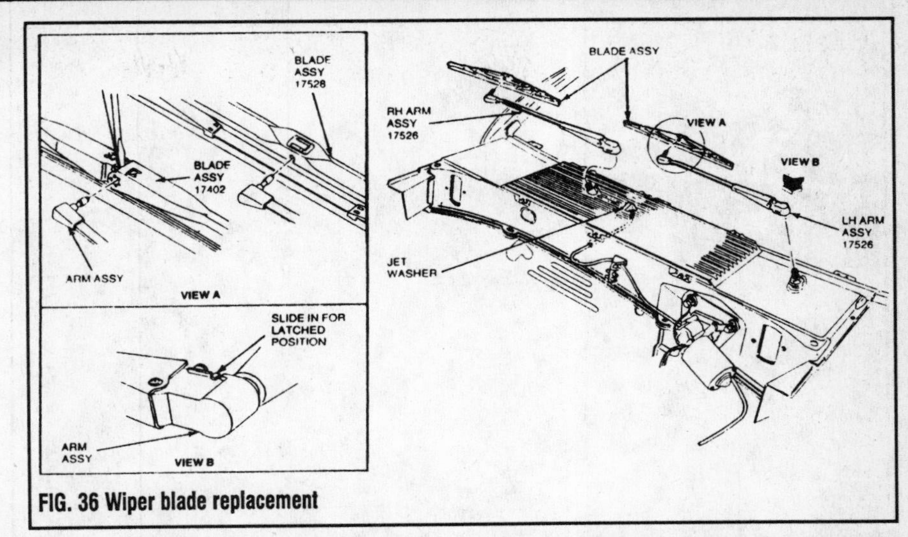

FIG. 36 Wiper blade replacement

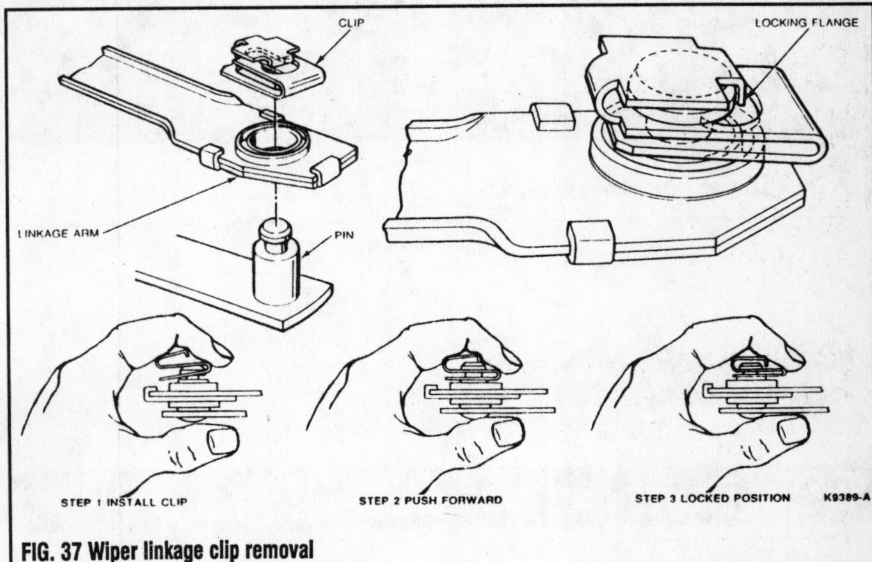

FIG. 37 Wiper linkage clip removal

Washer Reservoir and Pump Motor

REMOVAL & INSTALLATION

Reservoir

➧ SEE FIG. 40

1. Disconnect the wiring at the pump motor. Use a small screwdriver to unlock the connector tabs.
2. Remove the washer hose.
3. Remove the reservoir attaching screws or nuts and lift the assembly from the vehicle.

Motor/Impeller

➧ SEE FIG. 41

1. Remove the reservoir.
2. Using a small screwdriver, pry out the motor retaining ring.
3. Using a pliers, grip one edge of the electrical connector ring and pull the motor, seal and impeller from the reservoir.

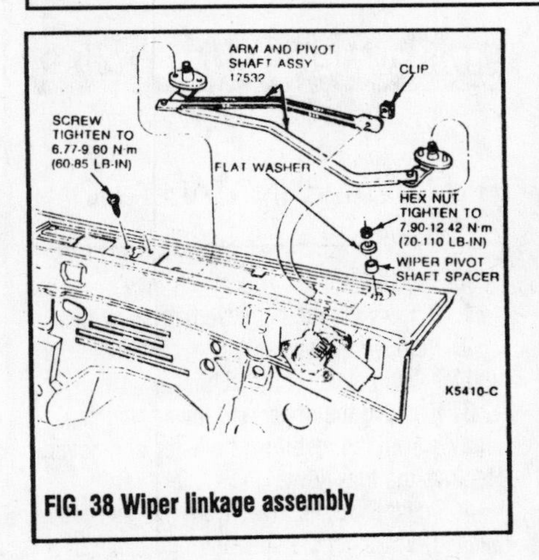

FIG. 38 Wiper linkage assembly

➡ **If the seal and impeller come apart from the motor, it can all be re-assembled.**

To install:

4. Take the time to clean out the reservoir before installing the motor.
5. Coat the seal with a dry lubricant, such as powdered graphite or spray Teflon®. This will aid assembly.
6. Align the small projection on the motor end cap with the slot in the reservoir and install the motor so that the seal seats against the bottom of the motor cavity.
7. Press the retaining ring into position. A 1 in., 12-point socket or length of 1 in. tubing, will do nicely as an installation tool.
8. Install the reservoir and connect the wiring.

➡ **It's not a good idea to run a new motor without filling the reservoir first. Dry-running will damage a new motor.**

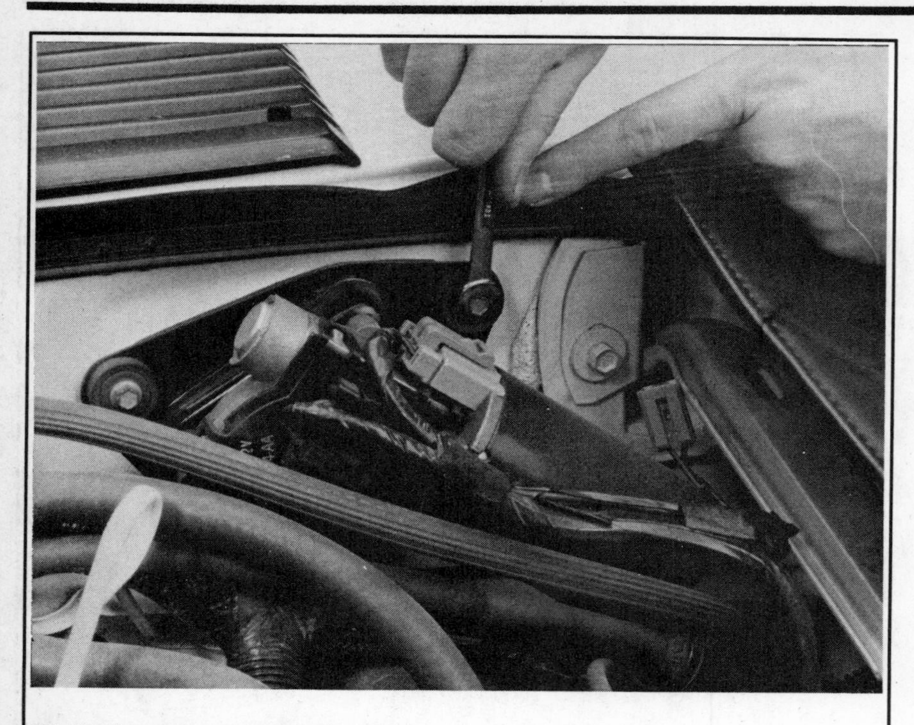
FIG. 39 Wiper motor assembly replacement

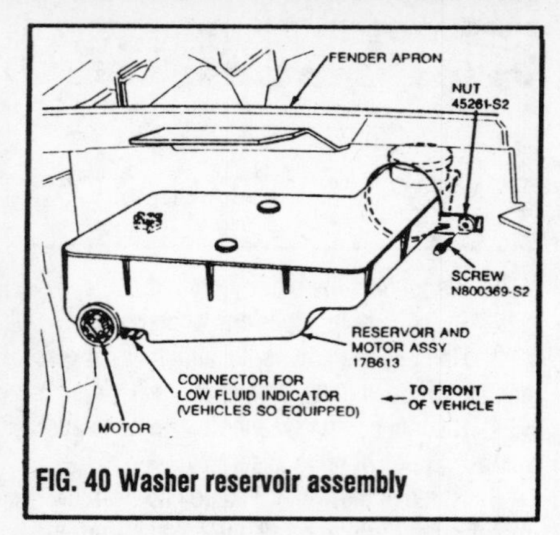

FIG. 40 Washer reservoir assembly

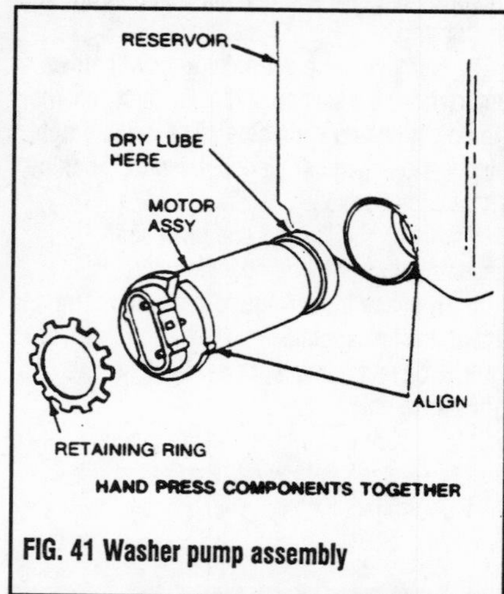

FIG. 41 Washer pump assembly

INSTRUMENTS AND SWITCHES

Precautions

Electronic modules, such as instrument clusters, powertrain controls and sound systems are sensitive to static electricity and can be damaged by static discharges which are below the levels that you can hear "snap" or detect on your skin. A detectable snap or shock of static electricity is in the 3,000 volt range. Some of these modules can be damaged by a charge of as little as 100 volts.

The following are some basic safeguards to avoid static electrical damage:

- Leave the replacement module in its original packing until you are ready to install it.
- Avoid touching the module connector pins
- Avoid placing the module on a non-conductive surface
- Use a commercially available static protection kit. These kits contain such things as grounding cords and conductive mats.

Instrument Cluster

REMOVAL & INSTALLATION

➧ SEE FIG. 42

1. Disconnect the negative battery cable.
2. Remove the switch assembly on the right and left-hand sides of the cluster.
3. Remove the upper and lower retaining screws from the instrument cluster trim cover. Remove the trim cover.
4. Pull the cluster away from the instrument panel. Disconnect the speedometer cable by pressing on the flat surface of the plastic connector located behind the instrument cluster.
5. Pull the cluster further away from the instrument panel. Disconnect the two printed circuit connectors from their receptacles in the cluster backplate. Remove the cluster.

To Install:

➡ If the gauges are being removed from the cluster assembly. DO NOT remove the gauge pointer; magnetic gauges cannot be recalibrated.

6. Apply a $^{3}/_{16}$ (4.6mm) bead of silicone damping grease in the drive hole of the speedometer head.
7. Connect the two cluster printed circuit connectors to the cluster backplate.
8. Position the instrument cluster to the instrument panel and install the retaining screws.
9. Install the instrument cluster trim panel and the retaining screws. Connect the negative battery cable.

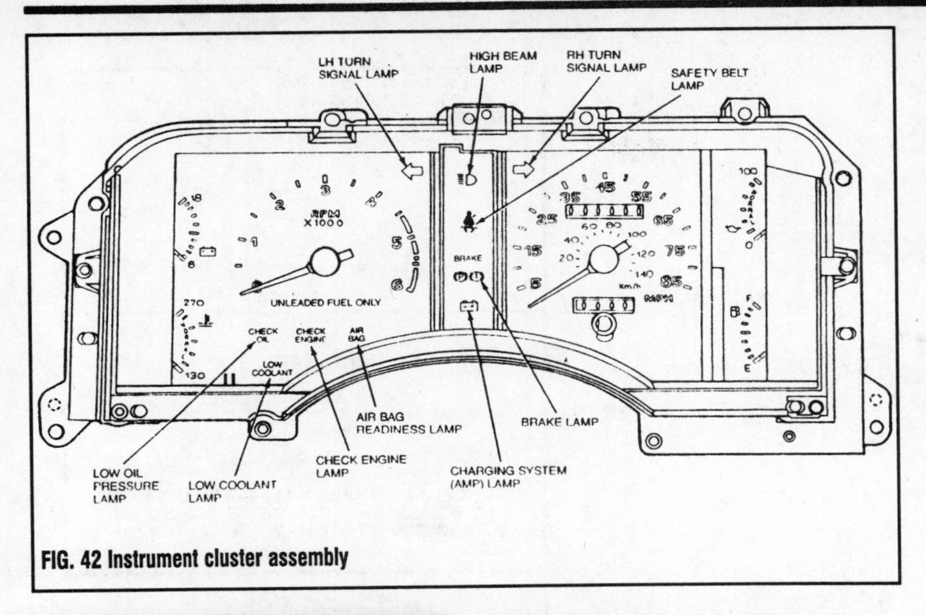

FIG. 42 Instrument cluster assembly

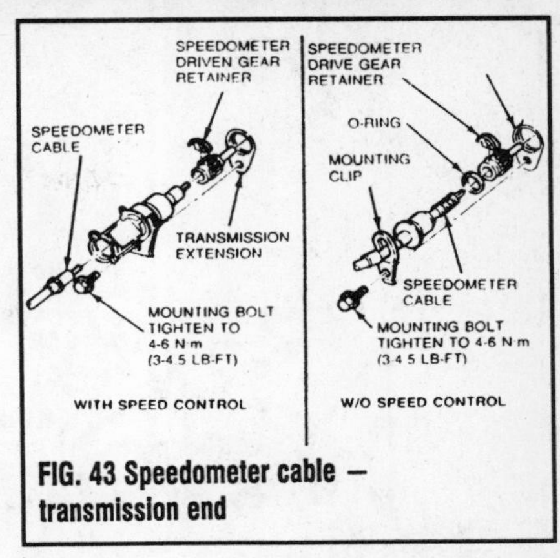

FIG. 43 Speedometer cable — transmission end

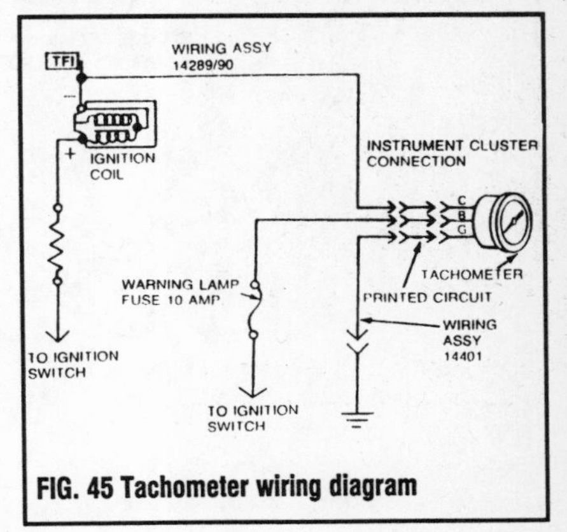

FIG. 45 Tachometer wiring diagram

Speedometer Head

REMOVAL & INSTALLATION

1. Remove the instrument cluster.
2. Disconnect the cable from the head.
3. Remove the lens and any surrounding trim.
4. Remove the 2 attaching screws.
5. Installation is the reverse of removal. Place a glob of silicone grease on the end of the cable core prior to connection.

Speedometer Cable Core

REMOVAL & INSTALLATION

➧ SEE FIGS. 43–44

1. Reach up behind the cluster and disconnect the cable by depressing the quick disconnect tab and pulling the cable away.
2. Remove the cable from the casing. If the cable is broken, raise the vehicle on a hoist and disconnect the cable from the transmission.

➡ On vehicles equipped with a transmission mounted speed sensor, remove the speedometer cable by pulling it out of the speed sensor. DO NOT attempt to remove the spring retainer clip with the speedometer in the sensor. To install the cable, snap it into the sensor.

3. Remove the cable from the casing.
4. To remove the casing from the vehicle pull it through the floor pan.
5. To replace the cable, slide the new cable into the casing and connect it at the transmission.
6. Route the cable through the floor pan and position the grommet in its groove in the floor.
7. Push the cable onto the speedometer head.

Tachometer

REMOVAL & INSTALLATION

➧ SEE FIG. 45

1. Disconnect the battery ground.
2. Remove the instrument cluster.
3. Remove the cluster mask and lens.
4. Remove the tachometer by prying the dial away from the cluster backplate. The tachometer is retained by clips.
5. Installation is the reverse of removal. Make sure the clips are properly seated.

Fuel Gauge

REMOVAL & INSTALLATION

➧ SEE FIG. 46

1. Disconnect the battery ground.
2. Remove the instrument cluster.
3. Remove the cluster mask and lens.
4. Remove the fuel/oil pressure gauge assembly.
5. Installation is the reverse of removal.

Sending Units

REMOVAL & INSTALLATION

Oil Pressure

➧ SEE FIGS. 47–48

1. Disconnect the wiring at the unit.
2. Unscrew the unit.
3. Coat the threads with electrically conductive sealer and screw the unit into place. The torque should be 10–18 ft. lbs.

Coolant Temperature

➧ SEE FIGS. 49–50

1. Remove the radiator cap to relieve any system pressure.
2. Disconnect the wiring at the unit.
3. Unscrew the unit.
4. Coat the threads with Teflon® tape or

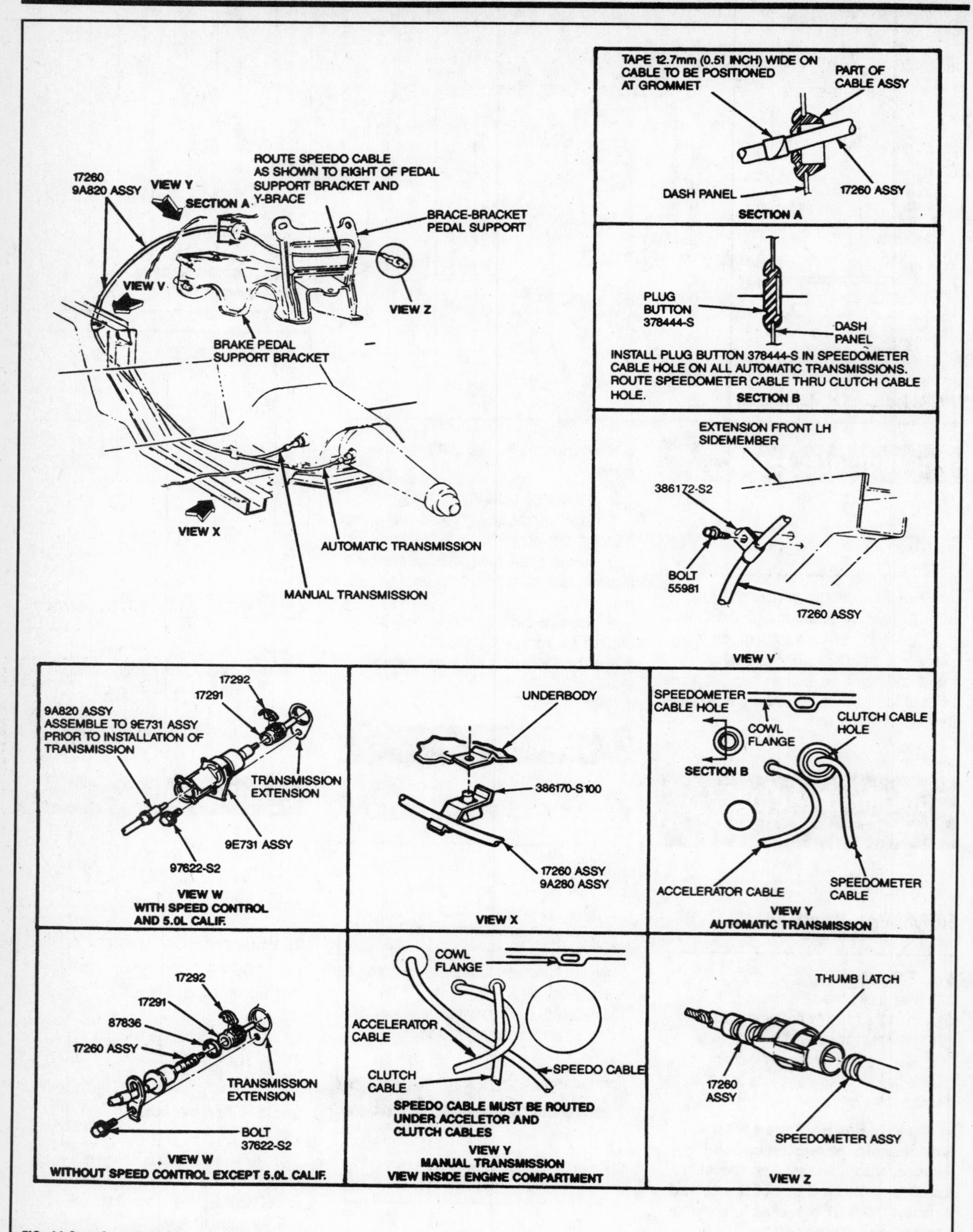

FIG. 44 Speedometer cable replacement

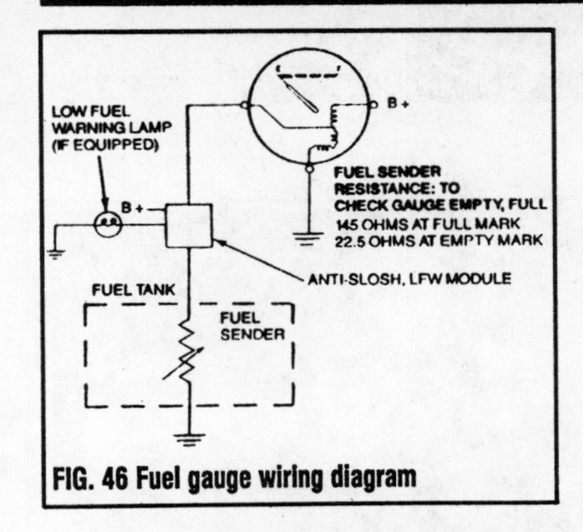

FIG. 46 Fuel gauge wiring diagram

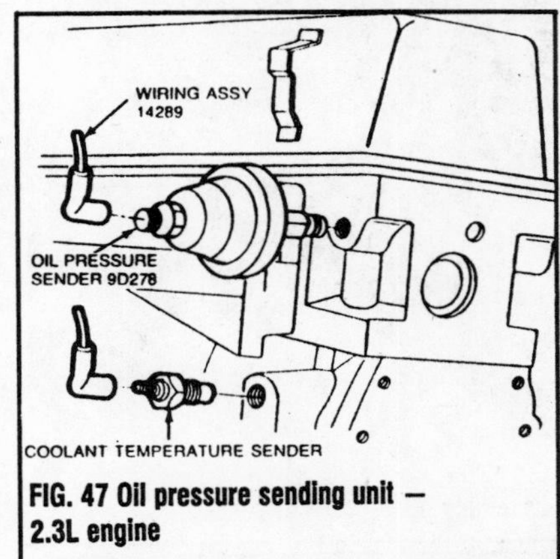

FIG. 47 Oil pressure sending unit — 2.3L engine

electrically conductive sealer and screw the unit into place. The torque should be 10–18 ft. lbs.

5. Replace any lost coolant.

Windshield Wiper Switch

REMOVAL & INSTALLATION

➧ SEE FIG. 51

1. Disconnect the negative battery cable.
2. Remove the steering column shroud attaching screws.
3. Grasp the top and bottom of the shroud and separate.
4. Remove the two wiper switch attaching bolts and remove the switch assembly. Push the wire connector off the wiper switch with a flat screwdriver.
5. To install, reverse the removal procedures.

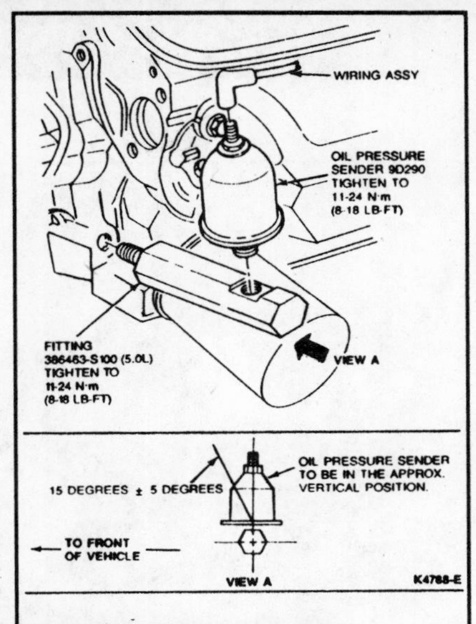

FIG. 48 Oil pressure sending unit — 5.0L engine

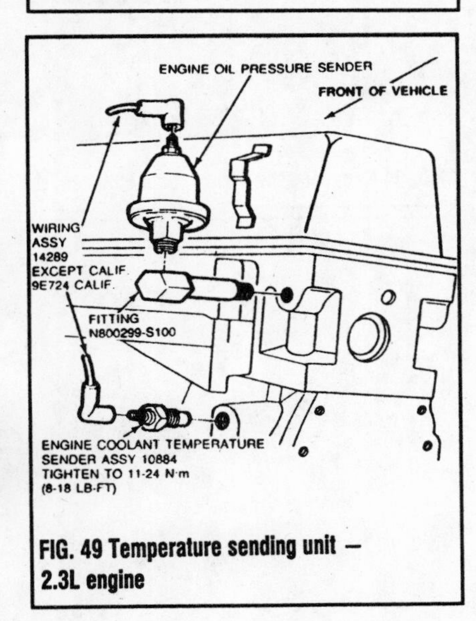

FIG. 49 Temperature sending unit — 2.3L engine

Headlight Switch

REMOVAL & INSTALLATION

➧ SEE FIG. 52

1. Disconnect the negative battery cable.
2. Disengage the locking tabs by pushing the tabs in with a small screwdriver and pulling on the paddles.

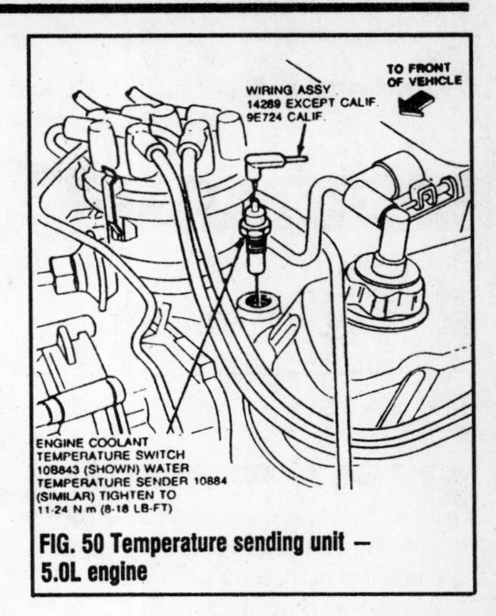

FIG. 50 Temperature sending unit — 5.0L engine

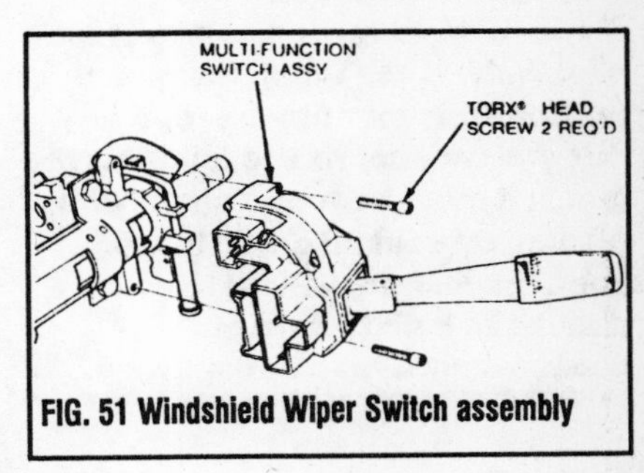

FIG. 51 Windshield Wiper Switch assembly

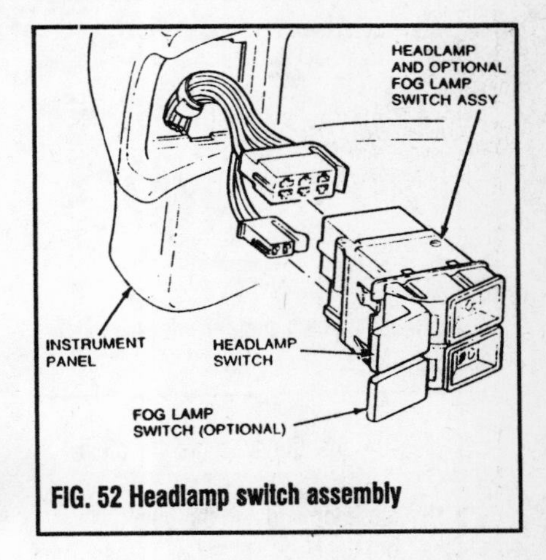

FIG. 52 Headlamp switch assembly

3. Using a suitable tool, pry the right hand side of the switch out of the instrument panel.
4. Pull the switch completely out of the opening and disconnect the two connectors.
5. To install, assemble the two connectors to the switch and insert the switch into the instrument panel opening until the locking tabs on both sides of the switch snap into place.

LIGHTING

Headlights

REMOVAL & INSTALLATION

➧ SEE FIGS. 53–55

CAUTION

The headlamp bulb contains high pressure halogen gas. The bulb may shatter if scratched or dropped! Hold the bulb by its plastic base only. If you touch the glass portion with your fingers, or if any dirt or oily deposits are found on the glass, it must be wiped clean with an alcohol soaked paper towel. Even the oil from your skin will cause the bulb to burn out prematurely due to hot-spotting.

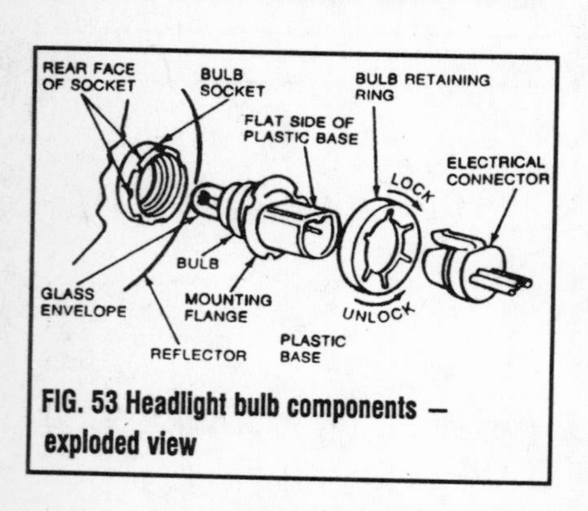

FIG. 53 Headlight bulb components — exploded view

1. Make sure that the headlight switch is **OFF**.
2. Raise the hood and find the bulb base protruding from the back of the headlamp assembly
3. Disconnect the wiring by grasping the connector and snapping it rearward firmly.
4. Rotate the bulb retaining ring counterclockwise (rear view) about 1/8 turn and slide it off the bulb base. Don't lose it; it's re-usable.
5. Carefully pull the bulb straight out of the headlamp assembly. Don't rotate it during removal.

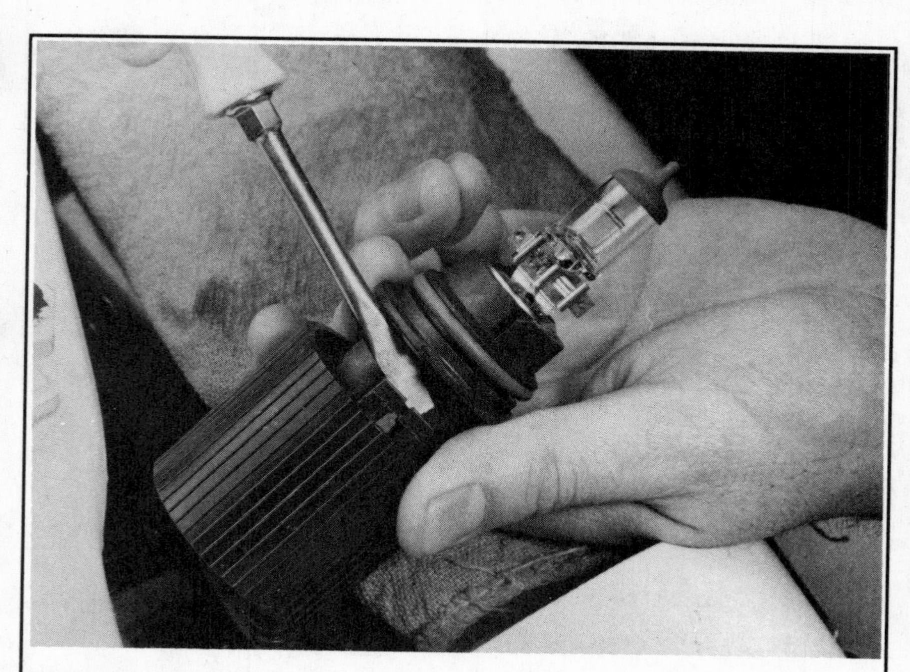

FIG. 54 Removing the wiring connector from the bulb assembly

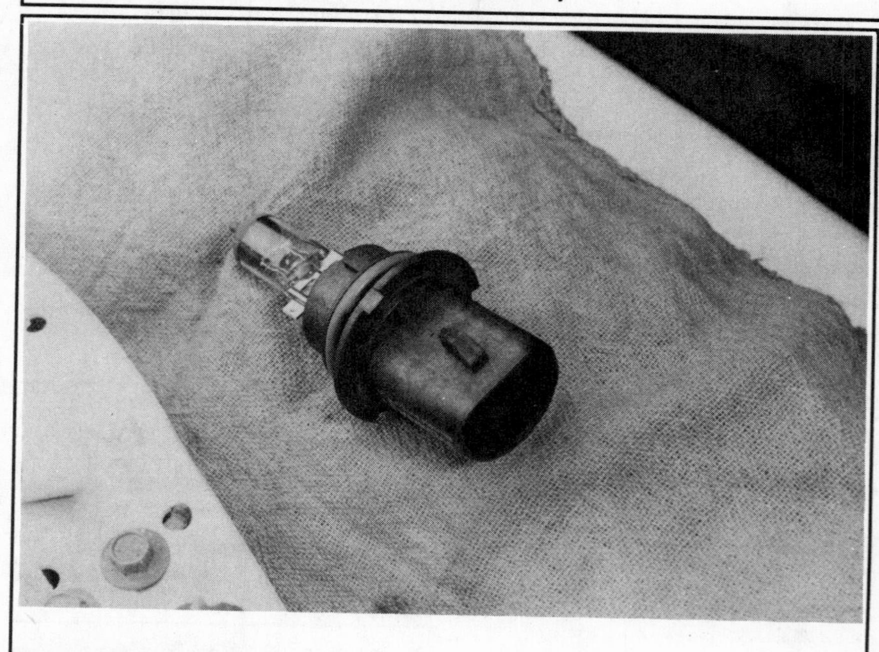

FIG. 55 Halogen bulb assembly

WARNING

Don't remove the old bulb until you are ready to immediately replace it! Leaving the headlamp assembly open, without a bulb, will allow foreign matter such as water, dirt, leaves, oil, etc. to enter the housing. This type of contamination will cut down on the amount and direction of light emitted, and eventually cause premature blow-out of the bulb.

To install:

6. With the flat side of the bulb base facing upward, insert it into the headlamp assembly. You may have to turn the bulb slightly to align the locating tabs. Once aligned, push the bulb firmly into place until the bulb base contacts the mounting flange in the socket.
7. Place the retaining ring over the bulb base, against the mounting flange and rotate it clockwise to lock it. It should lock against a definite stop when fully engaged.
8. Snap the electrical connector into place. A definite snap will be felt.
9. Turn the headlights on a check that everything works properly.

HEADLIGHT ADJUSTMENT

➡ **Before making any headlight adjustments, preform the following steps for preparation:**

1. Make sure all tires are properly inflated.
2. Take into consideration any faulty wheel alignment or improper rear axle tracking.
3. Make sure there is no load in the vehicle other than the driver.
4. Make sure all lenses are clean.

Each headlight is adjusted by means of two screws located at the 12 o'clock and 9 o'clock positions on the headlight underneath the trim ring. Always bring each beam into final position by turning the adjusting screws clockwise so that the headlight will be held against the tension springs when the operation is completed.

Parking Lamps

REMOVAL & INSTALLATION

➧ SEE FIG. 56

1. Remove the socket from the rear of the lamp housing.
2. Replace the bulb.
3. Installation is the reverse of removal.

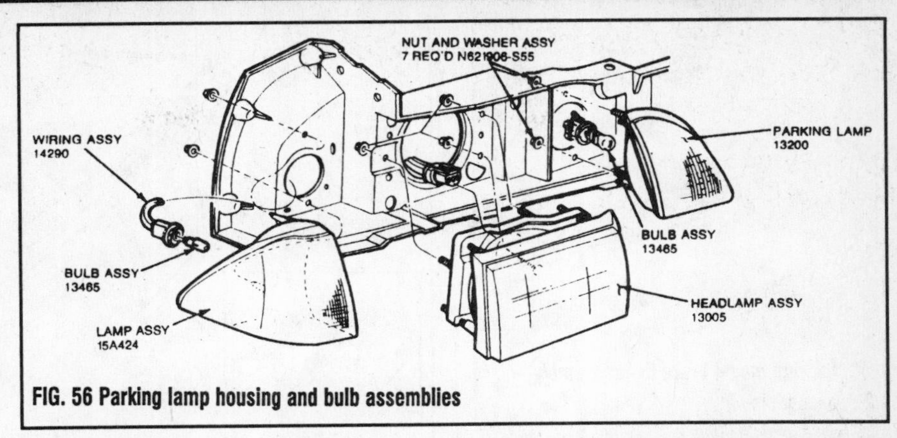

FIG. 56 Parking lamp housing and bulb assemblies

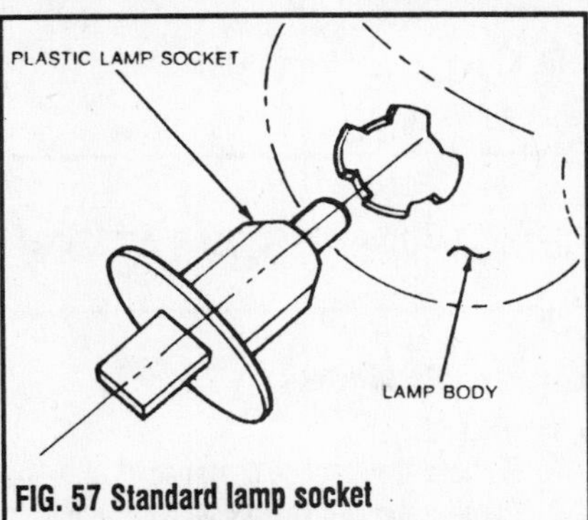

FIG. 57 Standard lamp socket

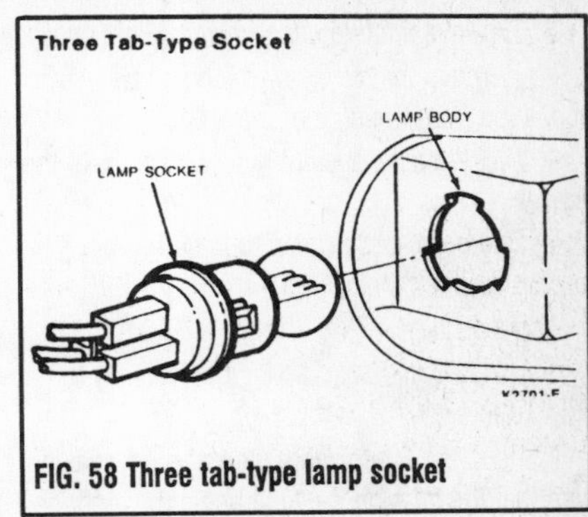

FIG. 58 Three tab-type lamp socket

Rear Lamps

REMOVAL & INSTALLATION

➧ SEE FIGS. 57–58

➡ **two types of rear socket are used; the standard plastic socket and the three tab-type socket.**

1. Remove the socket from the rear of the lamp housing by turning it counterclockwise to the stop.
2. Replace the bulb.
3. Installation is the reverse of removal. Index the smallest tab for locating the three locking tabs then press the socket into the lamp housing and rotate clockwise to the stop.

License Plate Lights

REMOVAL & INSTALLATION

➧ SEE FIG. 59

1. Open the trunk lid or vehicle hatch.
2. Remove the license lamp screws and lens.
3. Remove the socket and bulb, and twist the bulb out of the socket.
4. To install, reverse the removal procedure.

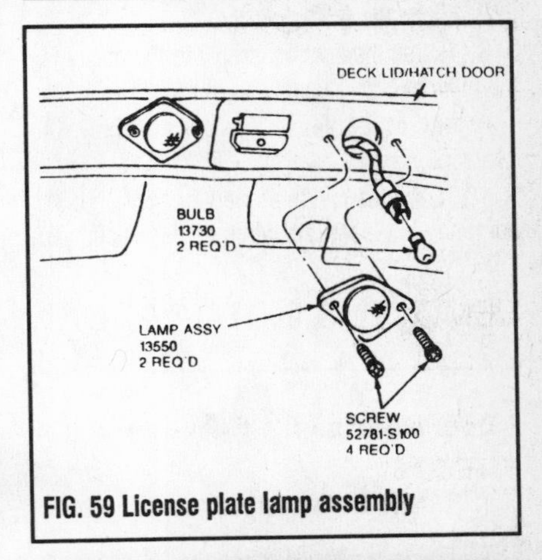

FIG. 59 License plate lamp assembly

High-Mount Brake Light

REMOVAL & INSTALLATION

2-Door Models

➧ SEE FIG. 60

1. Locate the wire to high-mount stoplight under package tray, from inside the luggage compartment. Pull the wire loose from plastic clip.

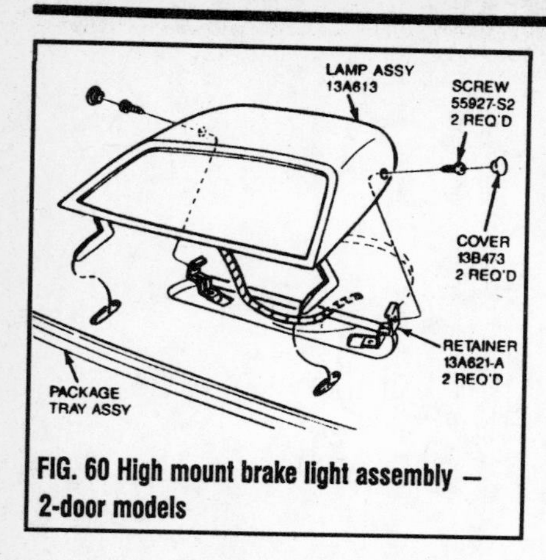

FIG. 60 High mount brake light assembly — 2-door models

2. Remove the two caps from the light cover.
3. Remove the screws, which can be accessed from the side of the light.
4. Pull the light assembly toward front of the vehicle.
5. The bulb sockets can then be removed by turning counterclockwise.

To Install:

6. Install sockets in the light by turning clockwise.
7. Push the light assembly rearward, making sure that the tabs on the light assembly engage into hole on the package tray.
8. Fasten the light assembly to the package tray with two screws.
9. Cover the holes on the light assembly with two caps.
10. From the inside of the luggage compartment insert the wires on the plastic clip.

REMOVAL & INSTALLATION

3-Door Models

➧ SEE FIG. 61

1. Remove the two screws from the light assembly.
2. The light can then be pulled rearward from spoiler.
3. Disengage the wiring harness "strain relief" clip by pulling straight out from light.
4. The bulb sockets can then be removed by gently twisting and pulling straight out of the light.

To Install:

5. Install the sockets in the light ensuring they are seated.
6. Insert the wiring harness "strain relief" clip.
7. Align the light mounting holes with U-nut in the spoiler.
8. Fasten the light to the spoiler with two screws.

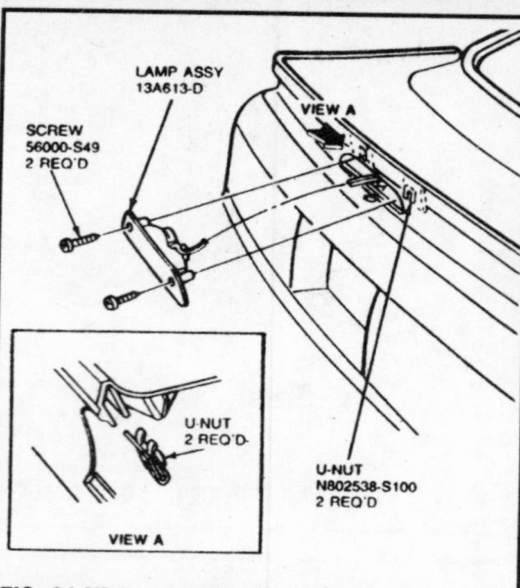

FIG. 61 High mount brake light assembly — 3-door models

REMOVAL & INSTALLATION

Convertible Models

➧ SEE FIG. 62

1. Remove the luggage crossbar.
2. Remove the two screws which can be accessed from the top of the light.
3. The light can then be lifted from the deck lid.
4. Disengage the wiring harness "strain relief" clip by pulling it straight out from the light.
5. The bulb sockets can then be remove by gently twisting and pulling it straight out of the light.

To Install:

6. Ensure the rubber seal is positioned properly on the light.
7. Install the sockets in the light ensuring they are seated.
8. Insert the wiring harness "strain relief" clip.
9. Align the light mounting holes with plastic nuts in the deck lid.
10. Fasten the light to the deck lid with two screws.
11. Install the luggage rack crossbar.

Fog Lights

REMOVAL & INSTALLATION

➧ SEE FIG. 63

➡ **The fog light assembly uses a halogen bulb and socket assembly.**

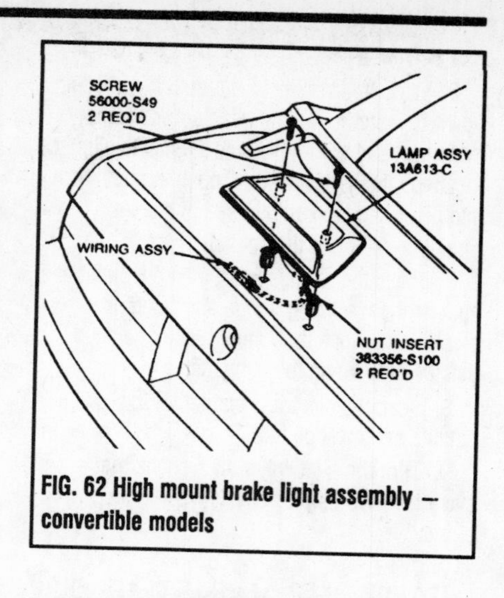

FIG. 62 High mount brake light assembly — convertible models

In the event of a bulb failure, perform the following procedures.

1. Lift up on the lens assembly retaining tab retaining the lens assembly to the light housing. Use caution to avoid dropping the lens.
2. Remove the lens assembly from the light housing and turn it to gain access to the rear of light body.
3. Disconnect the bulb wire lead from the pigtail connector.
4. Release the bulb socket retainer from the locking tab.
5. Remove the bulb and socket assembly from light body and pull the bulb directly out of the socket.

➡ **DO NOT touch the new bulb with bare hands. The stains will cause contamination of the quartz which may result in early failure of the light. Do not remove the protective plastic sleeve until the light is inserted into the socket. Ensure the circuit is not energized. If the quartz was inadvertently handled, it should be cleaned with a clean cloth moistened with alcohol before installation.**

To Install:

6. Insert the new bulb into the lens assembly and secure it with the retainer. Connect the wire lead to the pigtail connector.
7. Position the lens assembly right side up (as indicated on the lens) into the light housing.

➡ **There is a vertical tab on bottom of housing.**

8. Secure the lens assembly to the light housing with the retaining tab. Test the light for proper operation.

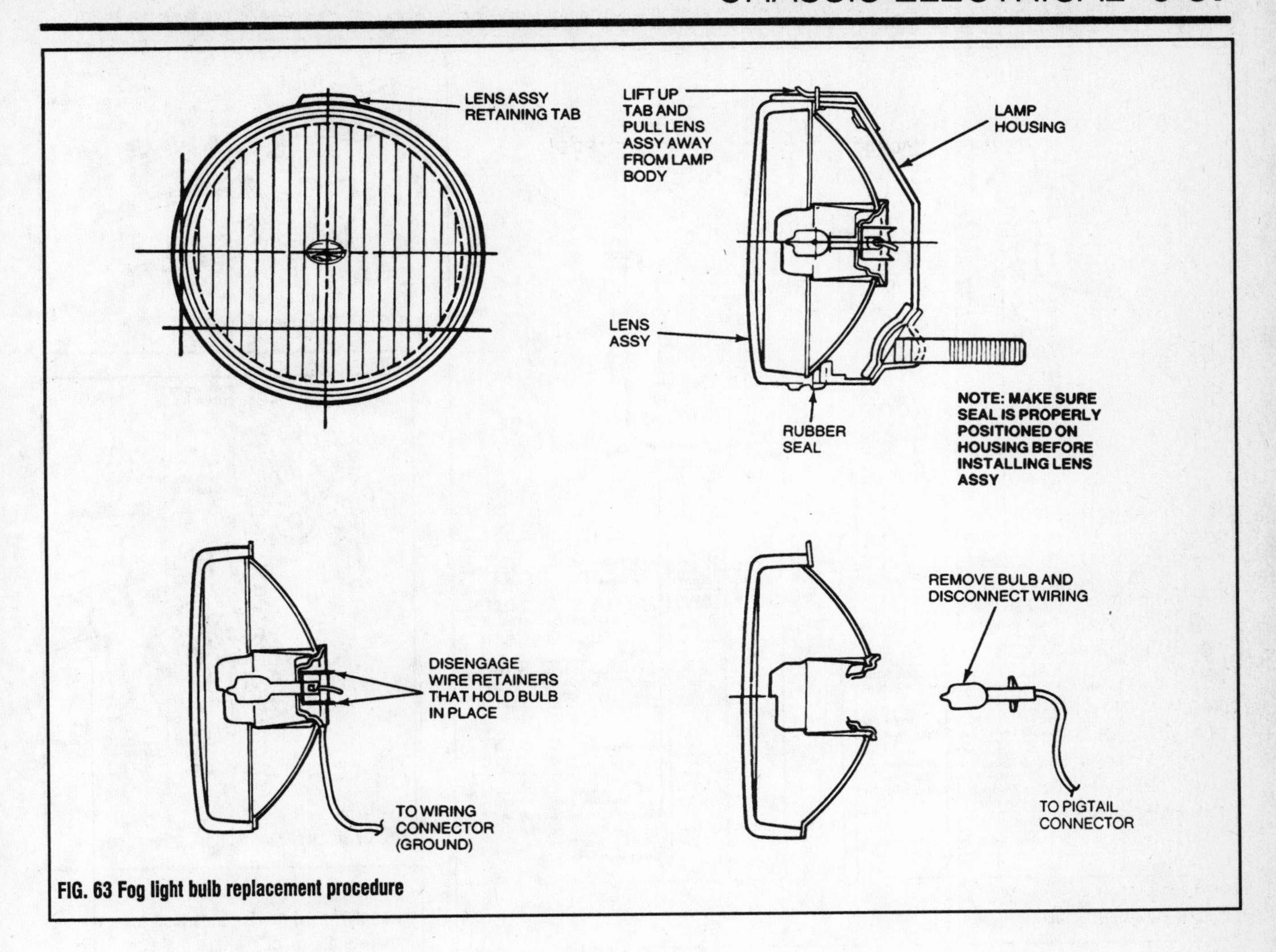

FIG. 63 Fog light bulb replacement procedure

CIRCUIT PROTECTION

Fuses

➧ SEE FIG. 64-64A

The fuse panel is located on the driver's side under the instrument panel.

Circuit Breakers

➧ SEE FIG. 65

Two circuits are protected by circuit breakers located in the fuse panel: the power windows (20 amp) and the windshield wiper circuit (8.25 amp). Three other circuits in the wiring harness are protected by circuit breakers: the headlights (22 amp) located in the headlight switch, the power seats and door locks (20 amp) located at the starter relay, and the convertible top (25 amp) located in the lower instrument panel reinforcement. The breakers are self-resetting.

Turn Signal and Hazard Flasher Locations

➧ SEE FIGS. 66–67

Two flasher units are used. One is used for the turn signal circuit and the other is for the hazard warning circuit. The turn signal flasher is mounted on the instrument panel reinforcement above the fuse panel. The hazard warning flasher is mounted in the fuse panel.

Fuse Link

➧ SEE FIG. 68

The fuse link is a short length of special, Hypalon (high temperature) insulated wire, integral with the engine compartment wiring harness and should not be confused with standard wire. It is several wire gauges smaller than the circuit which it protects. Under no circumstances should a fuse link replacement repair be made using a length of standard wire cut from bulk stock or from another wiring harness.

To repair any blown fuse link use the following procedure:

1. Determine which circuit is damaged, its location and the cause of the open fuse link. If the damaged fuse link is one of three fed by a

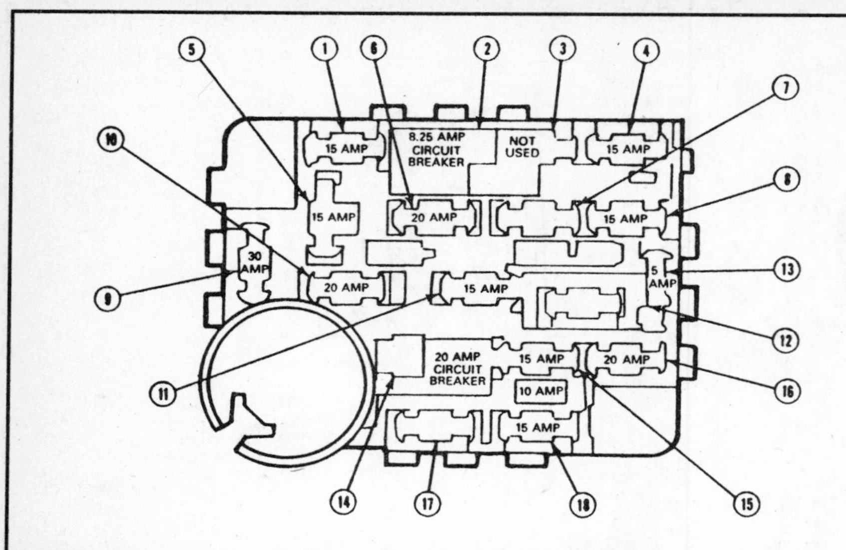

Cavity Number	Fuse Rating	Color	Circuit Protected
1	15 Amp	Lt. Blue	Stoplamps, Hazard Warning Lamps, Speed Control
2	8.25 Amp C. B.		Windshield Wiper, Windshield Washer Pump, Interval Wiper, Washer Fluid Level Indicator
3	Spare		Not Used
4	15 Amp	Red	Tail Lamps, Parking Lamps, Side Marker Lamps, Instrument Cluster Illumination Lamps, License Lamps
5	15 Amp	Lt. Blue	Turn Signal Lamps, Backup Lamps, Fluids Module, Heated Rear Window Relay
6	20 Amp	Yellow	A/C Clutch, Heated Rear Window Control, Luggage Compartment Lid Release, Speed Control Module, Clock/Radio Display, A/C Throttle Positioner, Day/Night Illumination Relay
7	Spare		Not Used
8	15 Amp	Lt. Blue	Courtesy Lamps, Key Warning Buzzer, Fuel Filler Door Release, Radio, Power Mirror

Cavity Number	Fuse Rating	Color	Circuit Protected
9	30 Amp	Lt. Green	Heater Blower Motor
10	20 Amp	Yellow	Flash-To-Pass, Low Oil Warning Relay
11	15 Amp	Lt. Blue	Radio, Tape Player, Premium Sound, Graphic Equalizer
12	Spare		Not Used
13	5 Amp	Tan	Instrument Cluster Illumination Lamps, Radio, Climate Control, Ash Receptacle Lamps, Floor "PRNDL" Lamp
14	20 Amp C. B.		Power Windows 15 Amp Fuse
15	15 Amp	Lt. Blue	Fog Lamps
16	20 Amp	Yellow	Horn, Cigar Lighter
17	Spare		Not Used
18	15 Amp	Lt. Blue	Warning Indicator Lamps, Throttle Solenoid Positioner, Low Fuel Module, Dual Timer Buzzer, Tachometer, Engine Idle Track Relay, Fluids Module/Display

FIG. 64 Fuse panel assembly and circuit locations

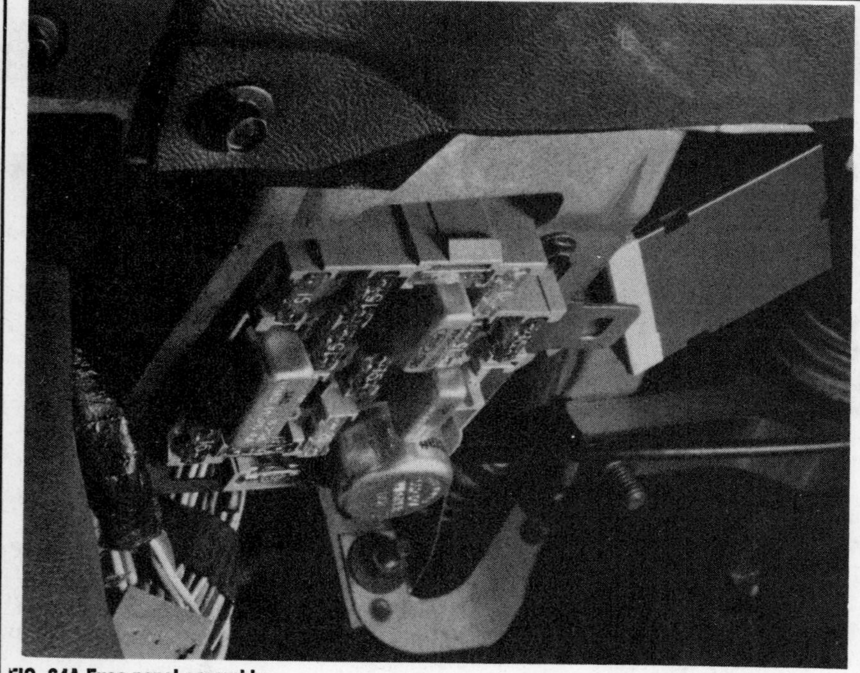

FIG. 64A Fuse panel assembly

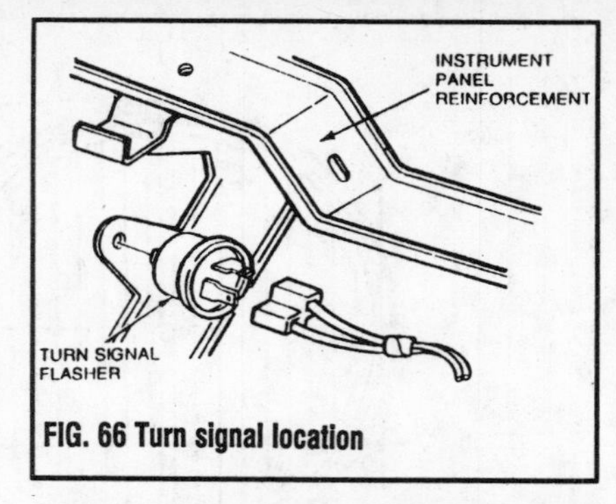

FIG. 66 Turn signal location

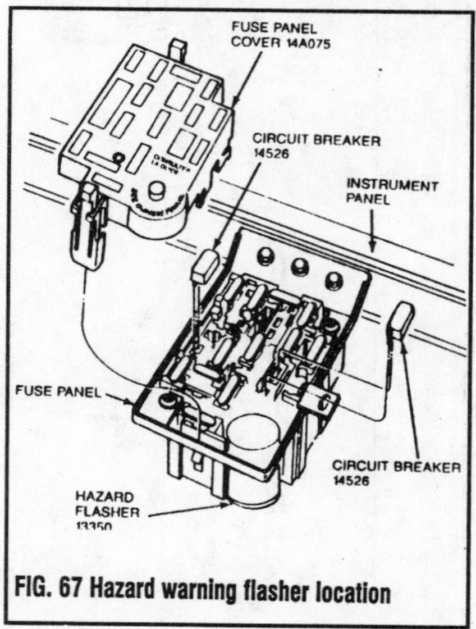

FIG. 67 Hazard warning flasher location

common No. 10 or 12 gauge feed wire, determine the specific affected circuit.

2. Disconnect the negative battery cable.
3. Cut the damaged fuse link from the wiring harness and discard it. If the fuse link is one of three circuits fed by a single feed wire, cut it out of the harness at each splice end and discard it.
4. Identify and procure the proper fuse link and butt connectors for attaching the fuse link to the harness.
5. To repair any fuse link in a 3-link group with one feed:
   a. After cutting the open link out of the harness, cut each of the remaining undamaged fuse links close to the feed wire weld.
   b. Strip approximately 1/2 in. (13mm) of insulation from the detached ends of the two good fuse links. Then insert two wire ends into one end of a butt connector and carefully push one stripped end of the replacement fuse link into the same end of the butt connector and crimp all three firmly together.

➡ **Care must be taken when fitting the three fuse links into the butt connector as the internal diameter**

Circuit	Circuit Protection and Rating	Location
Headlamps and High Beam Indicator	22 Amp CB	Integral with Lighting Switch
Heated Rear Window	16 GA Fuse Link	Engine Compartment
Power Windows, Power Seat, Power Door Locks	20 Amp. CB	Starter Motor Relay
Load Circuit	Fuse Link	In Harness
Engine Compartment Lamp	Fuse Link	In Harness
Convertible Top	25 Amp CB	Lower Instrument Panel-Reinforcement

FIG. 65 Fusible link and circuit breaker protected circuits

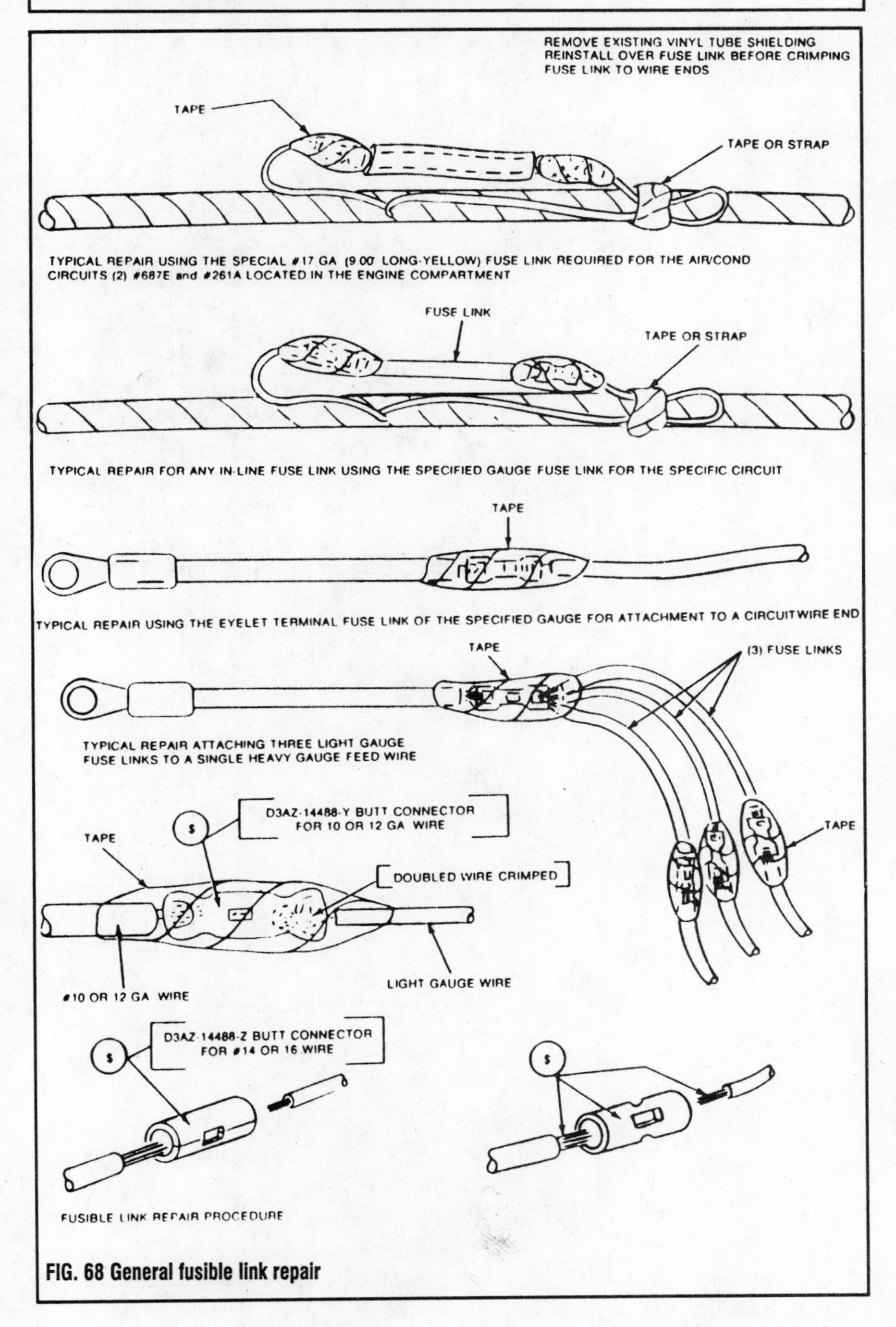

FIG. 68 General fusible link repair

is a snug it for three wires. Make sure to use a proper crimping tool. Pliers, side cutters, etc. will not apply the proper crimp to retain the wires and withstand a pull test.

c. After crimping the butt connector to the three fuse links, cut the weld portion from the feed wire and strip approximately 1/2 in. (13mm) of insulation from the cut end. Insert the stripped end into the open end of the butt connector and crimp very firmly.

d. To attach the remaining end of the replacement fuse link, strip approximately 1/2 in. (13mm) of insulation from the wire end of the circuit from which the blown fuse link was removed, and firmly crimp a butt connector or equivalent to the stripped wire. Then, insert the end of the replacement link into the other end of the butt connector and crimp firmly.

e. Using rosin core solder with a consistency of 60 percent tin and 40 percent lead, solder the connectors and the wires at the repairs and insulate with electrical tape.

6. To replace any fuse link on a single circuit in a harness, cut out the damaged portion, strip approximately 1/2 in. (13mm) of insulation from the two wire ends and attach the appropriate replacement fuse link to the stripped wire ends with two proper size butt connectors. Solder the connectors and wires and insulate the tape.

7. To repair any fuse link which has an eyelet terminal on one end such as the charging circuit, cut off the open fuse link behind the weld, strip approximately 1/2 in. (13mm) of insulation from the cut end and attach the appropriate new eyelet fuse link to the cut stripped wire with an appropriate size butt connector. Solder the connectors and wires at the repair and insulate with tape.

8. Connect the negative battery cable to the battery and test the system for proper operation.

➡ **Do not mistake a resistor wire for a fuse link. The resistor wire is generally longer and has print stating, "Resistor: don't cut or splice."**

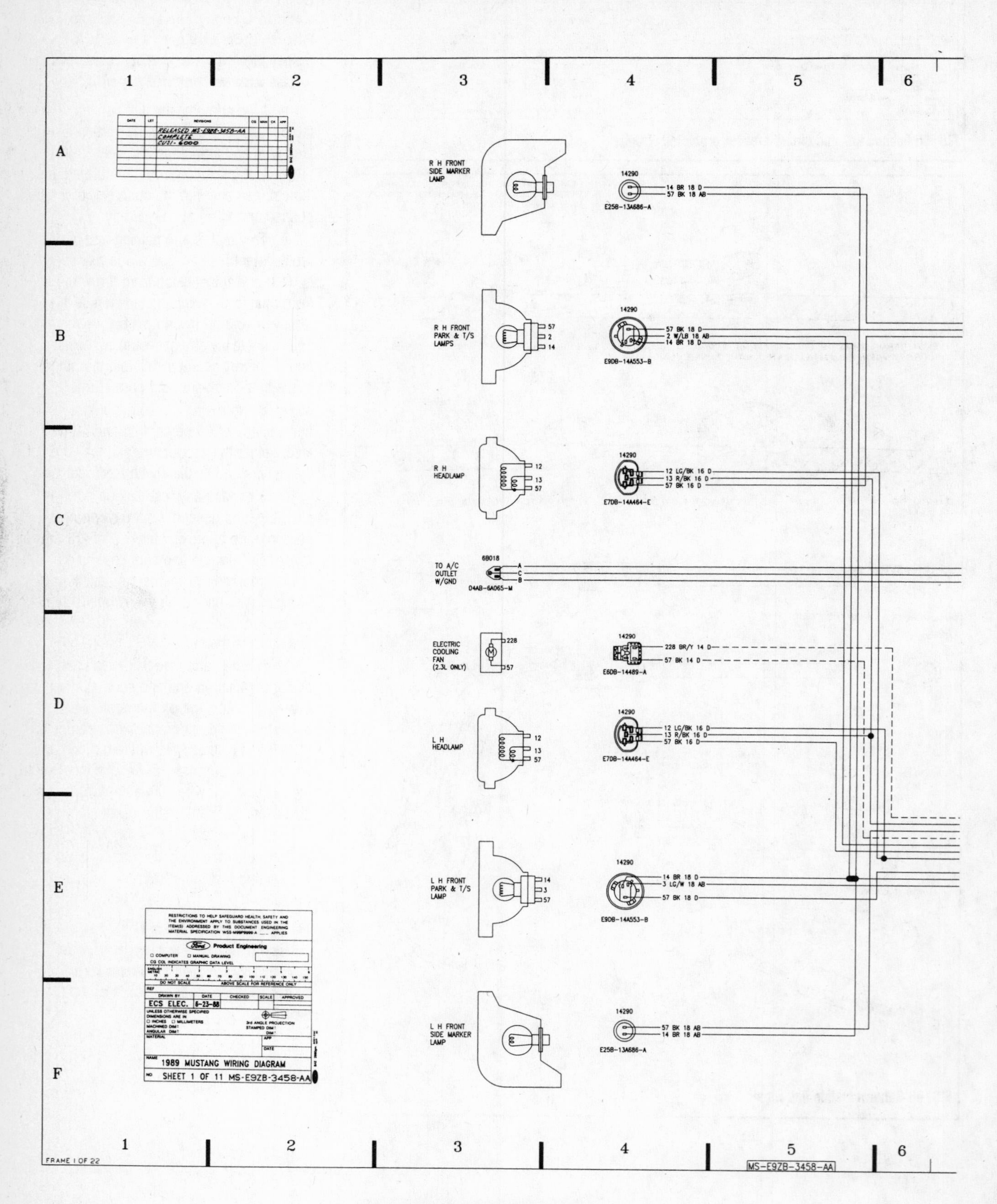

Wiring Diagram — 1989 Mustang 1 of 22

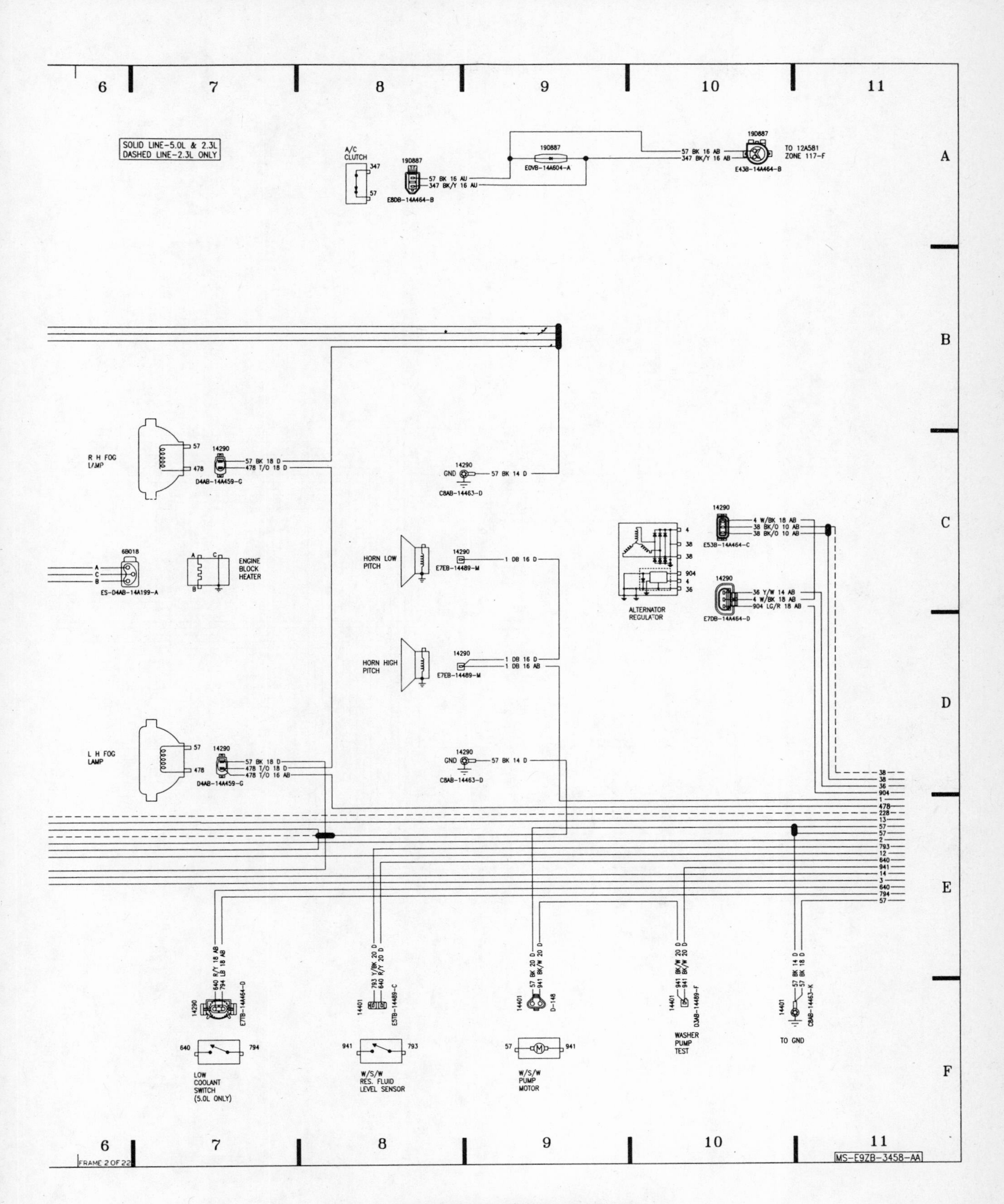

Wiring Diagram— 1989 Mustang 2 of 22

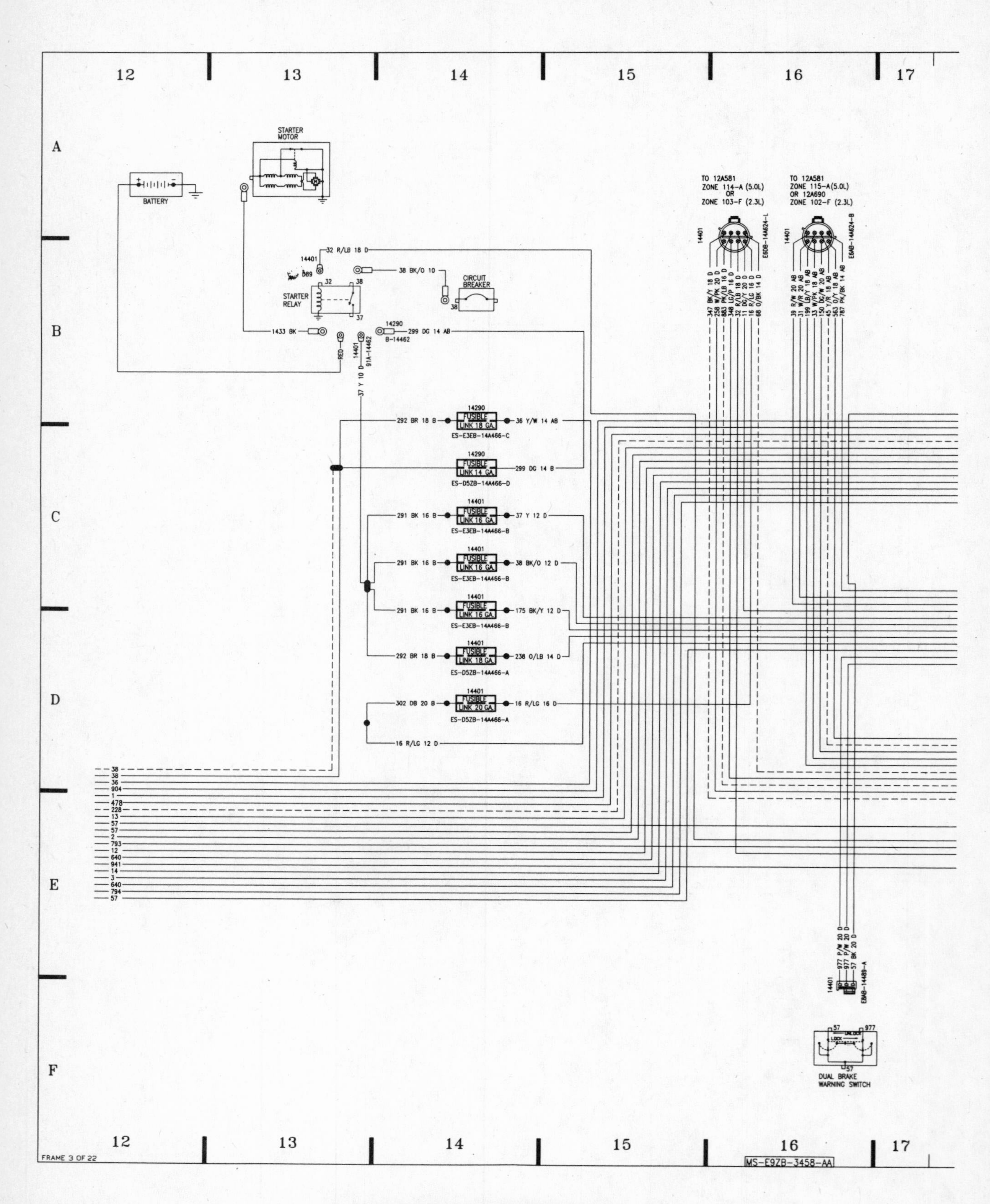
STARTER MOTOR
BATTERY
STARTER RELAY
CIRCUIT BREAKER
FUSIBLE LINK 18 GA
FUSIBLE LINK 14 GA
FUSIBLE LINK 16 GA
FUSIBLE LINK 20 GA
DUAL BRAKE WARNING SWITCH
FRAME 3 OF 22
MS-E9ZB-3458-AA

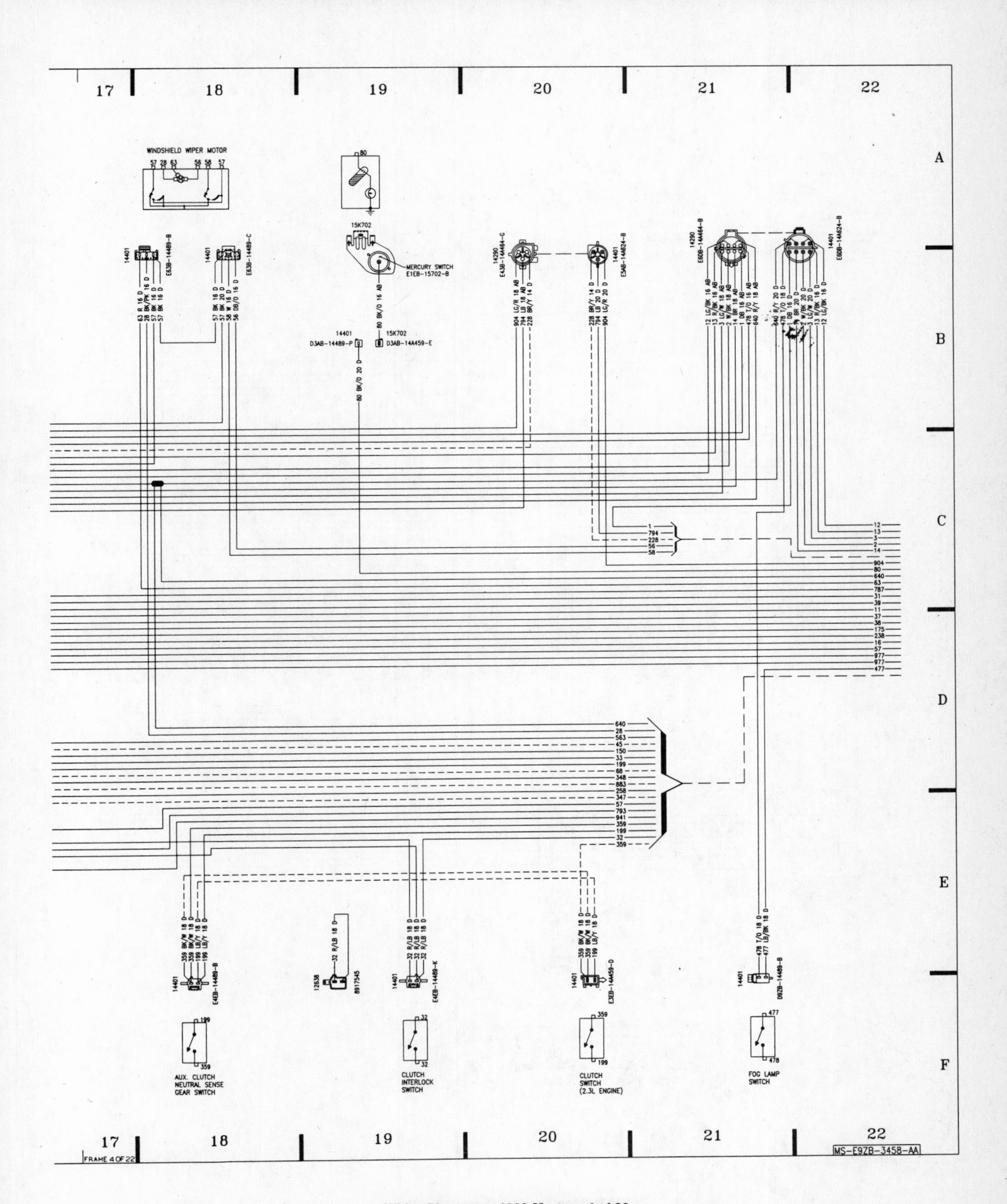

Wiring Diagram — 1989 Mustang 4 of 22

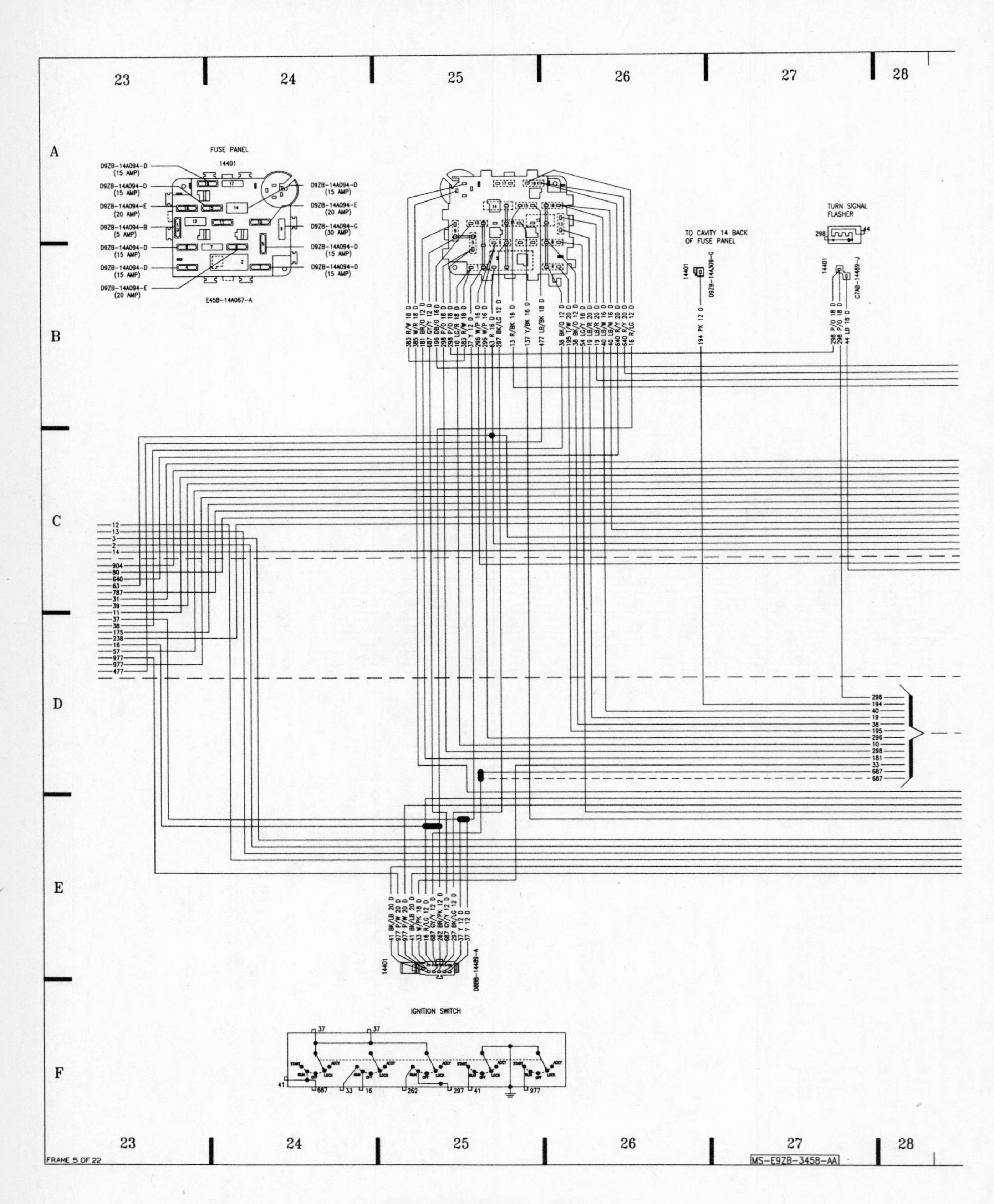
FUSE PANEL
14401
IGNITION SWITCH
TURN SIGNAL FLASHER
TO CAVITY 14 BACK OF FUSE PANEL
E45B-14A067-A
FRAME 5 OF 22
MS-E9ZB-3458-AA

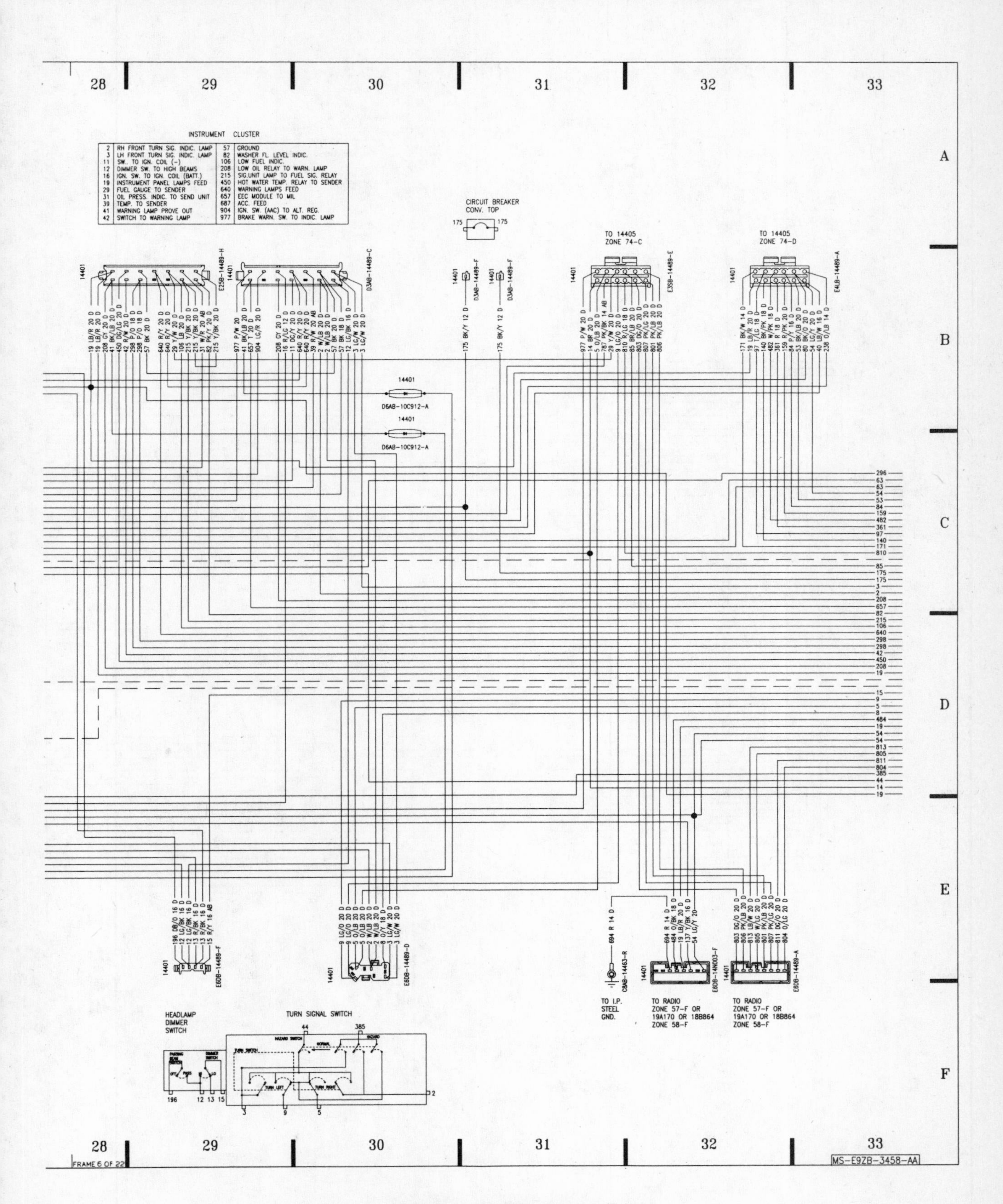
INSTRUMENT CLUSTER
2 RH FRONT TURN SIG. INDIC. LAMP
3 LH FRONT TURN SIG. INDIC. LAMP
11 SW.. TO IGN. COIL (-)
12 DIMMER SW. TO HIGH BEAMS
16 IGN. SW. TO IGN. COIL (BATT.)
19 INSTRUMENT PANEL LAMPS FEED
29 FUEL GAUGE TO SENDER
31 OIL PRESS. INDIC. TO SEND UNIT
39 TEMP. TO SENDER
41 WARNING LAMP PROVE OUT
42 SWITCH TO WARNING LAMP
57 GROUND
82 WASHER FL. LEVEL INDIC.
106 LOW FUEL INDIC.
208 LOW OIL RELAY TO WARN. LAMP
215 SIG.UNIT LAMP TO FUEL SIG. RELAY
450 HOT WATER TEMP. RELAY TO SENDER
640 WARNING LAMPS FEED
657 EEC MODULE TO MIL
687 ACC. FEED
904 IGN. SW. (AAC) TO ALT. REG.
977 BRAKE WARN. SW. TO INDIC. LAMP
CIRCUIT BREAKER CONV. TOP
TO 14405 ZONE 74-C
TO 14405 ZONE 74-D
D6AB-10C912-A
TO I.P. STEEL GND.
TO RADIO ZONE 57-F OR 19A170 OR 18B864 ZONE 58-F
HEADLAMP DIMMER SWITCH
TURN SIGNAL SWITCH
FRAME 6 OF 22
MS-E9ZB-3458-AA

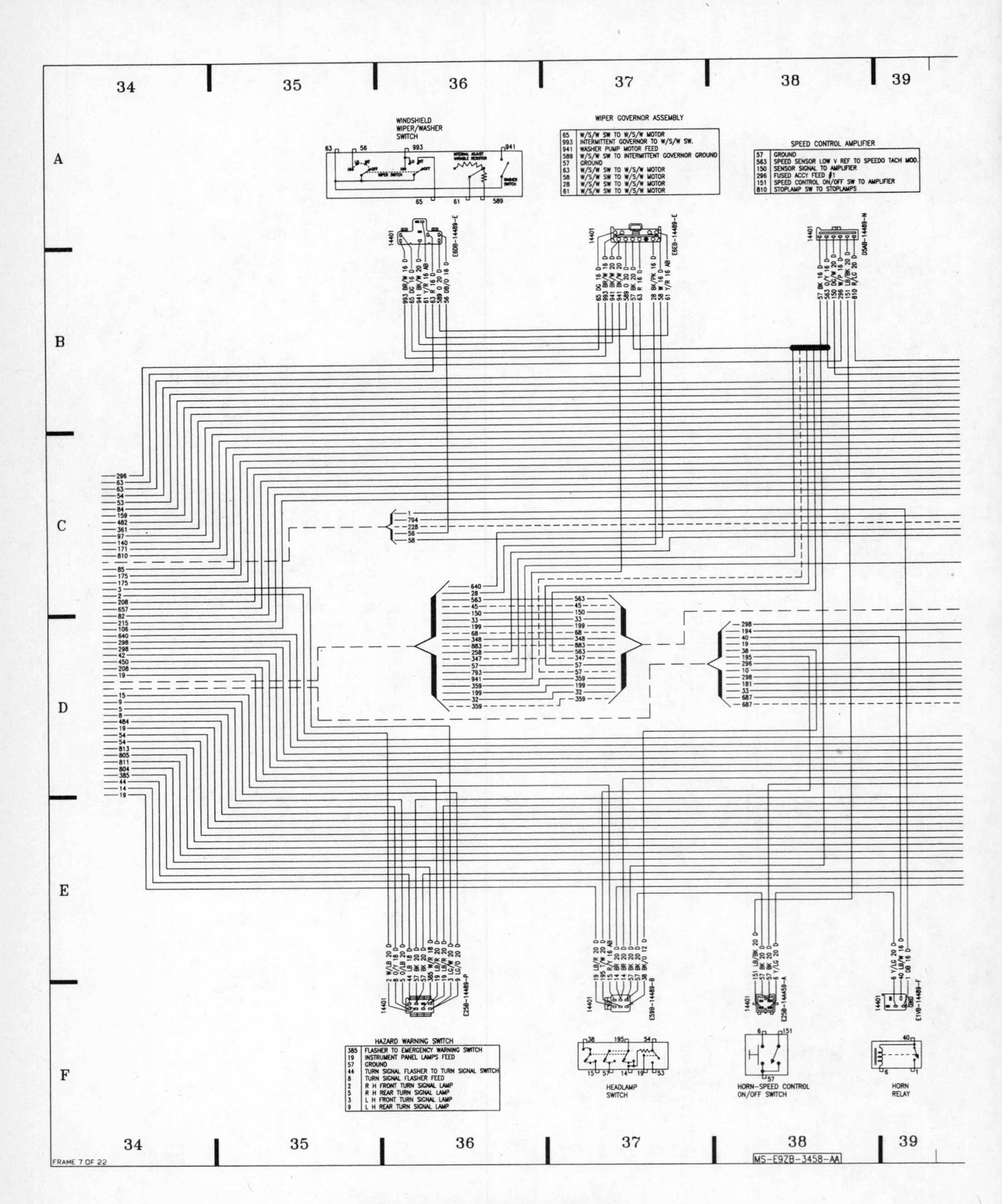

Wiring Diagram — 1989 Mustang 7 of 22

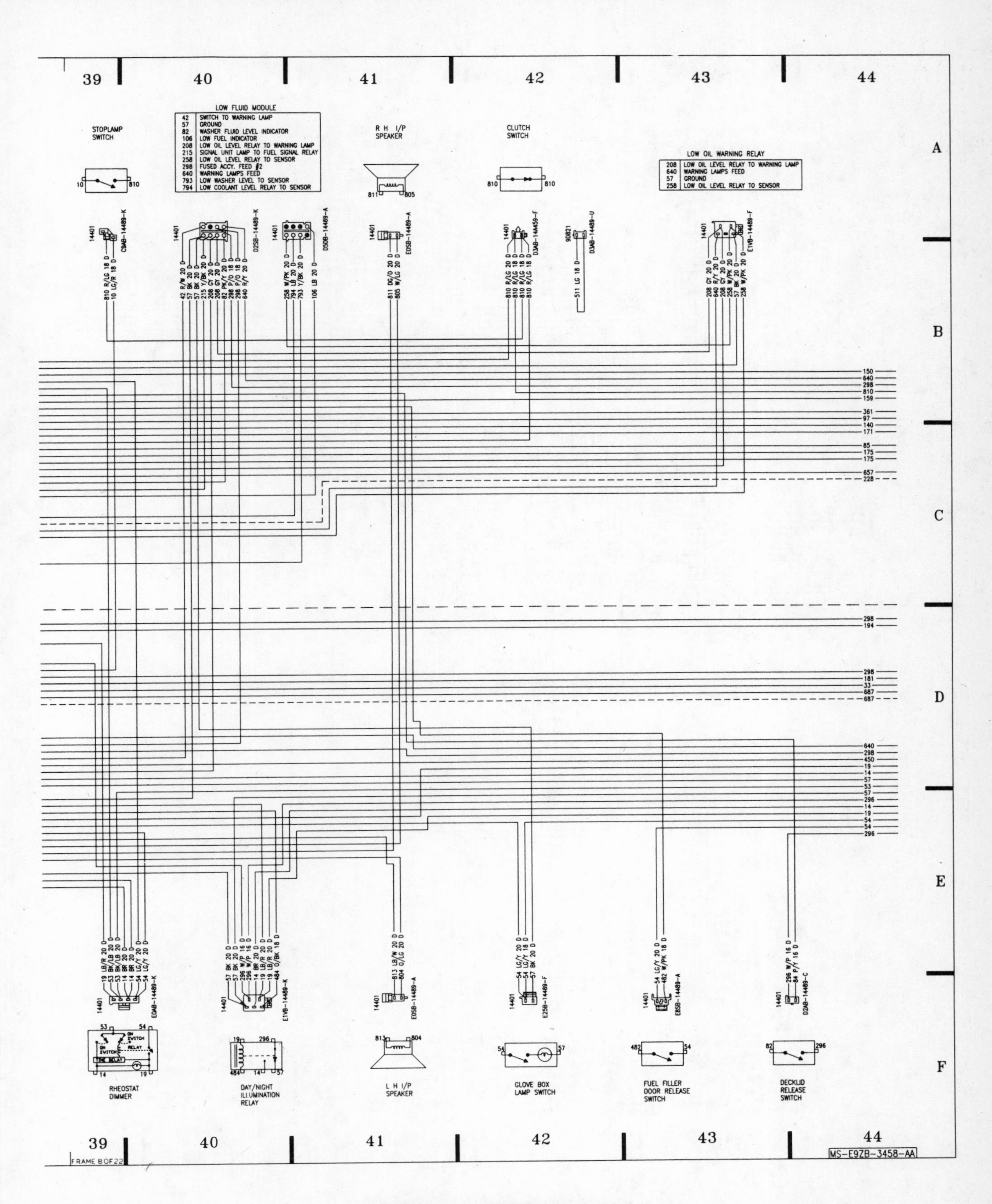
LOW FLUID MODULE
42 SWITCH TO WARNING LAMP
57 GROUND
82 WASHER FLUID LEVEL INDICATOR
106 LOW FUEL INDICATOR
208 LOW OIL LEVEL RELAY TO WARNING LAMP
215 SIGNAL UNIT LAMP TO FUEL SIGNAL RELAY
258 LOW OIL LEVEL RELAY TO SENSOR
298 FUSED ACCY. FEED #2
640 WARNING LAMPS FEED
793 LOW WASHER LEVEL TO SENSOR
794 LOW COOLANT LEVEL RELAY TO SENSOR
STOPLAMP SWITCH
R H I/P SPEAKER
CLUTCH SWITCH
LOW OIL WARNING RELAY
208 LOW OIL LEVEL RELAY TO WARNING LAMP
640 WARNING LAMPS FEED
57 GROUND
258 LOW OIL LEVEL RELAY TO SENSOR
RHEOSTAT DIMMER
DAY/NIGHT ILLUMINATION RELAY
L H I/P SPEAKER
GLOVE BOX LAMP SWITCH
FUEL FILLER DOOR RELEASE SWITCH
DECKLID RELEASE SWITCH
FRAME 8 OF 22
MS-E9ZB-3458-AA

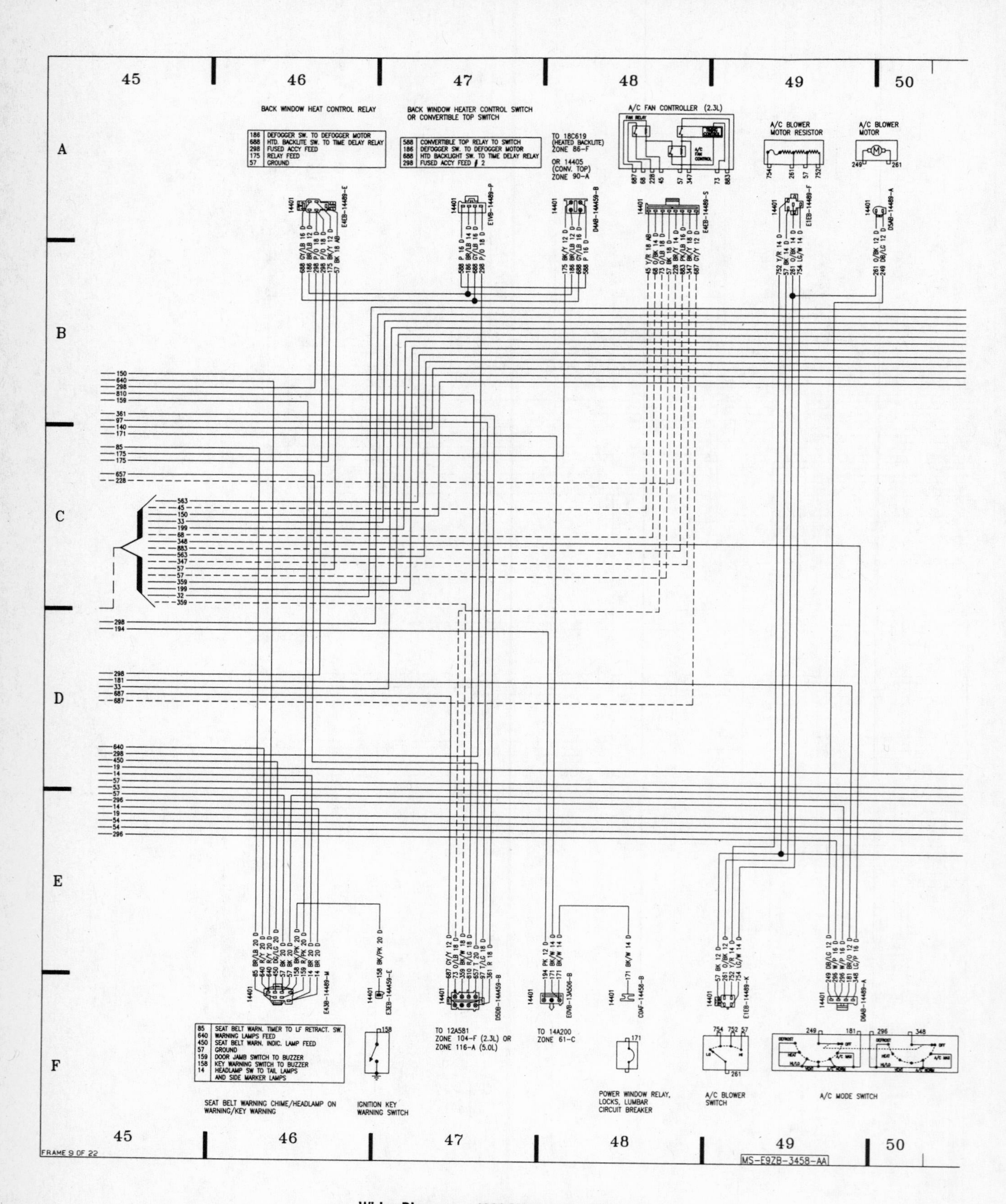

Wiring Diagram — 1989 Mustang 9 of 22

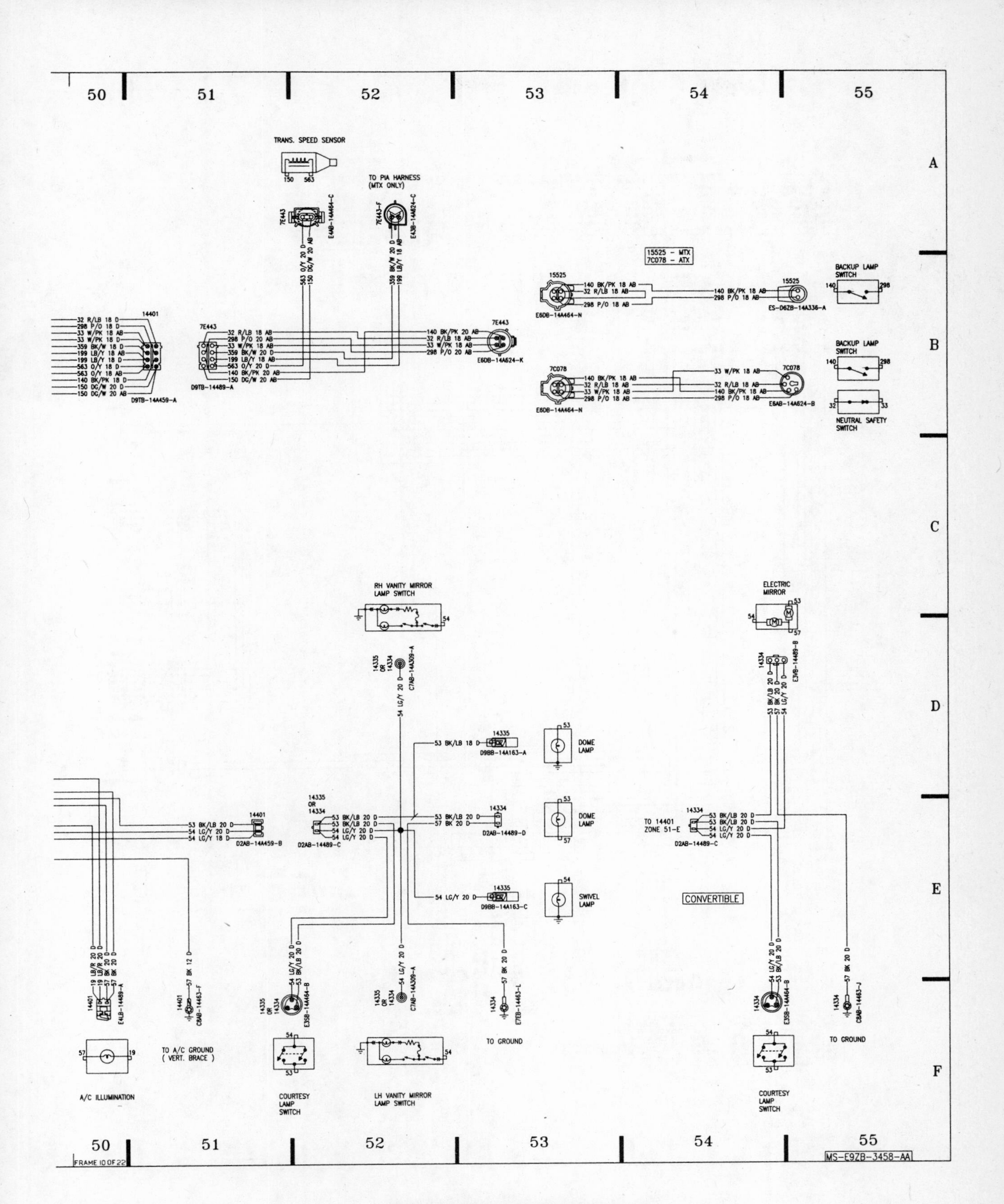
TRANS. SPEED SENSOR
TO PIA HARNESS (MTX ONLY)
15525 - MTX
7C078 - ATX
BACKUP LAMP SWITCH
NEUTRAL SAFETY SWITCH
RH VANITY MIRROR LAMP SWITCH
ELECTRIC MIRROR
DOME LAMP
SWIVEL LAMP
CONVERTIBLE
TO 14401 ZONE 51-E
A/C ILLUMINATION
TO A/C GROUND (VERT. BRACE)
COURTESY LAMP SWITCH
LH VANITY MIRROR LAMP SWITCH
TO GROUND
FRAME 10 OF 22
MS-E9ZB-3458-AA

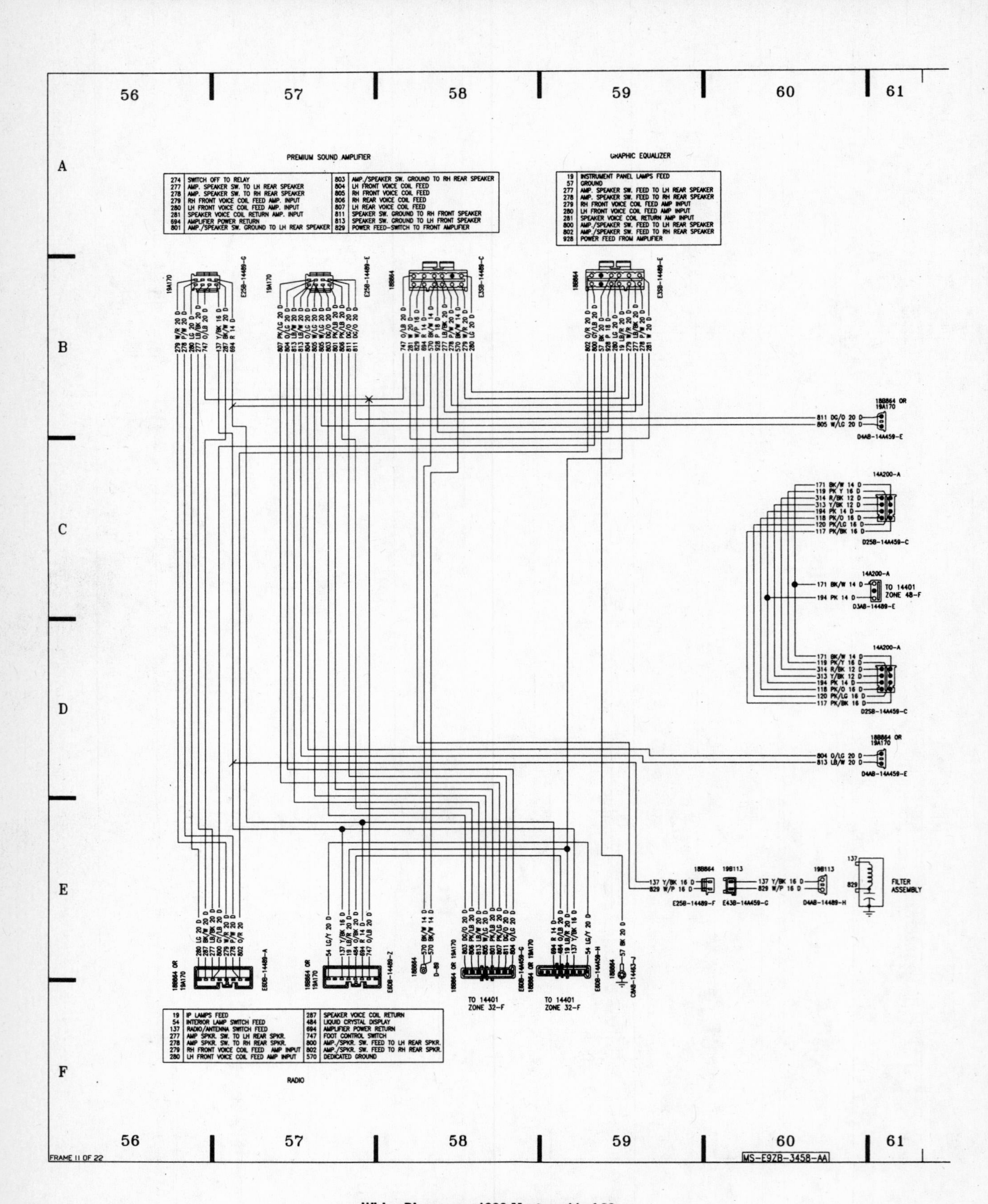
PREMIUM SOUND AMPLIFIER
274 SWITCH OFF TO RELAY
277 AMP. SPEAKER SW. TO LH REAR SPEAKER
278 AMP. SPEAKER SW. TO RH REAR SPEAKER
279 RH FRONT VOICE COIL FEED AMP. INPUT
280 LH FRONT VOICE COIL FEED AMP. INPUT
281 SPEAKER VOICE COIL RETURN AMP. INPUT
694 AMPLIFIER POWER RETURN
801 AMP./SPEAKER SW. GROUND TO LH REAR SPEAKER
803 AMP./SPEAKER SW. GROUND TO RH REAR SPEAKER
804 LH FRONT VOICE COIL FEED
805 RH FRONT VOICE COIL FEED
806 RH REAR VOICE COIL FEED
807 LH REAR VOICE COIL FEED
811 SPEAKER SW. GROUND TO RH FRONT SPEAKER
813 SPEAKER SW. GROUND TO LH FRONT SPEAKER
829 POWER FEED-SWITCH TO FRONT AMPLIFIER
GRAPHIC EQUALIZER
19 INSTRUMENT PANEL LAMPS FEED
57 GROUND
277 AMP. SPEAKER SW. FEED TO LH REAR SPEAKER
278 AMP. SPEAKER SW. FEED TO RH REAR SPEAKER
279 RH FRONT VOICE COIL FEED AMP INPUT
280 LH FRONT VOICE COIL FEED AMP INPUT
281 SPEAKER VOICE COIL RETURN AMP INPUT
800 AMP./SPEAKER SW. FEED TO LH REAR SPEAKER
802 AMP./SPEAKER SW. FEED TO RH REAR SPEAKER
928 POWER FEED FROM AMPLIFIER
RADIO
19 IP LAMPS FEED
54 INTERIOR LAMP SWITCH FEED
137 RADIO/ANTENNA SWITCH FEED
277 AMP SPKR. SW. TO LH REAR SPKR.
278 AMP SPKR. SW. TO RH REAR SPKR.
279 RH FRONT VOICE COIL FEED AMP INPUT
280 LH FRONT VOICE COIL FEED AMP INPUT
287 SPEAKER VOICE COIL RETURN
484 LIQUID CRYSTAL DISPLAY
694 AMPLIFIER POWER RETURN
747 FOOT CONTROL SWITCH
800 AMP./SPKR. SW. FEED TO LH REAR SPKR.
802 AMP./SPKR. SW. FEED TO RH REAR SPKR.
570 DEDICATED GROUND
FILTER ASSEMBLY
TO 14401 ZONE 48-F
TO 14401 ZONE 32-F
FRAME 11 OF 22
MS-E9ZB-3458-AA

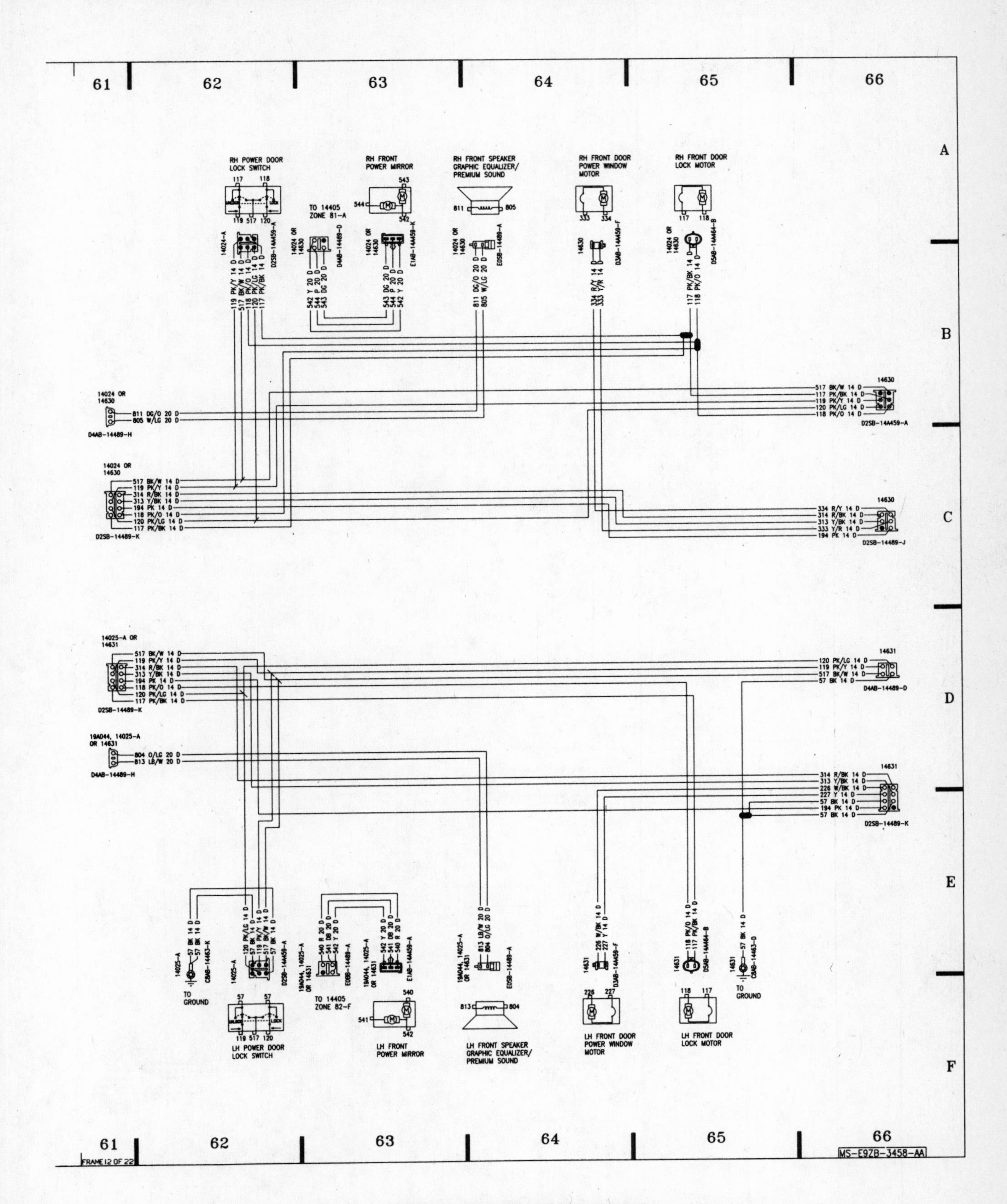

Wiring Diagram — 1989 Mustang 12 of 22

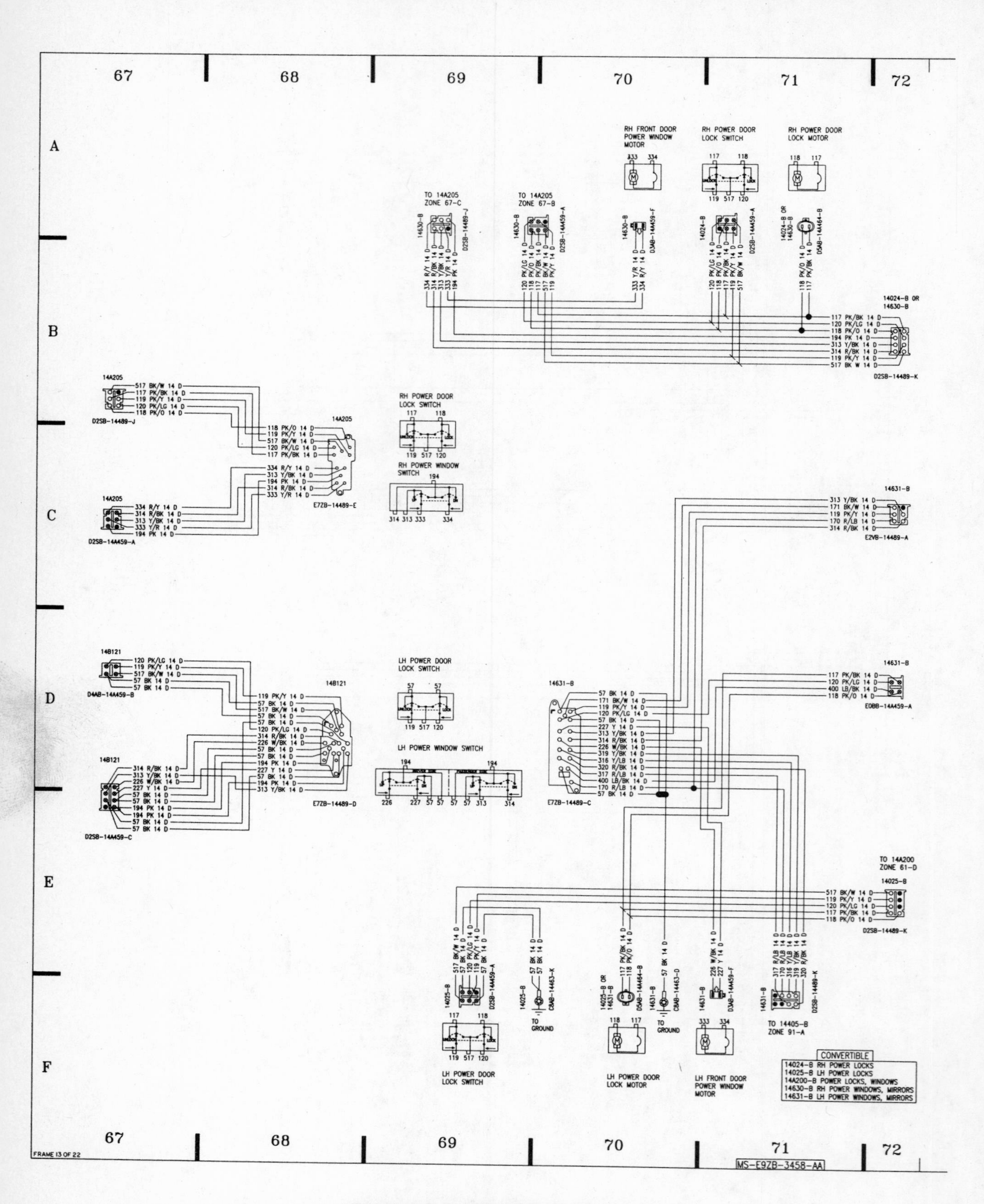

Wiring Diagram — 1989 Mustang 13 of 22

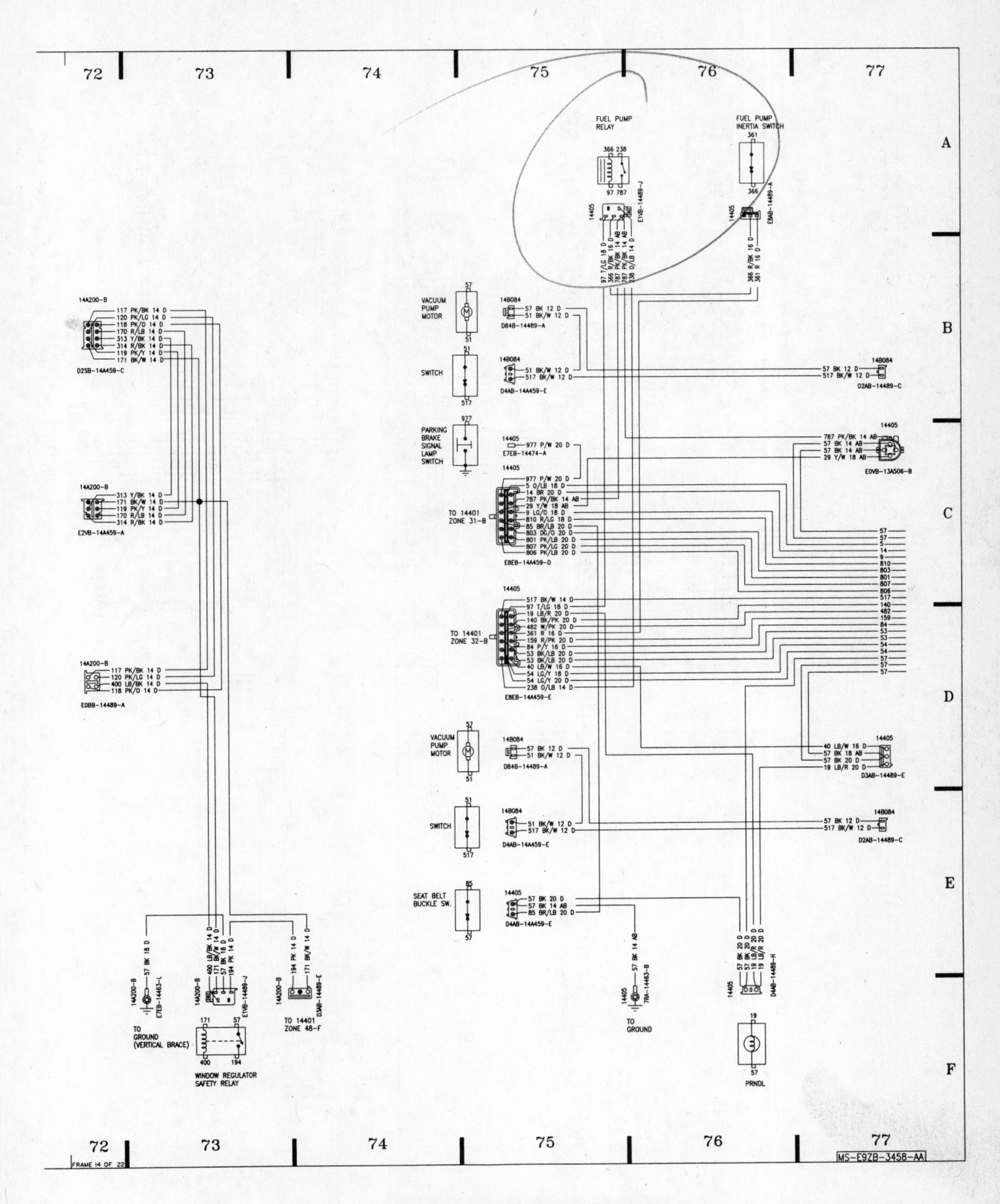

FUEL PUMP RELAY
FUEL PUMP INERTIA SWITCH
VACUUM PUMP MOTOR
SWITCH
PARKING BRAKE SIGNAL LAMP SWITCH
TO 14401 ZONE 31-B
TO 14401 ZONE 32-B
SEAT BELT BUCKLE SW.
TO GROUND (VERTICAL BRACE)
WINDOW REGULATOR SAFETY RELAY
TO 14401 ZONE 48-F
TO GROUND
PRNDL
FRAME 14 OF 22
MS-E9ZB-3458-AA

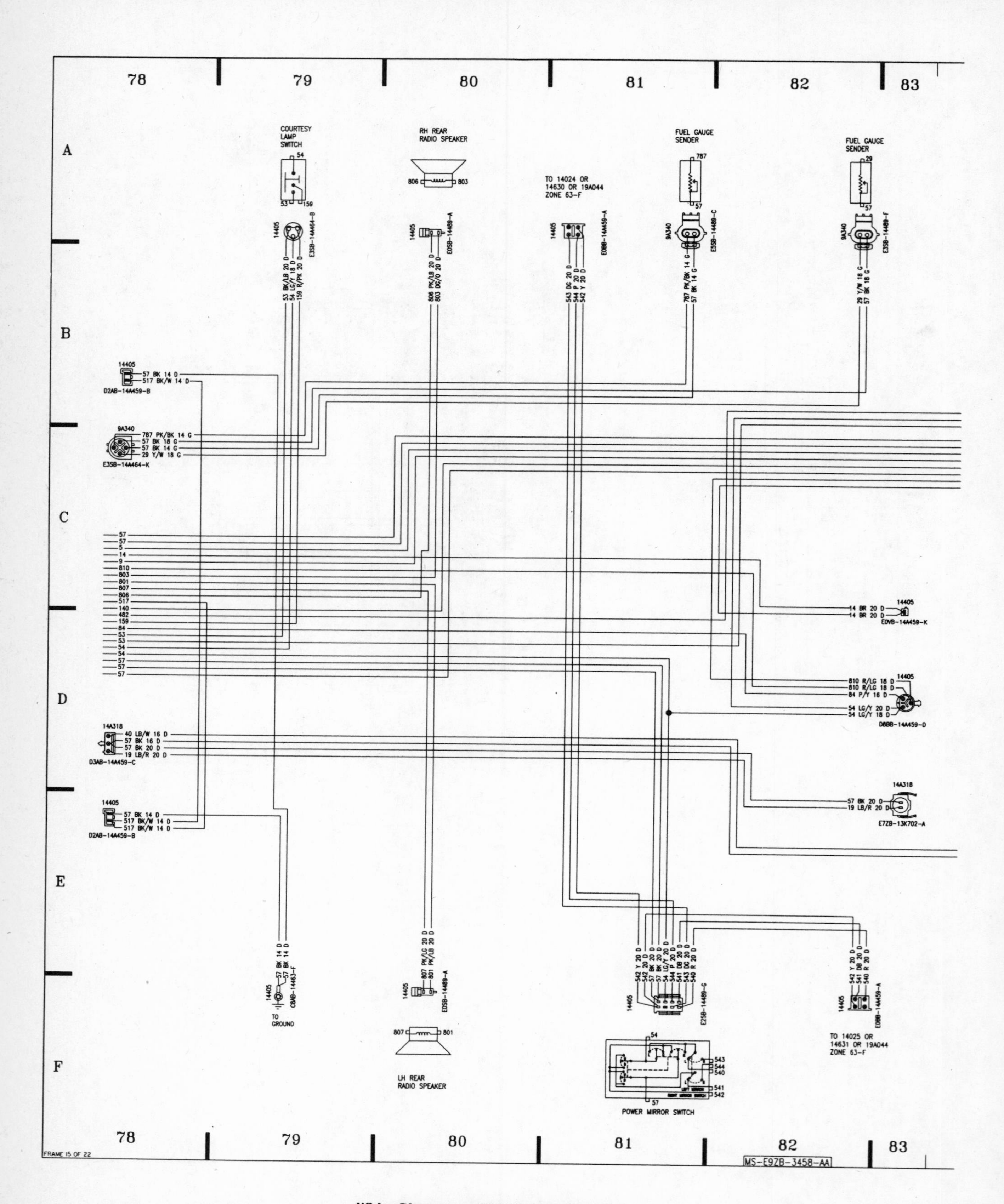

COURTESY LAMP SWITCH
RH REAR RADIO SPEAKER
TO 14024 OR 14630 OR 19A044 ZONE 63-F
FUEL GAUGE SENDER
FUEL GAUGE SENDER
LH REAR RADIO SPEAKER
POWER MIRROR SWITCH
TO 14025 OR 14631 OR 19A044 ZONE 63-F
TO GROUND
FRAME 15 OF 22
MS-E97B-3458-AA

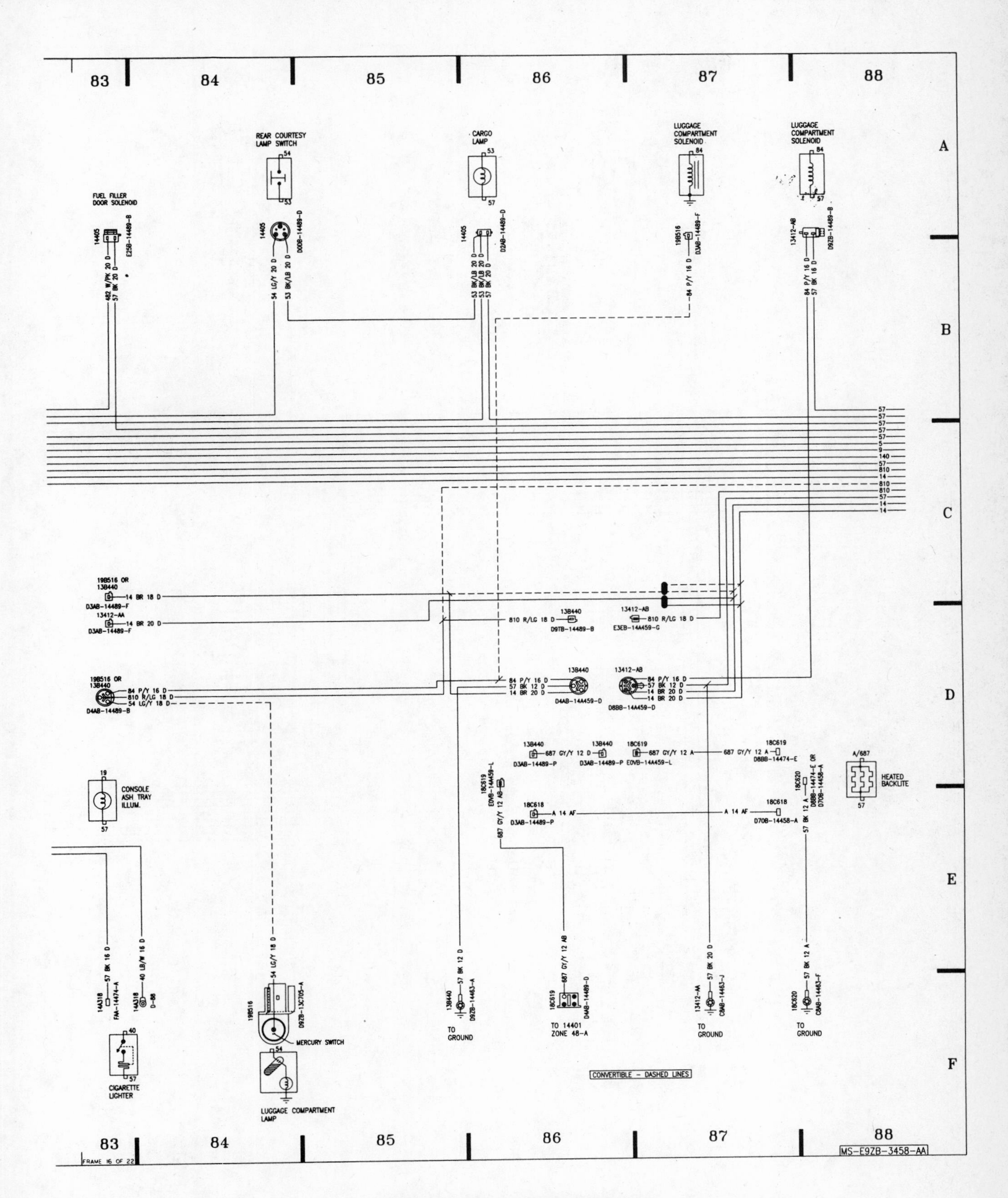
FUEL FILLER DOOR SOLENOID
REAR COURTESY LAMP SWITCH
CARGO LAMP
LUGGAGE COMPARTMENT SOLENOID
LUGGAGE COMPARTMENT SOLENOID
CONSOLE ASH TRAY ILLUM.
HEATED BACKLITE
MERCURY SWITCH
CIGARETTE LIGHTER
LUGGAGE COMPARTMENT LAMP
TO GROUND
TO 14401 ZONE 48-A
TO GROUND
TO GROUND
CONVERTIBLE - DASHED LINES
FRAME 16 OF 22
MS-E9ZB-3458-AA

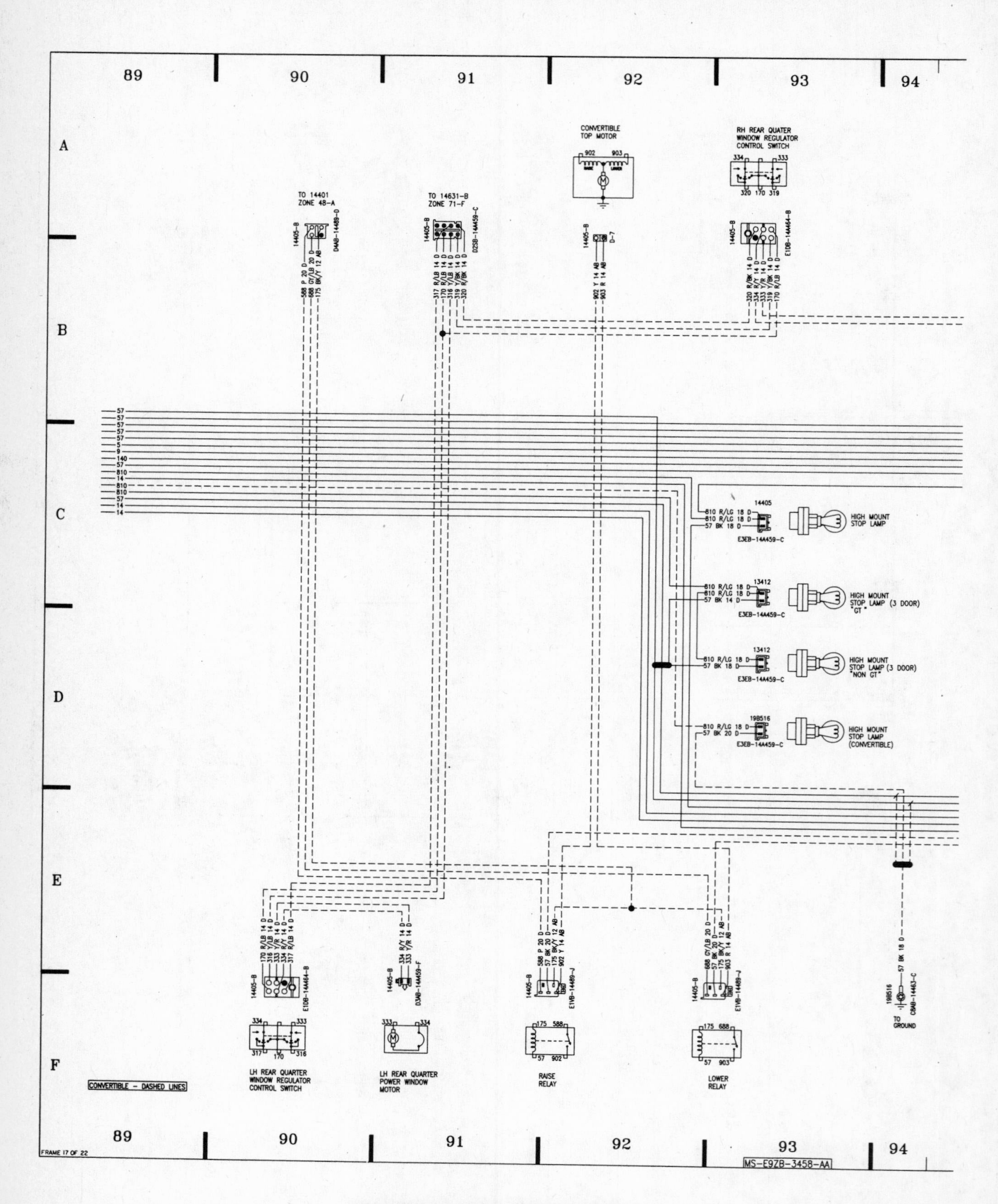

CONVERTIBLE TOP MOTOR
RH REAR QUATER WINDOW REGULATOR CONTROL SWITCH
TO 14401 ZONE 48-A
TO 14631-B ZONE 71-F
HIGH MOUNT STOP LAMP
HIGH MOUNT STOP LAMP (3 DOOR) "GT"
HIGH MOUNT STOP LAMP (3 DOOR) "NON GT"
HIGH MOUNT STOP LAMP (CONVERTIBLE)
LH REAR QUARTER WINDOW REGULATOR CONTROL SWITCH
LH REAR QUARTER POWER WINDOW MOTOR
RAISE RELAY
LOWER RELAY
TO GROUND
CONVERTIBLE - DASHED LINES
FRAME 17 OF 22
MS-E9ZB-3458-AA

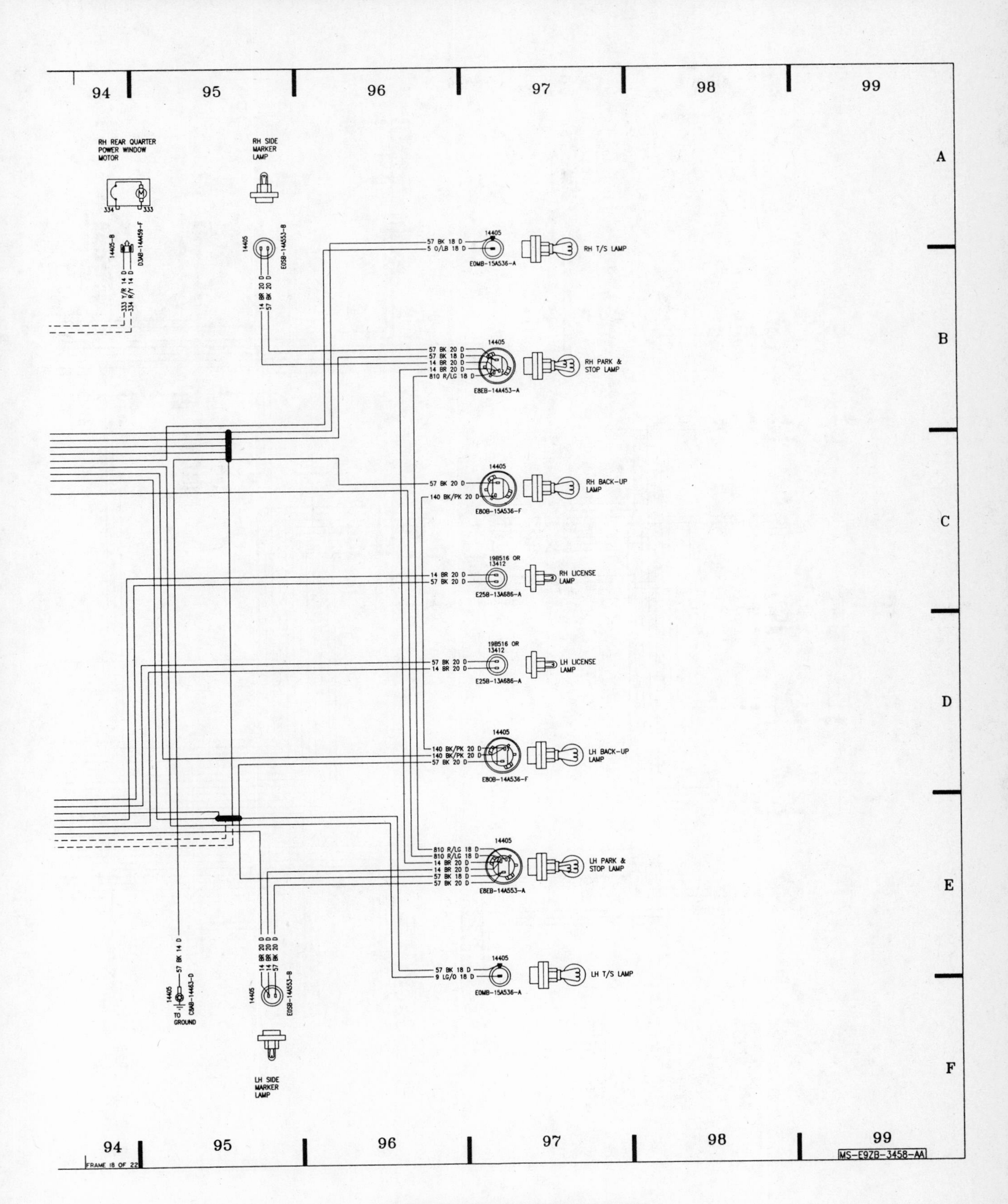
RH REAR QUARTER POWER WINDOW MOTOR
RH SIDE MARKER LAMP
RH T/S LAMP
RH PARK & STOP LAMP
RH BACK-UP LAMP
RH LICENSE LAMP
LH LICENSE LAMP
LH BACK-UP LAMP
LH PARK & STOP LAMP
LH T/S LAMP
LH SIDE MARKER LAMP
TO GROUND
FRAME 18 OF 22
MS-E9ZB-3458-AA

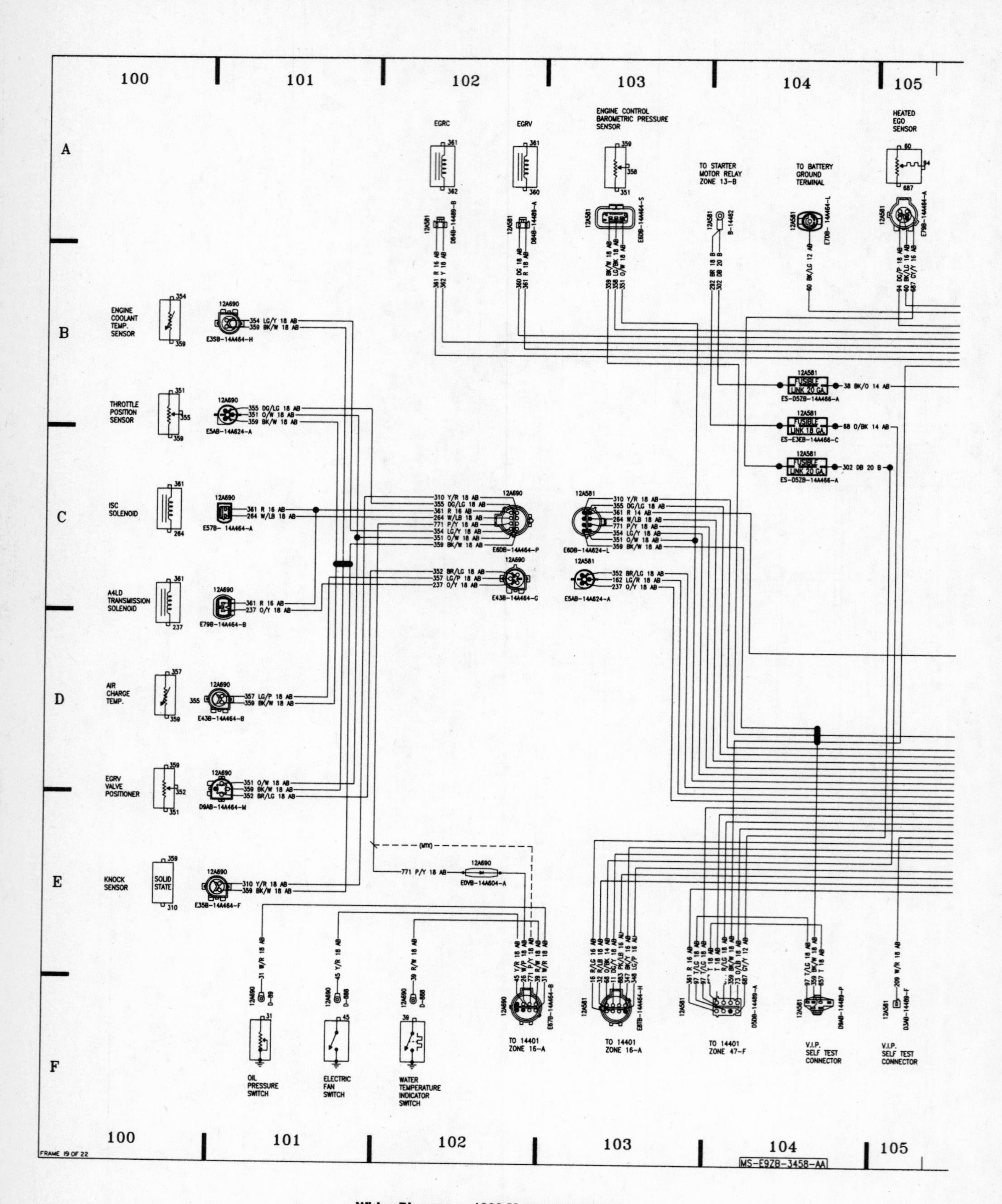

Wiring Diagram — 1989 Mustang 19 of 22

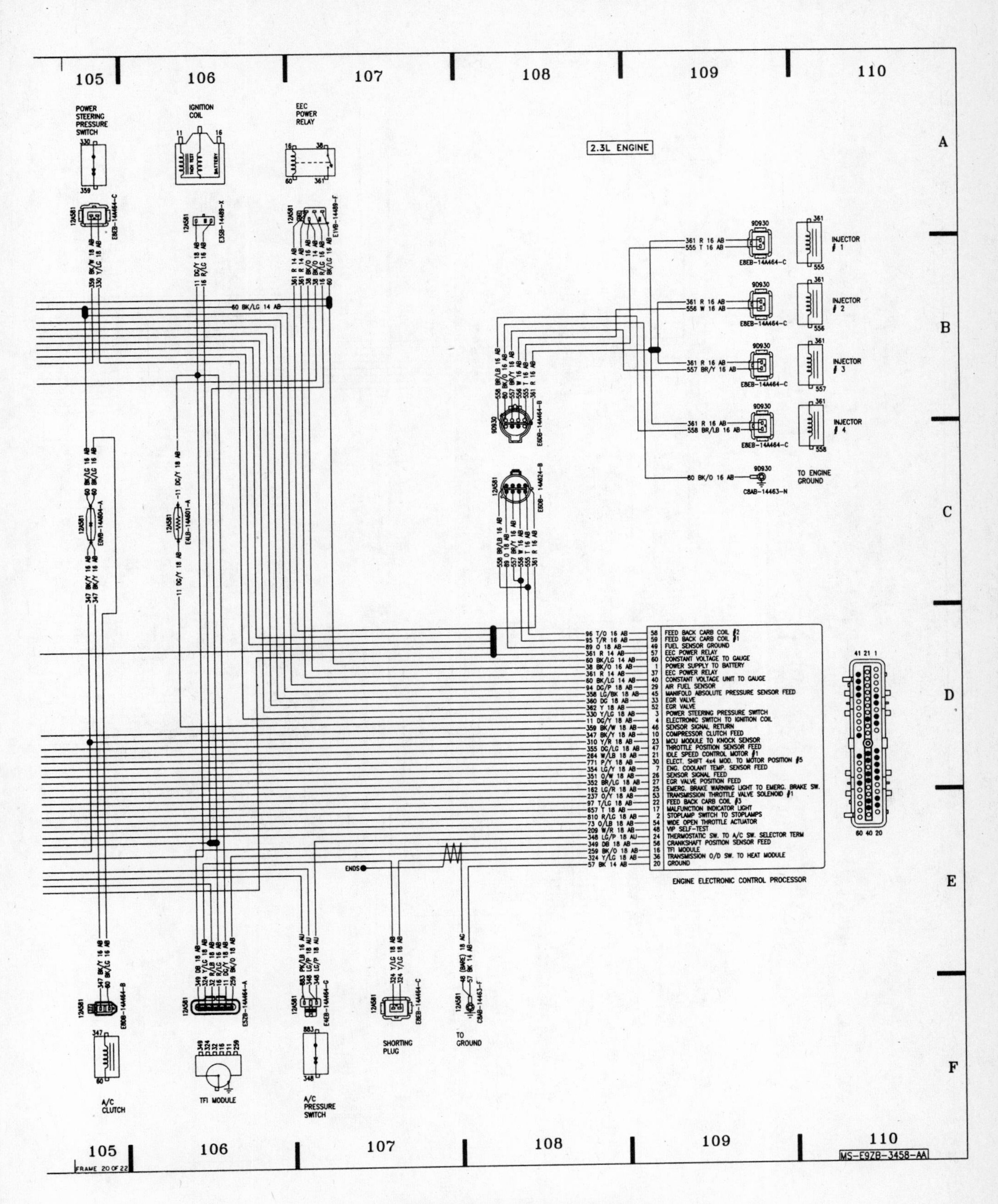
105
106
107
108
109
110
POWER STEERING PRESSURE SWITCH
IGNITION COIL
EEC POWER RELAY
2.3L ENGINE
INJECTOR #1
INJECTOR #2
INJECTOR #3
INJECTOR #4
TO ENGINE GROUND
A
B
C
D
E
F
96 T/O 16 AB 58 FEED BACK CARB COIL #2
95 T/R 16 AB 59 FEED BACK CARB COIL #1
89 O 18 AB 49 FUEL SENSOR GROUND
361 R 14 AB 57 EEC POWER RELAY
60 BK/LG 14 AB 60 CONSTANT VOLTAGE TO GAUGE
38 BK/O 16 AB 1 POWER SUPPLY TO BATTERY
361 R 14 AB 37 EEC POWER RELAY
60 BK/LG 14 AB 40 CONSTANT VOLTAGE UNIT TO GAUGE
94 DG/P 18 AB 29 AIR FUEL SENSOR
358 LG/BK 18 AB 45 MANIFOLD ABSOLUTE PRESSURE SENSOR FEED
360 DG 18 AB 33 EGR VALVE
362 Y 18 AB 52 EGR VALVE
330 Y/LG 18 AB 3 POWER STEERING PRESSURE SWITCH
11 DG/Y 18 AB 4 ELECTRONIC SWITCH TO IGNITION COIL
359 BK/W 18 AB 46 SENSOR SIGNAL RETURN
347 BK/Y 18 AB 10 COMPRESSOR CLUTCH FEED
310 Y/R 18 AB 23 MCU MODULE TO KNOCK SENSOR
355 DG/LG 18 AB 47 THROTTLE POSITION SENSOR FEED
264 W/LB 18 AB 21 IDLE SPEED CONTROL MOTOR #1
771 P/Y 18 AB 30 ELECT. SHIFT 4x4 MOD. TO MOTOR POSITION #5
354 LG/Y 18 AB 7 ENG. COOLANT TEMP. SENSOR FEED
351 O/W 18 AB 26 SENSOR SIGNAL FEED
352 BR/LG 18 AB 27 EGR VALVE POSITION FEED
162 LG/R 18 AB 25 EMERG. BRAKE WARNING LIGHT TO EMERG. BRAKE SW.
237 O/Y 18 AB 53 TRANSMISSION THROTTLE VALVE SOLENOID #1
97 T/LG 18 AB 22 FEED BACK CARB COIL #3
657 T 18 AB 17 MALFUNCTION INDICATOR LIGHT
810 R/LG 18 AB 2 STOPLAMP SWITCH TO STOPLAMPS
73 O/LB 18 AB 54 WIDE OPEN THROTTLE ACTUATOR
209 W/R 18 AB 48 VIP SELF-TEST
348 LG/P 18 AU 24 THERMOSTATIC SW. TO A/C SW. SELECTOR TERM
349 DB 18 AB 56 CRANKSHAFT POSITION SENSOR FEED
259 BK/O 18 AB 16 TFI MODULE
324 Y/LG 18 AB 36 TRANSMISSION O/D SW. TO HEAT MODULE
57 BK 14 AB 20 GROUND
ENGINE ELECTRONIC CONTROL PROCESSOR
ENDS
SHORTING PLUG
TO GROUND
A/C CLUTCH
TFI MODULE
A/C PRESSURE SWITCH
FRAME 20 OF 22
MS-E9ZB-3458-AA

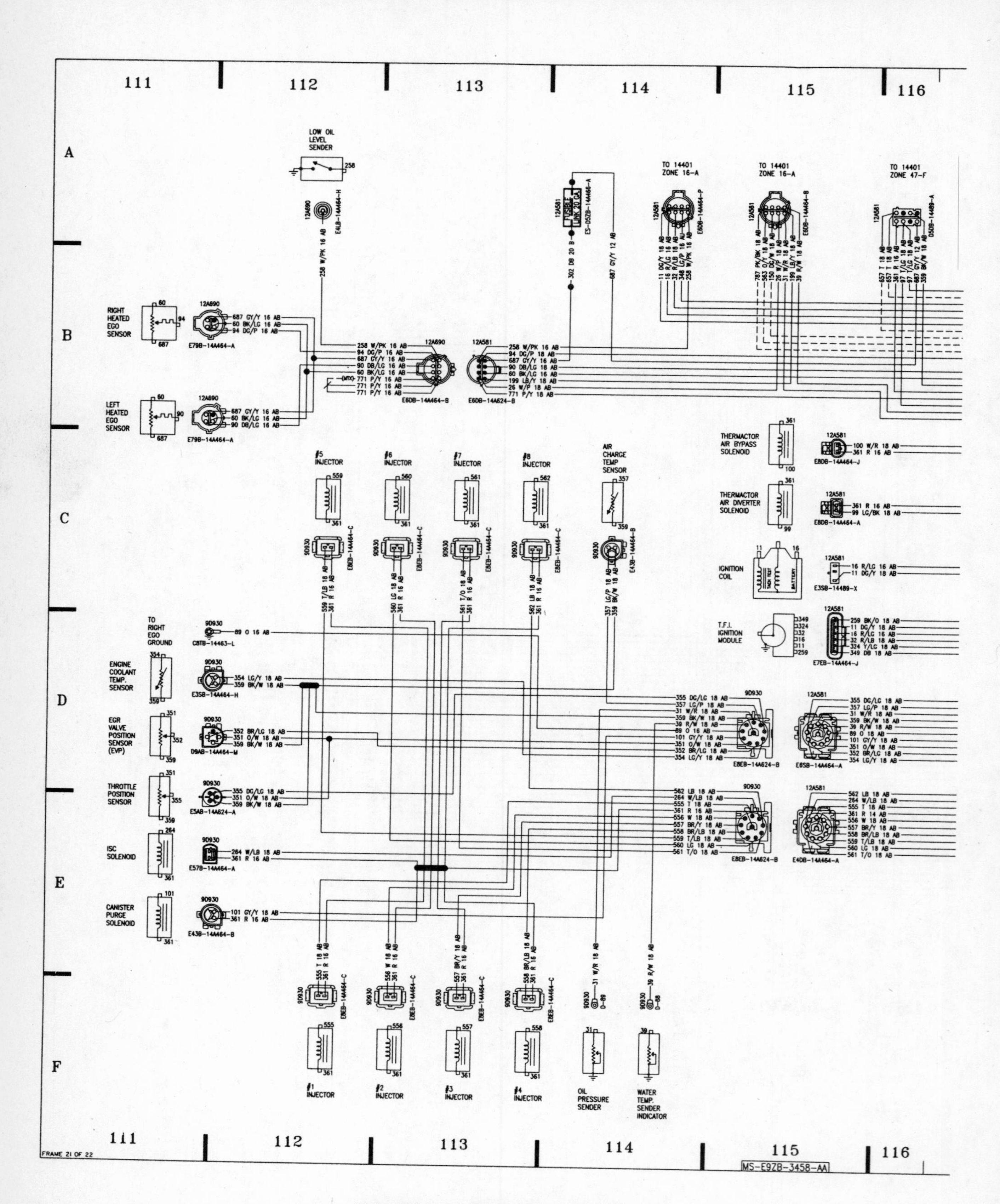
LOW OIL LEVEL SENDER
RIGHT HEATED EGO SENSOR
LEFT HEATED EGO SENSOR
#5 INJECTOR
#6 INJECTOR
#7 INJECTOR
#8 INJECTOR
AIR CHARGE TEMP SENSOR
THERMACTOR AIR BYPASS SOLENOID
THERMACTOR AIR DIVERTER SOLENOID
IGNITION COIL
T.F.I. IGNITION MODULE
TO RIGHT EGO GROUND
ENGINE COOLANT TEMP. SENSOR
EGR VALVE POSITION SENSOR (EVP)
THROTTLE POSITION SENSOR
ISC SOLENOID
CANISTER PURGE SOLENOID
#1 INJECTOR
#2 INJECTOR
#3 INJECTOR
#4 INJECTOR
OIL PRESSURE SENDER
WATER TEMP. SENDER INDICATOR
TO 14401 ZONE 16-A
TO 14401 ZONE 47-F
FRAME 21 OF 22
MS-E9ZB-3458-AA

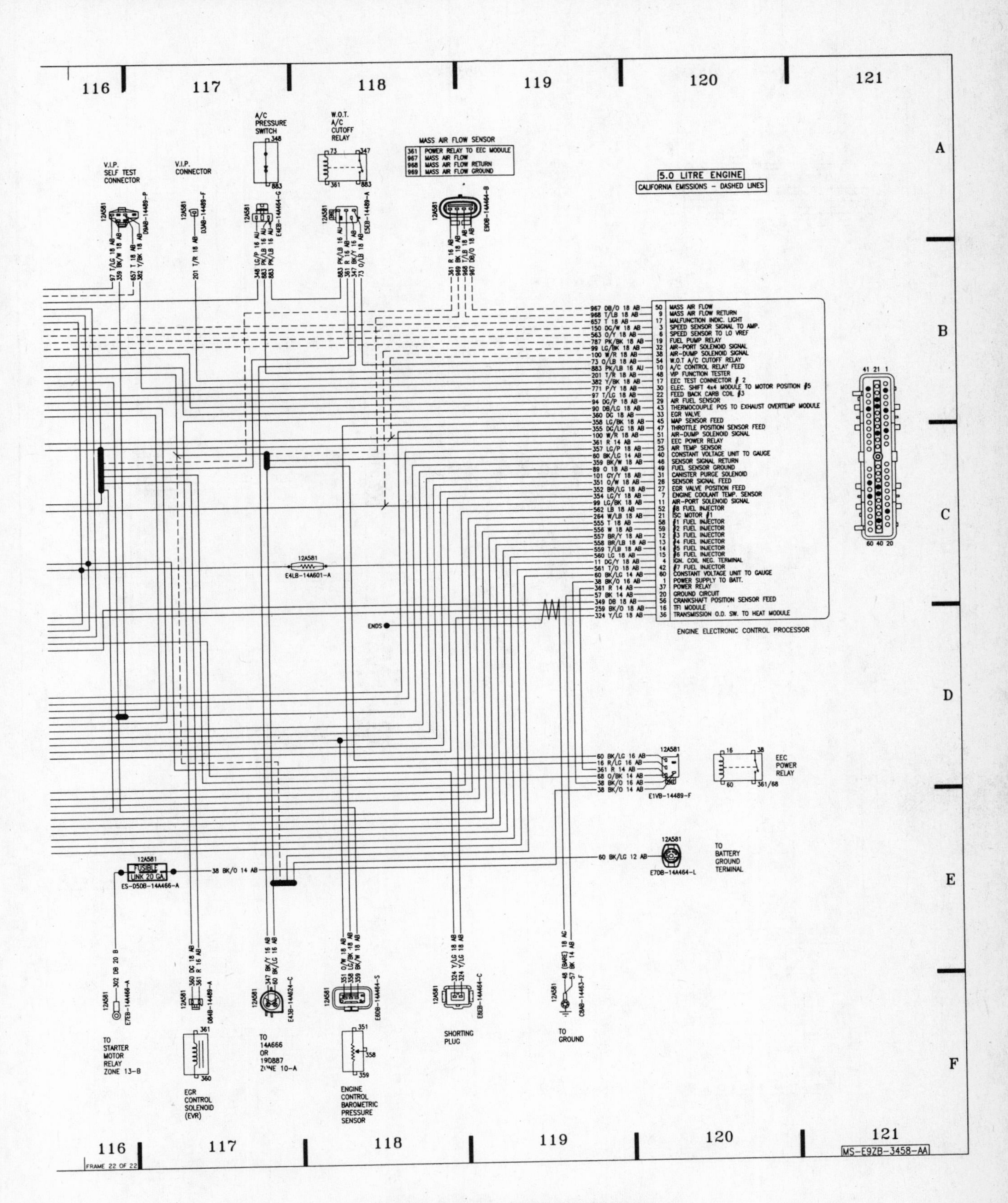
116
117
118
119
120
121
A/C PRESSURE SWITCH
W.O.T. A/C CUTOFF RELAY
MASS AIR FLOW SENSOR
361 POWER RELAY TO EEC MODULE
967 MASS AIR FLOW
968 MASS AIR FLOW RETURN
969 MASS AIR FLOW GROUND
V.I.P. SELF TEST CONNECTOR
V.I.P. CONNECTOR
5.0 LITRE ENGINE
CALIFORNIA EMISSIONS - DASHED LINES
967 DB/O 18 AB — 50 MASS AIR FLOW
968 T/LB 18 AB — 9 MASS AIR FLOW RETURN
657 T 18 AB — 17 MALFUNCTION INDIC. LIGHT
150 DG/W 18 AB — 3 SPEED SENSOR SIGNAL TO AMP.
563 O/Y 18 AB — 6 SPEED SENSOR TO LO VREF
787 PK/BK 18 AB — 19 FUEL PUMP RELAY
99 LG/BK 18 AB — 32 AIR-PORT SOLENOID SIGNAL
100 W/R 18 AB — 38 AIR-DUMP SOLENOID SIGNAL
73 O/LB 18 AB — 54 W.O.T A/C CUTOFF RELAY
883 PK/LB 16 AU — 10 A/C CONTROL RELAY FEED
201 T/R 18 AB — 48 VIP FUNCTION TESTER
382 Y/BK 18 AB — 17 EEC TEST CONNECTOR # 2
771 P/Y 18 AB — 30 ELEC. SHIFT 4x4 MODULE TO MOTOR POSITION #5
97 T/LG 18 AB — 22 FEED BACK CARB COIL #3
94 DG/P 18 AB — 29 AIR FUEL SENSOR
90 DB/LG 18 AB — 43 THERMOCOUPLE POS TO EXHAUST OVERTEMP MODULE
360 DG 18 AB — 33 EGR VALVE
358 LG/BK 18 AB — 45 MAP SENSOR FEED
355 DG/LG 18 AB — 47 THROTTLE POSITION SENSOR FEED
100 W/R 18 AB — 51 AIR-DUMP SOLENOID SIGNAL
361 R 14 AB — 57 EEC POWER RELAY
357 LG/P 18 AB — 25 AIR TEMP SENSOR
60 BK/LG 14 AB — 40 CONSTANT VOLTAGE UNIT TO GAUGE
359 BK/W 18 AB — 46 SENSOR SIGNAL RETURN
89 O 18 AB — 49 FUEL SENSOR GROUND
101 GY/Y 18 AB — 31 CANISTER PURGE SOLENOID
351 O/W 18 AB — 26 SENSOR SIGNAL FEED
352 BR/LG 18 AB — 27 EGR VALVE POSITION FEED
354 LG/Y 18 AB — 7 ENGINE COOLANT TEMP. SENSOR
99 LG/BK 18 AB — 11 AIR-PORT SOLENOID SIGNAL
562 LB 18 AB — 52 #8 FUEL INJECTOR
264 W/LB 18 AB — 21 ISC MOTOR #1
555 T 18 AB — 58 #1 FUEL INJECTOR
556 W 18 AB — 59 #2 FUEL INJECTOR
557 BR/Y 18 AB — 12 #3 FUEL INJECTOR
558 BR/LB 18 AB — 13 #4 FUEL INJECTOR
559 T/LB 18 AB — 14 #5 FUEL INJECTOR
560 LG 18 AB — 15 #6 FUEL INJECTOR
11 DG/Y 18 AB — 4 IGN. COIL NEG. TERMINAL
561 T/O 18 AB — 42 #7 FUEL INJECTOR
60 BK/LG 14 AB — 60 CONSTANT VOLTAGE UNIT TO GAUGE
38 BK/O 16 AB — 1 POWER SUPPLY TO BATT.
361 R 14 AB — 37 POWER RELAY
57 BK 14 AB — 20 GROUND CIRCUIT
349 DB 18 AB — 56 CRANKSHAFT POSITION SENSOR FEED
259 BK/O 18 AB — 16 TFI MODULE
324 Y/LG 18 AB — 36 TRANSMISSION O.D. SW. TO HEAT MODULE
ENGINE ELECTRONIC CONTROL PROCESSOR
ENDS
E4LB-14A601-A
60 BK/LG 16 AB
16 R/LG 16 AB
361 R 14 AB
68 O/BK 14 AB
38 BK/O 16 AB
38 BK/O 14 AB
E1VB-14489-F
EEC POWER RELAY
60 BK/LG 12 AB
E7DB-14A464-L
TO BATTERY GROUND TERMINAL
FUSIBLE LINK 20 GA.
ES-D5DB-14A466-A
38 BK/O 14 AB
TO STARTER MOTOR RELAY ZONE 13-B
EGR CONTROL SOLENOID (EVR)
TO 14A666 OR 19D887 ZONE 10-A
ENGINE CONTROL BAROMETRIC PRESSURE SENSOR
SHORTING PLUG
TO GROUND
A
B
C
D
E
F
FRAME 22 OF 22
MS-E9ZB-3458-AA

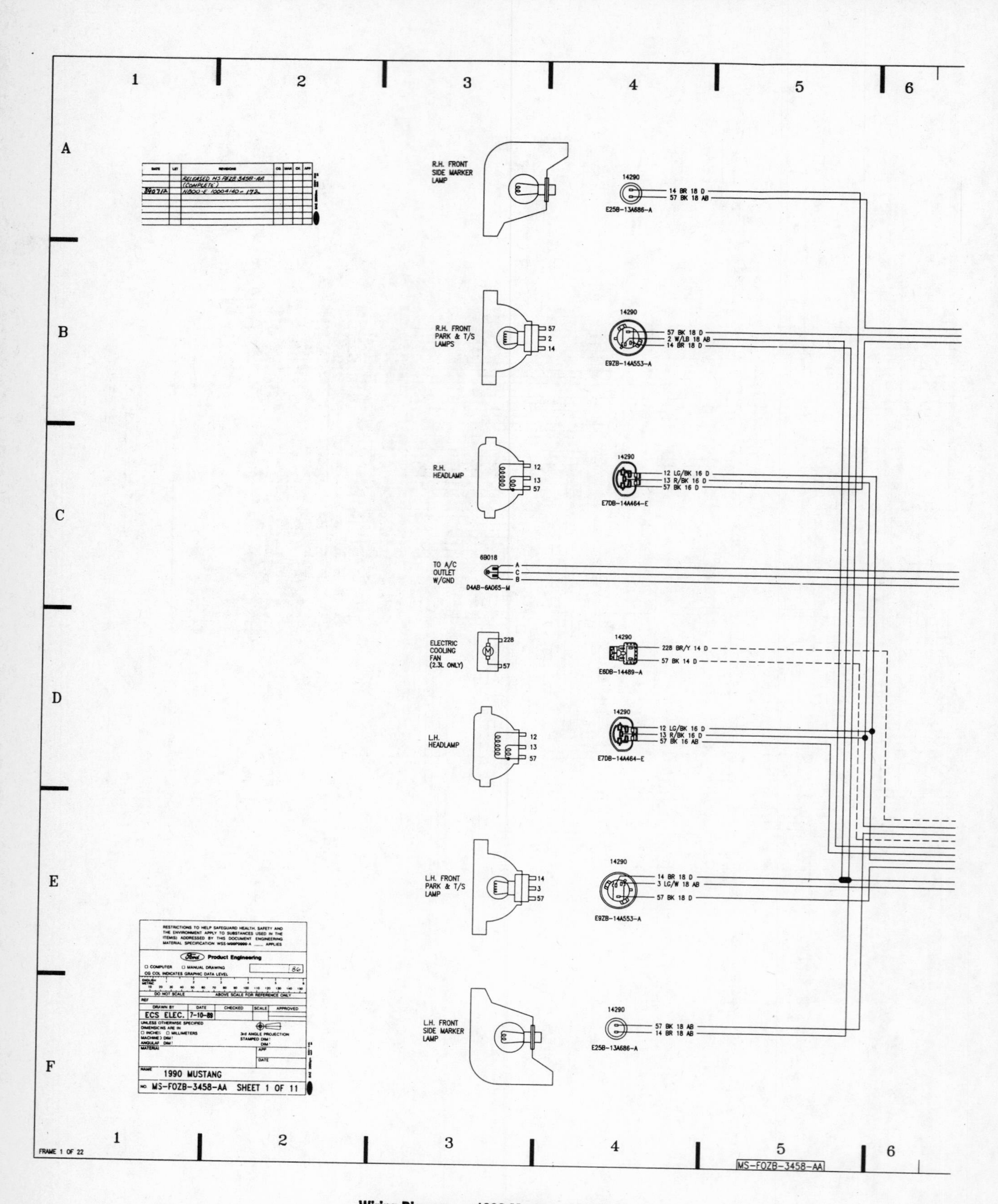
R.H. FRONT SIDE MARKER LAMP
14290
14 BR 18 D
57 BK 18 AB
E25B-13A686-A
R.H. FRONT PARK & T/S LAMPS
57
2
14
57 BK 18 D
2 W/LB 18 AB
14 BR 18 D
E9ZB-14A553-A
R.H. HEADLAMP
12
13
57
12 LG/BK 16 D
13 R/BK 16 D
57 BK 16 D
E7DB-14A464-E
TO A/C OUTLET W/GND
6B018
A
C
B
D4AB-6A065-M
ELECTRIC COOLING FAN (2.3L ONLY)
228
228 BR/Y 14 D
57 BK 14 D
E6DB-14489-A
L.H. HEADLAMP
57 BK 16 AB
L.H. FRONT PARK & T/S LAMP
3
3 LG/W 18 AB
L.H. FRONT SIDE MARKER LAMP
E25B-13A686-A
1990 MUSTANG
MS-FOZB-3458-AA SHEET 1 OF 11
ECS ELEC. 7-10-89
FRAME 1 OF 22
MS-FOZB-3458-AA

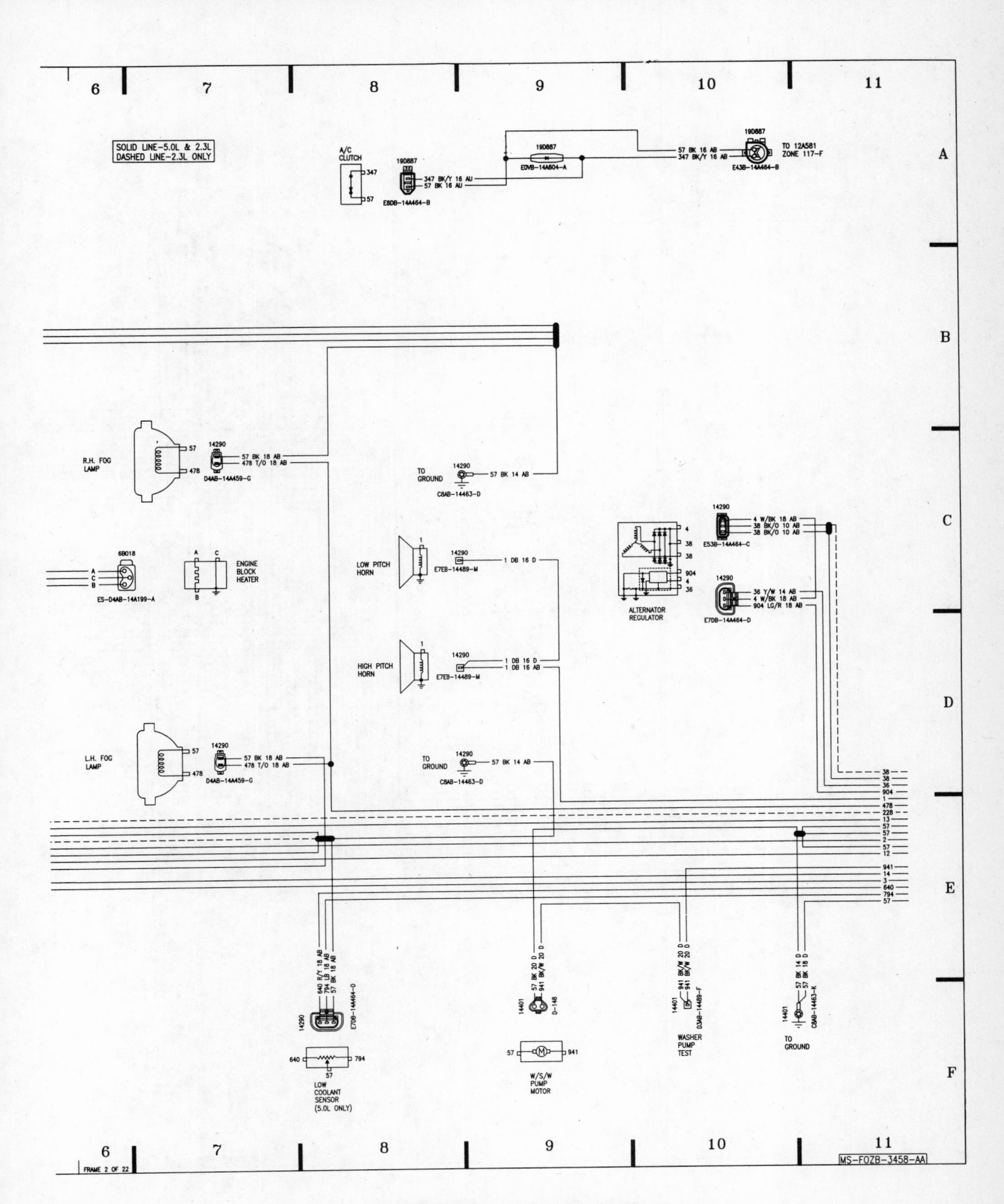
SOLID LINE-5.0L & 2.3L
DASHED LINE-2.3L ONLY
A/C CLUTCH
R.H. FOG LAMP
ENGINE BLOCK HEATER
LOW PITCH HORN
HIGH PITCH HORN
ALTERNATOR REGULATOR
L.H. FOG LAMP
TO GROUND
LOW COOLANT SENSOR (5.0L ONLY)
W/S/W PUMP MOTOR
WASHER PUMP TEST
FRAME 2 OF 22
MS-FOZB-3458-AA

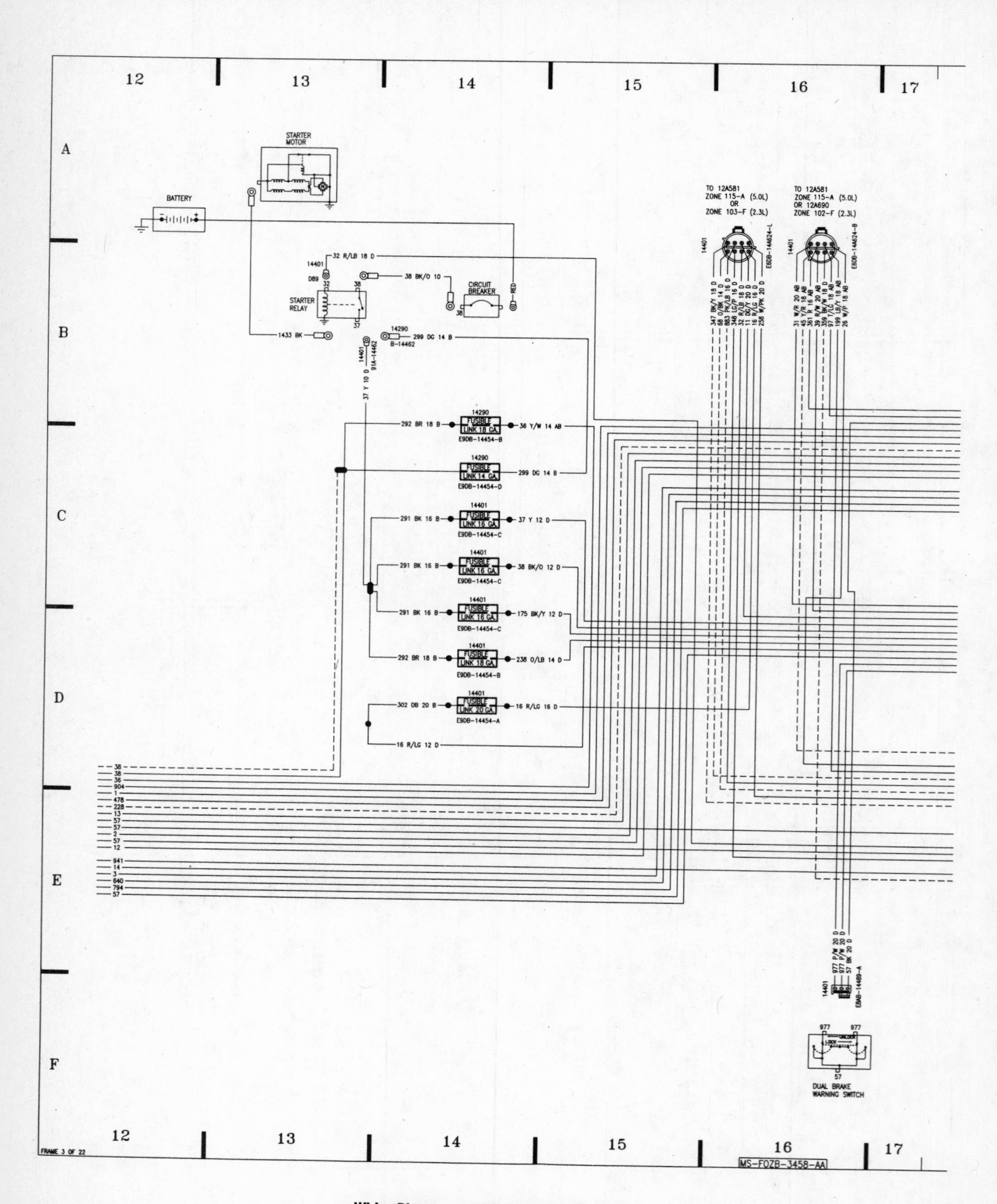
STARTER MOTOR
BATTERY
STARTER RELAY
CIRCUIT BREAKER
FUSIBLE LINK 18 GA
FUSIBLE LINK 14 GA
FUSIBLE LINK 16 GA
FUSIBLE LINK 20 GA
TO 12A581 ZONE 115-A (5.0L) OR ZONE 103-F (2.3L)
TO 12A581 ZONE 115-A (5.0L) OR 12A690 ZONE 102-F (2.3L)
DUAL BRAKE WARNING SWITCH
FRAME 3 OF 22
MS-F0ZB-3458-AA

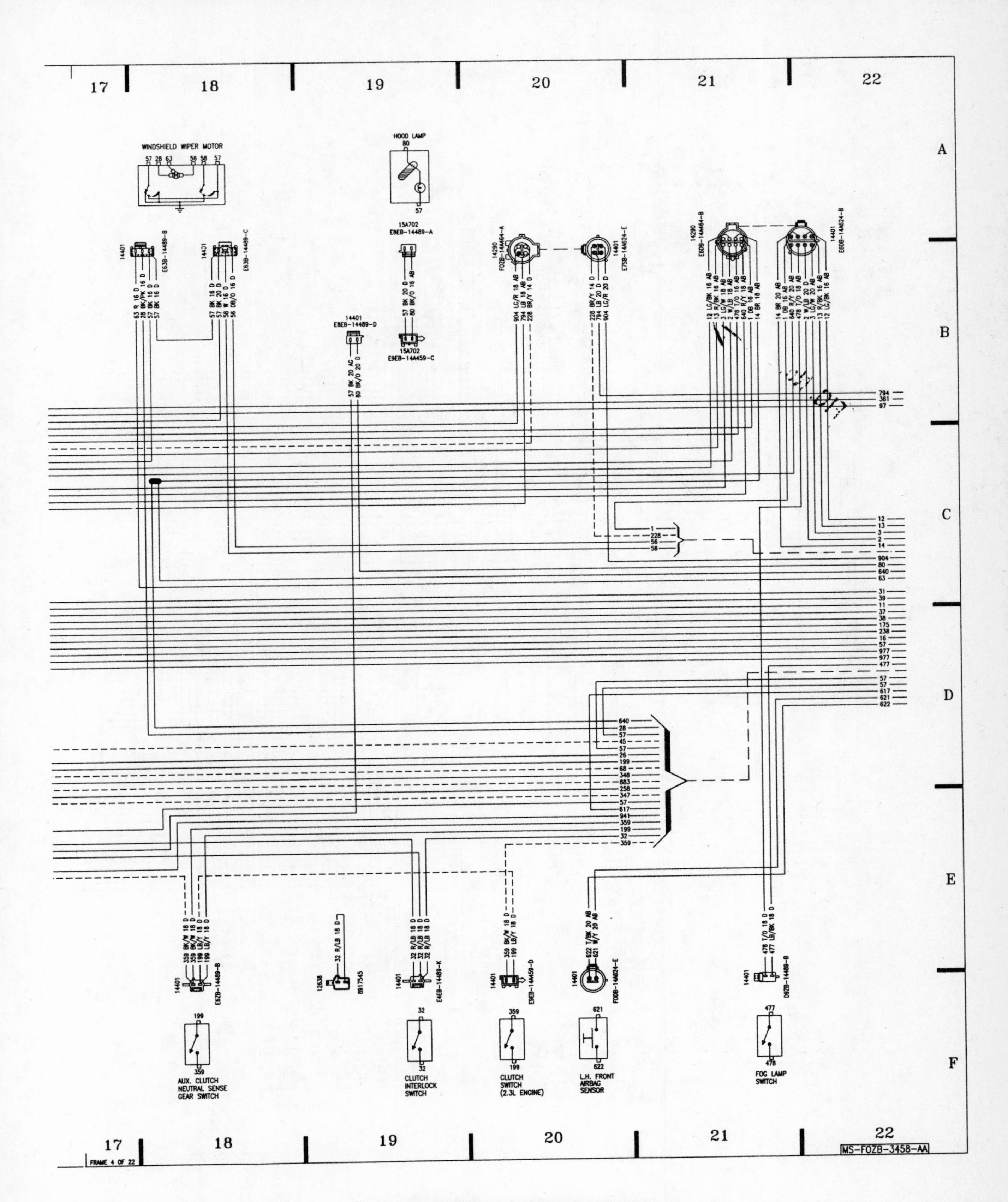

Wiring Diagram — 1990 Mustang 4 of 22

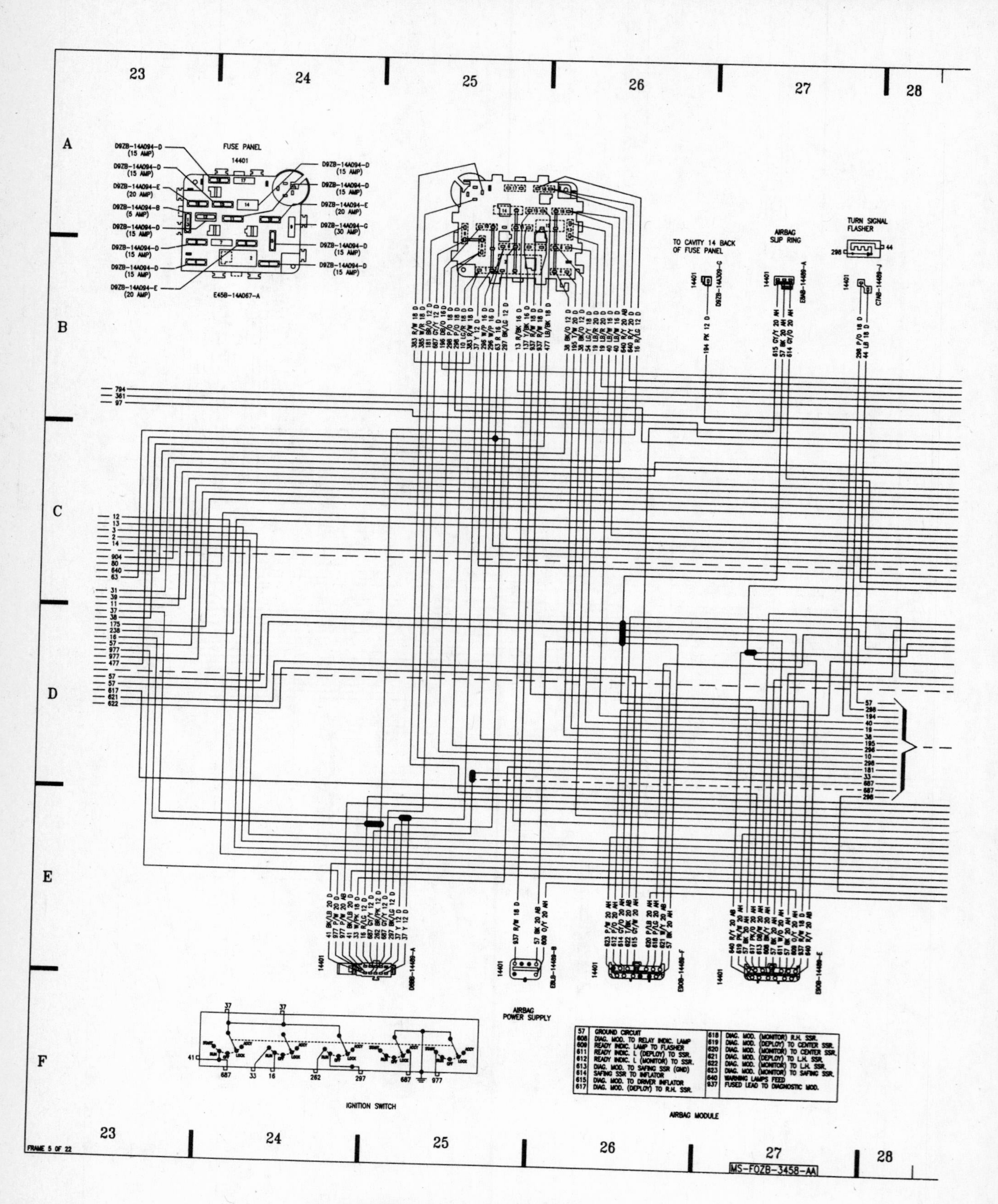

57	GROUND CIRCUIT	618	DIAG. MOD. (MONITOR) R.H. SSR.
608	DIAG. MOD. TO RELAY INDIC. LAMP	619	DIAG. MOD. (DEPLOY) TO CENTER SSR.
609	READY INDIC. LAMP TO FLASHER	620	DIAG. MOD. (MONITOR) TO CENTER SSR.
611	READY INDIC. L (DEPLOY) TO SSR.	621	DIAG. MOD. (DEPLOY) TO L.H. SSR.
612	READY INDIC. L (MONITOR) TO SSR.	622	DIAG. MOD. (MONITOR) TO L.H. SSR.
613	DIAG. MOD. TO SAFING SSR (GND)	623	DIAG. MOD. (MONITOR) TO SAFING SSR.
614	SAFING SSR TO INFLATOR	640	WARNING LAMPS FEED
615	DIAG. MOD. TO DRIVER INFLATOR	937	FUSED LEAD TO DIAGNOSTIC MOD.
617	DIAG. MOD. (DEPLOY) TO R.H. SSR.		

Wiring Diagram — 1990 Mustang 5 of 22

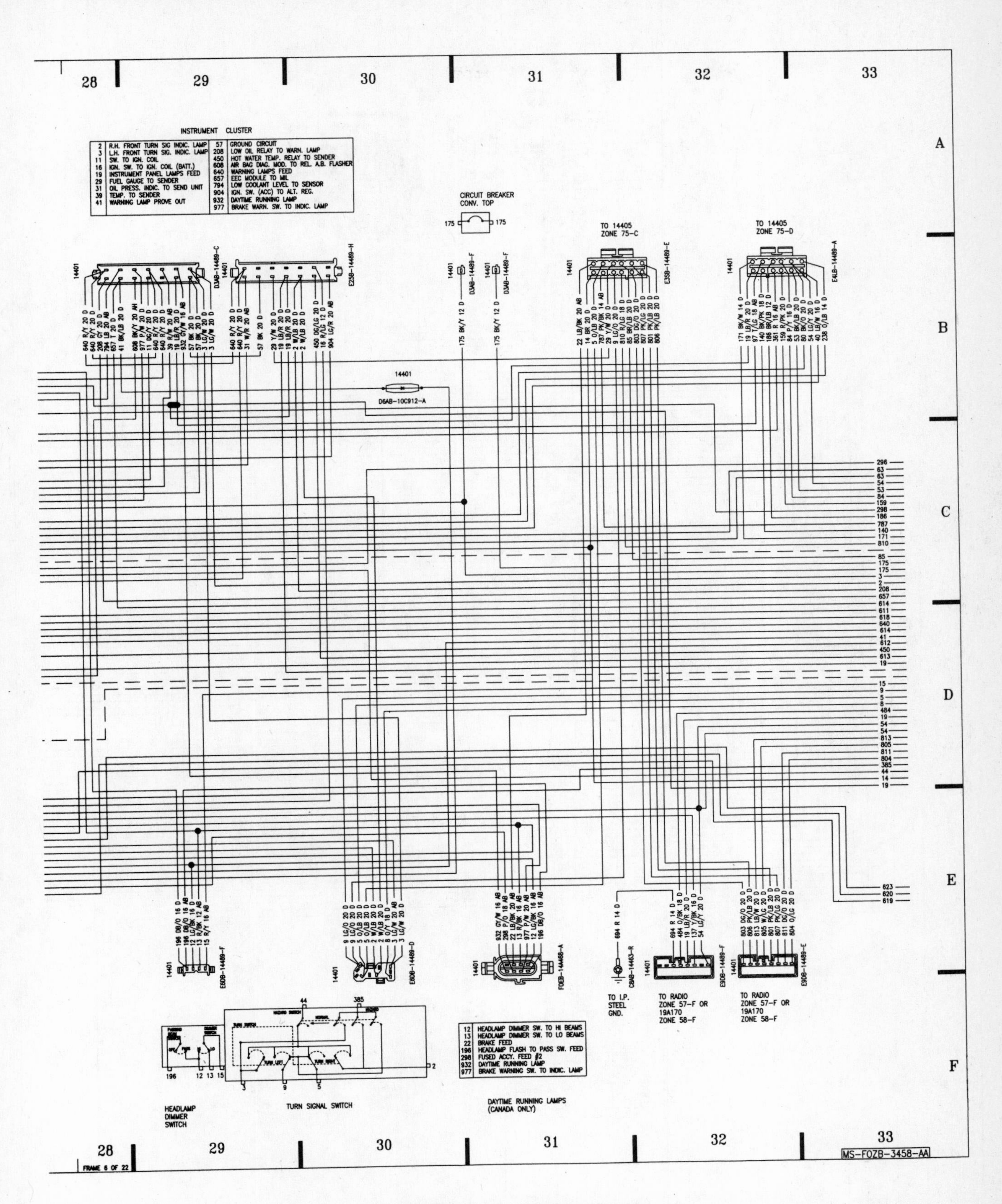

INSTRUMENT CLUSTER
2 R.H. FRONT TURN SIG INDIC. LAMP
3 L.H. FRONT TURN SIG. INDIC. LAMP
11 SW. TO IGN. COIL
16 IGN. SW. TO IGN. COIL (BATT.)
19 INSTRUMENT PANEL LAMPS FEED
29 FUEL GAUGE TO SENDER
31 OIL PRESS. INDIC. TO SEND UNIT
39 TEMP. TO SENDER
41 WARNING LAMP PROVE OUT
57 GROUND CIRCUIT
208 LOW OIL RELAY TO WARN. LAMP
450 HOT WATER TEMP. RELAY TO SENDER
608 AIR BAG DIAG. MOD. TO REL. A.B. FLASHER
640 WARNING LAMPS FEED
657 EEC MODULE TO MIL
794 LOW COOLANT LEVEL TO SENSOR
904 IGN. SW. (ACC) TO ALT. REG.
932 DAYTIME RUNNING LAMP
977 BRAKE WARN. SW. TO INDIC. LAMP
CIRCUIT BREAKER CONV. TOP
TO 14405 ZONE 75-C
TO 14405 ZONE 75-D
D6AB-10C912-A
TO I.P. STEEL GND.
TO RADIO ZONE 57-F OR 19A170 ZONE 58-F
TO RADIO ZONE 57-F OR 19A170 ZONE 58-F
12 HEADLAMP DIMMER SW. TO HI BEAMS
13 HEADLAMP DIMMER SW. TO LO BEAMS
22 BRAKE FEED
196 HEADLAMP FLASH TO PASS SW. FEED
298 FUSED ACCY. FEED #2
932 DAYTIME RUNNING LAMP
977 BRAKE WARNING SW. TO INDIC. LAMP
HEADLAMP DIMMER SWITCH
TURN SIGNAL SWITCH
DAYTIME RUNNING LAMPS (CANADA ONLY)
FRAME 6 OF 22
MS-F0ZB-3458-AA

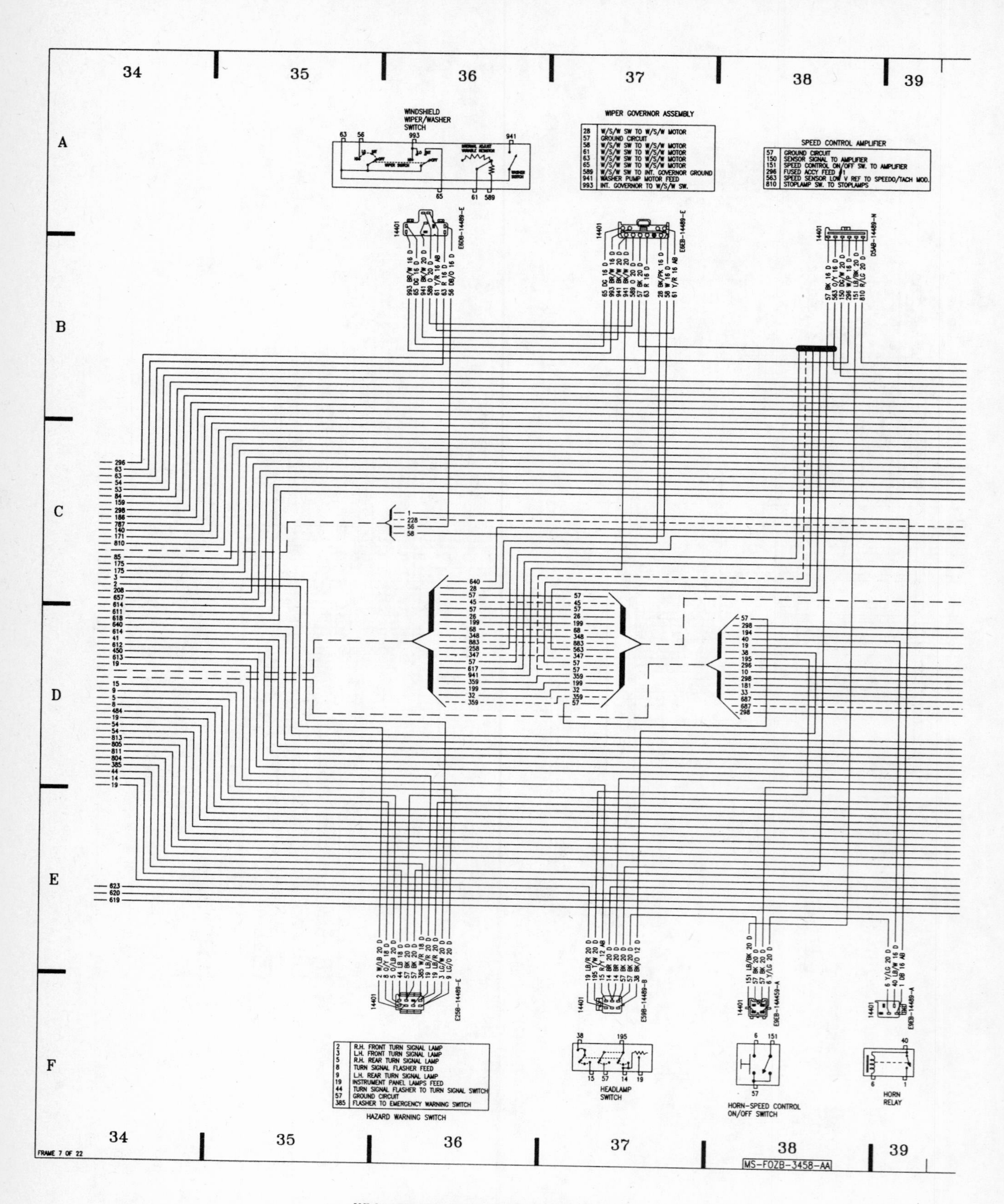

Wiring Diagram — 1990 Mustang 7 of 22

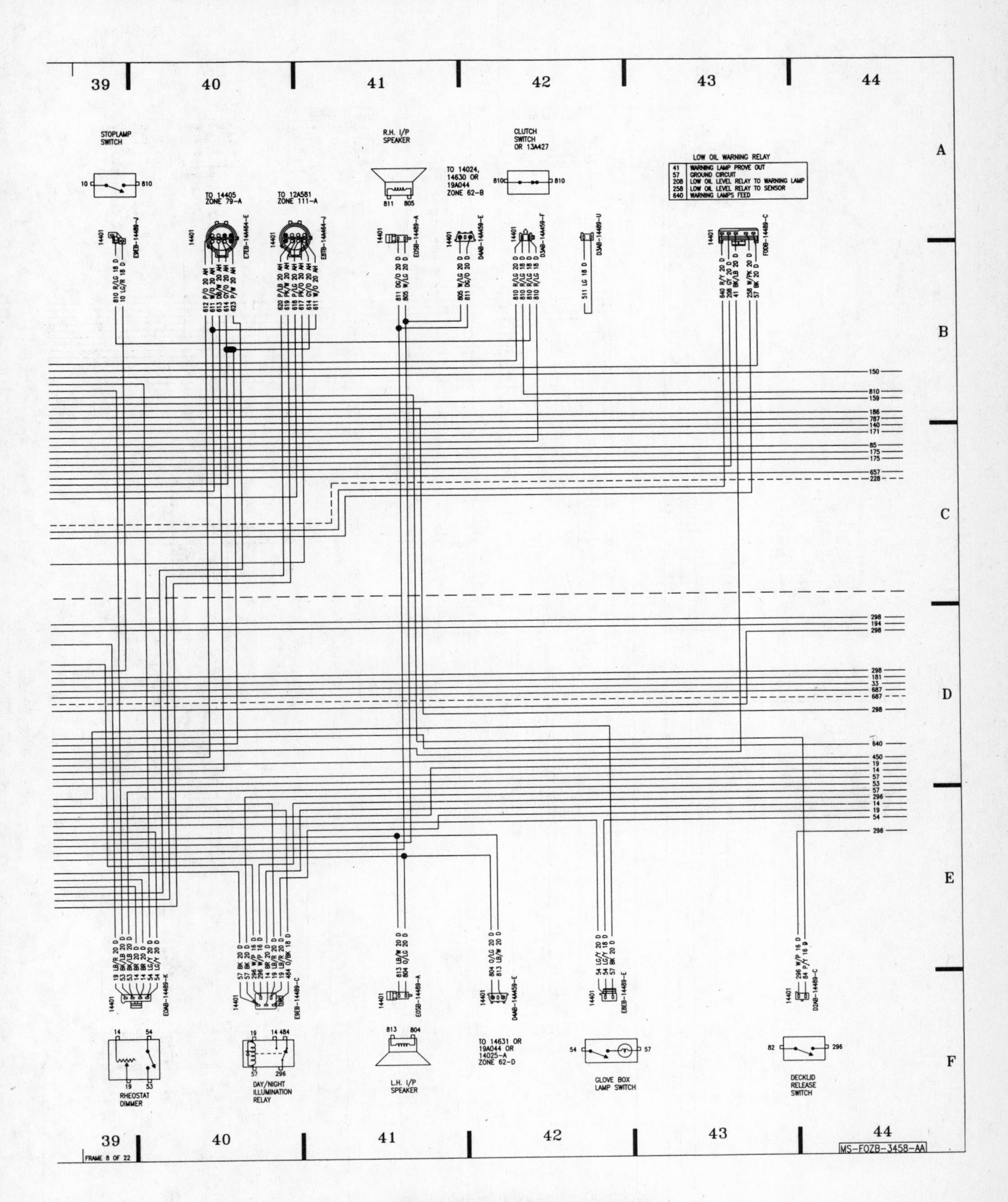
STOPLAMP SWITCH
R.H. I/P SPEAKER
CLUTCH SWITCH OR 13A427
LOW OIL WARNING RELAY
41 WARNING LAMP PROVE OUT
57 GROUND CIRCUIT
208 LOW OIL LEVEL RELAY TO WARNING LAMP
258 LOW OIL LEVEL RELAY TO SENSOR
640 WARNING LAMPS FEED
TO 14405 ZONE 79-A
TO 12A581 ZONE 111-A
TO 14024, 14630 OR 19A044 ZONE 62-B
RHEOSTAT DIMMER
DAY/NIGHT ILLUMINATION RELAY
L.H. I/P SPEAKER
TO 14631 OR 19A044 OR 14025-A ZONE 62-D
GLOVE BOX LAMP SWITCH
DECKLID RELEASE SWITCH
FRAME 8 OF 22
MS-F0ZB-3458-AA

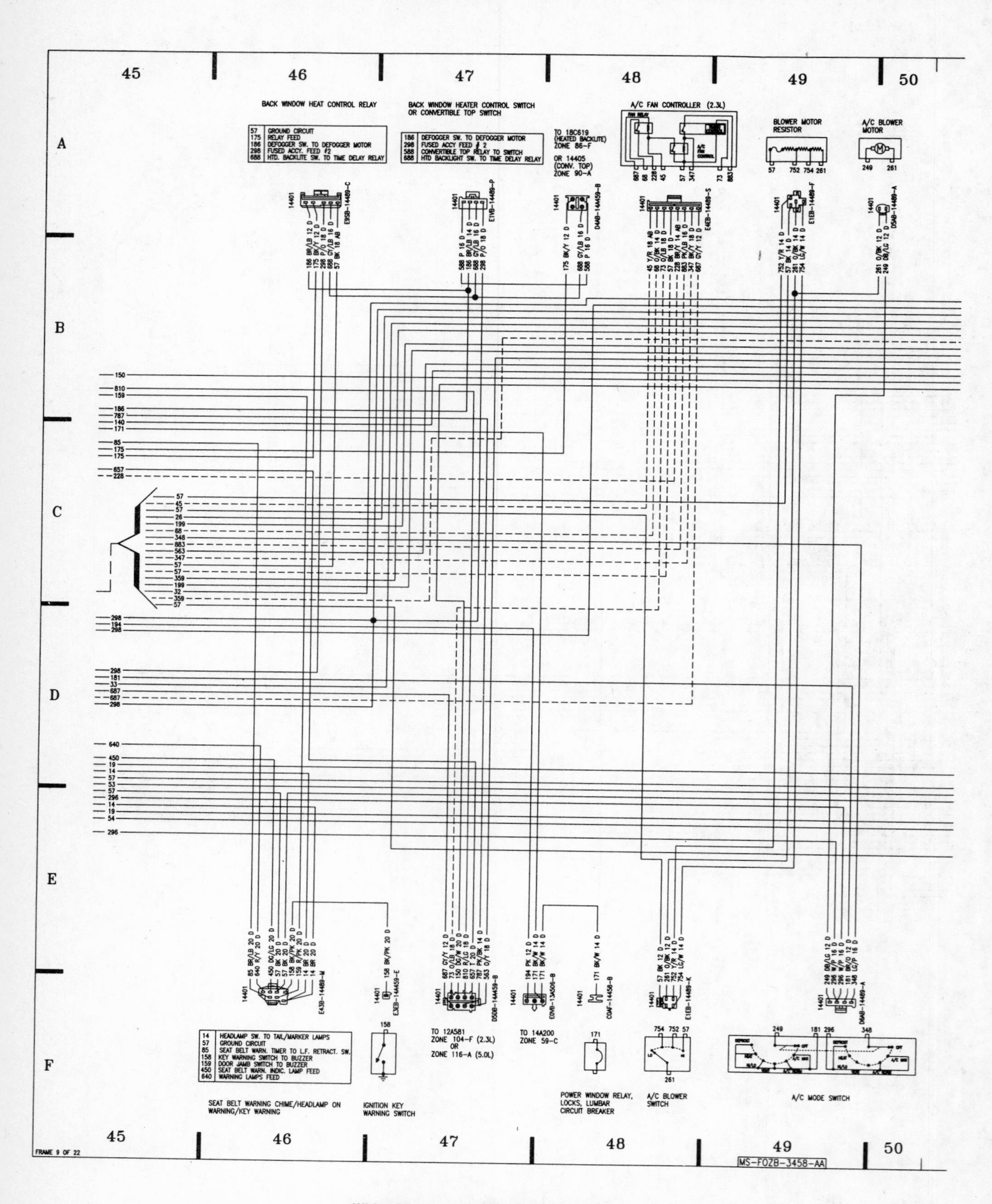

Wiring Diagram — 1990 Mustang 9 of 22

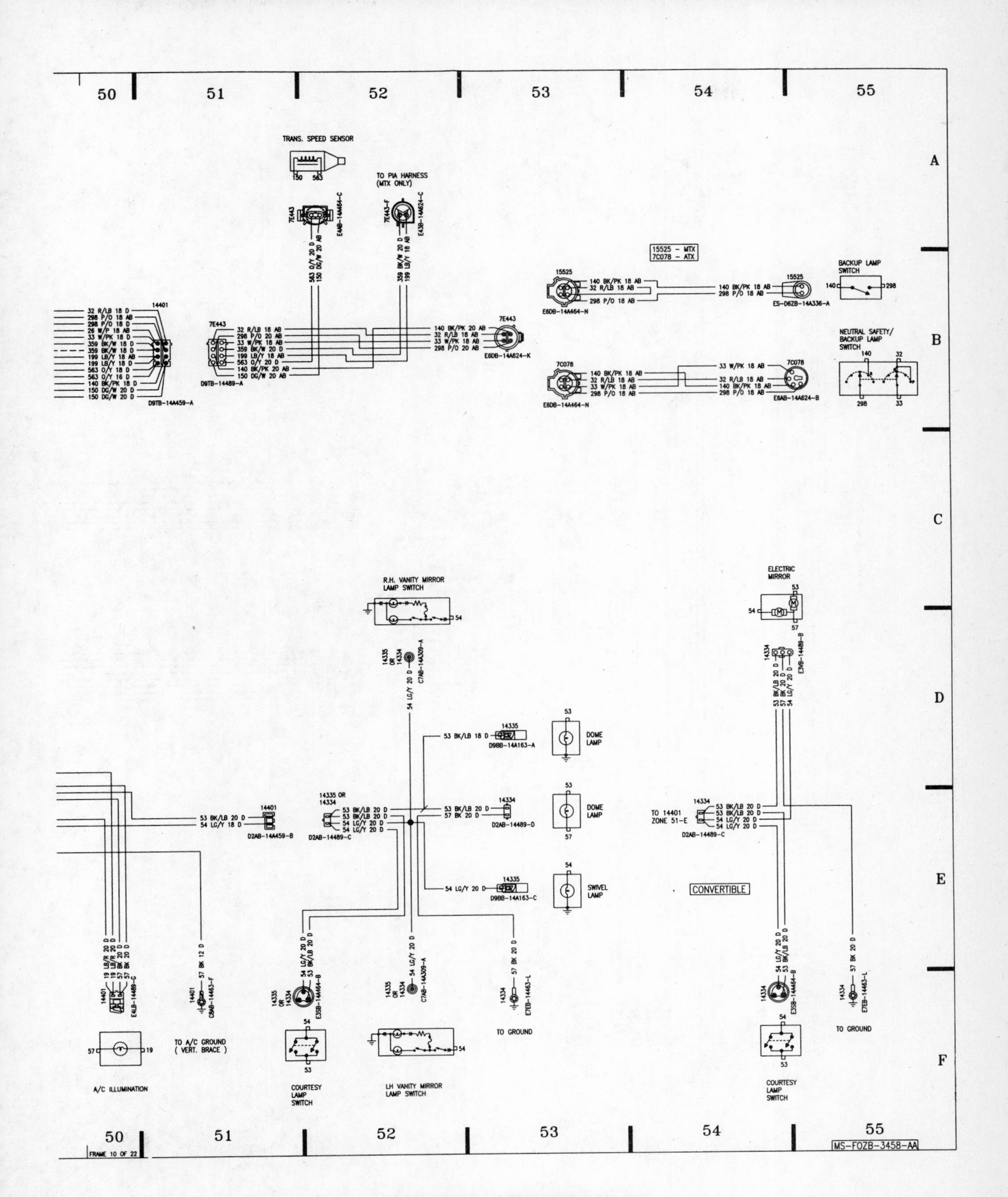

TRANS. SPEED SENSOR
TO PIA HARNESS (MTX ONLY)
15525 – MTX
7C078 – ATX
BACKUP LAMP SWITCH
NEUTRAL SAFETY/ BACKUP LAMP SWITCH
R.H. VANITY MIRROR LAMP SWITCH
ELECTRIC MIRROR
DOME LAMP
DOME LAMP
SWIVEL LAMP
CONVERTIBLE
TO 14401 ZONE 51–E
TO A/C GROUND (VERT. BRACE)
A/C ILLUMINATION
COURTESY LAMP SWITCH
LH VANITY MIRROR LAMP SWITCH
TO GROUND
TO GROUND
COURTESY LAMP SWITCH
FRAME 10 OF 22
MS–FOZB–3458–AA

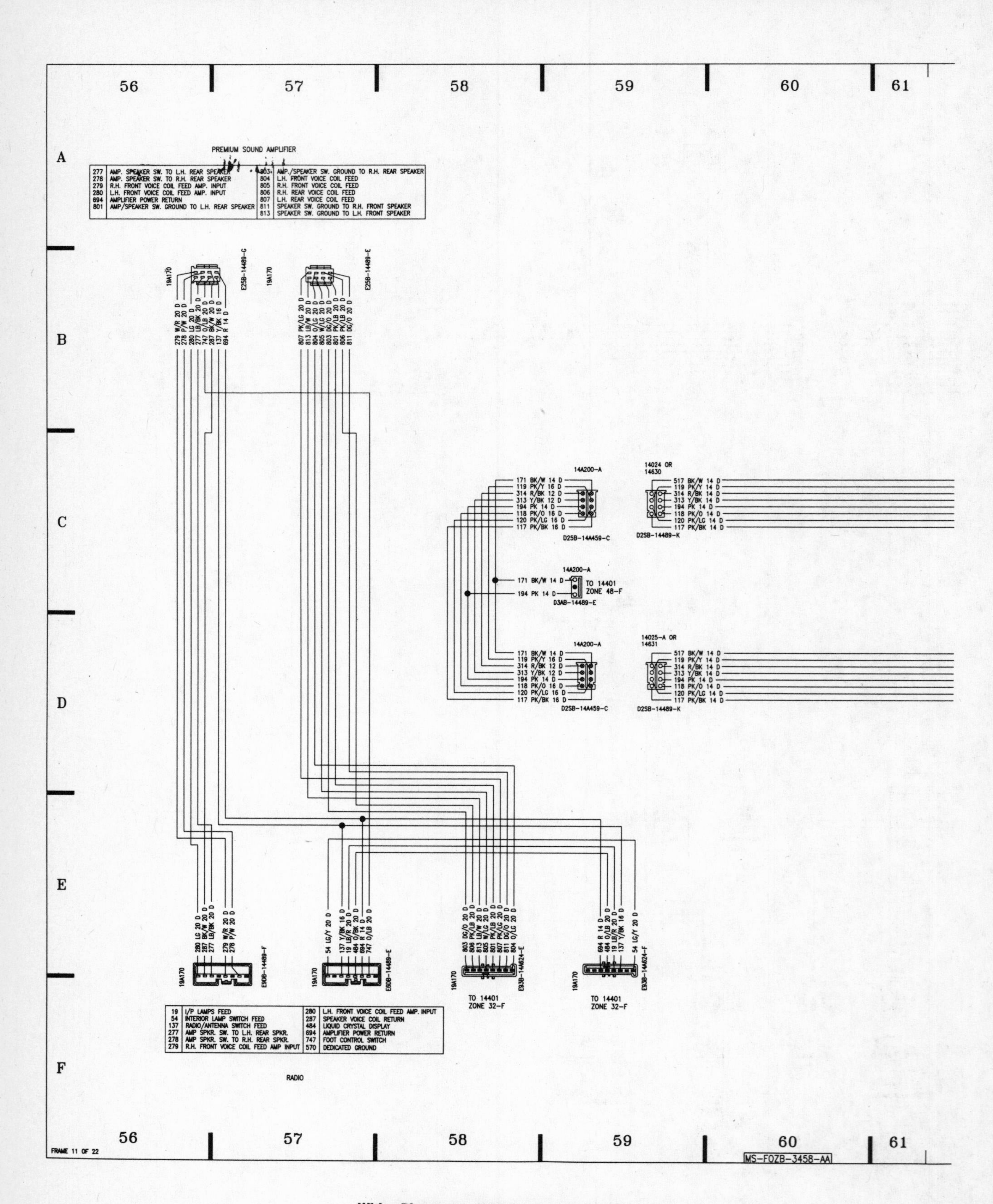

Wiring Diagram — 1990 Mustang 11 of 22

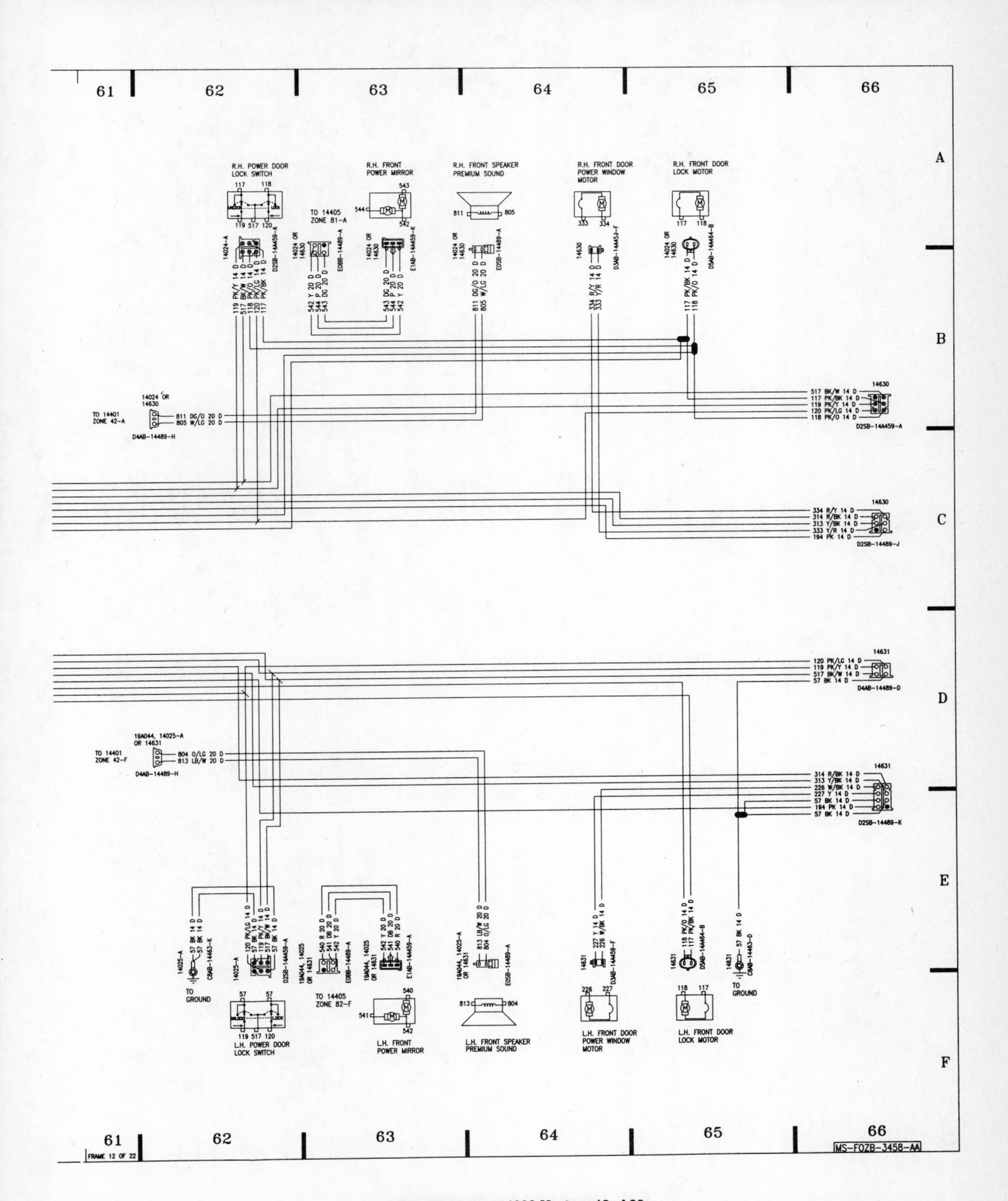
R.H. POWER DOOR LOCK SWITCH
R.H. FRONT POWER MIRROR
R.H. FRONT SPEAKER PREMIUM SOUND
R.H. FRONT DOOR POWER WINDOW MOTOR
R.H. FRONT DOOR LOCK MOTOR
TO 14405 ZONE 81-A
TO 14401 ZONE 42-A
TO 14401 ZONE 42-F
TO 14405 ZONE 82-F
TO GROUND
L.H. POWER DOOR LOCK SWITCH
L.H. FRONT POWER MIRROR
L.H. FRONT SPEAKER PREMIUM SOUND
L.H. FRONT DOOR POWER WINDOW MOTOR
L.H. FRONT DOOR LOCK MOTOR
FRAME 12 OF 22
MS-F0ZB-3458-AA

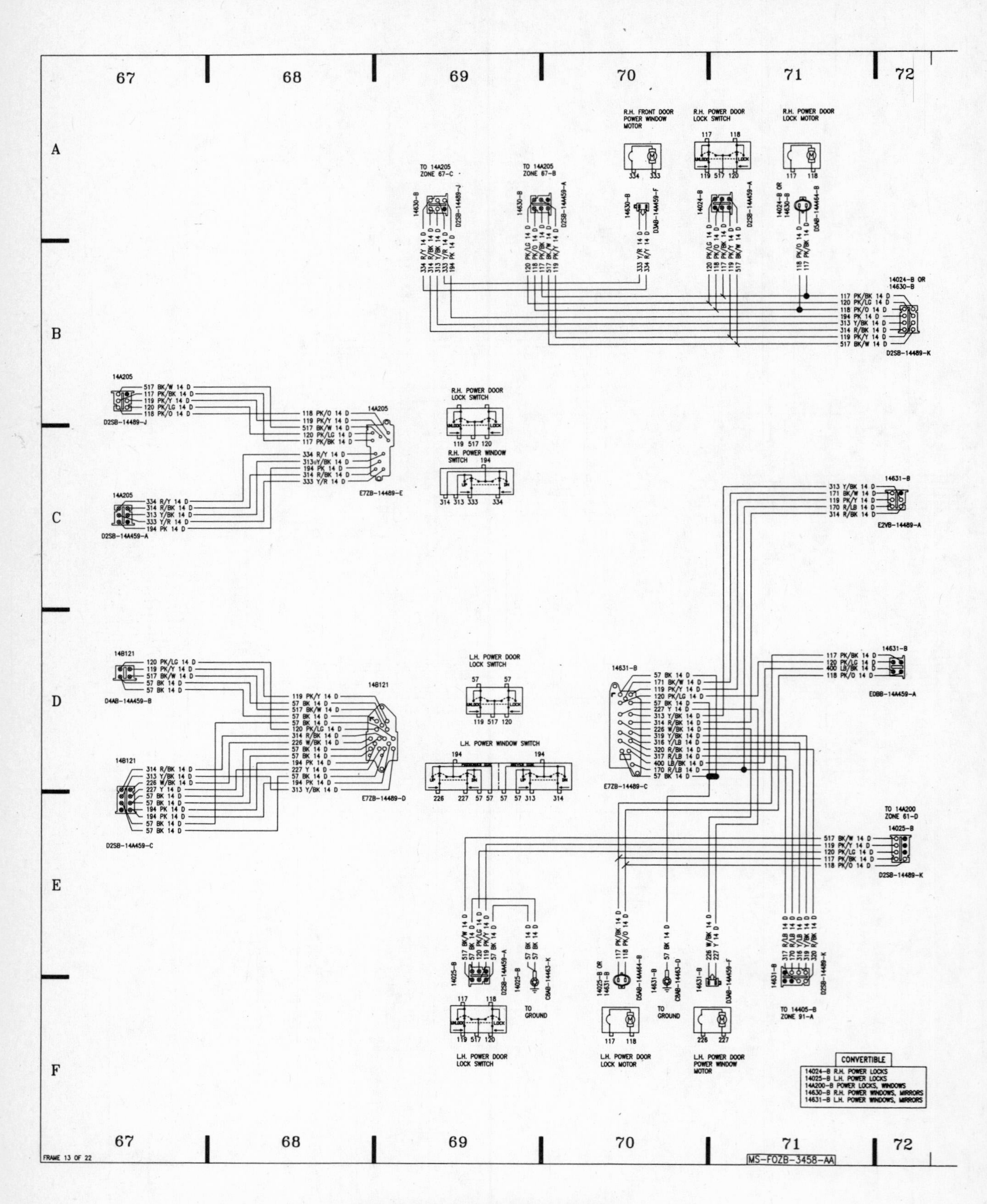

Wiring Diagram — 1990 Mustang 13 of 22

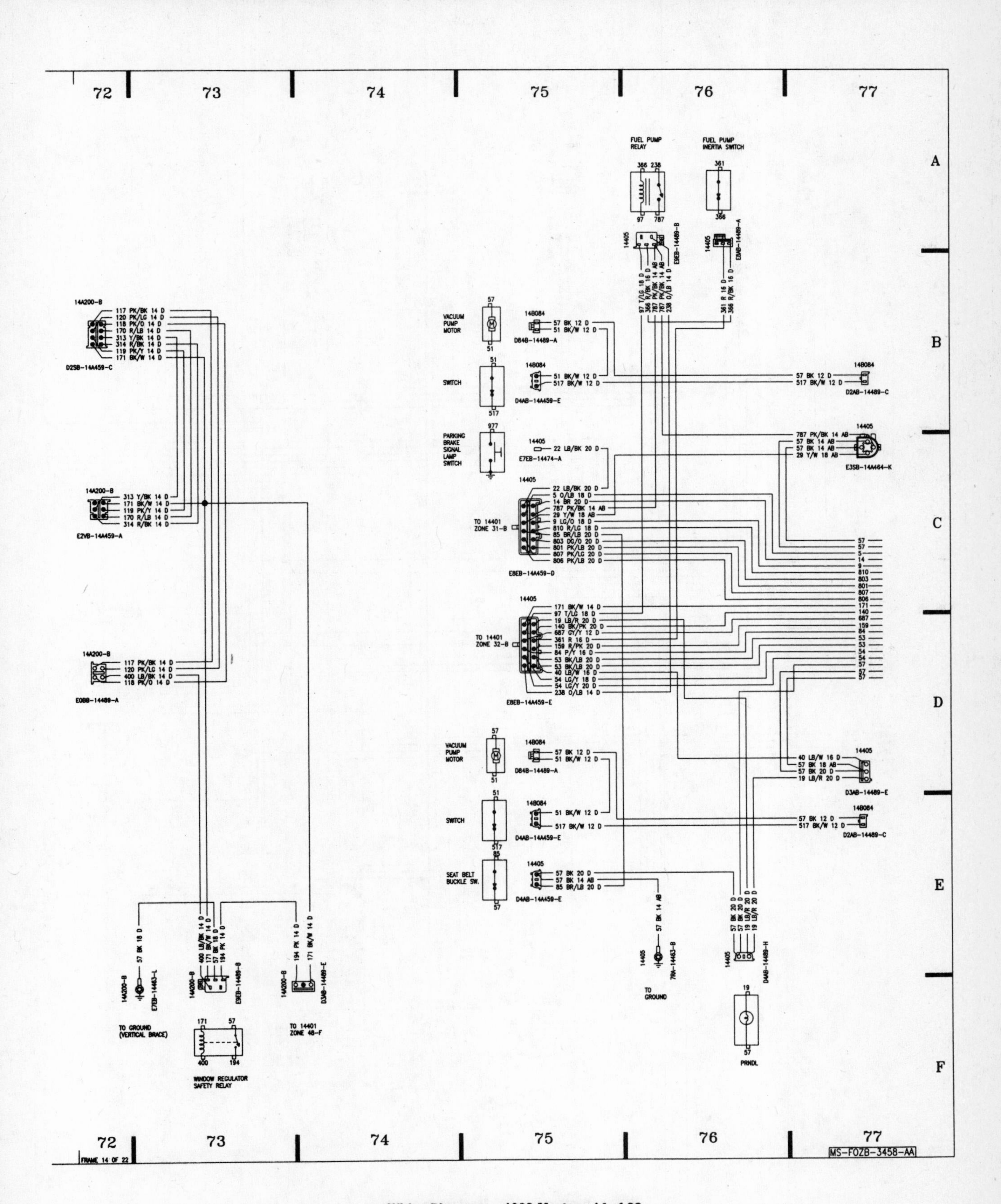

Wiring Diagram — 1990 Mustang 14 of 22

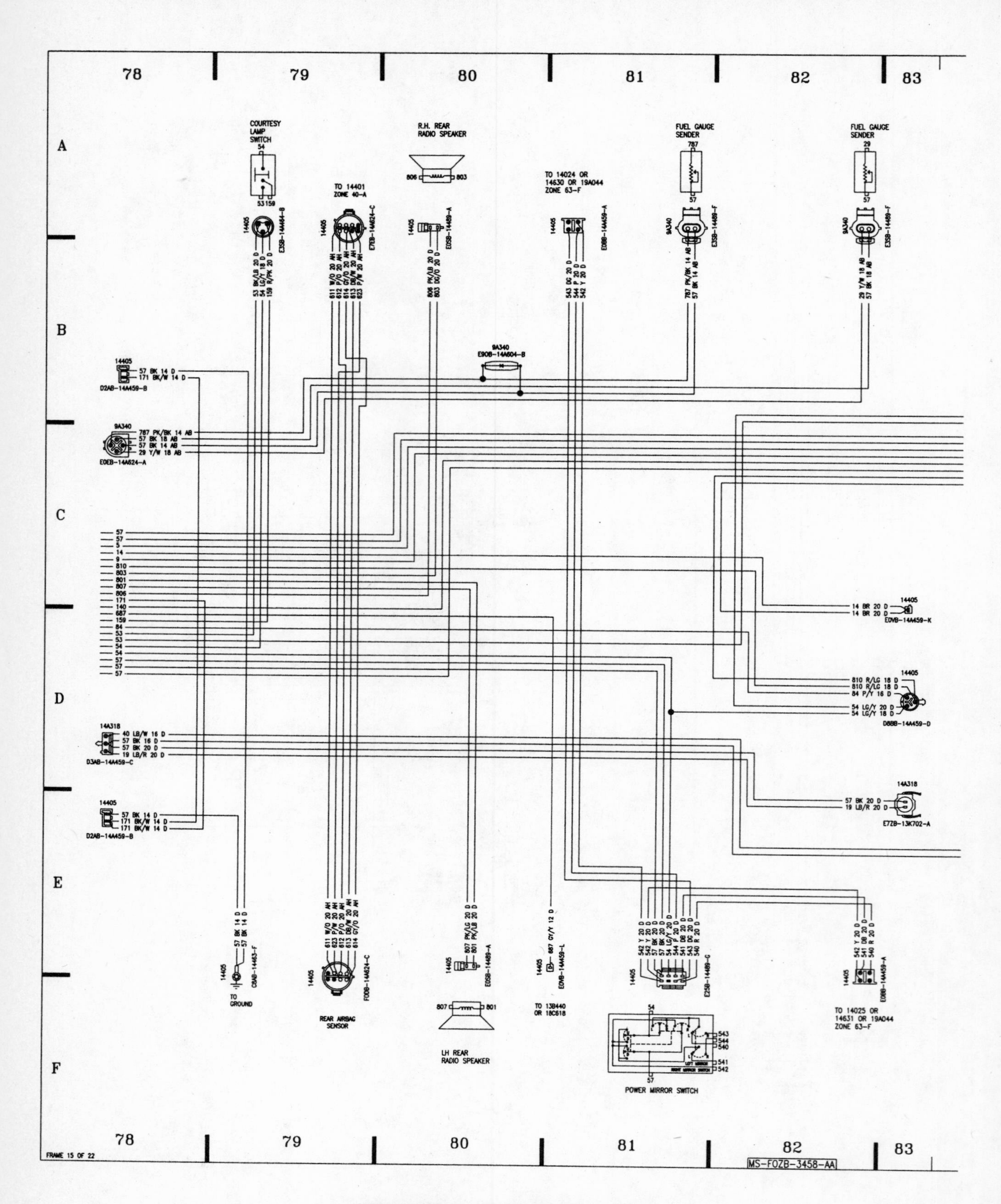

COURTESY LAMP SWITCH
R.H. REAR RADIO SPEAKER
FUEL GAUGE SENDER
FUEL GAUGE SENDER
TO 14401 ZONE 40-A
TO 14024 OR 14630 OR 19A044 ZONE 63-F
REAR AIRBAG SENSOR
LH REAR RADIO SPEAKER
TO GROUND
TO 13B440 OR 18C618
POWER MIRROR SWITCH
TO 14025 OR 14631 OR 19A044 ZONE 63-F
FRAME 15 OF 22
MS-F0ZB-3458-AA

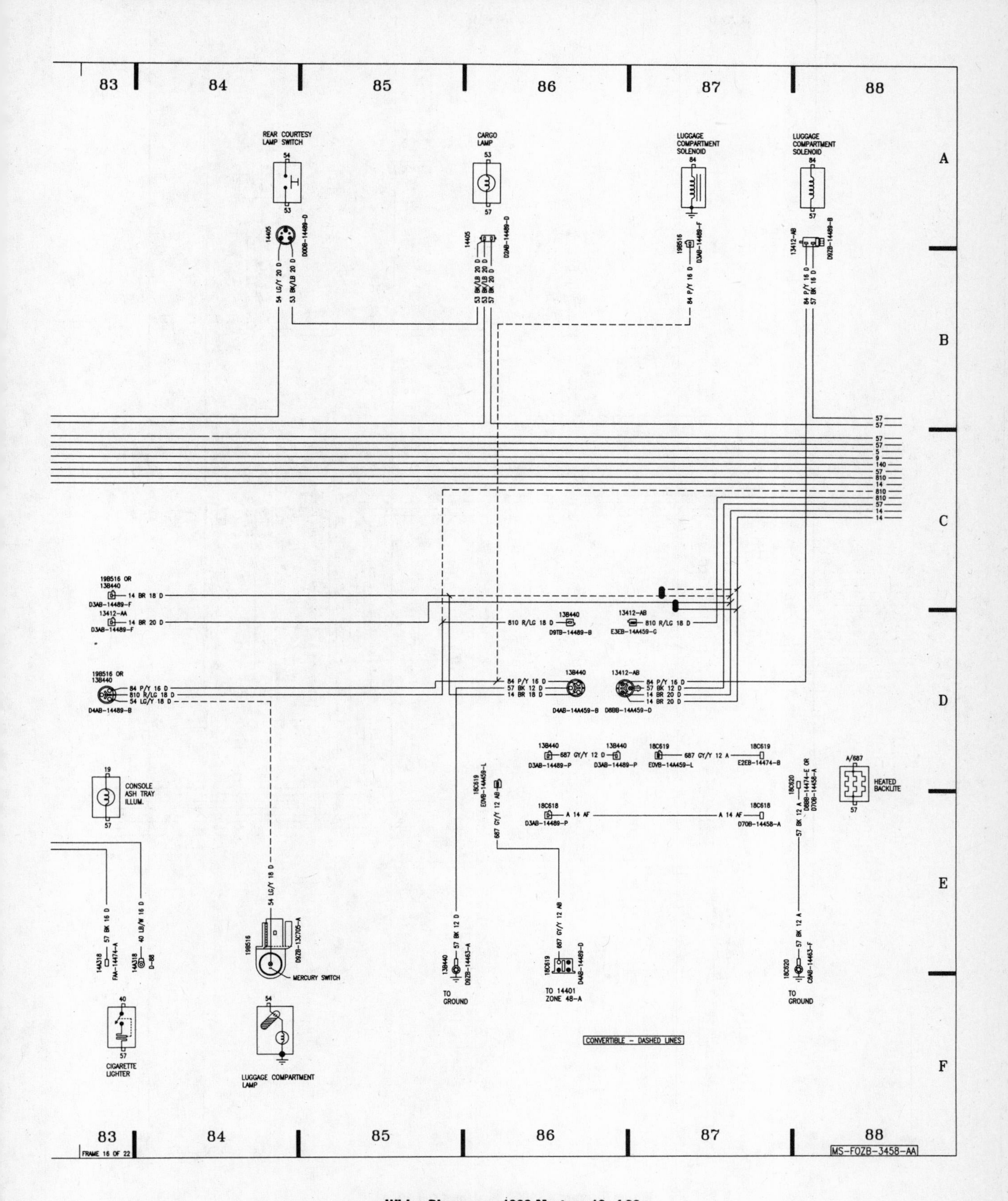
REAR COURTESY LAMP SWITCH
CARGO LAMP
LUGGAGE COMPARTMENT SOLENOID
LUGGAGE COMPARTMENT SOLENOID
CONSOLE ASH TRAY ILLUM.
MERCURY SWITCH
CIGARETTE LIGHTER
LUGGAGE COMPARTMENT LAMP
HEATED BACKLITE
TO GROUND
TO 14401 ZONE 48-A
TO GROUND
CONVERTIBLE - DASHED LINES
FRAME 16 OF 22
MS-F0ZB-3458-AA

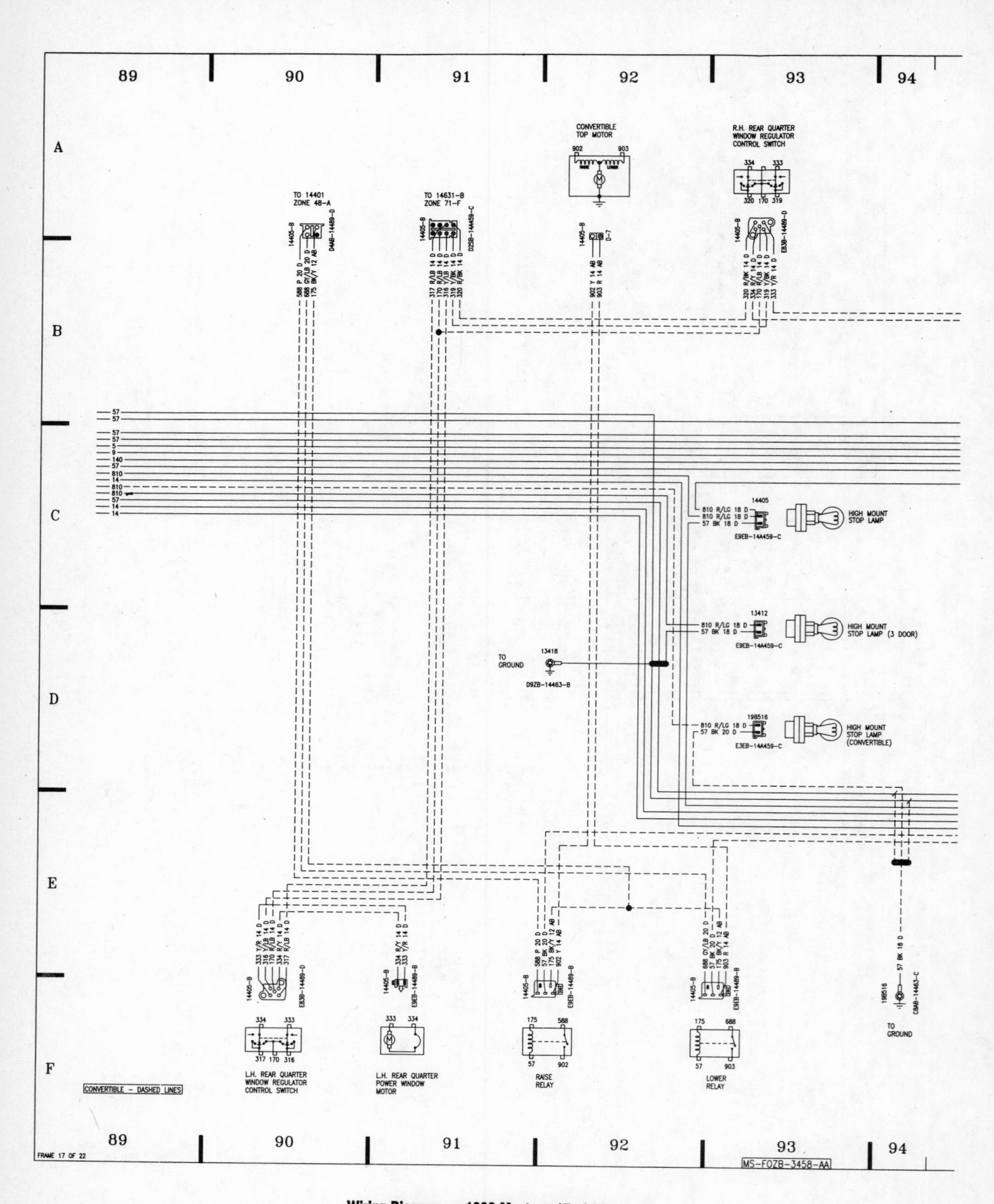

Wiring Diagram — 1990 Mustang 17 of 22

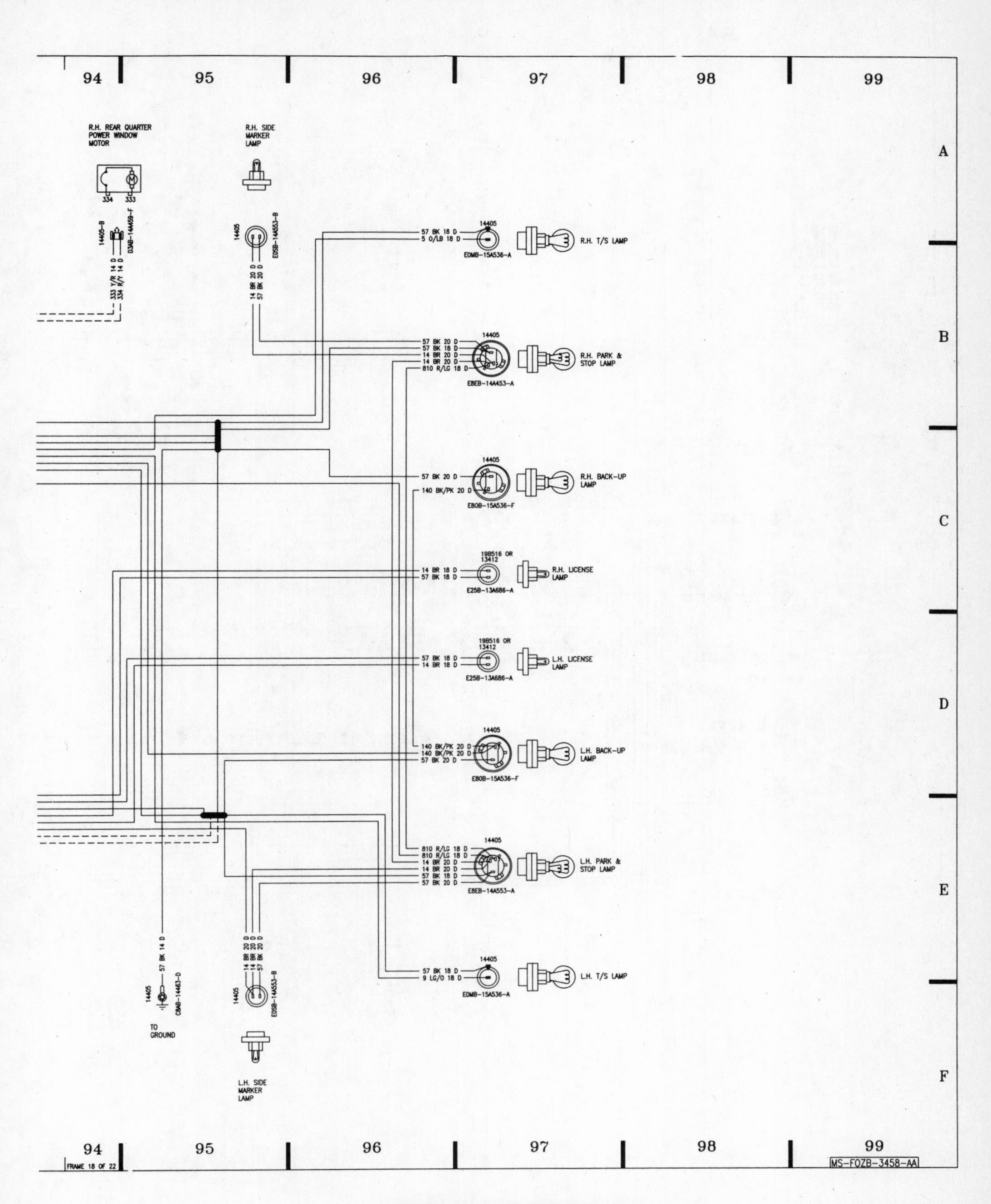

Wiring Diagram — 1990 Mustang 18 of 22

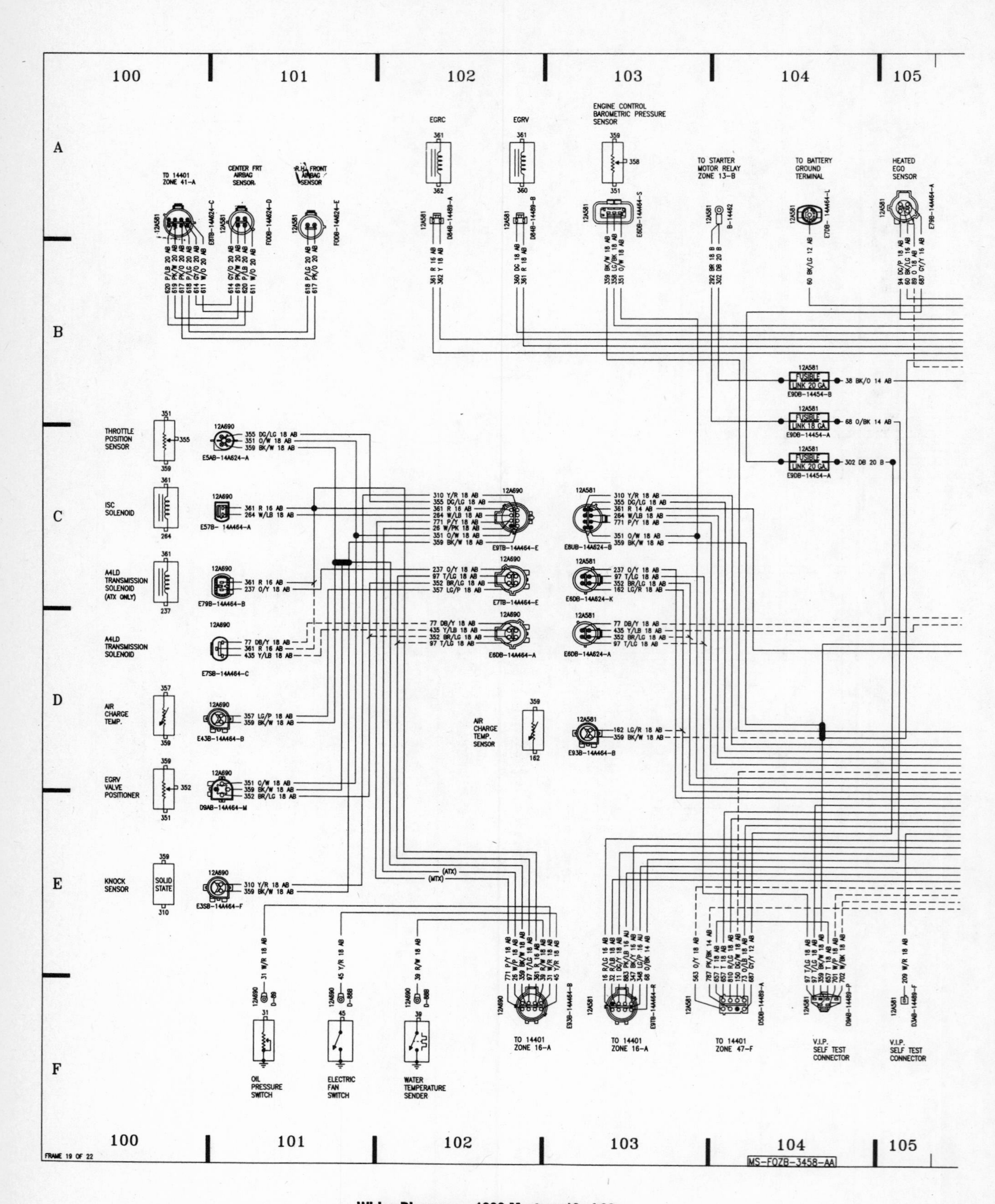

EGRC
EGRV
ENGINE CONTROL BAROMETRIC PRESSURE SENSOR
TO STARTER MOTOR RELAY ZONE 13-B
TO BATTERY GROUND TERMINAL
HEATED EGO SENSOR
CENTER FRT AIRBAG SENSOR
R.H. FRONT AIRBAG SENSOR
THROTTLE POSITION SENSOR
ISC SOLENOID
A4LD TRANSMISSION SOLENOID (ATX ONLY)
A4LD TRANSMISSION SOLENOID
AIR CHARGE TEMP.
AIR CHARGE TEMP. SENSOR
EGRV VALVE POSITIONER
KNOCK SENSOR
SOLID STATE
OIL PRESSURE SWITCH
ELECTRIC FAN SWITCH
WATER TEMPERATURE SENDER
TO 14401 ZONE 16-A
TO 14401 ZONE 47-F
V.I.P. SELF TEST CONNECTOR
FUSIBLE LINK 20 GA
FUSIBLE LINK 18 GA
FRAME 19 OF 22
MS-F0ZB-3458-AA

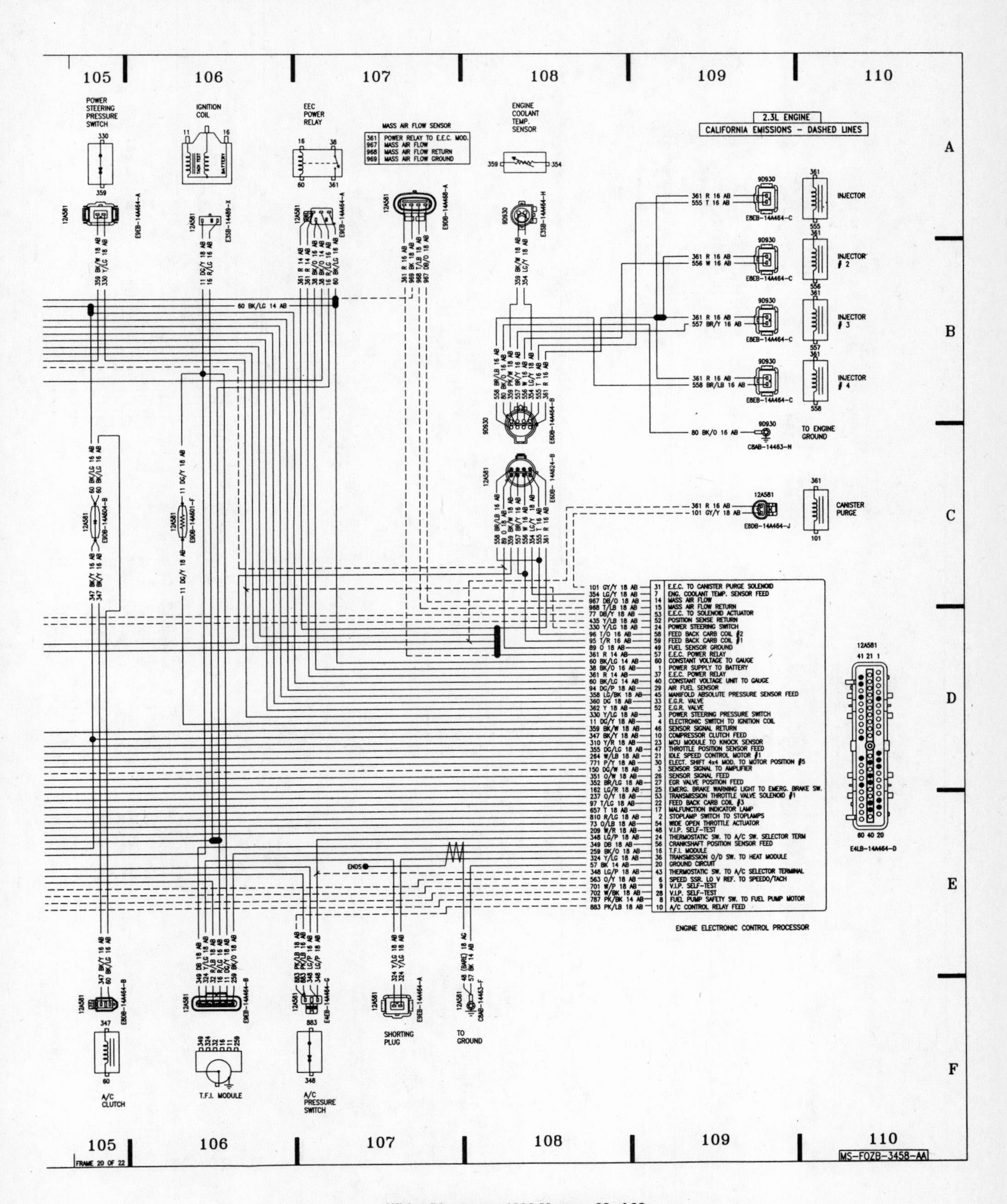

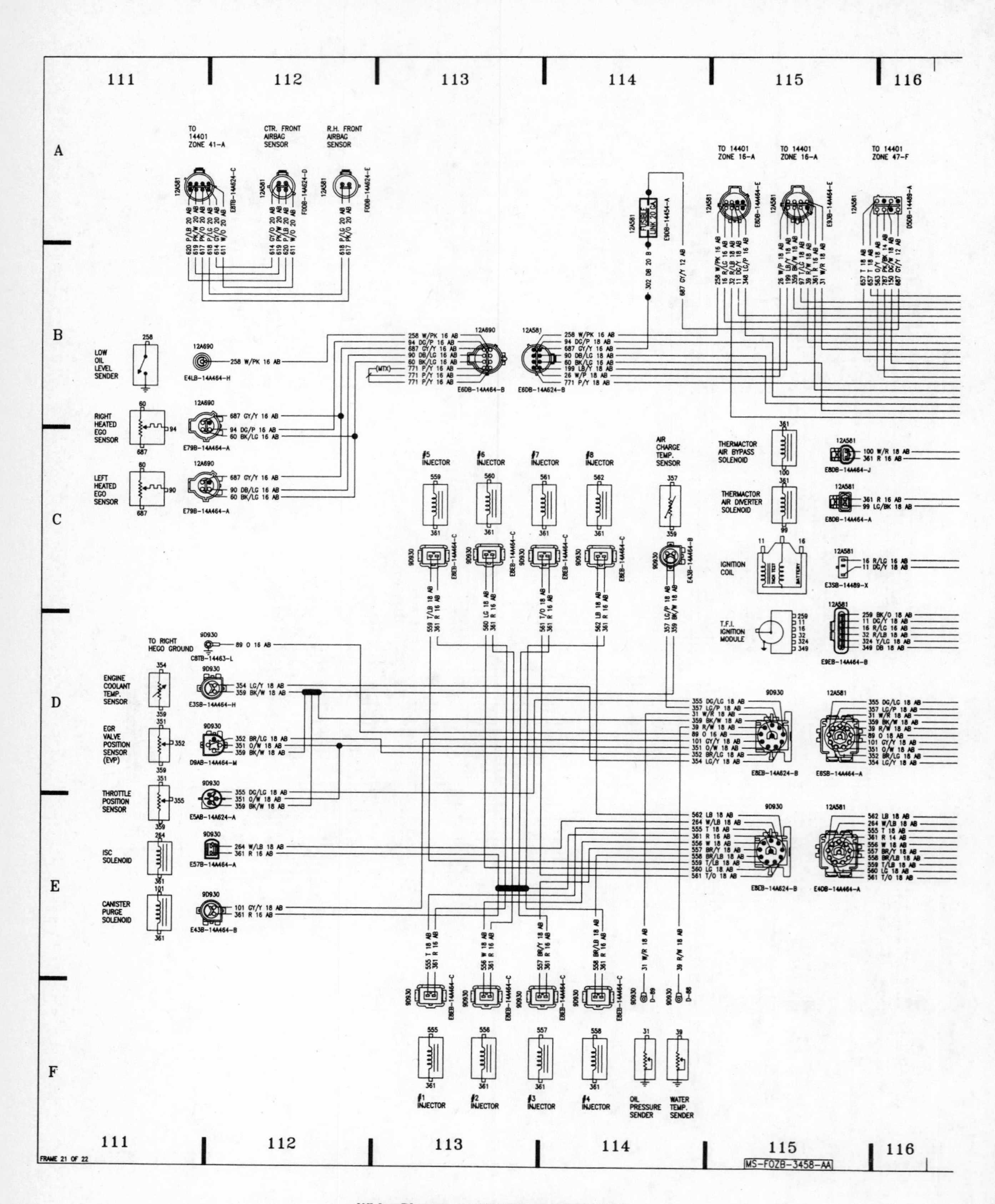

LOW OIL LEVEL SENDER
RIGHT HEATED EGO SENSOR
LEFT HEATED EGO SENSOR
ENGINE COOLANT TEMP. SENSOR
EGR VALVE POSITION SENSOR (EVP)
THROTTLE POSITION SENSOR
ISC SOLENOID
CANISTER PURGE SOLENOID
#5 INJECTOR
#6 INJECTOR
#7 INJECTOR
#8 INJECTOR
AIR CHARGE TEMP. SENSOR
THERMACTOR AIR BYPASS SOLENOID
THERMACTOR AIR DIVERTER SOLENOID
IGNITION COIL
T.F.I. IGNITION MODULE
#1 INJECTOR
#2 INJECTOR
#3 INJECTOR
#4 INJECTOR
OIL PRESSURE SENDER
WATER TEMP. SENDER
CTR. FRONT AIRBAG SENSOR
R.H. FRONT AIRBAG SENSOR
FUSIBLE LINK 20 GA.
TO RIGHT HEGO GROUND
FRAME 21 OF 22
MS-F0ZB-3458-AA

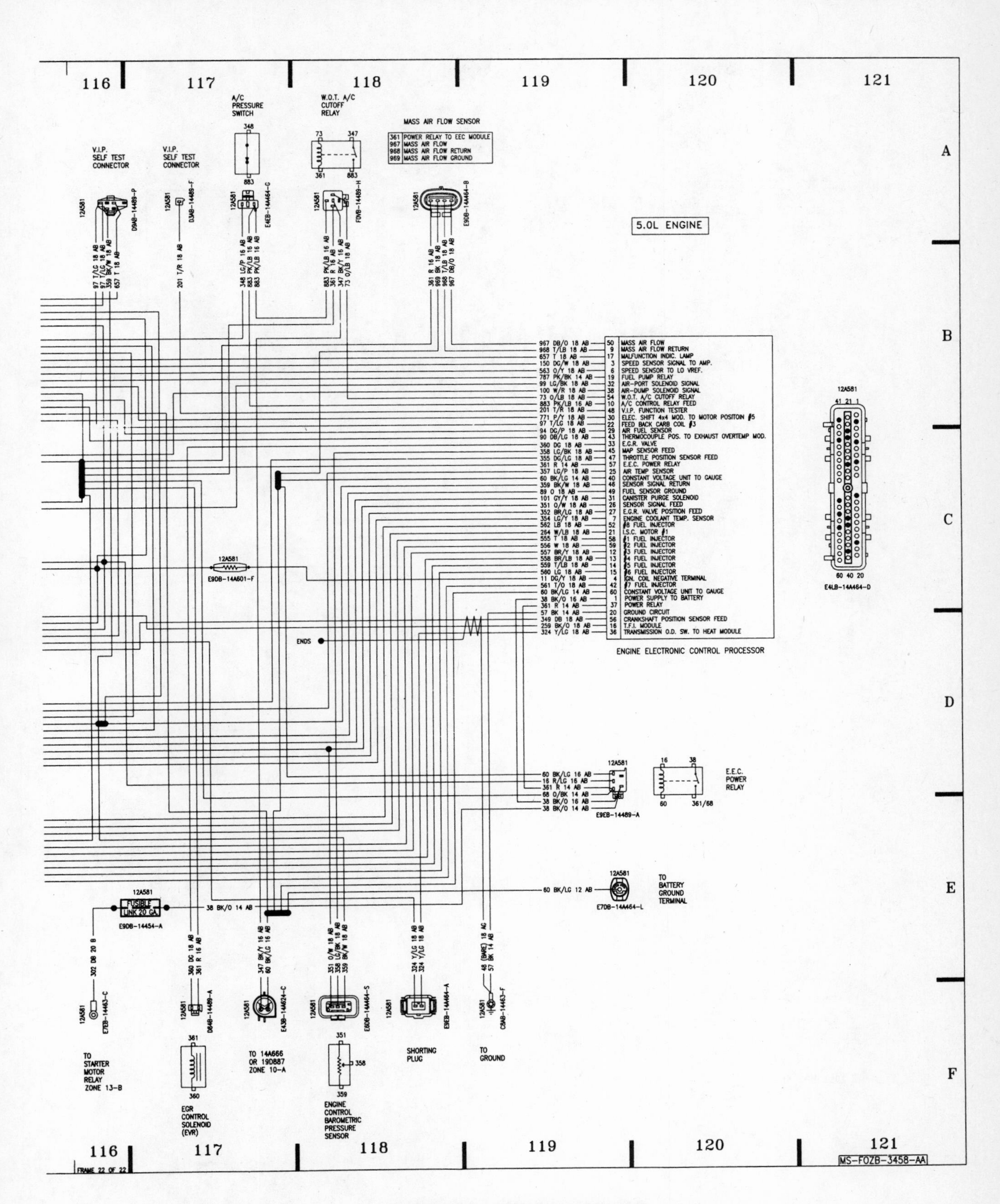

Wiring Diagram — 1990 Mustang 22 of 22

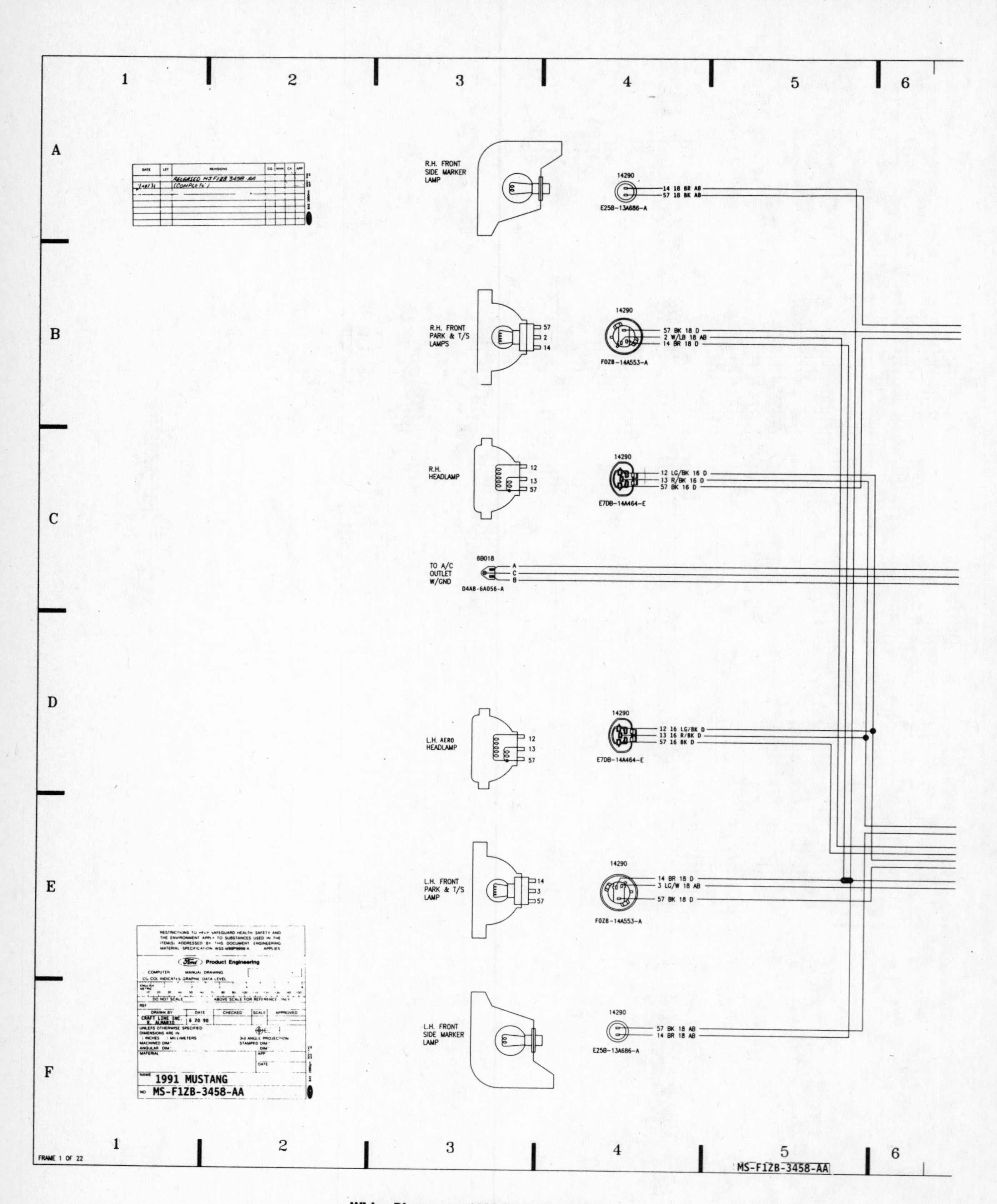

Wiring Diagram — 1991 Mustang 1 of 22

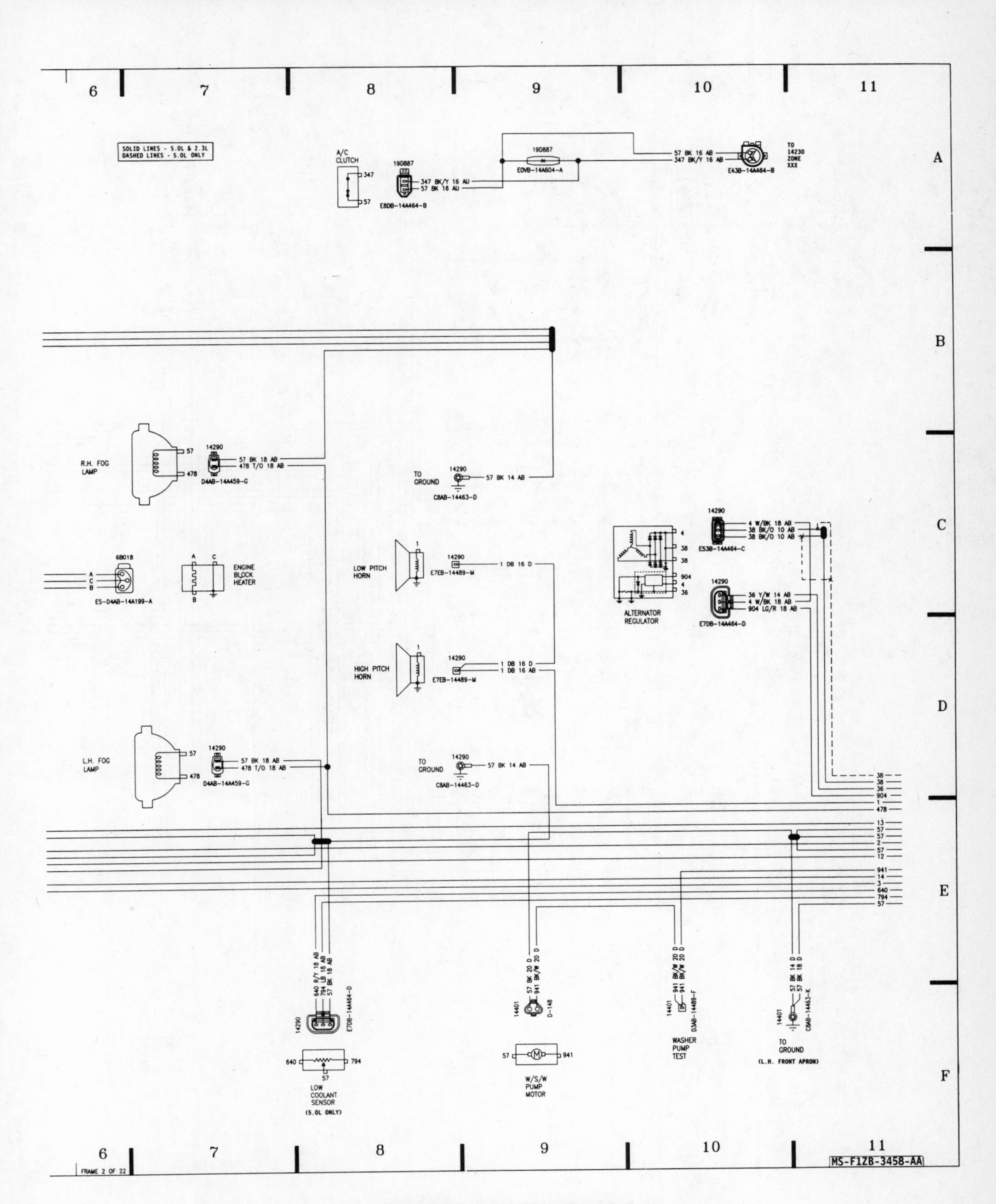
SOLID LINES - 5.0L & 2.3L
DASHED LINES - 5.0L ONLY
A/C CLUTCH
R.H. FOG LAMP
L.H. FOG LAMP
ENGINE BLOCK HEATER
LOW PITCH HORN
HIGH PITCH HORN
TO GROUND
ALTERNATOR REGULATOR
LOW COOLANT SENSOR
(5.0L ONLY)
W/S/W PUMP MOTOR
WASHER PUMP TEST
TO GROUND
(L.H. FRONT APRON)
FRAME 2 OF 22
MS-F1ZB-3458-AA

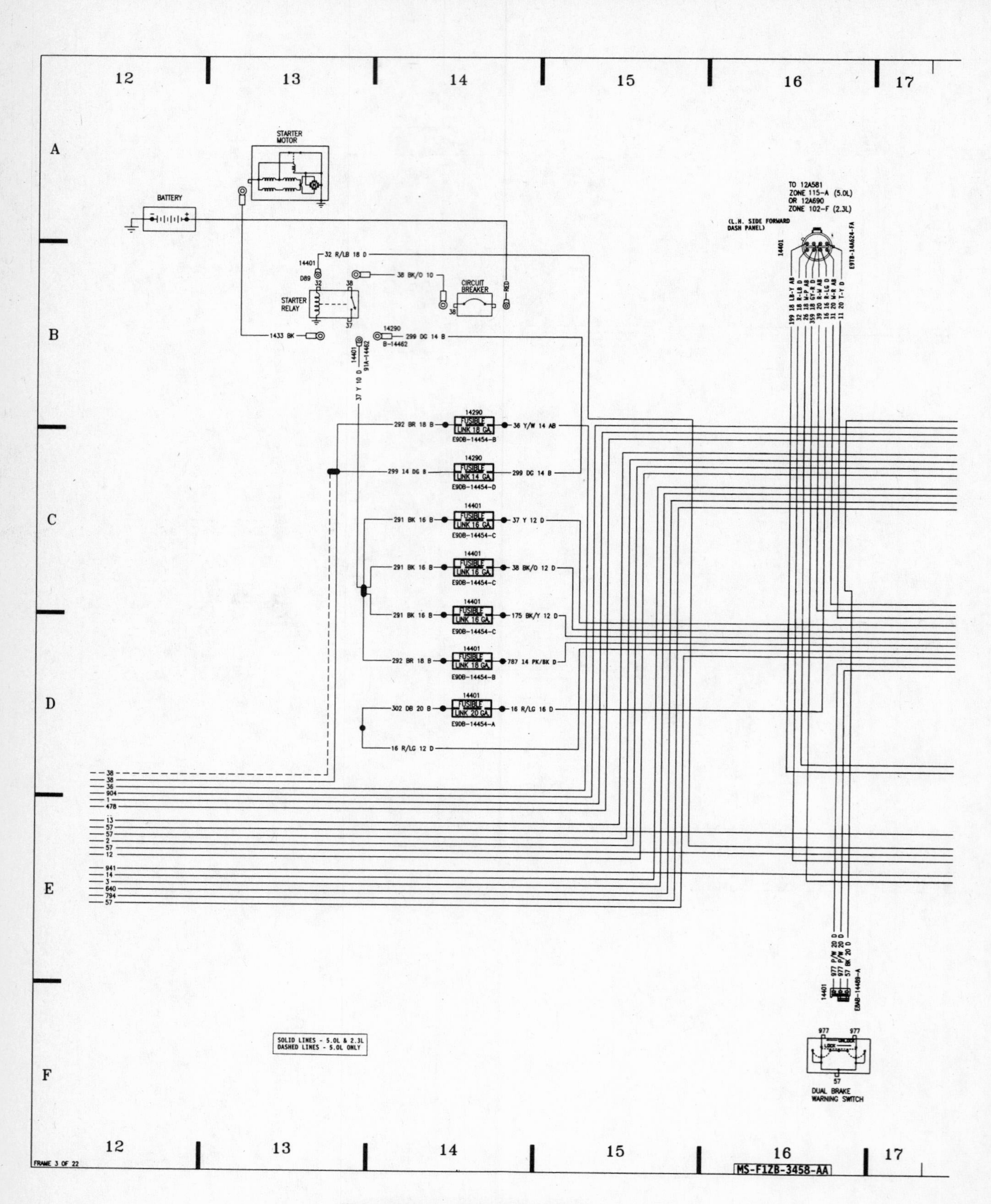
STARTER MOTOR
BATTERY
STARTER RELAY
CIRCUIT BREAKER
FUSIBLE LINK 18 GA
FUSIBLE LINK 14 GA
FUSIBLE LINK 16 GA
FUSIBLE LINK 20 GA
TO 12A581 ZONE 115-A (5.0L) OR 12A690 ZONE 102-F (2.3L)
(L.H. SIDE FORWARD DASH PANEL)
SOLID LINES - 5.0L & 2.3L
DASHED LINES - 5.0L ONLY
DUAL BRAKE WARNING SWITCH
FRAME 3 OF 22
MS-F1ZB-3458-AA

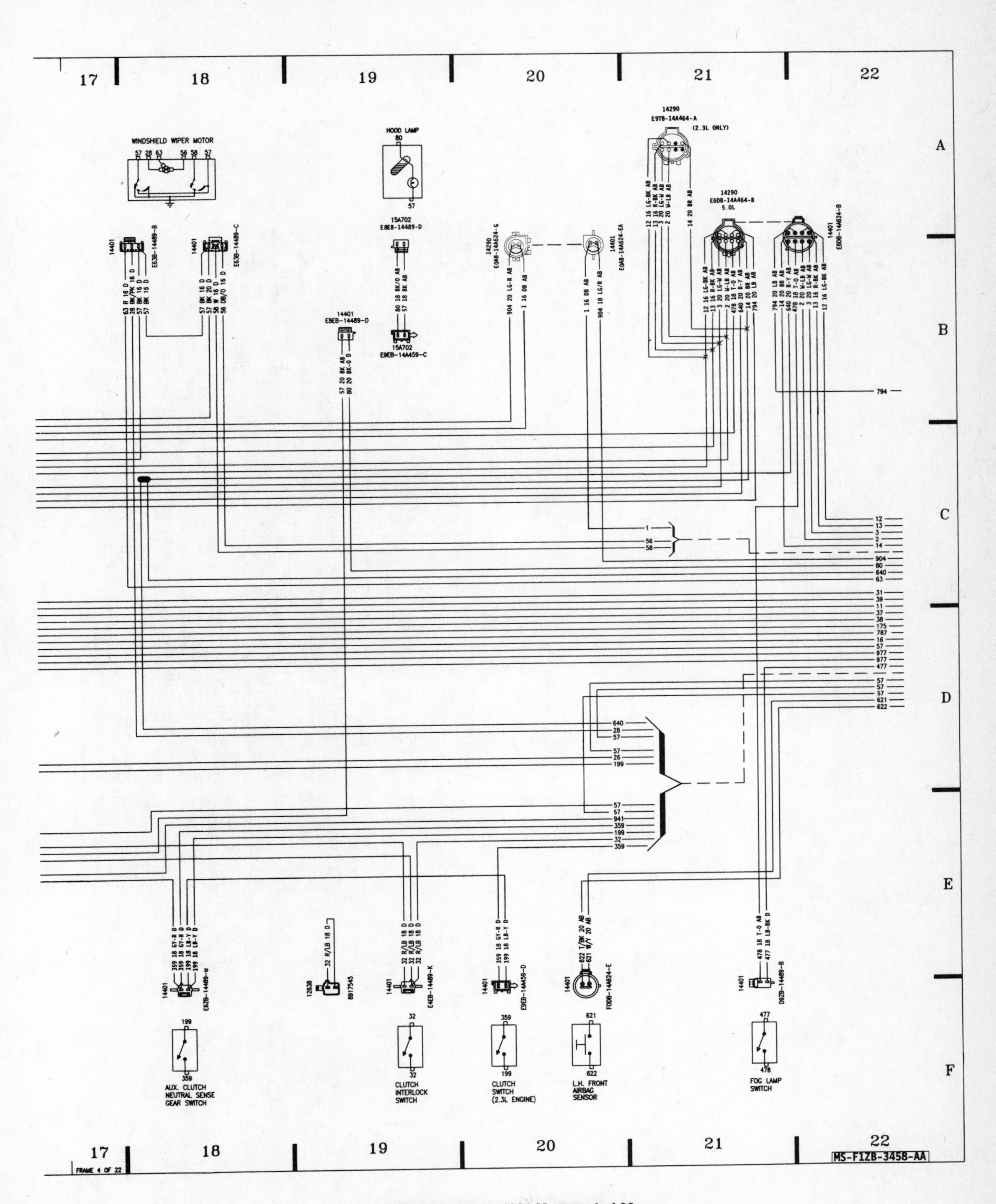

Wiring Diagram — 1991 Mustang 4 of 22

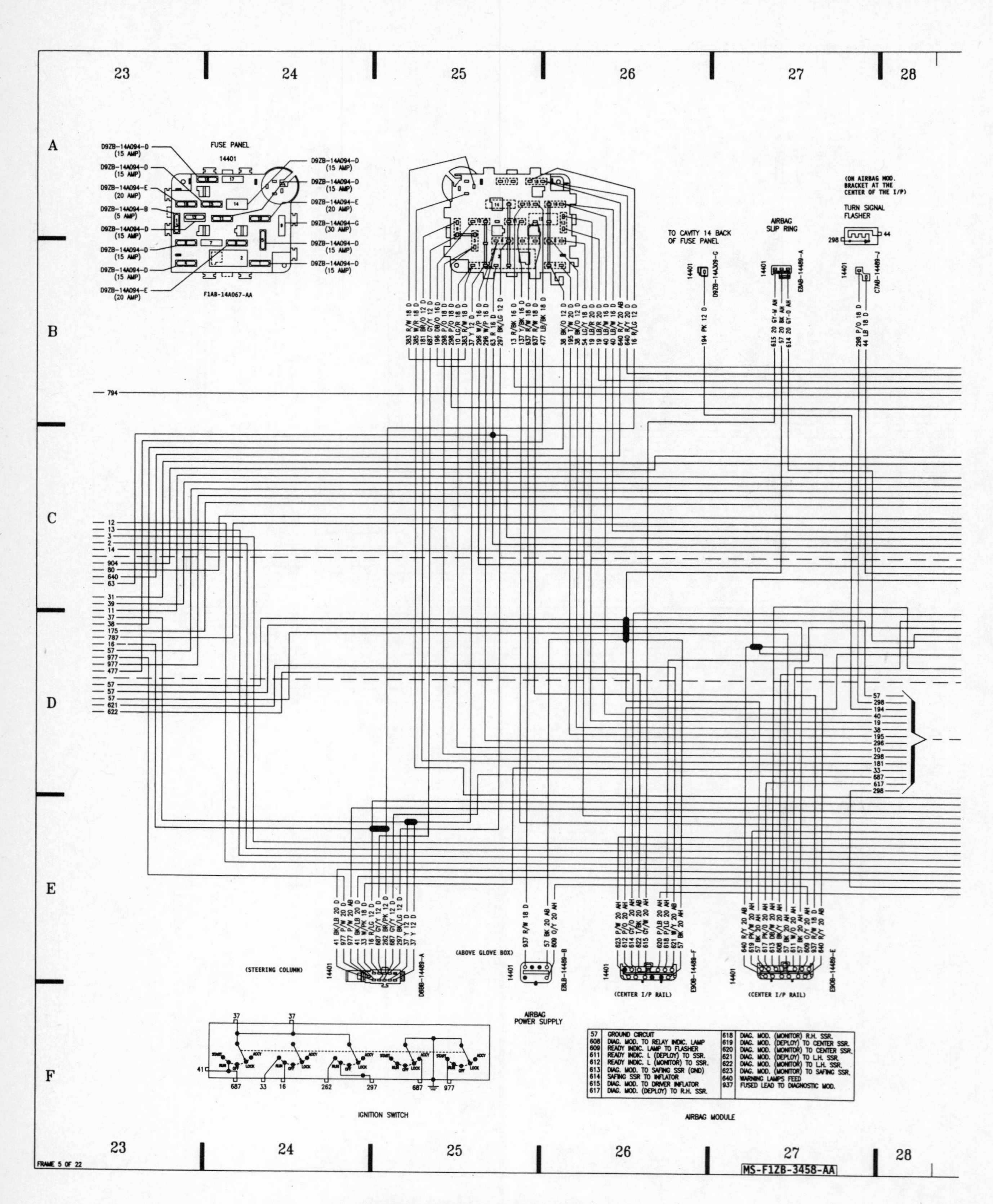

Wiring Diagram — 1991 Mustang 5 of 22

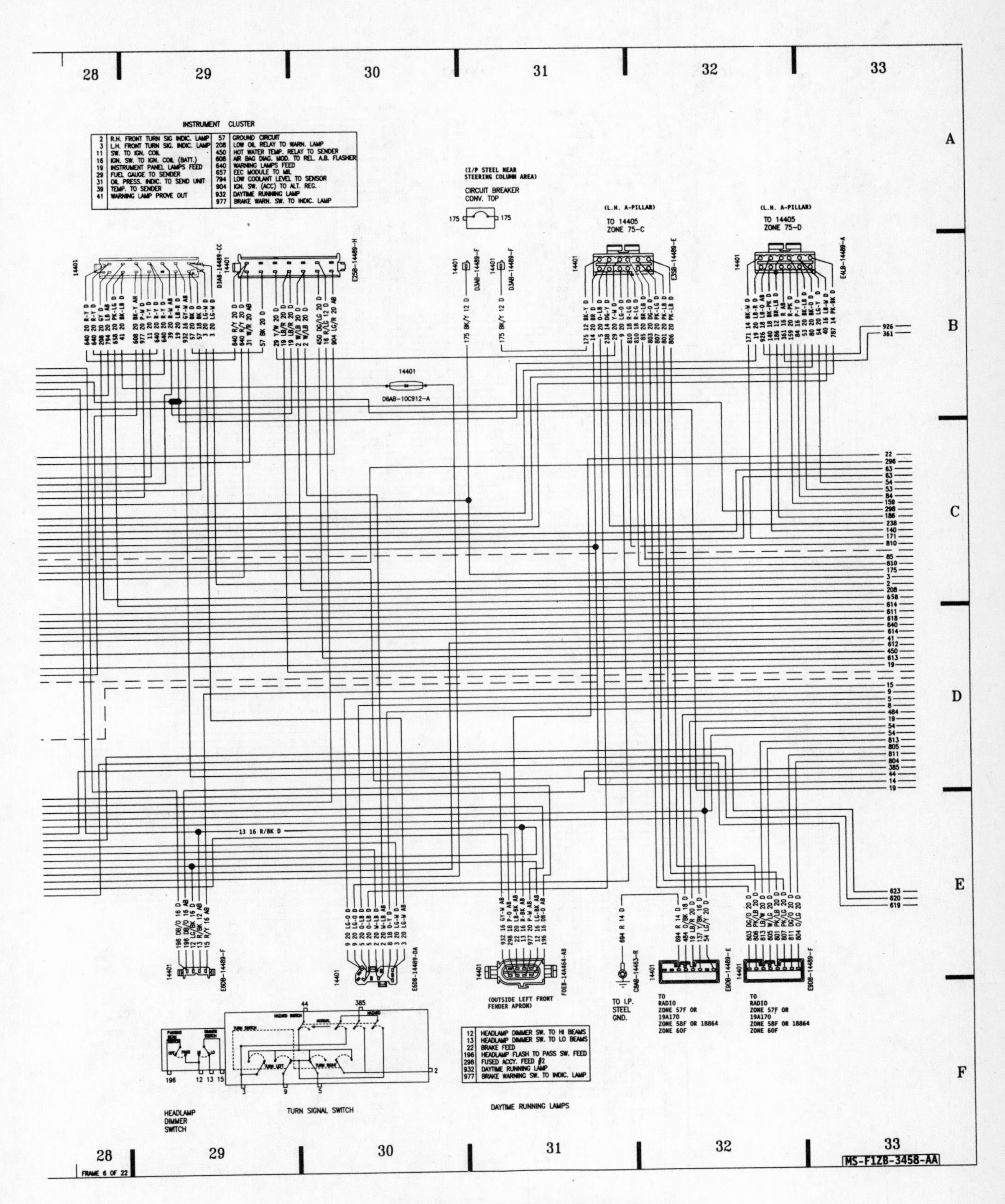
INSTRUMENT CLUSTER
2 R.H. FRONT TURN SIG INDIC. LAMP
3 L.H. FRONT TURN SIG. INDIC. LAMP
11 SW. TO IGN. COIL
16 IGN. SW. TO IGN. COIL (BATT.)
19 INSTRUMENT PANEL LAMPS FEED
29 FUEL GAUGE TO SENDER
31 OIL PRESS. INDIC. TO SEND UNIT
39 TEMP. TO SENDER
41 WARNING LAMP PROVE OUT
57 GROUND CIRCUIT
208 LOW OIL RELAY TO WARN. LAMP
450 HOT WATER TEMP. RELAY TO SENDER
608 AIR BAG DIAG. MOD. TO REL. A.B. FLASHER
640 WARNING LAMPS FEED
657 EEC MODULE TO MIL
794 LOW COOLANT LEVEL TO SENSOR
904 IGN. SW. (ACC) TO ALT. REG.
932 DAYTIME RUNNING LAMP
977 BRAKE WARN. SW. TO INDIC. LAMP
(I/P STEEL NEAR STEERING COLUMN AREA)
CIRCUIT BREAKER CONV. TOP
(L.H. A-PILLAR)
TO 14405 ZONE 75-C
TO 14405 ZONE 75-D
D6AB-10C912-A
13 16 R/BK D
(OUTSIDE LEFT FRONT FENDER APRON)
TO I.P. STEEL GND.
TO RADIO ZONE 57F OR 19A170 ZONE 58F OR 18864 ZONE 60F
12 HEADLAMP DIMMER SW. TO HI BEAMS
13 HEADLAMP DIMMER SW. TO LO BEAMS
22 BRAKE FEED
196 HEADLAMP FLASH TO PASS SW. FEED
298 FUSED ACCY. FEED #2
932 DAYTIME RUNNING LAMP
977 BRAKE WARNING SW. TO INDIC. LAMP
HEADLAMP DIMMER SWITCH
TURN SIGNAL SWITCH
DAYTIME RUNNING LAMPS
FRAME 6 OF 22
MS-F1ZB-3458-AA

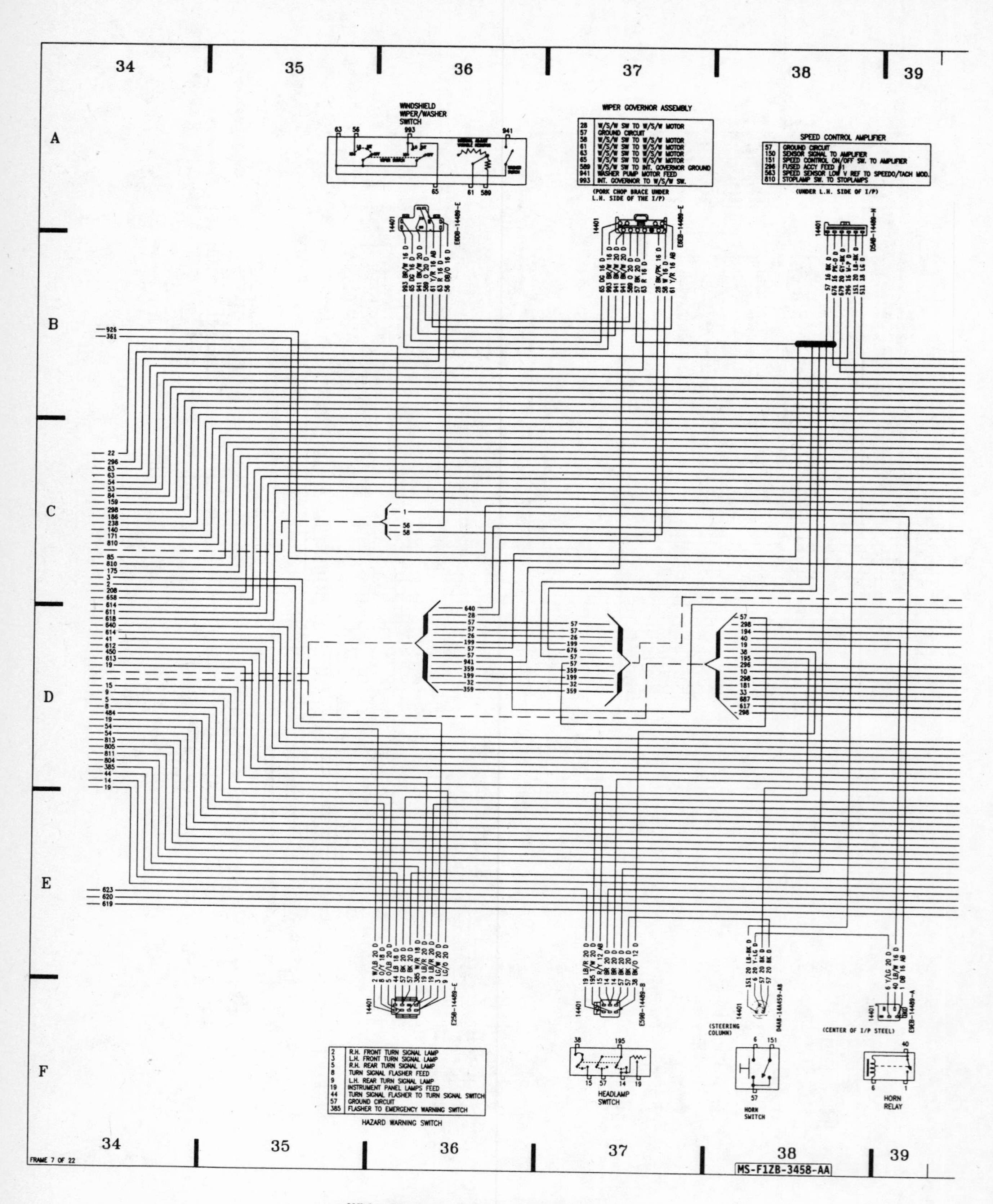

Wiring Diagram — 1991 Mustang 7 of 22

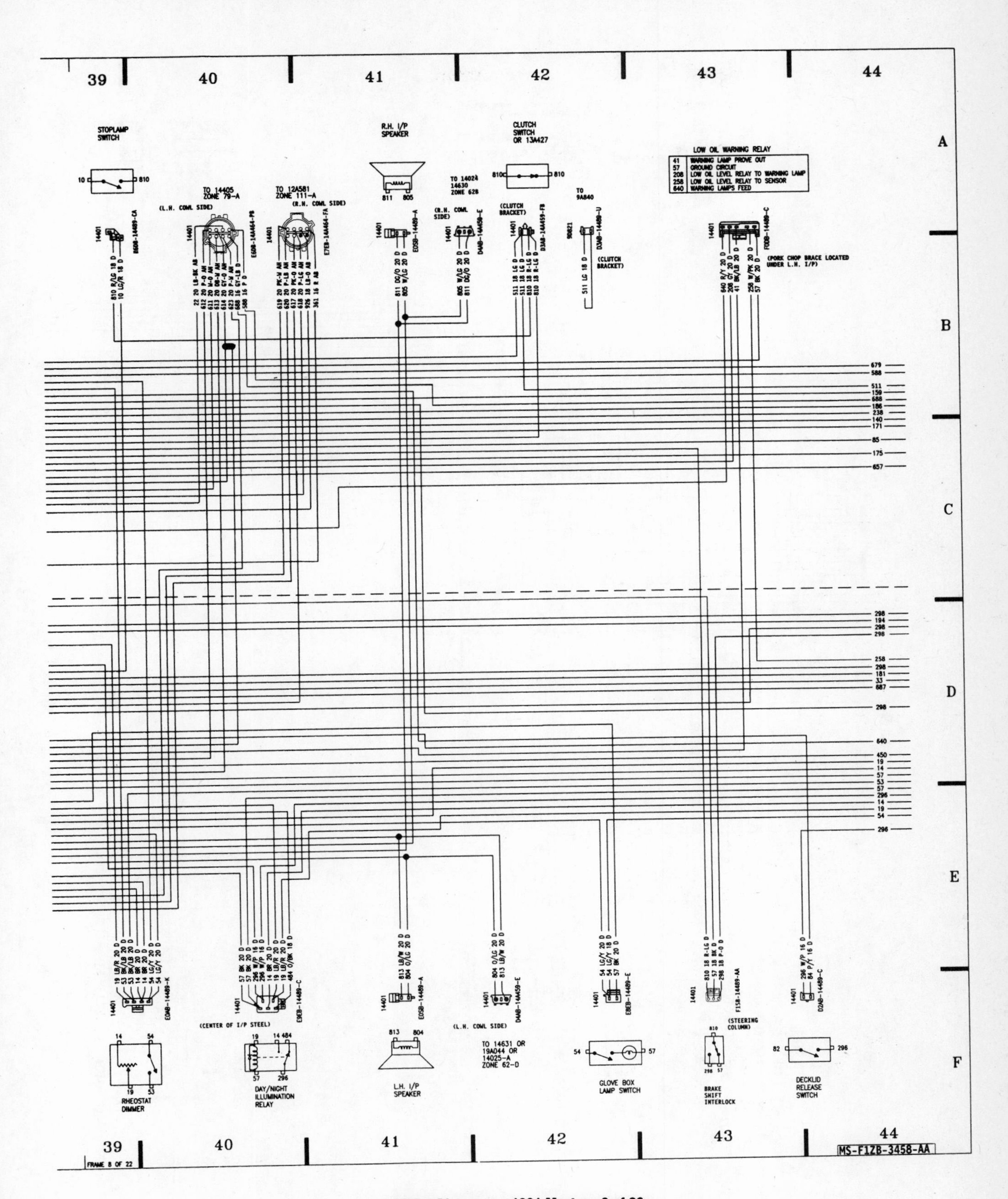
STOPLAMP SWITCH
R.H. I/P SPEAKER
CLUTCH SWITCH OR 13A427
LOW OIL WARNING RELAY
41 WARNING LAMP PROVE OUT
57 GROUND CIRCUIT
208 LOW OIL LEVEL RELAY TO WARNING LAMP
258 LOW OIL LEVEL RELAY TO SENSOR
640 WARNING LAMPS FEED
RHEOSTAT DIMMER
DAY/NIGHT ILLUMINATION RELAY
L.H. I/P SPEAKER
GLOVE BOX LAMP SWITCH
BRAKE SHIFT INTERLOCK
DECKLID RELEASE SWITCH
MS-F1ZB-3458-AA
FRAME 8 OF 22

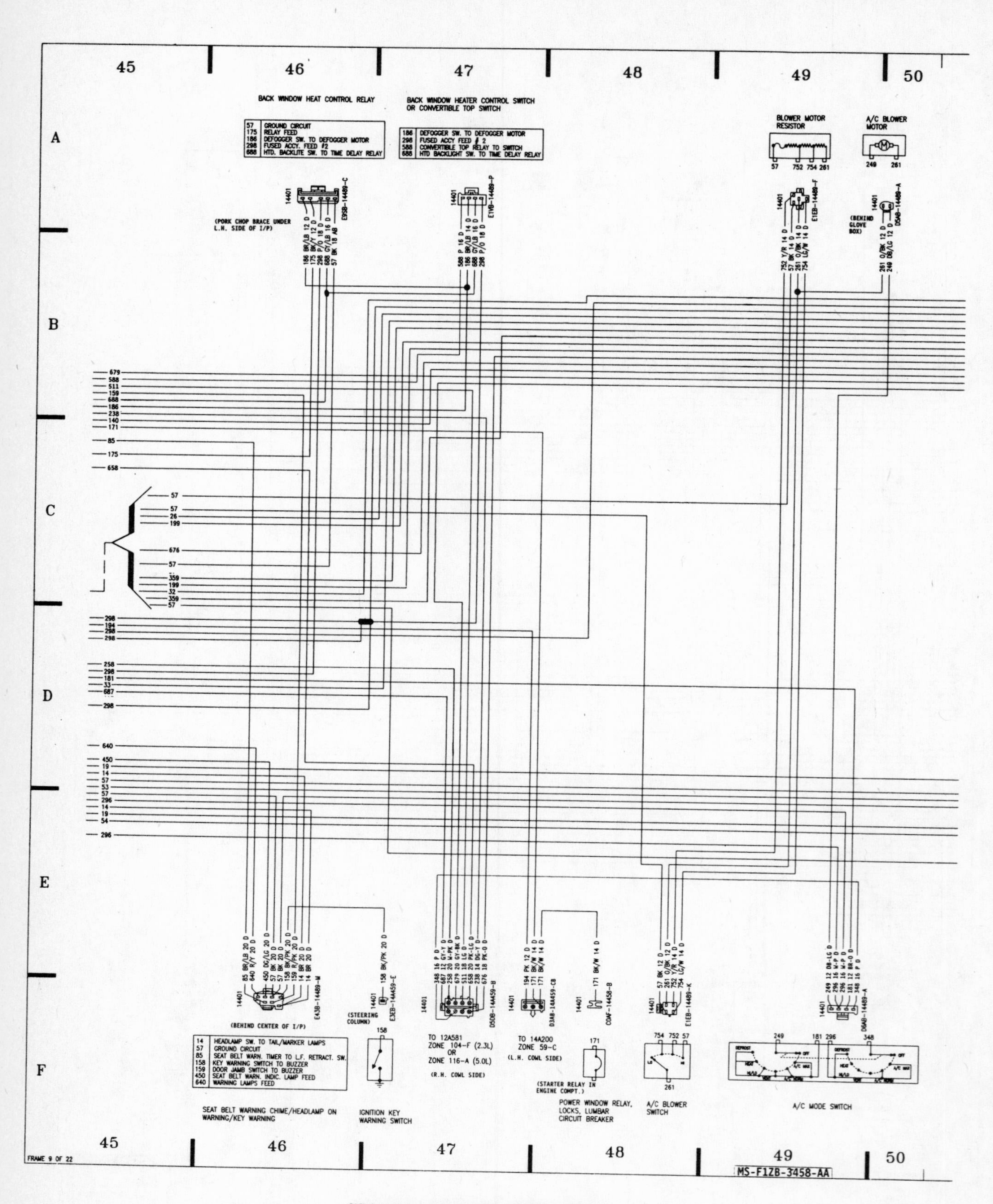

Wiring Diagram — 1991 Mustang 9 of 22

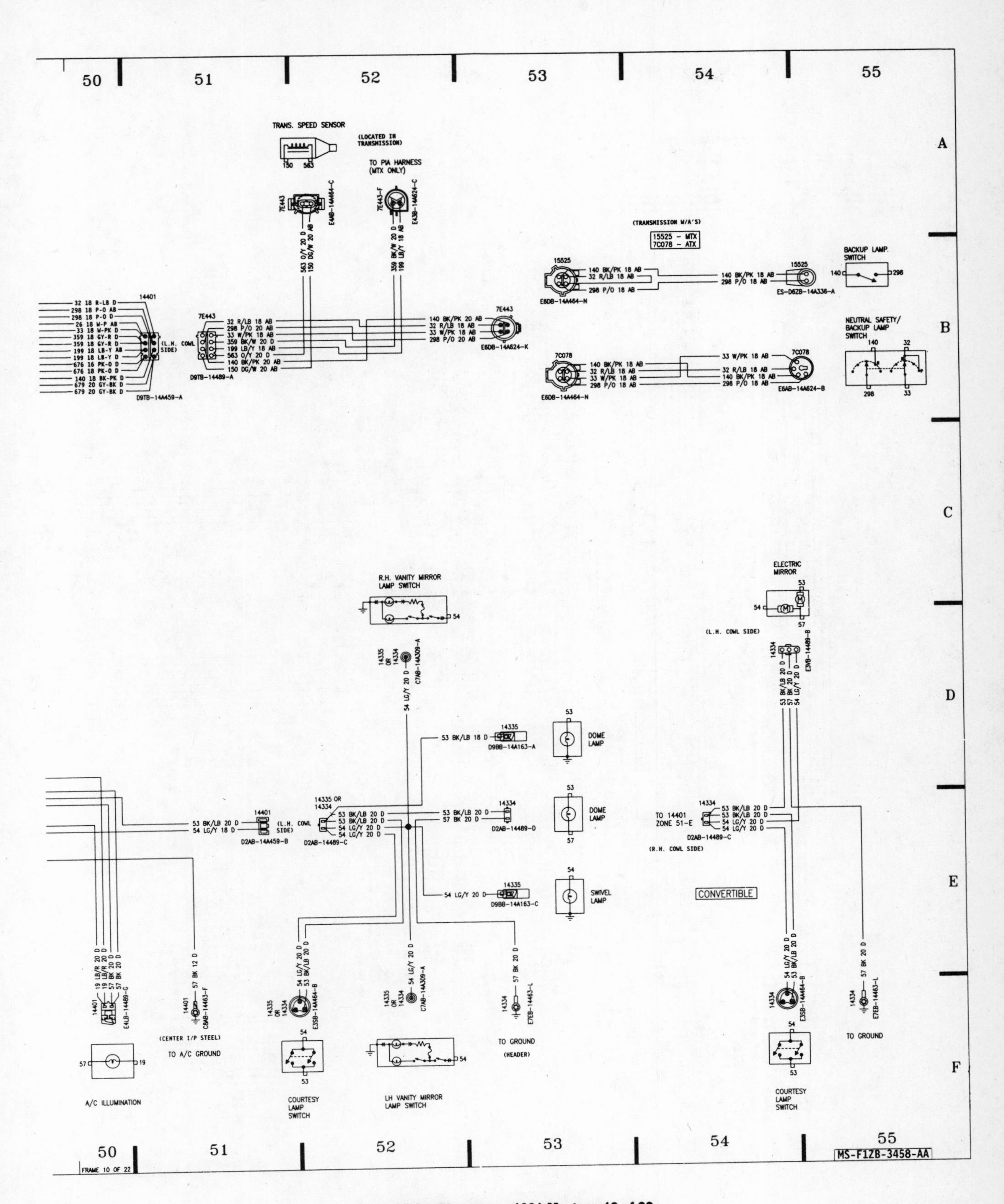

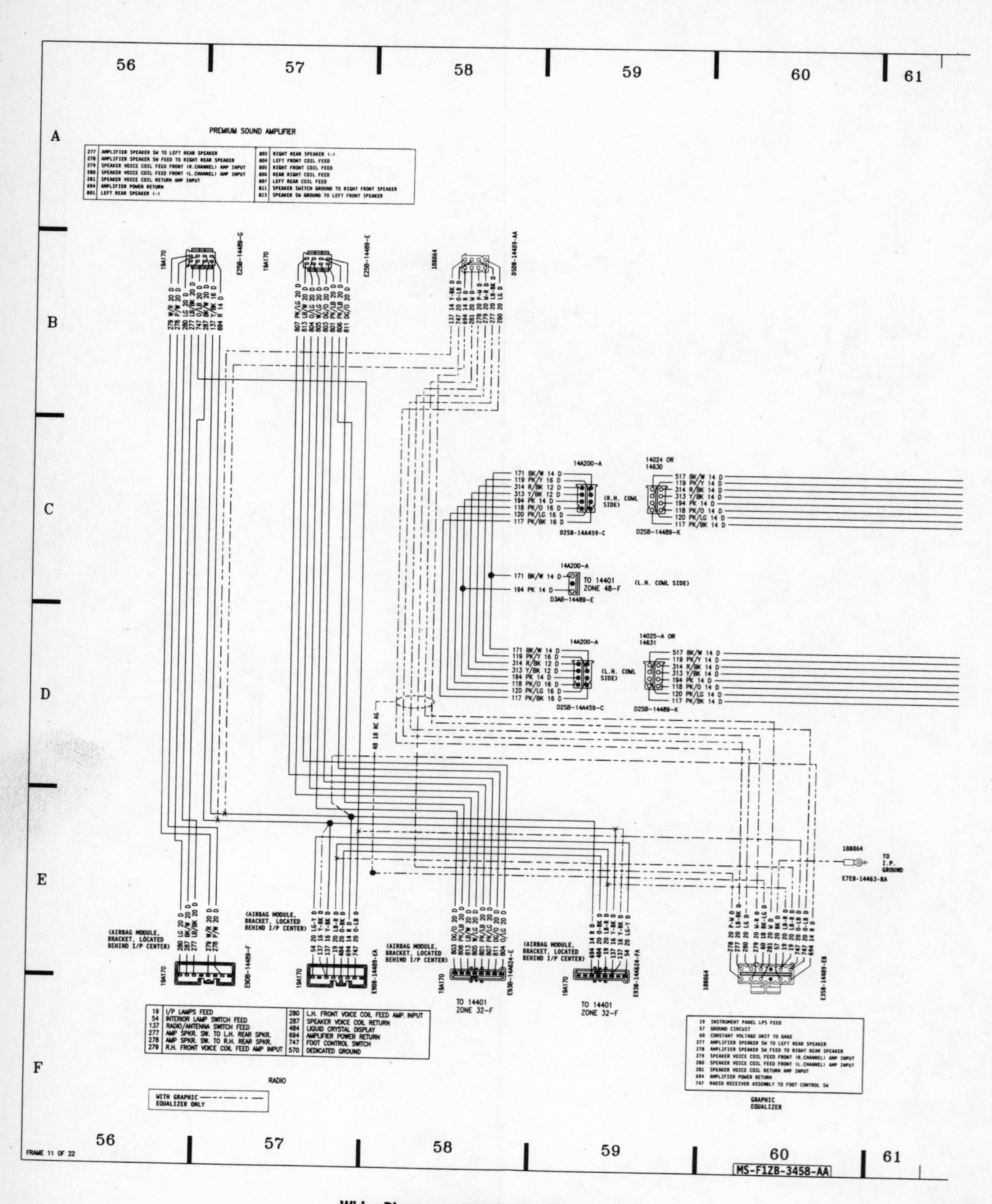
PREMIUM SOUND AMPLIFIER
277 AMPLIFIER SPEAKER SW TO LEFT REAR SPEAKER
278 AMPLIFIER SPEAKER SW FEED TO RIGHT REAR SPEAKER
279 SPEAKER VOICE COIL FEED FRONT (R.CHANNEL) AMP INPUT
280 SPEAKER VOICE COIL FEED FRONT (L.CHANNEL) AMP INPUT
281 SPEAKER VOICE COIL RETURN AMP INPUT
694 AMPLIFIER POWER RETURN
801 LEFT REAR SPEAKER (-)
803 RIGHT REAR SPEAKER (-)
804 LEFT FRONT COIL FEED
805 RIGHT FRONT COIL FEED
806 REAR RIGHT COIL FEED
807 LEFT REAR COIL FEED
811 SPEAKER SWITCH GROUND TO RIGHT FRONT SPEAKER
813 SPEAKER SW GROUND TO LEFT FRONT SPEAKER
19A170
E25B-14489-G
E25B-14489-E
18B864
D5OB-14489-AA
14A200-A
14024 OR 14630
(R.H. COWL SIDE)
D2SB-14A459-C
D2SB-14489-K
TO 14401 ZONE 48-F
(L.H. COWL SIDE)
D3AB-14489-E
14025-A OR 14631
48 18 NC AG
TO I.P. GROUND
E7EB-14463-RA
(AIRBAG MODULE, BRACKET, LOCATED BEHIND I/P CENTER)
E9DB-14489-F
E9DB-14489-EA
E93B-14A624-E
E93B-14A624-FA
TO 14401 ZONE 32-F
E35B-14489-EB
19 I/P LAMPS FEED
54 INTERIOR LAMP SWITCH FEED
137 RADIO/ANTENNA SWITCH FEED
277 AMP SPKR. SW. TO L.H. REAR SPKR.
278 AMP SPKR. SW. TO R.H. REAR SPKR.
279 R.H. FRONT VOICE COIL FEED AMP INPUT
280 L.H. FRONT VOICE COIL FEED AMP. INPUT
287 SPEAKER VOICE COIL RETURN
484 LIQUID CRYSTAL DISPLAY
694 AMPLIFIER POWER RETURN
747 FOOT CONTROL SWITCH
570 DEDICATED GROUND
RADIO
WITH GRAPHIC EQUALIZER ONLY
19 INSTRUMENT PANEL LPS FEED
57 GROUND CIRCUIT
60 CONSTANT VOLTAGE UNIT TO GAGE
277 AMPLIFIER SPEAKER SW TO LEFT REAR SPEAKER
278 AMPLIFIER SPEAKER SW FEED TO RIGHT REAR SPEAKER
279 SPEAKER VOICE COIL FEED FRONT (R.CHANNEL) AMP INPUT
280 SPEAKER VOICE COIL FEED FRONT (L.CHANNEL) AMP INPUT
281 SPEAKER VOICE COIL RETURN AMP INPUT
694 AMPLIFIER POWER RETURN
747 RADIO RECEIVER ASSEMBLY TO FOOT CONTROL SW
GRAPHIC EQUALIZER
FRAME 11 OF 22
MS-F1ZB-3458-AA

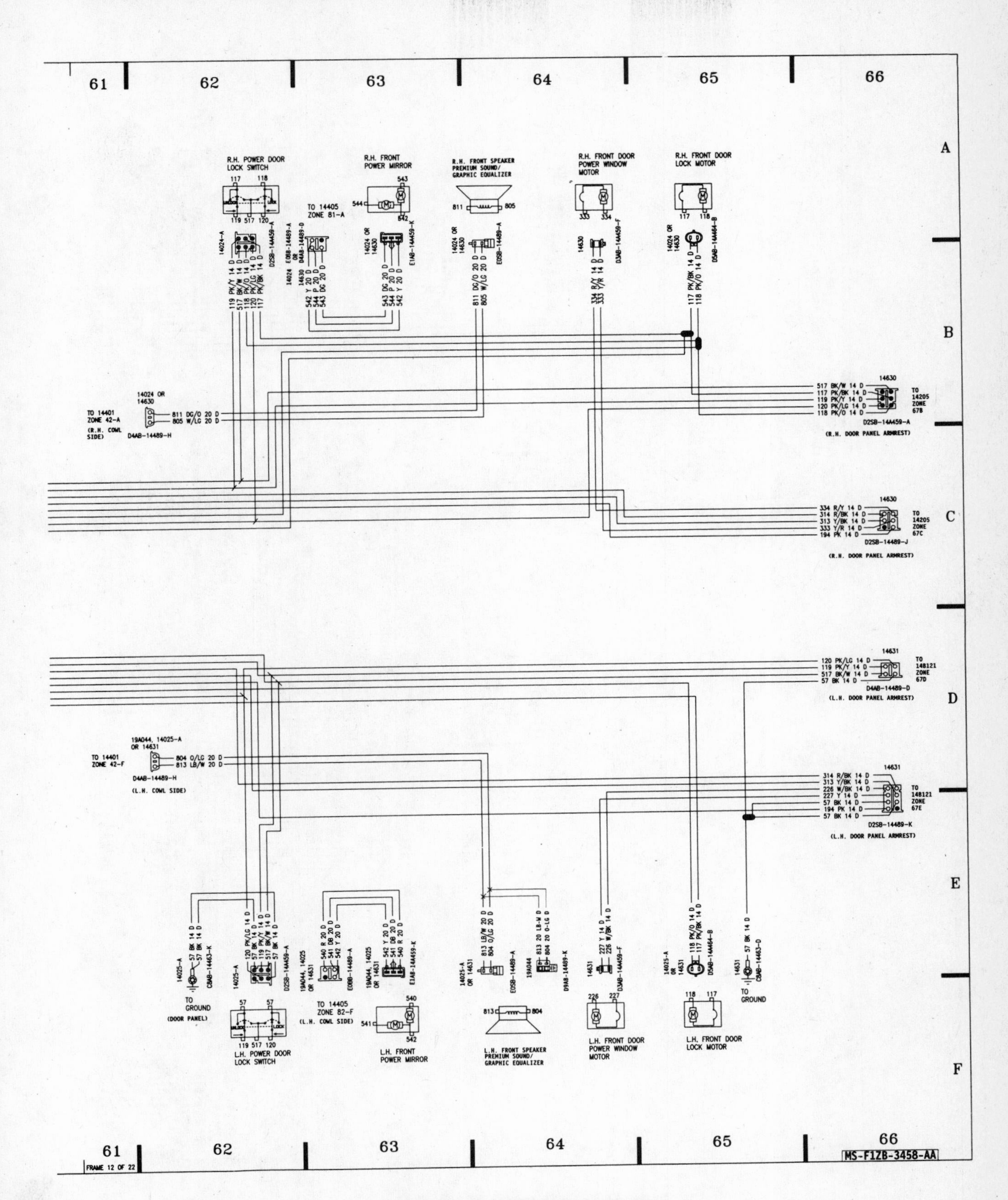
R.H. POWER DOOR LOCK SWITCH
R.H. FRONT POWER MIRROR
R.H. FRONT SPEAKER PREMIUM SOUND/ GRAPHIC EQUALIZER
R.H. FRONT DOOR POWER WINDOW MOTOR
R.H. FRONT DOOR LOCK MOTOR
TO 14405 ZONE 81-A
TO 14401 ZONE 42-A (R.H. COWL SIDE)
D4AB-14489-H
(R.H. DOOR PANEL ARMREST)
TO 14205 ZONE 67B
TO 14205 ZONE 67C
TO 148121 ZONE 67D
TO 148121 ZONE 67E
(L.H. DOOR PANEL ARMREST)
TO 14401 ZONE 42-F
(L.H. COWL SIDE)
TO GROUND (DOOR PANEL)
L.H. POWER DOOR LOCK SWITCH
TO 14405 ZONE 82-F (L.H. COWL SIDE)
L.H. FRONT POWER MIRROR
L.H. FRONT SPEAKER PREMIUM SOUND/ GRAPHIC EQUALIZER
L.H. FRONT DOOR POWER WINDOW MOTOR
L.H. FRONT DOOR LOCK MOTOR
TO GROUND
FRAME 12 OF 22
MS-F1ZB-3458-AA

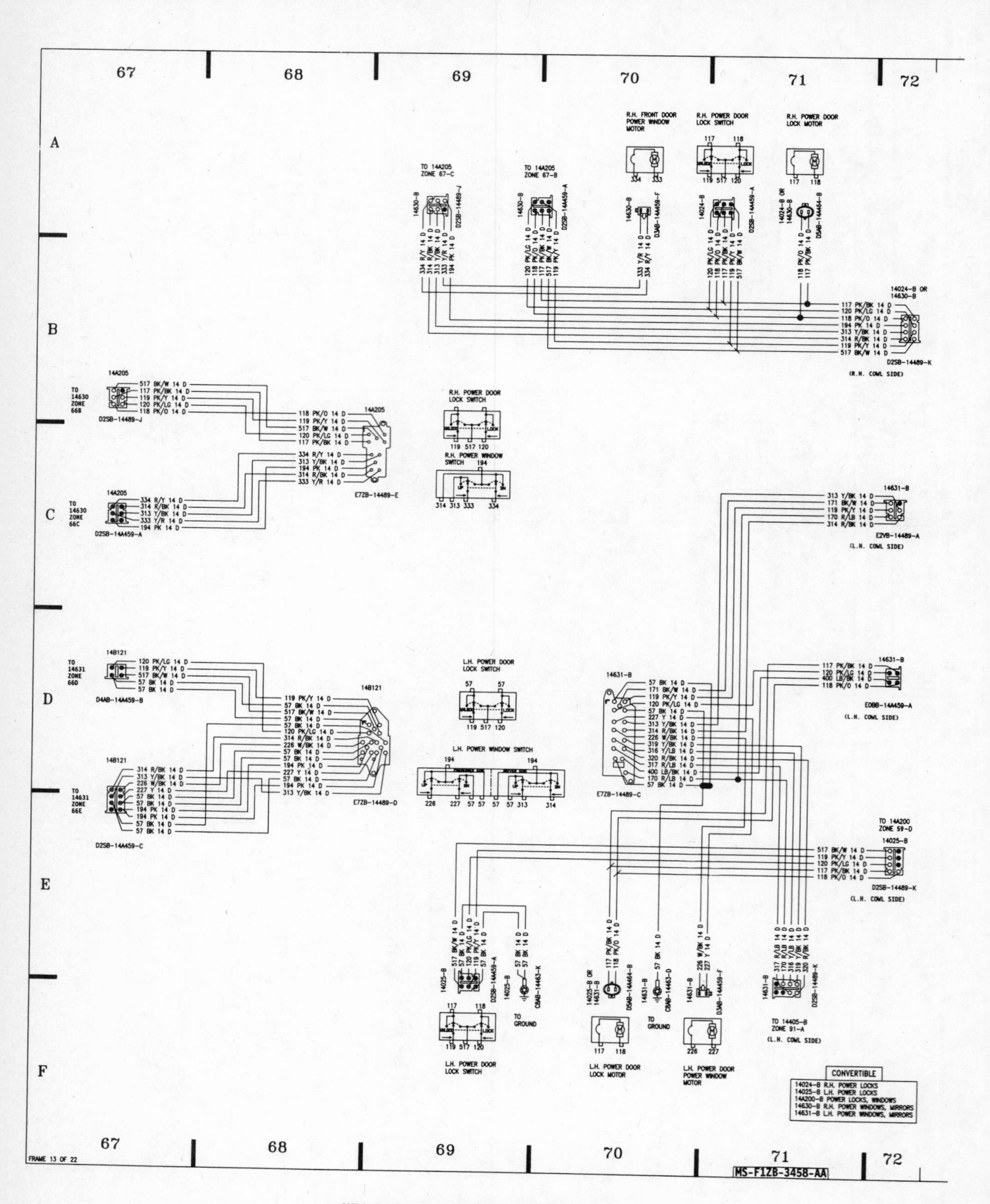

R.H. FRONT DOOR POWER WINDOW MOTOR
R.H. POWER DOOR LOCK SWITCH
R.H. POWER DOOR LOCK MOTOR
R.H. POWER WINDOW SWITCH
L.H. POWER DOOR LOCK SWITCH
L.H. POWER WINDOW SWITCH
L.H. POWER DOOR LOCK MOTOR
L.H. POWER DOOR POWER WINDOW MOTOR
(R.H. COWL SIDE)
(L.H. COWL SIDE)
CONVERTIBLE
14024-B R.H. POWER LOCKS
14025-B L.H. POWER LOCKS
14A200-B POWER LOCKS, WINDOWS
14630-B R.H. POWER WINDOWS, MIRRORS
14631-B L.H. POWER WINDOWS, MIRRORS
FRAME 13 OF 22
MS-F1ZB-3458-AA

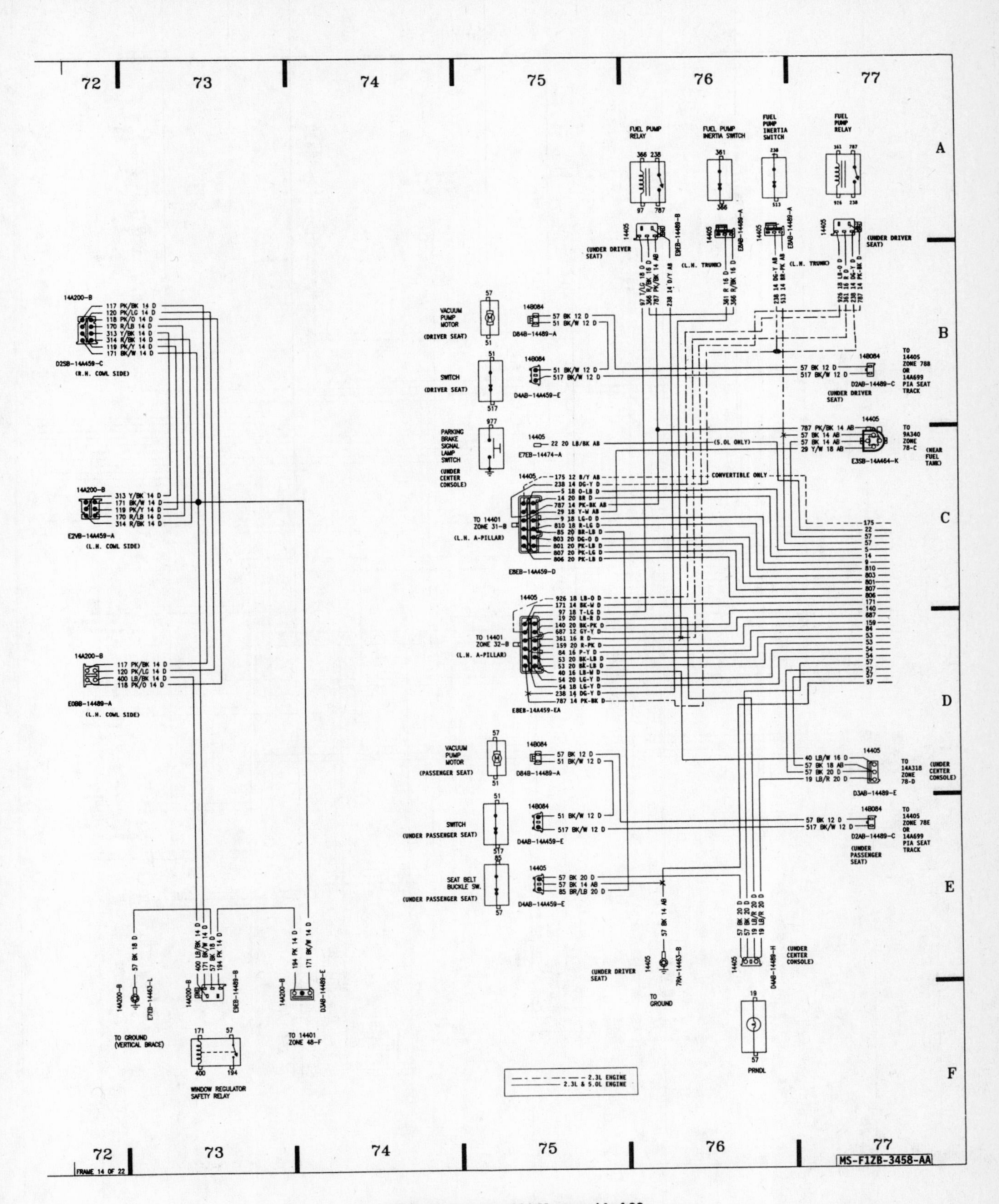

FUEL PUMP RELAY
FUEL PUMP INERTIA SWITCH
FUEL PUMP INERTIA SWITCH
FUEL PUMP RELAY
(UNDER DRIVER SEAT)
(L.H. TRUNK)
14A200-B
D2SB-14A459-C
(R.H. COWL SIDE)
E2VB-14A459-A
(L.H. COWL SIDE)
E0BB-14489-A
(L.H. COWL SIDE)
VACUUM PUMP MOTOR
(DRIVER SEAT)
D84B-14489-A
SWITCH
(DRIVER SEAT)
D4AB-14A459-E
PARKING BRAKE SIGNAL LAMP SWITCH
(UNDER CENTER CONSOLE)
E7EB-14474-A
(5.0L ONLY)
CONVERTIBLE ONLY
TO 14401 ZONE 31-B
(L.H. A-PILLAR)
E8EB-14A459-D
TO 14401 ZONE 32-B
(L.H. A-PILLAR)
E8EB-14A459-EA
TO 14405 ZONE 78B OR 14A699 PIA SEAT TRACK
D2AB-14489-C
(UNDER DRIVER SEAT)
TO 9A340 ZONE 78-C
(NEAR FUEL TANK)
E3SB-14A464-K
VACUUM PUMP MOTOR
(PASSENGER SEAT)
SWITCH
(UNDER PASSENGER SEAT)
SEAT BELT BUCKLE SW.
(UNDER PASSENGER SEAT)
TO 14A318 ZONE 78-D
(UNDER CENTER CONSOLE)
D3AB-14489-E
TO 14405 ZONE 78E OR 14A699 PIA SEAT TRACK
(UNDER PASSENGER SEAT)
TO GROUND (VERTICAL BRACE)
WINDOW REGULATOR SAFETY RELAY
TO 14401 ZONE 48-F
(UNDER DRIVER SEAT)
TO GROUND
(UNDER CENTER CONSOLE)
PRNDL
2.3L ENGINE
2.3L & 5.0L ENGINE
FRAME 14 OF 22
MS-F1ZB-3458-AA

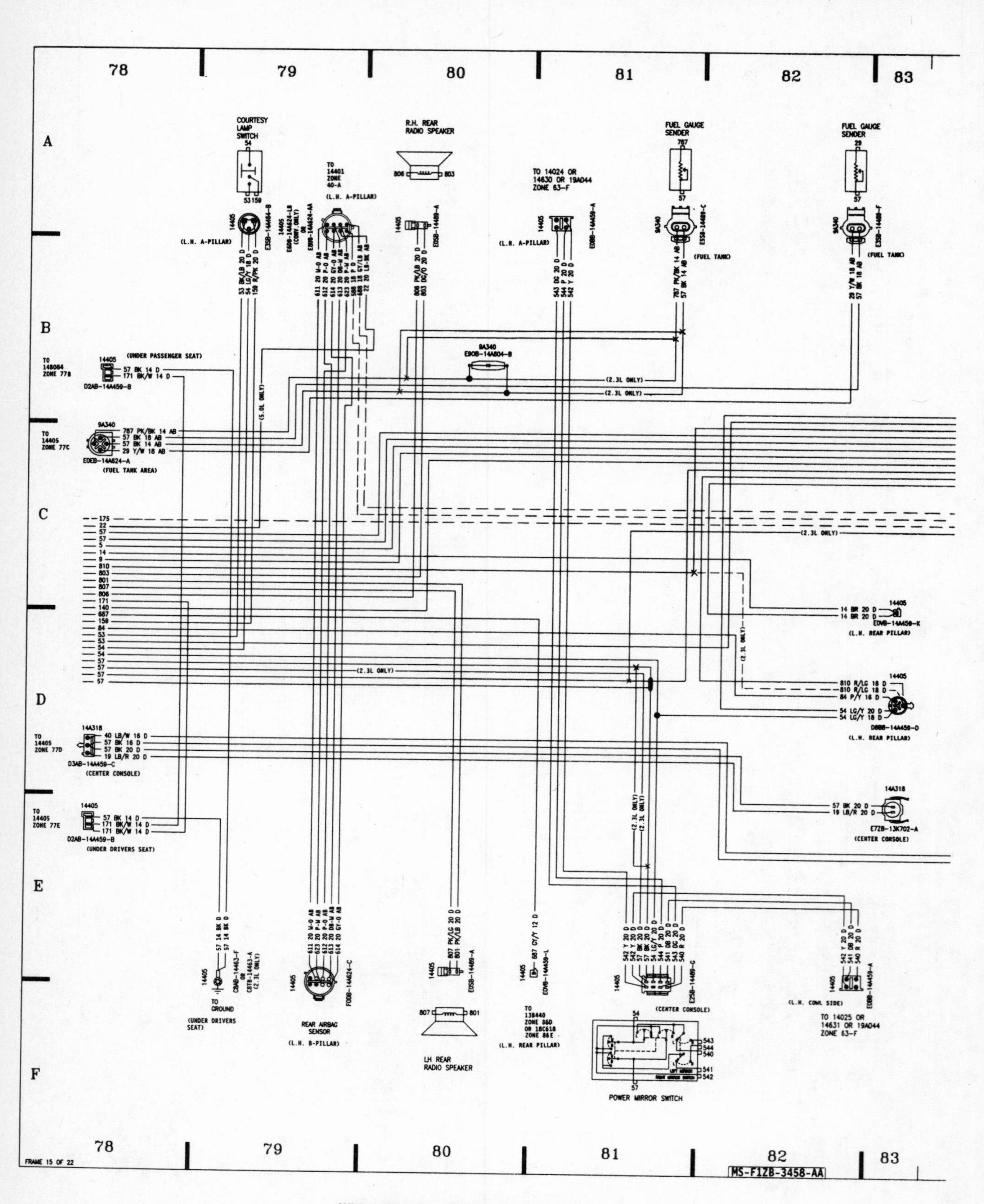

Wiring Diagram — 1991 Mustang 15 of 22

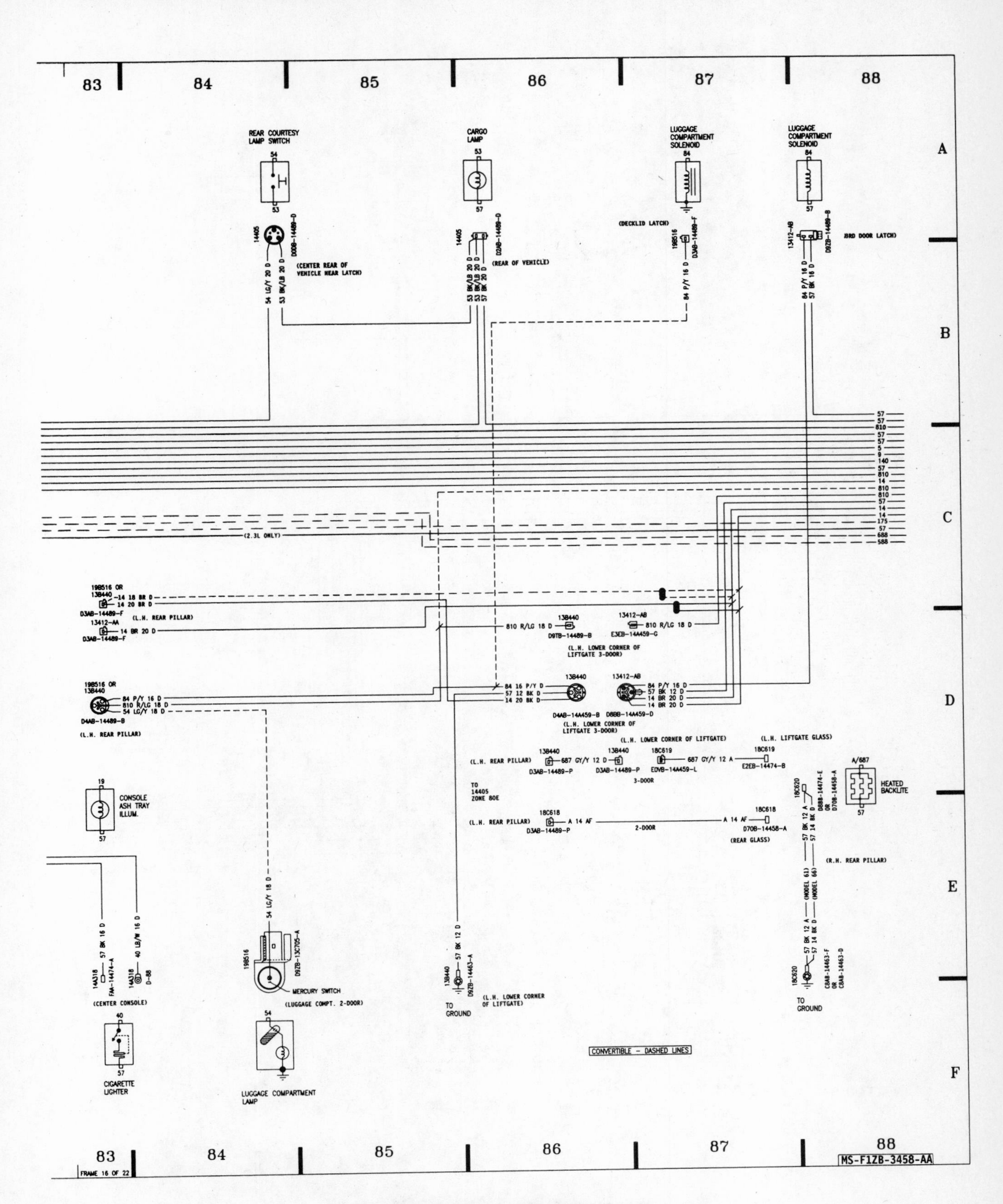
REAR COURTESY LAMP SWITCH
CARGO LAMP
LUGGAGE COMPARTMENT SOLENOID
(DECKLID LATCH)
(3RD DOOR LATCH)
(CENTER REAR OF VEHICLE NEAR LATCH)
(REAR OF VEHICLE)
(2.3L ONLY)
(L.H. REAR PILLAR)
(L.H. LOWER CORNER OF LIFTGATE 3-DOOR)
(L.H. LOWER CORNER OF LIFTGATE)
(L.H. LIFTGATE GLASS)
HEATED BACKLITE
CONSOLE ASH TRAY ILLUM.
2-DOOR
3-DOOR
(REAR GLASS)
(R.H. REAR PILLAR)
MERCURY SWITCH
(LUGGAGE COMPT. 2-DOOR)
(CENTER CONSOLE)
CIGARETTE LIGHTER
LUGGAGE COMPARTMENT LAMP
TO GROUND
CONVERTIBLE - DASHED LINES
FRAME 16 OF 22
MS-F1ZB-3458-AA

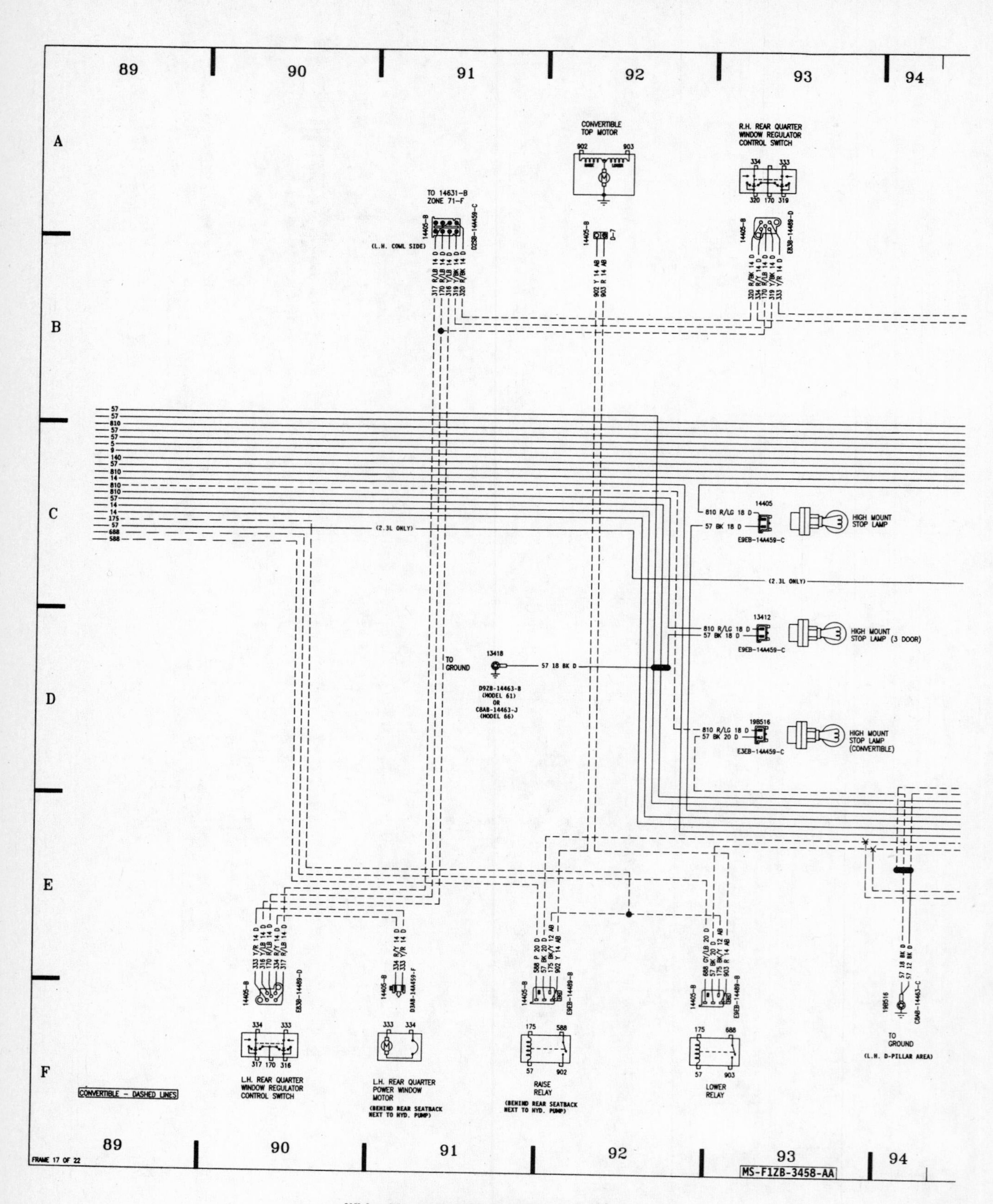
CONVERTIBLE TOP MOTOR
R.H. REAR QUARTER WINDOW REGULATOR CONTROL SWITCH
TO 14631-B ZONE 71-F
(L.H. COWL SIDE)
HIGH MOUNT STOP LAMP
HIGH MOUNT STOP LAMP (3 DOOR)
HIGH MOUNT STOP LAMP (CONVERTIBLE)
(2.3L ONLY)
TO GROUND
L.H. REAR QUARTER WINDOW REGULATOR CONTROL SWITCH
L.H. REAR QUARTER POWER WINDOW MOTOR
(BEHIND REAR SEATBACK NEXT TO HYD. PUMP)
RAISE RELAY
LOWER RELAY
TO GROUND (L.H. D-PILLAR AREA)
CONVERTIBLE - DASHED LINES
FRAME 17 OF 22
MS-F1ZB-3458-AA

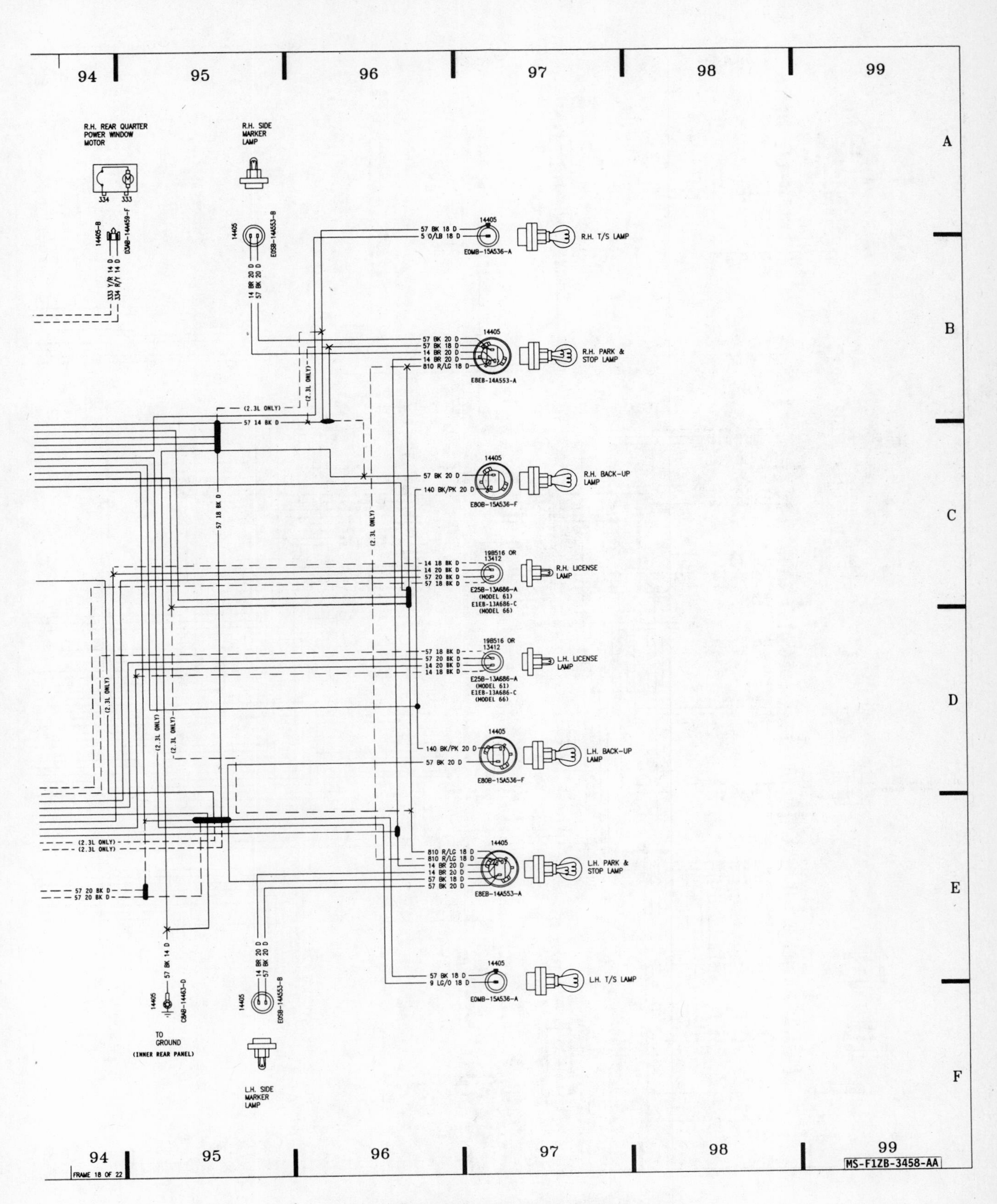

Wiring Diagram — 1991 Mustang 18 of 22

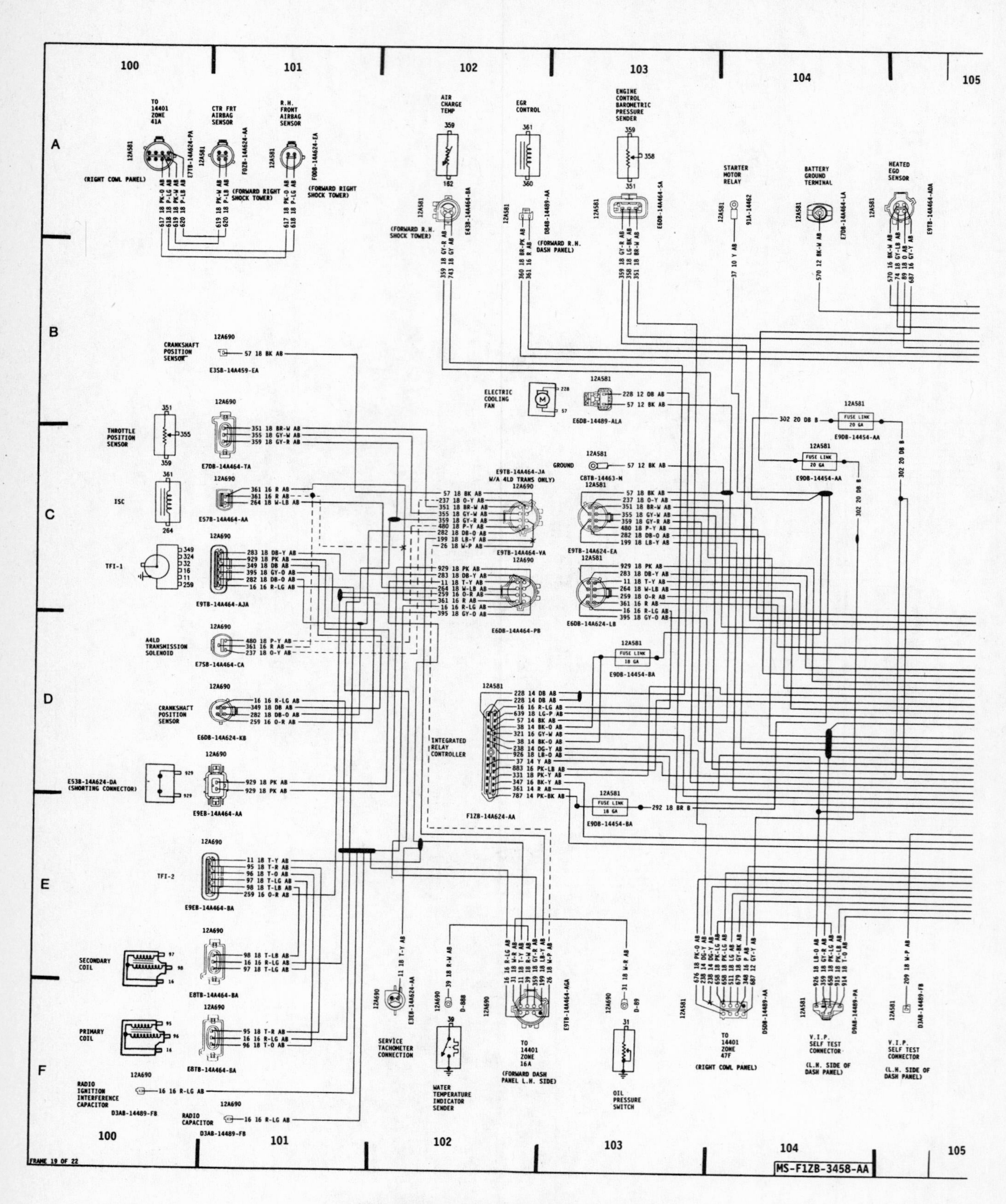

TO 14401 ZONE 41A
CTR FRT AIRBAG SENSOR
R.H. FRONT AIRBAG SENSOR
(RIGHT COWL PANEL)
(FORWARD RIGHT SHOCK TOWER)
AIR CHARGE TEMP
EGR CONTROL
ENGINE CONTROL BAROMETRIC PRESSURE SENDER
STARTER MOTOR RELAY
BATTERY GROUND TERMINAL
HEATED EGO SENSOR
(FORWARD R.H. SHOCK TOWER)
(FORWARD R.H. DASH PANEL)
CRANKSHAFT POSITION SENSOR
ELECTRIC COOLING FAN
THROTTLE POSITION SENSOR
ISC
TFI-1
A4LD TRANSMISSION SOLENOID
CRANKSHAFT POSITION SENSOR
INTEGRATED RELAY CONTROLLER
E53B-14A624-DA (SHORTING CONNECTOR)
TFI-2
SECONDARY COIL
PRIMARY COIL
RADIO IGNITION INTERFERENCE CAPACITOR
RADIO CAPACITOR
SERVICE TACHOMETER CONNECTION
WATER TEMPERATURE INDICATOR SENDER
TO 14401 ZONE 16A (FORWARD DASH PANEL L.H. SIDE)
OIL PRESSURE SWITCH
TO 14401 ZONE 47F (RIGHT COWL PANEL)
V.I.P. SELF TEST CONNECTOR (L.H. SIDE OF DASH PANEL)
FRAME 19 OF 22
MS-F1ZB-3458-AA

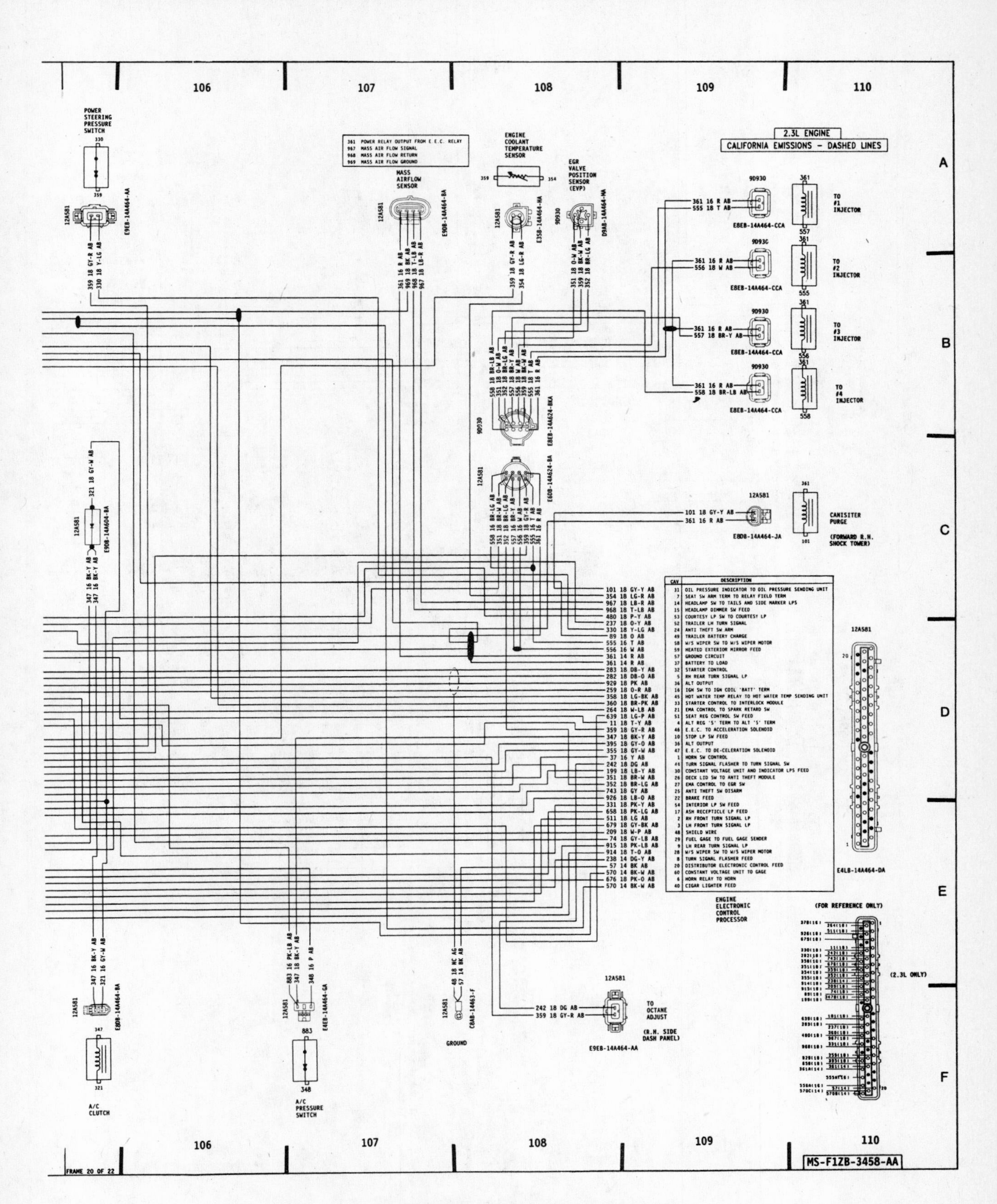

CAV	DESCRIPTION
31	OIL PRESSURE INDICATOR TO OIL PRESSURE SENDING UNIT
7	SEAT SW ARM TERM TO RELAY FIELD TERM
14	HEADLAMP SW TO TAILS AND SIDE MARKER LPS
15	HEADLAMP DIMMER SW FEED
53	COURTESY LP SW TO COURTESY LP
52	TRAILER LH TURN SIGNAL
24	ANTI THEFT SW ARM
49	TRAILER BATTERY CHARGE
58	W/S WIPER SW TO W/S WIPER MOTOR
59	HEATED EXTERIOR MIRROR FEED
57	GROUND CIRCUIT
37	BATTERY TO LOAD
32	STARTER CONTROL
5	RH REAR TURN SIGNAL LP
36	ALT OUTPUT
16	IGN SW TO IGN COIL 'BATT' TERM
45	HOT WATER TEMP RELAY TO HOT WATER TEMP SENDING UNIT
33	STARTER CONTROL TO INTERLOCK MODULE
21	EMA CONTROL TO SPARK RETARD SW
51	SEAT REG CONTROL SW FEED
4	ALT REG 'S' TERM TO ALT 'S' TERM
46	E.E.C. TO ACCELERATION SOLENOID
10	STOP LP SW FEED
36	ALT OUTPUT
47	E.E.C. TO DE-CELERATION SOLENOID
1	HORN SW CONTROL
44	TURN SIGNAL FLASHER TO TURN SIGNAL SW
30	CONSTANT VOLTAGE UNIT AND INDICATOR LPS FEED
26	DECK LID SW TO ANTI THEFT MODULE
27	EMA CONTROL TO EGR SW
25	ANTI THEFT SW DISARM
22	BRAKE FEED
54	INTERIOR LP SW FEED
17	ASH RECEPTICLE LP FEED
2	RH FRONT TURN SIGNAL LP
3	LH FRONT TURN SIGNAL LP
48	SHIELD WIRE
29	FUEL GAGE TO FUEL GAGE SENDER
9	LH REAR TURN SIGNAL LP
28	W/S WIPER SW TO W/S WIPER MOTOR
8	TURN SIGNAL FLASHER FEED
20	DISTRIBUTOR ELECTRONIC CONTROL FEED
60	CONSTANT VOLTAGE UNIT TO GAGE
6	HORN RELAY TO HORN
40	CIGAR LIGHTER FEED

Wiring Diagram — 1991 Mustang 20 of 22

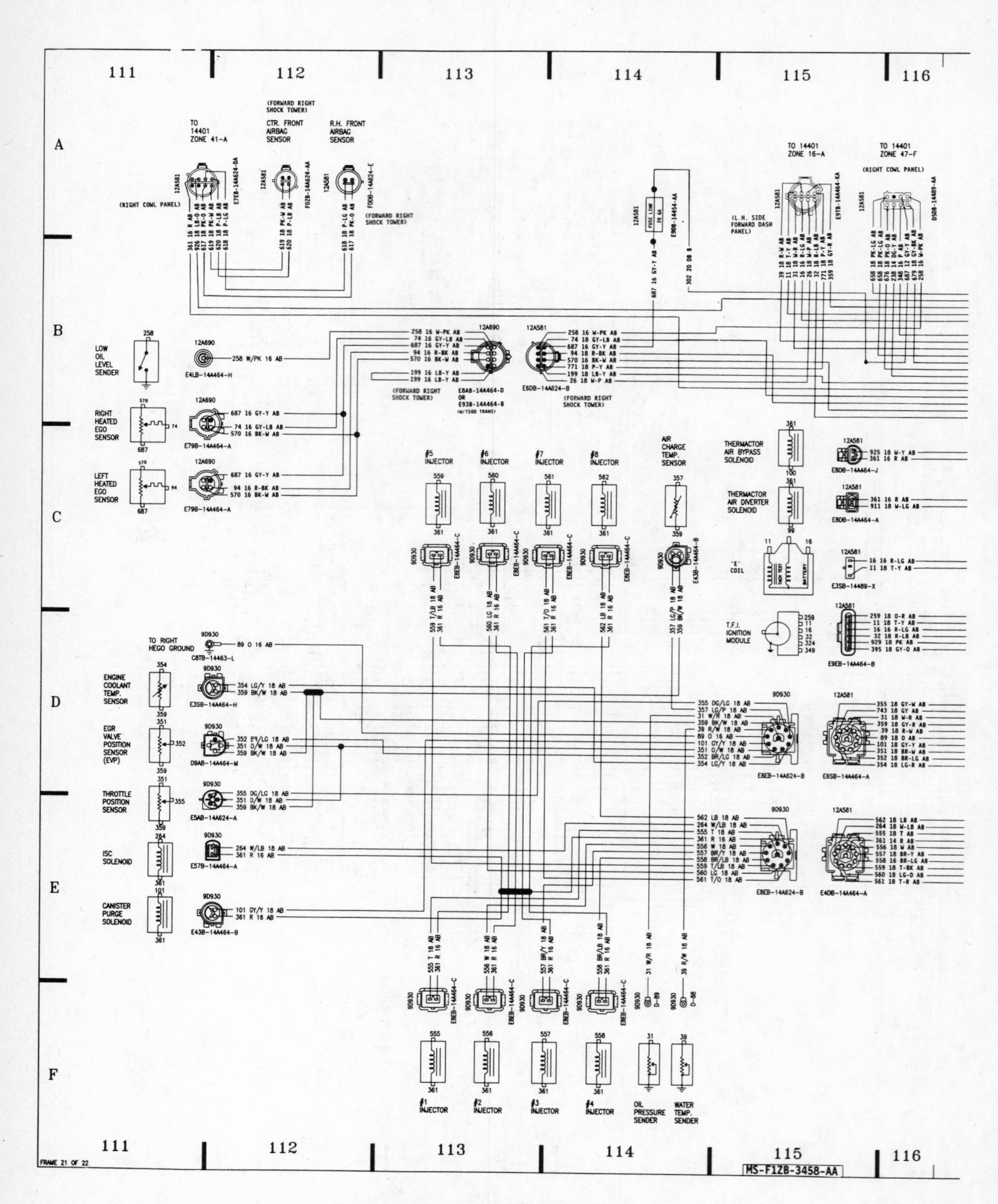

LOW OIL LEVEL SENDER
RIGHT HEATED EGO SENSOR
LEFT HEATED EGO SENSOR
TO RIGHT HEGO GROUND
ENGINE COOLANT TEMP. SENSOR
EGR VALVE POSITION SENSOR (EVP)
THROTTLE POSITION SENSOR
ISC SOLENOID
CANISTER PURGE SOLENOID
CTR. FRONT AIRBAG SENSOR
R.H. FRONT AIRBAG SENSOR
(FORWARD RIGHT SHOCK TOWER)
(RIGHT COWL PANEL)
#5 INJECTOR
#6 INJECTOR
#7 INJECTOR
#8 INJECTOR
AIR CHARGE TEMP. SENSOR
THERMACTOR AIR BYPASS SOLENOID
THERMACTOR AIR DIVERTER SOLENOID
'E' COIL
T.F.I. IGNITION MODULE
(L.H. SIDE FORWARD DASH PANEL)
#1 INJECTOR
#2 INJECTOR
#3 INJECTOR
#4 INJECTOR
OIL PRESSURE SENDER
WATER TEMP. SENDER
FRAME 21 OF 22
MS-F1ZB-3458-AA

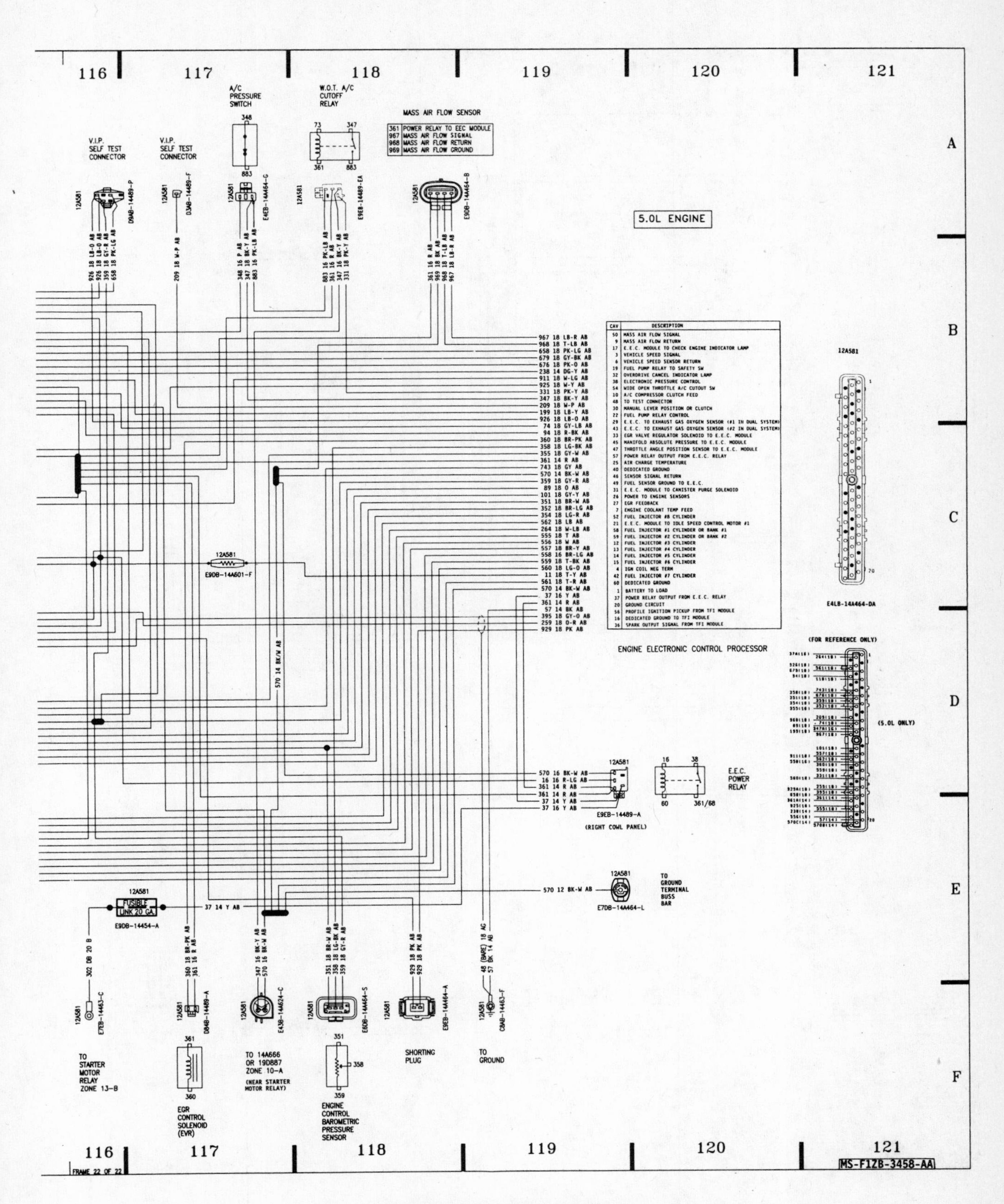

116
117
118
119
120
121
A
B
C
D
E
F
A/C PRESSURE SWITCH
W.O.T. A/C CUTOFF RELAY
MASS AIR FLOW SENSOR
361 POWER RELAY TO EEC MODULE
967 MASS AIR FLOW SIGNAL
968 MASS AIR FLOW RETURN
969 MASS AIR FLOW GROUND
V.I.P. SELF TEST CONNECTOR
V.I.P. SELF TEST CONNECTOR
5.0L ENGINE
967 18 LB-R AB
968 18 T-LB AB
658 18 PK-LG AB
679 18 GY-BK AB
676 18 PK-O AB
238 14 DG-Y AB
911 18 W-LG AB
925 18 W-Y AB
331 18 PK-Y AB
347 18 BK-Y AB
209 18 W-P AB
199 18 LB-Y AB
926 18 LB-O AB
74 18 GY-LB AB
94 18 R-BK AB
360 18 BR-PK AB
358 18 LG-BK AB
355 18 GY-W AB
361 14 R AB
743 18 GY AB
570 14 BK-W AB
359 18 GY-R AB
89 18 O AB
101 18 GY-Y AB
351 18 BR-W AB
352 18 BR-LG AB
354 18 LG-R AB
562 18 LB AB
264 18 W-LB AB
555 18 T AB
556 18 W AB
557 18 BR-Y AB
558 16 BR-LG AB
559 18 T-BK AB
560 18 LG-O AB
11 18 T-Y AB
561 18 T-R AB
570 14 BK-W AB
37 16 Y AB
361 14 R AB
57 14 BK AB
395 18 GY-O AB
259 18 O-R AB
929 18 PK AB
CAV DESCRIPTION
50 MASS AIR FLOW SIGNAL
9 MASS AIR FLOW RETURN
17 E.E.C. MODULE TO CHECK ENGINE INDICATOR LAMP
3 VEHICLE SPEED SIGNAL
6 VEHICLE SPEED SENSOR RETURN
19 FUEL PUMP RELAY TO SAFETY SW
32 OVERDRIVE CANCEL INDICATOR LAMP
38 ELECTRONIC PRESSURE CONTROL
54 WIDE OPEN THROTTLE A/C CUTOUT SW
10 A/C COMPRESSOR CLUTCH FEED
48 TO TEST CONNECTOR
30 MANUAL LEVER POSITION OR CLUTCH
22 FUEL PUMP RELAY CONTROL
29 E.E.C. TO EXHAUST GAS OXYGEN SENSOR (#1 IN DUAL SYSTEM)
43 E.E.C. TO EXHAUST GAS OXYGEN SENSOR (#2 IN DUAL SYSTEM)
33 EGR VALVE REGULATOR SOLENOID TO E.E.C. MODULE
45 MANIFOLD ABSOLUTE PRESSURE TO E.E.C. MODULE
47 THROTTLE ANGLE POSITION SENSOR TO E.E.C. MODULE
57 POWER RELAY OUTPUT FROM E.E.C. RELAY
25 AIR CHARGE TEMPERATURE
40 DEDICATED GROUND
46 SENSOR SIGNAL RETURN
49 FUEL SENSOR GROUND TO E.E.C.
31 E.E.C. MODULE TO CANISTER PURGE SOLENOID
26 POWER TO ENGINE SENSORS
27 EGR FEEDBACK
7 ENGINE COOLANT TEMP FEED
52 FUEL INJECTOR #8 CYLINDER
21 E.E.C. MODULE TO IDLE SPEED CONTROL MOTOR #1
58 FUEL INJECTOR #1 CYLINDER OR BANK #1
59 FUEL INJECTOR #2 CYLINDER OR BANK #2
12 FUEL INJECTOR #3 CYLINDER
13 FUEL INJECTOR #4 CYLINDER
14 FUEL INJECTOR #5 CYLINDER
15 FUEL INJECTOR #6 CYLINDER
4 IGN COIL NEG TERM
42 FUEL INJECTOR #7 CYLINDER
60 DEDICATED GROUND
1 BATTERY TO LOAD
37 POWER RELAY OUTPUT FROM E.E.C. RELAY
20 GROUND CIRCUIT
56 PROFILE IGNITION PICKUP FROM TFI MODULE
16 DEDICATED GROUND TO TFI MODULE
36 SPARK OUTPUT SIGNAL FROM TFI MODULE
ENGINE ELECTRONIC CONTROL PROCESSOR
12A581
E4LB-14A464-DA
(FOR REFERENCE ONLY)
(5.0L ONLY)
E9DB-14A601-F
570 14 BK/W AB
570 16 BK-W AB
16 16 R-LG AB
361 14 R AB
361 14 R AB
37 14 Y AB
37 16 Y AB
E9EB-14489-A
(RIGHT COWL PANEL)
E.E.C. POWER RELAY
570 12 BK-W AB
E7DB-14A464-L
TO GROUND TERMINAL BUSS BAR
FUSIBLE LINK 20 GA
E9DB-14454-A
37 14 Y AB
TO STARTER MOTOR RELAY ZONE 13-B
EGR CONTROL SOLENOID (EVR)
TO 14A666 OR 19D887 ZONE 10-A
(NEAR STARTER MOTOR RELAY)
ENGINE CONTROL BAROMETRIC PRESSURE SENSOR
SHORTING PLUG
TO GROUND
FRAME 22 OF 22
MS-F1ZB-3458-AA

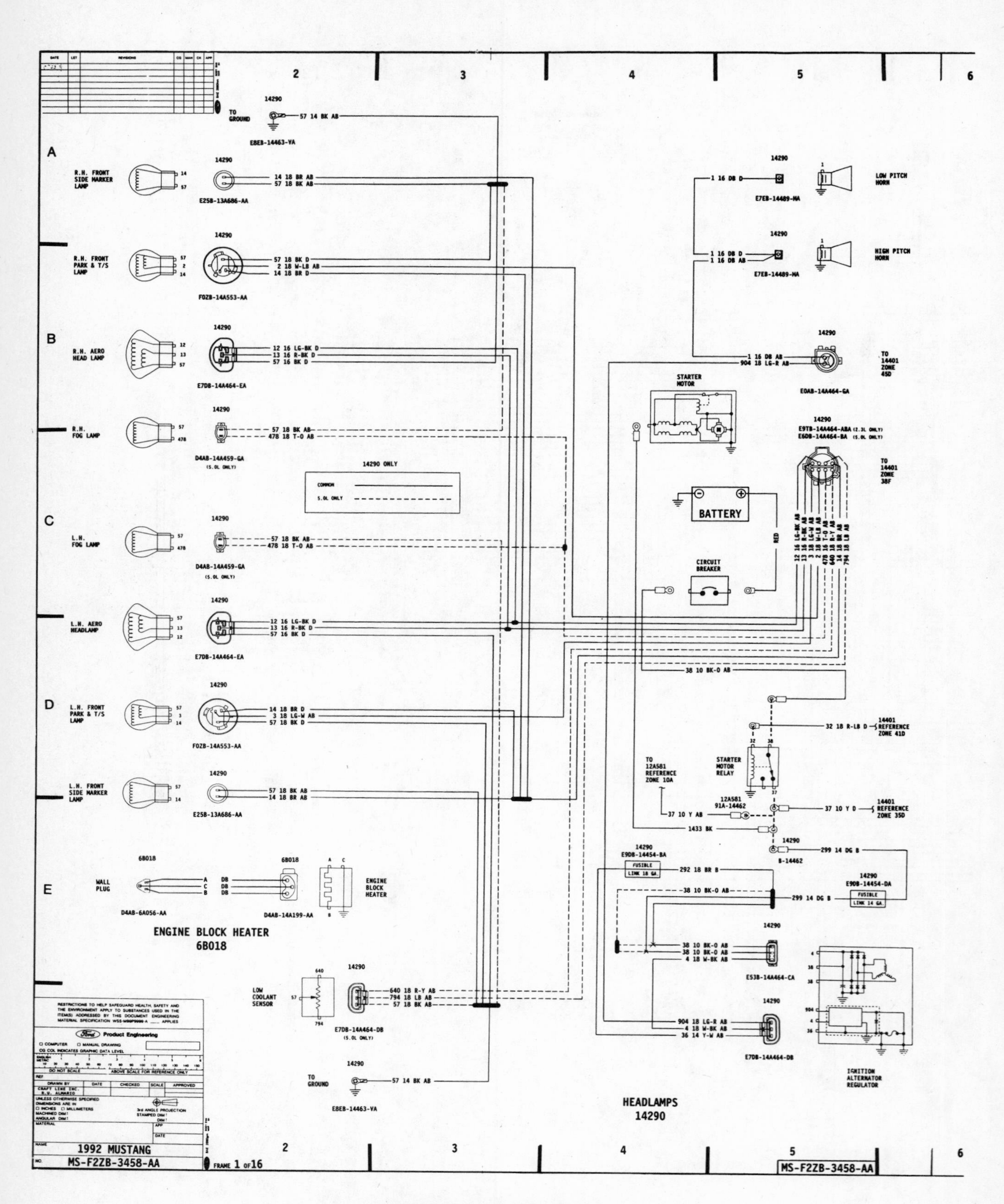
R.H. FRONT SIDE MARKER LAMP
R.H. FRONT PARK & T/S LAMP
R.H. AERO HEAD LAMP
R.H. FOG LAMP
L.H. FOG LAMP
L.H. AERO HEADLAMP
L.H. FRONT PARK & T/S LAMP
L.H. FRONT SIDE MARKER LAMP
LOW PITCH HORN
HIGH PITCH HORN
STARTER MOTOR
BATTERY
CIRCUIT BREAKER
STARTER MOTOR RELAY
ENGINE BLOCK HEATER
6B018
LOW COOLANT SENSOR
IGNITION ALTERNATOR REGULATOR
HEADLAMPS
14290
1992 MUSTANG
MS-F2ZB-3458-AA
FRAME 1 OF 16

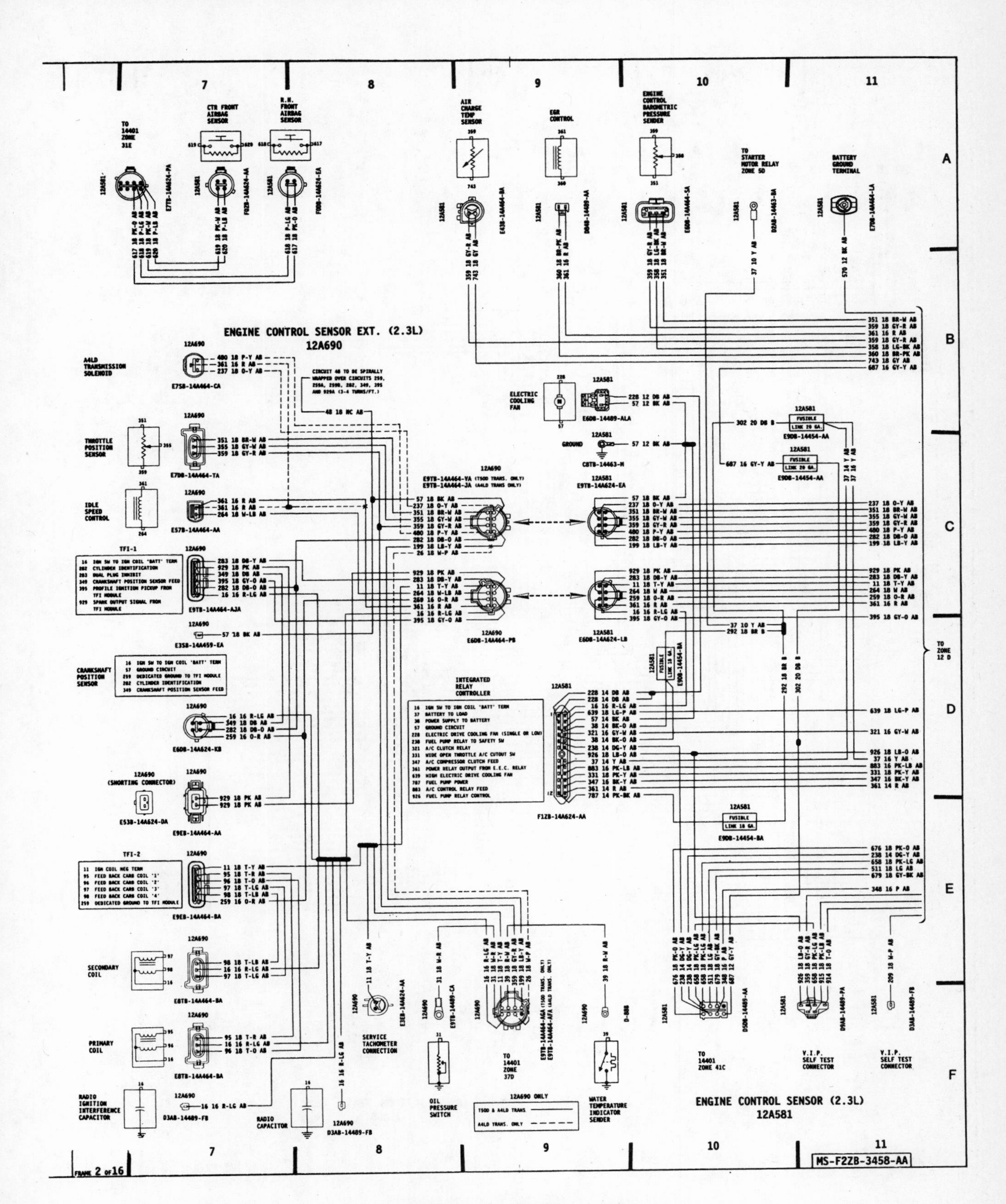
ENGINE CONTROL SENSOR EXT. (2.3L)
12A690
ENGINE CONTROL SENSOR (2.3L)
12A581
MS-F2ZB-3458-AA

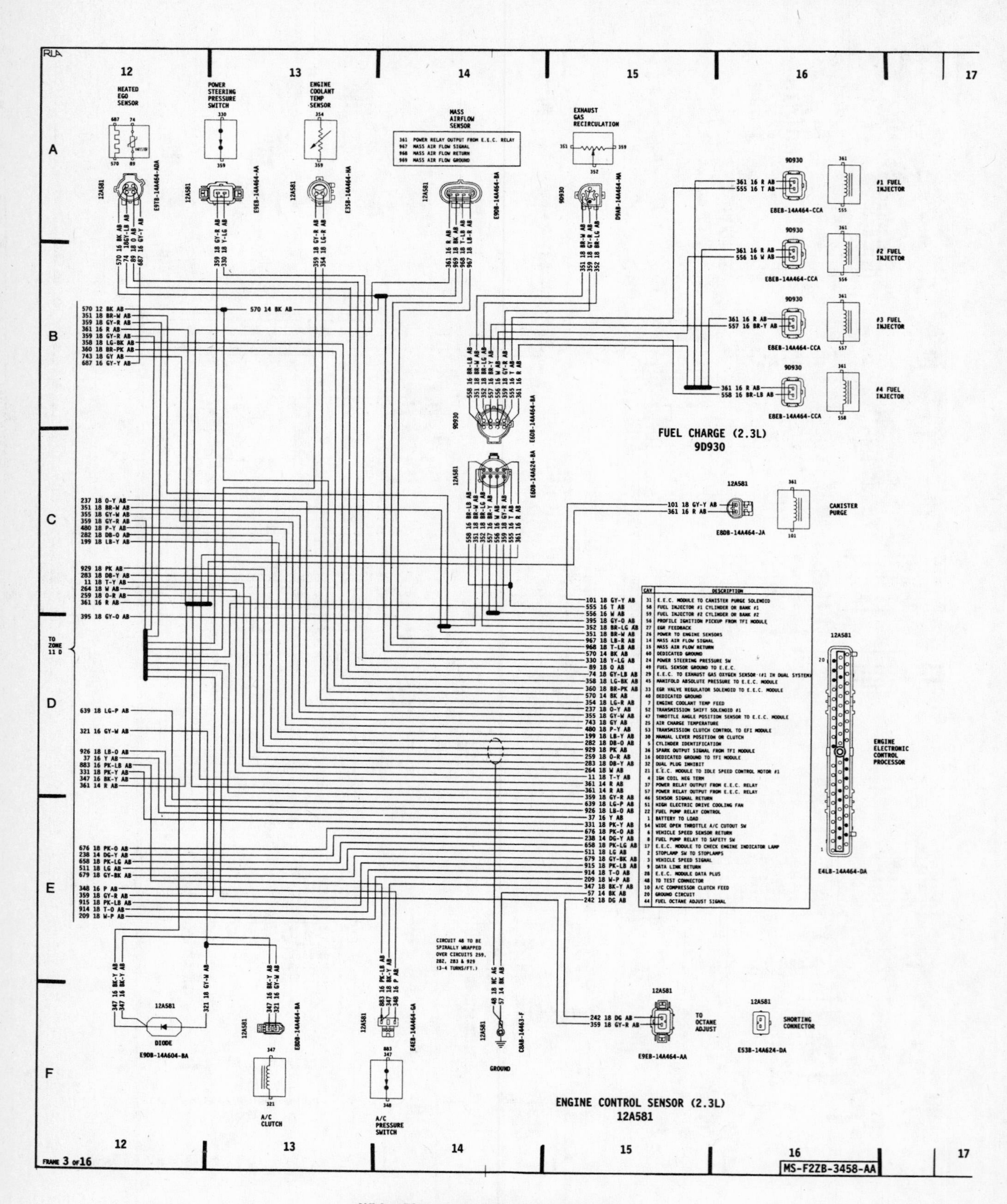

HEATED EGO SENSOR
POWER STEERING PRESSURE SWITCH
ENGINE COOLANT TEMP SENSOR
MASS AIRFLOW SENSOR
361 POWER RELAY OUTPUT FROM E.E.C. RELAY
967 MASS AIR FLOW SIGNAL
968 MASS AIR FLOW RETURN
969 MASS AIR FLOW GROUND
EXHAUST GAS RECIRCULATION
#1 FUEL INJECTOR
#2 FUEL INJECTOR
#3 FUEL INJECTOR
#4 FUEL INJECTOR
FUEL CHARGE (2.3L)
9D930
CANISTER PURGE
ENGINE ELECTRONIC CONTROL PROCESSOR
TO ZONE 11 D
CIRCUIT 48 TO BE SPIRALLY WRAPPED OVER CIRCUITS 259, 282, 283 & 929 (3-4 TURNS/FT.)
DIODE
A/C CLUTCH
A/C PRESSURE SWITCH
GROUND
TO OCTANE ADJUST
SHORTING CONNECTOR
ENGINE CONTROL SENSOR (2.3L)
12A581
FRAME 3 OF 16
MS-F2ZB-3458-AA

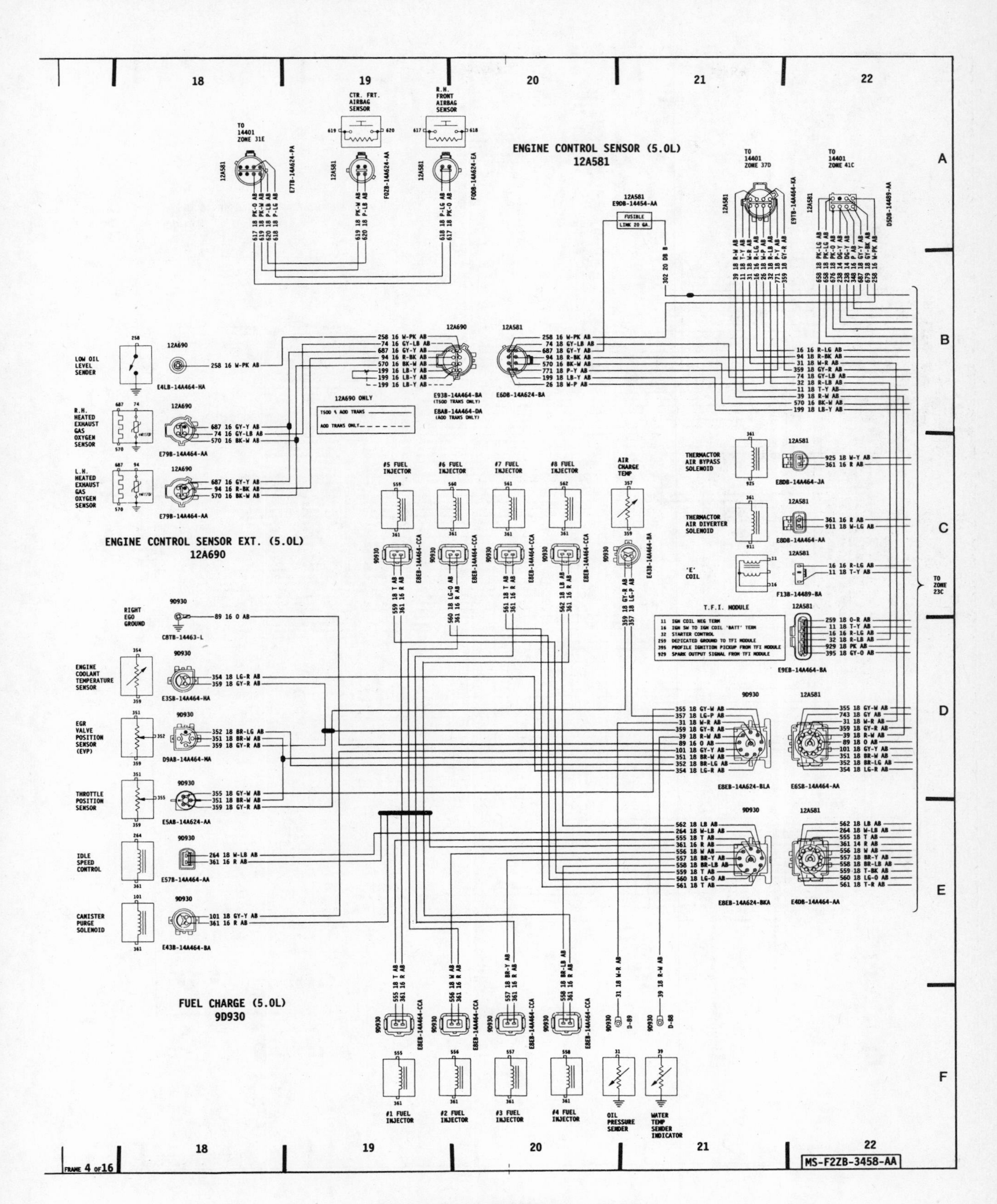
ENGINE CONTROL SENSOR (5.0L)
12A581
ENGINE CONTROL SENSOR EXT. (5.0L)
12A690
FUEL CHARGE (5.0L)
9D930
LOW OIL LEVEL SENDER
R.H. HEATED EXHAUST GAS OXYGEN SENSOR
L.H. HEATED EXHAUST GAS OXYGEN SENSOR
RIGHT EGO GROUND
ENGINE COOLANT TEMPERATURE SENSOR
EGR VALVE POSITION SENSOR (EVP)
THROTTLE POSITION SENSOR
IDLE SPEED CONTROL
CANISTER PURGE SOLENOID
CTR. FRT. AIRBAG SENSOR
R.H. FRONT AIRBAG SENSOR
#5 FUEL INJECTOR
#6 FUEL INJECTOR
#7 FUEL INJECTOR
#8 FUEL INJECTOR
AIR CHARGE TEMP
THERMACTOR AIR BYPASS SOLENOID
THERMACTOR AIR DIVERTER SOLENOID
'E' COIL
T.F.I. MODULE
#1 FUEL INJECTOR
#2 FUEL INJECTOR
#3 FUEL INJECTOR
#4 FUEL INJECTOR
OIL PRESSURE SENDER
WATER TEMP SENDER INDICATOR
TO ZONE 23C
FRAME 4 OF 16
MS-F2ZB-3458-AA

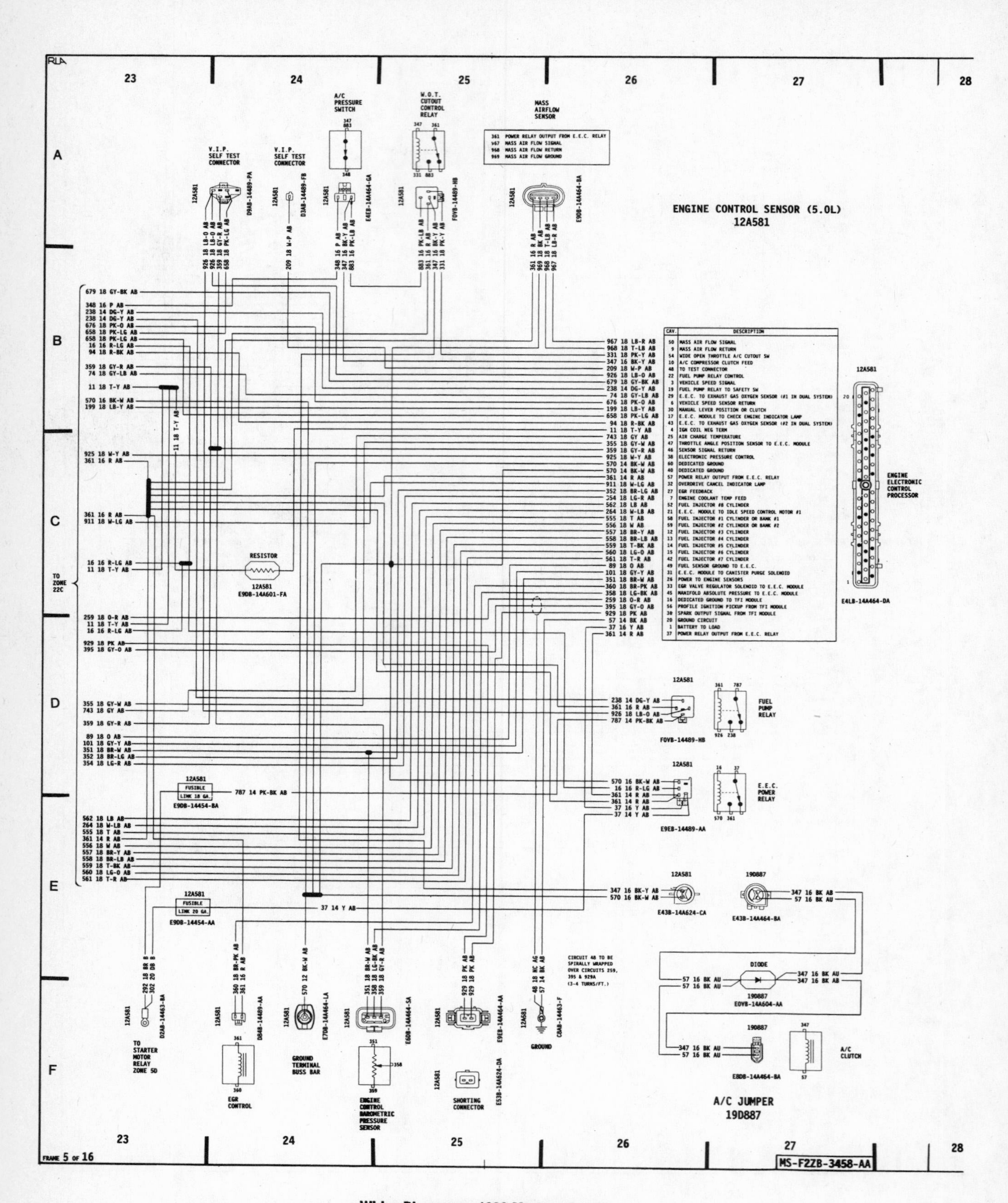
ENGINE CONTROL SENSOR (5.0L)
12A581
A/C PRESSURE SWITCH
W.O.T. CUTOUT CONTROL RELAY
MASS AIRFLOW SENSOR
361 POWER RELAY OUTPUT FROM E.E.C. RELAY
967 MASS AIR FLOW SIGNAL
968 MASS AIR FLOW RETURN
969 MASS AIR FLOW GROUND
V.I.P. SELF TEST CONNECTOR
CAV. DESCRIPTION
50 MASS AIR FLOW SIGNAL
9 MASS AIR FLOW RETURN
54 WIDE OPEN THROTTLE A/C CUTOUT SW
10 A/C COMPRESSOR CLUTCH FEED
48 TO TEST CONNECTOR
22 FUEL PUMP RELAY CONTROL
3 VEHICLE SPEED SIGNAL
19 FUEL PUMP RELAY TO SAFETY SW
29 E.E.C. TO EXHAUST GAS OXYGEN SENSOR (#1 IN DUAL SYSTEM)
6 VEHICLE SPEED SENSOR RETURN
30 MANUAL LEVER POSITION OR CLUTCH
17 E.E.C. MODULE TO CHECK ENGINE INDICATOR LAMP
43 E.E.C. TO EXHAUST GAS OXYGEN SENSOR (#2 IN DUAL SYSTEM)
4 IGN COIL NEG TERM
25 AIR CHARGE TEMPERATURE
47 THROTTLE ANGLE POSITION SENSOR TO E.E.C. MODULE
46 SENSOR SIGNAL RETURN
38 ELECTRONIC PRESSURE CONTROL
60 DEDICATED GROUND
40 DEDICATED GROUND
57 POWER RELAY OUTPUT FROM E.E.C. RELAY
32 OVERDRIVE CANCEL INDICATOR LAMP
27 EGR FEEDBACK
7 ENGINE COOLANT TEMP FEED
52 FUEL INJECTOR #8 CYLINDER
21 E.E.C. MODULE TO IDLE SPEED CONTROL MOTOR #1
58 FUEL INJECTOR #1 CYLINDER OR BANK #1
59 FUEL INJECTOR #2 CYLINDER OR BANK #2
12 FUEL INJECTOR #3 CYLINDER
13 FUEL INJECTOR #4 CYLINDER
14 FUEL INJECTOR #5 CYLINDER
15 FUEL INJECTOR #6 CYLINDER
42 FUEL INJECTOR #7 CYLINDER
49 FUEL SENSOR GROUND TO E.E.C.
31 E.E.C. MODULE TO CANISTER PURGE SOLENOID
26 POWER TO ENGINE SENSORS
33 EGR VALVE REGULATOR SOLENOID TO E.E.C. MODULE
45 MANIFOLD ABSOLUTE PRESSURE TO E.E.C. MODULE
16 DEDICATED GROUND TO TFI MODULE
56 PROFILE IGNITION PICKUP FROM TFI MODULE
36 SPARK OUTPUT SIGNAL FROM TFI MODULE
20 GROUND CIRCUIT
1 BATTERY TO LOAD
37 POWER RELAY OUTPUT FROM E.E.C. RELAY
ENGINE ELECTRONIC CONTROL PROCESSOR
E4LB-14A464-DA
TO ZONE 22C
RESISTOR
E9DB-14A601-FA
FUSIBLE LINK 18 GA.
E9DB-14454-BA
FUSIBLE LINK 20 GA.
E9DB-14454-AA
FUEL PUMP RELAY
F0VB-14489-HB
E.E.C. POWER RELAY
E9EB-14489-AA
E43B-14A624-CA
E43B-14A464-BA
DIODE
E0VB-14A604-AA
A/C CLUTCH
E8DB-14A464-BA
A/C JUMPER
19D887
CIRCUIT 48 TO BE SPIRALLY WRAPPED OVER CIRCUITS 259, 395 & 929A (3-4 TURNS/FT.)
TO STARTER MOTOR RELAY ZONE 5D
EGR CONTROL
GROUND TERMINAL BUSS BAR
ENGINE CONTROL BAROMETRIC PRESSURE SENSOR
SHORTING CONNECTOR
GROUND
FRAME 5 OF 16
MS-F2ZB-3458-AA

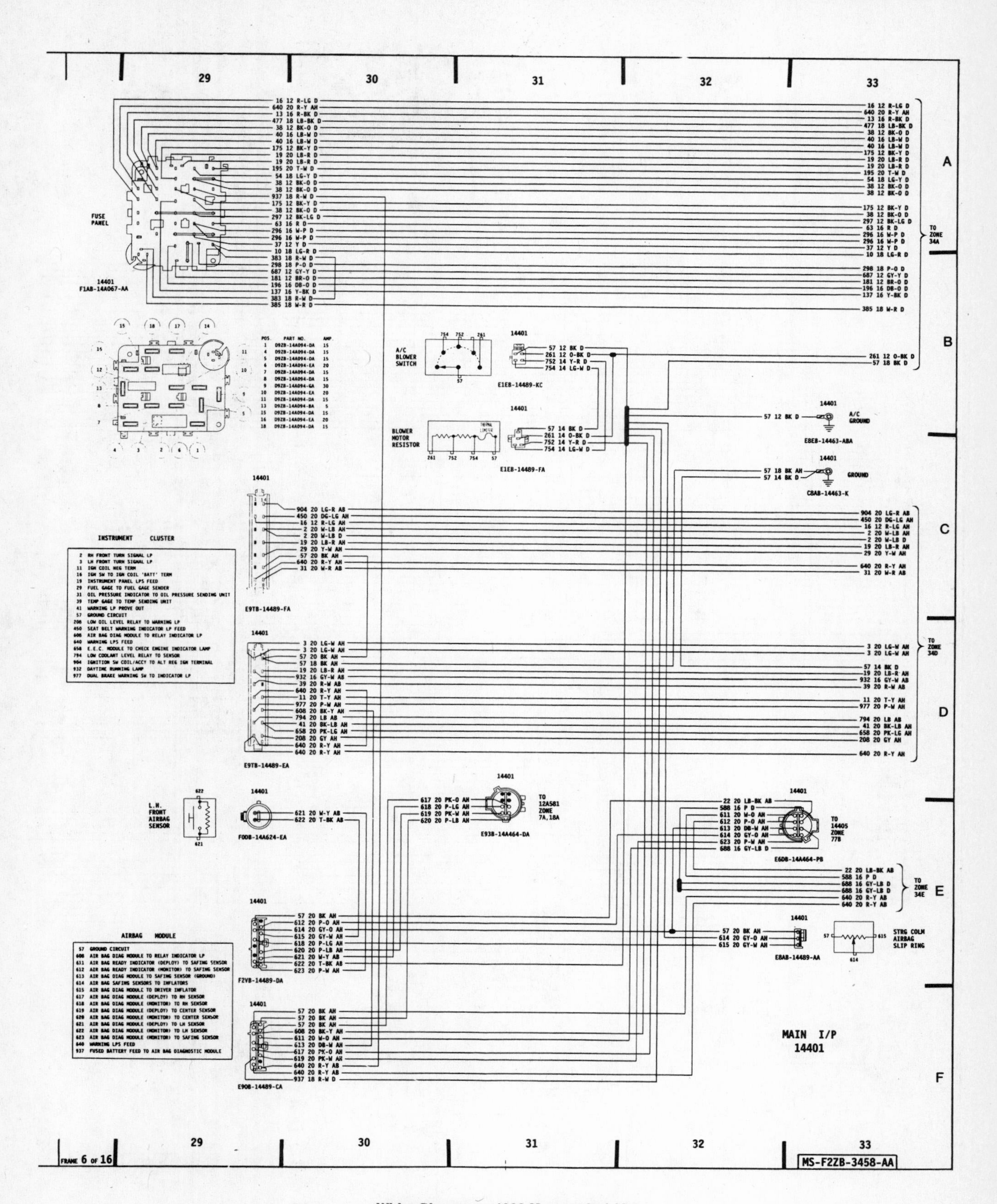

Wiring Diagram — 1992 Mustang 6 of 16

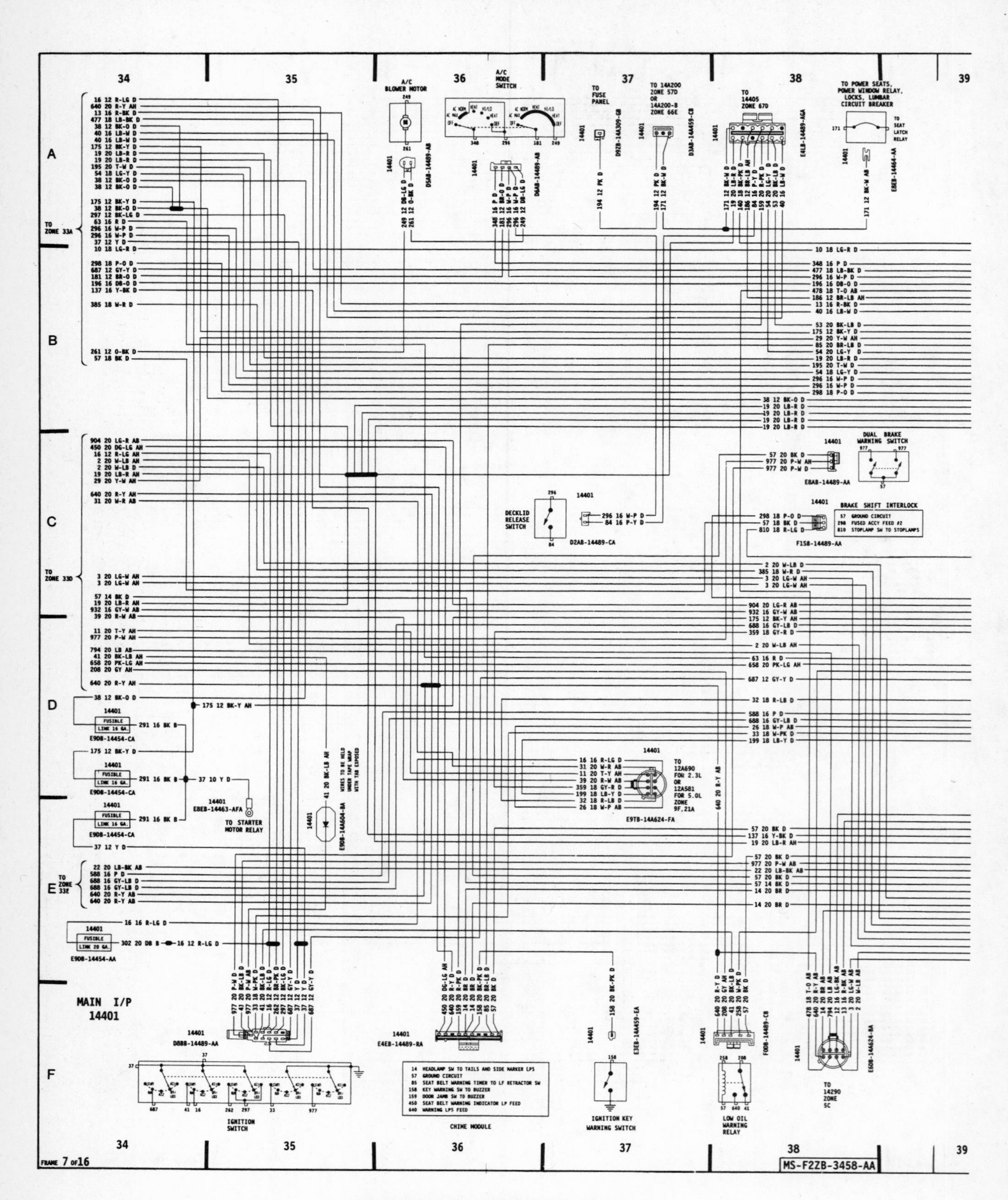

Wiring Diagram — 1992 Mustang 7 of 16

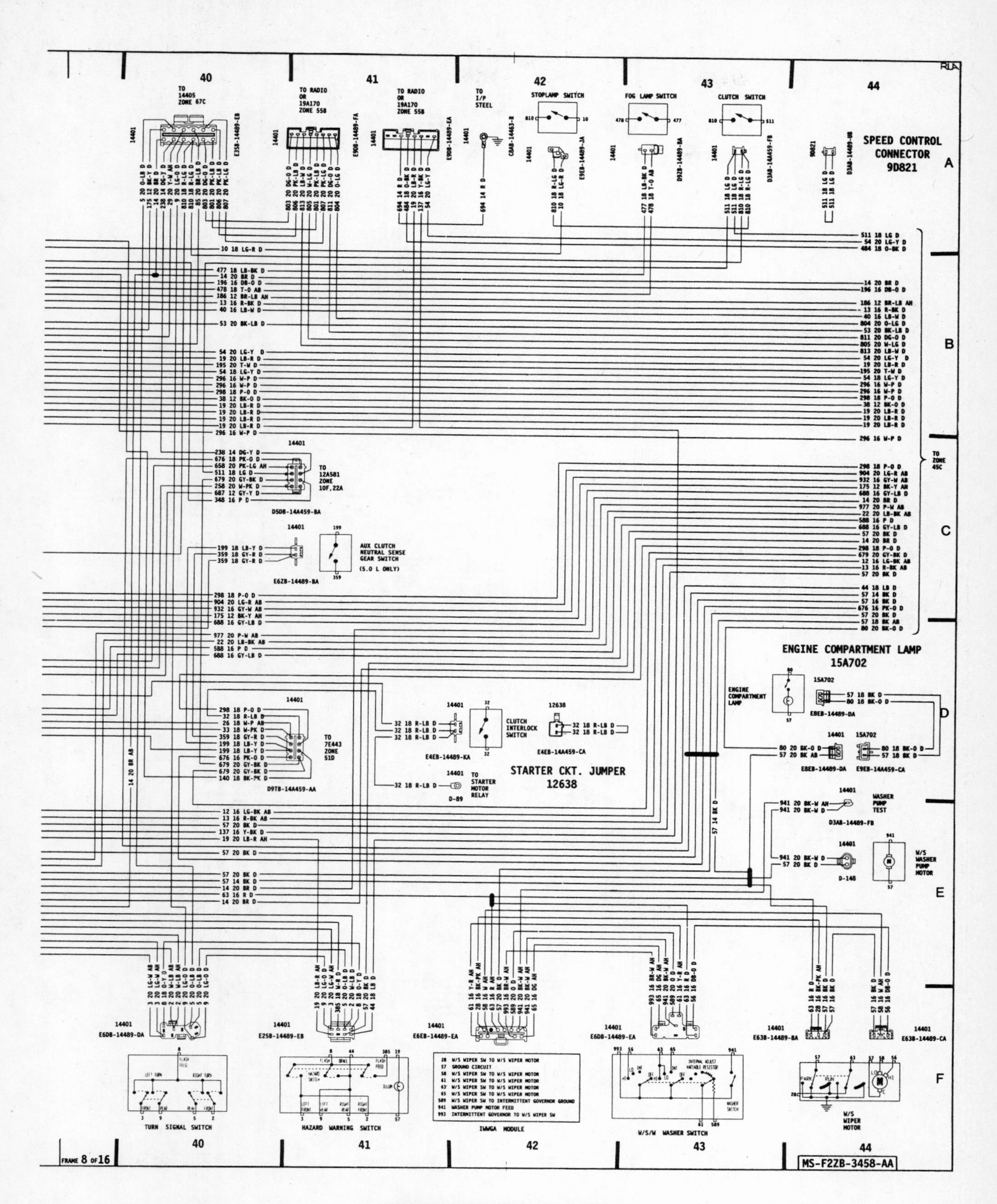

Wiring Diagram — 1992 Mustang 8 of 16

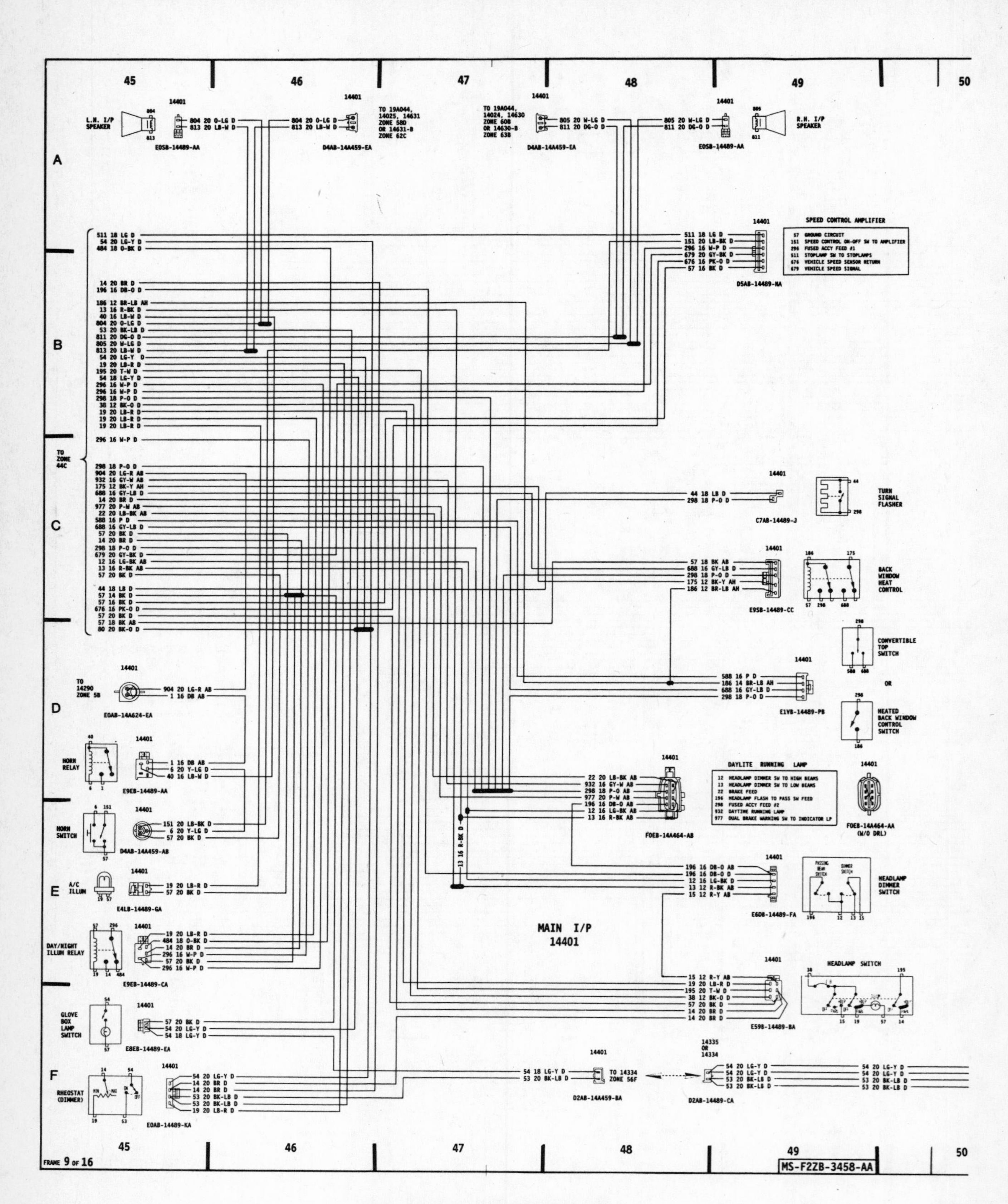

Wiring Diagram — 1992 Mustang 9 of 16

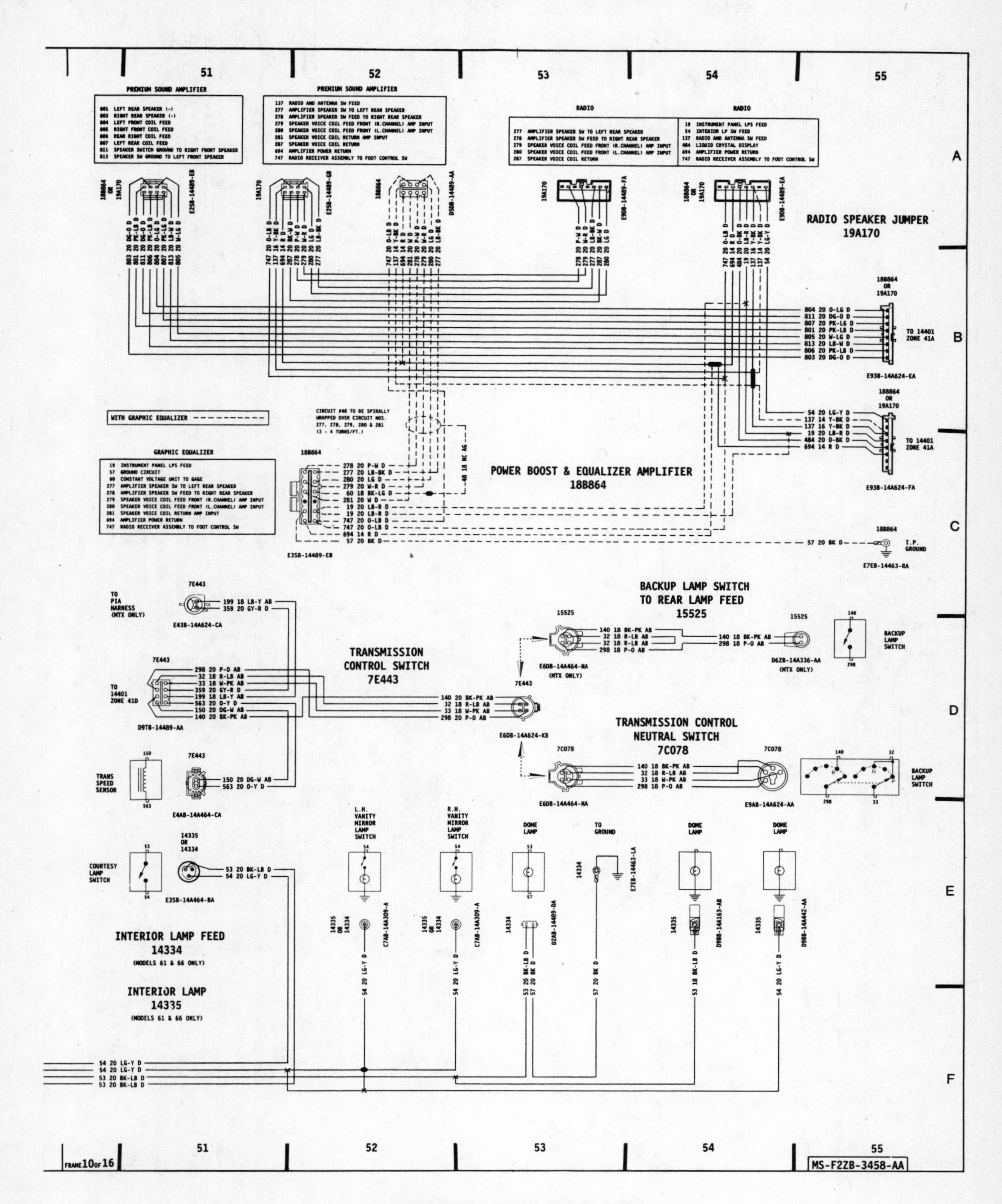

PREMIUM SOUND AMPLIFIER
RADIO
RADIO SPEAKER JUMPER 19A170
POWER BOOST & EQUALIZER AMPLIFIER 18B864
GRAPHIC EQUALIZER
WITH GRAPHIC EQUALIZER
BACKUP LAMP SWITCH TO REAR LAMP FEED 15525
TRANSMISSION CONTROL SWITCH 7E443
TRANSMISSION CONTROL NEUTRAL SWITCH 7C078
INTERIOR LAMP FEED 14334 (MODELS 61 & 66 ONLY)
INTERIOR LAMP 14335 (MODELS 61 & 66 ONLY)
COURTESY LAMP SWITCH
TRANS SPEED SENSOR
L.H. VANITY MIRROR LAMP SWITCH
R.H. VANITY MIRROR LAMP SWITCH
DOME LAMP
TO GROUND
BACKUP LAMP SWITCH
FRAME 10 OF 16
MS-F2ZB-3458-AA

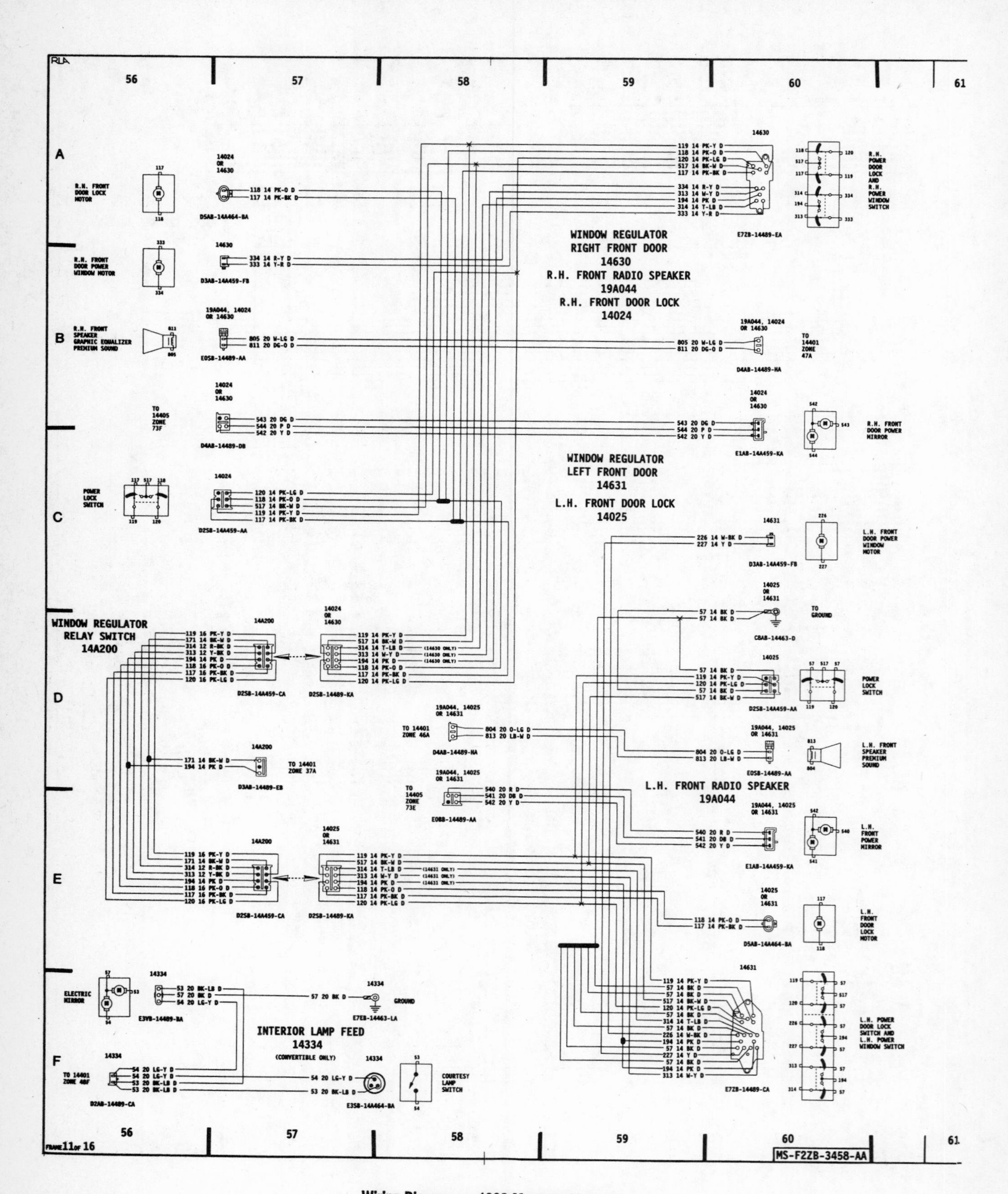
56
57
58
59
60
61
A
B
C
D
E
F
WINDOW REGULATOR
RIGHT FRONT DOOR
14630
R.H. FRONT RADIO SPEAKER
19A044
R.H. FRONT DOOR LOCK
14024
WINDOW REGULATOR
LEFT FRONT DOOR
14631
L.H. FRONT DOOR LOCK
14025
WINDOW REGULATOR
RELAY SWITCH
14A200
L.H. FRONT RADIO SPEAKER
19A044
INTERIOR LAMP FEED
14334
MS-F2ZB-3458-AA

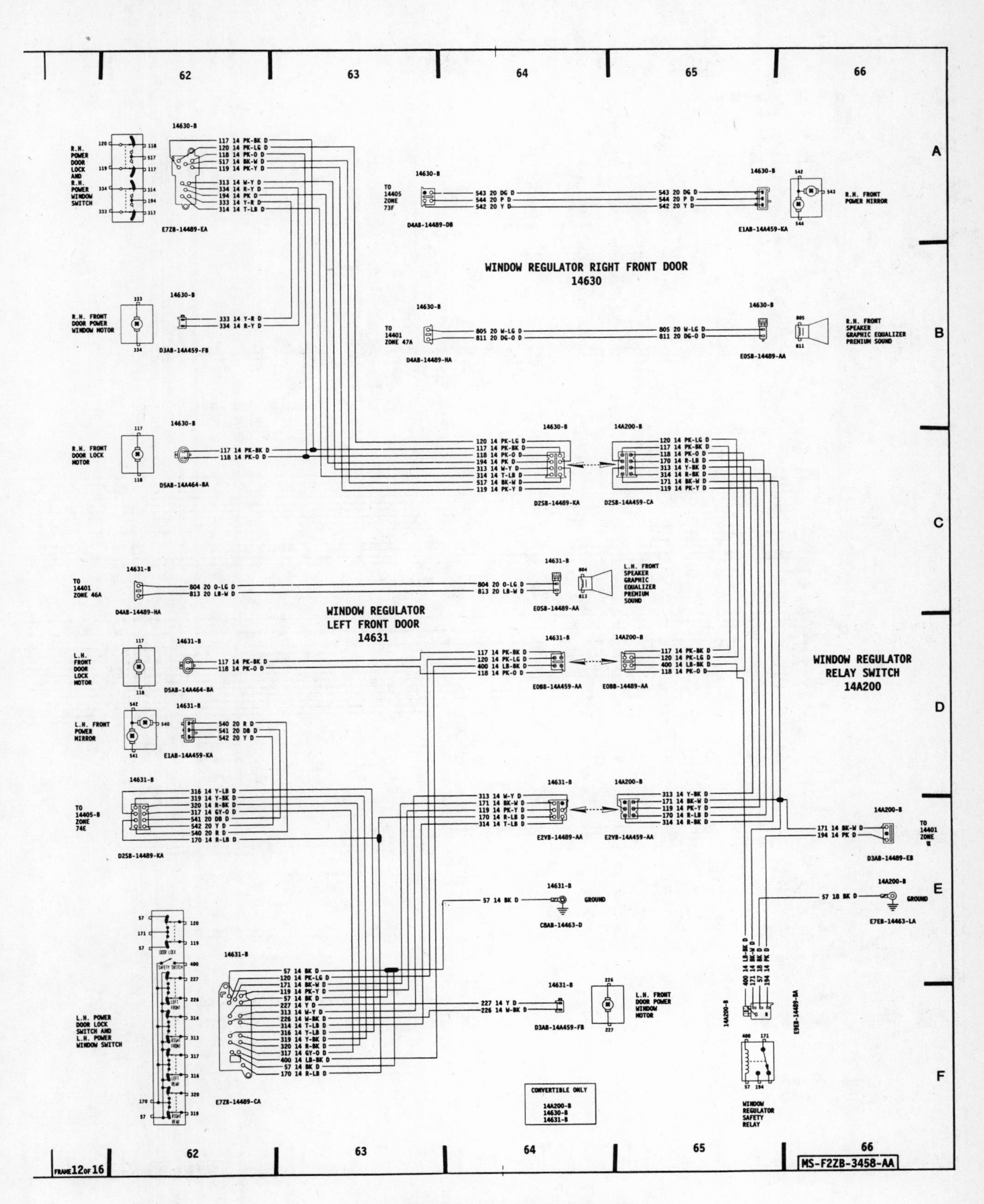

WINDOW REGULATOR RIGHT FRONT DOOR
14630
WINDOW REGULATOR
LEFT FRONT DOOR
14631
WINDOW REGULATOR
RELAY SWITCH
14A200
R.H. POWER DOOR LOCK AND R.H. POWER WINDOW SWITCH
R.H. FRONT DOOR POWER WINDOW MOTOR
R.H. FRONT DOOR LOCK MOTOR
R.H. FRONT POWER MIRROR
R.H. FRONT SPEAKER GRAPHIC EQUALIZER PREMIUM SOUND
L.H. FRONT SPEAKER GRAPHIC EQUALIZER PREMIUM SOUND
L.H. FRONT DOOR LOCK MOTOR
L.H. FRONT POWER MIRROR
L.H. POWER DOOR LOCK SWITCH AND L.H. POWER WINDOW SWITCH
L.H. FRONT DOOR POWER WINDOW MOTOR
WINDOW REGULATOR SAFETY RELAY
GROUND
CONVERTIBLE ONLY
14A200-B
14630-B
14631-B
FRAME 12 OF 16
MS-F2ZB-3458-AA

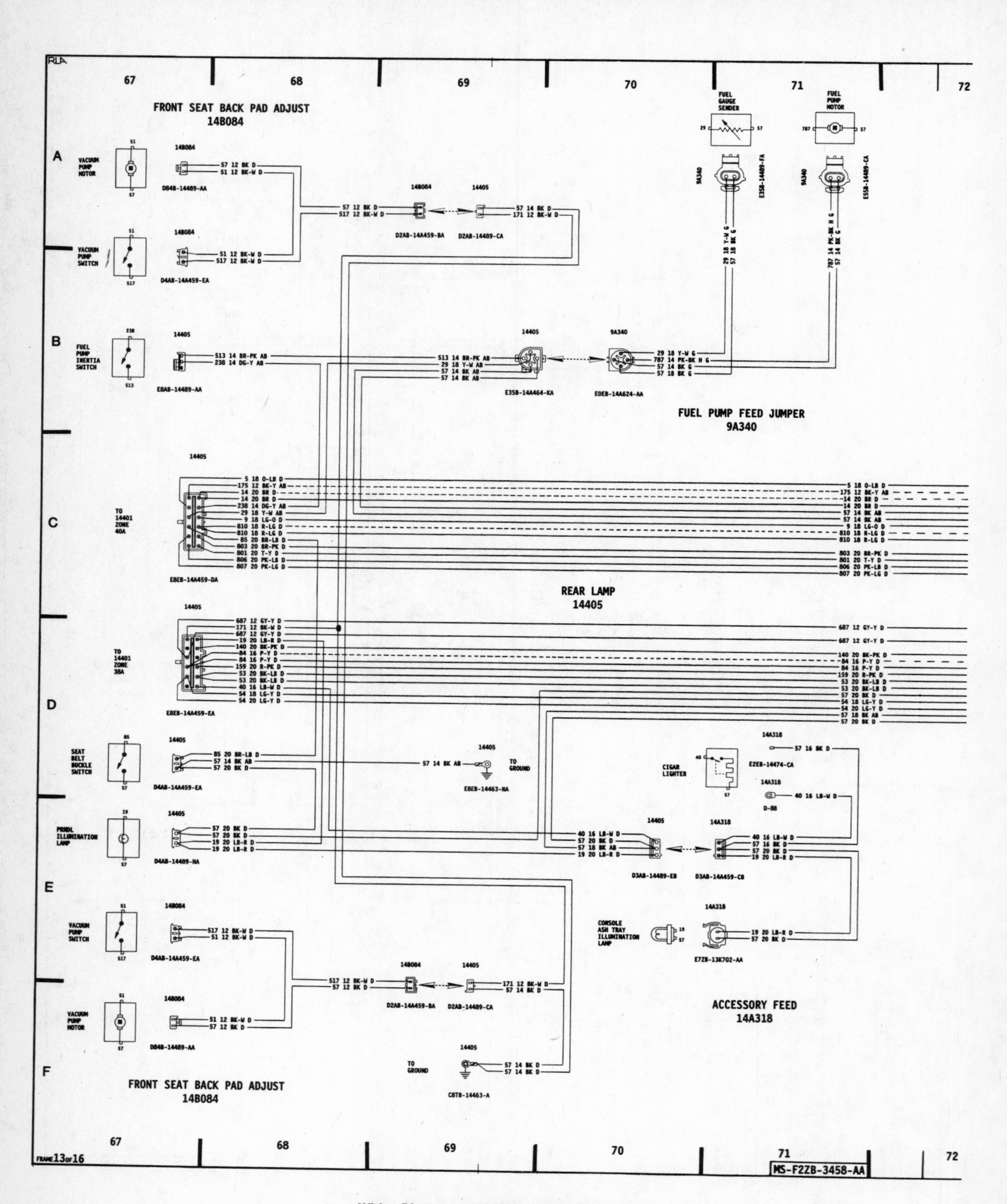
FRONT SEAT BACK PAD ADJUST
14B084
VACUUM PUMP MOTOR
VACUUM PUMP SWITCH
FUEL PUMP INERTIA SWITCH
FUEL GAUGE SENDER
FUEL PUMP MOTOR
FUEL PUMP FEED JUMPER
9A340
TO 14401 ZONE 40A
REAR LAMP
14405
TO 14401 ZONE 38A
SEAT BELT BUCKLE SWITCH
TO GROUND
CIGAR LIGHTER
PRNDL ILLUMINATION LAMP
CONSOLE ASH TRAY ILLUMINATION LAMP
ACCESSORY FEED
14A318
FRONT SEAT BACK PAD ADJUST
14B084
FRAME 13 OF 16
MS-F2ZB-3458-AA

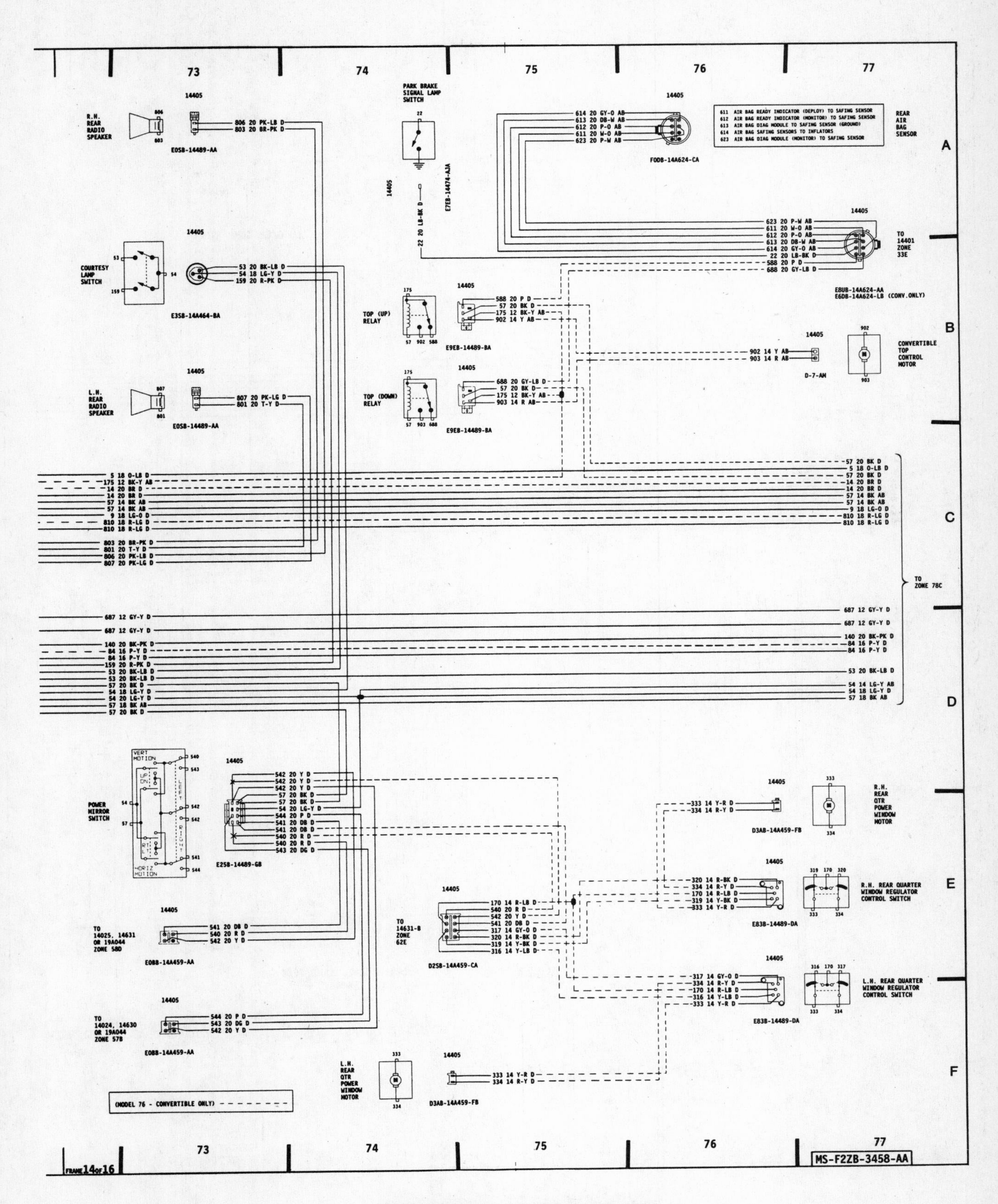

Wiring Diagram — 1992 Mustang 14 of 16

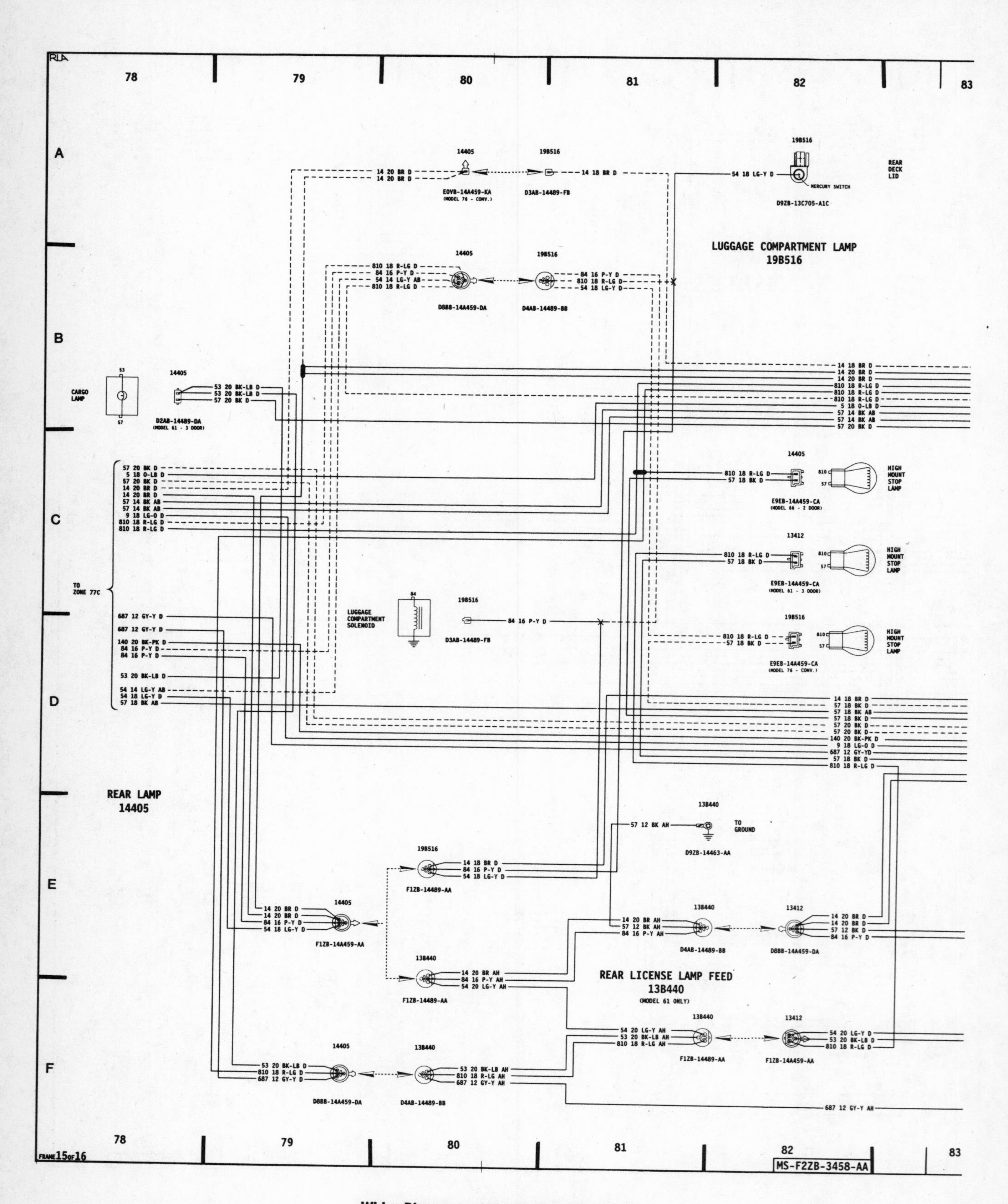

Wiring Diagram — 1992 Mustang 15 of 16

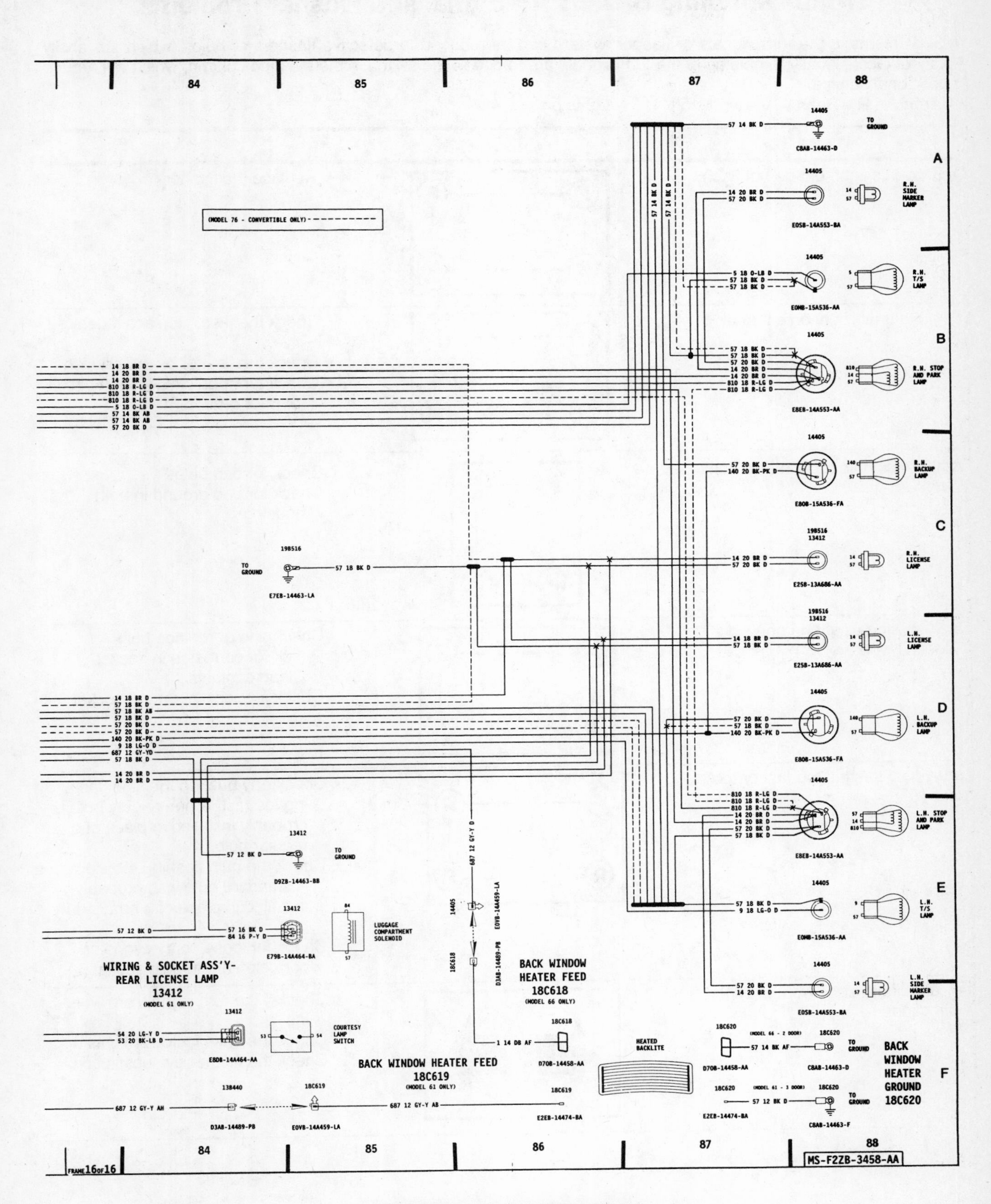

Wiring Diagram — 1992 Mustang 16 of 16

Troubleshooting Basic Turn Signal and Flasher Problems

Most problems in the turn signals or flasher system can be reduced to defective flashers or bulbs, which are easily replaced. Occasionally, problems in the turn signals are traced to the switch in the steering column, which will require professional service.

F = Front   R = Rear   ● = Lights off   o = Lights on

Problem	Solution
Turn signals light, but do not flash	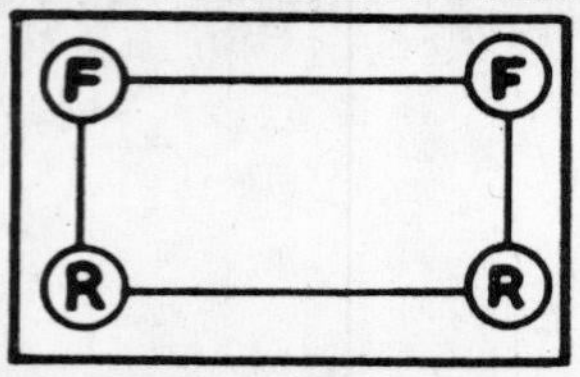 • Replace the flasher
No turn signals light on either side	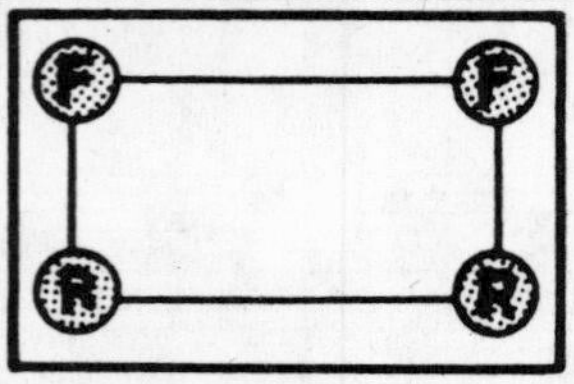 • Check the fuse. Replace if defective. • Check the flasher by substitution • Check for open circuit, short circuit or poor ground
Both turn signals on one side don't work	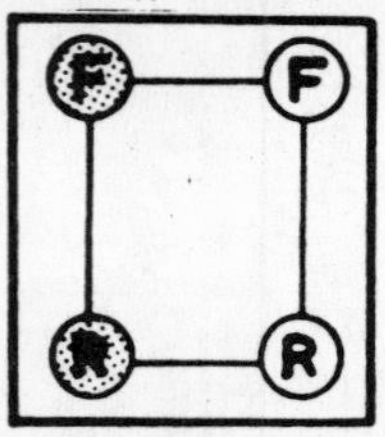 • Check for bad bulbs • Check for bad ground in both housings
One turn signal light on one side doesn't work	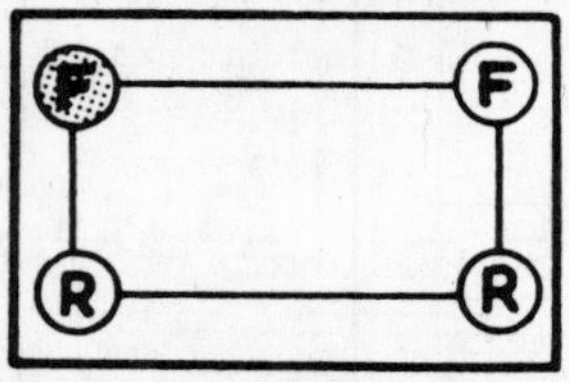 • Check and/or replace bulb • Check for corrosion in socket. Clean contacts. • Check for poor ground at socket
Turn signal flashes too fast or too slow	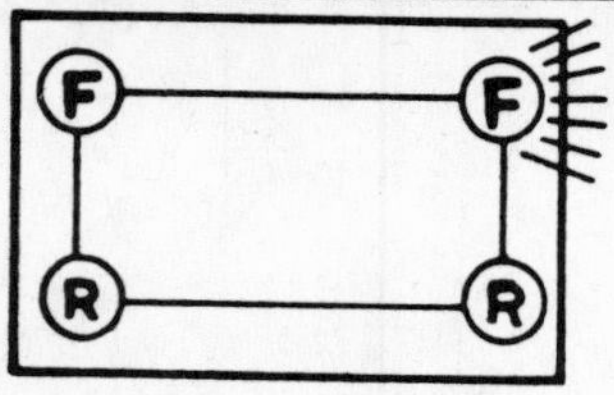 • Check any bulb on the side flashing too fast. A heavy-duty bulb is probably installed in place of a regular bulb. • Check the bulb flashing too slow. A standard bulb was probably installed in place of a heavy-duty bulb. • Check for loose connections or corrosion at the bulb socket
Indicator lights don't work in either direction	 • Check if the turn signals are working • Check the dash indicator lights • Check the flasher by substitution

Troubleshooting Basic Turn Signal and Flasher Problems

Most problems in the turn signals or flasher system can be reduced to defective flashers or bulbs, which are easily replaced. Occasionally, problems in the turn signals are traced to the switch in the steering column, which will require professional service.

F = Front R = Rear ● = Lights off ○ = Lights on

Problem		Solution
One indicator light doesn't light		• On systems with 1 dash indicator: See if the lights work on the same side. Often the filaments have been reversed in systems combining stoplights with taillights and turn signals. Check the flasher by substitution • On systems with 2 indicators: Check the bulbs on the same side Check the indicator light bulb Check the flasher by substitution

Troubleshooting Basic Lighting Problems

Problem	Cause	Solution
Lights		
One or more lights don't work, but others do	• Defective bulb(s) • Blown fuse(s) • Dirty fuse clips or light sockets • Poor ground circuit	• Replace bulb(s) • Replace fuse(s) • Clean connections • Run ground wire from light socket housing to car frame
Lights burn out quickly	• Incorrect voltage regulator setting or defective regulator • Poor battery/alternator connections	• Replace voltage regulator • Check battery/alternator connections
Lights go dim	• Low/discharged battery • Alternator not charging • Corroded sockets or connections • Low voltage output	• Check battery • Check drive belt tension; repair or replace alternator • Clean bulb and socket contacts and connections • Replace voltage regulator
Lights flicker	• Loose connection • Poor ground • Circuit breaker operating (short circuit)	• Tighten all connections • Run ground wire from light housing to car frame • Check connections and look for bare wires
Lights "flare"—Some flare is normal on acceleration—if excessive, see "Lights Burn Out Quickly"	• High voltage setting	• Replace voltage regulator
Lights glare—approaching drivers are blinded	• Lights adjusted too high • Rear springs or shocks sagging • Rear tires soft	• Have headlights aimed • Check rear springs/shocks • Check/correct rear tire pressure
Turn Signals		

Troubleshooting Basic Lighting Problems

Problem	Cause	Solution
Lights		
Turn signals don't work in either direction	• Blown fuse • Defective flasher • Loose connection	• Replace fuse • Replace flasher • Check/tighten all connections
Right (or left) turn signal only won't work	• Bulb burned out • Right (or left) indicator bulb burned out • Short circuit	• Replace bulb • Check/replace indicator bulb • Check/repair wiring
Flasher rate too slow or too fast	• Incorrect wattage bulb • Incorrect flasher	• Flasher bulb • Replace flasher (use a variable load flasher if you pull a trailer)
Indicator lights do not flash (burn steadily)	• Burned out bulb • Defective flasher	• Replace bulb • Replace flasher
Indicator lights do not light at all	• Burned out indicator bulb • Defective flasher	• Replace indicator bulb • Replace flasher

Troubleshooting Basic Dash Gauge Problems

Problem	Cause	Solution
Coolant Temperature Gauge		
Gauge reads erratically or not at all	• Loose or dirty connections • Defective sending unit • Defective gauge	• Clean/tighten connections • Bi-metal gauge: remove the wire from the sending unit. Ground the wire for an instant. If the gauge registers, replace the sending unit. • Magnetic gauge: disconnect the wire at the sending unit. With ignition ON gauge should register COLD. Ground the wire; gauge should register HOT.
Ammeter Gauge—Turn Headlights ON (do not start engine). Note reaction		
Ammeter shows charge Ammeter shows discharge Ammeter does not move	• Connections reversed on gauge • Ammeter is OK • Loose connections or faulty wiring • Defective gauge	• Reinstall connections • Nothing • Check/correct wiring • Replace gauge
Oil Pressure Gauge		
Gauge does not register or is inaccurate	• On mechanical gauge, Bourdon tube may be bent or kinked • Low oil pressure • Defective gauge	• Check tube for kinks or bends preventing oil from reaching the gauge • Remove sending unit. Idle the engine briefly. If no oil flows from sending unit hole, problem is in engine. • Remove the wire from the sending unit and ground it for an instant with the ignition ON. A good gauge will go to the top of the scale.

Troubleshooting Basic Dash Gauge Problems

Problem	Cause	Solution
Coolant Temperature Gauge		
Gauge does not register or is inaccurate	• Defective wiring	• Check the wiring to the gauge. If it's OK and the gauge doesn't register when grounded, replace the gauge.
	• Defective sending unit	• If the wiring is OK and the gauge functions when grounded, replace the sending unit
All Gauges		
All gauges do not operate	• Blown fuse • Defective instrument regulator	• Replace fuse • Replace instrument voltage regulator
All gauges read low or erratically	• Defective or dirty instrument voltage regulator	• Clean contacts or replace
All gauges pegged	• Loss of ground between instrument voltage regulator and car • Defective instrument regulator	• Check ground • Replace regulator

Troubleshooting Basic Dash Gauge Problems

Problem	Cause	Solution
Warning Lights		
Light(s) do not come on when ignition is ON, but engine is not started	• Defective bulb • Defective wire • Defective sending unit	• Replace bulb • Check wire from light to sending unit • Disconnect the wire from the sending unit and ground it. Replace the sending unit if the light comes on with the ignition ON.
Light comes on with engine running	• Problem in individual system • Defective sending unit	• Check system • Check sending unit (see above)

Troubleshooting the Heater

Problem	Cause	Solution
Blower motor will not turn at any speed	• Blown fuse • Loose connection • Defective ground • Faulty switch • Faulty motor • Faulty resistor	• Replace fuse • Inspect and tighten • Clean and tighten • Replace switch • Replace motor • Replace resistor
Blower motor turns at one speed only	• Faulty switch • Faulty resistor	• Replace switch • Replace resistor
Blower motor turns but does not circulate air	• Intake blocked • Fan not secured to the motor shaft	• Clean intake • Tighten security

Troubleshooting the Heater

Problem	Cause	Solution
Heater will not heat	• Coolant does not reach proper temperature • Heater core blocked internally • Heater core air-bound • Blend-air door not in proper position	• Check and replace thermostat if necessary • Flush or replace core if necessary • Purge air from core • Adjust cable
Heater will not defrost	• Control cable adjustment incorrect • Defroster hose damaged	• Adjust control cable • Replace defroster hose

Troubleshooting Basic Windshield Wiper Problems

Problem	Cause	Solution
Electric Wipers		
Wipers do not operate— Wiper motor heats up or hums	• Internal motor defect • Bent or damaged linkage • Arms improperly installed on linking pivots	• Replace motor • Repair or replace linkage • Position linkage in park and reinstall wiper arms
Electric Wipers		
Wipers do not operate— No current to motor	• Fuse or circuit breaker blown • Loose, open or broken wiring • Defective switch • Defective or corroded terminals • No ground circuit for motor or switch	• Replace fuse or circuit breaker • Repair wiring and connections • Replace switch • Replace or clean terminals • Repair ground circuits
Wipers do not operate— Motor runs	• Linkage disconnected or broken	• Connect wiper linkage or replace broken linkage
Vacuum Wipers		
Wipers do not operate	• Control switch or cable inoperative • Loss of engine vacuum to wiper motor (broken hoses, low engine vacuum, defective vacuum/fuel pump) • Linkage broken or disconnected • Defective wiper motor	• Repair or replace switch or cable • Check vacuum lines, engine vacuum and fuel pump • Repair linkage • Replace wiper motor
Wipers stop on engine acceleration	• Leaking vacuum hoses • Dry windshield • Oversize wiper blades • Defective vacuum/fuel pump	• Repair or replace hoses • Wet windshield with washers • Replace with proper size wiper blades • Replace pump

7
DRIVE TRAIN

UNDERSTANDING THE MANUAL TRANSMISSION AND CLUTCH

Because of the way an internal combustion engine breathes, it can produce torque, or twisting force, only within a narrow speed range. Most modern, overhead valve engines must turn at about 2,500 rpm to produce their peak torque. By 4,500 rpm they are producing so little torque that continued increases in engine speed produce no power increases.

The torque peak on overhead camshaft engines is, generally, much higher, but much narrower.

The manual transmission and clutch are employed to vary the relationship between engine speed and the speed of the wheels so that adequate engine power can be produced under all circumstances. The clutch allows engine torque to be applied to the transmission input shaft gradually, due to mechanical slippage. The car can, consequently, be started smoothly from a full stop.

The transmission changes the ratio between the rotating speeds of the engine and the wheels by the use of gears. 4-speed or 5-speed transmissions are most common. The lower gears allow full engine power to be applied to the rear wheels during acceleration at low speeds.

The clutch drive plate is a thin disc, the center of which is splined to the transmission input shaft. Both sides of the disc are covered with a layer of material which is similar to brake lining and which is capable of allowing slippage without roughness or excessive noise.

The clutch cover is bolted to the engine flywheel and incorporates a diaphragm spring which provides the pressure to engage the clutch. The cover also houses the pressure plate. The driven disc is sandwiched between the pressure plate and the smooth surface of the flywheel when the clutch pedal is released, thus forcing it to turn at the same speed as the engine crankshaft.

The transmission contains a mainshaft which passes all the way through the transmission, from the clutch to the driveshaft. This shaft is separated at one point, so that front and rear portions can turn at different speeds.

Power is transmitted by a countershaft in the lower gears and reverse. The gears of the countershaft mesh with gears on the mainshaft, allowing power to be carried from one to the other. All the countershaft gears are integral with that shaft, while several of the mainshaft gears can either rotate independently of the shaft or be locked to it. Shifting from one gear to the next causes one of the gears to be freed from rotating with the shaft and locks another to it. Gears are locked and unlocked by internal dog clutches which slide between the center of the gear and the shaft. The forward gears usually employ synchronizers; friction members which smoothly bring gear and shaft to the same speed before the toothed dog clutches are engaged.

The clutch is operating properly if:

1. It will stall the engine when released with the vehicle held stationary.
2. The shift lever can be moved freely between first and reverse gears when the vehicle is stationary and the clutch disengaged.

A clutch pedal free-play adjustment is incorporated in the linkage. If there is about 1-2 in. (25-50mm) of motion before the pedal begins to release the clutch, it is adjusted properly. Inadequate free-play wears all parts of the clutch releasing mechanisms and may cause slippage. Excessive free-play may cause inadequate release and hard shifting of gears.

Some clutches use a hydraulic system in place of mechanical linkage. If the clutch fails to release, fill the clutch master cylinder with fluid to the proper level and pump the clutch pedal to fill the system with fluid. Bleed the system in the same way as a brake system. If leaks are located, tighten loose connections or overhaul the master or slave cylinder as necessary.

MANUAL TRANSMISSION

Identification

All vehicles with manual transmission covered in this manual use the Ford T50D 5-speed overdrive manual transmission

Shift Lever and Boot Assembly

REMOVAL & INSTALLATION

➧ SEE FIG. 1

1. Remove the shift knob by rotating it counterclockwise.
2. Remove the console trim and lift the boot over the shift lever.

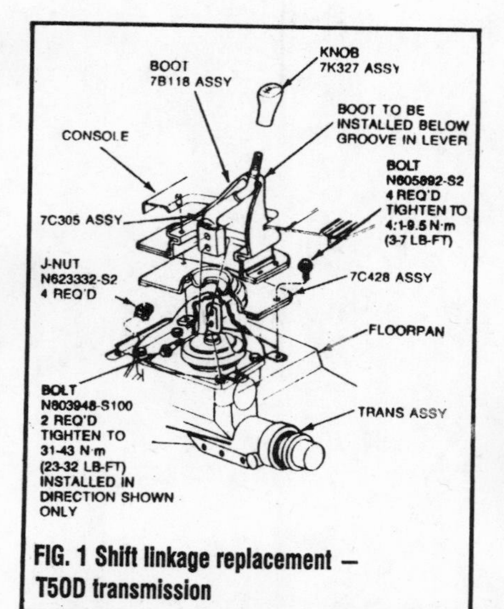

FIG. 1 Shift linkage replacement – T50D transmission

3. Remove the two bolts retaining the shift lever to the transmission.

To Install:

4. Install the two bolts into the shift lever and tighten to 23–32 ft. lbs. (31–43Nm).

➡ Shift lever bolts must only be in one direction, from the left hand side of the shift lever.

5. Install the console trim and the shift boot.
6. Install the shift knob by rotating clockwise until tension is felt. Then, rotate an additional 180 degrees to align the graphics on the shift knob.

Back-Up Light Switch

REMOVAL & INSTALLATION

1. Place the shift lever in neutral.

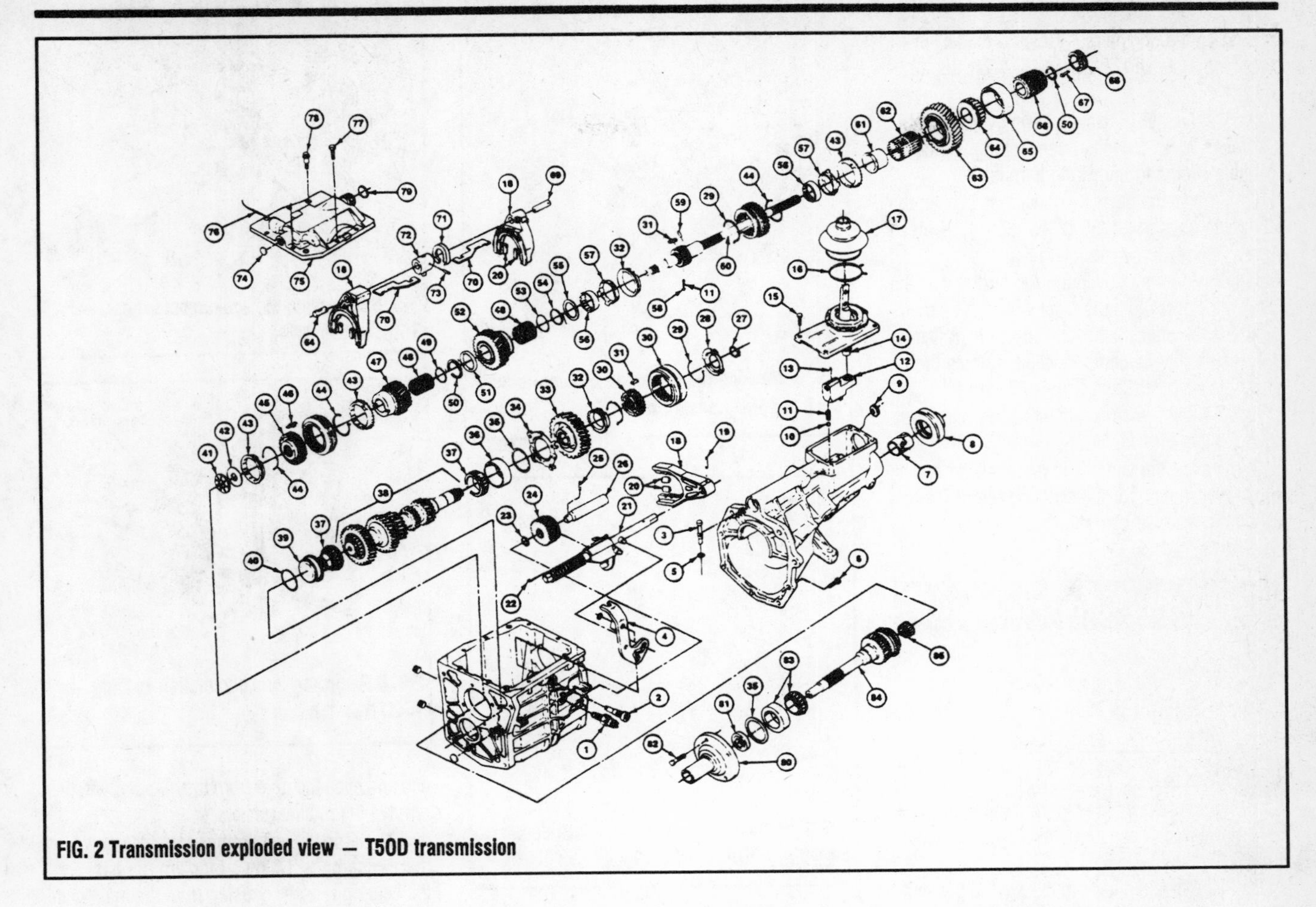

FIG. 2 Transmission exploded view — T50D transmission

2. Raise and support the car on jackstands.
3. Unplug the electrical connector at the switch.
4. Unscrew the switch from the transmission extension housing.
5. Screw the new switch into place and tighten it to 60 inch lbs.
6. Connect the wiring.

Transmission

REMOVAL & INSTALLATION

1. Disconnect the negative battery cable.
2. Raise and support the vehicle safely.
3. Matchmark the driveshaft for reasembly. Disconnect the driveshaft from the rear U-joint flange. Slide the driveshaft off the transmission output shaft and install an extension housing seal installation tool into the extension housing to prevent lubricant from leaking.
4. Remove the bolts and remove the catalytic converter.
5. Remove the 2 nuts attaching the rear transmission support to the crossmember. Remove the bolts.
6. Using a suitable jack, support the engine and transmission.
7. Remove the 2 nuts from the crossmember bolts. Remove the bolts, raise the jack slightly and remove the crossmember.
8. Lower the transmission to expose the 2 bolts securing the shift handle to the shift tower. Remove the 2 nuts and bolts and remove the shift handle.
9. Disconnect the wiring harness from the backup lamp switch. On the 5.0L engine, disconnect the neutral sensing switch.
10. Remove the bolt from the speedometer cable retainer and remove the speedometer driven gear from the transmission.
11. Remove the 4 bolts that secure the transmission to the flywheel housing.
12. Remove the transmission and jack rearward until the transmission input shaft clears the flywheel housing. If necessary lower the engine enough to obtain clearance for removing the transmission.

➡ Do not depress the clutch while the transmission is removed.

To install:

13. Make sure the mounting surface of the transmission and flywheel housing are clean and free of dirt, paint and burrs.
14. Install 2 guide pins in the flywheel housing lower mounting bolt holes. Raise the transmission and move forward on the guide pins until the input shaft splines enter the clutch hub splines and the case is positioned against the flywheel housing.
15. Install the 2 upper transmission-to-flywheel housing mounting bolts snug and remove the 2 guide pins. Install the 2 lower mounting bolts and tighten all the bolts to 45–65 ft. lbs. (61–88 Nm).
16. Raise the transmission with a jack until the shift handle can be secured to the shift tower. Install and tighten the attaching bolts and washers to 23–32 ft. lbs. (31–43 Nm).
17. Connect the speedometer cable to the extension housing and tighten the attaching screw to 36–54 inch lbs. (48–68 Nm).
18. Raise the rear of the transmission with the jack and install the transmission support. Install and tighten the attaching bolts to 36–50 ft. lbs. (48–68 Nm).
19. With the transmission extension housing

resting on the engine rear support, install the attaching bolts and tighten to 25–35 ft. lbs. (38–48 Nm).

20. Connect the backup lamp switch wiring harness. On 5.0L engine, connect the neutral sensing switch to the wiring harness.

21. Install the catalytic converter. Tighten the attaching bolts to 20–30 ft. lbs. (27–41 Nm).

22. Remove the extension housing installation tool and slide the forward end of the driveshaft over the transmission output shaft. Connect the driveshaft to the rear U-joint flange. Make sure the matchmarks align. Tighten the U-bolt nuts to 42–57 ft. lbs. (56–77 Nm).

23. Fill the transmission with the proper type and quantity of fluid.

24. Lower the vehicle. Check the shift and crossover motion for full shift engagement and smooth crossover operation.

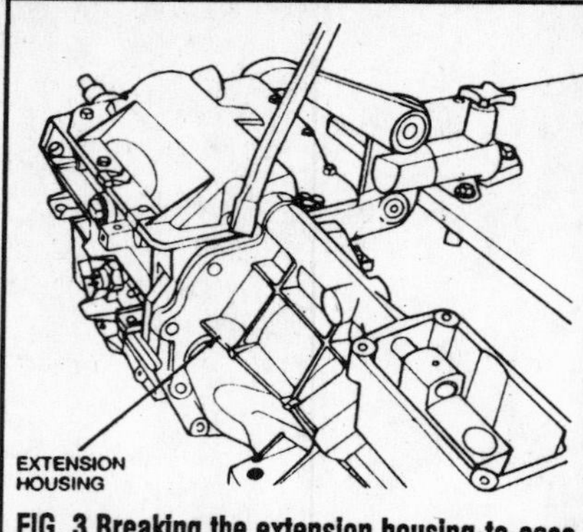

FIG. 3 Breaking the extension housing-to-case seal — T50D transmission

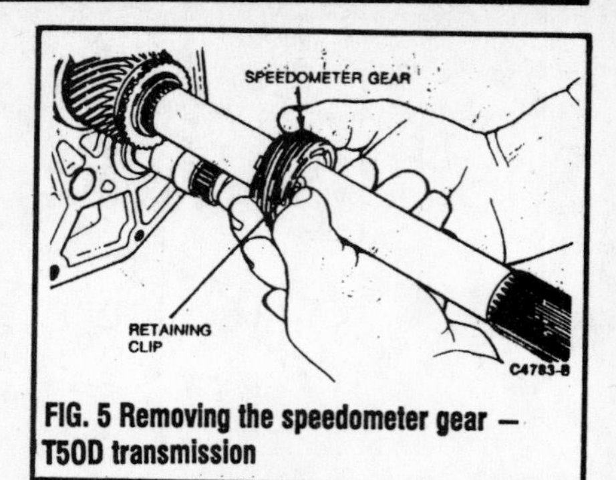

FIG. 5 Removing the speedometer gear — T50D transmission

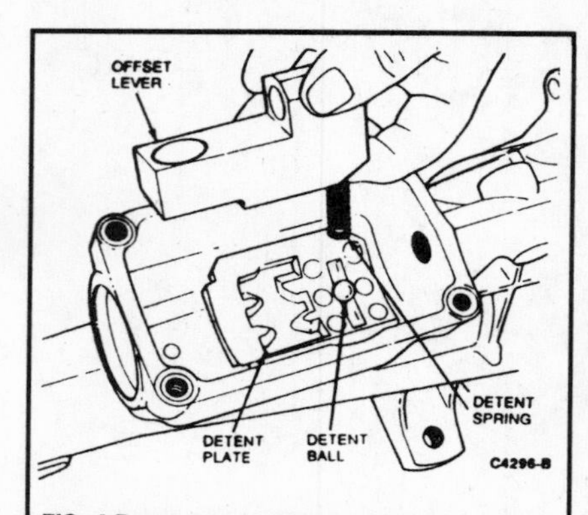

FIG. 4 Removing the offset lever and detent spring — T50D transmission

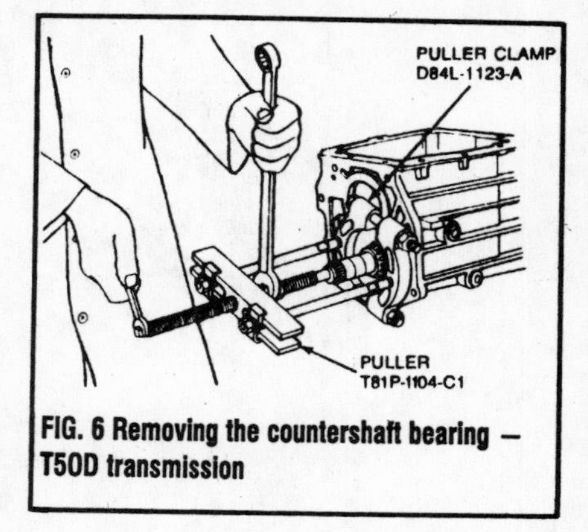

FIG. 6 Removing the countershaft bearing — T50D transmission

Transmission Overhaul

DISASSEMBLY

SEE FIGS. 2–6

1. Remove the drain plug from the lower right side of the main case and drain any excess oil from the transmission.

2. Place the shift lever in the Neutral position, then remove the turret cover-to-transmission bolts.

3. Using a medium pry bar, pry the turret cover from the extension housing.

4. Using a $^3/_{16}$ in. (5mm) pin punch and a hammer, remove the offset lever-to-shifter shaft roll pin, then the damper sleeve.

WARNING

If the extension housing is bolted in place, do not attempt to remove the offset lever; a lug, (located at the bottom of the offset lever) meshing with the detent plate, prevents rearward movement of the offset lever.

5. Remove the extension housing-to-main case bolts. Using a medium pry bar, pry the extension housing (break the seal) from the main case. Remove the extension housing/offset lever assembly by sliding it rearward.

6. Remove the offset lever, the roll pin, the detent spring/ball from the extension housing detent plate.

7. Remove the shift cover-to-main case bolts. Using a medium pry bar, pry the shift cover from the main case, then lift it slightly and slide it towards the filler plug side of the transmission. When the shift forks clear groove in the 5th/Reverse shift lever, continue lifting the cover.

8. Using a pair of needle-nose pliers, remove the 5th/Reverse shift lever-to-lever pivot pin C-clip.

9. Using the T50 Torx® Driver tool, remove the 5th/Reverse shift lever pivot pin but do not remove the 5th/Reverse shift lever. Remove the backup lamp switch.

10. Using a pair of snapring pliers, remove the 5th gear synchronizer snapring/spacer from the rear of the countershaft.

11. Remove the 5th gear, the synchronizer, the shift fork and the shift rail by gripping the components as an assembly and pulling them rearward from the main case.

➡ To disengage the 5th/Reverse shift rail, work it until it is free of the shift rail.

12. To remove the speedometer gear, press downward on the speedometer gear retaining clip and slide the gear from the output shaft, then remove the retaining clip.

13. Remove the front bearing retainer-to-main case bolts. Using a medium pry bar, pry the bearing retainer housing from the main case.

14. To remove the input shaft, rotate it until the flat on the clutch teeth aligns with the countershaft, then pull it from the main case; be careful not to drop the roller bearings, the thrust bearing or the race from the rear of the input shaft.

15. Remove the 4th gear blocking ring from the 3rd/4th synchronizer.

16. Pull the output shaft rearward, until the 1st gear stops against the case, then remove the output shaft bearing race.

➡ If the race sticks, work the shaft back and forth until it is free.

17. Tilt the output shaft so that the gear/synchronizer assembly end may be lifted up and out of the main case.

18. From the main case, remove the 5th/Reverse shift fork, the Reverse shift fork and the inhibitor spring.

19. Using a $^3/_{16}$ in. (5mm) pin punch and a hammer, drive the roll pin from the Reverse idler shaft.

20. Through the back of the main case, slide out the Reverse idler shaft, then remove the Reverse idler gear and the over travel rubber stop.

21. Using a hammer and a punch or chisel,

flatten the countershaft retainer tabs (all four corners). Remove the countershaft retainer-to-main case bolts, the retainer, the shims and the bearing race.

➡ **If the race sticks, work the shaft back and forth until it is free.**

22. Using the Puller tool No. T81P-1104-C1 and the Puller Clamp tool No. D84L-1123-A, press the bearing from the rear of the countershaft.

23. Move the countershaft rearward, tilt the assembly upward and remove it from the case.

24. Clean all of the parts in solvent and inspect for damage or wear; replace the parts as necessary. Remove the front bearing from the countershaft.

ASSEMBLY

➧ SEE FIGS. 7–12

1. Using an arbor press and the Bearing Installation tool No. T57L-4621-B, press a new bearing onto the front of the countershaft, then position the countershaft in the main case.

2. Using an arbor press and the Bearing Installation tool No. T83P-7025-AH, press the rear bearing onto the countershaft.

WARNING

When pressing the rear bearing onto the countershaft, place two pieces of 1/4 in. (6mm) bar stock inside the main case, between the countergear front and the main case to support it. During installation, if the countershaft is not properly supported, permanent distortion/damage may result to the main case.

3. Install the rear bearing race onto the countershaft. Install the countershaft bearing retainer and torque the retainer-to-main case bolts to 10-15 ft. lbs.

➡ **Initially, when installing the countershaft bearing retainer, do not use any shims.**

4. Using a Dial Indicator and the Bracket tool No. D78P-4201-F, measure the countergear end play; it should be 0.001-0.005 in. (0.0254-0.127mm). If the end play is excessive, remove the countershaft bearing retainer and install shims.

5. After reinstalling the countershaft bearing retainer, bend the retaining tabs over the mounting bolts.

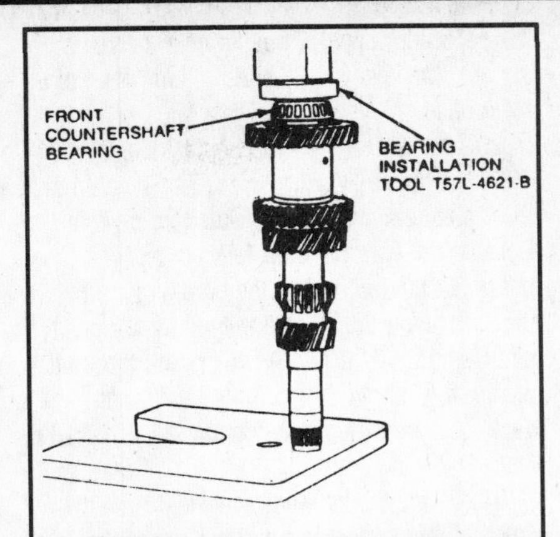

FIG. 7 Installing the front countershaft bearing T50D transmission

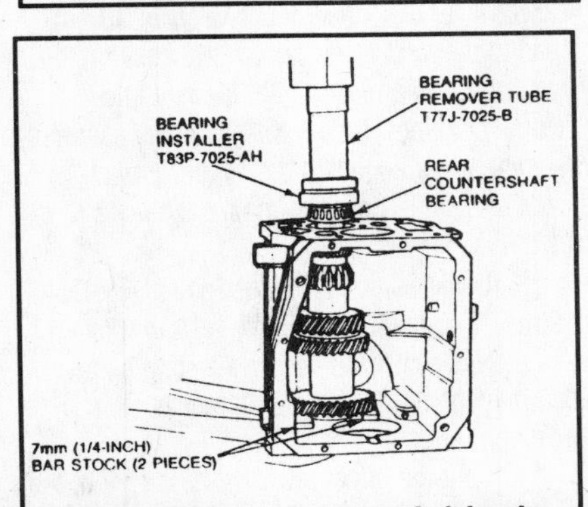

FIG. 8 Installing the rear countershaft bearing - T50D transmission

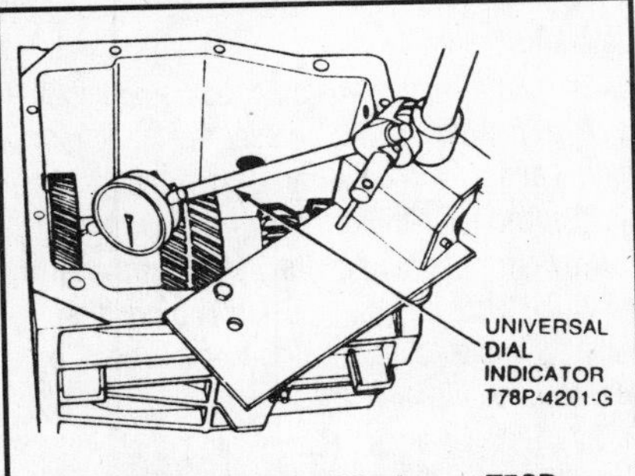

FIG. 9 Dial indicator positioning — T50D transmission

6. Install the Reverse idler gear in the main case with the shift lever groove facing the rear of the case, then the Reverse idler shaft and the rubber over travel stop.

7. Using a 3/16 in. (5mm) pin punch, drive the Reverse idler shaft roll pin into the idler shaft to secure the shaft.

8. Position the Reverse shifting fork and the 5th/Reverse shifting lever into the main case.

9. Install the output shaft assembly into the main case.

10. Using Polyethylene Grease No. D0AZ-19584-A, or equivalent, coat the input shaft roller bearings (place the bearings into the input shaft), the thrust bearing and the bearing race.

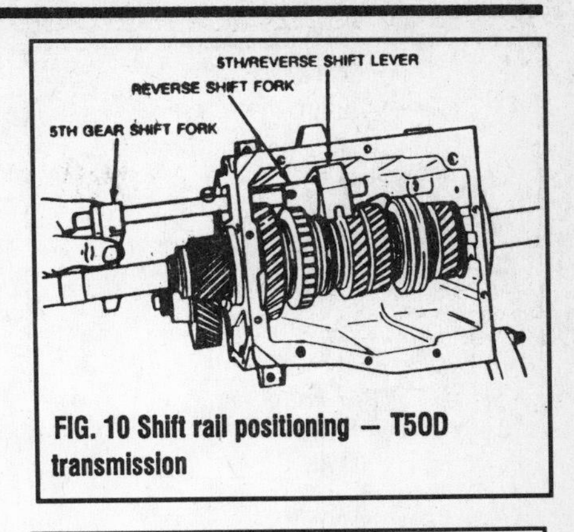

FIG. 10 Shift rail positioning — T50D transmission

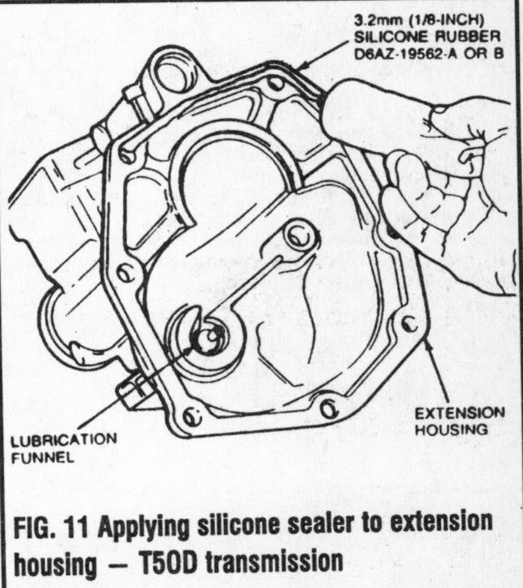

FIG. 11 Applying silicone sealer to extension housing — T50D transmission

11. Install the 4th gear blocking ring; align the blocking ring notches with the inserts of the 3rd/4th synchronizer.

12. To install the input shaft, align the flat on the synchronizer teeth the with the countershaft, then install the input shaft.

13. Install the input shaft bearing race into the input shaft bearing retainer; do not install the shims. Install the bearing retainer (inner notch facing upwards) onto the main case; do not use sealant. Torque the bearing retainer-to-main case bolts to 11-20 ft. lbs.

14. Install the output shaft rear bearing race; if necessary, tap the bearing into place using a plastic tipped hammer.

15. Install the 5th gear onto the countershaft. Install the shifting rail/5th gear shifting fork assembly into the main case.

➡ **When installing the shifting rail/5th gear shifting fork assembly, align the shift rail fork and slide the rail through the fork, stop after the rod passes through the fork.**

16. Place the shift lever return spring in the main case and slide the shifting rail through it; the long end of the spring MUST face the rear of the main case.

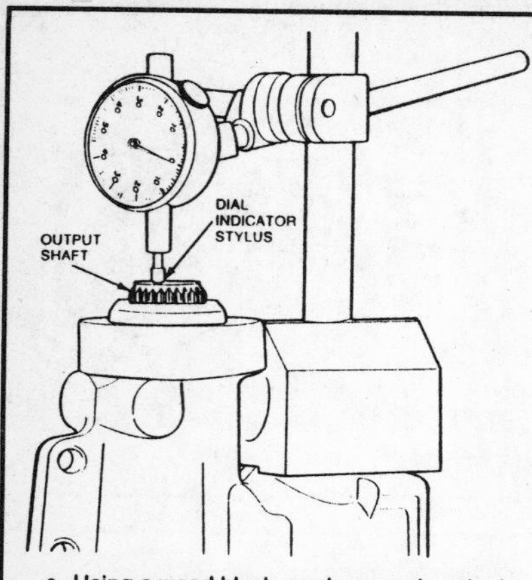

- Using a wood block, push upward on the input shaft and note the dial indicator reading.

NOTE: A shim must be installed that is the thickness of the dial indicator reading. This will provide zero end play.

CAUTION: Although zero end play is the ideal end play specification, a plus or minus .050mm (.002 inch) is an acceptable tolerance. Do not overload the bearings with too thick a shim.

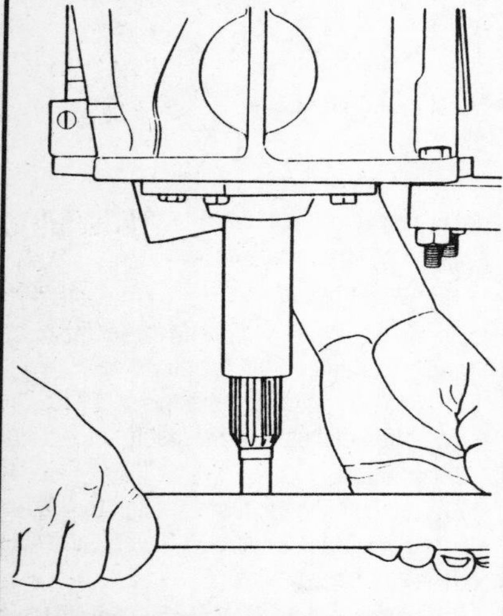

FIG. 12 Checking output shaft endplay — T50D transmission

17. Install the blocking ring and the 5th gear synchronizer into the 5th gear shifting fork, then slide the fork rail assembly into position.
18. Using a pair of snapring pliers, install the 5th gear synchronizer retainer and snapring.
19. Using a pair of needle nose pliers, connect the lever return spring to the front of the main case.
20. Apply Teflon® Pipe Sealant No. D8AZ-19554-A, or equivalent, to the 5th/Reverse shift lever pivot pin and the back-up light switch. Position the Reverse shift fork pin and the 5th gear shift rail pin so that they are engaged with the shift lever, then install the shift lever pivot pin. Using the T50 Torx® Driver tool, torque the pivot pin-to-shift lever to 23-32 ft. lbs. and the back-up light switch-to-transmission to 12-18 ft. lbs.
21. Install the speedometer gear onto the output shaft; make sure that the retainer clip engages a hole in the output shaft.
22. Using Silicone Rubber Sealant No. D6AZ-19562-A, or equivalent, apply a 1/8 in. (3mm) bead to the shift cover assembly. Position the synchronizers and the shifting cover into the Neutral positions, then install the cover assembly (shifting forks engaging the synchronizers). Torque the shift cover-to-main case bolts to 6-11 ft. lbs.
23. Using Silicone Rubber Sealant No. D6AZ-19562-A, or equivalent, apply a 1/8 in. (3mm) bead to the extension housing mating surface and the lubrication funnel in the extension housing.
24. Coat the offset lever's detent spring, the detent and the detent ball (place in the Neutral position) with petroleum jelly; position and install these parts into the extension housing (be sure to position the offset lever with the spring over the detent ball.
25. Install the extension housing and shift lever to the main case; be sure the lubrication funnel engages into the 5th gear synchronizer.
26. To install the offset lever, push it downward to compress the detent spring and to push the lever and the housing into position. Install the extension housing-to-main case bolts and torque to 20-45 ft. lbs.
27. Using a 3/16 in. (5mm) pin punch, drive the roll pin into the offset lever-to-shifter shaft hole. Install the damper sleeve into the offset lever.
28. To measure the output shaft end play, perform the following procedures:
    a. Position the transmission so that the extension is facing upwards.
    b. Using a Dial Indicator, secure it to the extension housing and position it so that it rides on the end of the output shaft.
    c. Rotate the input and the output shafts, then zero the dial indicator.
    d. Using a wooden block, push upwards on the input shaft and note the dial indicator reading.

WARNING

A shim must be installed that is the thickness of the dial indicator reading, which will provide a zero end play. DO NOT overload the bearings with too thick a shim; a ± 0.002 in. (± 0.050mm) is acceptable.

29. Place the transmission on a level surface and remove the input bearing retainer and the bearing race from the retainer; install the shim under the bearing race.
30. Using Silicone Rubber Sealant No. D6AZ-19562-A, or equivalent, apply a 1/8 in. (3mm) bead to the bearing retainer, install the retainer and check the end play.

➡ When applying sealant to the bearing retainer, sealant must not cover the notch on the inner edge of the retainer. Be sure to position the retainer with the inner notch facing upwards.

31. Using Silicone Rubber Sealant No. D6AZ-19562-A, or equivalent, apply a 1/8 in. (3mm) bead to the turret cover. Place the cover onto the extension housing and torque the bolts to 11-15 ft. lbs.
32. Install and torque the drain plug-to-main case to 15-30 ft. lbs.

CLUTCH

➧ SEE FIG. 13

CAUTION

The clutch driven disc contains asbestos, which has been determined to be a cancer causing agent. Never clean clutch surfaces with compressed air! Avoid inhaling any dust from any clutch surface! When cleaning clutch surfaces, use a commercially available brake cleaning fluid.

Self-Adjusting Clutch

➧ SEE FIG. 14

The free play in the clutch is adjusted by a built in mechanism that allows the clutch controls to be self-adjusted during normal operation.

The self-adjusting feature should be checked every 5,000 miles. This is accomplished by insuring that the clutch pedal travels to the top of its upward position. Grasp the clutch pedal with your hand or put your foot under the clutch pedal, pull up on the pedal until it stops. Very little effort is required (about 10 lbs.) During the application of upward pressure, a click may be heard which means an adjustment was necessary and has been accomplished.

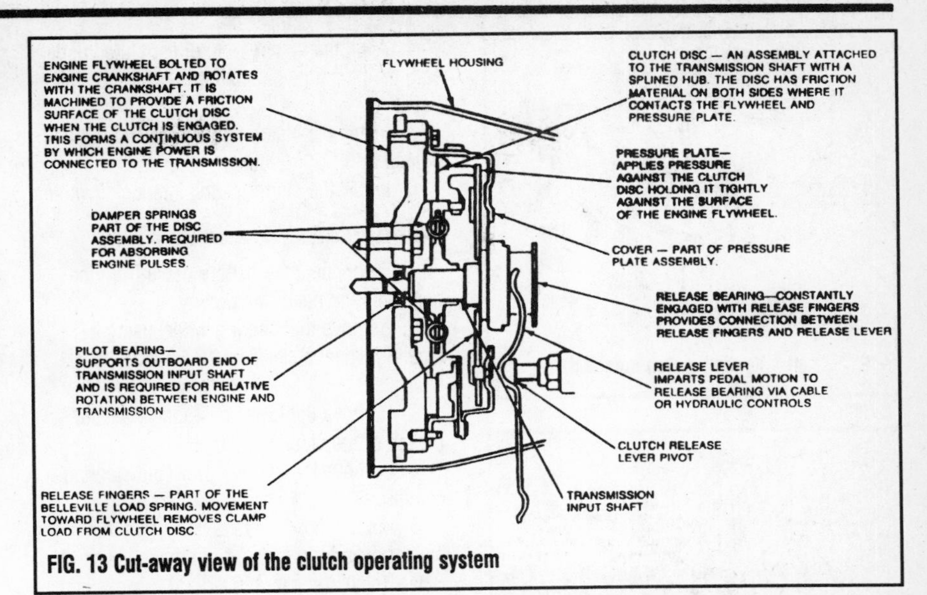

FIG. 13 Cut-away view of the clutch operating system

COMPONENTS

➧ SEE FIG. 15

The self-adjusting clutch control mechanism is automatically adjusted by a device on the clutch pedal. The system consists of a spring loaded gear quadrant, a spring loaded pawl, and a clutch cable which is spring loaded to preload the clutch release lever bearing to compensate for movement of the release lever, as the clutch disc wears. The spring loaded pawl located at the top of the clutch pedal, engages the gear quadrant when the clutch pedal is depressed and pulls the cable through its continuously adjusted stroke. Clutch cable adjustments are not required because of this feature.

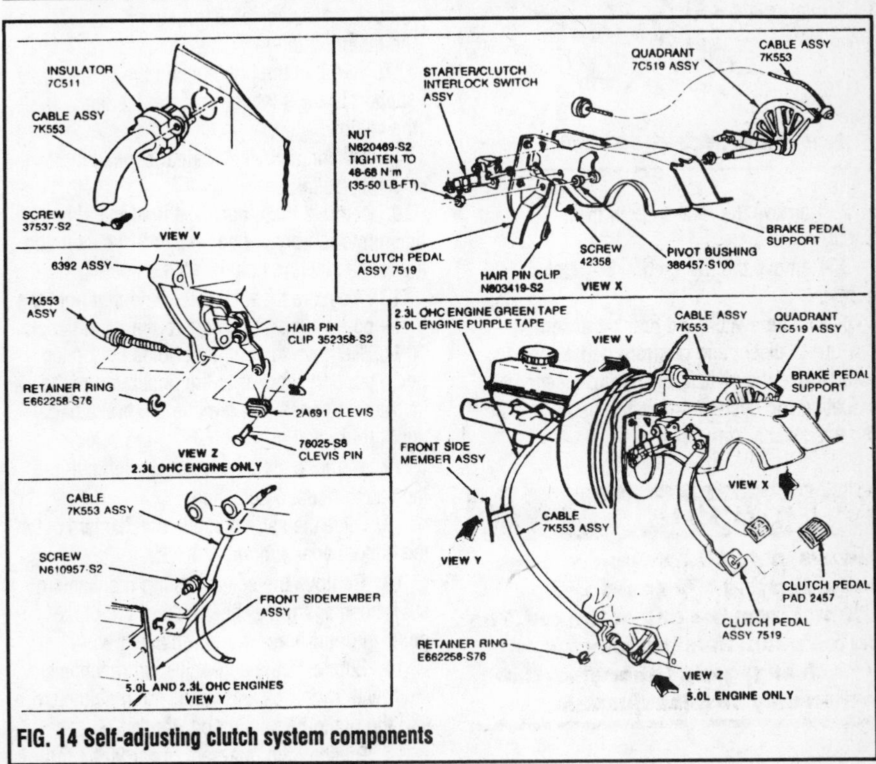

FIG. 14 Self-adjusting clutch system components

STARTER/CLUTCH INTERLOCK SWITCH

➧ SEE FIG. 16

The starter/clutch switch is designed to prevent starting the engine unless the clutch pedal is fully depressed. The switch is connected between the ignition switch and the starter motor relay coil and maintains an open circuit with the clutch pedal up (clutch engaged).

The switch is designed to self-adjust automatically the first time the clutch pedal is pressed to the floor. The self-adjuster consists of a two-piece clip snapped together over a serrated rod. When the plunger or rod is extended, the clip bottoms out on the switch body and allows the rod to ratchet over the serrations to a position determined by the clutch pedal travel limit. In this way, the switch is set to close the starter circuit when the clutch is pressed all the way to the floor (clutch disengaged).

Testing Continuity

1. Disconnect inline wiring connector at jumper harness.
2. Using a test lamp or continuity tester, check that switch is open with clutch pedal up (clutch engaged), and closed at approximately 1 in. (25mm) from the clutch pedal full down position (clutch disengaged).
3. If switch does not operate, check to see if the self-adjusting clip is out of position on the rod. It should be near the end of the rod.
4. If the self-adjusting clip is out of position, remove and reposition the clip to about 1 in. (25mm) from the end of the rod.
5. Reset the switch by pressing the clutch pedal to the floor.
6. Repeat Step 2. If switch is damaged, replace it.

REMOVAL & INSTALLATION

Starter/Clutch Interlock Switch

1. Disconnect the wiring connector.

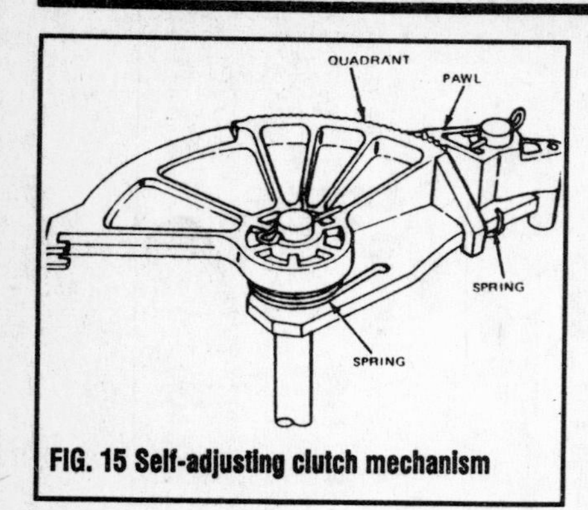

FIG. 15 Self-adjusting clutch mechanism

FIG. 16 Starter/clutch interlock switch

2. Remove the retaining pin from the clutch pedal.
3. Remove the switch bracket attaching screw.
4. Lift the switch and bracket assembly upward to disengage tab from pedal support.
5. Move the switch outward to disengage actuating rod eyelet from clutch pedal pin and remove switch from vehicle.

WARNING

Always install the switch with the self-adjusting clip about 1 in. (25mm) from the end of the rod. The clutch pedal must be fully up (clutch engaged). Otherwise, the switch may be misadjusted.

6. Place the eyelet end of the rod onto the pivot pin.
7. Swing the switch assembly around to line up hole in the mounting boss with the hole in the bracket.
8. Install the attaching screw.
9. Replace the retaining pin in the pivot pin.
10. Connect the wiring connector.

Clutch Pedal Assembly

1. Remove the starter/clutch interlock switch.
2. Remove the clutch pedal attaching nut.
3. Pull the clutch pedal off the clutch pedal shaft.
4. Align the square hole of the clutch pedal with the clutch pedal shaft and push the clutch pedal on.
5. Install the clutch pedal attaching nut and tighten to 32-50 ft. lbs.
6. Install the starter/clutch interlock switch.

Self-Adjusting Assembly

1. Disconnect the battery cable from the negative terminal of the battery.
2. Remove the steering wheel using a steering wheel puller Tool T67L-3600-A or equivalent.
3. Remove the lower dash panel section to the left of the steering column.
4. Remove the shrouds from the steering column.
5. Disconnect the brake lamp switch and the master cylinder pushrod from the brake pedal.
6. Rotate the clutch quadrant forward and unhook the clutch cable from the quadrant. Allow the quadrant to slowly swing rearward.
7. Remove the bolt holding the brake pedal support bracket lateral brace to the left side of the vehicle.
8. Disconnect all electrical connectors to the steering column.
9. Remove the 4 nuts that hold the steering column to the brake pedal support bracket and lower the steering column to the floor.
10. Remove the 4 booster nuts that hold the brake pedal support bracket to the dash panel.
11. Remove the bolt that holds the brake pedal support bracket to the underside of the instrument panel, and remove the brake pedal support bracket assembly from the vehicle.
12. Remove the clutch pedal shaft nut and the clutch pedal as outlined.
13. Slide the self-adjusting mechanism out of the brake pedal support bracket.
14. Remove the self-adjusting mechanism shaft bushings from either side of the brake pedal support bracket and replace if worn.
15. Lubricate the self-adjusting mechanism shaft with motor oil and install the mechanism into the brake pedal support bracket.
16. Position the quadrant towards the top of the vehicle. Align the flats on the shaft with the flats in the clutch pedal assembly, and install the retaining nuts. Tighten to 32-50 ft. lbs.
17. Position the brake pedal support bracket assembly beneath the instrument panel aligning the four holes with the studs in the dash panel. Install the four nuts loosely. Install the bolt through the support bracket into the instrument panel and tighten to 13-25 ft. lbs.
18. Tighten the four booster nuts that hold the brake pedal support bracket to the dash panel to 13-25 ft. lbs.
19. Connect the brake lamp switch and the master cylinder pushrod to the brake pedal.
20. Attach the clutch cable to the quadrant.
21. Position the steering column onto the four studs in the support bracket and start the four nuts.
22. Connect the steering column electrical connectors.
23. Install the steering column shrouds.
24. Install the brake pedal support lateral brace.
25. Tighten the steering column attaching nuts to 20-37 ft. lbs.
26. Install the lower dash panel section.
27. Install the steering wheel.
28. Connect the battery cable to the negative terminal on the battery.
29. Check the steering column for proper operation.
30. Depress the clutch pedal several times to adjust cable.

Quadrant Pawl, Self-Adjusting

1. Remove the self-adjusting mechanism.
2. Remove the two hairpin clips that hold the pawl and quadrant on the shaft assembly.
3. Remove the quadrant and quadrant spring.
4. Remove the pawl spring.
5. Remove the pawl.
6. Lubricate the pawl and quadrant pivot shafts with M1C75B or equivalent grease.
7. Install pawl. Position the teeth of the pawl toward the long shaft, and the spring hole at the end of the arm. Do not position the spring hole beneath the arm.
8. Insert the straight portion of the spring into the hole, with the coil up.
9. Keeping the straight portion in the hole rotate the spring 180 degrees to the left and slide the coiled portion of the spring over the boss.
10. Hook the bend portion of the spring under the arm.
11. Install the retainer clip on opposite side of spring.
12. Place the quadrant spring on the shaft with the bent portion of the spring in the hole in the arm.
13. Place the lubricated quadrant on the shaft aligning the projection at the bottom of the quadrant to a position beneath the arm of the shaft assembly. Push the pawl up so the bottom tooth of the pawl meshes with bottom tooth of quadrant.
14. Install the quadrant retainer pin.
15. Grasp the straight end of the quadrant spring with pliers and position behind the ear of the quadrant.
16. Install the self-adjusting mechanism.
17. Install the clutch pedal assembly.

Clutch Cable Assembly

1. Lift the clutch pedal to its upward most position to disengage the pawl and quadrant.

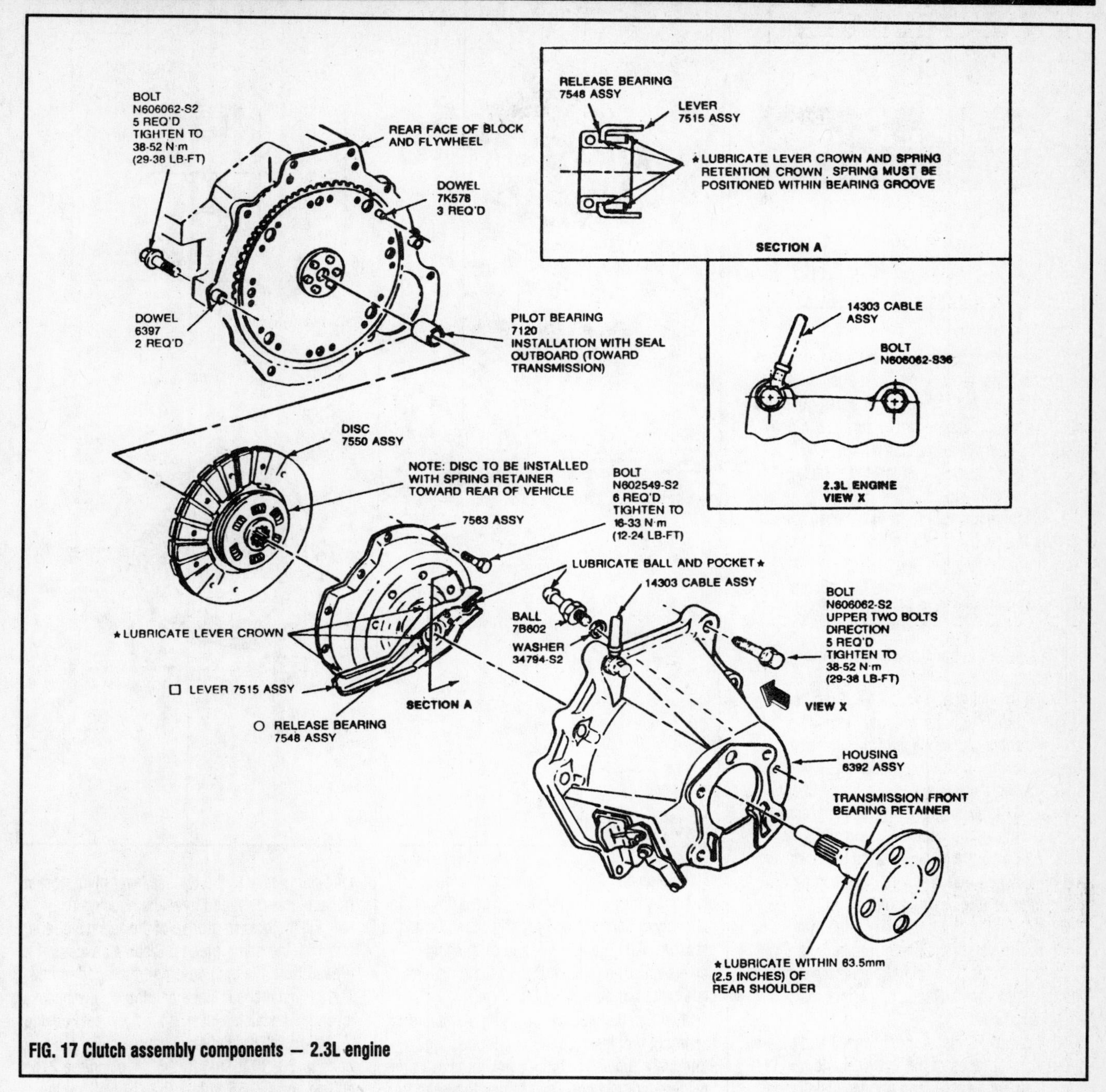

FIG. 17 Clutch assembly components — 2.3L engine

Push the quadrant forward, unhook the cable from the quadrant and allow to slowly swing rearward.

2. Open the hood and remove the screw that holds the cable assembly isolator to the dash panel.

3. Pull the cable through the dash panel and into the engine compartment. On 5.0L engines, remove cable bracket screw from fender apron.

4. Raise the vehicle and safely support on jackstands.

5. Remove the dust cover from the bell housing.

6. Remove the clip retainer holding the cable assembly to the bell housing.

7. Slide the ball on the end of the cable assembly through the hole in the clutch release lever and remove the cable.

8. Remove the dash panel isolator from the cable.

9. Install the dash panel isolator on the cable assembly.

10. Insert the cable through the hole in the bell housing and through the hole in the clutch release lever. Slide the ball on the end of the cable assembly away from the hole in the clutch release lever.

11. Install the clip retainer that holds the cable assembly to the bell housing.

12. Install the dust shield on the bell housing.

13. Push the cable assembly into the engine compartment and lower the vehicle. On 5.0L engines, install cable bracket screw in fender apron.

14. Push the cable assembly into the hole in the dash panel and secure the isolator with a screw.

15. Install the cable assembly by lifting the clutch pedal to disengage the pawl and quadrant, the, pushing the quadrant forward, hook the end of the cable over the rear of the quadrant.

16. Depress clutch pedal several times to adjust cable.

Driven Disc and Pressure Plate

REMOVAL & INSTALLATION

➧ SEE FIGS. 17–18

1. Disconnect the negative battery cable. Lift the clutch pedal to its uppermost position to disengage the pawl and quadrant. Push quadrant forward, unhook cable from quadrant and allow quadrant to slowly swing rearward.
2. Raise and safely support the vehicle. Remove the dust shield.
3. Disconnect cable from the release lever. Remove the retaining clip and remove the clutch cable from the flywheel housing.
4. Remove starter and bolts that secure engine rear plate to front lower part of flywheel housing.
5. Remove the transmission, then the flywheel housing.
6. Remove clutch release lever from housing by pulling it through the window in housing until retainer spring is disengaged from pivot. Remove release bearing from release lever.
7. Loosen the pressure plate cover attaching bolts evenly to release spring tension gradually and avoid distorting cover. If same pressure plate and cover are to be installed, mark cover and flywheel so that pressure plate can be installed in its original position.
8. Inspect the flywheel for scoring, cracks or other damage and machine or replace, as necessary. Inspect the pilot bearing for damage and free movement. Replace, as necessary.

To install:

9. If removed, install the flywheel. Make sure the mating surfaces of the flywheel and the crankshaft flange are clean prior to installation. Tighten the flywheel bolts to 56–64 ft. lbs. (73–87 Nm) on 2.3L engines and 75–85 ft. lbs. (102–115 Nm) on 5.0L engine.

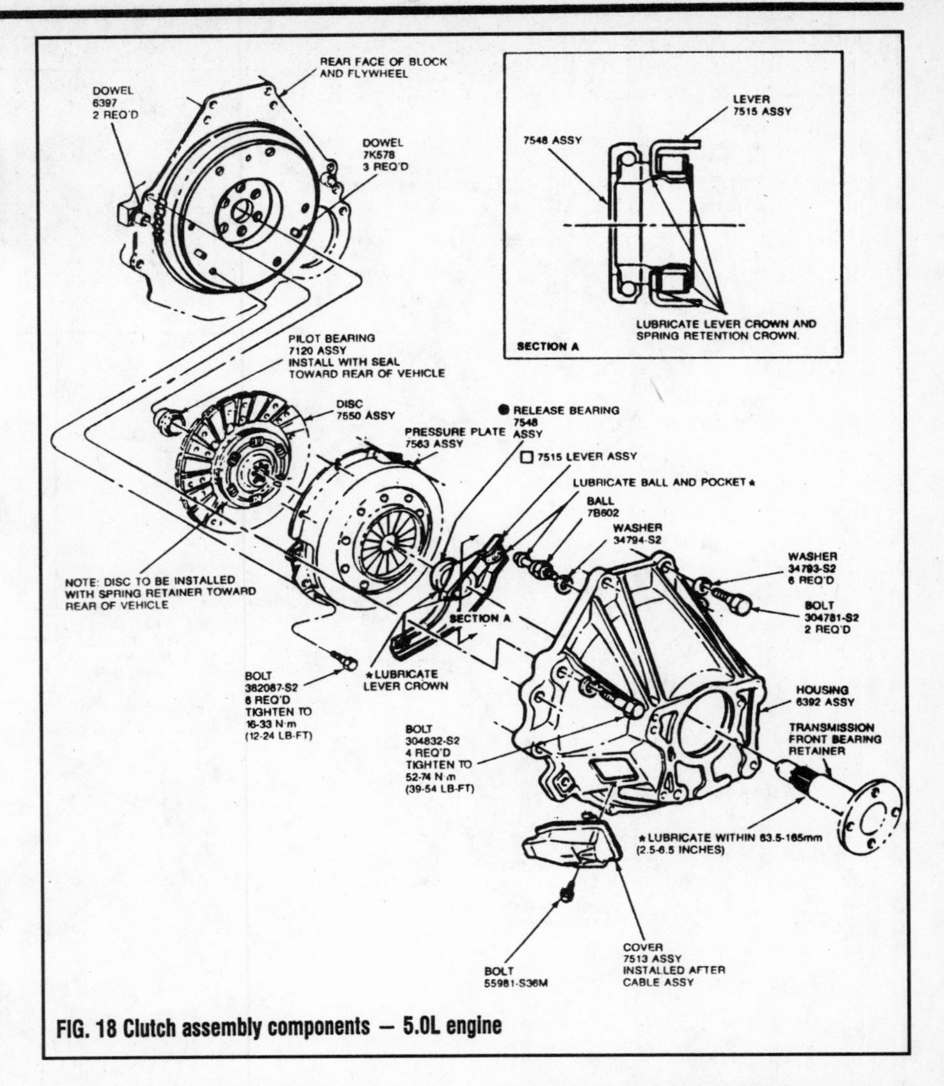

FIG. 18 Clutch assembly components — 5.0L engine

10. Position the clutch disc and pressure plate assembly on the flywheel. The 3 dowel pins on the flywheel must be properly aligned with the pressure plate. Bent, damaged or missing dowels must be replaced. Start the pressure plate bolts but do not tighten them.
11. Align the clutch disc using a suitable alignment tool inserted in the pilot bearing. Alternately tighten the bolts a few turns at a time, until they are all tight. Final torque the bolts to 12–24 ft. lbs. (17–32 Nm). Remove the alignment tool.
12. Apply a light coating of multi-purpose long-life grease to the release lever pivot pocket, release lever fork and flywheel housing pivot ball. Fill the grease groove of the release bearing hub with the same grease. Clean all excess grease from the inside bore of the bearing hub.
13. Install the release bearing on the release lever and install the lever in the flywheel housing.
14. Install the flywheel housing. Tighten the bolts to 28–38 ft. lbs. (38–52 Nm) on the 2.3L engine and 38–55 ft. lbs. (52–74 Nm) on the 5.0L engine.
15. Install the remaining components in the reverse order of removal.

Automatic Transmission

Understanding Automatic Transmissions

The automatic transmission allows engine torque and power to be transmitted to the rear wheels within a narrow range of engine operating speeds. The transmission will allow the engine to turn fast enough to produce plenty of power and torque at very low speeds, while keeping it at a sensible rpm at high vehicle speeds. The transmission performs this job entirely without driver assistance. The transmission uses a light fluid as the medium for the transmission of power. This fluid also works in the operation of various hydraulic control circuits and as a lubricant. Because the transmission fluid performs all of these three functions, trouble within the unit can easily travel from one part to another. For this reason, and because of the complexity and unusual operating principles of the transmission, a very sound understanding of the basic principles of operation will simplify troubleshooting.

THE TORQUE CONVERTER

The torque converter replaces the conventional clutch. It has three functions:

1. It allows the engine to idle with the vehicle at a standstill, even with the transmission in gear.
2. It allows the transmission to shift from range to range smoothly, without requiring that the driver close the throttle during the shift.
3. It multiplies engine torque to an increasing extent as vehicle speed drops and throttle opening is increased. This has the effect of making the transmission more responsive and reduces the amount of shifting required.

The torque converter is a metal case which is shaped like a sphere that has been flattened on opposite sides. It is bolted to the rear end of the engine's crankshaft. Generally, the entire metal case rotates at engine speed and serves as the engine's flywheel.

The case contains three sets of blades. One set is attached directly to the case. This set forms the torus or pump. Another set is directly connected to the output shaft, and forms the turbine. The third set is mounted on a hub which, in turn, is mounted on a stationary shaft through a one-way clutch. This third set is known as the stator.

A pump, which is driven by the converter hub at engine speed, keeps the torque converter full of transmission fluid at all times. Fluid flows continuously through the unit to provide cooling.

Under low speed acceleration, the torque converter functions as follows:

The torus is turning faster than the turbine. It picks up fluid at the center of the converter and, through centrifugal force, slings it outward. Since the outer edge of the converter moves faster than the portions at the center, the fluid picks up speed.

The fluid then enters the outer edge of the turbine blades. It then travels back toward the center of the converter case along the turbine blades. In impinging upon the turbine blades, the fluid loses the energy picked up in the torus.

If the fluid were now to immediately be returned directly into the torus, both halves of the converter would have to turn at approximately the same speed at all times, and torque input and output would both be the same.

In flowing through the torus and turbine, the fluid picks up two types of flow, or flow in two separate directions. It flows through the turbine blades, and it spins with the engine. The stator, whose blades are stationary when the vehicle is being accelerated at low speeds, converts one type of flow into another. Instead of allowing the fluid to flow straight back into the torus, the stator's curved blades turn the fluid almost 90 degrees toward the direction of rotation of the engine. Thus the fluid does not flow as fast toward the torus, but is already spinning when the torus picks it up. This has the effect of allowing the torus to turn much faster than the turbine. This difference in speed may be compared to the difference in speed between the smaller and larger gears in any gear train. The result is that engine power output is higher, and engine torque is multiplied.

As the speed of the turbine increases, the fluid spins faster and faster in the direction of engine rotation. As a result, the ability of the stator to redirect the fluid flow is reduced. Under cruising conditions, the stator is eventually forced to rotate on its one-way clutch in the direction of engine rotation. Under these conditions, the torque converter begins to behave almost like a solid shaft, with the torus and turbine speeds being almost equal.

THE PLANETARY GEARBOX

The ability of the torque converter to multiply engine torque is limited. Also, the unit tends to be more efficient when the turbine is rotating at relatively high speeds. Therefore, a planetary gearbox is used to carry the power output of the turbine to the driveshaft.

Planetary gears function very similarly to conventional transmission gears. However, their construction is different in that three elements make up one gear system, and, in that all three elements are different from one another. The three elements are: an outer gear that is shaped like a hoop, with teeth cut into the inner surface; a sun gear, mounted on a shaft and located at the very center of the outer gear; and a set of three planet gears, held by pins in a ring-like planet carrier, meshing with both the sun gear and the outer gear. Either the outer gear or the sun gear may be held stationary, providing more than one possible torque multiplication factor for each set of gears. Also, if all three gears are forced to rotate at the same speed, the gearset forms, in effect, a solid shaft.

Most modern automatics use the planetary gears to provide either a single reduction ratio of about 1.8:1, or two reduction gears: a low of about 2.5:1, and an intermediate of about 1.5:1. Bands and clutches are used to hold various portions of the gearsets to the transmission case or to the shaft on which they are mounted. Shifting is accomplished, then, by changing the portion of each planetary gearset which is held to the transmission case or to the shaft.

THE SERVOS AND ACCUMULATORS

The servos are hydraulic pistons and cylinders. They resemble the hydraulic actuators used on many familiar machines, such as bulldozers. Hydraulic fluid enters the cylinder, under pressure, and forces the piston to move to engage the band or clutches.

The accumulators are used to cushion the engagement of the servos. The transmission fluid must pass through the accumulator on the way to the servo. The accumulator housing contains a thin piston which is sprung away from the discharge passage of the accumulator. When fluid passes through the accumulator on the way to the servo, it must move the piston against spring pressure, and this action smooths out the action of the servo.

THE HYDRAULIC CONTROL SYSTEM

The hydraulic pressure used to operate the servos comes from the main transmission oil pump. This fluid is channeled to the various servos through the shift valves. There is generally a manual shift valve which is operated by the transmission selector lever and an automatic shift valve for each automatic upshift the transmission provides: i.e., 2-speed automatics have a low/high shift valve, while 3-speeds have a 1-2 valve, and a 2-3 valve.

There are two pressures which effect the operation of these valves. One is the governor pressure which is affected by vehicle speed. The other is the modulator pressure which is affected by intake manifold vacuum or throttle position. Governor pressure rises with an increase in vehicle speed, and modulator pressure rises as the throttle is opened wider. By responding to these two pressures, the shift valves cause the upshift points to be delayed with increased throttle opening to make the best use of the engine's power output.

Most transmissions also make use of an auxiliary circuit for downshifting. This circuit may be actuated by the throttle linkage or the vacuum line which actuates the modulator, or by a cable or solenoid. It applies pressure to a special downshift surface on the shift valve or valves.

The transmission modulator also governs the line pressure, used to actuate the servos. In this way, the clutches and bands will be actuated with a force matching the torque output of the engine.

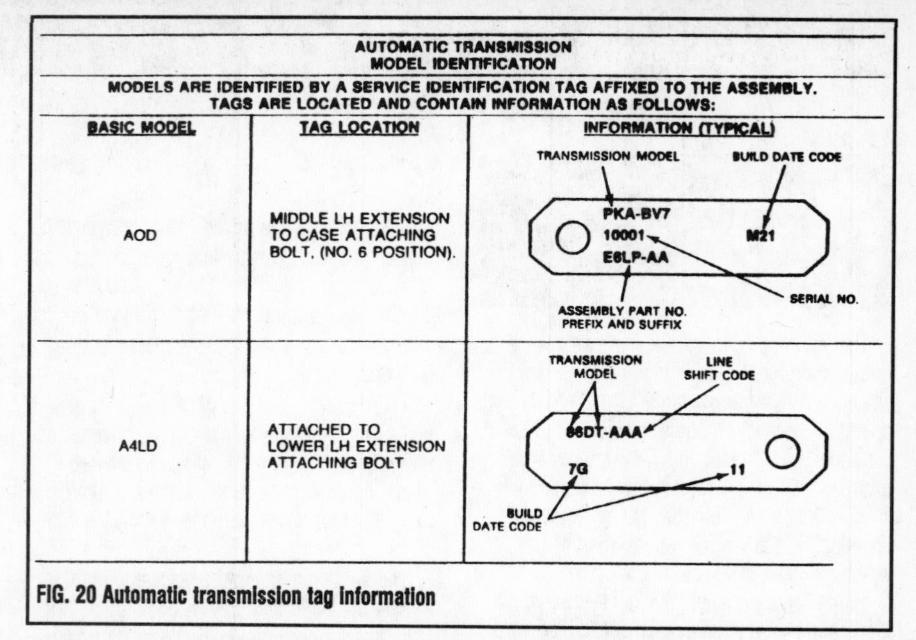
AUTOMATIC TRANSMISSION MODEL IDENTIFICATION

MODELS ARE IDENTIFIED BY A SERVICE IDENTIFICATION TAG AFFIXED TO THE ASSEMBLY. TAGS ARE LOCATED AND CONTAIN INFORMATION AS FOLLOWS:

BASIC MODEL	TAG LOCATION	INFORMATION (TYPICAL)
AOD	MIDDLE LH EXTENSION TO CASE ATTACHING BOLT. (NO. 6 POSITION).	TRANSMISSION MODEL: PKA-BV7; BUILD DATE CODE: M21; SERIAL NO.: 10001; ASSEMBLY PART NO. PREFIX AND SUFFIX: E6LP-AA
A4LD	ATTACHED TO LOWER LH EXTENSION ATTACHING BOLT	TRANSMISSION MODEL / LINE SHIFT CODE: 86DT-AAA; BUILD DATE CODE: 7G 11

FIG. 20 Automatic transmission tag information

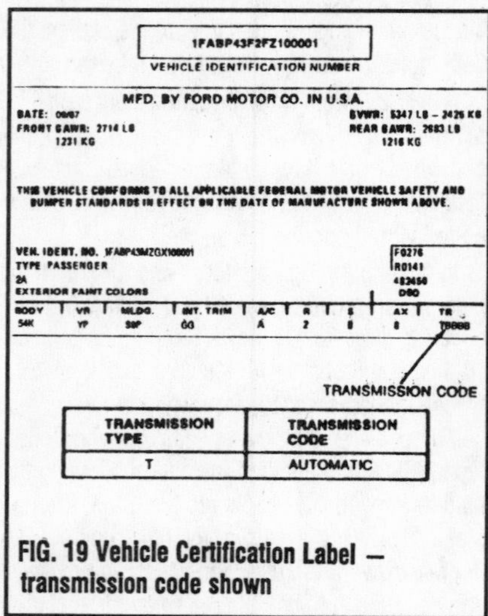

FIG. 19 Vehicle Certification Label — transmission code shown

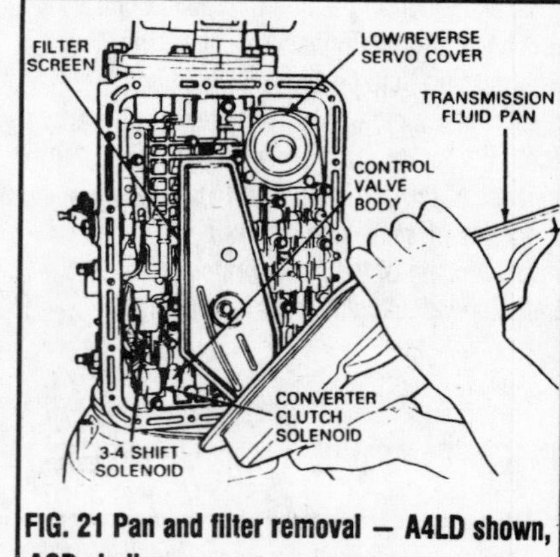

FIG. 21 Pan and filter removal — A4LD shown, AOD similar

Transmission Identification

➧ SEE FIGS. 19–20

All Mustangs are equipped with a Vehicle Certification Label, attached to the left side (driver's) door lock post.

The transmission code is located in the space marked TR on the label.

There are two different automatic transmissions used in the Mustang's covered in this manual, the "AOD" and the "A4LD" transmission.

Additional information is located on the tag attached to the transmission such as: service ID level, and build date.

Fluid Pan Removal and Filter Change

➡ **Refer to Section 1 for current fluid requirements.**

➧ SEE FIG. 21

1. Raise the car and support on jackstands.
2. Place a drain pan under the transmission.
3. Loosen the pan attaching bolts and drain the fluid from the transmission.
4. When the fluid has drained to the level of the pan flange, remove the remaining pan bolts working from the rear and both sides of the pan to allow it to drop and drain slowly.
5. When all of the fluid has drained, remove the pan and clean it thoroughly. discard the pan gasket.
6. Place a new gasket on the pan, and install the pan on the transmission. Tighten the attaching bolts to 12-16 ft. lbs.
7. Add three quarts of fluid to the transmission through the filler tube.
8. Lower the vehicle. Start the engine and move the gear selector through shift pattern. Allow the engine to reach normal operating temperature.
9. Check the transmission fluid. Add fluid, if necessary, to maintain correct level.

Adjustments

SHIFT LINKAGE ADJUSTMENT

Solid link type – 2.3L engine A4LD

➧ SEE FIG. 22

1. Place the transmission shift lever in OVERDRIVE for the A4LD transmission. Be certain that the selector is tight against the rearward OVERDRIVE stop.
2. Raise the vehicle and loosen the manual lever shift rod retaining nut. Move the transmission lever to the OVERDRIVE position. OVERDRIVE is the third detent from the full counterclockwise position.
3. With the transmission shift lever and transmission manual lever in position, tighten the retaining nut to 10-19 ft. lbs. (13–27Nm).
4. Check transmission operation for all selector lever detent positions.

Cable link type – 5.0L engine AOD

➧ SEE FIG. 23

1. Place the transmission shift lever in OVERDRIVE for the A4LD transmission. Be certain that the selector is tight against the rearward OVERDRIVE stop.
2. Raise the vehicle and loosen the manual lever shift cable retaining nut. Move the transmission lever to the OVERDRIVE position. OVERDRIVE is the third detent from the full counterclockwise position.
3. With the transmission shift lever and transmission manual lever in position, tighten the retaining nut to 10-19 ft. lbs. (13–27Nm).
4. Check transmission operation for all selector lever detent positions.

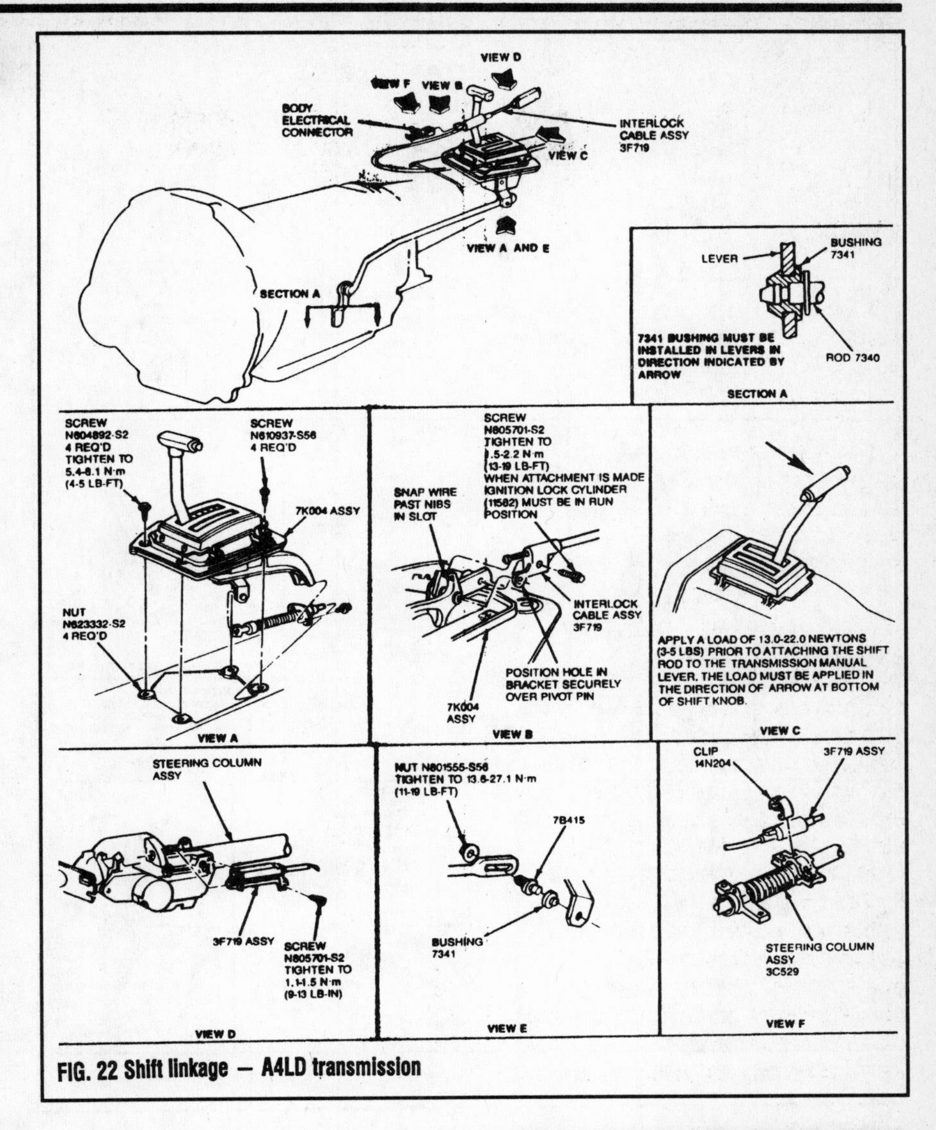

FIG. 22 Shift linkage – A4LD transmission

THROTTLE VALVE CABLE ADJUSTMENT

AOD Transmission

ADJUSTMENT WITH ENGINE OFF

➧ SEE FIG. 24

1. Set the parking brake and put the selector lever in **N**.
2. Remove the protective cover from the cable.
3. Make sure that the throttle lever is at the idle stop. If it isn't, check for binding or interference. NEVER ATTEMPT TO ADJUST THE IDLE STOP!
4. Make sure that the cable is free of sharp bends or is not rubbing on anything throughout its entire length.
5. Lubricate the TV lever ball stud with chassis lube.
6. Unlock the locking tab at the throttle body by prying with a small screwdriver.
7. Install a spring on the TV control lever, to hold it in the rearmost travel position. The spring must exert at least 10 lbs. of force on the lever.
8. Rotate the transmission outer TV lever 10-30° and slowly allow it to return.
9. Push down on the locking tab until flush.
10. Remove the retaining spring from the lever.

Neutral Start Switch

REMOVAL & INSTALLATION

➡ The neutral safety switch on the A4LD and AOD transmission is non-adjustable.

➧ SEE FIG. 25

1. Disconnect the battery ground.
2. Raise and support the car on jackstands.
3. Disconnect the switch harness by pushing the harness straight up off the switch with a long screwdriver underneath the rubber plug section.
4. Using special tool socket T74P-77247-A, or equivalent, on a ratchet extension at least 9 1/2 in. (241mm) long, unscrew the switch. Once the tool is on the switch, reach around the rear of the transmission over the extension housing.

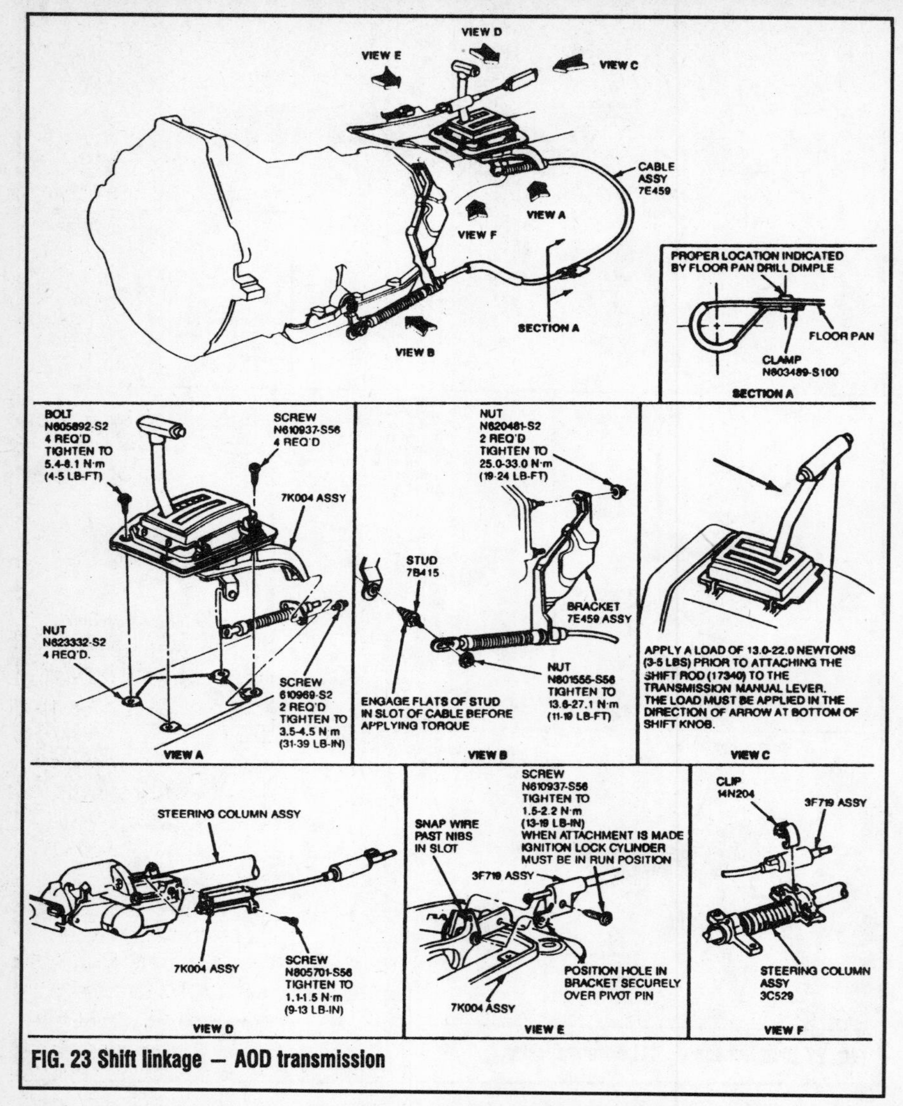

FIG. 23 Shift linkage — AOD transmission

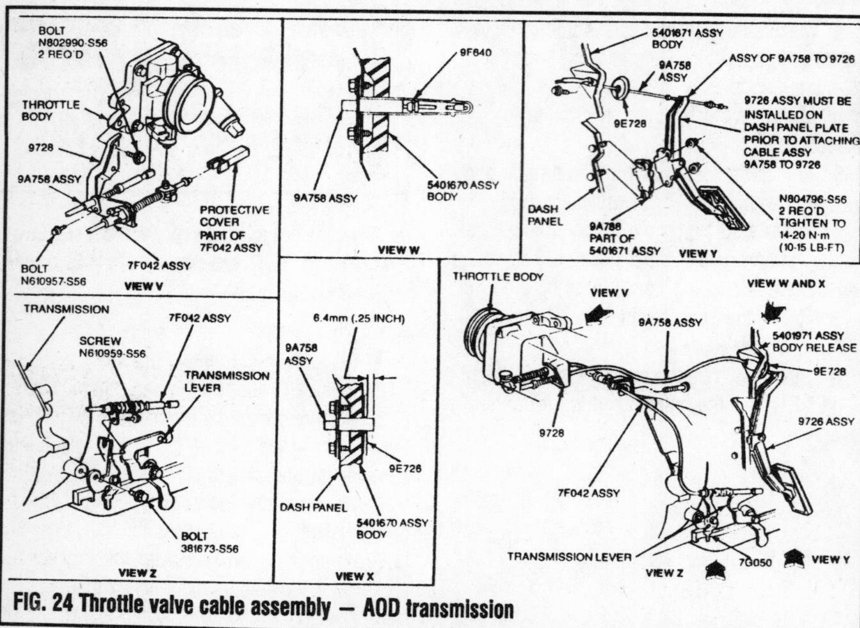

FIG. 24 Throttle valve cable assembly — AOD transmission

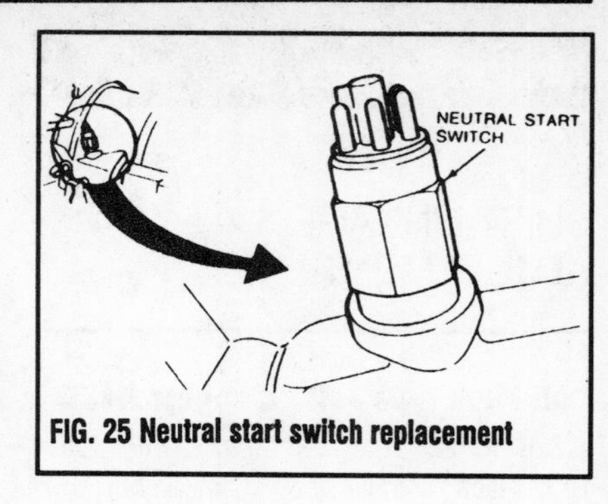

FIG. 25 Neutral start switch replacement

5. Installation is the reverse of removal. Use a new O-ring. Torque the switch to 11 ft. lbs.

Vacuum Modulator

REMOVAL & INSTALLATION

➧ SEE FIG. 26

1. Disconnect the hose from the vacuum diaphragm.
2. Remove the vacuum diaphragm retaining bolt and clamp. DO NOT pry or bend the clamp. Pull the vacuum diaphragm from the transmission case.
3. Remove the vacuum diaphragm control rod from the transmission case.

To Install:

4. Install the vacuum diaphragm control rod in the transmission case.
5. Push the vacuum diaphragm into the case. Secure the retaining clamp and bolt. Tight the bolt to 80–106 inch lbs. (9–12Nm).
6. Install the hose to the vacuum diaphragm.

Transmission

REMOVAL & INSTALLATION

1. Raise and safely support the vehicle.
2. Place the drain pan under the transmission fluid pan. Starting at the rear of the pan and working toward the front, loosen the attaching bolts and allow the fluid to drain. Finally removal all of the pan attaching bolts except two at the front, to allow the fluid to further drain. With fluid drained, install two bolts on the rear side of the pan to temporarily hold it in place.
3. Remove the converter drain plug access cover from the lower end of the converter housing.

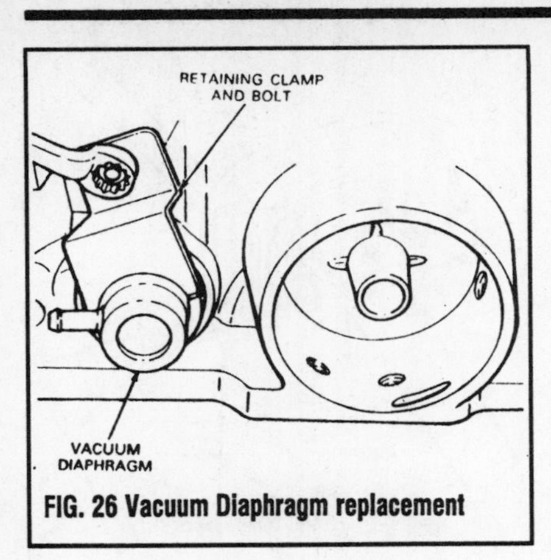

FIG. 26 Vacuum Diaphragm replacement

4. Remove the converter-to-flywheel attaching nuts. place a wrench on the crankshaft pulley attaching bolt to turn the converter to gain access to the nuts.

5. Place a drain pan under the converter to catch the fluid. With the wrench on the crankshaft pulley attaching bolts, turn the converter to gain access to the converter drain plug and remove the plug. After the fluid has been drained, reinstall the plug.

6. Disconnect the driveshaft from the rear axle and slide shaft rearward from the transmission. Install a seal installation tool in the extension housing to prevent fluid leakage.

7. Disconnect the cable from the terminal or the starter motor. Remove the three attaching bolts and remove the starter motor. Disconnect the neutral start switch wires at the plug connector.

8. Remove the rear mount-to-crossmember attaching bolts and the two crossmember-to-frame attaching bolts.

9. Remove the two engine rear support-to-extension housing attaching bolts.

10. Disconnect the TV linkage rod from the transmission TV lever. Disconnect the manual rod from the transmission manual lever at the transmission.

11. Remove the two bolts securing the bellcrank bracket to the converter housing.

12. Raise the transmission with a transmission jack to provide clearance to remove the crossmember. Remove the rear mount from the crossmember and remove the crossmember from the side supports.

13. Lower the transmission to gain access to the oil cooler lines.

14. Disconnect each oil line from the fittings on the transmission.

15. Disconnect the speedometer cable from the extension housing.

16. Remove the bolt that secures the transmission fluid filler tube to the cylinder block. Lift the filler tube and the dipstick from the transmission.

17. Secure the transmission to the jack with the chain.

18. Remove the converter housing-to-cylinder block attaching bolts.

19. Carefully move the transmission and converter assembly away from the engine and, at the same time, lower the jack to clear the underside of the vehicle.

20. Remove the converter and mount the transmission in a holding fixture.

21. Tighten the converter drain plug to 20-28 ft. lbs.

22. Position the converter on the transmission, making sure the converter drive flats are fully engaged in the pump gear by rotating the converter.

23. With the converter properly installed, place the transmission on the jack. Secure the transmission to the jack with a chain.

24. Rotate the converter until the studs and drain plug are in alignment with the holes in the flywheel.

WARNING

Lube the pilot bushing.

25. Align the yellow balancing marks on converter and flywheel on models with the 8-5.0L.

26. move the converter and transmission assembly forward into position, using care not to damage the flywheel and the converter pilot. The converter must rest squarely against the flywheel. This indicates that the converter pilot is not binding in the engine crankshaft.

27. Install and tighten the converter housing-to-engine attaching bolts to 40-50 ft. lbs. make sure that the vacuum tube retaining clips are properly positioned.

28. Remove the safety chain from around the transmission.

29. Install a new O-ring on the lower end of the transmission filler tube. Insert the tube in the transmission case and secure the tube to the engine with the attaching bolts.

30. Connect the speedometer cable to the extension housing.

31. Connect the oil cooler lines to the right side of the transmission case.

32. Position the crossmember on the side supports. Position the rear mount on the crossmember and install the attaching bolt and nut.

33. Secure the engine rear support to the extension housing and tighten the bolts to 35-40 ft. lbs.

34. Lower the transmission and remove the jack.

35. Secure the crossmember to the side supports with the attaching bolts and tighten them to 35-40 ft. lbs.

36. Position the bellcrank to the converter housing and install the two attaching bolts.

37. Connect the TV linkage rod to the transmission TV lever. Connect the manual linkage rod to the manual lever at the transmission.

38. Secure the converter-to-flywheel attaching nuts and tighten them to 20-30 ft. lbs.

39. Install the converter housing access cover and secure it with the attaching bolts.

40. Secure the starter motor in place with the attaching bolts. Connect the cable to the terminal on the starter. Connect the neutral start switch wires at the plug connector.

41. Connect the driveshaft to the rear axle.

42. Adjust the shift linkage as required.

43. Adjust throttle linkage.

44. Lower the vehicle.

45. Fill the transmission to the correct level. Start the engine and shift the transmission to all ranges, then recheck the fluid level.

DRIVELINE

Driveshaft and U-Joints

➧ SEE FIG. 27

The driveshaft is the means by which the power from the engine and transmission (in the front of the car) is transferred to the differential and rear axles, and finally to the rear wheels.

The driveshaft assembly incorporates two universal joints, one at each end, and a slip yoke at the front end of the assembly, which fits into the back of the transmission.

All driveshafts are balanced when installed in a car. It is therefore imperative that before applying undercoating to the chassis, the

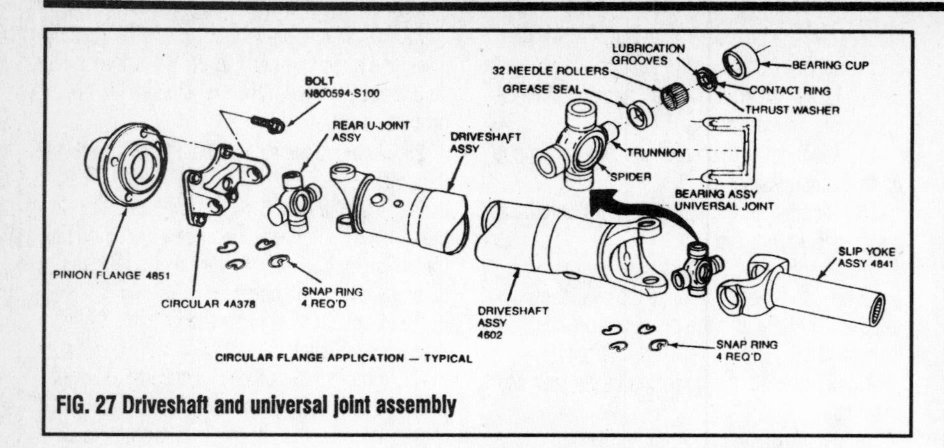

FIG. 27 Driveshaft and universal joint assembly

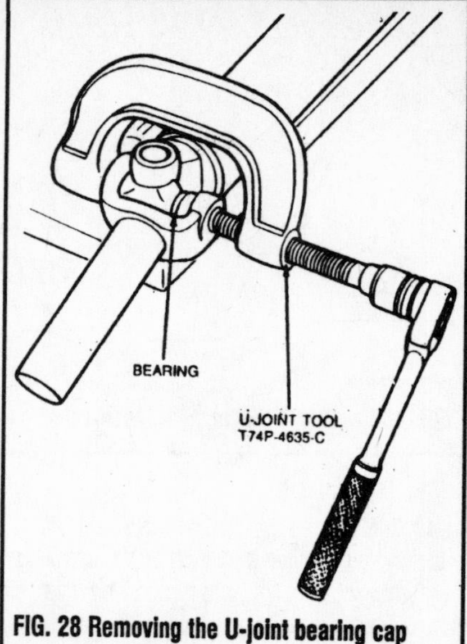

FIG. 28 Removing the U-joint bearing cap

driveshaft and universal joint assembly be completely covered to prevent the accidental application of undercoating to the surfaces, and the subsequent loss of balance.

DRIVESHAFT REMOVAL

The procedure for removing the driveshaft assembly, complete with universal joint and slip yoke, is as follows:

1. Mark the relationship of the rear dirveshaft yoke and the drive pinion flange of the axle. If the original yellow alignment marks are visible, there is not need for new marks. The purpose of this marking is to facilitate installation of the assembly in its exact original position, thereby maintaining proper balance.
2. Remove the four bolts or U-clamps which hold the rear universal joint to the pinion flange. Wrap tape around the loose bearing caps in order to prevent them from falling off the spider.
3. Pull the driveshaft toward the rear of the vehicle until the slip yoke clears the transmission housing and the seal. Plug the hole at the rear of the transmission housing or place a container under the opening to catch any fluid which might leak.

UNIVERSAL JOINT OVERHAUL

➧ SEE FIG. 28–29

1. Position the driveshaft assembly in a sturdy vise.
2. Remove the snaprings which retain the bearings in the slip yoke (front only) and in the driveshaft (front and rear).
3. Using a large vise or an arbor press and a socket smaller than the bearing cap on one side and a socket larger than the bearing cap on the other side, drive one of the bearings in toward the center of the universal joint, which will force the opposite bearing out.
4. As each bearing is forced far enough out of the universal joint assembly that it is accessible, grip it with a pair of pliers, and pull it from the driveshaft yoke. Drive the spider in the opposite direction in order to make the opposite bearing accessible, and pull it free with a pair of pliers. Use this procedure to remove all bearings from both universal joints.
5. After removing the bearings, lift the spider from the yoke.
6. Thoroughly clean all dirt and foreign matter from the yokes on both ends of the driveshaft.

WARNING

When installing new bearings in the yokes, it is advisable to use an arbor press. However, if this tool is not available, the bearings should be driven into position with extreme car, as a heavy jolt on the needle bearings can easily damage or misalign them, greatly shortening their lift and hampering their efficiency.

7. Start a new bearing into the yoke at the rear of the driveshaft.
8. Position a new spider in the rear yoke and press the new bearing ¹/₄ in. (6mm) below the outer surface of the yoke.
9. With the bearing in position, install a new snapring.
10. Start a new bearing into the opposite side of the yoke.
11. Press the bearing until the opposite bearing, which you have just installed, contacts the inner surface of the snapring.
12. Install a new snapring on the second bearing. It may be necessary to grind the surface of this second snapring.

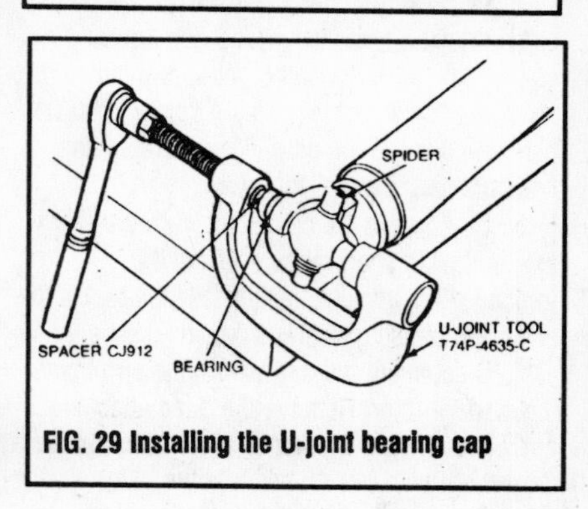

FIG. 29 Installing the U-joint bearing cap

13. Reposition the driveshaft in the vise, so that the front universal joint is accessible.
14. Install the new bearings, new spider, and new snaprings in the same manner as you did for the rear universal joint.
15. Position the slip yoke on the spider. Install new bearings, nylon thrust bearings, and snaprings.
16. Check both reassembled joints for freedom of movement. If misalignment of any part is causing a bind, a sharp rap on the side of the yoke with a brass hammer should seat the bearing needle and provide the desired freedom of movement. Care should be exercised to firmly support the shaft end during this operation, as well as to prevent blows to the bearings themselves. Under no circumstances should the driveshaft be installed in a car if there is any binding in the universal joints.

DRIVESHAFT INSTALLATION

1. Carefully inspect the rubber seal on the output shaft and the seal in end of the

transmission extension housing. Replace them if they are damaged.

2. Examine the lugs on the axle pinion flange and replace the flange if the lugs are shaved or distorted.

3. Coat the yoke spline with special-purpose lubricant. The Ford part number for this lubricant if B8A-19589-A.

4. Remove the plug from the rear of the transmission housing.

5. Insert the yoke into the transmission housing and onto the transmission output shaft. Make sure that the yoke assembly does not bottom on the output shaft with excessive force.

6. Locate the marks which you made on the rear driveshaft yoke and the pinion flange prior to removal of the driveshaft assembly. Install the driveshaft assembly with the marks properly aligned.

7. Install the U-bolts and nuts or bolts which attach the universal joint to the pinion flange. Torque the U-bolts nuts to 8-15 ft. lbs. Flange bolts are tighten to 70-95 ft. lbs.

REAR AXLE

Understanding Drive Axles

The drive axle is a special type of transmission that reduces the speed of the drive from the engine and transmission and divides the power to the wheels. Power enters the axle from the driveshaft via the companion flange. The flange is mounted on the drive pinion shaft. The drive pinion shaft and gear which carry the power into the differential turn at engine speed. The gear on the end of the pinion shaft drives a large ring gear the axis of rotation of which is 90 degrees away from the of the pinion. The pinion and gear reduce the gear ratio of the axle, and change the direction of rotation to turn the axle shafts which drive both wheels. The axle gear ratio is found by dividing the number of pinion gear teeth into the number of ring gear teeth.

The ring gear drives the differential case. The case provides the two mounting points for the ends of a pinion shaft on which are mounted two pinion gears. The pinion gears drive the two side gears, one of which is located on the inner end of each axle shaft.

By driving the axle shafts through the arrangement, the differential allows the outer drive wheel to turn faster than the inner drive wheel in a turn.

The main drive pinion and the side bearings, which bear the weight of the differential case, are shimmed to provide proper bearing preload, and to position the pinion and ring gears properly.

WARNING

The proper adjustment of the relationship of the ring and pinion gears is critical. It should be attempted only by those with extensive equipment and/or experience.

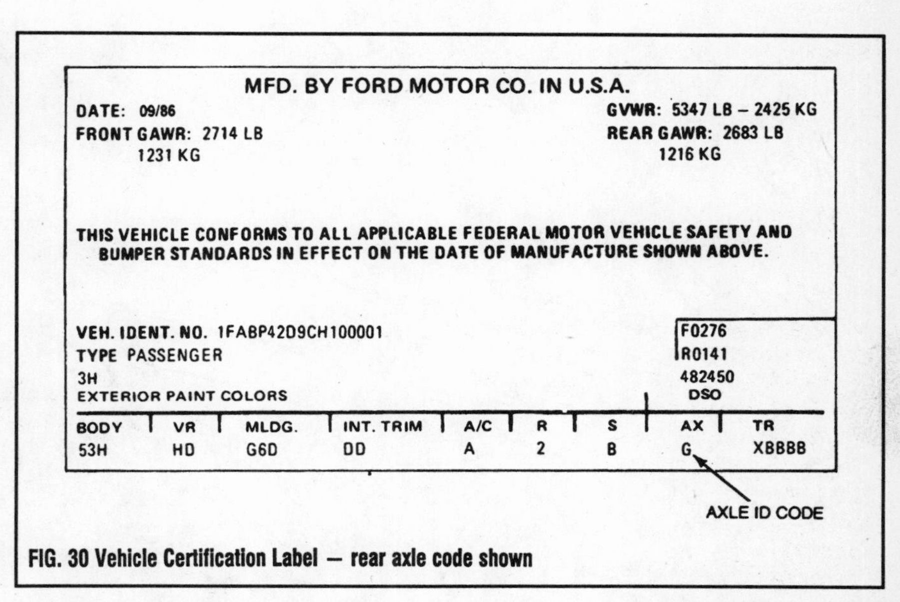

FIG. 30 Vehicle Certification Label — rear axle code shown

Limited-slip differentials include clutches which tend to link each axle shaft to the differential case. Clutches may be engaged either by spring action or by pressure produced by the torque on the axles during a turn. During turning on a dry pavement, the effects of the clutches are overcome, and each wheel turns at the required speed. When slippage occurs at either wheel, however, the clutches will transmit some of the power to the wheel which has the greater amount of traction. Because of the presence of clutches, limited-slip units require a special lubricant.

Determining Axle Ratio

The drive axle is said to have a certain axle ratio. This number (usually a whole number and a decimal fraction) is actually a comparison of the number of gear teeth on the ring gear and the pinion gear. For example, a 4.11 rear means that theoretically, there are 4.11 teeth on the ring gear and one tooth on the pinion gear or, put another way, the driveshaft must turn 4.11 times to turn

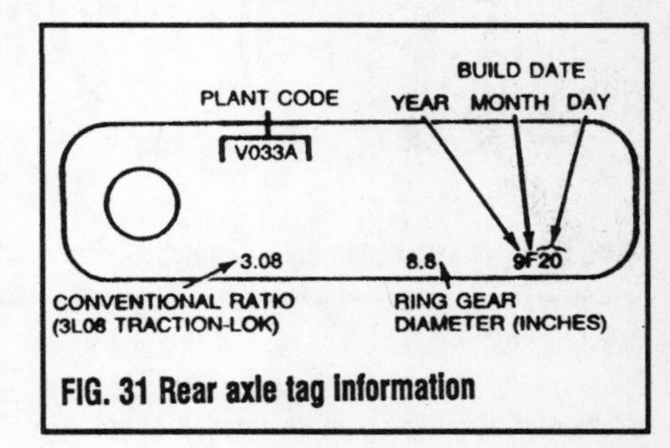

FIG. 31 Rear axle tag information

the wheels once. Actually, on a 4.11 rear, there might be 37 teeth on the ring gear and 9 teeth on the pinion gear. By dividing the number of teeth on the pinion gear into the number of teeth on the ring gear, the numerical axle ratio (4.11) is obtained. This also provides a good method of ascertaining exactly what axle ratio one is dealing with.

Another method of determining gear ratio is to jack up and support the car so that both rear wheels are off the ground. Make a chalk mark on the rear wheel and the driveshaft. Put the transmission in neutral. Turn the rear wheel one complete turn and count the number of turns that

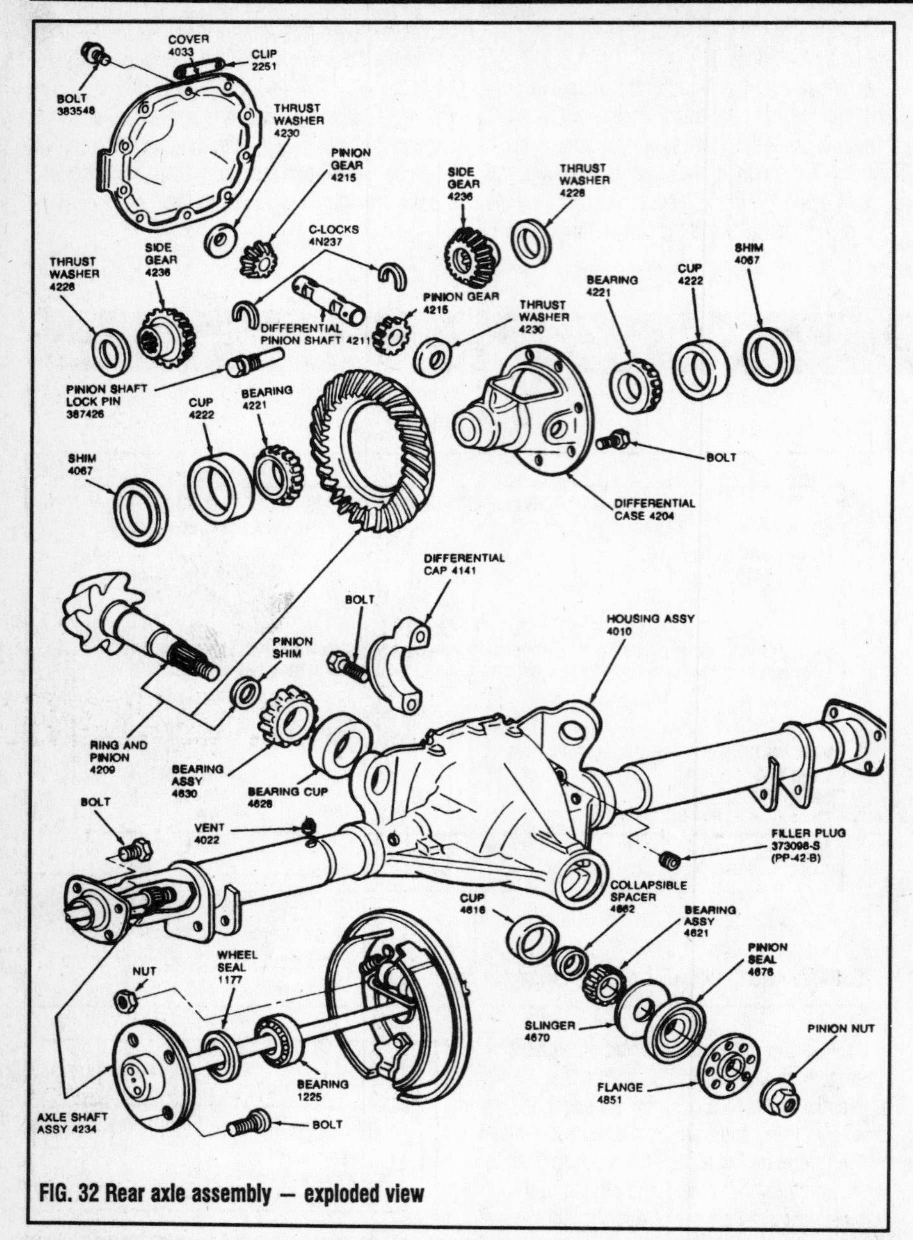

FIG. 32 Rear axle assembly — exploded view

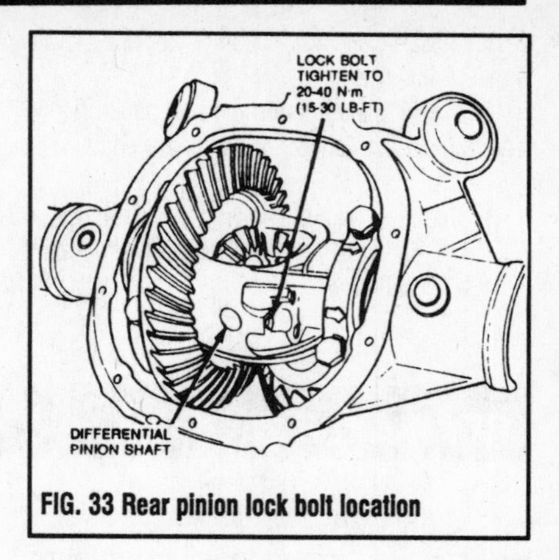

FIG. 33 Rear pinion lock bolt location

the driveshaft makes. The number of turns that the driveshaft makes in one complete revolution of the rear wheel is an approximation of the rear axle ratio.

Axle Identification

➧ SEE FIG. 30–32

All Mustangs are equipped with a Vehicle Certification Label, attached to the left side (driver's) door lock post.

The axle code is located in the space marked AX on the label.

There are two different rear axles used in the Mustang's covered in this manual, the "Integral Carrier 7.5 Inch Ring Gear" and the "Integral Carrier 8.8 Inch Ring Gear".

Additional information is located on the tag attached to the rear axle such as: ring gear size, gear ratio, and build date.

The "Drive Axle Application Chart" in section one will provide gear ratios from the code you obtain from the Vehicle Certification Label.

Rear Axle Shaft, Bearing and Seal

REMOVAL & INSTALLATION

➧ SEE FIGS. 33–34

1. Raise and safely support the vehicle. Remove wheel and tire assembly and remove brake drum or brake rotor.
2. If equipped, remove the anti-lock brake speed sensor.
3. Clean all dirt from the area of the carrier cover. Drain the axle lubricant by removing the housing cover.
4. Remove differential pinion shaft lock bolt and pinion shaft.
5. Push flanged end of axle shafts toward the center of the vehicle and remove the C-lock from button end of the axle shaft. Remove the axle shaft from the housing, being careful not to damage the oil seal.
6. Insert wheel bearing and seal replacer tool T85L–1225–AH or equivalent, in the bore and position it behind the bearing so the tangs on the tool engage the bearing outer race. Remove bearing and seal as a unit using an impact slide hammer.

To install:

7. Lubricate the new bearing with rear axle lubricant. Install the bearing into the housing bore using a suitable bearing installer.
8. Install a new axle seal using a seal installer.

➡ Check for the presence of an axle shaft O-ring on the spline end of the shaft and install, if not present.

9. Carefully slide the axle shaft into the axle housing, without damaging the bearing or seal assembly. Start the splines into the side gear and push firmly until the button end of the axle shaft can be seen in the differential case.
10. Install the C-lock on the button end of the axle shaft splines, then push the shaft outboard until the shaft splines engage and the C-lock seats in the counterbore of the differential side gear.
11. Insert the differential pinion shaft through the case and pinion gears, aligning the hole in the shaft with the lock bolt hole. Apply a suitable locking compound to the lock bolt and install in

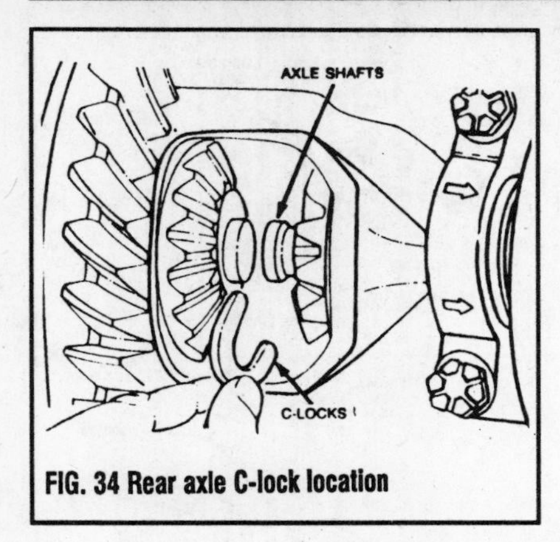

FIG. 34 Rear axle C-lock location

the case and pinion shaft. Tighten to 15–30 ft. lbs. (20–41 Nm).

12. Cover the inside of the differential case with a shop rag and clean the machined surface of the carrier and cover. Remove the shop rag.

13. Apply a bead of silicone sealer to the cover and install on the carrier. Tighten the bolts in a criss-cross pattern. Final torque the cover retaining bolts to 10–15 ft. lbs. (15–20 Nm).

14. Add rear axle lubricant to the carrier to a level $^1/_4$–$^9/_{16}$ in. (6–15mm) below the bottom of the fill hole. Install the filler plug and tighten to 15–30 ft. lbs. (20–41 Nm).

15. Install the anti-lock speed sensor, if equipped. Tighten the retaining bolt to 40–60 inch lbs. (4.5–6.8 Nm).

16. Install the brake calipers and rotors or the brake drums, as required. Install the wheel and tire assembly and lower the vehicle.

Pinion Seal

REMOVAL & INSTALLATION

➧ SEE FIGS. 35–36

1. Raise and safely support the vehicle. Matchmark the rear driveshaft yoke and the companion flange so they may be reassembled in the same way to maintain balance.

2. Disconnect the driveshaft from the rear axle companion flange, remove the driveshaft and remove the driveshaft from the extension housing. Plug the extension housing to prevent leakage.

3. Install an inch pound torque wrench on the pinion nut and record the torque required to maintain rotation of the pinion through several revolutions.

4. While holding the companion flange with holder tool No. T78P–4851–A or equivalent, remove the pinion nut.

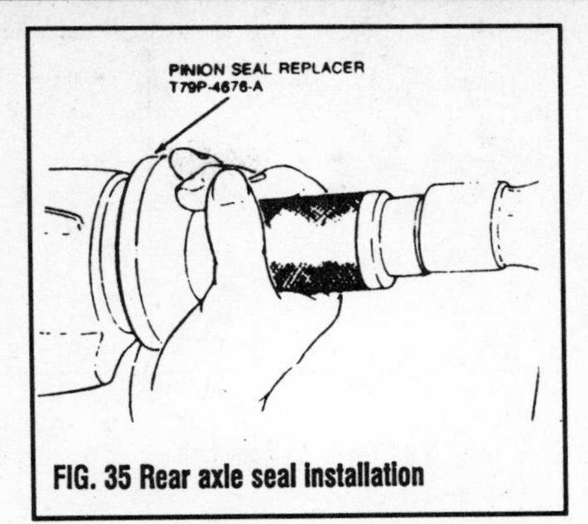

FIG. 35 Rear axle seal installation

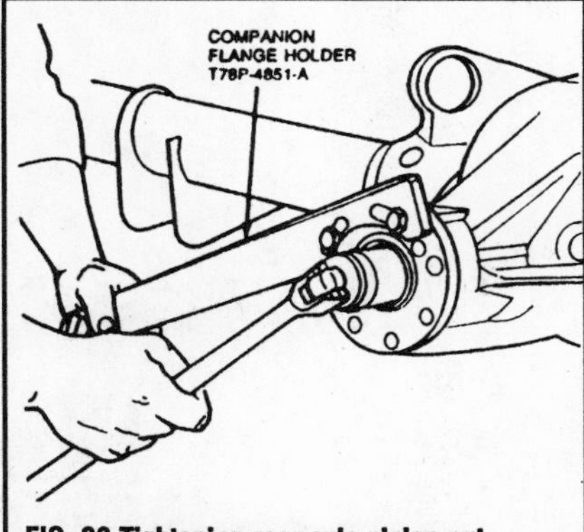

FIG. 36 Tightening rear axle pinion nut

5. Clean the area around the oil seal and place a pan under the seal.

6. Mark the companion flange in relation to the pinion shaft so the flange can be installed in the same position.

7. Remove the rear axle companion flange using tool No. T65L–4851–B or equivalent.

8. Pry the seal out of the housing using a prybar.

To install:

9. Clean the oil seal seat surface and install the seal in the carrier using seal replacer tool No. T79P–4676–A or equivalent. Apply lubricant to the lips of the seal.

10. Apply a small amount of lubricant to the companion flange splines, align the marks on the flange and and the pinion shaft and install the flange.

11. Install a new nut on the pinion shaft and apply lubricant on the washer side of the nut.

12. Hold the flange with the holder tool while tightening the nut. Rotate the pinion to ensure proper seating and take frequent pinion bearing torque preload readings until the original recorded preload reading is obtained.

13. If the original recorded preload is less than the minimum specification of 170 ft. lbs. (230 Nm) on the 7.5 in. diameter ring gear axle or 140 ft. lbs. (190 Nm) on the 8.8 in. diameter ring gear axle, tighten to specification. If the preload is higher than specification tighten to the original reading as recorded.

➡ Under no circumstances should the pinion nut be backed off to reduce preload. If reduced preload is required, a new collapsible pinion spacer and pinion nut should be installed.

14. Remove the plug from the transmission extension housing and install the front end of the driveshaft on the transmission output shaft.

15. Connect the rear end of the driveshaft to the axle companion flange, aligning the scribe marks and tighten the 4 bolts to 71–95 ft. lbs. (95–130 Nm).

16. Add lubricant to the axle until it is $^1/_4$–$^9/_{16}$ in. (6–15mm) below the bottom of the fill hole with the axle in operating position.

17. Make sure the axle vent is not plugged with debris.

Axle Housing

REMOVAL & INSTALLATION

➧ SEE FIGS. 37–38

1. Raise and safely support the vehicle. Position safety stands under the rear frame crossmember.

2. Remove the cover and drain the axle lubricant.

3. Remove the wheel and tire assemblies. Remove the brake drums or brake rotors.

4. Remove the lock bolt from the pinion shaft and remove the shaft.

5. Remove the anti-lock brake sensor before removing the axle shafts, if equipped.

6. Push the axle shafts inward to remove the C-locks and remove the axle shafts.

7. If necessary, remove the bolt attaching the brake junction block to rear cover.

8. Remove the brake lines from the clips and position out of the way.

9. If equipped with drum brakes, remove the 4 retaining nuts from each backing plate and wire the backing plate to the underbody.

10. Matchmark the driveshaft yoke and companion flange. Disconnect the driveshaft at the companion flange and wire it to the underbody.

11. Support the axle housing with jackstands. Disengage the brake line from the clips that retain the line to the axle housing.

12. Disconnect the axle vents from the rear axle housing.

➡ Some axle vents may be secured to the housing assembly through the brake junction block. At assembly, a thread lock/sealer must be applied to ensure retension.

13. If equipped with air springs, proceed as follows:

a. Disconnect the negative battery cable. Turn **OFF** the air suspension switch located in the trunk.

b. Remove the heat shield and spring retainer clip from the top of the air spring.

c. Disconnect the electrical connector and the air line from the air spring solenoid, located on the air spring.

d. Remove the solenoid clip and rotate the solenoid counterclockwise to the first stop.

e. Pull the solenoid straight out slowly to the second stop to bleed the air from the system.

⁂ CAUTION

Do not fully release the solenoid until the air is completely bled from the air spring, or personal injury may result.

f. After the air is fully bled from the system, rotate the solenoid counterclockwise to the 3rd stop and remove the solenoid from the solenoid housing.

g. Insert air spring removal tool T90P–5310–A or equivalent, between the axle tube and spring seat on the forward side of the axle.

h. Position the tool so that its flat end rests on the piston knob. Push downward, forcing the piston and retainer clip off the axle spring seat.

i. Remove the air spring.

14. Disconnect the lower shock absorber studs from the mounting brackets on the axle housing. If equipped, disconnect the quad shock from the quad shock bracket.

15. Disconnect the upper arms from the mountings on the axle housing ear brackets.

16. Lower the axle housing assembly unti the coil springs are released and lift out the coil springs.

17. Disconnect the suspension lower arms at the axle housing.

18. Lower the axle housing and remove it from the vehicle.

To install:

19. Position the axle housing under the vehicle and raise the axle with a hoist or jack. Connect the lower suspension arms to their mounting brackets on the axle housing. Do not tighten the bolts and nuts at this time.

20. Reposition the rear coil springs.

21. Raise the housing into position.

22. Connect the upper arms to the mounting ears on the housing. Tighten the nuts and bolts to 70–100 ft. lbs. (95–135 Nm). Tighten the lower arm bolts and nuts to 70–100 ft. lbs. (95–135 Nm)

23. Install the axle vent and the the brake line to the clips that retain the line to the axle housing. If equipped with air springs, install the solenoids and connect the air lines and electrical connectors.

24. If equipped with drum brakes, install the brake backing plates on the axle housing flanges.

25. Connect the lower shock absorber studs to the mounting bracket on the axle housing. If equipped, connect the quad shock to the quad shock bracket.

26. Connect the driveshaft to the companion flange and tighten the bolts and nuts to 70–95 ft. lbs. (95–130 Nm).

27. Slide the rear axle shafts into the housing until the splines enter the side gear. Push the axle shafts inward and install the C-lock at the end of each shaft spline. Pull the shafts outboard until the C-lock enters the recess in the side gears.

28. Install the pinion shaft and the pinion shaft lock bolt. Tighten to 15–30 ft. lbs. (20–41 Nm).

29. Install the anti-lock sensor, if equipped.

30. Install the rear brake drums or disc brake rotors and calipers.

31. Install the rear carrier cover using new silicone sealer. Tighten to 15 ft. lbs. (20 Nm).

32. Install the brake junction block on the

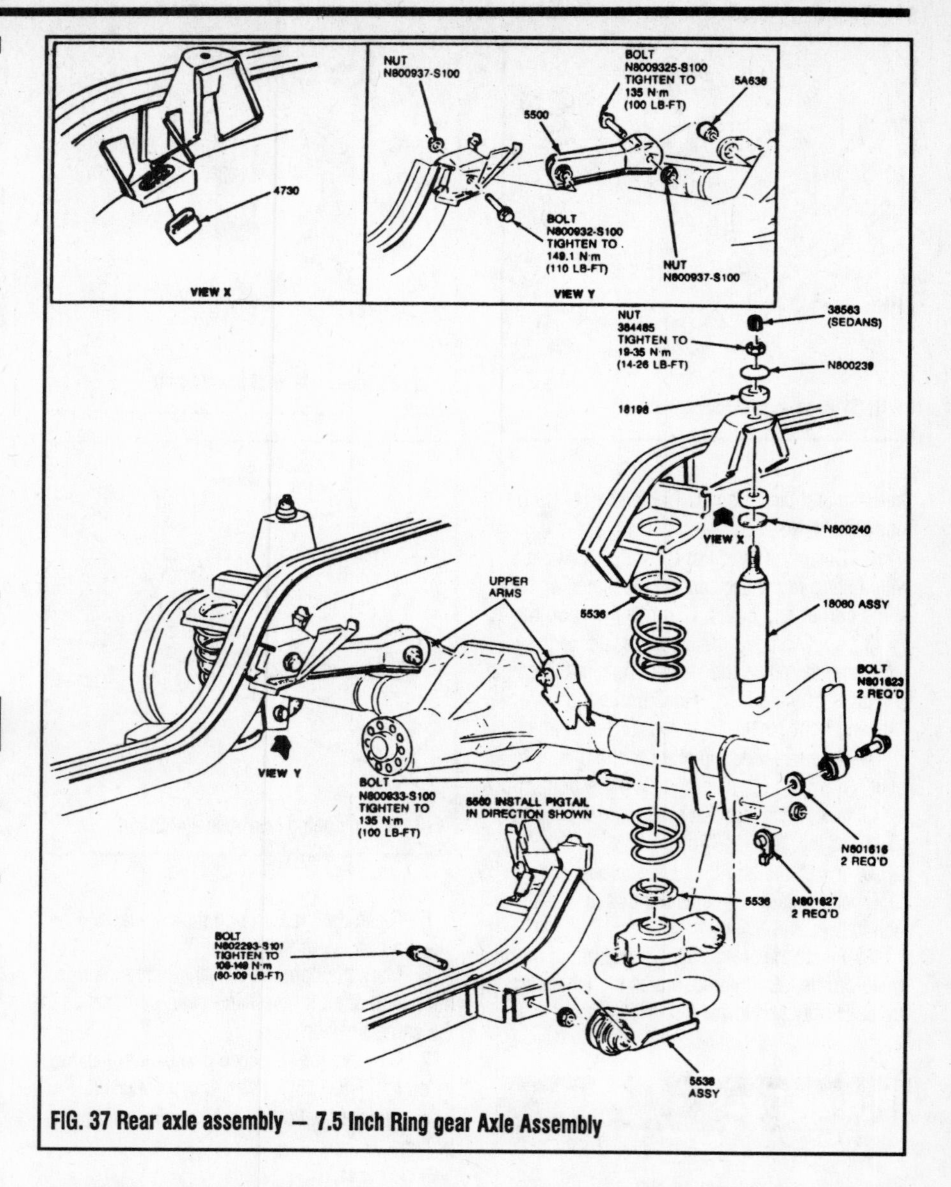

FIG. 37 Rear axle assembly — 7.5 Inch Ring gear Axle Assembly

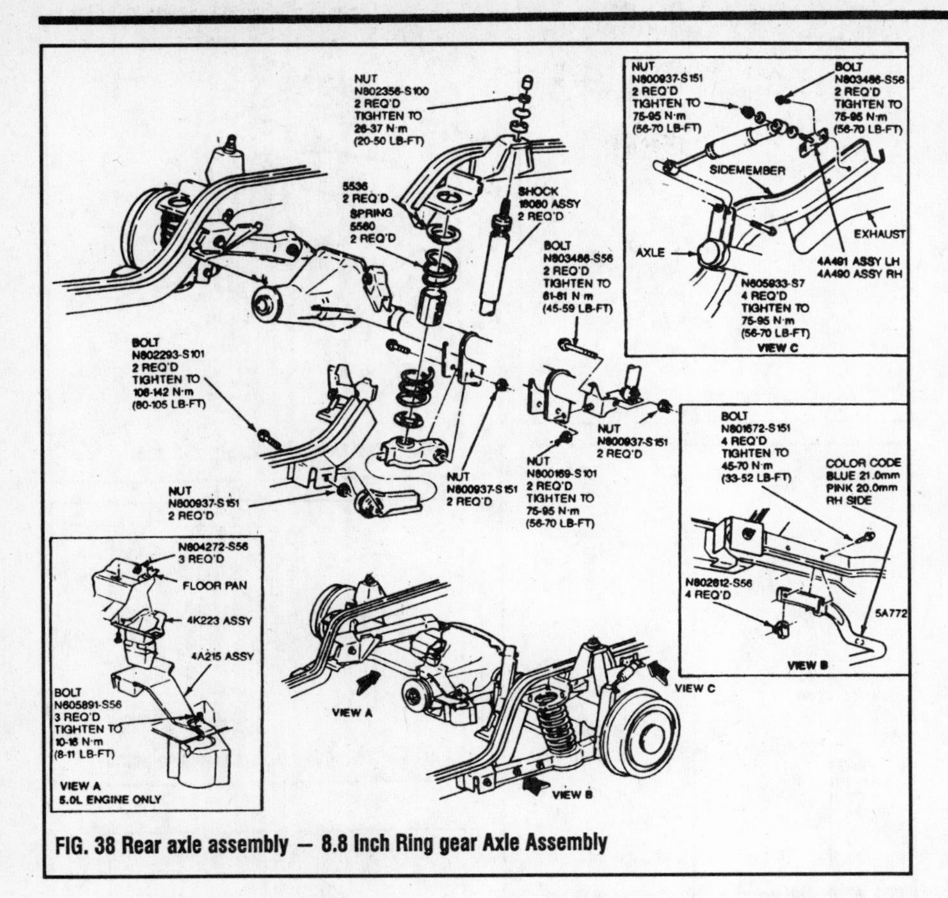

FIG. 38 Rear axle assembly — 8.8 Inch Ring gear Axle Assembly

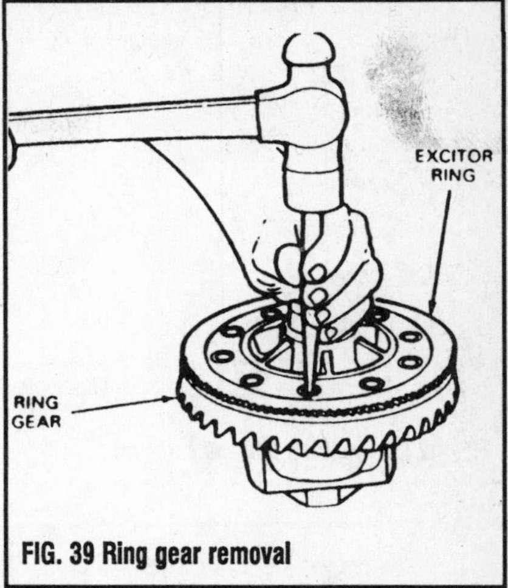

FIG. 39 Ring gear removal

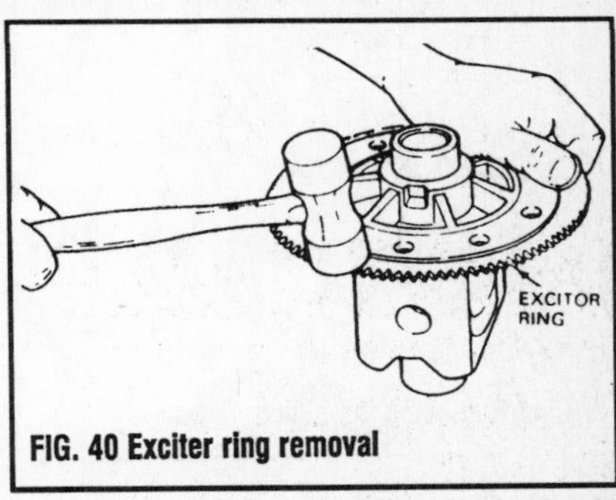

FIG. 40 Exciter ring removal

carrier cover and tighten to 10–18 ft. lbs. (14–24 Nm).

33. Fill the axle with lubricant to the bottom of the filler hole. Install the filler plug and tighten to 15–30 ft. lbs. (20-41 Nm).

34. Lower the vehicle.

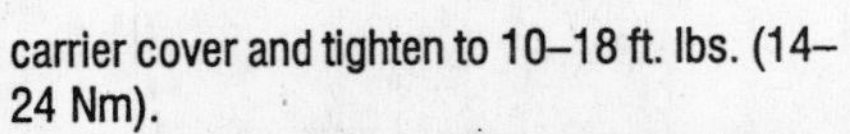

Conventional Differential Overhaul

DISASSEMBLY

➧ SEE FIGS. 39–43

Differential Carrier

1. Remove the cover and clean the lubricant from the internal parts.

2. Using a dial indicator, measure and record the ring gear backlash and the runout; the backlash should be 0.008–0.015 in. (0.20–0.38mm) and the runout should be less than 0.004 in. (0.10mm).

3. Mark 1 differential bearing cap to ensure it is installed its original position.

4. Loosen the differential bearing cap bolts.

5. Using a prybar, pry the differential carrier until the bearing cups and shims are loose in the bearing caps.

6. Remove the bearing caps and the differential assembly.

7. If necessary, remove ring gear-to-differential case bolts. Using a hammer and a punch, strike the alternate bolt holes around the ring gear to dislodge it from the differential.

8. If necessary, remove the exciter ring by striking it with a soft hammer.

9. Remove the pinion shaft lock bolt from the differential case. Remove the differential pinion shaft, the pinion gears and the thrust washers.

10. Remove the side gears and thrust washers.

11. Using a bearing puller tool, press the bearings from the differential carrier.

Pinion Gear

1. Remove the differential carrier assembly.

2. Using a companion flange holding tool, remove the companion flange nut.

3. Using a puller tool, press the companion flange from the pinion gear.

4. Using a soft hammer, drive the pinion gear from the housing.

5. Using a prybar, remove the pinion gear oil seal from the housing.

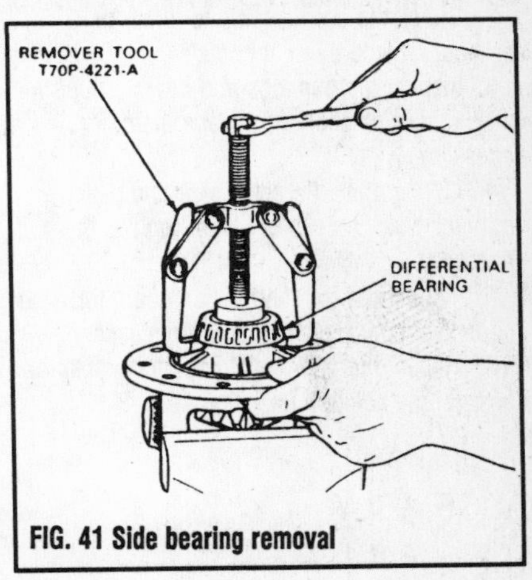

FIG. 41 Side bearing removal

6. Remove the oil slinger and the front pinion bearing.

7. Using a shop press, press the bearing cone from from the pinion gear.

8. Remove and record the shim from the pinion gear.

INSPECTION

1. Clean the differential components in solvent and use compressed air to dry them; do not use compressed air on the bearings, only shop towels.

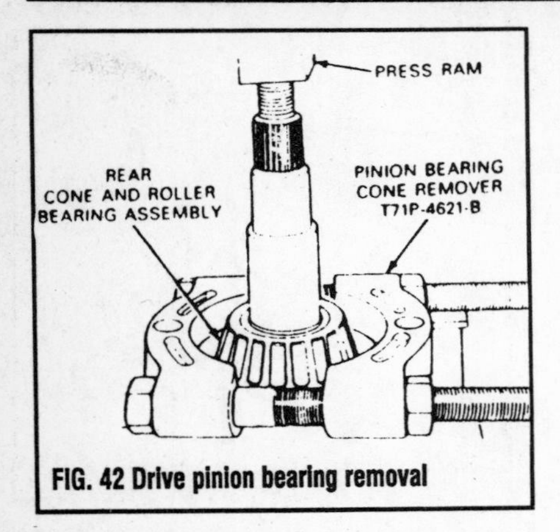

FIG. 42 Drive pinion bearing removal

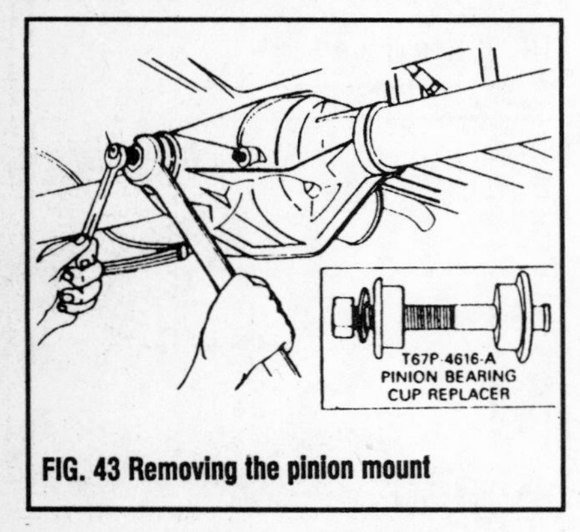

FIG. 43 Removing the pinion mount

2. Check the components for wear or damage; replace them, if necessary.

3. Inspect the bearings and bearing cups for wear, cracks or scoring; replace them, if necessary.

4. Inspect the differential side and pinion gears for wear, cracks or chips; replace them, if necessary.

5. Inspect the ring and pinion gears for wear and/or damage; replace them, if necessary.

6. Inspect the differential case for cracks or damage; replace it, if necessary.

ASSEMBLY

➧ SEE FIGS. 44–74

Pinion Gear

➡ **When replacing the ring and pinion gear, the correct shim thickness for the new gear set to be installed is determined by following procedure using a pinion depth gauge tool set.**

1. Assemble the appropriate aligning adapter, the gauge disc and gauge block to the screw.

2. Place the rear pinion bearing over the aligning tool and insert it into the rear portion of the bearing cup of the carrier. Place the front bearing into the front bearing cup and assemble the tool handle into the screw. Roll the assembly back and forth a few times to seat the bearings while tightening the tool handle, by hand, to 20 ft. lbs. (27 Nm).

➡ **The gauge block must be offset 45 degrees to obtain an accurate reading.**

3. Center the gauge tube into the differential bearing bore. Install the bearing caps and tighten the bolts to 70–85 ft. lbs. (96–115 Nm); be sure to install the caps with the triangles pointing outward.

4. Place the selected shim(s) on the pinion and press the pinion bearing cone and roller assembly until it is firmly seated on the shaft, using the pinion bearing cone replacer and the axle bearing/seal plate.

5. Place the collapsible spacer on the pinion stem against the pinion stem shoulder.

6. Install the front pinion bearing and oil slinger in the housing bore and install the pinion seal on the pinion seal replacer. Using a hammer, install the seal until it seats.

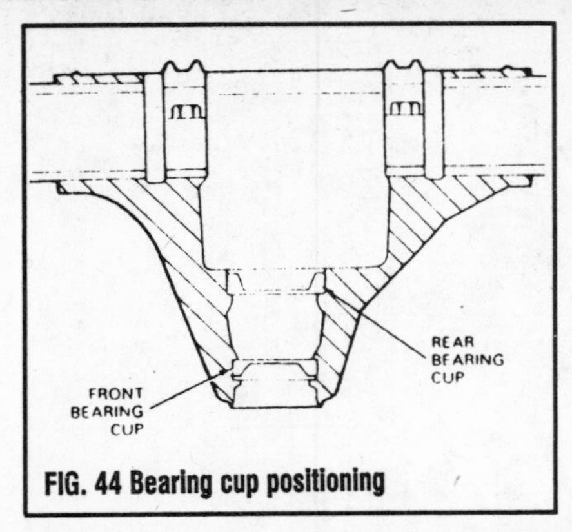

FIG. 44 Bearing cup positioning

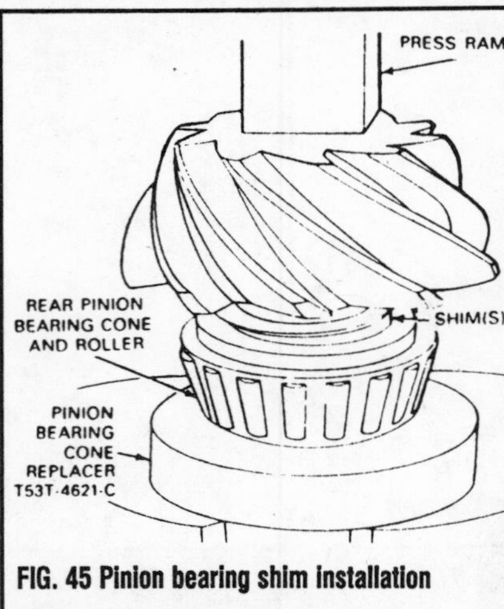

FIG. 45 Pinion bearing shim installation

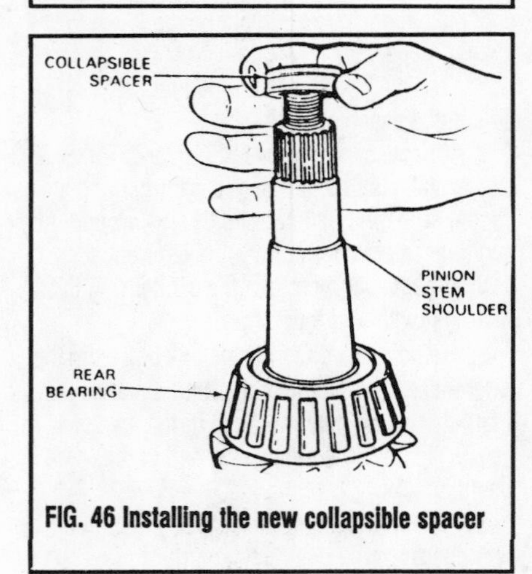

FIG. 46 Installing the new collapsible spacer

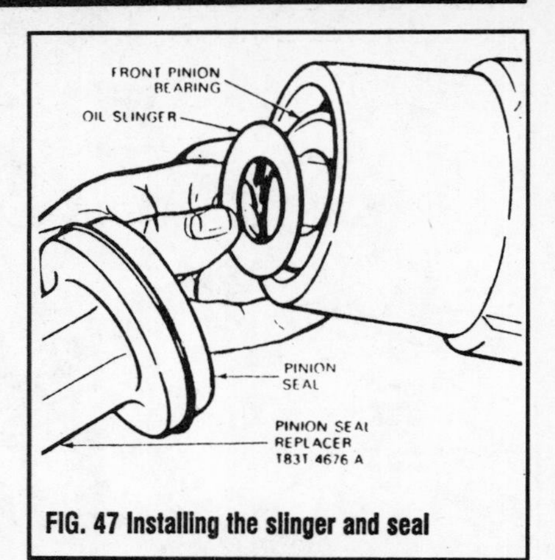

FIG. 47 Installing the slinger and seal

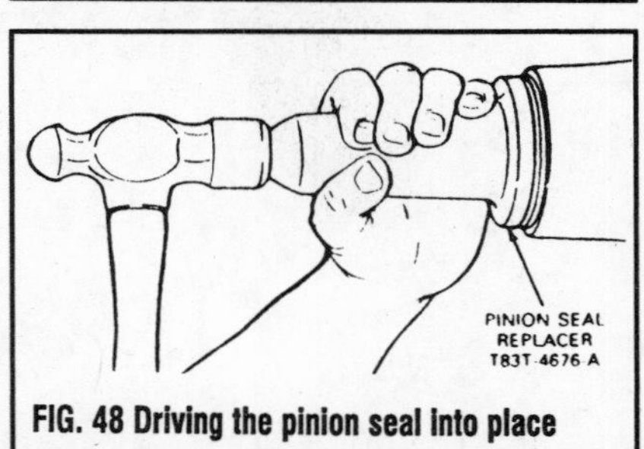

FIG. 48 Driving the pinion seal into place

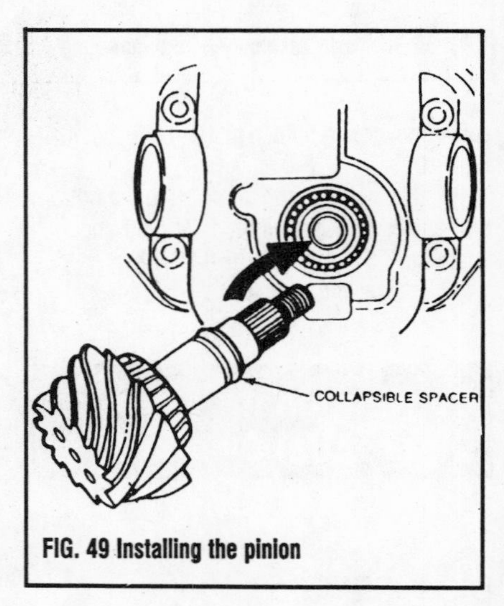

FIG. 49 Installing the pinion

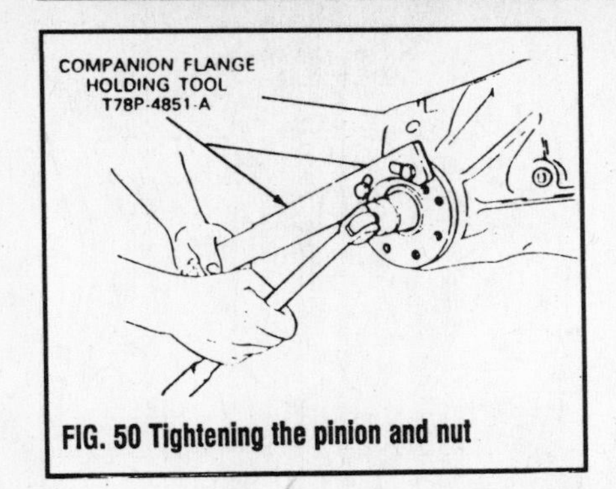

FIG. 50 Tightening the pinion and nut

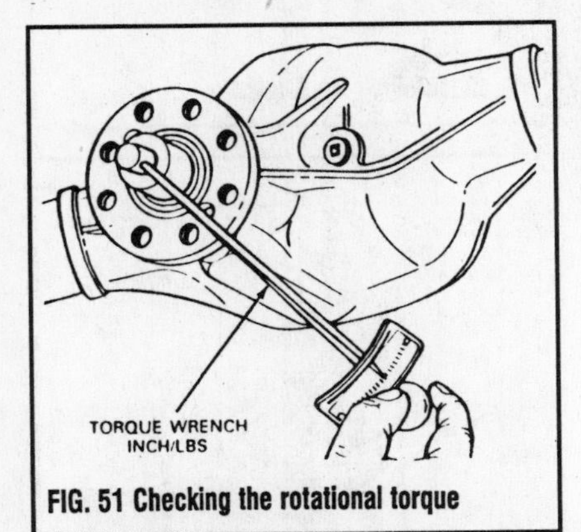

FIG. 51 Checking the rotational torque

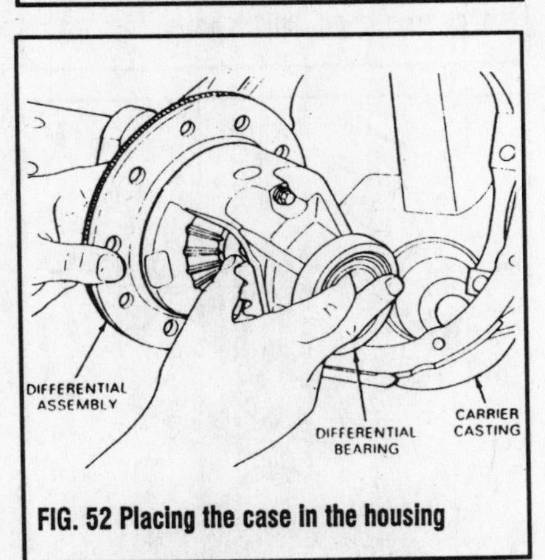

FIG. 52 Placing the case in the housing

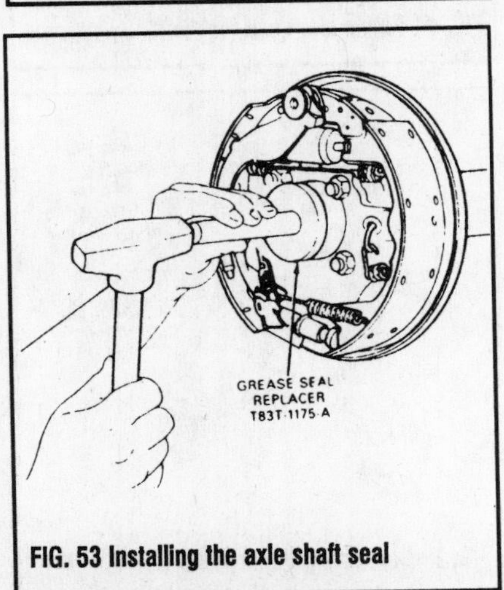

FIG. 53 Installing the axle shaft seal

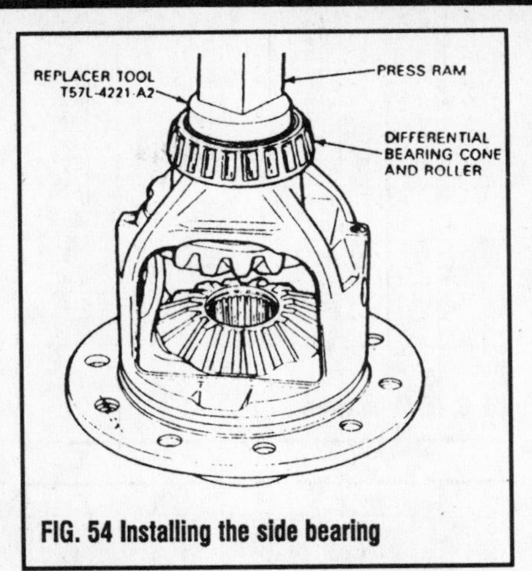

FIG. 54 Installing the side bearing

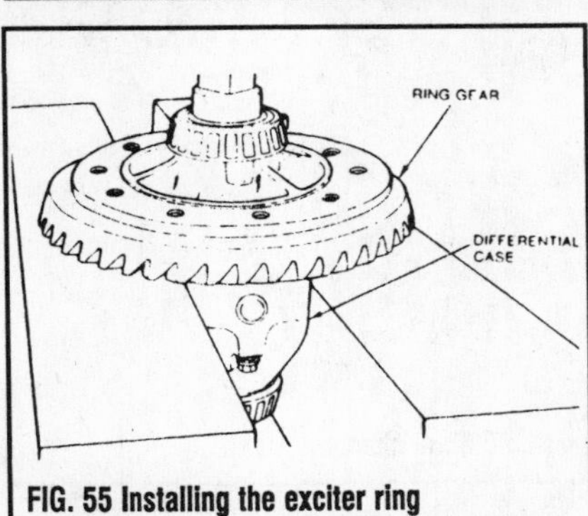

FIG. 55 Installing the exciter ring

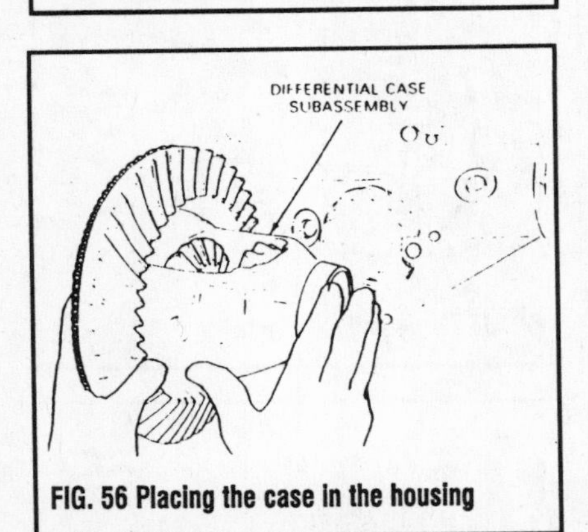

FIG. 56 Placing the case in the housing

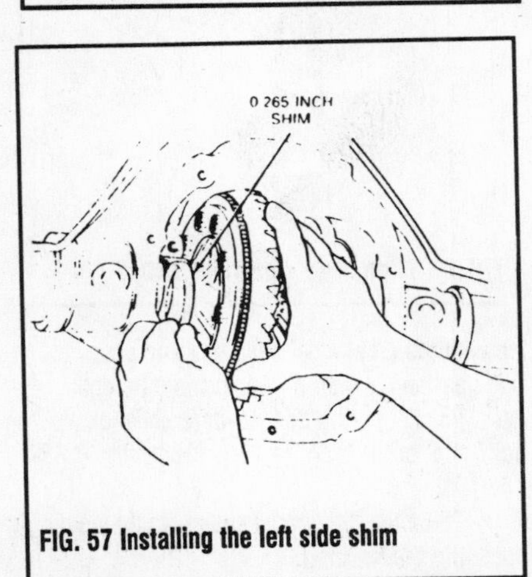

FIG. 57 Installing the left side shim

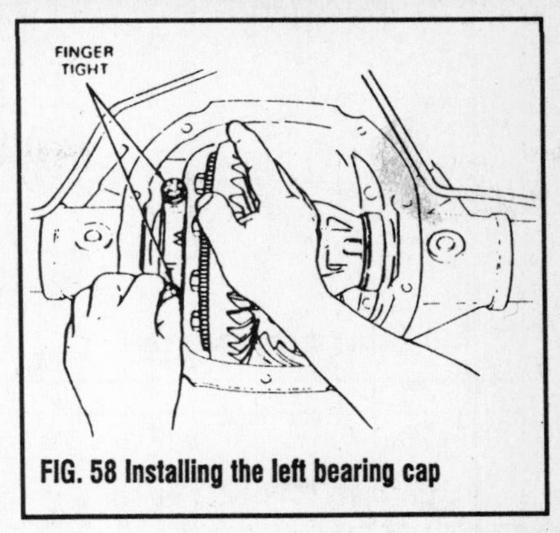

FIG. 58 Installing the left bearing cap

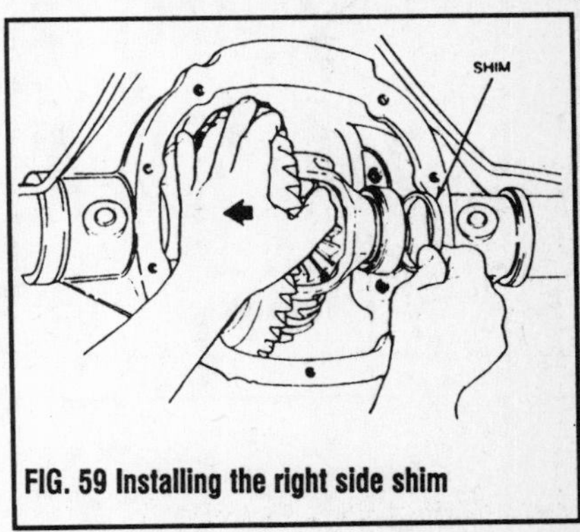

FIG. 59 Installing the right side shim

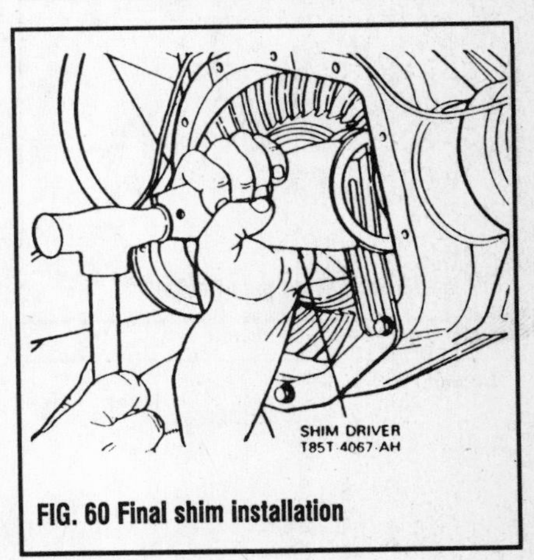

FIG. 60 Final shim installation

7. From the rear of the axle housing, install the drive pinion assembly into the housing pinion shaft bore.

8. Lubricate the pinion shaft splines and install the companion flange.

9. Using a companion flange holder tool, torque the pinion nut to 160 ft. lbs. (217 Nm); rotate the pinion gear, occasionally, to ensure proper bearing seating.

10. Using an inch pound torque wrench, frequently, measure the pinion bearing preload; it should be 8–14 inch lbs. for used bearings or 16–29 inch lbs. for new bearings.

➡ **If the preload is higher than the specification, tighten to the**

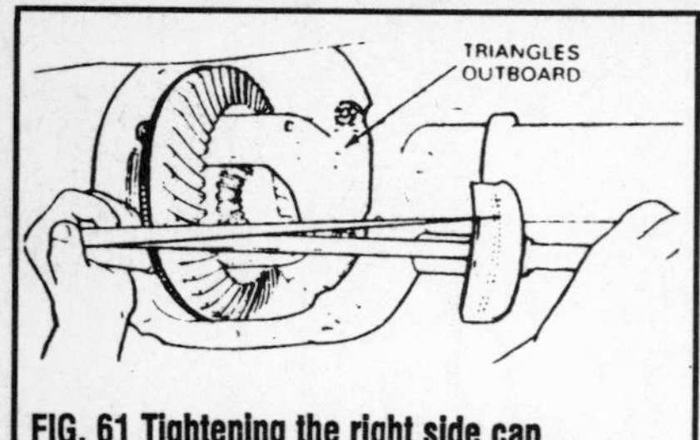

FIG. 61 Tightening the right side cap

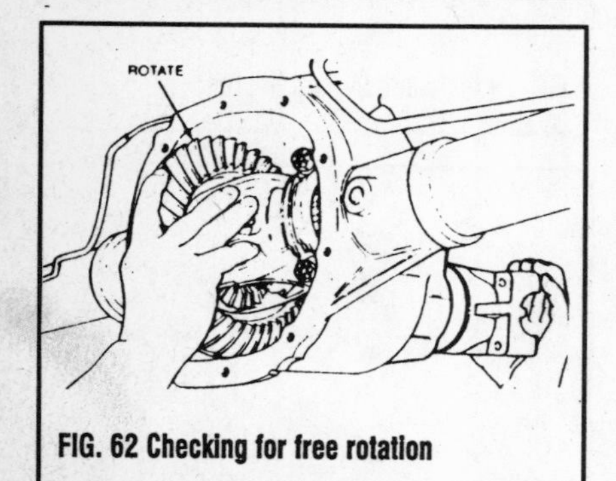

FIG. 62 Checking for free rotation

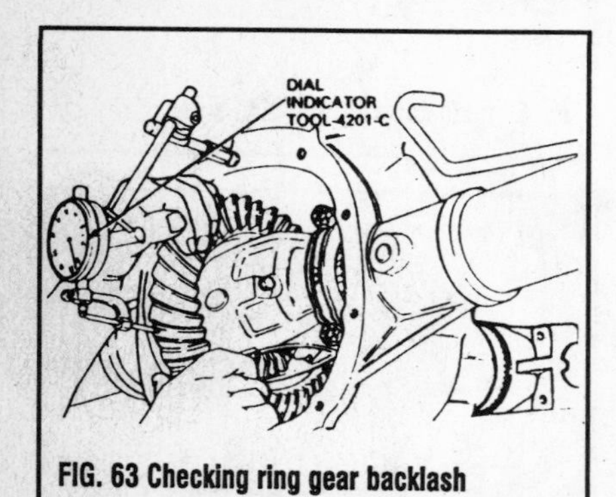

FIG. 63 Checking ring gear backlash

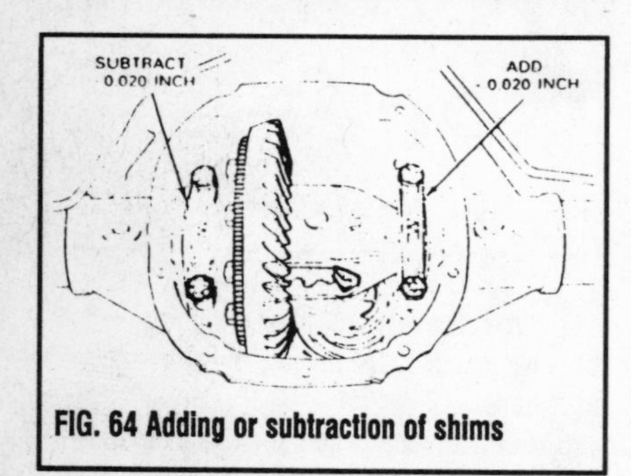

FIG. 64 Adding or subtraction of shims

original reading as recorded; never back off the pinion nut.

Differential Carrier

1. If the bearings were removed, use a shop press to press them onto the differential case.
2. Install the side gears and thrust washers. Install the pinion gears, the thrust washer, the pinion shaft and the pinion shaft lock bolt.
3. Using a shop press, align the excitor ring tab with the differential case slot and press the ring gear and exciter ring onto the differential case. Install the ring gear-to-differential case bolts and torque them to 100–120 ft. lbs. (135–162 Nm).
4. Place the differential case with the bearing cups into the housing.

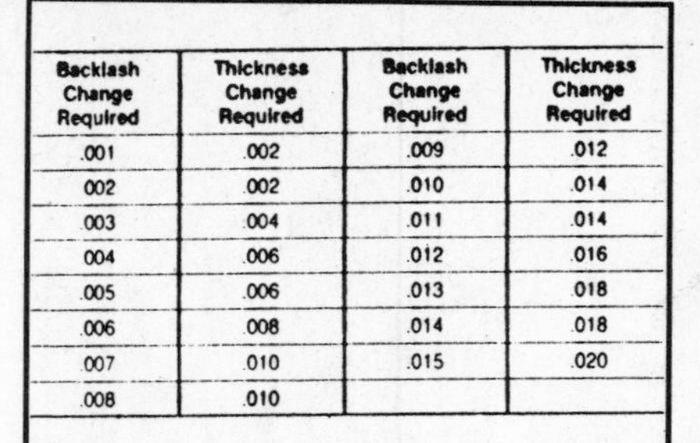

Backlash Change Required	Thickness Change Required	Backlash Change Required	Thickness Change Required
.001	.002	.009	.012
.002	.002	.010	.014
.003	.004	.011	.014
.004	.006	.012	.016
.005	.006	.013	.018
.006	.008	.014	.018
.007	.010	.015	.020
.008	.010		

FIG. 65 Shim chart

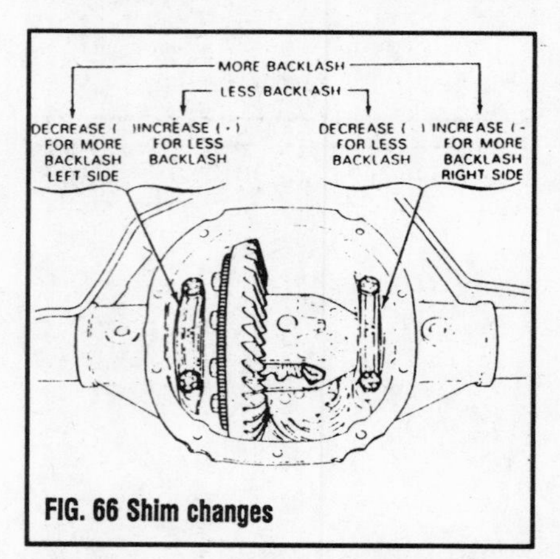

FIG. 66 Shim changes

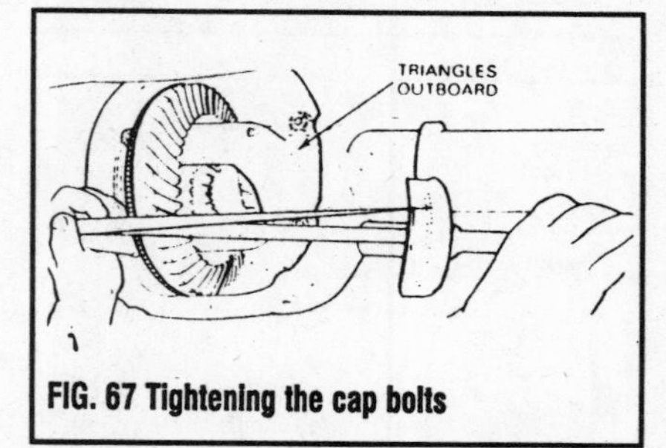

FIG. 67 Tightening the cap bolts

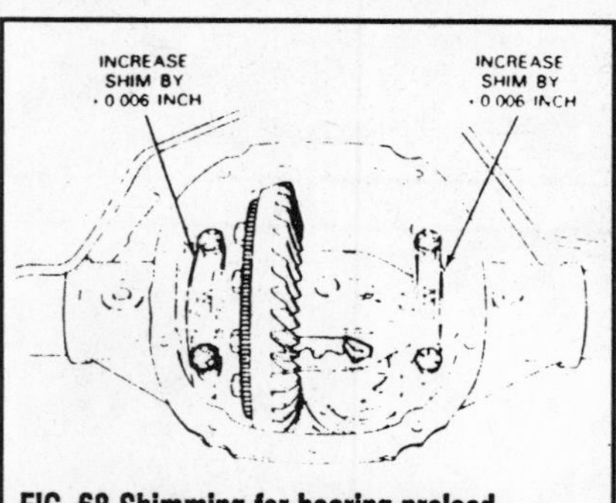

FIG. 68 Shimming for bearing preload

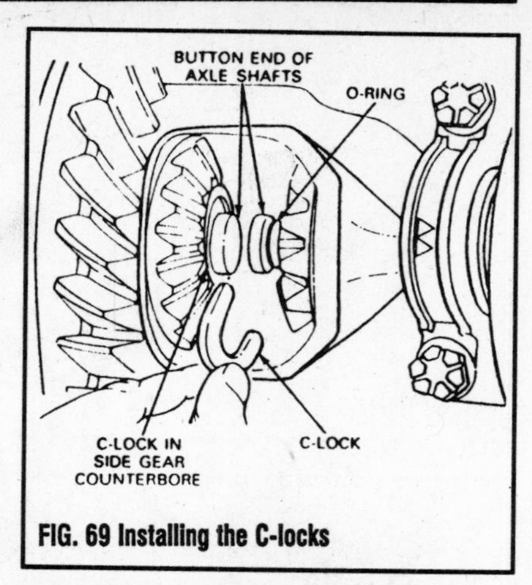

FIG. 69 Installing the C-locks

PINION
DIFFERENTIAL
GEARS

FIG. 70 Installing the pinion gears

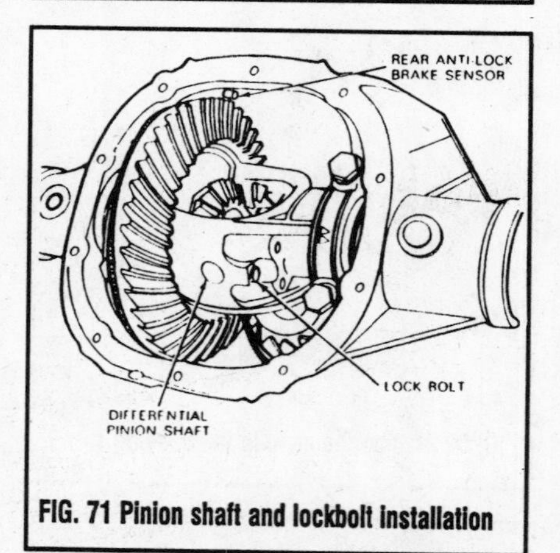

FIG. 71 Pinion shaft and lockbolt installation

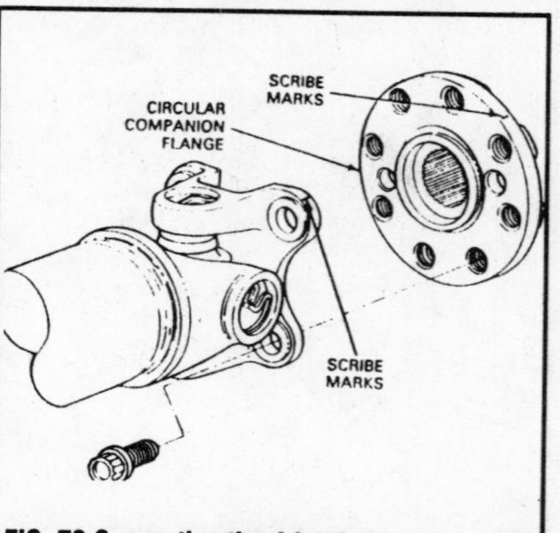

FIG. 72 Connecting the driveshaft

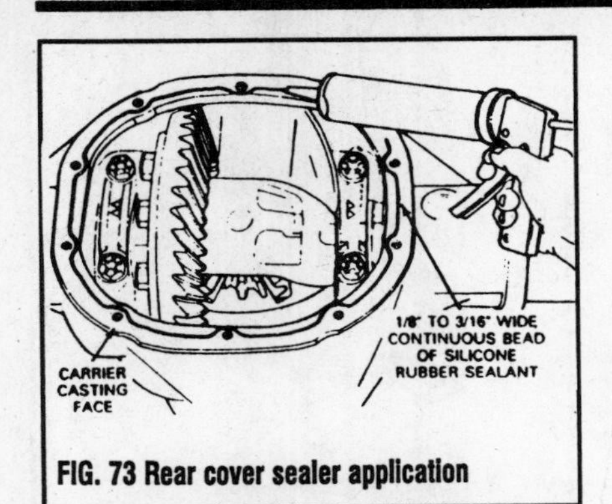

FIG. 73 Rear cover sealer application

5. On the left side, install a 0.265 in shim. Install the bearing cap and tighten the bolts finger tight.
6. On the right side, install progressively larger shims until the largest can be installed by hand. Install the bearing cap.
7. Torque the bearing cap-to-housing bolts to 70–85 ft. lbs. (95–115 Nm).
8. Rotate the assembly to ensure free rotation.
9. Adjust the ring gear backlash.

ADJUSTMENT

Ring Gear and Pinion Backlash

1. Using a dial indicator, measure the ring gear and pinion backlash; it should be 0.008–0.015 in. (0.20–0.38mm). If the backlash is 0.001–0.007 in. (0.0.025–0.178mm) or greater than 0.015 in. (0.38mm) proceed to Step 3. If the backlash is zero, proceed to Step 2.
2. If the backlash is zero, add 0.020 in. (0.50mm) shim(s) to the right side and subtract a 0.020 in. (0.50mm) shim(s) from the left side.
3. If the backlash is within specification, go to step 7. If the backlash is 0.001–0.007 in. (0.025–0.178mm) or greater than 0.015 in. (0.38mm), increase the thickness of a shim on 1 side and decrease the same thickness of another shim on the other side, until the backlash comes within range.
4. Install and torque the bearing cap bolts to 80–95 ft. lbs. (109–128 Nm).
5. Rotate the assembly several times to ensure proper seating.
6. Recheck the backlash, if it is not within specification, go to step 7.
7. Remove the bearing caps. Increase the shim sizes, on both sides by 0.006 in. (0.152mm); make sure the shims are fully seated and the assembly turns freely. Use a shim driver to install the shims.
8. Install the bearing caps and torque the bearing caps to 80–95 ft. lbs. (109–128 Nm). Recheck the backlash; if not to specification, repeat this entire procedure.

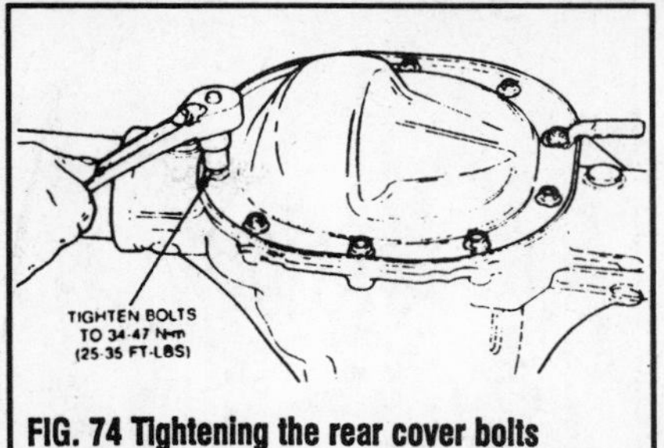

FIG. 74 Tightening the rear cover bolts

Limited Slip Differential Overhaul

➧ SEE FIG. 75

Other than the preload spring, pinion shaft and gears and the clutch packs, the overhaul of this unit is identical to that of the conventional differential.

For removal of these components, see the Adjustment procedures immediately following.

CLUTCH PACK PRELOAD ADJUSTMENT

➧ SEE FIGS. 76–91

This adjustment can be made with the unit in the truck. The axle shafts, however, must be removed completely.

1. Using a punch, drive the S-shaped preload spring half way out of the differential case.
2. Rotate the case 180°. Hold the preload spring with a pliers and tap the spring until it is removed from the differential.

CAUTION

Be careful! The preload spring is under tension!

3. Using gear rotator tool T80P-4205-A, or its equivalent, rotate the pinion gears until the can be removed from the differential. A 12 in.

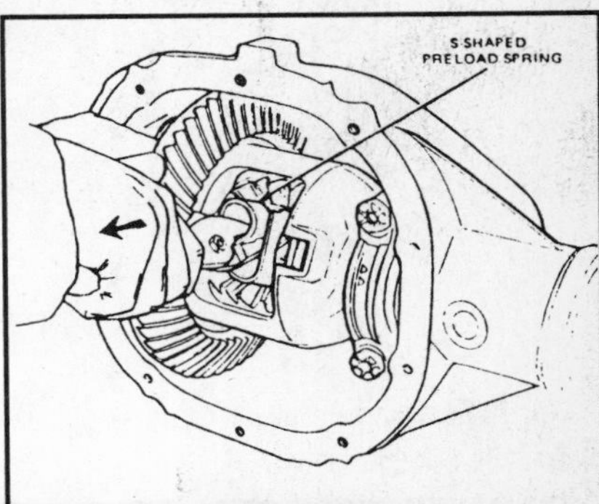

FIG. 76 Removing the preload spring — Traction-Lok differential

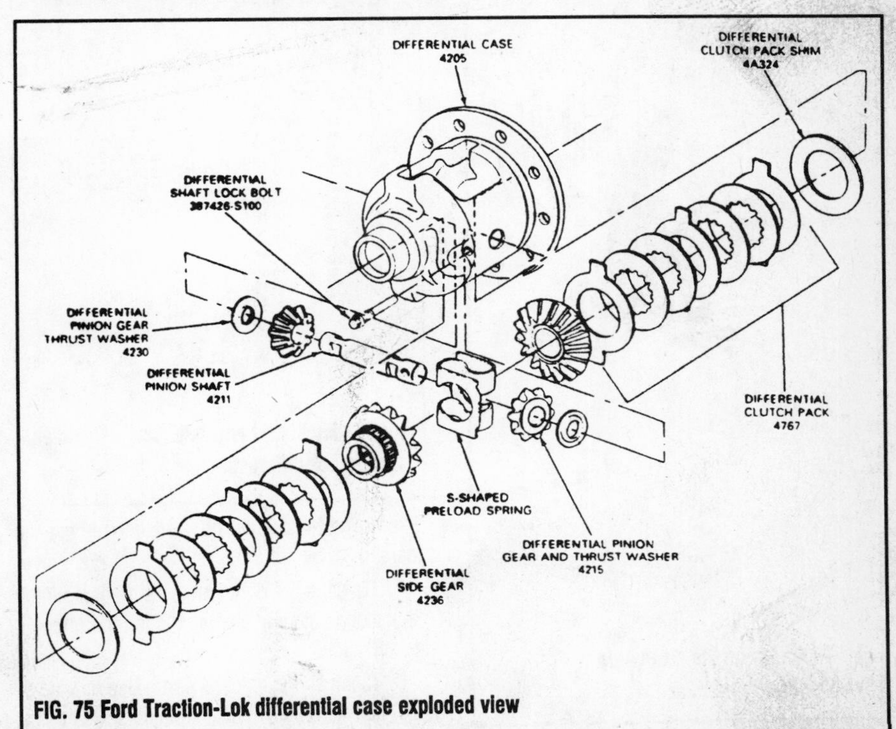

FIG. 75 Ford Traction-Lok differential case exploded view

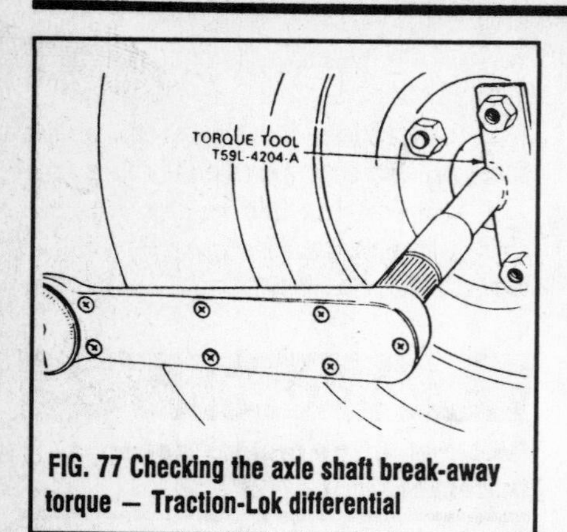

FIG. 77 Checking the axle shaft break-away torque — Traction-Lok differential

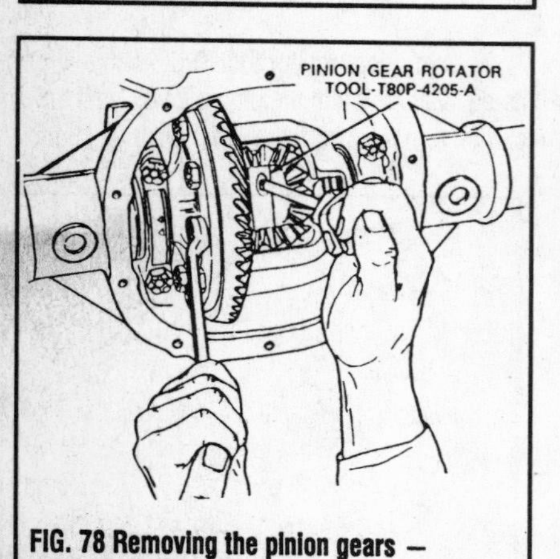

FIG. 78 Removing the pinion gears — Traction-Lok differential

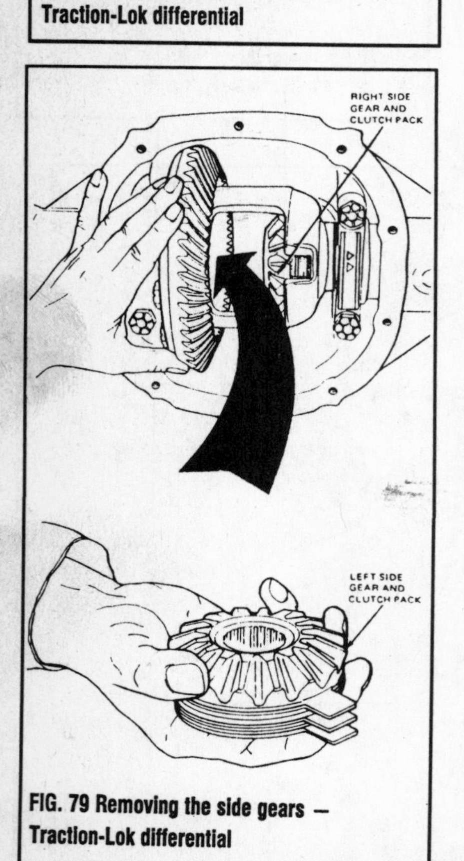

FIG. 79 Removing the side gears — Traction-Lok differential

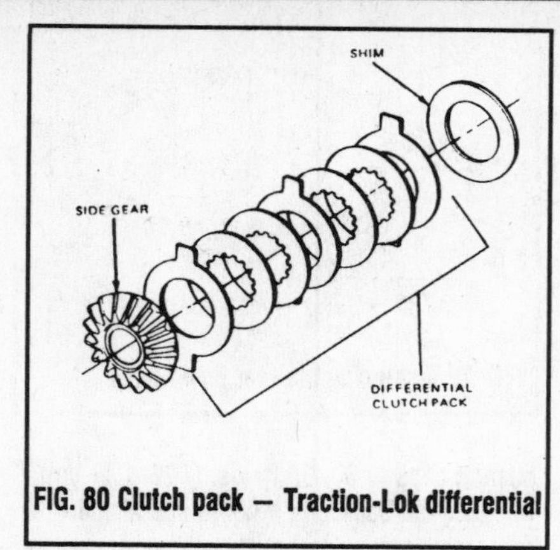

FIG. 80 Clutch pack — Traction-Lok differential

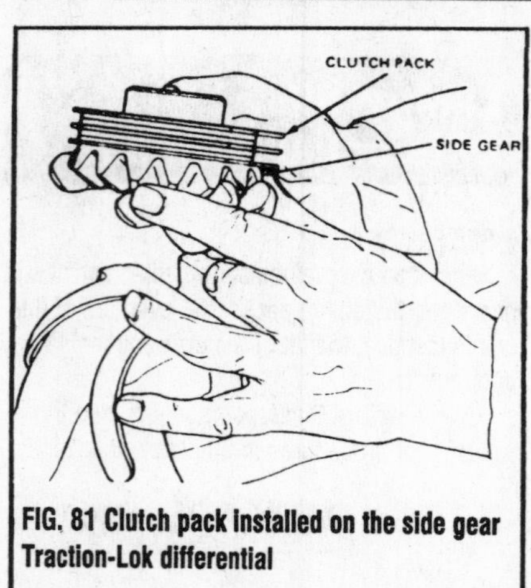

FIG. 81 Clutch pack installed on the side gear Traction-Lok differential

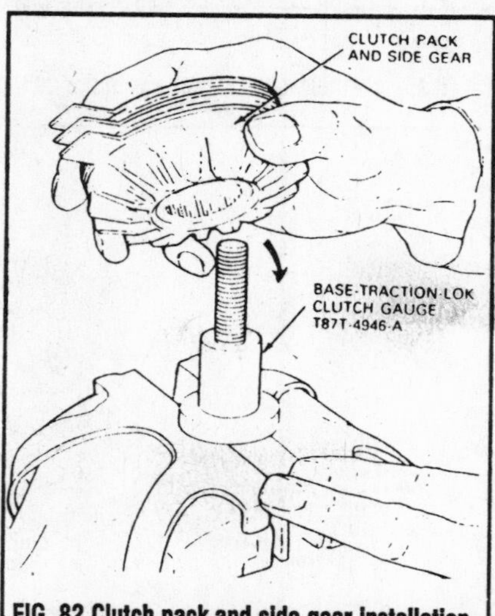

FIG. 82 Clutch pack and side gear installation Traction-Lok differential

(305mm) extension will be needed to remove the gears.

4. Remove the right and left side gear clutch pack along with any shims, and tag them for identification.

5. Clean and inspect all parts. Use only acid-free, non-flammable cleaning agents. Use a lint-

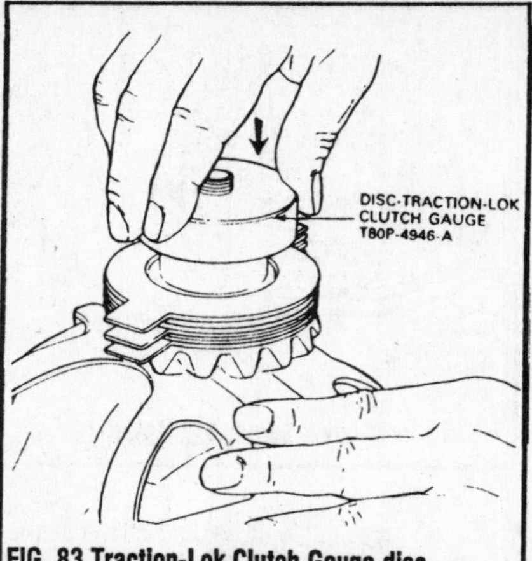

FIG. 83 Traction-Lok Clutch Gauge disc installed — Traction-Lok differential

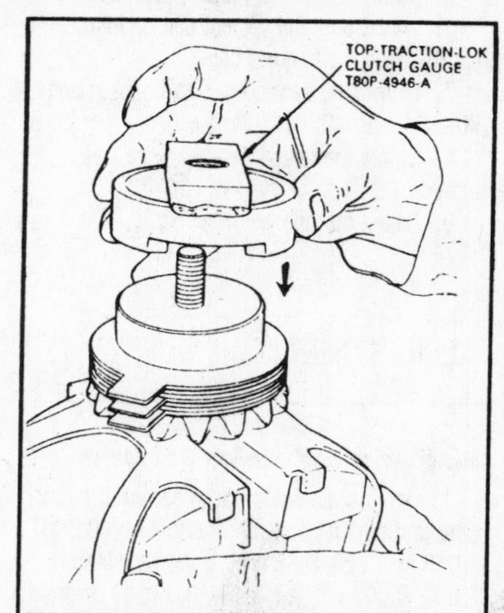

FIG. 84 Traction-Lok Clutch Gauge top installation — Traction-Lok differential

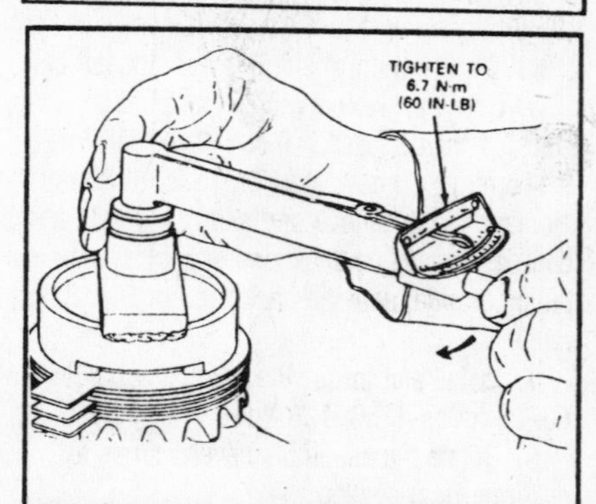

FIG. 85 Traction-Lok Clutch Gauge top nut installation — Traction-Lok differential

free cloth to wipe the parts dry. Replace any damaged parts.

6. Assemble the clutch packs — without the shims — on their side gears. Coat all friction

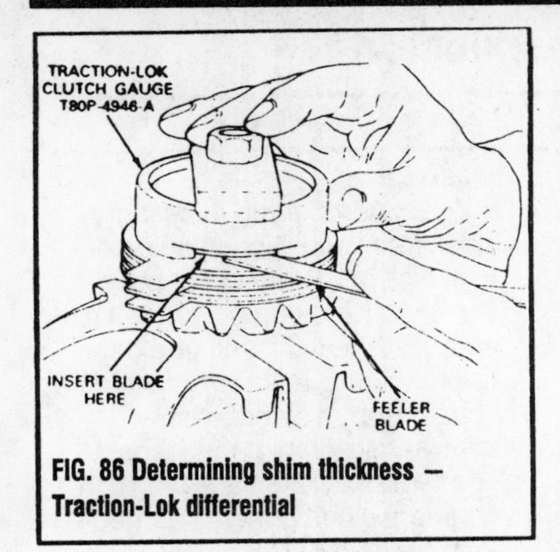

FIG. 86 Determining shim thickness — Traction-Lok differential

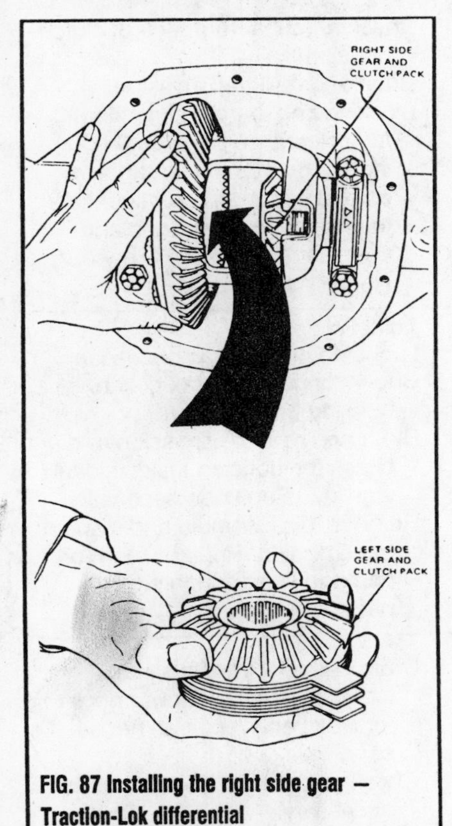

FIG. 87 Installing the right side gear — Traction-Lok differential

plates with limited slip lubricant, such as Ford Additive Friction Modifier, or equivalent meeting EST-M2C118-A specifications.

7. Place the base portion of the Traction-Lok Clutch Gauge T87T-4946-A, or equivalent, in a vise. Install the clutch pack and side gear — without the shims — over the base.

8. Install the tool's disc over the base and on top of the clutch pack.

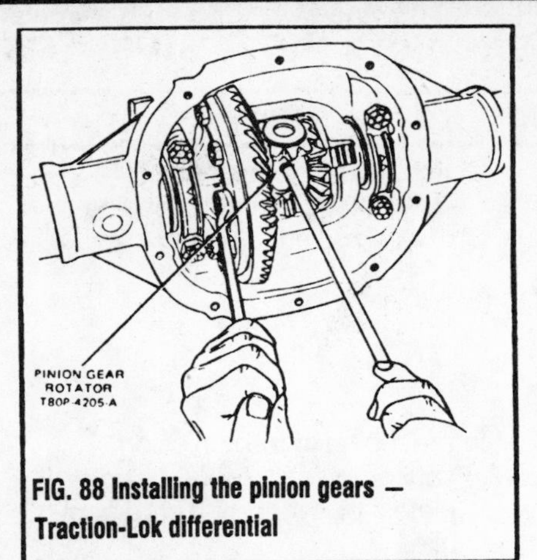

FIG. 88 Installing the pinion gears — Traction-Lok differential

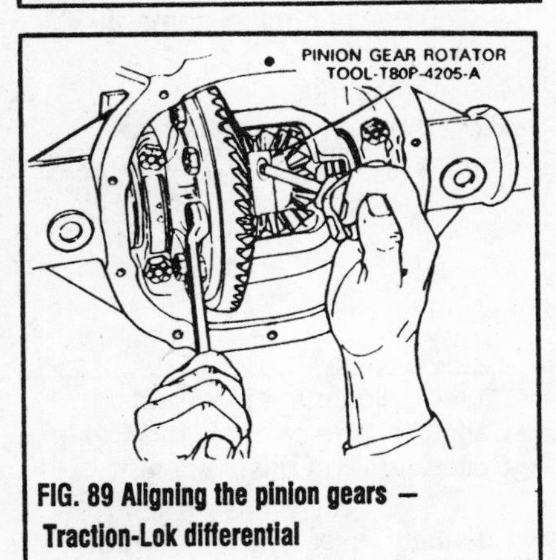

FIG. 89 Aligning the pinion gears — Traction-Lok differential

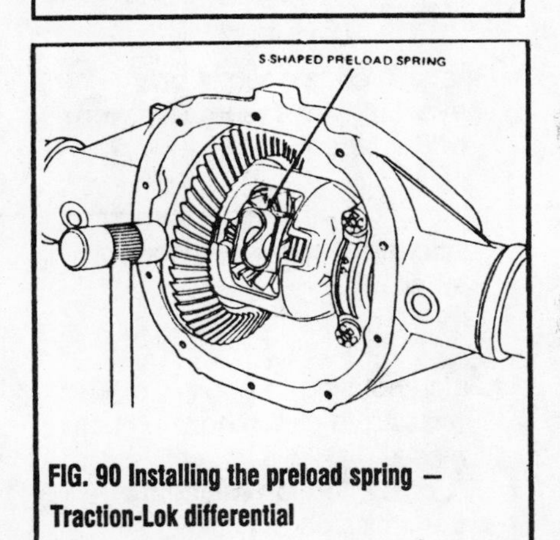

FIG. 90 Installing the preload spring — Traction-Lok differential

9. Install the top portion of the tool over the disc and base stud.

10. Install the tool's nut and torque it to 60 inch lbs.

11. Using feeler gauges, determine the distance between the tool and the clutch pack. This will be the thickness of the necessary shim.

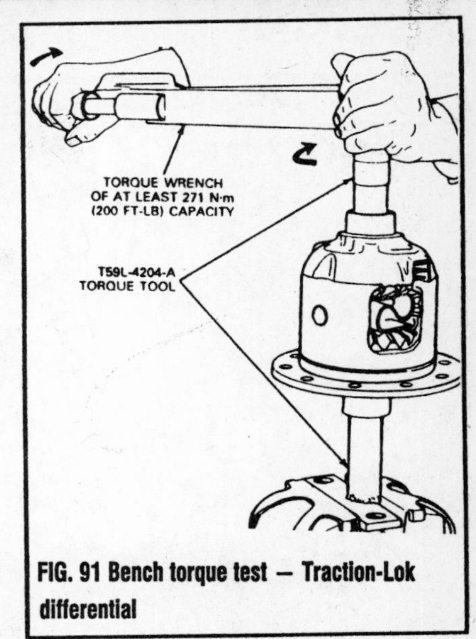

FIG. 91 Bench torque test — Traction-Lok differential

12. Install the right side gear, clutch pack and new shim into the differential case. Repeat this for the left side.

13. Place the pinion gears and thrust washers 180° apart on the side gears. Install tool T80P-4205-A, or equivalent. A 12 in. (305mm) extension should be used to install the gears.

14. Rotate the tool until the pinion gears are aligned with the pinion shaft hole. Remove the tool.

15. Hold the S-shaped preload spring at the differential case window, and, with a plastic mallet, drive the spring into position. Check the spring for damage.

BENCH TORQUE TEST

This test must be made any time the differential has been removed, or adjustments have been made.

Using the locker tools in set T59L-4204-A, or equivalent, check the torque required to rotate one side gear while the other is held stationary.

The initial breakaway torque, if original clutch plates are used, should be at least 20 ft. lbs.

The rotating torque needed to keep the side gear turning, with new plates, will fluctuate.

Troubleshooting the Manual Transmission

Problem	Cause	Solution
Transmission shifts hard	• Clutch adjustment incorrect	• Adjust clutch
	• Clutch linkage or cable binding	• Lubricate or repair as necessary
	• Shift rail binding	• Check for mispositioned selector arm roll pin, loose cover bolts, worn shift rail bores, worn shift rail, distorted oil seal, or extension housing not aligned with case. Repair as necessary.
	• Internal bind in transmission caused by shift forks, selector plates, or synchronizer assemblies	• Remove, dissemble and inspect transmission. Replace worn or damaged components as necessary.
	• Clutch housing misalignment	• Check runout at rear face of clutch housing
	• Incorrect lubricant	• Drain and refill transmission
	• Block rings and/or cone seats worn	• Blocking ring to gear clutch tooth face clearance must be 0.030 inch or greater. If clearance is correct it may still be necessary to inspect blocking rings and cone seats for excessive wear. Repair as necessary.
Gear clash when shifting from one gear to another	• Clutch adjustment incorrect	• Adjust clutch
	• Clutch linkage or cable binding	• Lubricate or repair as necessary
	• Clutch housing misalignment	• Check runout at rear of clutch housing
	• Lubricant level low or incorrect lubricant	• Drain and refill transmission and check for lubricant leaks if level was low. Repair as necessary.
	• Gearshift components, or synchronizer assemblies worn or damaged	• Remove, disassemble and inspect transmission. Replace worn or damaged components as necessary.
Transmission noisy	• Lubricant level low or incorrect lubricant	• Drain and refill transmission. If lubricant level was low, check for leaks and repair as necessary.
	• Clutch housing-to-engine, or transmission-to-clutch housing bolts loose	• Check and correct bolt torque as necessary
	• Dirt, chips, foreign material in transmission	• Drain, flush, and refill transmission
	• Gearshift mechanism, transmission gears, or bearing components worn or damaged	• Remove, disassemble and inspect transmission. Replace worn or damaged components as necessary.
	• Clutch housing misalignment	• Check runout at rear face of clutch housing

Troubleshooting the Manual Transmission

Problem	Cause	Solution
Jumps out of gear	• Clutch housing misalignment	• Check runout at rear face of clutch housing
	• Gearshift lever loose	• Check lever for worn fork. Tighten loose attaching bolts.
	• Offset lever nylon insert worn or lever attaching nut loose	• Remove gearshift lever and check for loose offset lever nut or worn insert. Repair or replace as necessary.
	• Gearshift mechanism, shift forks, selector plates, interlock plate, selector arm, shift rail, detent plugs, springs or shift cover worn or damaged	• Remove, disassemble and inspect transmission cover assembly. Replace worn or damaged components as necessary.
	• Clutch shaft or roller bearings worn or damaged	• Replace clutch shaft or roller bearings as necessary
Jumps out of gear (cont.)	• Gear teeth worn or tapered, synchronizer assemblies worn or damaged, excessive end play caused by worn thrust washers or output shaft gears	• Remove, disassemble, and inspect transmission. Replace worn or damaged components as necessary.
	• Pilot bushing worn	• Replace pilot bushing
Will not shift into one gear	• Gearshift selector plates, interlock plate, or selector arm, worn, damaged, or incorrectly assembled	• Remove, disassemble, and inspect transmission cover assembly. Repair or replace components as necessary.
	• Shift rail detent plunger worn, spring broken, or plug loose	• Tighten plug or replace worn or damaged components as necessary
	• Gearshift lever worn or damaged	• Replace gearshift lever
	• Synchronizer sleeves or hubs, damaged or worn	• Remove, disassemble and inspect transmission. Replace worn or damaged components.
Locked in one gear—cannot be shifted out	• Shift rail(s) worn or broken, shifter fork bent, setscrew loose, center detent plug missing or worn	• Inspect and replace worn or damaged parts
	• Broken gear teeth on countershaft gear, clutch shaft, or reverse idler gear	• Inspect and replace damaged part
	Gearshift lever broken or worn, shift mechanism in cover incorrectly assembled or broken, worn damaged gear train components	• Disassemble transmission. Replace damaged parts or assemble correctly.

Troubleshooting Basic Clutch Problems

Problem	Cause
Excessive clutch noise	Throwout bearing noises are more audible at the lower end of pedal travel. The usual causes are: • Riding the clutch • Too little pedal free-play • Lack of bearing lubrication A bad clutch shaft pilot bearing will make a high pitched squeal, when the clutch is disengaged and the transmission is in gear or within the first 2″ of pedal travel. The bearing must be replaced. Noise from the clutch linkage is a clicking or snapping that can be heard or felt as the pedal is moved completely up or down. This usually requires lubrication. Transmitted engine noises are amplified by the clutch housing and heard in the passenger compartment. They are usually the result of insufficient pedal free-play and can be changed by manipulating the clutch pedal.
Clutch slips (the car does not move as it should when the clutch is engaged)	This is usually most noticeable when pulling away from a standing start. A severe test is to start the engine, apply the brakes, shift into high gear and SLOWLY release the clutch pedal. A healthy clutch will stall the engine. If it slips it may be due to: • A worn pressure plate or clutch plate • Oil soaked clutch plate • Insufficient pedal free-play
Clutch drags or fails to release	The clutch disc and some transmission gears spin briefly after clutch disengagement. Under normal conditions in average temperatures, 3 seconds is maximum spin-time. Failure to release properly can be caused by: • Too light transmission lubricant or low lubricant level • Improperly adjusted clutch linkage
Low clutch life	Low clutch life is usually a result of poor driving habits or heavy duty use. Riding the clutch, pulling heavy loads, holding the car on a grade with the clutch instead of the brakes and rapid clutch engagement all contribute to low clutch life.

Troubleshooting Basic Automatic Transmission Problems

Problem	Cause	Solution
Fluid leakage	• Defective pan gasket • Loose filler tube • Loose extension housing to transmission case • Converter housing area leakage	• Replace gasket or tighten pan bolts • Tighten tube nut • Tighten bolts • Have transmission checked professionally
Fluid flows out the oil filler tube	• High fluid level • Breather vent clogged • Clogged oil filter or screen • Internal fluid leakage	• Check and correct fluid level • Open breather vent • Replace filter or clean screen (change fluid also) • Have transmission checked professionally

Troubleshooting Basic Automatic Transmission Problems

Problem	Cause	Solution
Transmission overheats (this is usually accompanied by a strong burned odor to the fluid)	• Low fluid level • Fluid cooler lines clogged • Heavy pulling or hauling with insufficient cooling • Faulty oil pump, internal slippage	• Check and correct fluid level • Drain and refill transmission. If this doesn't cure the problem, have cooler lines cleared or replaced. • Install a transmission oil cooler • Have transmission checked professionally
Buzzing or whining noise	• Low fluid level • Defective torque converter, scored gears	• Check and correct fluid level • Have transmission checked professionally
No forward or reverse gears or slippage in one or more gears	• Low fluid level • Defective vacuum or linkage controls, internal clutch or band failure	• Check and correct fluid level • Have unit checked professionally
Delayed or erratic shift	• Low fluid level • Broken vacuum lines • Internal malfunction	• Check and correct fluid level • Repair or replace lines • Have transmission checked professionally

Lockup Torque Converter Service Diagnosis

Problem	Cause	Solution
No lockup	• Faulty oil pump • Sticking governor valve • Valve body malfunction (a) Stuck switch valve (b) Stuck lockup valve (c) Stuck fail-safe valve • Failed locking clutch • Leaking turbine hub seal • Faulty input shaft or seal ring	• Replace oil pump • Repair or replace as necessary • Repair or replace valve body or its internal components as necessary • Replace torque converter • Replace torque converter • Repair or replace as necessary
Will not unlock	• Sticking governor valve • Valve body malfunction (a) Stuck switch valve (b) Stuck lockup valve (c) Stuck fail-safe valve	• Repair or replace as necessary • Repair or replace valve body or its internal components as necessary
Stays locked up at too low a speed in direct	• Sticking governor valve • Valve body malfunction (a) Stuck switch valve (b) Stuck lockup valve (c) Stuck fail-safe valve	• Repair or replace as necessary • Repair or replace valve body or its internal components as necessary
Locks up or drags in low or second	• Faulty oil pump • Valve body malfunction (a) Stuck switch valve (b) Stuck fail-safe valve	• Replace oil pump • Repair or replace valve body or its internal components as necessary

Lockup Torque Converter Service Diagnosis

Problem	Cause	Solution
Sluggish or stalls in reverse	• Faulty oil pump • Plugged cooler, cooler lines or fittings • Valve body malfunction (a) Stuck switch valve (b) Faulty input shaft or seal ring	• Replace oil pump as necessary • Flush or replace cooler and flush lines and fittings • Repair or replace valve body or its internal components as necessary
Loud chatter during lockup engagement (cold)	• Faulty torque converter • Failed locking clutch • Leaking turbine hub seal	• Replace torque converter • Replace torque converter • Replace torque converter
Vibration or shudder during lockup engagement	• Faulty oil pump • Valve body malfunction • Faulty torque converter • Engine needs tune-up	• Repair or replace oil pump as necessary • Repair or replace valve body or its internal components as necessary • Replace torque converter • Tune engine
Vibration after lockup engagement	• Faulty torque converter • Exhaust system strikes underbody • Engine needs tune-up • Throttle linkage misadjusted	• Replace torque converter • Align exhaust system • Tune engine • Adjust throttle linkage

Lockup Torque Converter Service Diagnosis

Problem	Cause	Solution
Vibration when revved in neutral	• Torque converter out of balance	• Replace torque converter
Overheating: oil blows out of dip stick tube or pump seal	• Plugged cooler, cooler lines or fittings • Stuck switch valve	• Flush or replace cooler and flush lines and fittings • Repair switch valve in valve body or replace valve body
Shudder after lockup engagement	• Faulty oil pump • Plugged cooler, cooler lines or fittings • Valve body malfunction • Faulty torque converter • Fail locking clutch • Exhaust system strikes underbody • Engine needs tune-up • Throttle linkage misadjusted	• Replace oil pump • Flush or replace cooler and flush lines and fittings • Repair or replace valve body or its internal components as necessary • Replace torque converter • Replace torque converter • Align exhaust system • Tune engine • Adjust throttle linkage

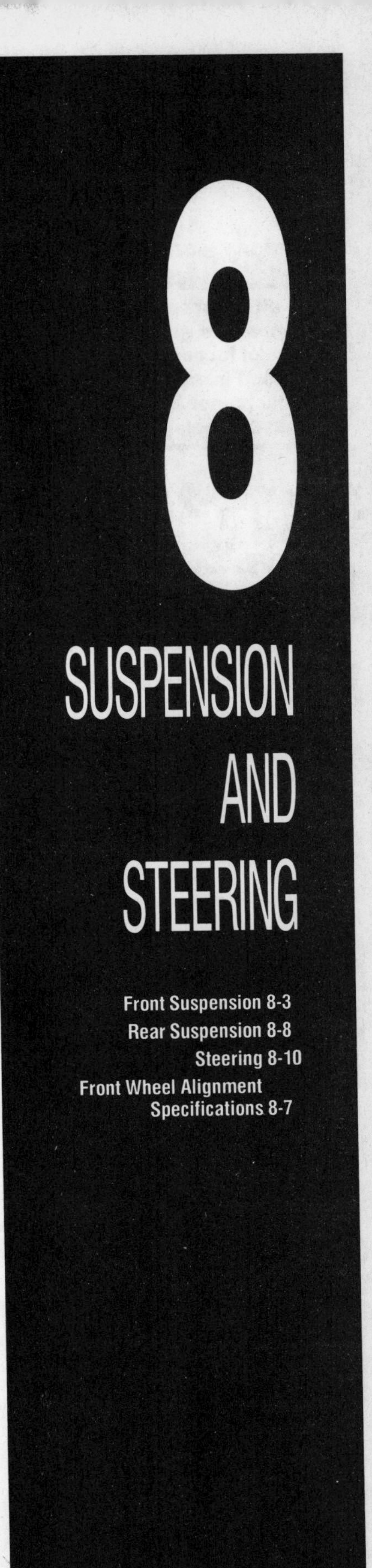

8 SUSPENSION AND STEERING

WHEELS

CAUTION

Some aftermarket wheels may not be compatible with these vehicles. The use of incompatible wheels may result in equipment failure and possible personal injury! Use only approved wheels!

Front or Rear Wheels

REMOVAL & INSTALLATION

SEE FIG. 1

1. Set the parking brake and block the opposite wheel.
2. On vehicles with an automatic transmission, place the selector lever in **P**. On vehicles with a manual transmission, place the transmission in reverse.
3. If equipped, remove the wheel cover.
4. Break loose the lug nuts.
5. Raise the vehicle until the tire is clear of the ground.
6. Remove the lug nuts and remove the wheel.

To install:

7. Clean the wheel lugs and brake drum or hub of all foreign material.
8. Position the wheel on the hub or drum and hand-tighten the lug nuts. Make sure that the coned ends face inward.
9. Using the lug wrench, tighten all the lugs, in a criss-cross fashion until they are snug.
10. Lower the vehicle. Tighten the nuts, in the sequence shown, to 100 ft. lbs.

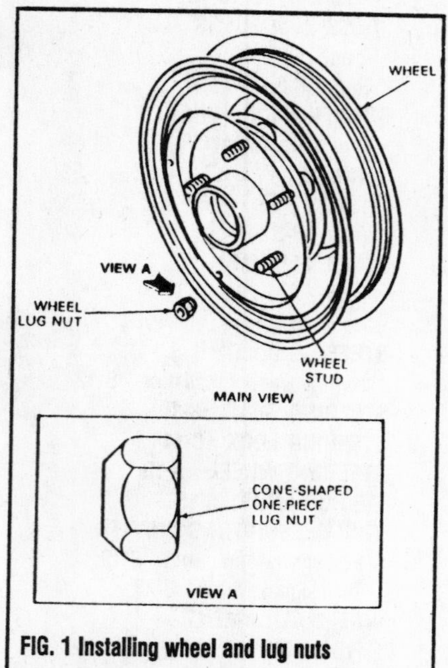

FIG. 1 Installing wheel and lug nuts

Wheel Lug Nut Stud

REPLACEMENT

Front Wheels

USING A PRESS

SEE FIG. 2

1. Remove the wheel.

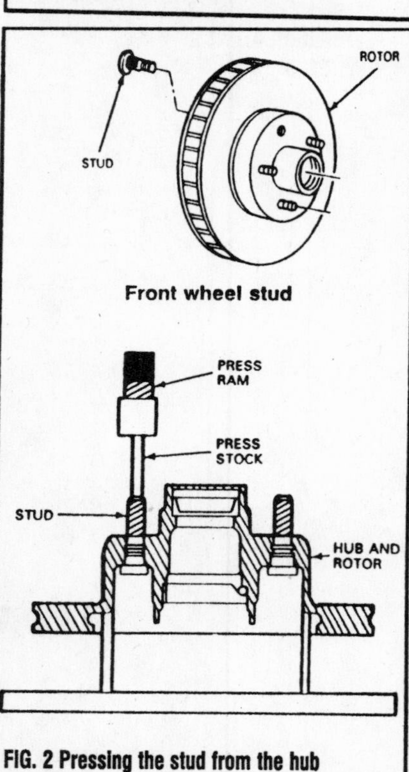

FIG. 2 Pressing the stud from the hub

2. Place the hub/rotor assembly in a press, supported by the hub surface. NEVER rest the assembly on the rotor!
3. Press the stud from the hub.

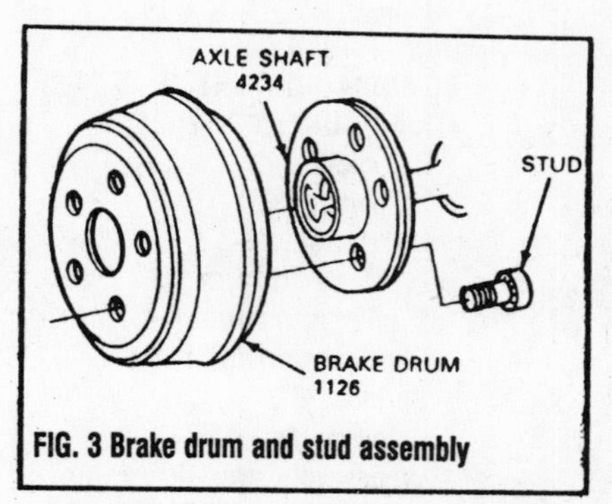

FIG. 3 Brake drum and stud assembly

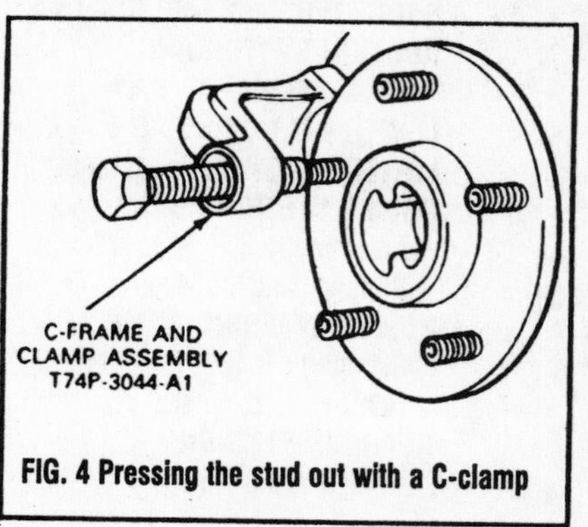

FIG. 4 Pressing the stud out with a C-clamp

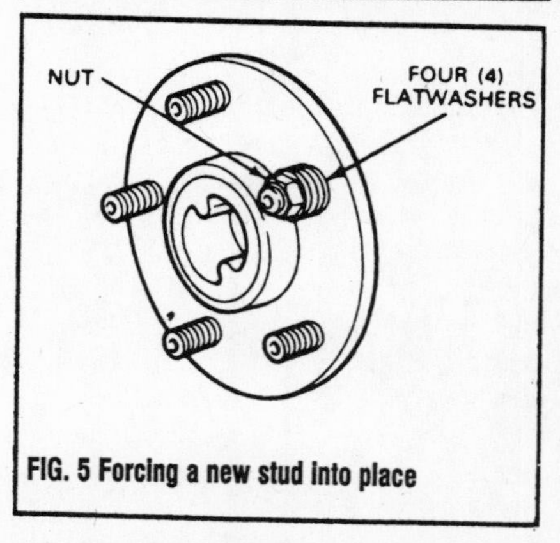

FIG. 5 Forcing a new stud into place

4. Position the new stud in the hub and align the serrations. Make sure it is square and press it into place.

USING A HAMMER AND DRIVER

1. Remove the wheel.
2. Support the hub/rotor assembly on a flat, hard surface, resting the assembly on the hub. NEVER rest the assembly on the rotor!
3. Position a driver, such as a drift or broad punch, on the outer end of the stud and drive it from the hub.
4. Turn the assembly over, coat the serrations of the new stud with liquid soap, position the stud in the hole, aligning the

serrations, and, using the drift and hammer, drive it into place until fully seated.

Rear Wheels

➧ SEE FIGS. 3–5

1. Remove the wheel.
2. Remove the drum or rotor from the axle shaft or hub studs.
3. Using a large C-clamp and socket, press the stud from the drum or rotor.
4. Coat the serrated part of the stud with liquid soap and place it in the hole. Align the serrations.
5. Place 3 or 4 flat washers on the outer end of the stud and thread a lug nut on the stud with the flat side against the washers. Tighten the lug nut until the stud is drawn all the way in.

Do not use an impact wrench!

FRONT SUSPENSION

➧ SEE FIG. 6

Coil Springs

WARNING

Always use extreme caution when working with coil springs. make sure the vehicle is supported sufficiently.

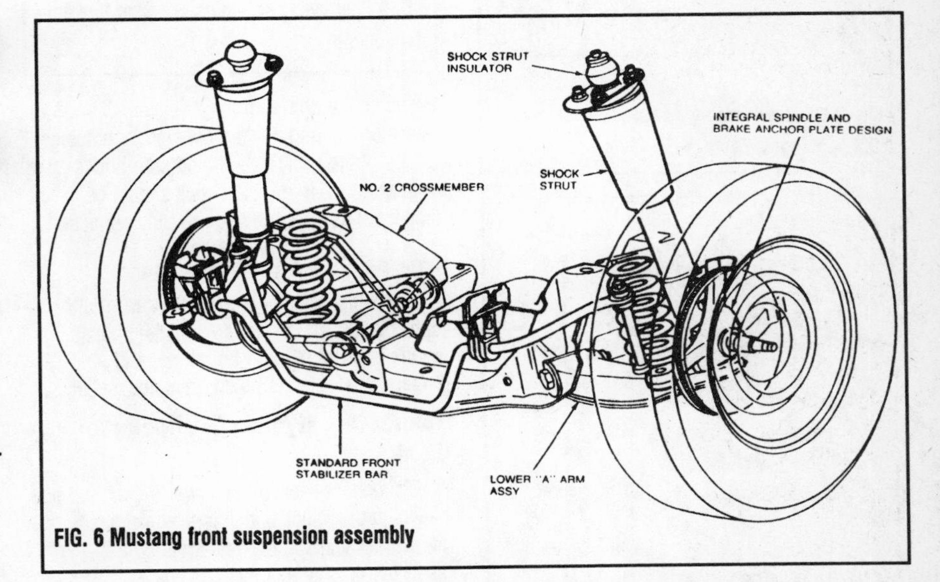

FIG. 6 Mustang front suspension assembly

REMOVAL & INSTALLATION

➧ SEE FIGS. 7–9

1. Raise and safely support the vehicle, allowing the control arms to hang free.
2. Remove the wheel and tire assembly and the brake caliper. Suspend the caliper with a length of wire; do not let the caliper hang by the brake hose.
3. Disconnect the tie rod end from the steering spindle and disconnect the stabilizer link from the lower arm.
4. Remove the steering gear bolts, if necessary and position the gear so the suspension arm bolt can be removed.
5. On all vehicles except Mustang with 5.0L engine, use spring compressor tool T82P–5310–A or equivalent to place the upper plate in position into the spring pocket cavity on the crossmember. The hooks on the plate should be facing the center of the vehicle.
6. On Mustang with 5.0L engine, use spring compressor tool D78P–5310–A or equivalent, to install a plate between the coils near the toe of the spring. Mark the location of the upper plate on the coils for installation.
7. Install the compression rod into the lower arm spring pocket hole, through the coil spring and into the upper plate.
8. Install the lower plate, lower ball nut, thrust washer and bearing and forcing nut onto the compression rod. Tighten the forcing nut until a drag on the nut is felt.
9. Remove the suspension arm-to-crossmember nuts and bolts. The compressor tool forcing nut may have to be tightened or loosened for easy bolt removal.
10. Loosen the compression rod forcing nut until spring tension is relieved and remove the forcing nut. Remove the compression rod and coil spring.

To Install:

11. Place the insulator on top of the spring. Position the spring into the lower arm pocket. Make sure the spring pigtail is positioned between the 2 holes in the lower arm spring pocket.
12. Position the spring into the upper spring seat in the crossmember.
13. On all except Mustang with 5.0L engine, insert the compression rod through the control arm and spring, then hook it to the upper plate. The upper plate is installed with the hooks facing the center of the vehicle.
14. On Mustang with 5.0L engine, install the upper plate between the coils in the location marked during removal.
15. Install the lower plate, ball nut, thrust washer and bearing and forcing nut onto the compression rod.
16. Tighten the forcing nut, position the lower arm into the crossmember and install new lower arm-to-crossmember bolts and nuts. Do not tighten at this time.
17. Remove the spring compressor tool from the vehicle. Raise the suspension arm to a normal attitude position with a jack. Tighten the lower arm-to-crossmember attaching nuts to 110–150 ft. lbs. (149–203 Nm). Remove the jack.
18. Install the steering gear-to-crossmember bolts and nuts, if removed. Hold the bolts and tighten the nuts to 90–100 ft. lbs. (122–135 Nm).
19. Connect the stabilizer bar link to the lower suspension arm. Tighten the attaching nut to 8–12 ft. lbs. (12–16 Nm).
20. Position the tie rod into the steering spindle and install the retaining nut. Tighten the

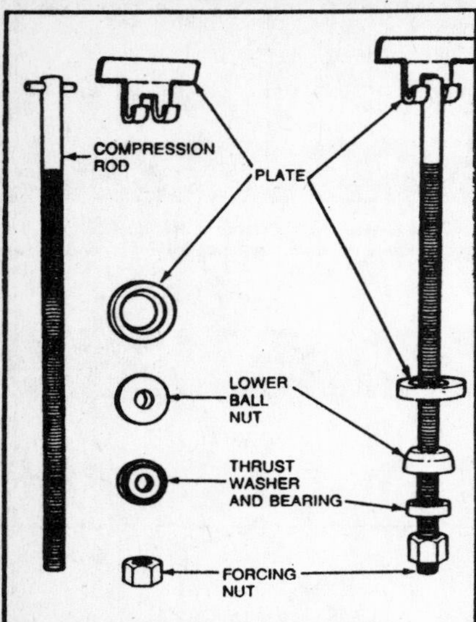

FIG. 7 Front coil spring compressor tool — 2.3L engine vehicles

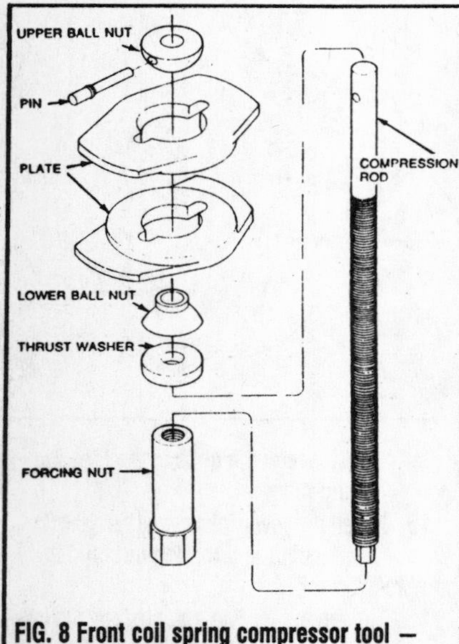

FIG. 8 Front coil spring compressor tool — 5.0L engine vehicles

nut to 35 ft. lbs. (47 Nm) and continue tightening the nut to align the next castellation with the hole in the stud. Install a new cotter pin.

21. Install the brake caliper and the wheel and tire assembly. Lower the vehicle.

MacPherson Strut

TESTING

Bounce Test

Each shock absorber can be tested by bouncing the corner of the vehicle until maximum up and down movement is obtained. Let go of the vehicle. It should stop bouncing in 1–2 bounces. If not, the shock should be inspected for damage and possibly replaced.

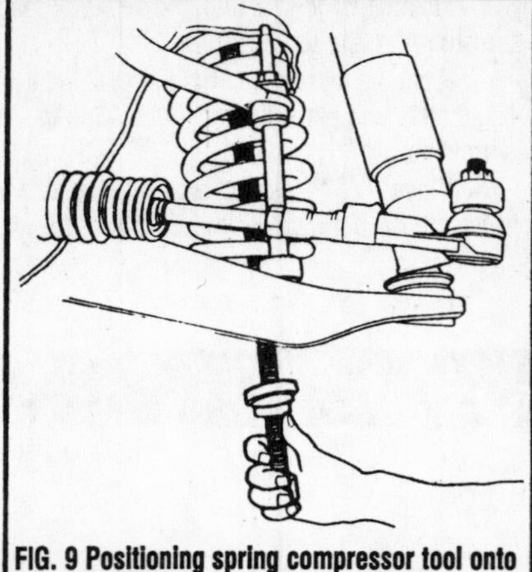
FIG. 9 Positioning spring compressor tool onto the vehicle

Inspect the Shock Mounts

Check the shock mountings for worn or defective grommets, loose mounting nuts, interference or missing bump stops. If no apparent defects are noted, continue testing.

Inspecting Hydraulic Shocks for Leaks

Disconnect each shock lower mount and pull down on the shock until it is fully extended. inspect for leaks in the seal area. Shock absorber fluid is very thin and has a characteristic odor and dark brown color. Don't confuse the glossy paint on some shocks with leaking fluid. A slight trace of fluid is a normal condition; they are designed to seep a certain amount of fluid past the seals for lubrication. If you are in doubt as to whether the fluid on the shock is coming from the shock itself or from some other source, wipe the seal area clean and manually operate the shock (see the following procedure). Fluid will appear if the unit is leaking.

Manually Operating the Shocks

It may be necessary to fabricate a holding fixture for certain types of shock absorbers. If a suspected problem is in the front shocks, disconnect both front shock lower mountings.

Grip the lower end of the shock and pull down (rebound stroke) and then push up (compression stroke). The control arms will limit the movement of front shocks during the compression stroke. Compare the rebound resistance of both shocks and compare the compression resistance. Usually any shock showing a noticeable difference will be the one at fault.

If the shock has internal noises, extend the shock fully then exert an extra pull. If a small additional movement is felt, this usually means a loose piston and the shock should be replaced. Other noises that are cause for replacing shocks are a squeal after a full stroke in both directions, a clicking noise on fast reverse and a lag at reversal near mid-stroke.

REMOVAL & INSTALLATION

➧ SEE FIG. 10

1. Disconnect the negative battery cable.
2. Place the ignition switch in the **UNLOCKED** position to permit free movement of the front wheels.
3. Raise the vehicle by the lower control arms until the wheels are just off the ground. From the engine compartment, remove and discard the 3 upper mount retaining nuts. Do not remove the pop-rivet holding the camber plate position.
4. Continue to raise the front of the vehicle by the lower control arms and position safety stands under the frame jacking pads, rearward of the wheels.
5. Remove the wheel and tire assembly and remove the brake caliper. Support the caliper with a length of wire; do not let the caliper hang by the brake hose.
6. Remove the 2 lower nuts that attach the strut to the spindle, leaving the bolts in place. Carefully remove both spindle-to-strut bolts, push the bracket free of the spindle and remove the strut.
7. Compress the strut to clear the upper mount of the body mounting pad. Remove the upper mount and jounce bumper, if necessary.

To install:

8. Install the upper mount and jounce bumper, if removed.
9. Position the 3 upper mount studs into the body mounting pad and camber plate and start 3 new nuts.
10. Compress the strut and position into the spindle. Install 2 new lower retaining bolts and hand start the nuts. Remove the suspension load from the control arms by lowering the vehicle. Tighten the lower retaining nuts to 140–200 ft. lbs. (190–271 Nm).
11. Raise the suspension control arms and tighten the upper mount retaining nuts to 40–55 ft. lbs. (54–75 Nm).
12. Install the brake caliper and the wheel and tire assembly.
13. Lower the vehicle to the ground and check the front end alignment.

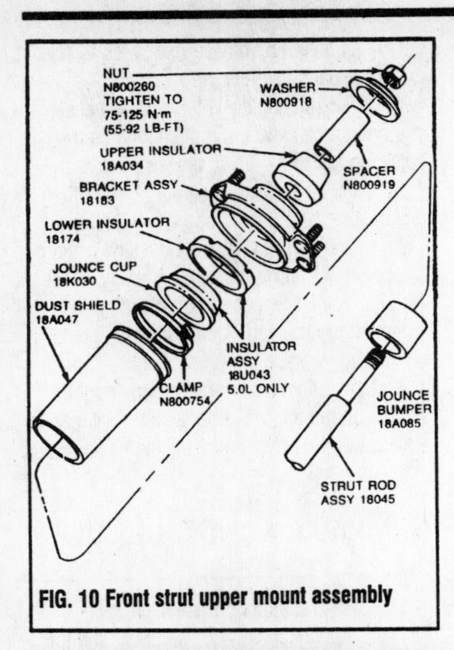

FIG. 10 Front strut upper mount assembly

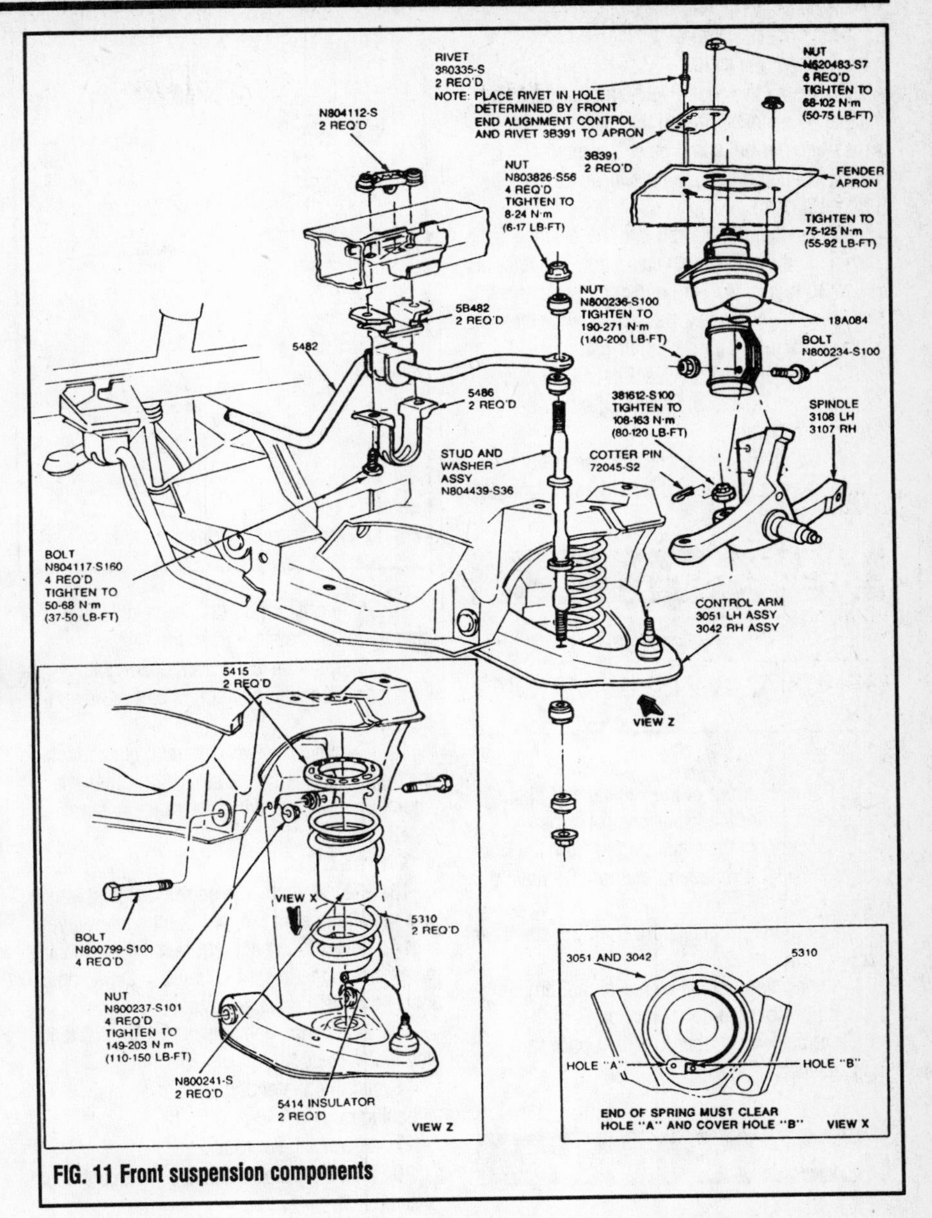

FIG. 11 Front suspension components

Lower Ball Joints

INSPECTION

1. Support the vehicle in normal driving position with ball joints loaded.
2. Wipe the wear indicator and ball joint cover checking surface clean.
3. The checking surface should project outside the cover. If the checking surface is inside the cover, replace the lower arm assembly.

REMOVAL & INSTALLATION

The ball joint is an integral part of the lower control arm. If the ball joint is defective, the entire lower control arm must be replaced.

Lower Control Arm

REMOVAL & INSTALLATION

➧ SEE FIG. 11

1. Raise and safely support the vehicle. Allow the control arms to hang free.
2. Remove the wheel and tire assembly.
3. If necessary, remove the brake caliper and suspend with a length of wire; do not let the caliper hang by the brake hose. Remove the brake rotor and dust shield.
4. Disconnect the tie rod end from the steering spindle. Disconnect the stabilizer bar link from the lower arm.
5. Remove the steering gear bolts and lower the gear out of the way to provide clearance, if necessary, for suspension arm bolt removal.
6. Remove the cotter pin and loosen the lower ball joint stud nut 1–2 turns. Do not remove the nut at this time. Tap the spindle boss sharply to relieve the stud pressure.
7. Install a suitable spring compressor and compress the spring so it is free in the seat.
8. Remove and discard the ball joint nut and raise he entire strut and spindle assembly. Wire out of the way to obtain working room.
9. Remove and discard the suspension arm-to-crossmember nuts and bolts. Remove the lower control arm and, except Mark VII, remove the coil spring.

To install:

10. Position the coil spring into the lower arm pocket. Make sure the spring pigtail is positioned between the 2 holes in the pocket.
11. Position the lower arm to the crossmember and install new arm-to-crossmember bolts and nuts. Do not tighten at this time.
12. Raise the control arm with a jack to a normal attitude position and remove the spring compressor.
13. With the jack in place, tighten the lower arm-to-crossmember attaching nuts to 110–150 ft. lbs. (149–203 Nm).
14. Tighten the ball joint stud nut to 100–120

ft. lbs. (136–163 Nm) and install a new cotter pin. Remove the jack.

15. Install the dust shield, rotor and brake caliper, if removed. Install the steering gear-to-crossmember bolts and nuts, if removed. Hold the bolts and tighten the nuts to 90–100 ft. lbs. (122–136 Nm).

16. Position the tie rod into the steering spindle and install the retaining nut. Tighten the nut to 35 ft. lbs. (47 Nm) and continue tightening the nut to align the next castellation with the hole in the stud. Install a new cotter pin.

17. Connect the stabilizer bar link to the lower control arm. Tighten the retaining nut to 9–12 ft. lbs. (12–16 Nm).

18. Install the wheel and tire assembly and lower the vehicle. Check the front end alignment.

Stabilizer Bar

REMOVAL & INSTALLATION

➧ SEE FIG. 11

1. Raise the front of the vehicle and place jackstands under the lower control arms.

2. Disconnect the stabilizer bar from the links and the insulator mounting clamps. Remove the stabilizer bar.

3. Cut the worn insulators from the stabilizer bar.

4. Installation is the reverse of the removal procedure. Coat the necessary parts of the stabilizer bar with rubber lubricant prior to installation.

Spindle

REMOVAL & INSTALLATION

➧ SEE FIG. 11

1. Raise and support the front end on jackstands under the frame.

2. Remove the wheels.

3. Remove the calipers and suspend them out of the way.

4. Remove the hub and rotor assemblies.

5. Remove the rotor dust shields.

6. Unbolt the stabilizer links from the control arms.

7. Using a separator, disconnect the tie rod ends from the spindle.

8. Remove the cotter pin and loosen the ball joint stud nut a few turns. Don't remove it at this time!

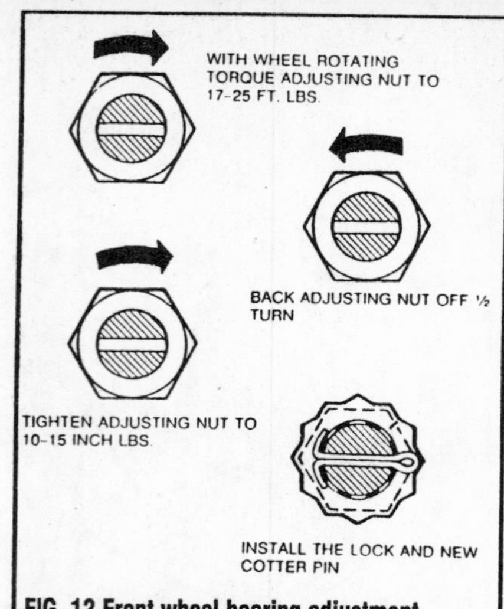

FIG. 12 Front wheel bearing adjustment

9. Using a hammer, tap the spindle boss sharply to relieve stud pressure.

10. Support the lower control arm with a floor jack, compress the coil spring and remove the stud nut.

11. Remove the two bolts and nuts attaching the spindle to the shock strut. Compress the shock strut until working clearance is obtained.

12. Remove the spindle.

To install:

13. Place the spindle on the ball joint stud, and install the stud nut, but don't tighten it yet.

14. Lower the shock strut until the attaching holes are aligned with the holes in the spindle. Install two new bolts and nuts.

15. Tighten the ball stud nut to 80-120 ft. lbs. and install the cotter pin.

16. Torque the shock strut-to-spindle attaching nuts to 140-200 ft. lbs.

17. Lower the floor jack.

18. Install the stabilizer links. Torque the nuts to 6-17 ft. lbs.

19. Attach the tie rod ends and torque the nuts to 35-47 ft. lbs.

20. The remainder of installation is the reverse of removal.

Front Wheel Bearings

ADJUSTMENT

➧ SEE FIG. 12

1. Raise and safely support the front of the vehicle.

2. Remove the wheel cover and grease cap.

3. Remove the cotter pin and nut retainer.

4. Loosen the adjusting nut 3 turns and rock the wheel back and forth a few times to release the brake pads from the rotor.

5. While rotating the wheel and hub assembly in a counterclockwise direction, tighten the adjusting nut to 17–25 ft. lbs. (23–34 Nm).

6. Back off the adjusting nut 1/2 turn, then retighten to 10–28 inch lbs.(1.1–3.2 Nm).

7. Install the nut retainer and a new cotter pin. Check the wheel rotation. If it is noisy or rough, the bearings either need to be cleaned and repacked or replaced. After adjustment is completed, replace the grease cap.

8. Lower the vehicle. Before driving the vehicle, pump the brake pedal several times to restore normal brake pedal travel.

REMOVAL & INSTALLATION

1. Raise and support the vehicle safely. Remove the wheel and tire assembly and the caliper. Suspend the caliper with a length of wire; do not let it hang from the brake hose.

2. Pry off the dust cap. Tap out and discard the cotter pin. Remove the nut retainer.

3. Being careful not to drop the outer bearing, pull off the brake disc and wheel hub assembly.

4. Remove the inner grease seal using a prybar. Remove the inner wheel bearing.

5. Clean the wheel bearings with solvent and inspect them for pits, scratches and excessive wear. Wipe all the old grease from the hub and inspect the bearing races. If either bearings or races are damaged, the bearing races must be removed and the bearings and races replaced as an assembly.

6. If the bearings are to be replaced, drive out the races from the hub using a brass drift.

7. Make sure the spindle, hub and bearing assemblies are clean prior to installation.

To install:

8. If the bearing races were removed, install new ones using a suitable bearing race installer. Pack the bearings with a bearing packer. If a packer is not available, work as much grease as possible between the rollers and cages.

9. Coat the inner surface of the hub and bearing races with grease.

10. Install the inner bearing in the hub. Being careful not to distort it, install the oil seal with its lip facing the bearing. Drive the seal in until its outer edge is even with the edge of the hub. Lubricate the lip of the seal with grease.

11. Install the hub/disc assembly on the spindle, being careful not to damage the oil seal.

12. Install the outer bearing, washer and spindle nut. Install the caliper and the wheel and tire assembly and adjust the bearings.

WHEEL ALIGNMENT

The caster and camber are set at the factory and cannot be changed. Only the toe is adjustable.

TOE ADJUSTMENT

➧ SEE FIG. 13

Toe is the difference in width (distance), between the front and rear inside edges of the front tires.

1. Turn the steering wheel, from left to right, several times and center.

➡ **If car has power steering, start the engine before centering the steering wheel.**

2. Secure the centered steering wheel with a steering wheel holder, or any device that will keep it centered.
3. Release the tie rod end bellows clamps so the bellows will not twist while adjustment is made. Loosen the jam nuts on the tie rod ends. Adjust the left and right connector sleeves until each wheel has one-half of the desired toe setting.

FIG. 13 Toe adjustment on tie rod end

4. After the adjustment has been made, tighten the jam nuts and secure the bellows clamps. Release the steering wheel lock and check for steering wheel center. Readjust, if necessary until steering wheel is centered and toe is within specs.

WHEEL ALIGNMENT

			Caster		Camber			Steering
Year	**Model**		**Range (deg.)**	**Preferred Setting (deg.)**	**Range (deg.)**	**Preferred Setting (deg.)**	**Toe-in (in.)**	**Axis Inclination (deg.)**
1989	Mustang	Exc. 5.0L GT	1 5/32P–1 29/32P	1 3/16P	27/32N–21/32P	3/32N	1/16–5/16	15 23/32
		5.0L GT	1/2P–2P	1 9/32P	5/7N–29/32P	5/32P	1/16–5/16	15 23/32
1990	Mustang	Exc. 5.0L GT	1 5/32P–2 5/8P	1 29/32P	1/4N–1/4P	1/2N	1/4N–0	15 23/32
		5.0L GT	1 5/32P–2 5/8P	1 29/32P	1 3/8N–1/8P	5/8N	1/4N–0	15 23/32
1991	Mustang	Exc. 5.0L GT	1 5/32P–2 5/8P	1 29/32P	1/4N–1/4P	1/2N	1/4N–0	15 23/32
		5.0L GT	1 5/32P–2 5/8P	1 29/32P	1 3/8N–1/8P	5/8N	1/4N–0	15 23/32
1992	Mustang	Exc. 5.0L GT	1 5/32P–2 5/8P	1 29/32P	1/4N–1/4P	1/2N	1/4N–0	15 23/32
		5.0L GT	1 5/32P–2 5/8P	1 29/32P	1 3/8N–1/8P	5/8N	1/4N–0	15 23/32

P—Positive
N—Negative

REAR SUSPENSION

➧ SEE FIG. 14

Coil Springs

REMOVAL & INSTALLATION

➧ SEE FIG. 15

1. Raise and safely support the vehicle. Support the body at the rear body crossmember.
2. Remove the stabilizer bar, if equipped.
3. Support the axle with a suitable jack.
4. Place another jack under the lower arm axle pivot bolt. Remove and discard the bolt and nut. Lower the jack slowly until the coil spring load is relieved.
5. Remove the coil spring and insulator from the vehicle.

To install:

6. Place the upper spring insulator on top of the spring. Place the lower spring insulator on the lower arm.
7. Position the coil spring on the lower arm spring seat, so the pigtail on the lower arm is at the rear of the vehicle and pointing toward the left side of the vehicle.
8. Slowly raise the jack until the arm is in position. Insert a new rear pivot bolt and nut with the nut facing outward. Do not tighten at this time.
9. Raise the axle to curb height. Tighten the lower arm-to-axle pivot bolt to 70–100 ft. lbs. (95–135 Nm).
10. Install the stabilizer bar, if equipped. Remove the crossmember supports and lower the vehicle.

Shock Absorbers

TESTING

Bounce Test

Each shock absorber can be tested by bouncing the corner of the vehicle until maximum up and down movement is obtained. Let go of the vehicle. It should stop bouncing in 1–2 bounces. If not, the shock should be inspected for damage and possibly replaced.

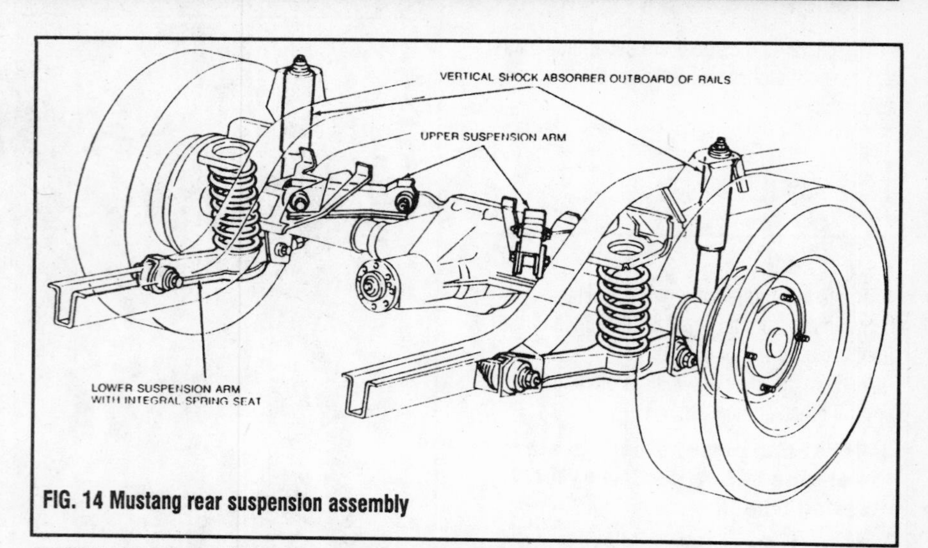

FIG. 14 Mustang rear suspension assembly

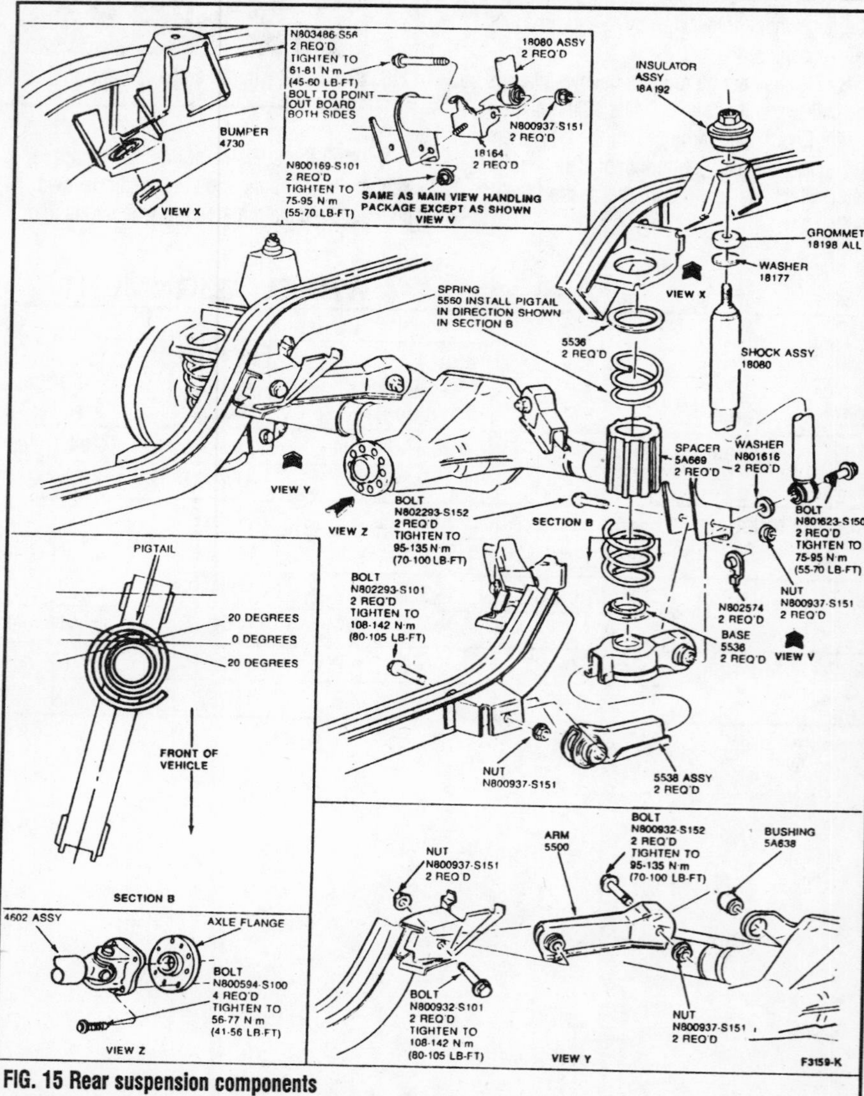

FIG. 15 Rear suspension components

Inspect the Shock Mounts

Check the shock mountings for worn or defective grommets, loose mounting nuts, interference or missing bump stops. If no apparent defects are noted, continue testing.

Inspecting Hydraulic Shocks for Leaks

Disconnect each shock lower mount and pull down on the shock until it is fully extended. inspect for leaks in the seal area. Shock absorber fluid is very thin and has a characteristic odor and dark brown color. Don't confuse the glossy paint on some shocks with leaking fluid. A slight trace of fluid is a normal condition; they are designed to seep a certain amount of fluid past the seals for lubrication. If you are in doubt as to whether the fluid on the shock is coming from the shock itself or from some other source, wipe the seal area clean and manually operate the shock (see the following procedure). Fluid will appear if the unit is leaking.

Manually Operating the Shocks

It may be necessary to fabricate a holding fixture for certain types of shock absorbers. If a suspected problem is in the front shocks, disconnect both front shock lower mountings.

Grip the lower end of the shock and pull down (rebound stroke) and then push up (compression stroke). The control arms will limit the movement of front shocks during the compression stroke. Compare the rebound resistance of both shocks and compare the compression resistance. Usually any shock showing a noticeable difference will be the one at fault.

If the shock has internal noises, extend the shock fully then exert an extra pull. If a small additional movement is felt, this usually means a loose piston and the shock should be replaced. Other noises that are cause for replacing shocks are a squeal after a full stroke in both directions, a clicking noise on fast reverse and a lag at reversal near mid-stroke.

REMOVAL & INSTALLATION

➧ SEE FIG. 15–16

1. Raise the vehicle and support it by the rear axle housing. Open the luggage compartment. On Mustang 3-door, open the hatch back door.
2. Remove the trim panels, as necessary, to gain access to the shock absorber. Remove the shock absorber retaining nut washer and insulator.
3. Remove the shock absorber bolt washer and nut at the lower arm and remove the shock absorber.

FIG. 16 Rear shock absorber attaching bolt is a Torx® bolt

➡ Vehicles are equipped with gas pressurized shock absorbers which will extend unassisted.

To install:

4. Prime the new shock absorber as follows:
   a. With the shock absorber right side up, extend it fully.
   b. Turn the shock upside down and fully compress it.
   c. Repeat the previous 2 steps at least 3 times to make sure any trapped air has been expelled.
5. Place the inner washer and insulator on the upper retaining stud and position the stud through the shock tower mounting hole.
6. Attach the lower end of the shock absorber with the retaining bolt and nut. Tighten the bolt to 55–70 ft. lbs. (75–95Nm).
7. Install the upper insulator, washer and retaining nut and tighten to 27 ft. lbs. (37–47 Nm).
8. Lower the vehicle.

Rear Control Arms

REMOVAL & INSTALLATION

➧ SEE FIG. 15

Upper Arm

➡ If 1 arm needs to be replaced, replace the other arm also.

1. Raise and safely support the vehicle at the rear crossmember.
2. Remove and discard the upper arm pivot bolts and nuts and remove the control arm.

To install:

3. Place the upper arm into the bracket of the body side rail. Install a new pivot bolt and nut with the nut facing outboard. Do not tighten at this time.
4. Using a jack, raise the suspension until the upper arm-to-axle pivot hole is in position with the hole in the axle bushing. Install a new pivot bolt and nut with the nut facing inboard.
5. Raise the suspension to curb height. Tighten the front upper arm bolt to 70–100 ft. lbs. (95–135 Nm) and the rear upper arm bolt to 80–105 ft. lbs. (108–142 Nm).
6. Remove the supports and lower the vehicle.

Lower Arm

➡ If 1 arm needs to be replaced, replace the other arm also.

1. Raise and safely support the vehicle at the rear crossmember.
2. Remove the stabilizer bar, if equipped.
3. Place a jack under the lower arm-to-axle pivot bolt. Remove and discard the bolt and nut. Lower the jack slowly until the coil spring can be removed.
4. Remove and discard the lower arm-to-frame pivot bolt and nut. Remove the lower arm.

To install:

5. Position the lower arm assembly into the front arm bracket. Install a new pivot bolt and nut with the nut facing outwards. Do not tighten at this time.
6. Position the coil spring on the lower arm spring seat, so the pigtail on the lower arm is at the rear of the vehicle and pointing toward the left side of the vehicle.
7. Slowly raise the jack until the arm is in position. Insert a new rear pivot bolt and nut with the nut facing outward. Do not tighten at this time.
8. Raise the axle to curb height. Tighten the lower arm front bolt to 80–105 ft. lbs. (108–142

Nm) and the rear bolt to 70–100 ft. lbs. (95–135 Nm).

9. Install the stabilizer bar, if equipped. Remove the crossmember supports and lower the vehicle.

Sway Bar

REMOVAL & INSTALLATION

➧ SEE FIG. 16

1. Raise and support the front end on jackstands.
2. Disconnect the stabilizer bar from the links, or the links from the lower arm.

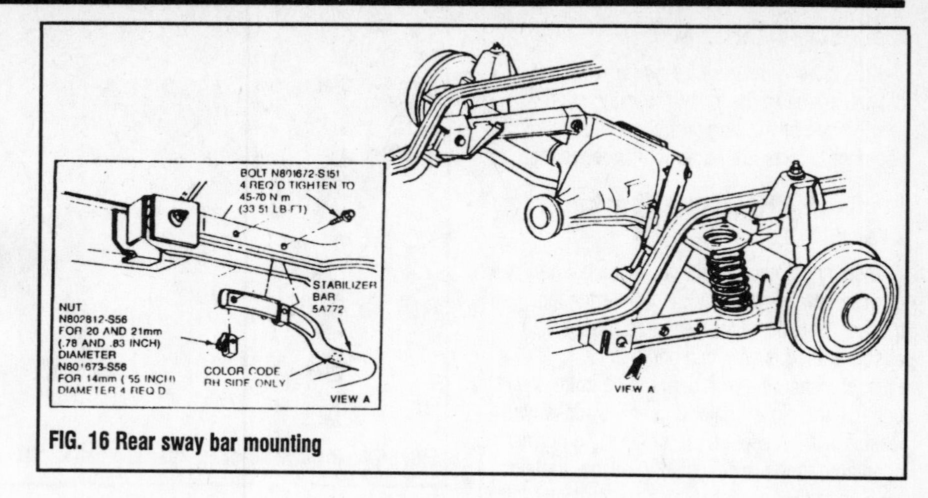

FIG. 16 Rear sway bar mounting

3. Disconnect the bar from the retaining clamps.
4. Installation is the reverse of removal. Torque the bar fasteners to 33–51 ft. lbs. (45–70Nm).

STEERING

Air Bag

DISARMING

1. Disconnect the negative battery cable and the backup power supply.

➡ The backup power supply allows air bag deployment if the battery or battery cables are damaged in an accident before the crash sensors close. The power supply is a capacitor that will leak down in approximately 15 minutes after the battery is disconnected or in 1 minute if the battery positive cable is grounded. The backup power supply must be disconnected before any air bag related service is performed.

2. Remove the 4 nut and washer assemblies retaining the driver air bag module to the steering wheel.
3. Disconnect the driver air bag module connector and attach a jumper wire to the air bag terminals on the clockspring.
4. If equipped with a passenger air bag, open the glove compartment and rotate all the way down, past the stops. Disconnect the passenger air bag connector and attach a jumper wire to the air bag terminals on the wiring harness side of the passenger air bag module connector.

Steering Wheel

**** CAUTION**

If equipped with an air bag, the negative battery cable and air bag backup power supply must be disconnected, before working on the system. Failure to do so may result in deployment of the air bag and possible personal injury.

REMOVAL & INSTALLATION

With Air Bag

➧ SEE FIG. 17

1. Center the front wheels to the straight ahead position.
2. Disconnect the negative battery cable and air bag backup power supply.

➡ The backup power supply allows air bag deployment if the battery or battery cables are damaged in an accident before the crash sensors close. The power supply is a capacitor that will leak down in approximately 15 minutes after the battery is disconnected or in 1 minute if the battery positive cable is grounded. The backup power supply must be disconnected before any air bag related service is performed.

3. Remove the 4 air bag module retaining nuts and lift the module off the steering wheel. Disconnect the electrical connector from the air bag module and remove the module.

**** CAUTION**

When carrying a live air bag, make sure the bag and trim cover are pointed away from the body. In the unlikely event of an accidental deployment, the bag will then deploy with minimal chance of injury. In addition, when placing a live air bag on a bench or other surface, always face the bag and trim cover up, away from the surface. This will reduce the motion of the air bag if it is accidentally deployed.

4. Disconnect the cruise control wire harness from the steering wheel, if equipped.

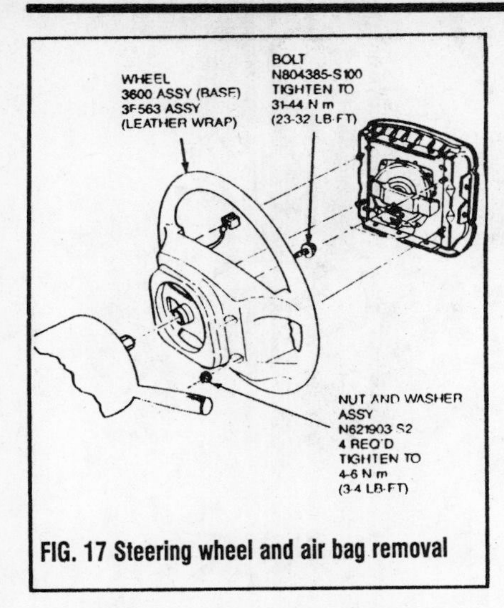

FIG. 17 Steering wheel and air bag removal

5. Remove and discard the steering wheel bolt. Remove the steering wheel using a suitable puller. Route the contact assembly wire harness through the steering wheel as the wheel is lifted off the shaft.

➡ **Do not use a knock-off type steering wheel puller or strike the retaining bolt with a hammer. This could cause damage to the steering shaft bearing.**

To install:

6. Make sure the front wheels are in the straight ahead position.

7. Route the contact assembly wire harness through the steering wheel opening at the 3 o'clock position and install the steering wheel on the steering shaft. The steering wheel and shaft alignment marks should be aligned. Make sure the air bag contact wire is not pinched.

8. Install a new steering wheel retaining bolt and tighten to 23–33 ft. lbs. (31–48 Nm).

9. If equipped, connect the cruise control wire harness to the wheel and snap the connector assembly into the steering wheel clip. Make sure the wiring does not get trapped between the steering wheel and contact assembly.

10. Connect the air bag wire harness to the air bag module and install the module to the steering wheel. Tighten the module retaining nuts to 3–4 ft. lbs. (4–6 Nm).

11. Connect the air bag backup power supply and negative battery cable. Verify the air bag warning indicator.

Without Air Bag

1. Disconnect the negative battery cable.

2. Remove the horn pad and cover assembly. Disconnect the horn electrical connector.

3. Disconnect the cruise control switch electrical connector, if equipped.

4. Remove and discard the steering wheel bolt. Remove the steering wheel using a suitable puller.

➡ **Do not use a knock-off type steering wheel puller or strike the retaining bolt with a hammer. This could cause damage to the steering shaft bearing.**

To install:

5. Align the index marks on the steering wheel and shaft and install the steering wheel.

6. Install a new steering wheel retaining bolt and tighten to 30 ft. lbs. (41 Nm).

7. Connect the cruise control electrical connector, if equipped.

8. Connect the horn electrical connector and install the horn pad and cover.

9. Connect the negative battery cable.

Combination Switch

The combination switch incorporates the turn signal, dimmer and wiper switch functions on the Mustang.

REMOVAL & INSTALLATION

➧ SEE FIG. 18

1. Disconnect the negative battery cable.

2. Remove the shroud retaining screws and remove the upper and lower shrouds.

3. Remove the switch retaining screws and lift up the switch assembly.

4. With the wiring connectors exposed, carefully lift the connector retainer tabs and disconnect the connectors.

5. Installation is the reverse of the removal procedure. Tighten the attaching bolts to 18–26 ft. lbs. (2–3Nm).

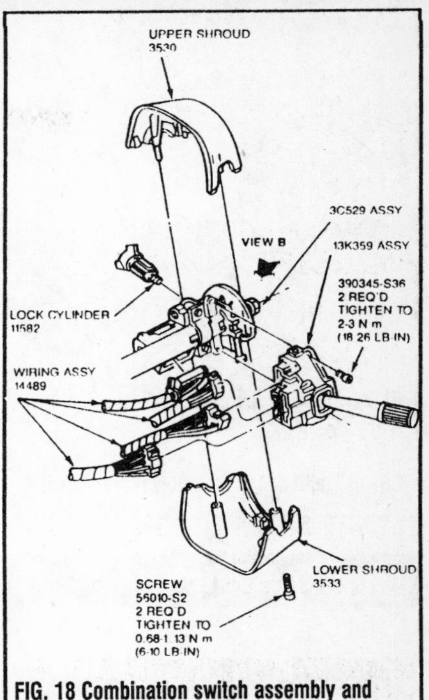

FIG. 18 Combination switch assembly and related parts

Ignition Lock

REMOVAL & INSTALLATION

➧ SEE FIG. 19

1. Disconnect the negative battery cable and, if equipped, the air bag backup power supply.

2. On the Mustang equipped with tilt column, remove the upper extension shroud by unsnapping the shroud retaining clip at the 9 o'clock position.

3. Remove the trim shroud halves by removing the attaching screws. Remove the electrical connector from the key warning switch.

4. Turn the ignition to the **RUN** position.

5. Place a 1/8 in. (3mm) diameter wire pin or small drift punch in the hole in the casting surrounding the lock cylinder and depress the retaining pin while pulling out on the lock cylinder to remove it from the column housing.

To install:

6. To install the lock cylinder, turn it to the **RUN** position and depress the retaining pin. Insert the lock cylinder into its housing in the lock cylinder casting.

7. Make sure that the cylinder is fully seated and aligned in the interlocking washer before turning the key to the **OFF** position. This action will permit the cylinder retaining pin to extend into the hole in the lock cylinder housing.

8. Using the ignition key, rotate the cylinder to ensure the correct mechanical operation in all positions.

9. Check for proper start in **P** or **N**. Also make sure that the start circuit cannot be actuated in **D** or **R** positions and that the column is locked in the **LOCK** position.

10. Connect the key warning buzzer electrical connector and install the trim shrouds, if required.

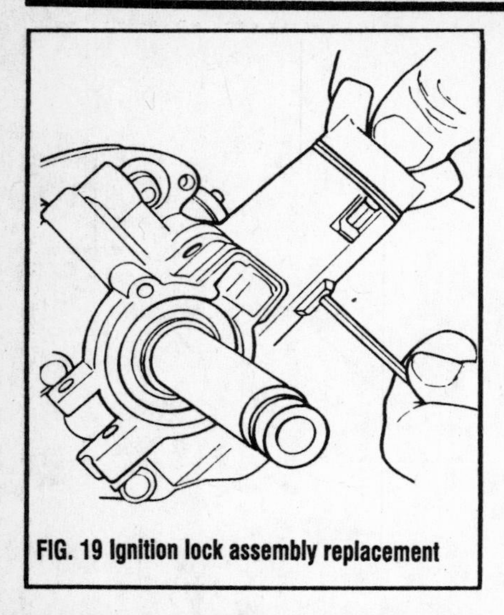

FIG. 19 Ignition lock assembly replacement

Ignition Switch

REMOVAL & INSTALLATION

1. Disconnect the negative battery cable.
2. On all vehicles with tilt column, remove the upper extension shroud by unsnapping the shroud from the retaining clips at the 9 o'clock position.
3. Remove the steering column shroud.
4. Disconnect the electrical connector from the ignition switch.
5. Rotate the ignition key lock cylinder to the **RUN** position.
6. Remove the 2 screws attaching the ignition switch.
7. Disengage the ignition switch from the actuator pin and remove the switch.

To install:

8. Adjust the new ignition switch by sliding the carrier to the **RUN** position.
9. Check to ensure that the ignition key lock cylinder is in the **RUN** position. The **RUN** position is achieved by rotating the key lock cylinder approximately 90 degrees from the **LOCK** position.
10. Install the ignition switch onto the actuator pin.
11. Align the switch mounting holes and install the attaching screws. Tighten the screws to 50–69 inch lbs. (5.6–7.9 Nm).
12. Connect the electrical connector to the ignition switch.
13. Connect the negative battery cable. Check the ignition switch for proper function in **START** and **ACC** positions. Make sure the column is locked in the **LOCK** position.
14. Install the remaining components in the reverse order of removal.

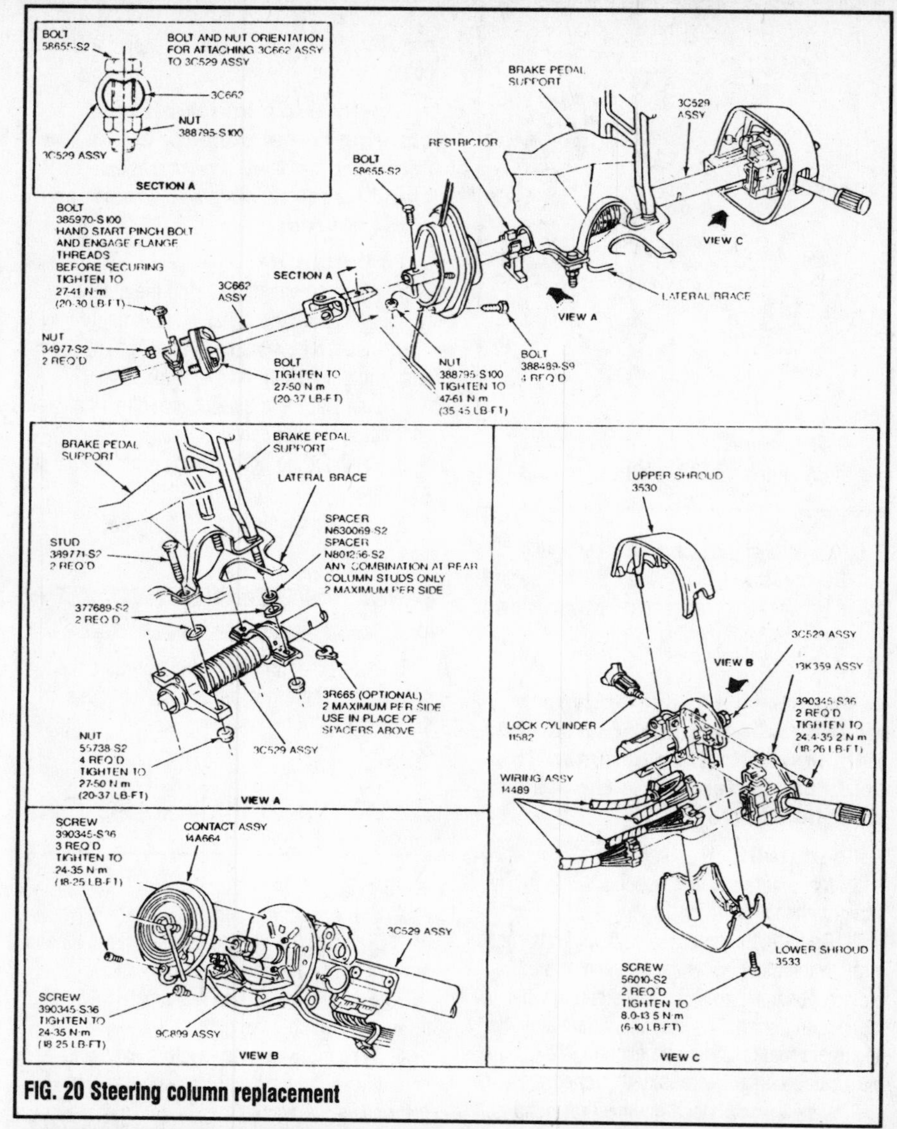

FIG. 20 Steering column replacement

Steering Column

REMOVAL & INSTALLATION

➧ SEE FIGS. 20–21

1. Disconnect the negative battery cable. If equipped with an air bag, disconnect the air bag backup power supply.

➡ **If equipped with an air bag, do not remove the steering column wheel and air bag module as an assembly unless the column is locked or the steering shaft is secured to keep it from turning. This will avoid damage to the clockspring assembly.**

2. Remove the 2 nuts that attach the flexible coupling to the flange on the steering input shaft. Disengage the safety strap and bolt assembly from the flexible coupling. If equipped with column shift, disconnect the transmission shift rod from the transmission control selector lever.
3. If equipped with column shift, remove the shift linkage grommet and replace with new using shift linkage insulator tool T67P–7341–A or equivalent.
4. Remove the steering wheel and the steering column trim shrouds.
5. Remove the steering column cover and hood release mechanism directly under the column.
6. Disconnect the electrical connectors to the steering column switches.
7. If equipped with column shift, loosen the 4 nuts holding the column to the brake pedal support, allowing the column to be lowered enough for access to the shift indicator lever and cable assembly.

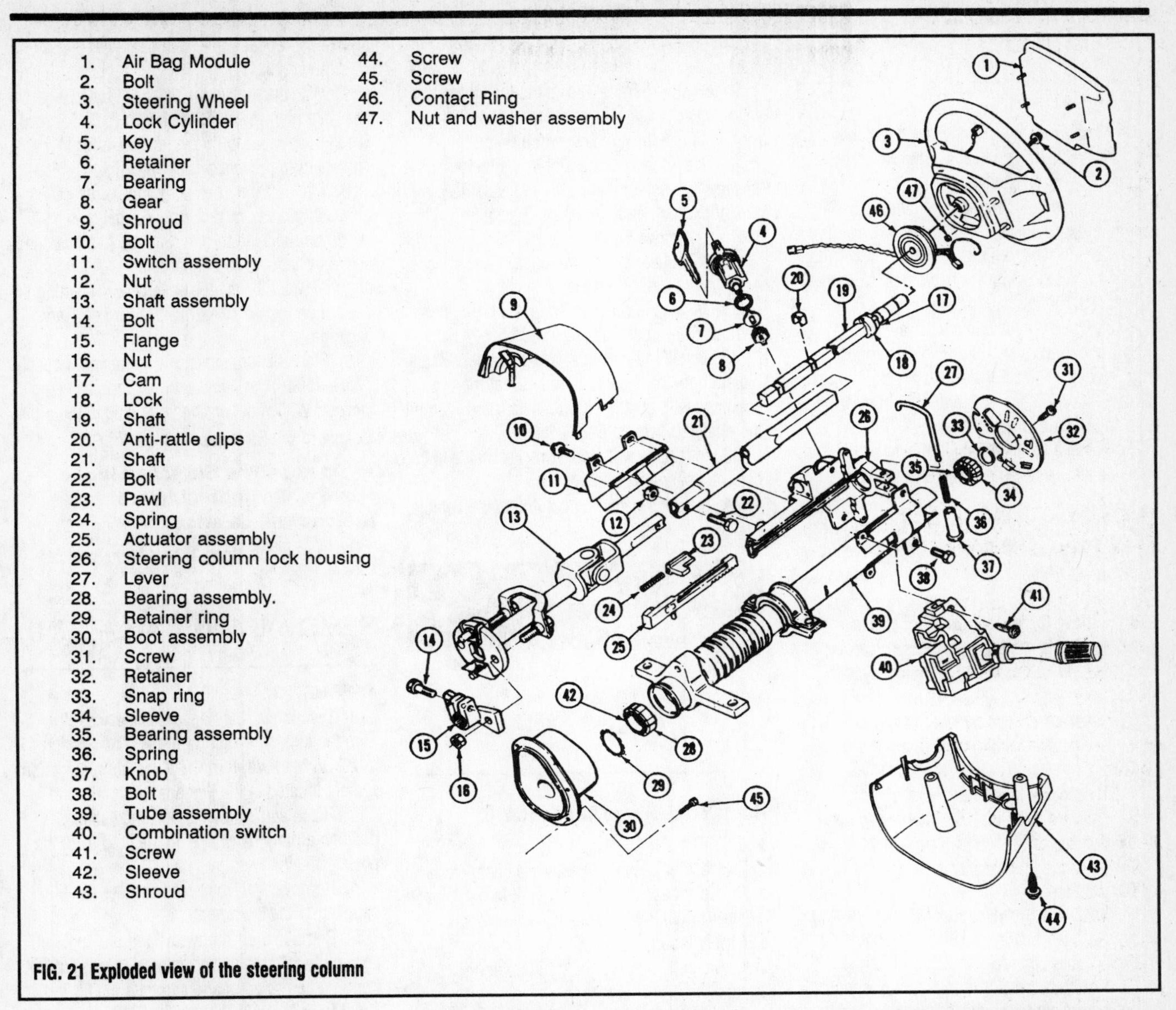

FIG. 21 Exploded view of the steering column

➡ **Be careful not to lower the column too far, so the plastic lever or cable is not damaged due to the weight of the column.**

8. If equipped with column shift, reach between the steering column and instrument panel and gently lift the shift indicator cable off the cleat on the shift indicator lever. Remove the shift indicator cable clamp from the steering column tube.

9. Remove the 4 screws that attach the dust boot to the dash panel.

10. Remove the 4 attaching nuts holding the column to the brake pedal support. Lower the column to clear the 4 mounting bolts and pull the column out, so the U-joint assembly will pass through the clearance hole in the dash panel.

To install:

11. Install the steering column by inserting the U-joint assembly through the opening in the dash panel. Be careful not to damage the column during installation.

12. Align the 4 bolts on the brake pedal support with the mounting holes on the column collar and bracket. If equipped with floor shift, attach the nuts and tighten to 20–37 ft. lbs. (27–50 Nm). On column shift, attach the nuts loosely, so the column will hang with a clearance between the column and instrument panel.

13. Connect the electrical connectors to the steering column switches.

14. Engage the safety strap and bolt assembly to the flange on the steering gear input shaft. Install the 2 nuts that attach the steering column lower shaft and U-joint assembly to the flange on the steering gear input shaft. Tighten the nuts to 20–37 ft. lbs. (27–50 Nm).

➡ **The safety strap must be properly positioned to prevent metal-to-metal contact after tightening the nuts. The flexible coupling must not be distorted when the nuts are tightened. Pry the steering shaft up or down with a suitable pry bar to achieve plus or minus 1/8 in. (3mm) coupling insulator flatness.**

15. Engage the dust boot at the base of the steering column to the dash panel opening. Install the 4 screws that attach the dust boot to the dash panel.

16. Install the steering wheel and the trim shrouds.

17. Install the hood release mechanism and steering column cover beneath the steering column.

18. Connect the negative battery cable. Check the steering column for proper operation.

DISASSEMBLY

➧ SEE FIG. 21

1. Disconnect the negative battery cable. Remove the steering wheel. Remove the steering column from the vehicle.
2. Remove the lower U-joint, spring, sensor ring and bushing.
3. Remove the turn signal canceling cam by pushing upward using a suitable tool. Note the direction of the flush surface.
4. Remove the ignition switch assembly. Remove the upper snapring and coil spring.
5. Remove the steel sleeve and ring. Remove the shift control assembly and shift control bracket, column shift vehicles.
6. Remove the shift cable bracket on column shift vehicles. Using a drift tap lock actuator cam pivot pin loose.
7. Remove the plastic bearing retainer from the lock cylinder bore. Remove the metal bearing from the lock cylinder bore. Remove the ignition lock gear.
8. Remove the two tilt pivot bolts. Use caution as the tilt spring will release when the bolts are removed. Remove the lock cylinder housing.
9. Remove the steering shaft from the column assembly. Remove the column lock actuator. Remove the lower bearing and mounting bracket.
10. Remove the tilt position lever arm pivot pin using a drift. Remove the lever lock arms and springs.

To install:

11. Install the steering shaft into the housing. Install the lower bearing and column mounting bracket. Tighten the screws to 5in-i8 ft. lbs.
12. Install the sensor ring, bushing, spring and flex coupling to the steering shaft. Tighten the pinch bolt to 29in-i41 ft. lbs.
13. Position the lock actuator assembly in the housing. Position the actuator cam in the lock housing and install the cam pivot pin. Be sure the pin is flush with the housing.
14. Install one tilt lever spring and arm into the housing. Install the outer lever spring and arm with the pivot pin. Tap the pin in place while driving out the drift.
15. Support the housing in a vise and drive the pin flush with the housing. Position the two nuts or spacers to hold the tilt lock arms away from the housing.
16. Position the tilt spring on the lock housing. With an assistant, install the lock housing and pivot bolts. Torque them to 14in-i20 ft. lbs. Lube the pivot bolts with grease before installation.
17. Install the steel sleeve and ring gear over the steering shaft and onto the upper bearing.
18. Install the spring and a new snapring on the top side of the spring using a 3/4 in. × 2 1/4 in. (19mm × 57mm) PVC pipe. Install the turn signal cancel cam, flush surface UP.
19. Install the ignition switch. Align the pin from the switch with the slot in the lock/column assembly. Position the slot in the assembly with the index mark on the casting. Torque the retaining screws 5in-i8 inch lbs.
20. Install the ignition lock gear. Coat the gear with grease before installation.
21. Install the metal bearing. Coat the metal bearing with grease before installation.
22. Install the plastic bearing retainer. Install the shift control tube assembly. Coat the bushings with grease prior to installation. Torque the retaining screws 5–8 ft. lbs.
23. Install the shift cable bracket on the lower column bearing assembly.
24. Install the steering column in the vehicle.

Power Steering Rack and Pinion

ADJUSTMENT

Rack Yoke Plug Clearance

➧ SEE FIG. 22

The rack yoke plug clearance adjustment is not a normal service adjustment. It is only required when the input shaft and valve assembly is removed.

1. Clean the exterior of the steering rack thoroughly.
2. Install 2 long bolts and washers through the bushings and attach the rack to bench mounted holding fixture T57L–500–B or equivalent.
3. Do not remove the external pressure lines, unless they are leaking or damaged. If the lines are removed, install new seals. If the lines are damaged, they must be replaced.
4. Drain the power steering fluid by rotating the input shaft lock-to-lock twice using pinion shaft torque adapter tool T74P–3504–R or equivalent. Cover the ports on the valve housing with a shop cloth while draining the gear to avoid possible oil spray.
5. Insert an inch pound torque wrench with a maximum capacity of 30–60 inch lbs. (3.39–6.77 Nm) into the pinion shaft torque adapter tool. Position the adapter and wrench on the input shaft splines.
6. Loosen the yoke plug locknut with pinion housing locknut wrench T78P–3504–H or equivalent. Loosen the yoke plug with a 3/4 in. socket wrench.
7. With the rack at the center of travel, tighten the yoke plug to 45–50 inch lbs. (5.0–5.6 Nm). Clean the threads of the yoke plug prior to tightening to prevent a false reading.
8. Back off the yoke plug approximately 1/8 turn, 44 degrees minimum to 54 degrees maximum until the torque required to initiate and sustain rotation of the input shaft is 7–18 inch lbs. (0.79–2.03 Nm) for base power steering or 7–24 inch lbs. (0.79–2.71 Nm) for handling package.
9. Place pinion housing yoke locknut wrench T78P–3504–H or equivalent, on the yoke plug locknut. While holding the yoke plug, tighten the locknut to 44–66 ft. lbs. (60–89 Nm).

➡ **Do not allow the yoke plug to move while tightening or the preload will be affected.**

10. Install the steering rack in the vehicle.

REMOVAL & INSTALLATION

➧ SEE FIG. 23

1. Disconnect the negative battery cable. Turn the ignition switch to the **RUN** position.
2. On Mark VII, turn **OFF** the air suspension switch, located in the luggage compartment.
3. Raise and safely support the vehicle. Position a drain pan to catch the fluid from the power steering lines.
4. Remove the 1 bolt retaining the flexible coupling to the input shaft.
5. Remove the front wheel and tire assemblies. Remove the cotter pins and nuts from the tie rod ends and separate the tie rod studs from the spindles.

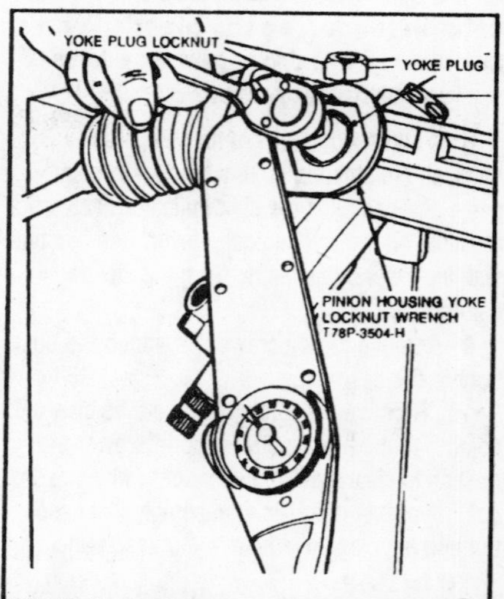

FIG. 22 Rack yoke plug clearance adjustment

6. Remove the 2 nuts, insulator washers and bolts retaining the steering rack to the crossmember. Remove the front rubber insulators.

7. Position the rack to allow access to the hydraulic lines and disconnect the lines.

8. Remove the steering rack.

To install:

9. Install new plastic seals on the hydraulic line fittings.

10. Install the rack on the mounting spikes and install the hydraulic lines. Tighten the fittings to 10–15 ft. lbs. (14–20 Nm) on 1989 vehicles and 20–25 ft. lbs. (27–33 Nm) on 1990–92 vehicles.

➡ The hoses are designed to swivel when properly tightened. Do not attempt to eliminate looseness by over-tightening the fittings.

11. Install the front rubber insulators. Make sure all rubber insulators are pushed completely inside the gear housing before installing the mounting bolts.

12. Insert the input shaft into the flexible coupling. Install the mounting bolts, insulator washers and nuts. Tighten the nuts to 30–40 ft. lbs. (41–54 Nm) while holding the bolts. Install and tighten the flexible coupling bolt to 20–30 ft. lbs. (28–40 Nm).

13. Connect the tie rod ends to the spindle arms and install the retaining nuts. Tighten to 35–47 ft. lbs. (48–63 Nm). After tightening, tighten the nuts to their nearest cotter pin castellation and install 2 new cotter pins.

14. Lower the vehicle. Turn the ignition switch to **OFF** and connect the negative battery cable.

15. Fill the power steering system with the proper type and quantity of fluid. If the tie rod ends were loosened, check and adjust the front end alignment.

FIG. 23 Power rack and pinion assembly

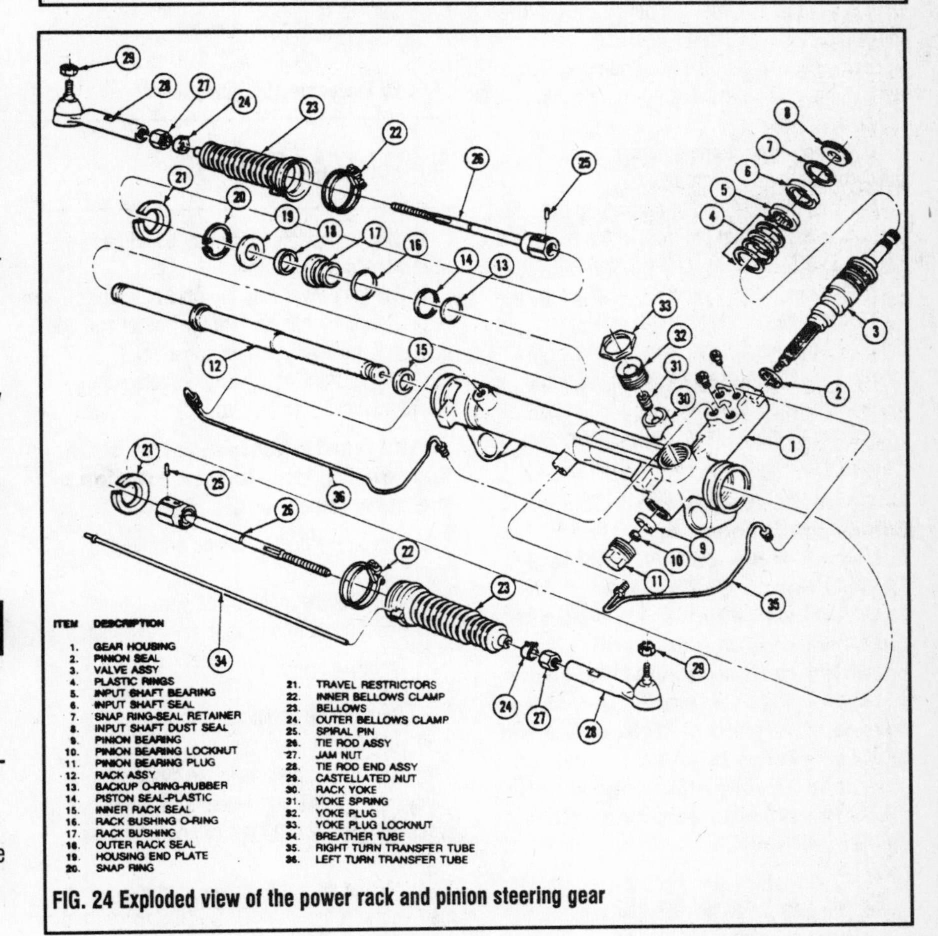

FIG. 24 Exploded view of the power rack and pinion steering gear

Tie Rod Ends

REMOVAL & INSTALLATION

➧ SEE FIG. 24

1. Raise and safely support the vehicle.

2. Remove the cotter pin and nut from the tie rod end ball stud. Disconnect the tie rod end from the spindle using ball stud remover tool 3290-D or equivalent.

3. Holding the tie rod end with a wrench, loosen the tie rod jam nut. Grip the tie rod end with pliers and remove the assembly from the tie rod, but first note the depth to which the tie rod was located by using the jam nut as a marker.

To install:

4. Clean the tie rod threads.

5. Thread the new tie rod end onto the tie rod to the same depth as the removed tie rod end.

6. Place the tie rod end ball stud into the spindle and install the nut. Make sure the front wheels are in the straight ahead position.

7. Tighten the nut to 35 ft. lbs. (48 Nm) and

continue tightening the nut to align the next castellation of the nut with the cotter pin hole in the stud. Install a new cotter pin.

8. Set the toe to specification. Tighten the jam nut to 35–50 ft. lbs. (48–68 Nm).

Power Steering Pump

REMOVAL & INSTALLATION

➧ SEE FIG. 25

1. Disconnect the negative battery cable.
2. Disconnect the fluid return hose at the reservoir and drain the fluid into a container.
3. Remove the pressure hose from the pump and, if necessary, drain the fluid into a container. Do not remove the fitting from the pump.
4. Remove the pump mounting bracket. Disconnect the belt from the pulley and remove the pump.
5. On engines with the fixed pump system, remove the pulley before removing the pump.

To install:

6. On non-fixed pump systems, install the pulley on the pump, if removed.
7. Place the pump on the mounting bracket and install the bolts at the front of the pump. Tighten to 30–45 ft. lbs. (40–62 Nm).
8. On fixed pump systems, install the pulley.
9. Place the belt on the pump pulley and adjust the tension, if necessary.
10. Install the pressure hose to the pump fitting. Tighten the tube nut with a tube nut wrench rather than with an open-end wrench. Tighten to 20–25 ft. lbs. (27–34 Nm).

➡ Do not overtighten this fitting. Swivel and/or end play of the fitting is normal and does not indicate a loose fitting. Over-tightening the tube nut can collapse the tube nut wall, resulting in a leak and requiring replacement of the entire pressure hose assembly. Use of an open-end wrench to tighten the nut can deform the tube nut hex which may result in improper torque and may make further servicing of the system difficult.

11. Connect the return hose to the pump and tighten the clamp. Fill the reservoir with the proper type and quantity of fluid.

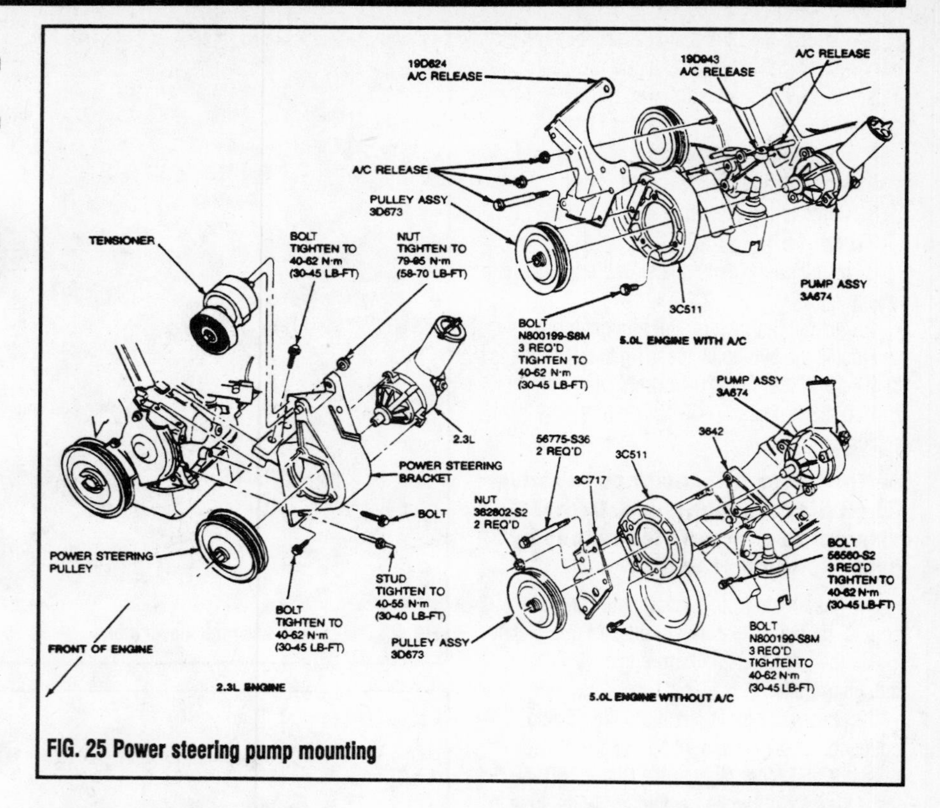

FIG. 25 Power steering pump mounting

SYSTEM BLEEDING

1. Disconnect the ignition coil and raise the front wheels off the floor.
2. Fill the power steering fluid reservoir.
3. Crank the engine with the starter and add fluid until the level remains constant.
4. While cranking the engine, rotate the steering wheel from lock-to-lock.

➡ The front wheels must be off the floor during lock-to-lock rotation of the steering wheel.

5. Check the fluid level and add fluid, if necessary.
6. Connect the ignition coil wire. Start the engine and allow it to run for several minutes.
7. Rotate the steering wheel from lock-to-lock.
8. Shut off the engine and check the fluid level. Add fluid, if necessary.
9. If air is still present in the system, purge the system of air using power steering pump air evacuator tool 021–00014 or equivalent, as follows:
    a. Make sure the power steering pump reservoir is full to the COLD FULL mark on the dipstick.
    b. Tightly insert the rubber stopper of the air evacuator assembly into the pump reservoir fill neck.
    c. Apply 15 in. Hg maximum vacuum on the pump reservoir for a minimum of 3 minutes with the engine idling. As air purges from the system, vacuum will fall off. Maintain adequate vacuum with the vacuum source.
    d. Release the vacuum and remove the vacuum source. Fill the reservoir to the COLD FULL mark.
    e. With the engine idling, apply 15 in. Hg vacuum to the pump reservoir. Slowly cycle the steering wheel from lock-to-lock every 30 seconds for approximately 5 minutes. Do not hold the steering wheel on the stops while cycling. Maintain adequate vacuum with the vacuum source as the air purges.
    f. Release the vacuum and remove the vacuum source. Fill the reservoir to the COLD FULL mark.
    g. Start the engine and cycle the steering wheel. Check for oil leaks at all connections. In severe cases of aeration, it may be necessary to repeat Steps 9b–9f.

Troubleshooting the Turn Signal Switch

Problem	Cause	Solution
Turn signal will not cancel	• Loose switch mounting screws • Switch or anchor bosses broken • Broken, missing or out of position detent, or cancelling spring	• Tighten screws • Replace switch • Reposition springs or replace switch as required
Turn signal difficult to operate	• Turn signal lever loose • Switch yoke broken or distorted • Loose or misplaced springs • Foreign parts and/or materials in switch • Switch mounted loosely	• Tighten mounting screws • Replace switch • Reposition springs or replace switch • Remove foreign parts and/or material • Tighten mounting screws
Turn signal will not indicate lane change	• Broken lane change pressure pad or spring hanger • Broken, missing or misplaced lane change spring • Jammed wires	• Replace switch • Replace or reposition as required • Loosen mounting screws, reposition wires and retighten screws
Turn signal will not stay in turn position	• Foreign material or loose parts impeding movement of switch yoke • Defective switch	• Remove material and/or parts • Replace switch
Hazard switch cannot be pulled out	• Foreign material between hazard support cancelling leg and yoke	• Remove foreign material. No foreign material impeding function of hazard switch—replace turn signal switch.
No turn signal lights	• Inoperative turn signal flasher • Defective or blown fuse • Loose chassis to column harness connector • Disconnect column to chassis connector. Connect new switch to chassis and operate switch by hand. If vehicle lights now operate normally, signal switch is inoperative • If vehicle lights do not operate, check chassis wiring for opens, grounds, etc.	• Replace turn signal flasher • Replace fuse • Connect securely • Replace signal switch • Repair chassis wiring as required

Troubleshooting the Turn Signal Switch (cont.)

Problem	Cause	Solution
Instrument panel turn indicator lights on but not flashing	• Burned out or damaged front or rear turn signal bulb	• Replace bulb
	• If vehicle lights do not operate, check light sockets for high resistance connections, the chassis wiring for opens, grounds, etc.	• Repair chassis wiring as required
	• Inoperative flasher	• Replace flasher
	• Loose chassis to column harness connection	• Connect securely
	• Inoperative turn signal switch	• Replace turn signal switch
	• To determine if turn signal switch is defective, substitute new switch into circuit and operate switch by hand. If the vehicle's lights operate normally, signal switch is inoperative.	• Replace turn signal switch
Stop light not on when turn indicated	• Loose column to chassis connection	• Connect securely
	• Disconnect column to chassis connector. Connect new switch into system without removing old.	• Replace signal switch
Stop light not on when turn indicated (cont.)	Operate switch by hand. If brake lights work with switch in the turn position, signal switch is defective.	
	• If brake lights do not work, check connector to stop light sockets for grounds, opens, etc.	• Repair connector to stop light circuits using service manual as guide
Turn indicator panel lights not flashing	• Burned out bulbs	• Replace bulbs
	• High resistance to ground at bulb socket	• Replace socket
	• Opens, ground in wiring harness from front turn signal bulb socket to indicator lights	• Locate and repair as required
Turn signal lights flash very slowly	• High resistance ground at light sockets	• Repair high resistance grounds at light sockets
	• Incorrect capacity turn signal flasher or bulb	• Replace turn signal flasher or bulb
	• If flashing rate is still extremely slow, check chassis wiring harness from the connector to light sockets for high resistance	• Locate and repair as required
	• Loose chassis to column harness connection	• Connect securely
	• Disconnect column to chassis connector. Connect new switch into system without removing old. Operate switch by hand. If flashing occurs at normal rate, the signal switch is defective.	• Replace turn signal switch

Troubleshooting the Turn Signal Switch (cont.)

Problem	Cause	Solution
Hazard signal lights will not flash—turn signal functions normally	• Blow fuse • Inoperative hazard warning flasher • Loose chassis-to-column harness connection • Disconnect column to chassis connector. Connect new switch into system without removing old. Depress the hazard warning lights. If they now work normally, turn signal switch is defective. • If lights do not flash, check wiring harness "K" lead for open between hazard flasher and connector. If open, fuse block is defective	• Replace fuse • Replace hazard warning flasher in fuse panel • Conect securely • Replace turn signal switch • Repair or replace brown wire or connector as required

Troubleshooting the Power Steering Pump

Problem	Cause	Solution
Chirp noise in steering pump	• Loose belt	• Adjust belt tension to specification
Belt squeal (particularly noticeable at full wheel travel and stand still parking)	• Loose belt	• Adjust belt tension to specification
Growl noise in steering pump	• Excessive back pressure in hoses or steering gear caused by restriction	• Locate restriction and correct. Replace part if necessary.
Growl noise in steering pump (particularly noticeable at stand still parking)	• Scored pressure plates, thrust plate or rotor • Extreme wear of cam ring	• Replace parts and flush system • Replace parts
Groan noise in steering pump	• Low oil level • Air in the oil. Poor pressure hose connection.	• Fill reservoir to proper level • Tighten connector to specified torque. Bleed system by operating steering from right to left—full turn.
Rattle noise in steering pump	• Vanes not installed properly • Vanes sticking in rotor slots	• Install properly • Free up by removing burrs, varnish, or dirt
Swish noise in steering pump	• Defective flow control valve	• Replace part
Whine noise in steering pump	• Pump shaft bearing scored	• Replace housing and shaft. Flush system.
Hard steering or lack of assist	• Loose pump belt • Low oil level in reservoir **NOTE:** Low oil level will also result in excessive pump noise • Steering gear to column misalignment	• Adjust belt tension to specification • Fill to proper level. If excessively low, check all lines and joints for evidence of external leakage. Tighten loose connectors. • Align steering column

Troubleshooting the Power Steering Pump (cont.)

Problem	Cause	Solution
Hard steering or lack of assist	• Lower coupling flange rubbing against steering gear adjuster plug • Tires not properly inflated	• Loosen pinch bolt and assemble properly • Inflate to recommended pressure
Foaming milky power steering fluid, low fluid level and possible low pressure	• Air in the fluid, and loss of fluid due to internal pump leakage causing overflow	• Check for leaks and correct. Bleed system. Extremely cold temperatures will cause system aeriation should the oil level be low. If oil level is correct and pump still foams, remove pump from vehicle and separate reservoir from body. Check welsh plug and body for cracks. If plug is loose or body is cracked, replace body.
Low pump pressure	• Flow control valve stuck or inoperative • Pressure plate not flat against cam ring	• Remove burrs or dirt or replace. Flush system. • Correct
Momentary increase in effort when turning wheel fast to right or left	• Low oil level in pump • Pump belt slipping • High internal leakage	• Add power steering fluid as required • Tighten or replace belt • Check pump pressure. (See pressure test)
Steering wheel surges or jerks when turning with engine running especially during parking	• Low oil level • Loose pump belt • Steering linkage hitting engine oil pan at full turn • Insufficient pump pressure	• Fill as required • Adjust tension to specification • Correct clearance • Check pump pressure. (See pressure test). Replace flow control valve if defective.
Steering wheel surges or jerks when turning with engine running especially during parking (cont.)	• Sticking flow control valve	• Inspect for varnish or damage, replace if necessary
Excessive wheel kickback or loose steering	• Air in system	• Add oil to pump reservoir and bleed by operating steering. Check hose connectors for proper torque and adjust as required.
Low pump pressure	• Extreme wear of cam ring • Scored pressure plate, thrust plate, or rotor • Vanes not installed properly • Vanes sticking in rotor slots • Cracked or broken thrust or pressure plate	• Replace parts. Flush system. • Replace parts. Flush system. • Install properly • Freeup by removing burrs, varnish, or dirt • Replace part

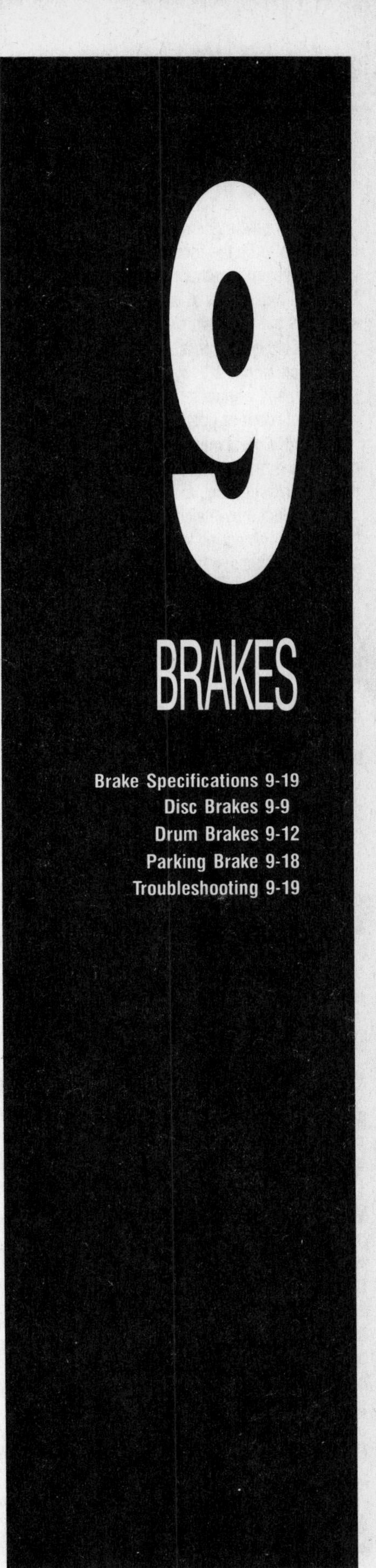

9 BRAKES

BRAKE OPERATING SYSTEM

The hydraulic system transports the power required to force the frictional surfaces of the braking system together from the pedal to the individual brake units at each wheel. A hydraulic system is used for two reasons.

First, fluid under pressure can be carried to all parts of an automobile by small pipes and flexible hoses without taking up a significant amount of room or posing routing problems.

Second, a great mechanical advantage can be given to the brake pedal end of the system, and the foot pressure required to actuate the brakes can be reduced by making the surface area of the master cylinder pistons smaller than that of any of the pistons in the wheel cylinders or calipers.

The master cylinder consists of a fluid reservoir and a double cylinder and piston assembly. Double type master cylinders are designed to separate the front and rear braking systems hydraulically in case of a leak.

Steel lines carry the brake fluid to a point on the vehicle's frame near each of the vehicle's wheels. The fluid is then carried to the calipers and wheel cylinders by flexible tubes and steel lines in order to allow for suspension and steering movements.

In drum brake systems, each wheel cylinder contains two pistons, one at either end, which push outward in opposite directions.

In disc brake systems, the cylinders are part of the calipers. The cylinders are used to force the brake pads against the disc.

All pistons employ some type of seal, usually made of rubber, to minimize fluid leakage. A rubber dust boot seals the outer end of the cylinder against dust and dirt. The boot fits around the outer end of the piston on disc brake calipers, and around the brake actuating rod on wheel cylinders.

The hydraulic system operates as follows: When at rest, the entire system, from the piston(s) in the master cylinder to those in the wheel cylinders or calipers, is full of brake fluid. Upon application of the brake pedal, fluid trapped in front of the master cylinder piston(s) is forced through the lines to the wheel cylinders. Here, it forces the pistons outward, in the case of drum brakes, and inward toward the disc, in the case of disc brakes. The motion of the pistons is opposed by return springs mounted outside the cylinders in drum brakes, and by spring seals, in disc brakes.

Upon release of the brake pedal, a spring located inside the master cylinder immediately returns the master cylinder pistons to the normal position. The pistons contain check valves and the master cylinder has compensating ports drilled in it. These are uncovered as the pistons reach their normal position. The piston check valves allow fluid to flow toward the wheel cylinders or calipers as the pistons withdraw. Then, as the return springs force the brake pads or shoes into the released position, the excess fluid reservoir through the compensating ports. It is during the time the pedal is in the released position that any fluid that has leaked out of the system will be replaced through the compensating ports.

Dual circuit master cylinders employ two pistons, located one behind the other, in the same cylinder. The primary piston is actuated directly by mechanical linkage from the brake pedal through the power booster. The secondary piston is actuated by fluid trapped between the two pistons. If a leak develops in front of the secondary piston, it moves forward until it bottoms against the front of the master cylinder, and the fluid trapped between the pistons will operate the rear brakes. If the rear brakes develop a leak, the primary piston will move forward until direct contact with the secondary piston takes place, and it will force the secondary piston to actuate the front brakes. In either case, the brake pedal moves farther when the brakes are applied, and less braking power is available.

All dual circuit systems use a switch to warn the driver when only half of the brake system is operational. This switch is located in a valve body which is mounted on the firewall or the frame below the master cylinder. A hydraulic piston receives pressure from both circuits, each circuit's pressure being applied to one end of the piston. When the pressures are in balance, the piston remains stationary. When one circuit has a leak, however, the greater pressure in that circuit during application of the brakes will push the piston to one side, closing the switch and activating the brake warning light.

In disc brake systems, this valve body also contains a metering valve and, in some cases, a proportioning valve. The metering valve keeps pressure from traveling to the disc brakes on the front wheels until the brake shoes on the rear wheels have contacted the drums, ensuring that the front brakes will never be used alone. The proportioning valve controls the pressure to the rear brakes to lessen the chance of rear wheel lock-up during very hard braking.

Warning lights may be tested by depressing the brake pedal and holding it while opening one of the wheel cylinder bleeder screws. If this does not cause the light to go on, substitute a new lamp, make continuity checks, and, finally, replace the switch as necessary.

The hydraulic system may be checked for leaks by applying pressure to the pedal gradually and steadily. If the pedal sinks very slowly to the floor, the system has a leak. This is not to be confused with a springy or spongy feel due to the compression of air within the lines. If the system leaks, there will be a gradual change in the position of the pedal with a constant pressure.

Check for leaks along all lines and at wheel cylinders. If no external leaks are apparent, the problem is inside the master cylinder.

Disc Brakes

BASIC OPERATING PRINCIPLES

Instead of the traditional expanding brakes that press outward against a circular drum, disc brake systems utilize a disc (rotor) with brake pads positioned on either side of it. Braking effect is achieved in a manner similar to the way you would squeeze a spinning phonograph record between your fingers. The disc (rotor) is a casting with cooling fins between the two braking surfaces. This enables air to circulate between the braking surfaces making them less sensitive to heat buildup and more resistant to fade. Dirt and water do not affect braking action since contaminants are thrown off by the centrifugal action of the rotor or scraped off the by the pads. Also, the equal clamping action of the two brake pads tends to ensure uniform, straight line stops. Disc brakes are inherently self-adjusting.

There are three general types of disc brake:

1. A fixed caliper.
2. A floating caliper.
3. A sliding caliper.

The fixed caliper design uses two pistons mounted on either side of the rotor (in each side of the caliper). The caliper is mounted rigidly and does not move.

The sliding and floating designs are quite similar. In fact, these two types are often lumped together. In both designs, the pad on the inside of the rotor is moved into contact with the rotor by hydraulic force. The caliper, which is not held in a fixed position, moves slightly, bringing the outside pad into contact with the rotor. There are various methods of attaching floating calipers. Some pivot at the bottom or top, and some slide on mounting bolts. In any event, the end result is the same.

All the cars covered in this book employ the sliding caliper design.

Drum Brakes

BASIC OPERATING PRINCIPLES

➧ SEE FIG. 1,2

Drum brakes employ two brake shoes mounted on a stationary backing plate. These shoes are positioned inside a circular drum which rotates with the wheel assembly. The shoes are held in place by springs. This allows them to slide toward the drums (when they are applied) while keeping the linings and drums in alignment. The shoes are actuated by a wheel cylinder which is mounted at the top of the backing plate. When the brakes are applied, hydraulic pressure forces the wheel cylinder's actuating links outward. Since these links bear directly against the top of the brake shoes, the tops of the shoes are then forced against the inner side of the drum. This action forces the bottoms of the two shoes to contact the brake drum by rotating the entire assembly slightly (known as servo action). When pressure within the wheel cylinder is relaxed, return springs pull the shoes back away from the drum.

Most modern drum brakes are designed to self-adjust themselves during application when the vehicle is moving in reverse. This motion causes both shoes to rotate very slightly with the drum, rocking an adjusting lever, thereby causing rotation of the adjusting screw.

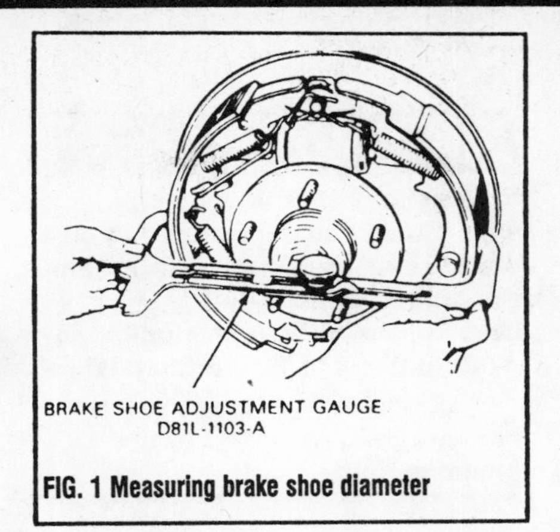

FIG. 1 Measuring brake shoe diameter

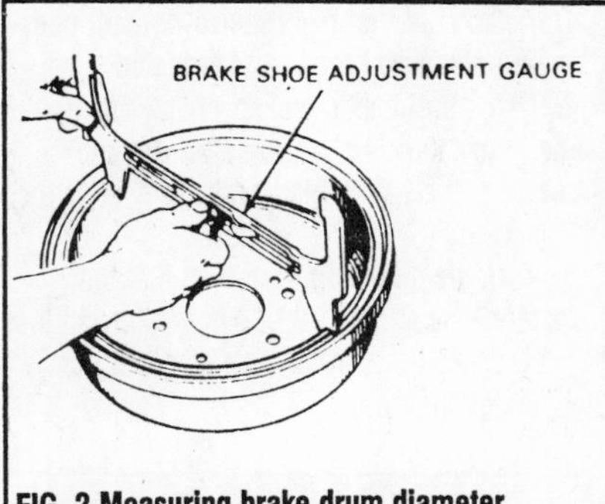

FIG. 2 Measuring brake drum diameter

Power Boosters

Power brakes operate just as non-power brake systems except in the actuation of the master cylinder pistons. A vacuum diaphragm is located on the front of the master cylinder and assists the driver in applying the brakes, reducing both the effort and travel he must put into moving the brake pedal.

The vacuum diaphragm housing is connected to the intake manifold by a vacuum hose. A check valve is placed at the point where the hose enters the diaphragm housing, so that during periods of low manifold vacuum brake assist vacuum will not be lost.

Depressing the brake pedal closes off the vacuum source and allows atmospheric pressure to enter on one side of the diaphragm. This causes the master cylinder pistons to move and apply the brakes. When the brake pedal is released, vacuum is applied to both sides of the diaphragm, and return springs return the diaphragm and master cylinder pistons to the released position. If the vacuum fails, the brake pedal rod will butt against the end of the master cylinder actuating rod, and direct mechanical application will occur as the pedal is depressed.

The hydraulic and mechanical problems that apply to conventional brake systems also apply to power brakes, and should be checked for if the tests below do not reveal the problem.

Test for a system vacuum leak as described below:

1. Operate the engine at idle without touching the brake pedal for at least one minute.
2. Turn off the engine, and wait one minute.
3. Test for the presence of assist vacuum by depressing the brake pedal and releasing it several times. Light application will produce less and less pedal travel, if vacuum was present. If there is no vacuum, air is leaking into the system somewhere.

Test for system operation as follows:

1. Pump the brake pedal (with engine off) until the supply vacuum is entirely gone.
2. Put a light, steady pressure on the pedal.
3. Start the engine, and operate it at idle. If the system is operating, the brake pedal should fall toward the floor if constant pressure is maintained on the pedal.

Power brake systems may be tested for hydraulic leaks just as ordinary systems are tested.

BRAKE SYSTEM

⁂ WARNING

Clean, high quality brake fluid is essential to the safe and proper operation of the brake system. You should always buy the highest quality brake fluid that is available. If the brake fluid becomes contaminated, drain and flush the system and fill the master cylinder with new fluid. Never reuse any brake fluid. Any brake fluid that is removed from the system should be discarded.

Adjustments

DRUM BRAKES

➧ SEE FIG. 3

The drum brakes are self-adjusting and require a manual adjustment only after the brake

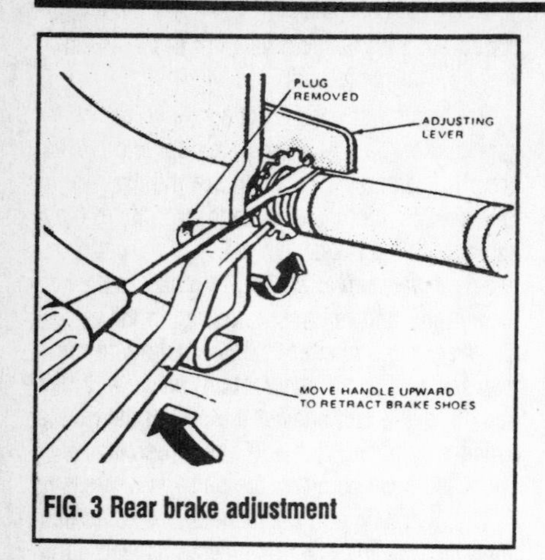

FIG. 3 Rear brake adjustment

shoes have been replaced, or when the length of the adjusting screw has been changed while performing some other service operation, as i.e., taking off brake drums.

To adjust the brakes, follow the procedures given below:

Drum Installed

1. Raise and support the rear end on jackstands.
2. Remove the rubber plug from the adjusting slot on the backing plate.
3. Insert a brake adjusting spoon into the slot and engage the lowest possible tooth on the starwheel. Move the end of the brake spoon downward to move the starwheel upward and expand the adjusting screw. Repeat this operation until the brakes lock the wheels.
4. Insert a small screwdriver or piece of firm wire (coat hanger wire) into the adjusting slot and push the automatic adjusting lever out and free of the starwheel on the adjusting screw and hold it there.
5. Engage the topmost tooth possible on the starwheel with the brake adjusting spoon. Move the end of the adjusting spoon upward to move the adjusting screw starwheel downward and contract the adjusting screw. Back off the adjusting screw starwheel until the wheel spins freely with a minimum of drag. Keep track of the number of turns that the starwheel is backed off, or the number of strokes taken with the brake adjusting spoon.
6. Repeat this operation for the other side. When backing off the brakes on the other side, the starwheel adjuster must be backed off the same number of turns to prevent side-to-side brake pull.
7. When the brakes are adjusted make several stops while backing the vehicle, to equalize the brakes at both of the wheels.
8. Remove the safety stands and lower the vehicle. Road test the vehicle.

Drum Removed

CAUTION

Brake shoes contain asbestos, which has been determined to be a cancer causing agent. Never clean the brake surfaces with compressed air! Avoid inhaling any dust from any brake surface! When cleaning brake surfaces, use a commercially available brake cleaning fluid.

1. Make sure that the shoe-to-contact pad areas are clean and properly lubricated.
2. Using and inside caliper check the inside diameter of the drum. Measure across the diameter of the assembled brake shoes, at their widest point.
3. Turn the adjusting screw so that the diameter of the shoes is 0.030 in. (0.76mm) less than the brake drum inner diameter.
4. Install the drum.

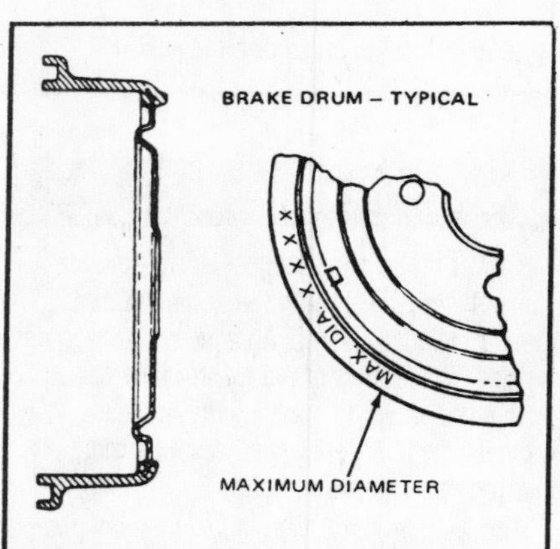

FIG. 4 Brake drum maximum diameter location

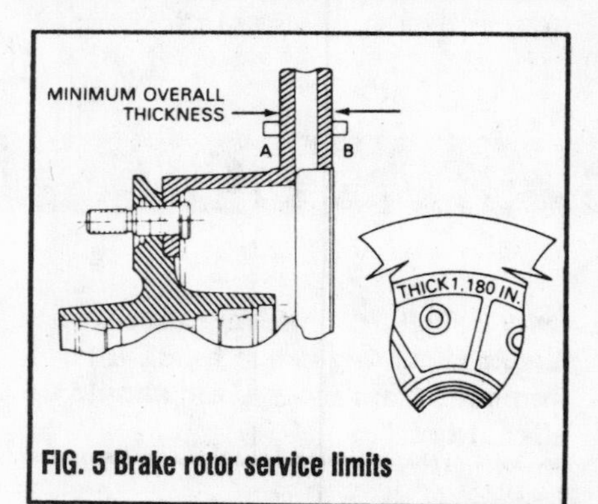

FIG. 5 Brake rotor service limits

Stoplight Switch

REMOVAL & INSTALLATION

➧ SEE FIG. 6

1. Disconnect the negative battery cable.
2. Disconnect the wire harness at the connector from the switch. The locking tab on the connector must be lifted before the connector can be removed.
3. Remove the hairpin retainer, slide the stoplight switch, the pushrod and the nylon washers and bushings away from the pedal and remove the switch.

➡ **Since the switch side plate nearest the brake pedal is slotted, it is not necessary to remove the brake master cylinder pushrod and 1 washer from the brake pedal pin. To install:**

4. Position the switch so the U-shaped side is nearest the pedal and directly over/under the pin. Then slide the switch down/up trapping the master cylinder pushrod and black bushing between the switch side plates. Push the switch and pushrod assembly firmly toward the brake pedal arm. Assemble the outside white plastic washer to the pin and install the hairpin retainer to trap the whole assembly.
5. Assemble the wire harness connector to the switch. Check the switch for proper operation.

➡ **The stoplight switch wire harness must be long enough to travel with the switch during full pedal stroke. If wire length is insufficient, reroute the harness or service, as required.**

Brake Pedal

REMOVAL & INSTALLATION

➧ SEE FIG. 6

1. Remove the brake light switch.
2. Slide the pushrod and spacer from the pedal pin.
3. If the vehicle is equipped with speed control, you can leave the speed control bracket in place.

4a. On vehicles with manual transmission:

a. Disconnect the clutch cable from the clutch pedal.

b. Remove the pedal and bushings.

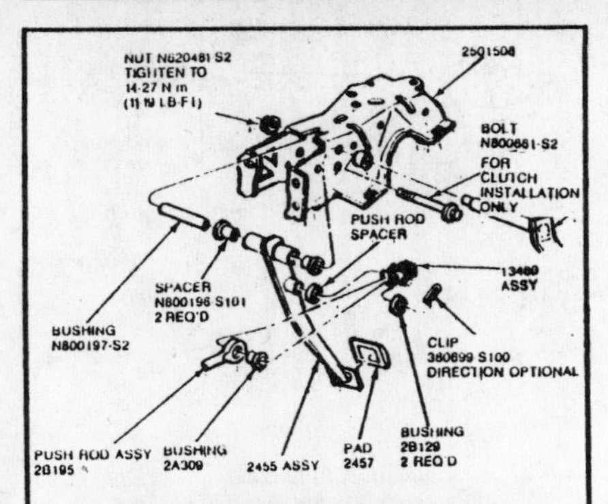
FIG. 6 Brake pedal assembly and related components

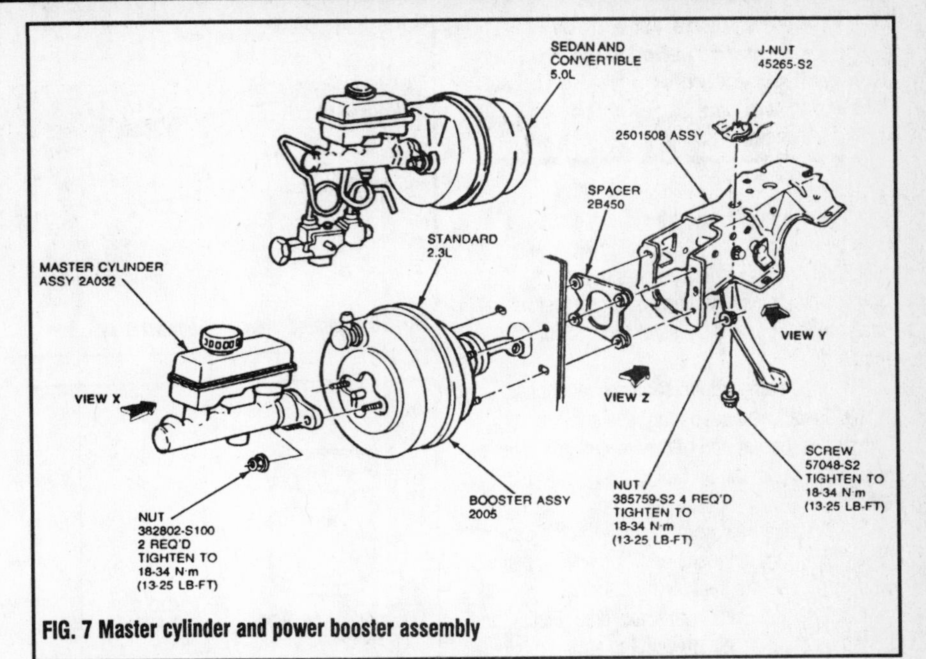
FIG. 7 Master cylinder and power booster assembly

4b. On vehicles with automatic transmission:

a. Remove the spring retainer and bushing from the brake pedal shaft.

b. From the other end, pull out the shaft and remove the pedal.

c. Remove the bushings and spring washer from the pedal.

To install:

5a. On vehicles with automatic transmission:

a. Install the bushings and spring washer from the pedal.

b. Install the pedal.

c. Install the spring retainer and bushing on the brake pedal shaft.

5b. On vehicles with manual transmission:

a. Install the pedal and bushings.

b. Reposition the clutch pedal.

c. Install the clutch cable.

6. Slide the pushrod and spacer on the pedal pin.

7. Install the brake light switch.

Master Cylinder

REMOVAL & INSTALLATION

➧ SEE FIG. 7

1. Disconnect the negative battery cable.
2. Remove the brake lines from the primary and secondary outlet ports of the master cylinder.
3. Disconnect the brake warning indicator connector.
4. Remove the nuts attaching master cylinder to the brake booster assembly.
5. Slide the master cylinder forward and upward from the vehicle.

To install:

6. Before installation, bench bleed the new master cylinder as follows:

a. Mount the new master cylinder in a holding fixture. Be careful not to damage the housing.

b. Fill the master cylinder reservoir with brake fluid.

c. Using a suitable tool inserted into the booster pushrod cavity, push the master cylinder piston in slowly. Place a suitable container under the master cylinder to catch the fluid being expelled from the outlet ports.

d. Place a finger tightly over each outlet port and allow the master cylinder piston to return.

e. Repeat the procedure until clear fluid only is expelled from the master cylinder. Plug the outlet ports and remove the master cylinder from the holding fixture.

7. Mount the master cylinder on the booster.
8. Attach the brake fluid lines to the master cylinder.
9. Connect the brake warning indicator switch connector.
10. Bleed the system. Operate the brakes several times, then check for external hydraulic leaks.

OVERHAUL

➧ SEE FIGS. 8–10

The most important thing to remember when rebuilding the master cylinder is cleanliness. Work in clean surroundings with clean tools and clean cloths or paper for drying purposes. Have plenty of clean alcohol and brake fluid on hand to clean and lubricate the internal components. There are service repair kits available for overhauling the master cylinder.

1. Remove the master cylinder from the vehicle and drain the brake fluid.
2. Using a large screwdriver, pry the reservoir off the master cylinder.
3. Mount the cylinder in a vise so that the outlets are up then remove the seal from the hub.
4. Remove the proportioning valve from the master cylinder.
5. Remove the stopscrew from the bottom of the master cylinder.
6. Depress the primary piston and remove the snapring from the rear of the bore.
7. Remove the secondary piston assembly using compressed air. Cover the bore opening with a cloth to prevent damage to the piston.
8. Using compressed air in the outlet port at the blind end and plugging the other port, remove the primary piston.
9. Clean metal parts in brake fluid and discard the rubber parts.
10. Inspect the bore for damage or wear, and check the pistons for damage and proper clearance in the bore.

✲✲ CAUTION

DO NOT HONE THE CYLINDER BORE! If the bore is pitted or scored deeply, the master cylinder assembly must be replaced. If any evidence of contamination exist in

the master cylinder, the entire hydraulic system should be flushed and refilled with clean brake fluid. Blow out the passages with compressed air.

11. If the master cylinder is not damaged, it may be serviced with a rebuilding kit. The rebuilding kit may contain secondary and primary piston assemblies instead of just rubber seals. In this case, seal installation is not required.
12. Clean all parts in isopropyl alcohol.
13. Install new secondary seals in the two grooves in the flat end of the front piston. The lips of the seals will be facing away from each other.
14. Install a new primary seal and the seal protector on the opposite end of the front piston with the lips of the seal facing outward.
15. Coat the seals with brake fluid. Install the spring on the front piston with the spring retainer in the primary seal.
16. Insert the piston assembly, spring end first, into the bore and use a wooden rod to seat it.
17. Coat the rear piston seals with brake fluid and install them into the piston grooves with the lips facing the spring end.
18. Assemble the spring onto the piston and install the assembly into the bore spring first. Install the snapring.
19. Hold the piston train at the bottom of the bore and install the stopscrew. Install a new seal on the hub.

➡ **Whenever the reservoir or master cylinder is replaced, new reservoir grommets should be used.**

20. Coat the new grommet with clean brake fluid and insert them into the master cylinder. Bench-bleed the cylinder or install and bleed the cylinder on the car.
21. Press the reservoir into place. A snap should be felt, indicating that the reservoir is properly positioned.

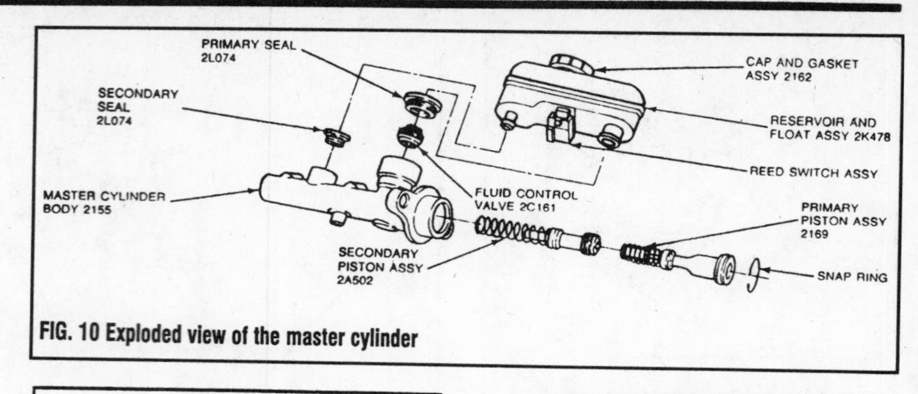

FIG. 10 Exploded view of the master cylinder

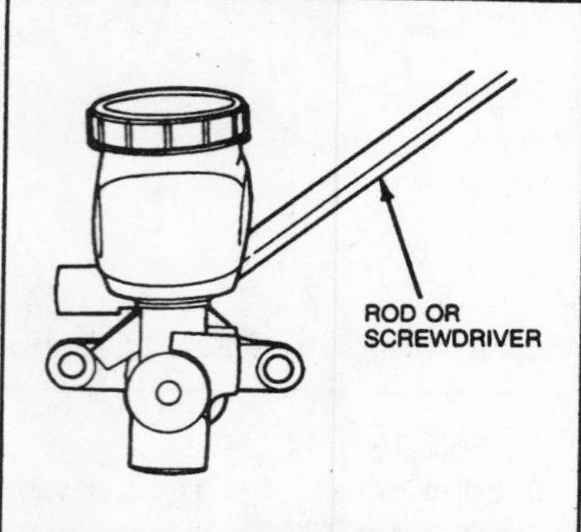

FIG. 8 Removing master cylinder reservoir

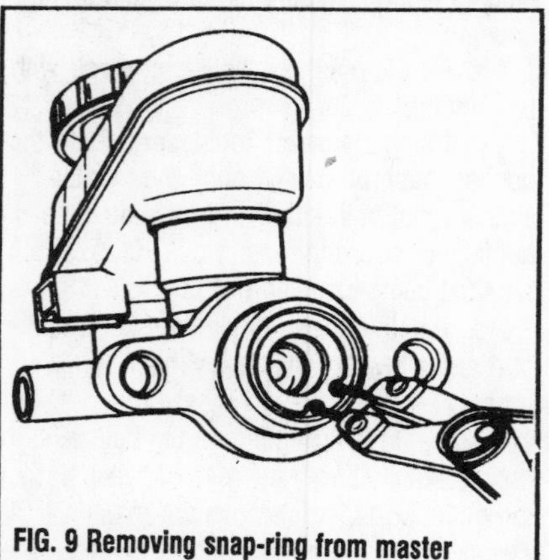

FIG. 9 Removing snap-ring from master cylinder

Pressure Differential Valve

REMOVAL & INSTALLATION

1. Disconnect the electrical leads from the valve.
2. Unscrew the valve from the master cylinder.
3. Install the valve in the reverse order of removal.
4. Bleed the master cylinder.

Power Brake Booster

REMOVAL & INSTALLATION

➧ SEE FIG. 7

1. Disconnect the negative battery cable. Remove the air cleaner.
2. Remove the cruise control actuator cable and cruise control servo.
3. On Mustang equipped with the 2.3L engine, perform the following:
   a. Relieve the fuel system pressure.
   b. Disconnect the accelerator cable from the throttle body. Remove the screw that secures the accelerator cable to the accelerator shaft bracket and remove the cable from the bracket.
   c. Remove the screws that secure the accelerator shaft bracket to the manifold and rotate the bracket toward the engine. Remove the horn.
   d. Disconnect the 2 manifold injector connectors located near the oil dipstick retaining bracket. Disconnect the 2 fuel hoses to the fuel supply manifold.
   e. Remove the 3 bolts holding the oil dipstick bracket to the upper intake manifold. Remove the dipstick and bracket.
   f. Remove the windshield wiper motor and remove the vacuum hoses directly over the brake booster at the dash panel vacuum tee.
   g. Remove the bolt holding the clutch cable stand, move the bracket to the side rail at the fender inner panel.
   h. If equipped with cruise control, move the cruise control cable aside to clear the booster.
4. Disconnect the manifold vacuum hose from the booster check valve.
5. Disconnect the brake lines from the master cylinder, remove the master cylinder-to-booster retaining nuts and remove the master cylinder.
6. Working inside the vehicle below the instrument panel, remove the stoplight switch connector. Remove the switch retaining pin and slide the switch off the brake pedal pin just far enough for the outer arm to clear the pin, then remove the switch. Be careful not to damage the switch.
7. Remove the booster-to-dash panel attaching nuts. If necessary, remove the cowl top intrusion bolt.
8. On Mustang equipped with cruise

control, remove and set aside the control amplifier which is mounted to the lower outboard booster stud.

9. Slide the booster pushrod, washers and bushing off the brake pedal pin. Remove the booster.

10. Installation is the reverse of the removal procedure. Tighten the booster-to-dash panel attaching nuts and the master cylinder attaching nuts to 13–25 ft. lbs. (18–34 Nm). If the brake lines were disconnected, bleed the brake system.

BRAKE BOOSTER PUSHROD ADJUSTMENT

➧ SEE FIGS. 11–12

The pushrod has an adjustment screw to maintain the correct relationship between the booster control valve plunger and the master cylinder piston. If the plunger is too long it will prevent the master cylinder piston from completely releasing hydraulic pressure, causing the brakes to drag. If the plunger is too short it will cause excessive pedal travel and an undesirable clunk in the booster area. Remove the master cylinder for access to the booster pushrod.

To check the adjustment of the screw, fabricate a gauge (from cardboard, following the dimensions in the illustration) and place it against the master cylinder mounting surface of the booster body. Adjust the pushrod screw by turning it until the end of the screw just touches the inner edge of the slot in the gauge. Install the master cylinder and bleed the system.

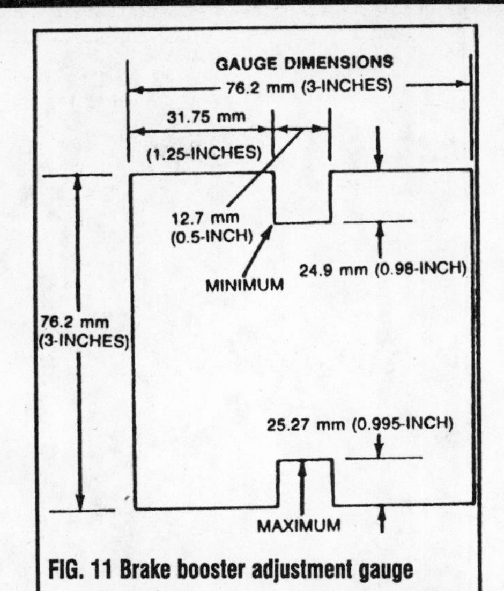

FIG. 11 Brake booster adjustment gauge

FIG. 12 Brake booster pushrod adjustment

Brake Hoses and Lines

HYDRAULIC BRAKE LINE CHECK

➧ SEE FIG. 13

The hydraulic brake lines and brake linings are to be inspected at the recommended intervals in the maintenance schedule. Follow the steel tubing from the master cylinder to the flexible hose fitting at each wheel. If a section of the tubing is found to be damaged, replace the entire section with tubing of the same type, size, shape, and length.

✱✱ CAUTION

Copper tubing should never be used in the brake system! Use only SAE J526 or J527 steel tubing.

When installing a new section of brake tubing, flush clean brake fluid or denatured alcohol through to remove any dirt or foreign material from the line. Be sure to flare both ends to provide sound, leak-proof connections.

✱✱ CAUTION

Double-flare the lines! Never single-flare a brake line!

When bending the tubing to fit the underbody contours, be careful not to kink or crack the line. Torque all hydraulic connections to 10–15 lbs.

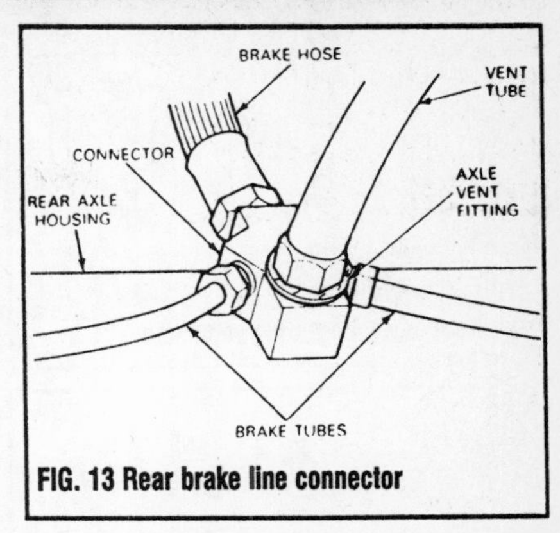

FIG. 13 Rear brake line connector

Check the flexible brake hoses that connect the steel tubing to each wheel cylinder. Replace the hose if it shows any signs of softening, cracking, or other damage. When installing a new front brake hose, position the hose to avoid contact with other chassis parts. Place a new copper gasket over the hose fitting and thread the hose assembly into the front wheel cylinder. A new rear brake hose must be positioned clear of the exhaust pipe or shock absorber. Thread the hose into the rear brake tube connector. When installing either a new front or rear brake hose, engage the opposite end of the hose to the bracket on the frame. Install the horseshoe type retaining clip and connect the tube to the hose with the tube fitting nut.

Always bleed the system after hose or line replacement. Before bleeding, make sure that the master cylinder is topped up with high temperature, extra heavy duty fluid of at least SAE 70R3 (DOT 3) quality.

FLARING A BRAKE LINE

Using a Split-Die Type Flaring Tool

➧ SEE FIG. 14–15

1. Using a tubing cutter, cut the required length of line.
2. Square the end with a file and chamfer the end.
3. Place the tube in the proper size die hole and position it so that it is flush with the die face. Lock the line with the wing nut.
4. The punches with most tools are marked to identify the sequence of flaring. Such marks are usually Op.1 and Op.2 or something similar.
5. Slide the Op.1 punch into position and tighten the screw to form a single flare.
6. Remove the punch and position the Op.2 punch. Tighten the screw to form the double flare.

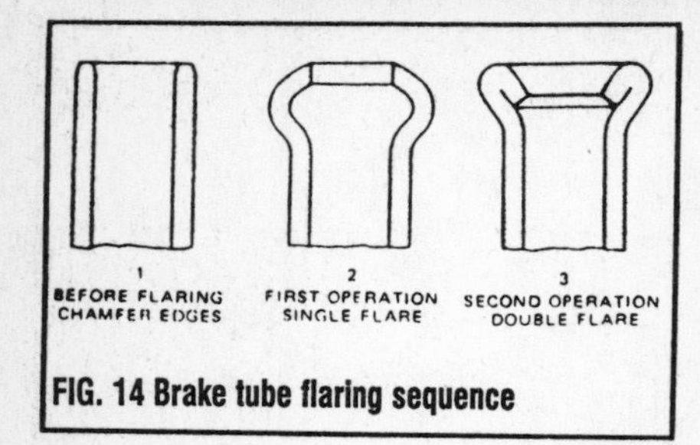

FIG. 14 Brake tube flaring sequence

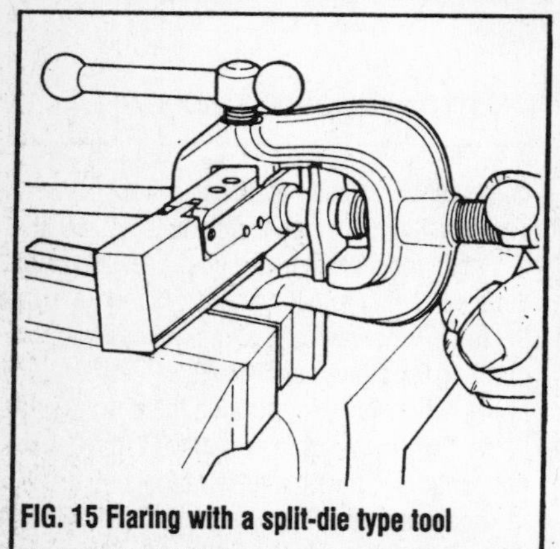
FIG. 15 Flaring with a split-die type tool

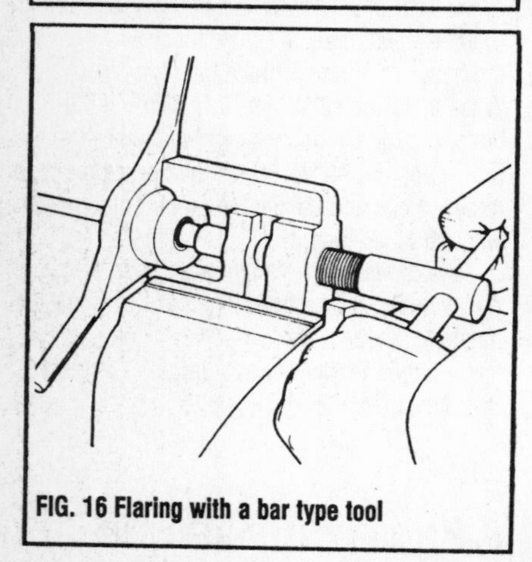
FIG. 16 Flaring with a bar type tool

7. Remove the punch and release the line from the die.
8. Inspect the finished flare for cracks or uneven flare form. If the flare is not perfect, cut it off and re-flare the end.

Using a Flaring Bar Type Tool

➧ SEE FIG. 16

1. Using a tubing cutter, cut the required length of line.
2. Square the end with a file and chamfer the end.
3. Insert the tube into the proper size hole in the bar, until the end of the tube sticks out as far as the thickness of the adapter above the bar, or, depending on the tool, even with the bar face.
4. Fit the adapter onto the tube and slide the bar into the yoke. Lock the bar in position with the tube beneath the yoke screw.
5. Tighten the yoke screw and form the single flare.
6. Release the yoke screw and remove the adapter.
7. Install the second adapter and form the double flare.
8. Release the screw and remove the tube. Check the flare for cracks or uneven flaring. If the flare isn't perfect, cut it off and re-flare the line.

FIG. 17 Brake system bleeding

Brake System Bleeding

➧ SEE FIG. 17

1. Clean all dirt from the master cylinder filler cap.
2. If the master cylinder is known or suspected to have air in the bore, it must be bled before any of the wheel cylinders or calipers. To bleed the master cylinder, loosen the upper secondary left front outlet fitting approximately 3/4 turn. Have an assistant depress the brake pedal slowly through it's full travel. Close the outlet fitting and let the pedal return slowly to the fully released position. Wait 5 seconds and then repeat the operation until all air bubbles disappear.
3. Repeat Step 2 with the right-hand front outlet fitting.
4. Continue to bleed the brake system by removing the rubber dust cap from the wheel cylinder bleeder fitting or caliper fitting at the right-hand rear of the vehicle. Place a suitable box wrench on the bleeder fitting and attach a rubber drain tube to the fitting. The end of the tube should fit snugly around the bleeder fitting. Submerge the other end of the tube in a container partially filled with clean brake fluid and loosen the fitting 3/4 turn.
5. Have an assistant push the brake pedal down slowly through it's full travel. Close the bleeder fitting and allow the pedal to slowly return to it's full release position. Wait 5 seconds and repeat the procedure until no bubbles appear at the submerged end of the bleeder tube. Secure the bleeder fitting and remove the bleeder tube. Install the rubber dust cap on the bleeder fitting.
6. Repeat the procedure in Steps 4 and 5 in the following sequence: left front, left rear and right front. Refill the master cylinder reservoir after each wheel cylinder or caliper has been bled and install the master cylinder cover and gasket. When brake bleeding is completed, the fluid level should be filled to the maximum level indicated on the reservoir.
7. Always make sure the disc brake pistons are returned to their normal positions by depressing the brake pedal several times until normal pedal travel is established. If the pedal feels spongy, repeat the bleeding procedure.

FRONT DISC BRAKES

CAUTION

Brake shoes contain asbestos, which has been determined to be a cancer causing agent. Never clean the brake surfaces with compressed air! Avoid inhaling any dust from any brake surface! When cleaning brake surfaces, use a commercially available brake cleaning fluid.

Brake Pads

REMOVAL & INSTALLATION

SEE FIG. 18–25

1. Remove and discard half the brake fluid from the master cylinder.
2. Raise and safely support vehicle. Remove the front wheel and tire assemblies.
3. Remove the caliper locating pins and remove the caliper from the anchor plate and rotor, but do not disconnect the brake hose.
4. Remove the outer brake pad from the caliper assembly and remove the inner brake pad from the caliper piston.
5. Inspect the disc brake rotor for scoring and wear. Replace or machine, as necessary.
6. Suspend the caliper inside the fender housing with a length of wire. Do not let the caliper hang by the brake hose.

To install:

7. Use a large C-clamp to push the caliper piston back into its bore.
8. Install the inner brake pad, then the outer brake pad, making sure the clips are engaged in the caliper piston.
9. Install the caliper and the wheel and tire assembly. Lower the vehicle.
10. Pump the brake pedal prior to moving the vehicle to seat the brake pads. Refill the master cylinder.

FIG. 18 Spray caliper bolts with a brake cleaner to ease removal

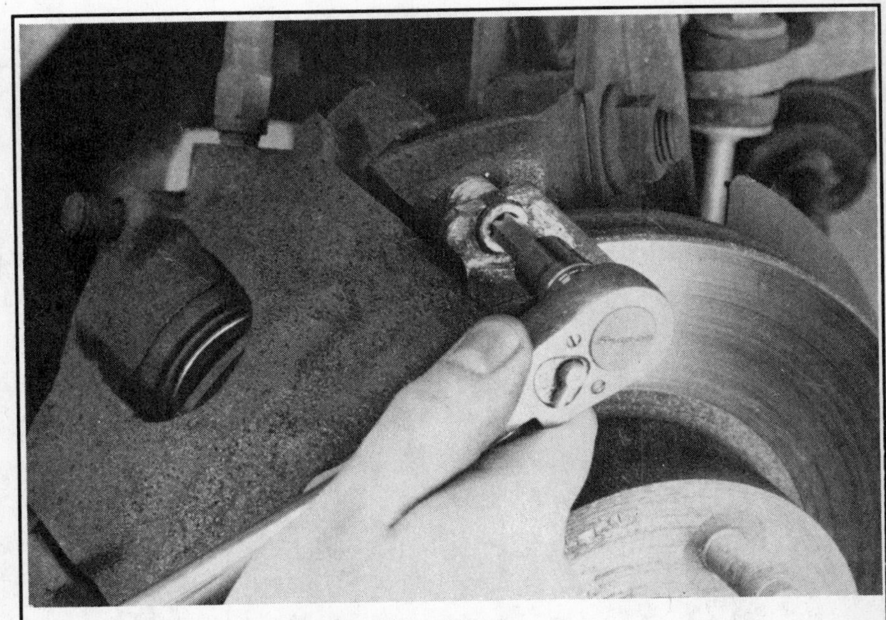

FIG. 19 Loosening caliper locating pins with special tool

INSPECTION

Remove the brake pads as described above and measure the thickness of the lining. If the lining at any point on the pad assembly is less than 0.125in. (3.175mm), thick (from shoe surface), or there is evidence of the lining being contaminated by brake fluid or oil, replace the brake pad.

Brake Caliper

REMOVAL & INSTALLATION

SEE FIG. 26

1. Raise and safely support the vehicle.

FIG. 20 Removing caliper locating pins

FIG. 21 Removing the caliper assembly

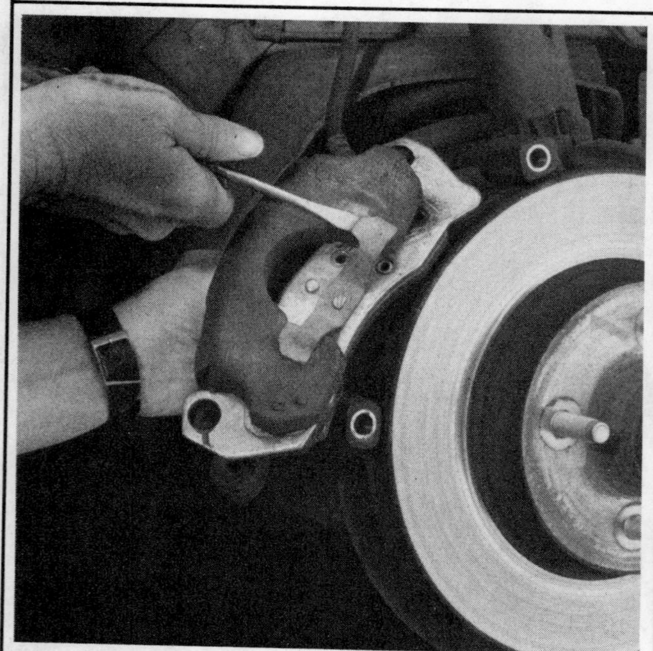

FIG. 22 Removing the outer brake pad

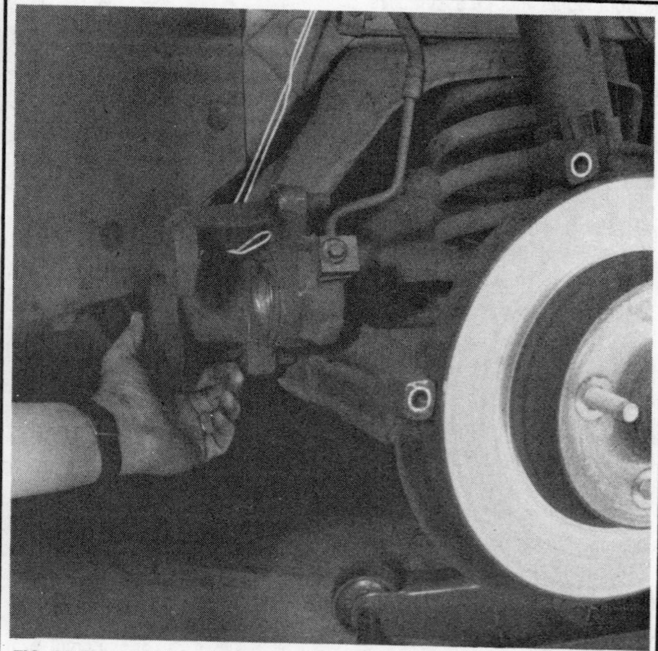

FIG. 23 Suspend the caliper assembly from the fender

FIG. 24 Pushing the caliper piston back into the bore

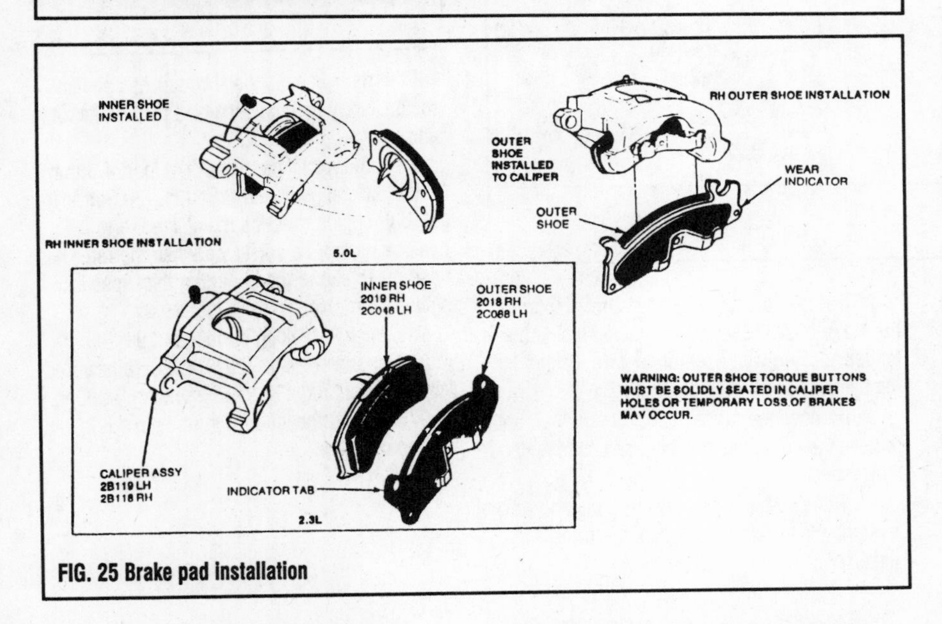

FIG. 25 Brake pad installation

Remove the front wheel and tire assembly.

2. If equipped with a hollow brake hose retaining bolt, remove the bolt and plug the brake hose.

3. If the brake line screws into the caliper, loosen the brake line fitting that connects the brake hose to the brake line at the frame bracket. Remove the retaining clip from the hose and bracket and disengage the hose from the bracket. Unscrew the hose from the caliper.

4. Remove the caliper locating pins and remove the caliper. If removing both calipers, mark the right and left sides so they may be reinstalled correctly.

To install:

5. Install the caliper over the rotor with the outer brake shoe against the rotor's braking surface.

6. Lubricate the inside of the locating pin insulators with silicone dialectic grease. Install the caliper locating pins and tighten to 45–65 ft. lbs. (61–88 Nm).

7. If equipped with a hollow brake hose retaining bolt, install new copper washers on each side of the brake hose fitting outlet and install the bolt, through the hose fitting and into the caliper. Tighten the bolt to 25 ft. lbs. (34 Nm).

8. If the brake hose screws into the caliper, thread the brake hose into the caliper and tighten to 20–30 ft. lbs. (28–41 Nm).

➡ This is a special self-sealing fitting that does not require a gasket. When the hose is correctly tightened, there should be 1 or 2 threads of the fitting still showing at the caliper. It is not necessary for the hose fitting to be flush with the caliper for sealing, so do not over-tighten.

9. Bleed the brake system, install the wheel and tire assembly and lower the vehicle.

10. To position the brake pads, apply the brake pedal several times before moving the vehicle.

OVERHAUL

➧ SEE FIG. 27

1. Clean the outside of the caliper in alcohol after removing it from the vehicle and removing the brake pads.

2. Drain the caliper through the inlet port.

3. Roll some thick shop cloths or rags and place them between the piston and the outer legs of the caliper.

4. Apply compressed air to the caliper inlet port until the piston comes out of the caliper bore. Use low air pressure to avoid having the piston pop out too rapidly and possible causing injury.

5. If the piston becomes cocked in the cylinder bore and will not come out, remove the air pressure and tap the piston with a soft hammer to try and straighten it. Do not use a sharp tool or pry the piston out of the bore. Reapply the air pressure.

6. Remove the boot from the piston and seal from the caliper cylinder bore.

7. Clean the piston and caliper in alcohol.

8. Lubricate the piston seal with clean brake fluid, and position the seal in the groove in the cylinder bore.

9. Coat the outside of the piston and both of the beads of dust boot with clean brake fluid. Insert the piston through the dust boot until the boot is around the bottom (closed end) of the piston.

10. Hold the piston and dust boot directly above the caliper cylinder bore, and use your fingers to work the bead of dust boot into the groove near the top of the cylinder bore.

11. After the bead is seated in the groove, press straight down on the piston until it bottoms in the bore. Be careful not to cock the piston in the bore. Be careful not to cock the piston in the bore. Use a C-clamp with a block of wood inserted between the clamp and the piston to bottom the piston, if necessary.

12. Install the brake pads and install the caliper. Bleed the brake hydraulic system and

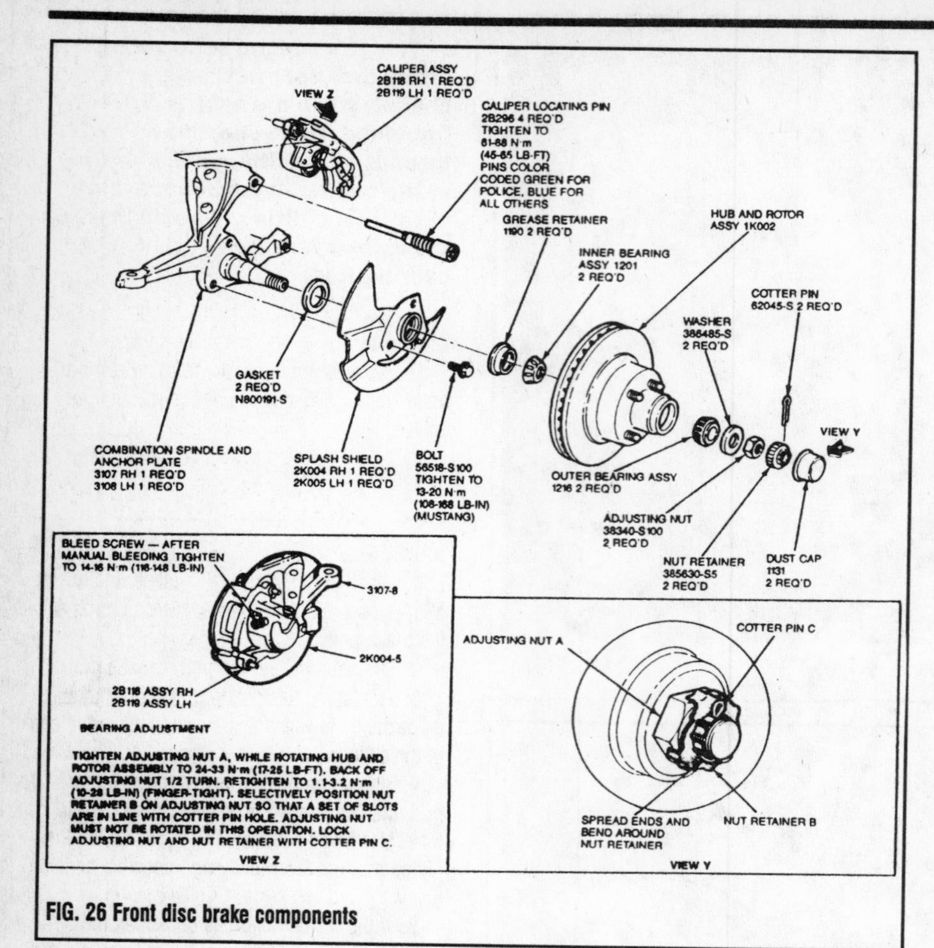

FIG. 26 Front disc brake components

recenter the pressure differential valve. Do not drive the vehicle until a firm brake pedal is obtained.

Brake Rotor

REMOVAL & INSTALLATION

➧ SEE FIG. 26

1. Raise and safely support the vehicle. Remove the wheel and tire assembly.
2. Remove the caliper, but do not disconnect the brake hose. Suspend the caliper inside the fender housing with a length of wire. Do not let the caliper hang by the brake hose.
3. Remove the grease cap from the hub and remove the cotter pin, nut lock, adjusting nut and flatwasher.
4. Remove the outer bearing cone and roller assembly and remove the hub and rotor assembly.

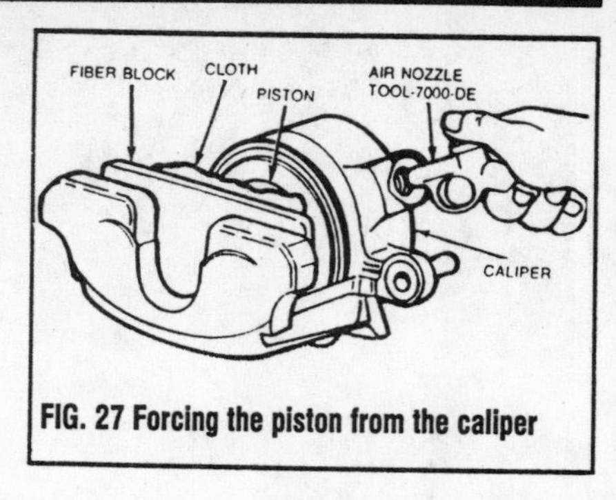

FIG. 27 Forcing the piston from the caliper

5. Inspect the rotor for scoring and wear. Replace or machine as necessary. If machining, observe the minimum thickness specification.
6. Installation is the reverse of removal. Make sure the grease in the rotor is clean and adequate. Adjust the wheel bearings.

INSPECTION

If the rotor is deeply scarred or has shallow cracks, it may be refinished on a disc brake rotor lathe.

A minimum thickness of 0.972 in. (24.68mm) for the 5.0L engine, and 0.810 in. (20.5mm) for the 2.3L engine must be maintained. If the damage cannot be corrected when the rotor has been machined to the minimum thickness it should be replaced.

The finished braking surfaces of the rotor must be parallel within 0.007 in. (0.18mm) and lateral run-out must not be more than 0.003 in. (0.076mm) on the inboard surface in a 5 in. (127mm) radius.

REAR DRUM BRAKES

⁂ CAUTION

Brake shoes contain asbestos, which has been determined to be a cancer causing agent. Never clean the brake surfaces with compressed air! Avoid inhaling any dust from any brake surface! When cleaning brake surfaces, use a commercially available brake cleaning fluid.

Rear Brake Drum

REMOVAL & INSTALLATION

1. Raise and safely support the vehicle.

2. Remove the wheel and tire assembly.
3. Remove the drum retaining nuts and remove the brake drum.

➡ **If the drum will not come off, pry the rubber plug from the backing plate. Insert a narrow rod through the hole in the backing plate and disengage the adjusting lever from the adjusting screw. While holding the adjustment lever away from the screw, back off the adjusting screw with a brake adjusting tool.**

4. Inspect the brake drum for scoring and wear. Replace or machine as necessary. If machining, observe the maximum diameter specification.
5. Installation is the reverse of removal.

INSPECTION

Check that there are no cracks or chips in the braking surface. Excessive bluing indicates overheating and a replacement drum is needed. The drum can be machined to remove minor damage and to establish a rounded braking surface on a warped drum. Never exceed the maximum oversize of the drum when machining the braking surface. The maximum inside diameter is stamped on the rim of the drum.

FIG. 28 Remove the brake drum

Rear Brake Shoes

REMOVAL & INSTALLATION

➧ SEE FIGS. 28–34

1. Raise and safely support the vehicle. Remove the rear wheel and tire assemblies. Remove the brake drum.
2. Remove the shoe-to-anchor springs and unhook the cable eye from the anchor pin. Remove the anchor pin plate.
3. Remove the shoe hold-down springs, shoes, adjusting screw, pivot nut, socket and automatic adjustment parts.
4. Remove the parking brake link, spring and retainer. Disconnect the parking brake cable from the parking brake lever.
5. After removing the rear brake secondary shoe, disassemble the parking brake lever from the shoe by removing the retaining clip and spring washer.

To install:

➧ SEE FIGS. 35–43

6. Before installing the rear brake shoes, assemble the parking brake lever to the secondary shoe and secure it with the spring washer and retaining clip.
7. Apply a light coating of caliper slide grease at the points where the brake shoes contact the backing plate. Be careful not to get any lubricant on the brake linings.
8. Position the brake shoes on the backing plate. The primary shoe with the short lining faces the front of the vehicle, the secondary to the rear. Secure the assembly with the hold-down springs. Install the parking brake link, spring and retainer. Back-off the parking brake adjustment, then connect the parking brake cable to the parking brake lever.
9. Install the anchor pin plate on the anchor pin. Place the cable eye over the anchor pin with the crimped side toward the drum. Install the primary shoe to the anchor pin.
10. Install the cable guide on the secondary shoe web with the flanged hole fitted into the hole in the secondary shoe web. Thread the cable around the cable guide groove.

➡ **The cable must be positioned in the groove and not between the guide and the shoe web.**

11. Install the secondary shoe-to-anchor spring. Make sure the cable eye is not cocked or binding on the anchor pin when installed. All parts should be flat on the anchor pin.
12. Apply a thin coat of lubricant to the threads and the socket end of the adjusting screw. Turn the adjusting screw into the adjusting pivot nut to the limit of the threads, then back-off $^{1}/_{2}$ turn.

➡ **Make sure the socket end of the adjusting screw is stamped with an R or L, indicating the right or left side of the vehicle. The adjusting screw assemblies must be installed on the correct side for proper brake shoe adjustment.**

13. Place the adjusting socket on the screw and install the assembly between the shoe ends with the adjusting screw toothed wheel nearest the secondary shoe.
14. Hook the cable hook into the hole in the adjusting lever. The adjusting levers are stamped with an **R** or **L** to indicate their installation on the right or left side.
15. Position the hooked end of the adjuster spring completely into the large hole in the primary shoe web. Connect the loop end of the spring to the adjuster lever hole.
16. Pull the adjuster lever, cable and automatic adjuster spring down and toward the rear, engaging the pivot hook in the large hole of the secondary shoe web.
17. Make sure the upper ends of the brake shoes are seated against the anchor pin and the shoes are centered on the backing plate.
18. Adjust the brakes using brake adjustment gauge D81L–1103–A or equivalent.
19. Install the brake drum, wheel and tire assemblies and lower the vehicle.
20. Apply the brakes several times while backing up the vehicle. After each stop, the vehicle must be moved forward.

FIG. 29 Clean brake drum with brake cleaner

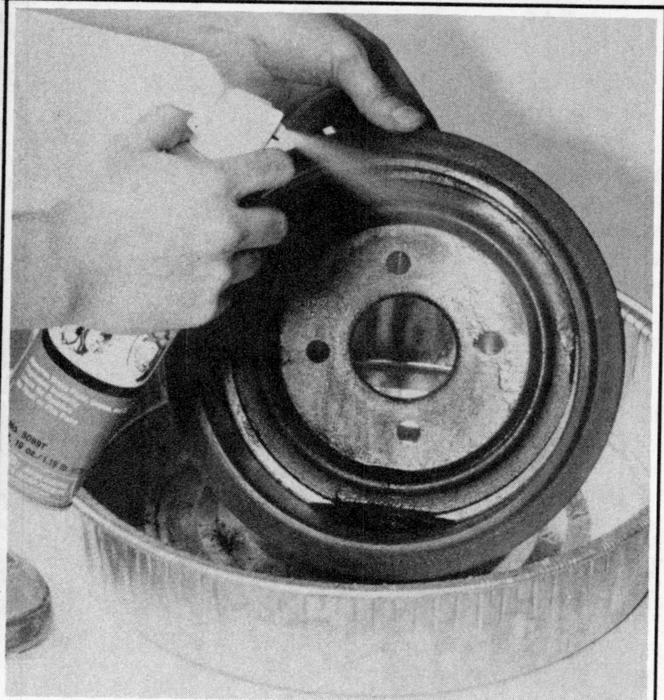

FIG. 30 Clean all brake components with brake cleaner

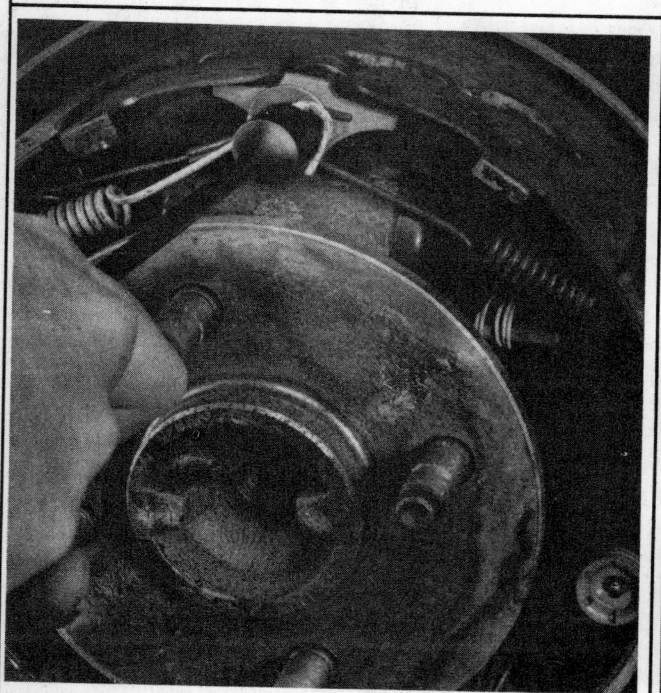

FIG. 31 Remove brake return springs with special tool

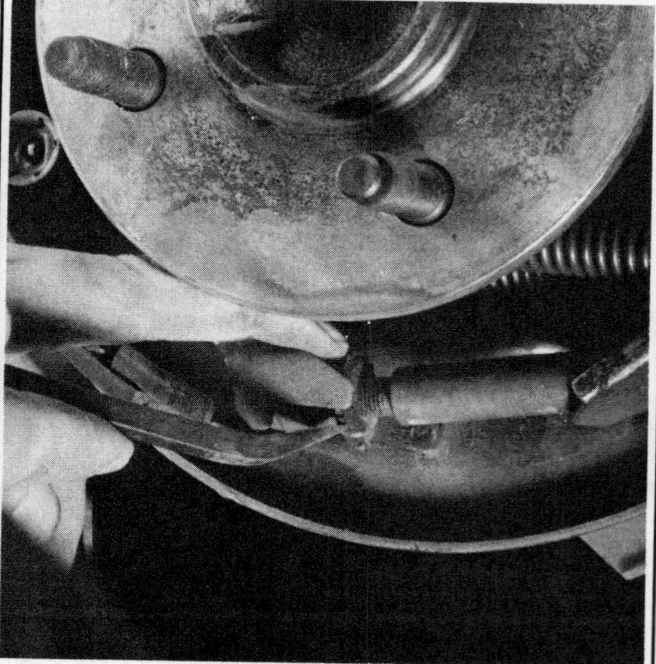

FIG. 32 Loosen brake adjuster screw

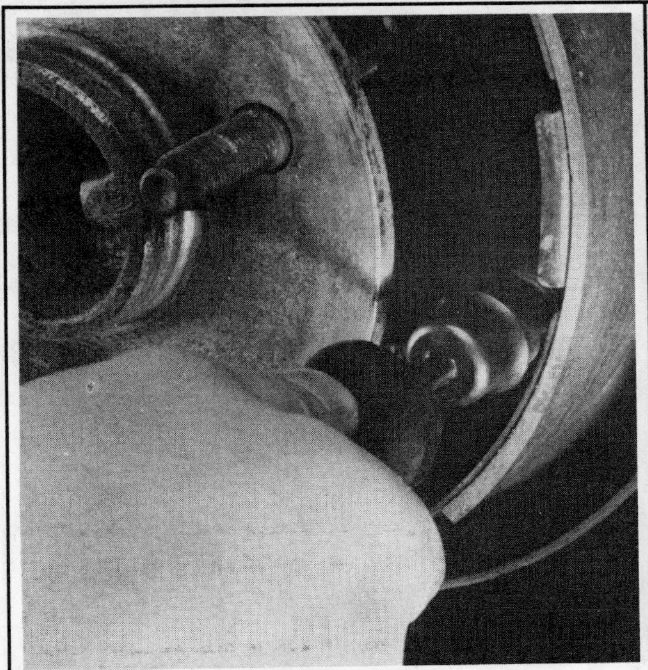

FIG. 33 Remove brake shoe hold-down springs with special tool

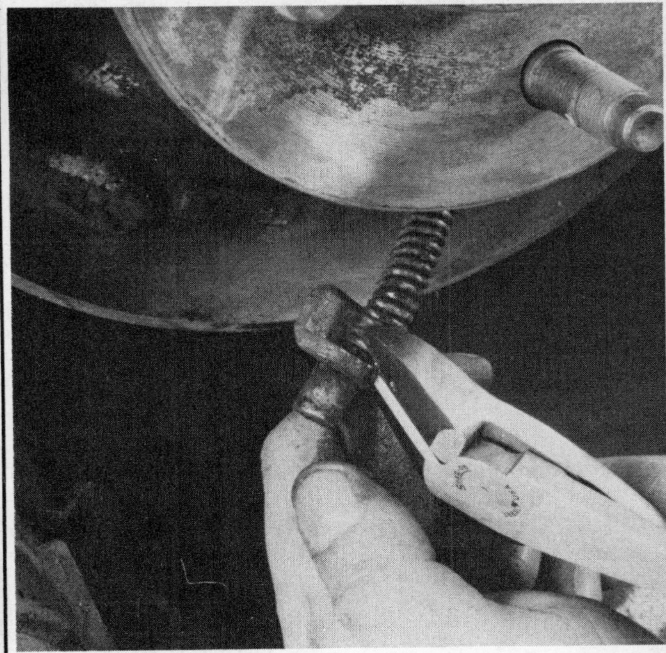

FIG. 34 Remove parking brake lever

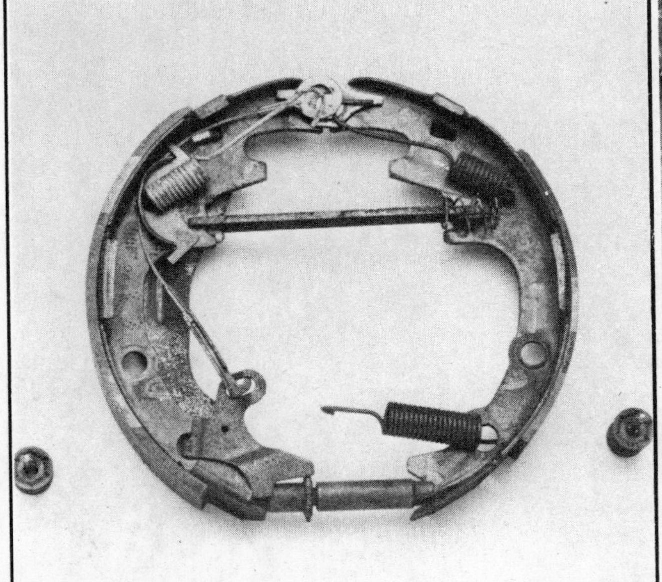

FIG. 35 Assemble and inspect all brake components

FIG. 36 Grease backing plate shoe contact points

FIG. 37 Install parking brake lever

FIG. 38 Install brake shoe hold-down springs

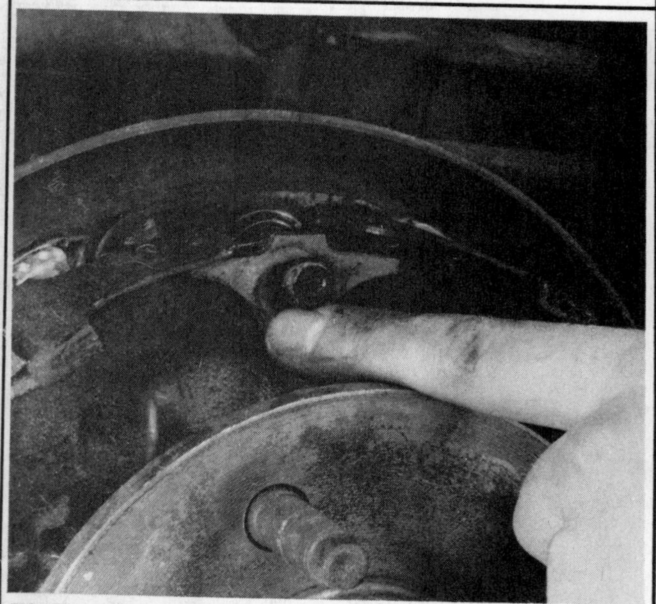

FIG. 39 Install brake shoe retainer

FIG. 40 Install primary brake shoe return spring

FIG. 41 Install secondary brake shoe return spring

FIG. 42 Install brake adjuster spring

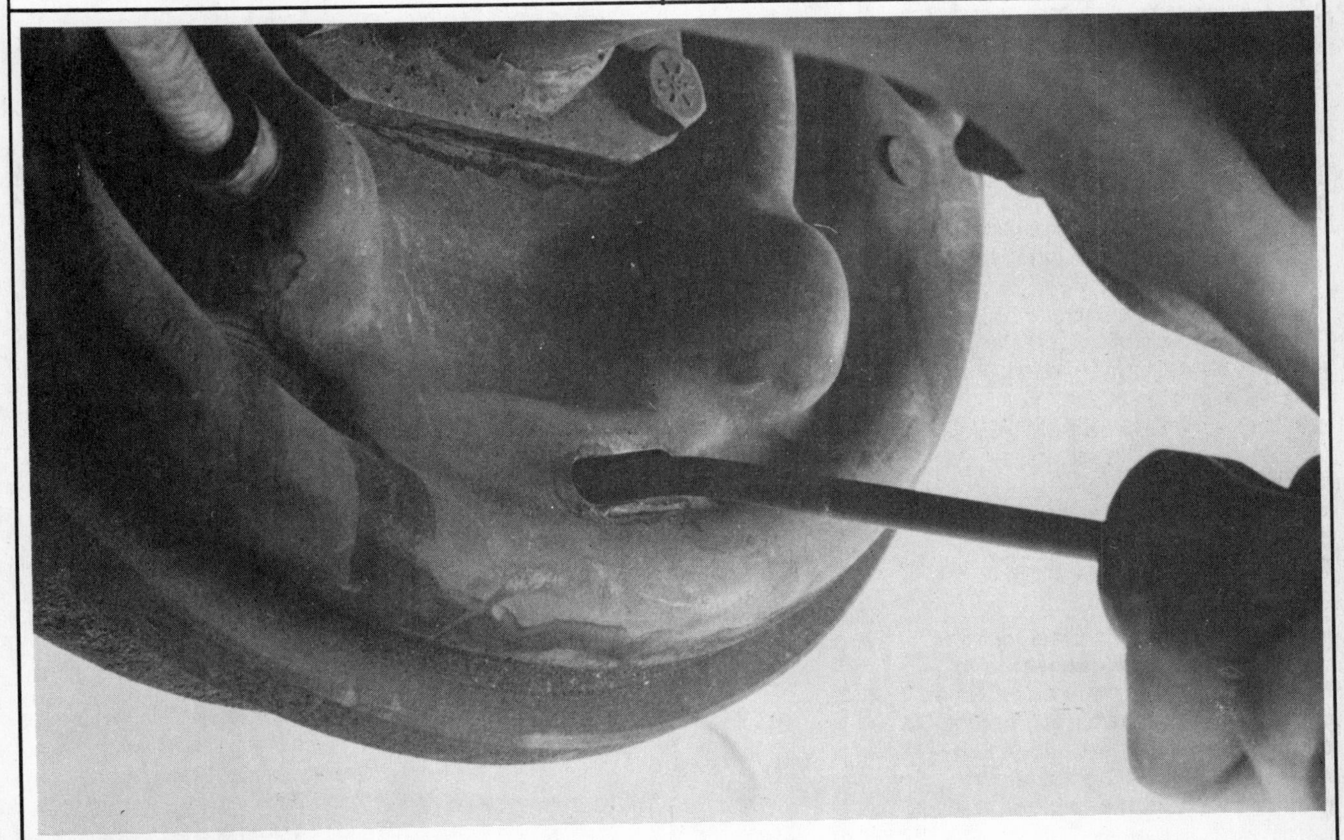
FIG. 43 Adjust brake shoes

Wheel Cylinder

REMOVAL & INSTALLATION

1. Remove the wheel and tire assembly and the brake drum.
2. Remove the brake shoe assembly.
3. Disconnect the brake line from the wheel cylinder at the backing plate.
4. Remove the wheel cylinder attaching bolts and remove the wheel cylinder.
5. Installation is the reverse of the removal procedure. Tighten the wheel cylinder attaching bolts to 10–20 ft. lbs. (14–28 Nm).
6. Bleed the brake system.

PARKING BRAKE

Parking Brake Cable

ADJUSTMENT

➧ SEE FIG. 44

1. Make sure the parking brake is fully released.
2. Place the transmission in **N**. Raise and safely support the vehicle.
3. Tighten the adjusting nut against the cable equalizer, causing a rear wheel brake drag. Loosen the adjusting nut until the rear brakes are fully released. There should be no brake drag.
4. Lower the vehicle and check the operation of the parking brake.

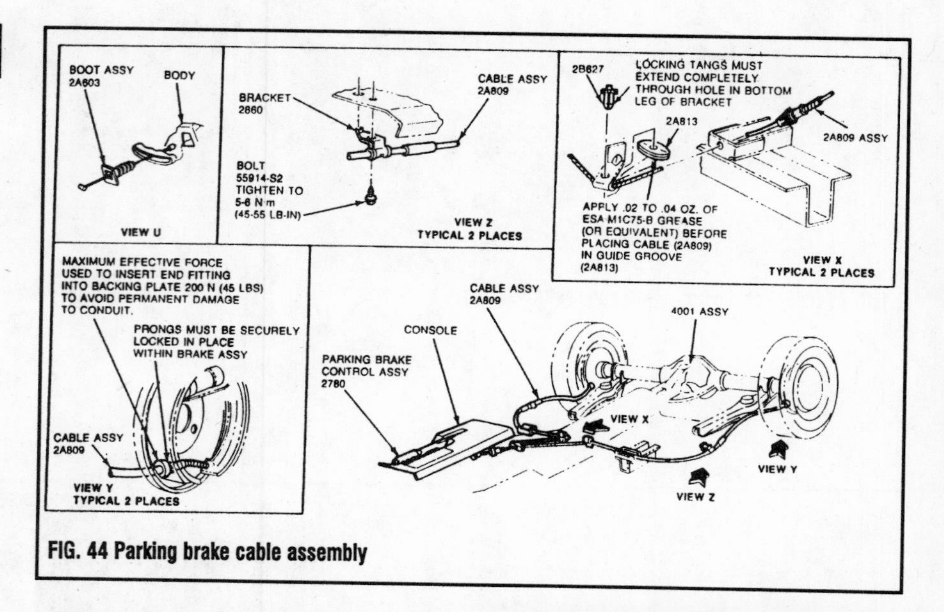

FIG. 44 Parking brake cable assembly

REMOVAL & INSTALLATION

➧ SEE FIG. 44

1. Place the parking brake control in the released position. Release the cable tension as follows:
   a. Remove the floor console.
   b. With an assistant inside the vehicle, raise and safely support the vehicle.
   c. Have another assistant pull the equalizer rearward approximately 1–2$1/2$ in. (25–64mm) to rotate the self-adjuster reel backward.
   d. Insert a steel lockpin through the holes in the lever and control assembly. This locks the ratchet wheel in the cable released position.

➡ Do not remove the steel lockpin until the cables are connected to the equalizer. Pin removal releases the tension in the ratchet wheel causing the spring to unwind and release tension. If the pin is removed without the cables attached, the entire assembly must be removed to reset the spring tension.

2. Raise and safely support the vehicle. Remove the rear cables from the equalizer.
3. Remove the cable snap fitting from the body. Remove the retaining clip that attaches the cable to the underbody.
4. Remove the wheel and tire assemblies and the brake drums.
5. Remove the self adjuster springs and remove the cable retainers from the backing plates.
6. Disconnect the cable ends from the parking brake levers, compress the cable retainer prongs and pull the cable ends from the backing plates.
7. Installation is the reverse of the removal procedure. Adjust the parking brake.

Parking Brake Lever

REMOVAL & INSTALLATION

➧ SEE FIGS. 45–46

1. Place the parking brake control in the released position. Release the cable tension as follows:
   a. Remove the floor console.
   b. With an assistant inside the vehicle, raise and safely support the vehicle.
   c. Have another assistant pull the equalizer rearward approximately 1–2$1/2$ in. (25–64mm) to rotate the self-adjuster reel backward.
   d. Insert a steel lockpin through the holes in the lever and control assembly. This locks the ratchet wheel in the cable released position.

➡ Do not remove the steel lockpin until the cables are connected to the equalizer. Pin removal releases the tension in the ratchet wheel causing the spring to unwind and release tension. If the pin is removed without the cables attached, the entire assembly must be removed to reset the spring tension.

2. Disconnect the equalizer from the control.
3. Remove the bolts attaching the parking brake lever to the floorpan.

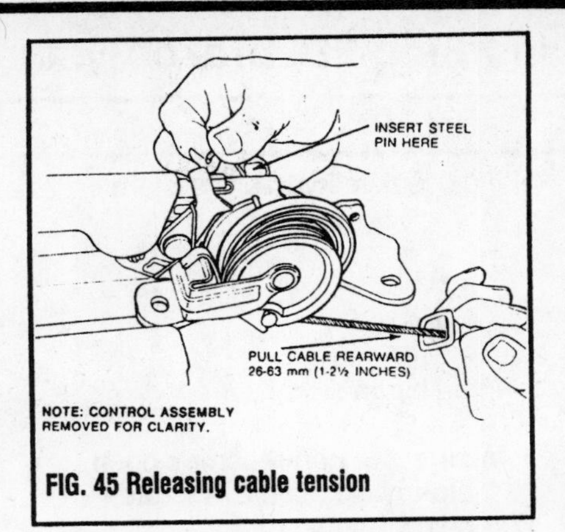

FIG. 45 Releasing cable tension

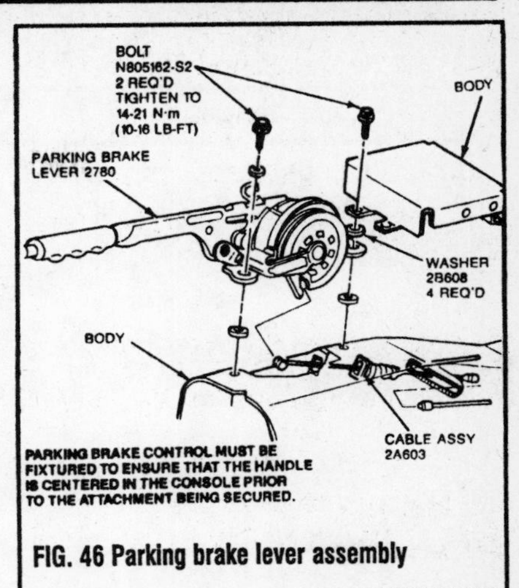

FIG. 46 Parking brake lever assembly

To Install:

4. Route the cable and equalizer around the control assembly pulley. Install the cable anchor pin in the pivot hole in the ratchet.
5. Connect the rear cable to the equalizer.
6. With the cable attached, position the control assembly on the floorpan. Install and tighten the bolts to 10–16 ft. lbs. (12–21Nm).
7. Remove the lockpin from the control assembly to reset the tension.
8. Apply the parking brake several times. Check that the parking brakes apply and the **BRAKE** indicator lights when the brakes are applied.
9. Install the console.

BRAKE SPECIFICATIONS

All measurements in inches unless noted

			Brake Disc			Brake Drum Diameter			Minimum Lining Thickness	
Year	Model	Master Cylinder Bore	Original Thickness	Minimum Thickness	Maximum Runout	Original Inside Diameter	Max. Wear Limit	Maximum Machine Diameter	Front	Rear
1989	Mustang	①	③	②	0.003	9.00	—	9.060	0.125	0.030
1990	Mustang	①	③	②	0.003	9.00	—	9.060	0.125	0.030
1991	Mustang	①	③	②	0.003	9.00	—	9.060	0.125	0.030
1992	Mustang	①	③	②	0.003	9.00	—	9.060	0.125	0.030

① 2.3L engine—0.827
5.0L engine—1.125
② 2.3L engine—0.810
5.0L engine—0.972
③ 2.3L engine—0.870
5.0L engine—1.030

Troubleshooting the Brake System

Problem	Cause	Solution
Low brake pedal (excessive pedal travel required for braking action.)	• Excessive clearance between rear linings and drums caused by inoperative automatic adjusters	• Make 10 to 15 alternate forward and reverse brake stops to adjust brakes. If brake pedal does not come up, repair or replace adjuster parts as necessary.
	• Worn rear brakelining	• Inspect and replace lining if worn beyond minimum thickness specification
	• Bent, distorted brakeshoes, front or rear	• Replace brakeshoes in axle sets
	• Air in hydraulic system	• Remove air from system. Refer to Brake Bleeding.

Troubleshooting the Brake System (cont.)

Problem	Cause	Solution
Low brake pedal (pedal may go to floor with steady pressure applied.)	• Fluid leak in hydraulic system	• Fill master cylinder to fill line; have helper apply brakes and check calipers, wheel cylinders, differential valve tubes, hoses and fittings for leaks. Repair or replace as necessary.
	• Air in hydraulic system	• Remove air from system. Refer to Brake Bleeding.
	• Incorrect or non-recommended brake fluid (fluid evaporates at below normal temp).	• Flush hydraulic system with clean brake fluid. Refill with correct-type fluid.
	• Master cylinder piston seals worn, or master cylinder bore is scored, worn or corroded	• Repair or replace master cylinder
Low brake pedal (pedal goes to floor on first application—o.k. on subsequent applications.)	• Disc brake pads sticking on abutment surfaces of anchor plate. Caused by a build-up of dirt, rust, or corrosion on abutment surfaces	• Clean abutment surfaces
Fading brake pedal (pedal height decreases with steady pressure applied.)	• Fluid leak in hydraulic system	• Fill master cylinder reservoirs to fill mark, have helper apply brakes, check calipers, wheel cylinders, differential valve, tubes, hoses, and fittings for fluid leaks. Repair or replace parts as necessary.
	• Master cylinder piston seals worn, or master cylinder bore is scored, worn or corroded	• Repair or replace master cylinder
Spongy brake pedal (pedal has abnormally soft, springy, spongy feel when depressed.)	• Air in hydraulic system	• Remove air from system. Refer to Brake Bleeding.
	• Brakeshoes bent or distorted	• Replace brakeshoes
	• Brakelining not yet seated with drums and rotors	• Burnish brakes
	• Rear drum brakes not properly adjusted	• Adjust brakes

Troubleshooting the Brake System (cont.)

Problem	Cause	Solution
Decreasing brake pedal travel (pedal travel required for braking action decreases and may be accompanied by a hard pedal.)	• Caliper or wheel cylinder pistons sticking or seized	• Repair or replace the calipers, or wheel cylinders
	• Master cylinder compensator ports blocked (preventing fluid return to reservoirs) or pistons sticking or seized in master cylinder bore	• Repair or replace the master cylinder
	• Power brake unit binding internally	• Test unit according to the following procedure: (a) Shift transmission into neutral and start engine (b) Increase engine speed to 1500 rpm, close throttle and fully depress brake pedal (c) Slow release brake pedal and stop engine (d) Have helper remove vacuum check valve and hose from power unit. Observe for backward movement of brake pedal. (e) If the pedal moves backward, the power unit has an internal bind—replace power unit
Grabbing brakes (severe reaction to brake pedal pressure.)	• Brakelining(s) contaminated by grease or brake fluid	• Determine and correct cause of contamination and replace brakeshoes in axle sets
	• Parking brake cables incorrectly adjusted or seized	• Adjust cables. Replace seized cables.
	• Incorrect brakelining or lining loose on brakeshoes	• Replace brakeshoes in axle sets
	• Caliper anchor plate bolts loose	• Tighten bolts
	• Rear brakeshoes binding on support plate ledges	• Clean and lubricate ledges. Replace support plate(s) if ledges are deeply grooved. Do not attempt to smooth ledges by grinding.
	• Incorrect or missing power brake reaction disc	• Install correct disc
	• Rear brake support plates loose	• Tighten mounting bolts
Chatter or shudder when brakes are applied (pedal pulsation and roughness may also occur.)	• Brakeshoes distorted, bent, contaminated, or worn	• Replace brakeshoes in axle sets
	• Caliper anchor plate or support plate loose	• Tighten mounting bolts
	• Excessive thickness variation of rotor(s)	• Refinish or replace rotors in axle sets
Noisy brakes (squealing, clicking, scraping sound when brakes are applied.)	• Bent, broken, distorted brakeshoes	• Replace brakeshoes in axle sets
	• Excessive rust on outer edge of rotor braking surface	• Remove rust

Troubleshooting the Brake System (cont.)

Problem	Cause	Solution
Hard brake pedal (excessive pedal pressure required to stop vehicle. May be accompanied by brake fade.)	• Loose or leaking power brake unit vacuum hose	• Tighten connections or replace leaking hose
	• Incorrect or poor quality brake-lining	• Replace with lining in axle sets
	• Bent, broken, distorted brakeshoes	• Replace brakeshoes
	• Calipers binding or dragging on mounting pins. Rear brakeshoes dragging on support plate.	• Replace mounting pins and bushings. Clean rust or burrs from rear brake support plate ledges and lubricate ledges with molydisulfide grease. **NOTE:** If ledges are deeply grooved or scored, do not attempt to sand or grind them smooth—replace support plate.
	• Caliper, wheel cylinder, or master cylinder pistons sticking or seized	• Repair or replace parts as necessary
	• Power brake unit vacuum check valve malfunction	• Test valve according to the following procedure: (a) Start engine, increase engine speed to 1500 rpm, close throttle and immediately stop engine (b) Wait at least 90 seconds then depress brake pedal (c) If brakes are not vacuum assisted for 2 or more applications, check valve is faulty
	• Power brake unit has internal bind	• Test unit according to the following procedure: (a) With engine stopped, apply brakes several times to exhaust all vacuum in system (b) Shift transmission into neutral, depress brake pedal and start engine (c) If pedal height decreases with foot pressure and less pressure is required to hold pedal in applied position, power unit vacuum system is operating normally. Test power unit. If power unit exhibits a bind condition, replace the power unit.

Troubleshooting the Brake System (cont.)

Problem	Cause	Solution
Hard brake pedal (excessive pedal pressure required to stop vehicle. May be accompanied by brake fade.)	• Master cylinder compensator ports (at bottom of reservoirs) blocked by dirt, scale, rust, or have small burrs (blocked ports prevent fluid return to reservoirs).	• Repair or replace master cylinder **CAUTION:** Do not attempt to clean blocked ports with wire, pencils, or similar implements. Use compressed air only.
	• Brake hoses, tubes, fittings clogged or restricted	• Use compressed air to check or unclog parts. Replace any damaged parts.
	• Brake fluid contaminated with improper fluids (motor oil, transmission fluid, causing rubber components to swell and stick in bores	• Replace all rubber components, combination valve and hoses. Flush entire brake system with DOT 3 brake fluid or equivalent.
	• Low engine vacuum	• Adjust or repair engine
Dragging brakes (slow or incomplete release of brakes)	• Brake pedal binding at pivot	• Loosen and lubricate
	• Power brake unit has internal bind	• Inspect for internal bind. Replace unit if internal bind exists.
	• Parking brake cables incorrrectly adjusted or seized	• Adjust cables. Replace seized cables.
	• Rear brakeshoe return springs weak or broken	• Replace return springs. Replace brakeshoe if necessary in axle sets.
	• Automatic adjusters malfunctioning	• Repair or replace adjuster parts as required
	• Caliper, wheel cylinder or master cylinder pistons sticking or seized	• Repair or replace parts as necessary
	• Master cylinder compensating ports blocked (fluid does not return to reservoirs).	• Use compressed air to clear ports. Do not use wire, pencils, or similar objects to open blocked ports.
Vehicle moves to one side when brakes are applied	• Incorrect front tire pressure	• Inflate to recommended cold (reduced load) inflation pressure
	• Worn or damaged wheel bearings	• Replace worn or damaged bearings
	• Brakelining on one side contaminated	• Determine and correct cause of contamination and replace brakelining in axle sets
	• Brakeshoes on one side bent, distorted, or lining loose on shoe	• Replace brakeshoes in axle sets
	• Support plate bent or loose on one side	• Tighten or replace support plate
	• Brakelining not yet seated with drums or rotors	• Burnish brakelining
	• Caliper anchor plate loose on one side	• Tighten anchor plate bolts
	• Caliper piston sticking or seized	• Repair or replace caliper
	• Brakelinings water soaked	• Drive vehicle with brakes lightly applied to dry linings
	• Loose suspension component attaching or mounting bolts	• Tighten suspension bolts. Replace worn suspension components.
	• Brake combination valve failure	• Replace combination valve

Troubleshooting the Brake System (cont.)

Problem	Cause	Solution
Noisy brakes (squealing, clicking, scraping sound when brakes are applied.) (cont.)	• Brakelining worn out—shoes contacting drum of rotor	• Replace brakeshoes and lining in axle sets. Refinish or replace drums or rotors.
	• Broken or loose holdown or return springs	• Replace parts as necessary
	• Rough or dry drum brake support plate ledges	• Lubricate support plate ledges
	• Cracked, grooved, or scored rotor(s) or drum(s)	• Replace rotor(s) or drum(s). Replace brakeshoes and lining in axle sets if necessary.
	• Incorrect brakelining and/or shoes (front or rear).	• Install specified shoe and lining assemblies
Pulsating brake pedal	• Out of round drums or excessive lateral runout in disc brake rotor(s)	• Refinish or replace drums, re-index rotors or replace

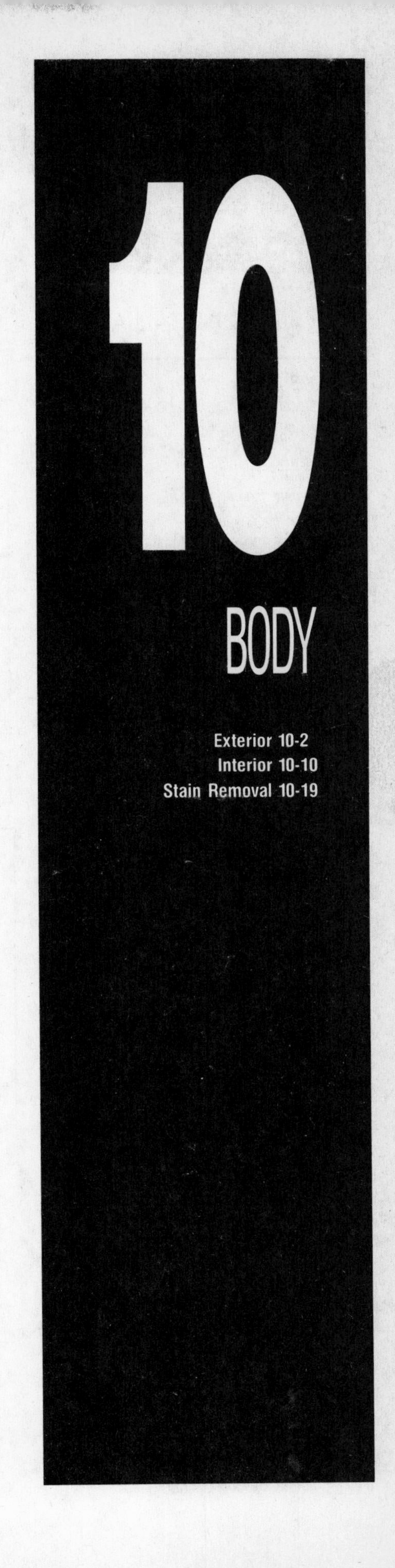

10 BODY

EXTERIOR

Doors

REMOVAL & INSTALLATION

➧ SEE FIGS. 1–3

1. Remove the door trim panel.
2. Remove the watershield, and, if a new door is being installed, save all the molding clips and moldings.
3. Remove the wiring harness, actuator and speakers.
4. If a new door is being installed, remove all window and lock components.
5. Support the door and unbolt the hinges from the door.
6. Installation is the reverse of removal. New holes may have to be drilled in a replacement door for the trim. Align the door and tighten the hinge bolts securely.

ADJUSTMENT

➧ SEE FIGS. 1–4

➡ **Loosen the hinge-to-door bolts for lateral adjustment only. Loosen the hinge-to-body bolts for both lateral and vertical adjustment.**

1. Determine which hinge bolts are to be loosened and back them out just enough to allow movement.
2. To move the door safely, use a padded pry bar. When the door is in the proper position, tighten the bolts to 24 ft. lbs. and check the door operation. There should be no binding or interference when the door is closed and opened.
3. Door closing adjustment can also be affected by the position of the lock striker plate. Loosen the striker plate bolts and move the striker plate just enough to permit proper closing and locking of the door.

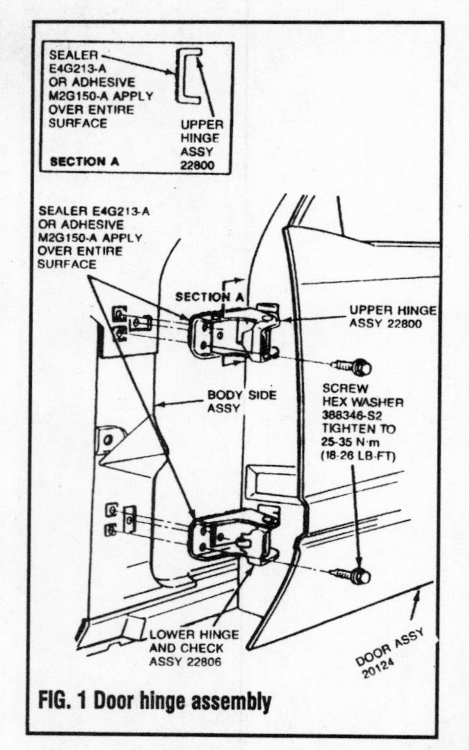

FIG. 1 Door hinge assembly

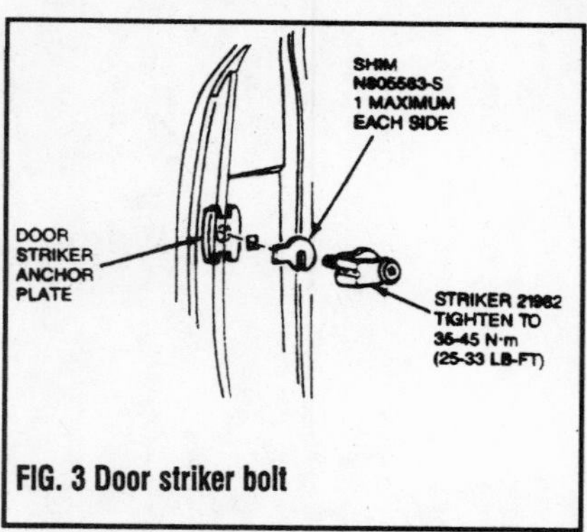

FIG. 3 Door striker bolt

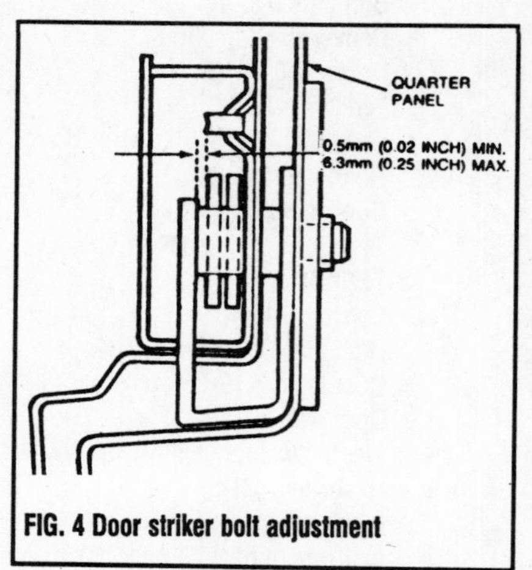

FIG. 4 Door striker bolt adjustment

Hood

REMOVAL & INSTALLATION

➧ SEE FIG. 5–6

1. Open and support the hood.
2. Matchmark the hood-to-hinge positions.
3. Have an assistant support the hood while you remove the hinge-to-hood bolts.
4. You and your assistant can remove the hood.
5. Installation is the reverse of removal.

ALIGNMENT

➧ SEE FIG. 6

1. Side-to-side and fore-aft adjustments can be made by loosening the hood-to-hinge attachment bolts and positioning the hood as necessary.
2. Hood vertical fit can be adjusted by raising or lowering the hinge-to-fender reinforcement bolts.
3. To ensure a snug fit of the hood against the rear hood bumpers, it may be necessary to rotate the hinge around the 3 attaching bolts.

Hood Latch Control Cable

REMOVAL & INSTALLATION

➧ SEE FIG. 5

1. From inside the vehicle, release the hood.
2. Remove the two bolts retaining the latch to the upper radiator support.
3. Remove the screw retaining the cable end retainer to the latch assembly.
4. Disengage the cable by rotating it out of the latch return spring.
5. To facilitate installing the cable, fasten a length of fishing line about 8 ft. long to latch the end of the cable.
6. From the inside vehicle, unseat the sealing grommet from the cowl side, remove the cable mounting bracket attaching screws and carefully pull the cable assembly out. Do not pull the "fish line" out.

To Install:

7. Using the previously installed fish line, pull the new cable assembly through the retaining wall, seat the grommet securely, and install the cable mounting bracket attaching screws.

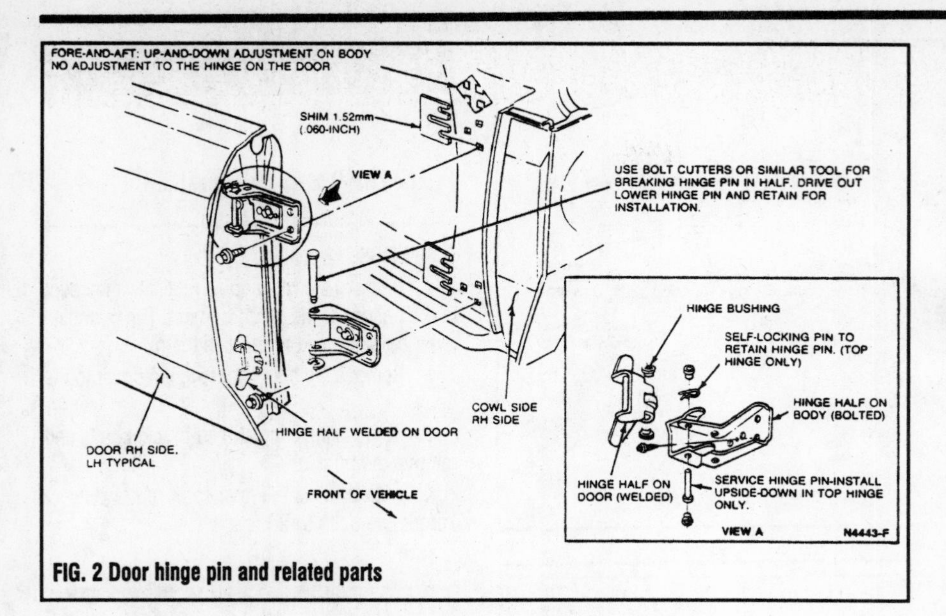

FIG. 2 Door hinge pin and related parts

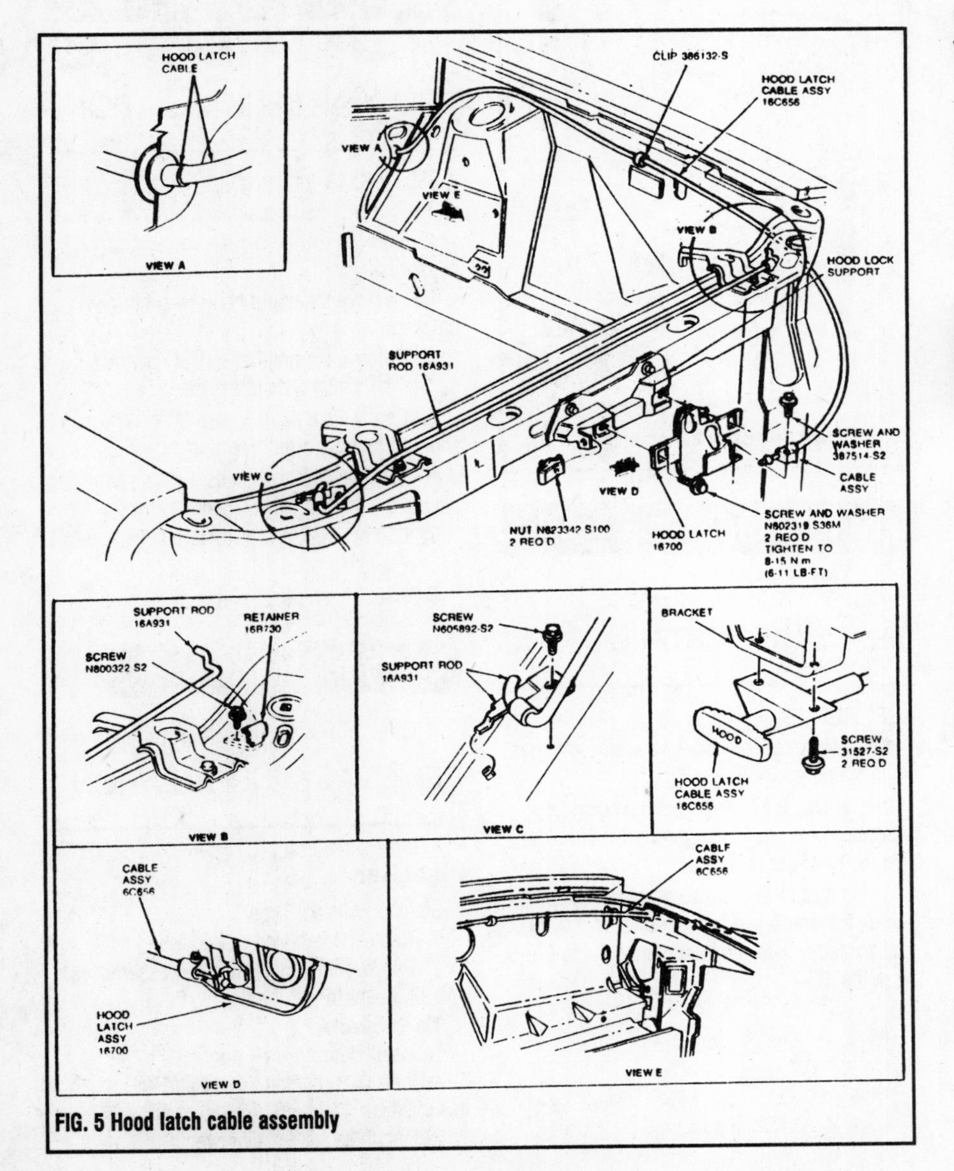

FIG. 5 Hood latch cable assembly

8. Thread the terminal end of cable into the hood latch return spring.
9. Route the cable through the V-slot on the latch and install the cable end retaining screw.
10. Check the hood latch cable release operation before the closing hood. Adjust if necessary.

Hood latch

REMOVAL & INSTALLATION

➧ SEE FIG. 5

1. From inside the vehicle release the hood.
2. Remove the two bolts retaining the latch to the upper radiator support.
3. Remove the two bolts retaining the hood latch assembly-to-radiator support and remove the latch.

To Install:

1. Engage the hood latch to the control cable and position the hood latch to the radiator support.
2. Install the two attaching bolts.
3. Adjust the hood latch and torque the attaching bolts to 7–10 ft. lbs.

Trunk Lid/Hatch Door

REMOVAL & INSTALLATION

➧ SEE FIG. 7–9

The trunk lid can be removed by removing the hinge-to-trunk lid bolts and sliding the trunk lid off of the hinges. The gas struts used to support the trunk lid must be removed with the lid in the fully open position.

ALIGNMENT

1. Fore-aft fit may be adjusted by loosening the hinge-to-lid bolts and positioning the lid as necessary.

2. Vertical fit can be adjusted by adding or deleting shims located between the the hinges and trunk lid.

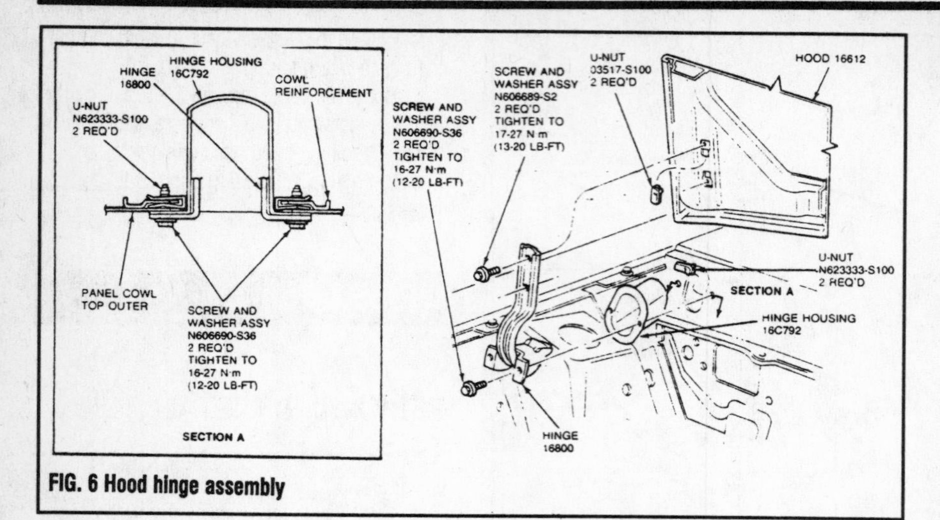

FIG. 6 Hood hinge assembly

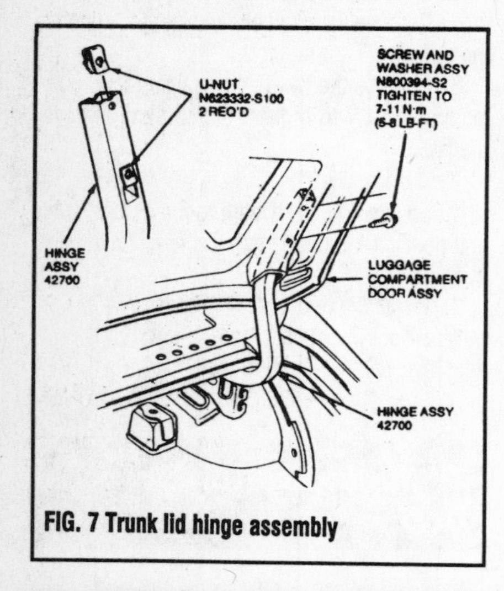

FIG. 7 Trunk lid hinge assembly

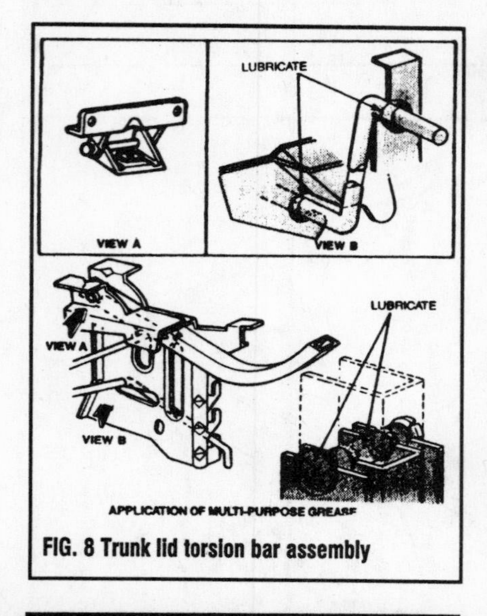

FIG. 8 Trunk lid torsion bar assembly

Liftgate Support Cylinder

REMOVAL & INSTALLATION

➧ SEE FIG. 9

1. Open the liftgate and temporarily support it.
2. The lift cylinder end fitting is a spring-clip design and removal is accomplished by sliding a small screwdriver under it and prying up to remove it from the ball stud.
3. Remove the support cylinder.

To Install:

4. Install each cylinder to the C-pillar and the liftgate bracket ball socket by pushing the cylinder's locking wedge onto the socket.
5. Close the liftgate. Check the support cylinder operation.

Front Bumper

REMOVAL & INSTALLATION

➧ SEE FIGS. 10–11

1. The bumper is removed by removing the 8 attaching bolts.
2. Squeeze the bumper pad retaining tabs with pliers and push them through their holes until the pads are removed.
3. Installation is the reverse of removal. Torque the bracket-to-bumper bolts to 14–20 ft. lbs.; the isolator-to-reinforcement bolts to 25–38 ft. lbs.

Rear Bumpers

REMOVAL & INSTALLATION

➧ SEE FIGS. 12–13

1. Remove the 6 pushpins retaining the cover to the bottom of the reinforcement. The pushpins must be destroyed to remove them.
2. Remove the nuts attaching the cover to the body.
3. Remove the 6 isolator-to-bracket bolts and remove the bumper.
4. Installation is the reverse of removal. Torque the bolts to 20 ft. lbs.

Grille

REMOVAL & INSTALLATION

➧ SEE FIGS. 10–11

1. Remove the license plate bolts or rivets.
2. Remove the lower grille-to-radiator support screws.
3. Remove the upper grille-to-support brackets.
4. Remove the grille-to-fender nuts and detach the reinforcement assembly.
5. Remove the headlamp side marker, park and turn lamps from the reinforcement.
6. Remove the 2 pushnuts per side attaching the lower corner reinforcement assembly.
7. Drill out the grille-to-reinforcement rivets and remove the grille.
8. Installation is the reverse of removal.

Outside Mirror

REMOVAL & INSTALLATION

Right Hand

Standard Manual Type

1. Remove the inside sail cover.
2. Remove the nut and washer assemblies and lift the mirror off the door.

To Install:

3. Install the mirror on door.
4. Install and tighten the nut and washer assemblies.
5. Install the inside sail cover.

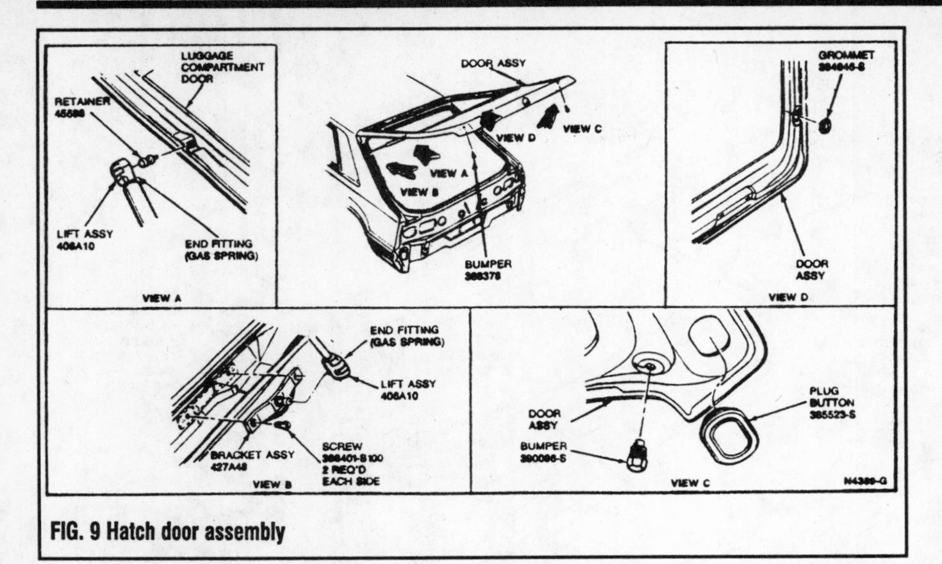

FIG. 9 Hatch door assembly

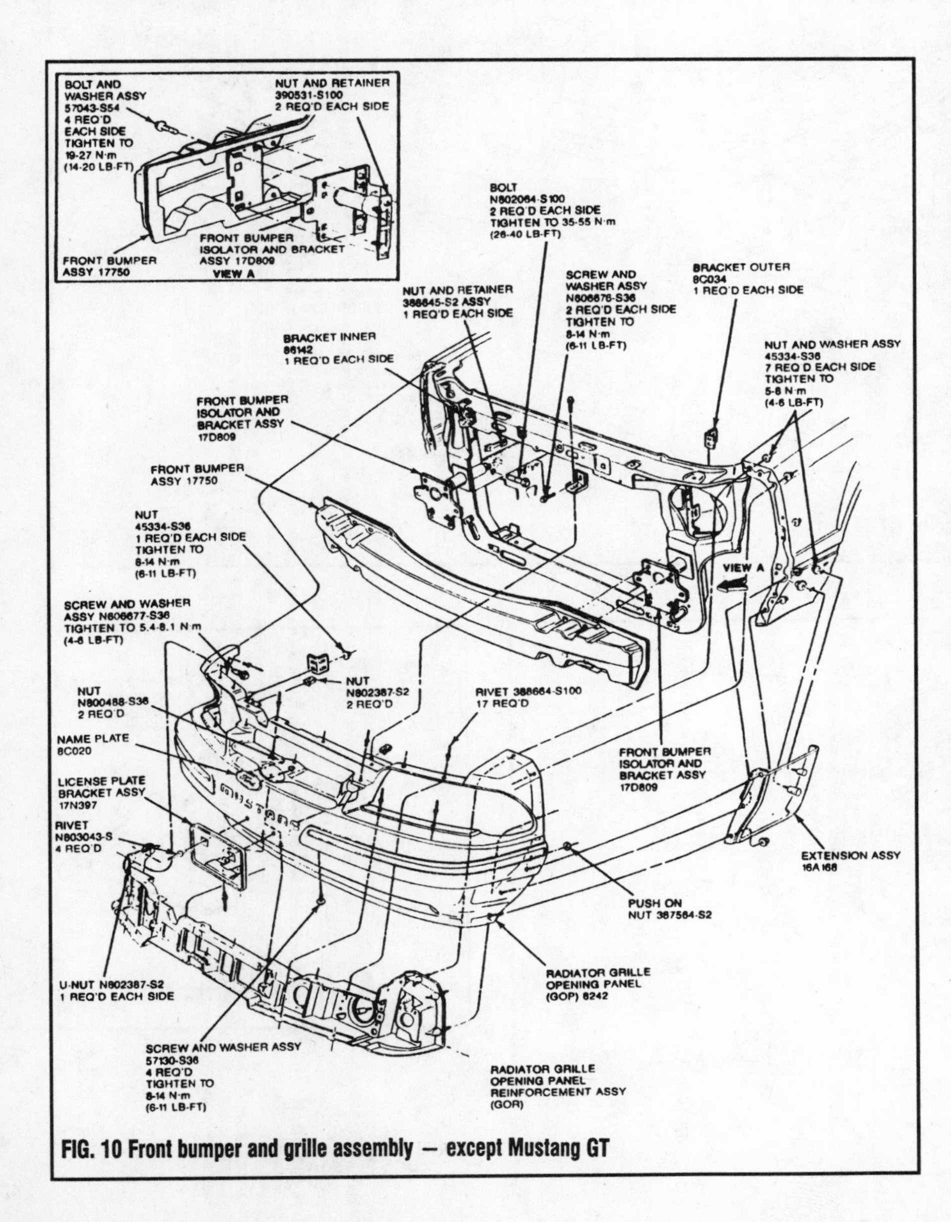

FIG. 10 Front bumper and grille assembly — except Mustang GT

Left Hand Remote Control

➧ SEE FIG. 14

1. Pull the nob assembly to remove it from the control shaft.
2. Remove the interior sail cover retainer screw and remove the cover.
3. Loosen the setscrew retaining control assembly to the sail cover.
4. Remove the mirror attaching nuts, washers and grommet. Remove the mirror and the control assembly.

To install:

5. Seat the grommet in the outer door panel and position the mirror to the door. Install the attaching nuts and washer and tighten to 25–39 inch lbs.
6. Route the control mechanism through the door and position to the sail trim panel. Tighten the setscrew to 2–6 inch lbs.
7. Position the sail cover to the door and install the retaining screw.
8. Position the rubber knob onto the control shaft and push to install.

Power Outside Mirrors

➧ SEE FIG. 15

➡ Outside mirrors that are frozen must be thawed prior to adjustment. Do not attempt to free-up the mirror by pressing the glass assembly.

1. Disconnect the negative (–) battery cable.
2. Remove the one screw retaining the mirror mounting hole cover and remove the cover.
3. Remove the door trim panel.
4. Disconnect the mirror assembly wiring connector. Remove the necessary wiring guides.
5. Remove the three mirror retaining nuts on the sail mirrors, two on door mirrors. Remove the mirror while guiding the wiring and connector through hole in the door.

To install:

6. Install the mirror assembly by routing the connector and wiring through the hole in the door. Attach with the three retaining nuts on the sail mirrors, two on the door mirrors. Tighten the retaining nuts.
7. Connect the mirror wiring connector and install the wiring guides.
8. Replace the mirror mounting hole cover and install one screw.
9. Replace the door trim panel.
10. Connect the negative (–) battery cable.

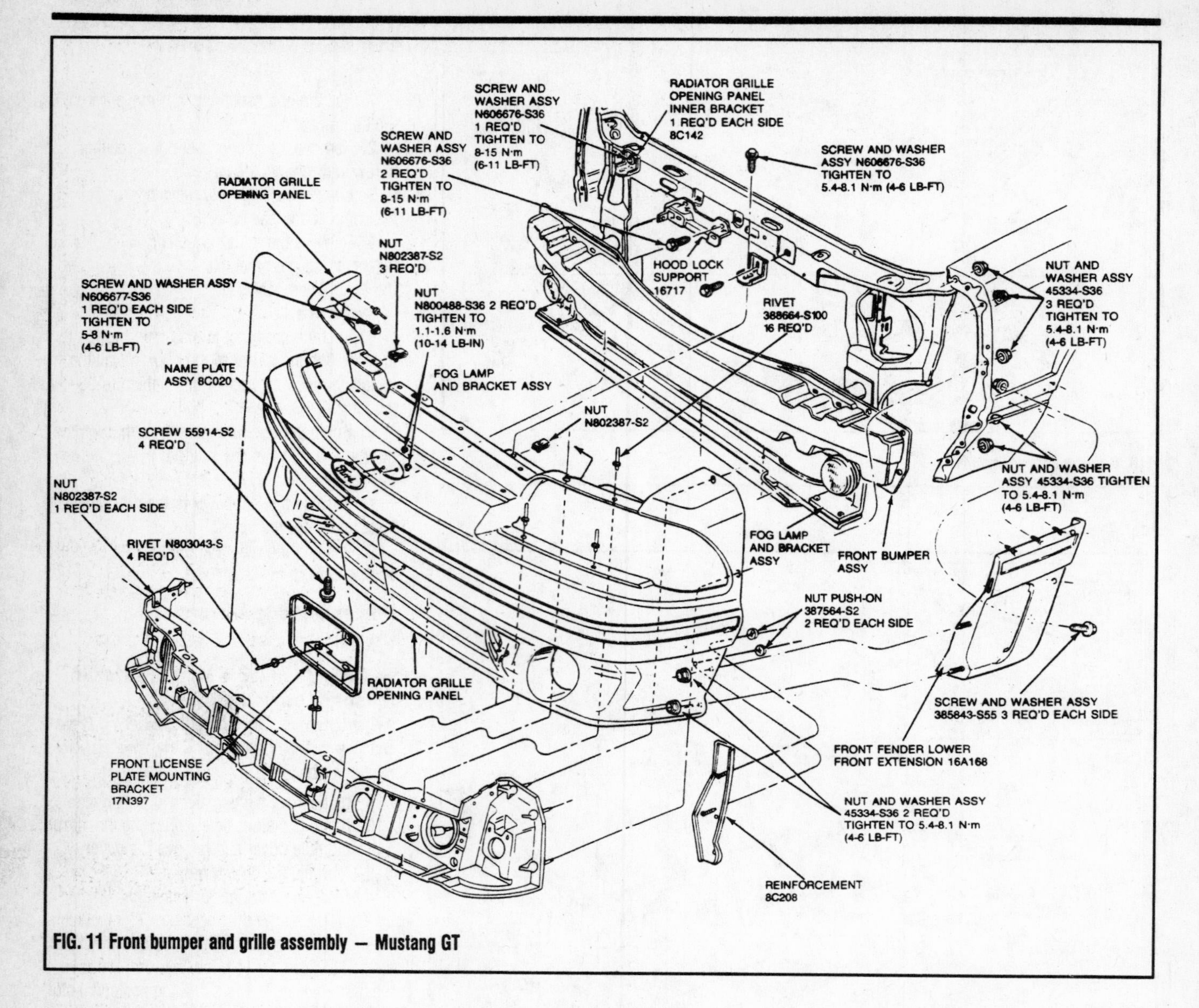

FIG. 11 Front bumper and grille assembly — Mustang GT

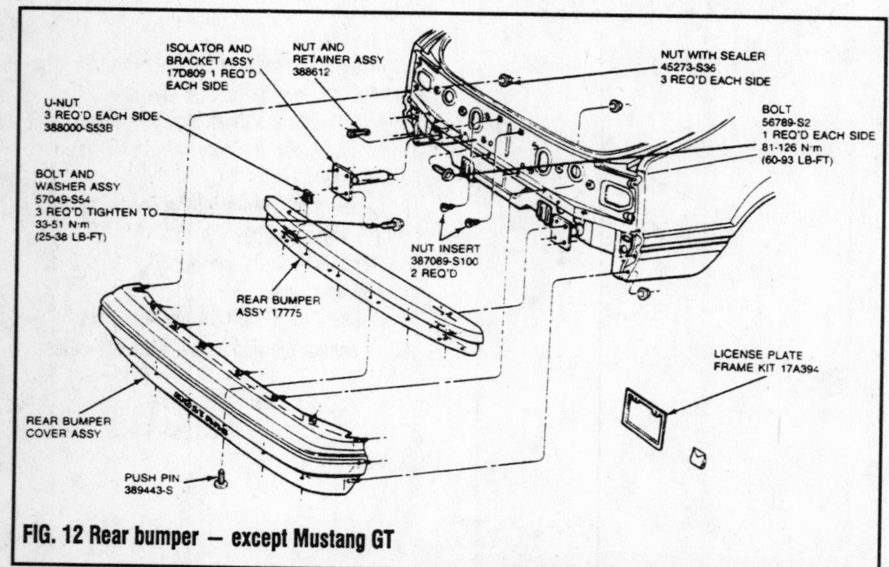

FIG. 12 Rear bumper — except Mustang GT

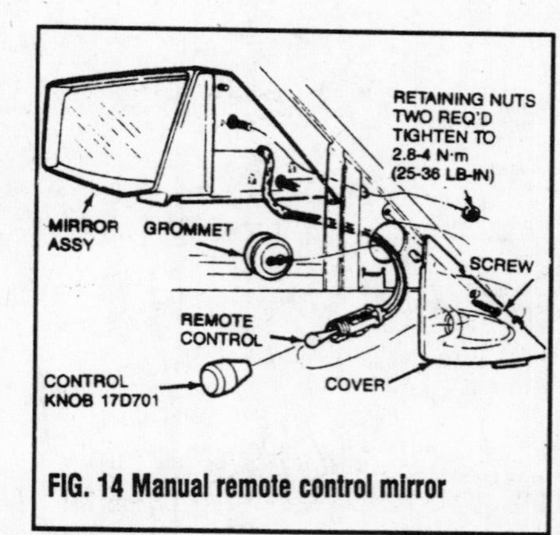

FIG. 14 Manual remote control mirror

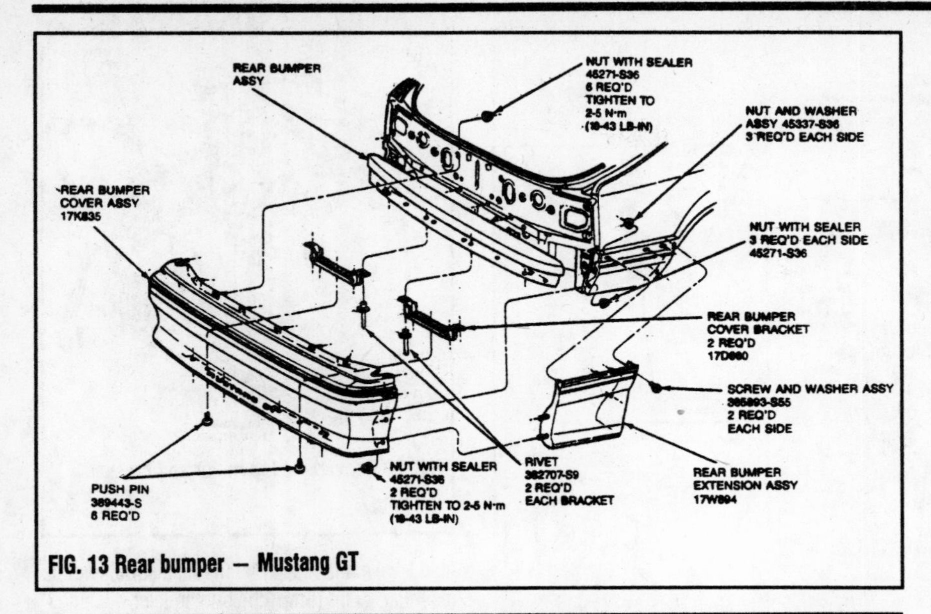

FIG. 13 Rear bumper — Mustang GT

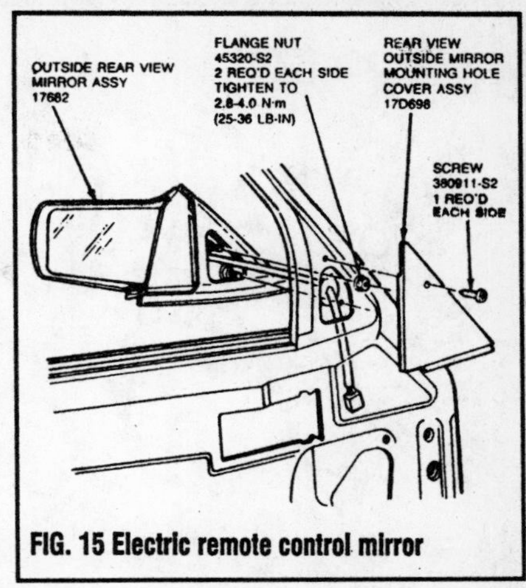

FIG. 15 Electric remote control mirror

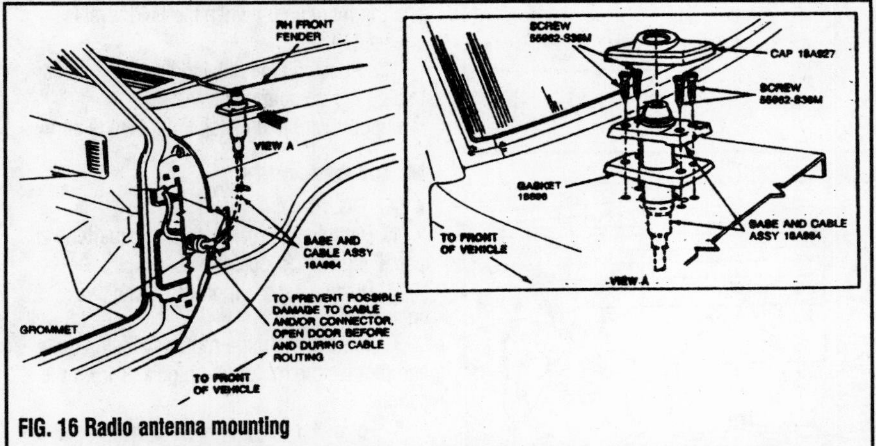

FIG. 16 Radio antenna mounting

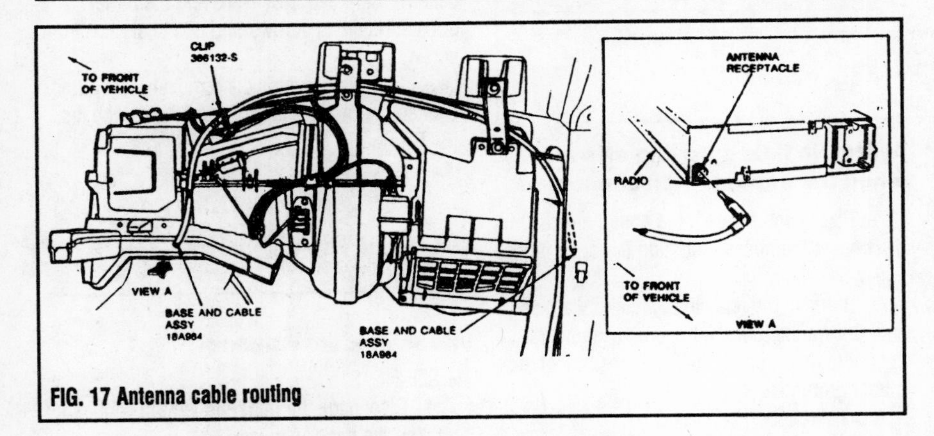

FIG. 17 Antenna cable routing

Antenna

REMOVAL & INSTALLATION

➧ SEE FIGS. 16–17

1. Remove the radio and disconnect the antenna cable at the radio by pulling it straight out of the set.
2. Remove the right hand cowl side trim panel.
3. Remove the antenna cable clip holding the antenna cable to the heater plenum.
4. Remove the antenna cap and antenna base retaining screws and pull the cable through the holes in the door hinge pillar and fender and remove the antenna assembly.

To Install:

5. With the right hand door open position the antenna assembly in the fender opening, put the gasket in position on the antenna and install the antenna base to the fender.
6. Pull the antenna lead through the door hinge pillar opening. Seat the grommet by pulling the cable from inside the vehicle.
7. Route the cable behind the heater plenum and attach the locating clip. Connect the lead to the rear of the radio.
8. Install the right side cowl trim panel. Install the radio.

Fenders

REMOVAL & INSTALLATION

➧ SEE FIG. 18

1. Remove the front bumper assembly.
2. Remove the screws retaining the grille opening reinforcement panel to the fender. Remove the screws retaining the upper front fender mounting bracket to the fender.
3. Remove the screws retaining the front fender mounting bracket to the fender. Remove the screws retaining the lower rear fender to the side of the body.
4. Remove the bolts retaining the upper and lower front fender. Remove the three retaining bolts from the catwalk area of the fender apron. Remove the fender from the vehicle.
5. Installation is the reverse of the removal procedure.

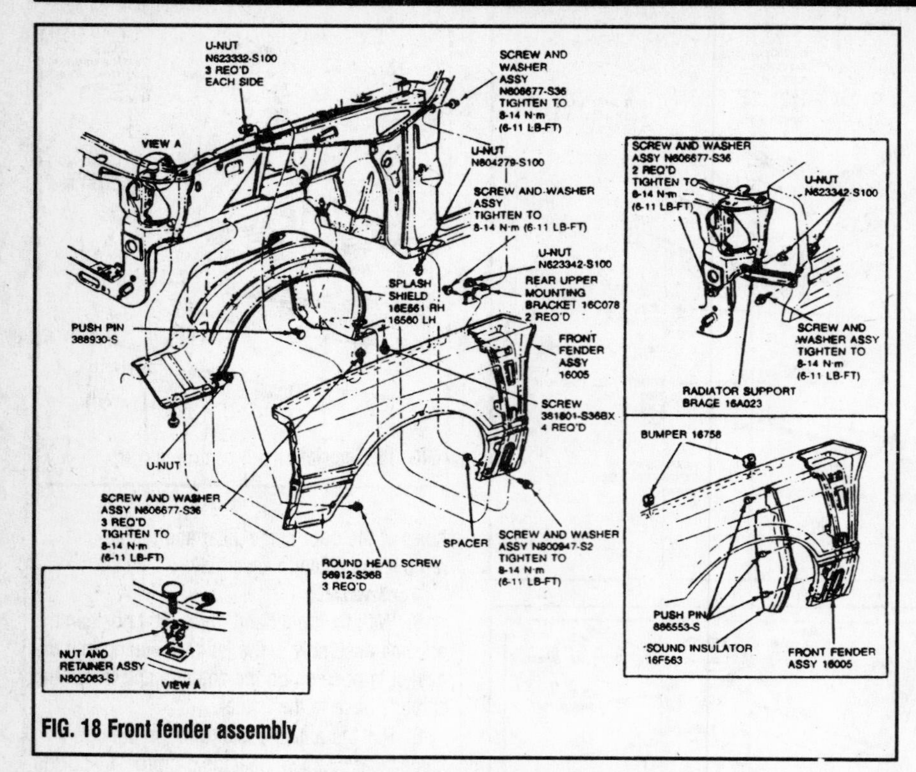

FIG. 18 Front fender assembly

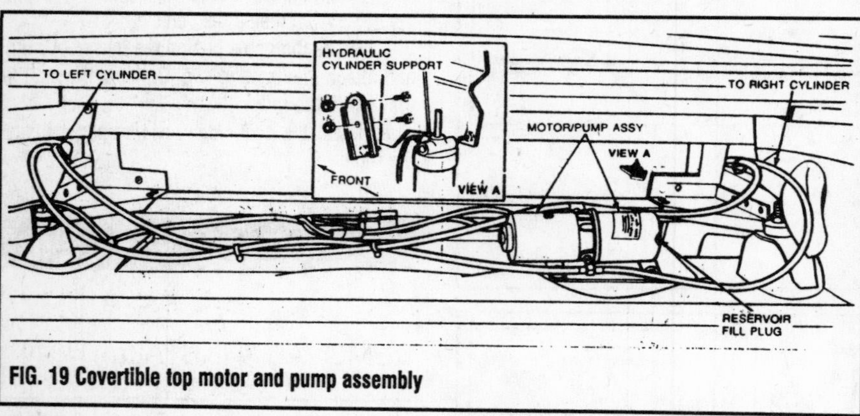

FIG. 19 Covertible top motor and pump assembly

Convertible Top

MOTOR REPLACEMENT

Motor and Pump

➧ SEE FIG. 19

1. Open the top to the fully raised position.
2. Disconnect the negative battery cable.
3. Disconnect the wires at the pump motor.
4. Vent the reservoir by removing the filler plug, and reinstall the filler plug. The plug is located on the top of the reservoir.

➡ **The reservoir must be vented to equalize pressure. This will minimizes the possibility of hydraulic fluid spraying all over when the hoses are disconnected.**

5. Place rags beneath the hose connections, disconnect the hoses, and plug the open fittings and lines.
6. Remove the retaining nuts and washers and remove the motor and pump assembly from the floorpan. Be careful not to loose any of the rubber grommets.

To Install:

7. Remove any plugs from the lines and fittings, and connect the lines to the pump.
8. Install the assembly on the floorpan. Check that all rubber grommets are in the proper position.
9. Connect the wires at the pump motor.
10. Connect the negative battery cable.
11. Operate the top assembly 2–3 times to bleed any air from the system, and check the fluid level in the reservoir. The fluid should not be less that 1/4 in. (6mm) below the filler opening. The top must be up when the fluid level is checked.
12. Position the retaining clips on the rear folding top compartment support wire and install the retaining screws.

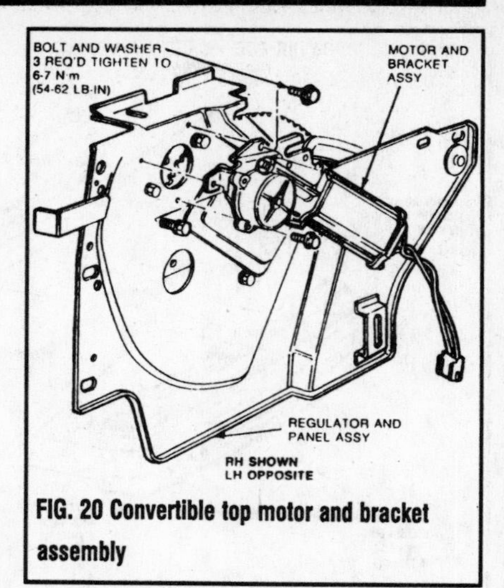

FIG. 20 Convertible top motor and bracket assembly

Motor Assembly

➧ SEE FIG. 20

1. Remove the rear seat back, quarter trim panel and front shoulder belt retractor.
2. Disconnect the motor-to-switch connector.
3. Remove the three motor attaching bolts and remove the motor and regulator from the panel.
4. For motor service, remove the bracket and seal from the motor by removing the three bracket retaining screws and one seal retaining screw.
5. To install, reverse the removal procedures. Tighten the motor assembly to 54–62 inch lbs. (6–7Nm).

TOP REPLACEMENT

Rear Window Valance

➧ SEE FIG. 21

1. Disengage the right and left top latches by rotating the handles inward.
2. Raise the top and put 4–6 inch blocks under the front of the top.

➡ **This will relieve tension on the rear window.**

3. Disconnect the negative battery cable.
4. Remove the retaining screws and unclip the sling at the quarter and rear belt areas. Lower the sling to the storage area.

5. Remove the nut and washers from all the tacking strips on the quarter side panels and at the bottom of the rear valance assembly. Remove all the tacking strips.

6. Unzip the rear window and put aside.

7. Carefully remove the staples from the tacking strip and the rear window valance and save the tacking strip.

8. Remove the caps and screws from the No. 4 bow.

9. Slide the transverse roof molding out of the transverse carrier roof molding.

10. Remove the retaining staples from the transverse roof carrier molding.

11. Position the left and right quarter sides forward.

12. Remove the staples from the zipper on the No. 4 bow and discard the zipper

To install:

13. Position the old valance on the new valance and trace the square cutout holes onto the new valance.

14. Staple the tacking strip to the new valance using a staple gun and 1/4 in. × 5/16 in. (6mm × 8mm) staples.

➡ Staples should be shot in a row with a minimal gap of 1/4 in. (6mm) between each staple.

15. Measure end to end on the No. 4 bow and mark the centerline with a grease pencil.

16. Measure end to end on the valance and mark the centerline with a grease pencil on the zipper of the valance.

17. Position the zipper of the valance on the No. 4 bow, aligning the centerline of the valance with the centerline of the No. 4 bow.

18. Staple the zipper portion of the valance to the No. 4 bow using a staple gun and 1/4 × 1/2 in. (6mm × 13mm) staples.

➡ Staples should be shot in a row with a minimal gap of 1/4 in. (6mm) between each staple.

19. Trim off excess material on the zipper.

20. Move the right and left quarter sides back into proper position.

21. Position the roof transverse carrier molding above the convertible top material on the No. 4 bow and staple using a staple gun and 1/4 in. × 1/2 in. (6mm × 13mm) staples.

➡ Use RTV sealer on the top of the transverse roof carrier to prevent leaks where stapled.

22. Slide the transverse roof molding into the transverse roof carrier molding, and attach with roof molding caps and screws.

23. Install the rear window valance in the proper position and zip.

24. Finger-tighten the nut and washer assemblies for the tacking strips on the quarter side panels and at the bottom of the rear window assembly.

25. Remove the blocks used to relieve tension and open the latches, lower the top into position and lock the latches closed.

26. Adjust and secure all tacking strips on the quarter side panels and at the bottom of the rear window valance assembly.

➡ Be certain that the rear window valance has no visible sags or buckles.

27. Cling on the sling well and attach with the retaining screws.

28. Connect the negative battery cable.

29. Raise and lower the top and confirm proper operation.

Top Stay Pad

➧ SEE FIGS. 22–23

1. Remove the top, quarter stay pad and rear window valance.

2. Remove the stay pad staples at bow No. 1 and bow No. 2.

3. Separate the inside and outside flaps, remove the needle pad and reinforcement pad.

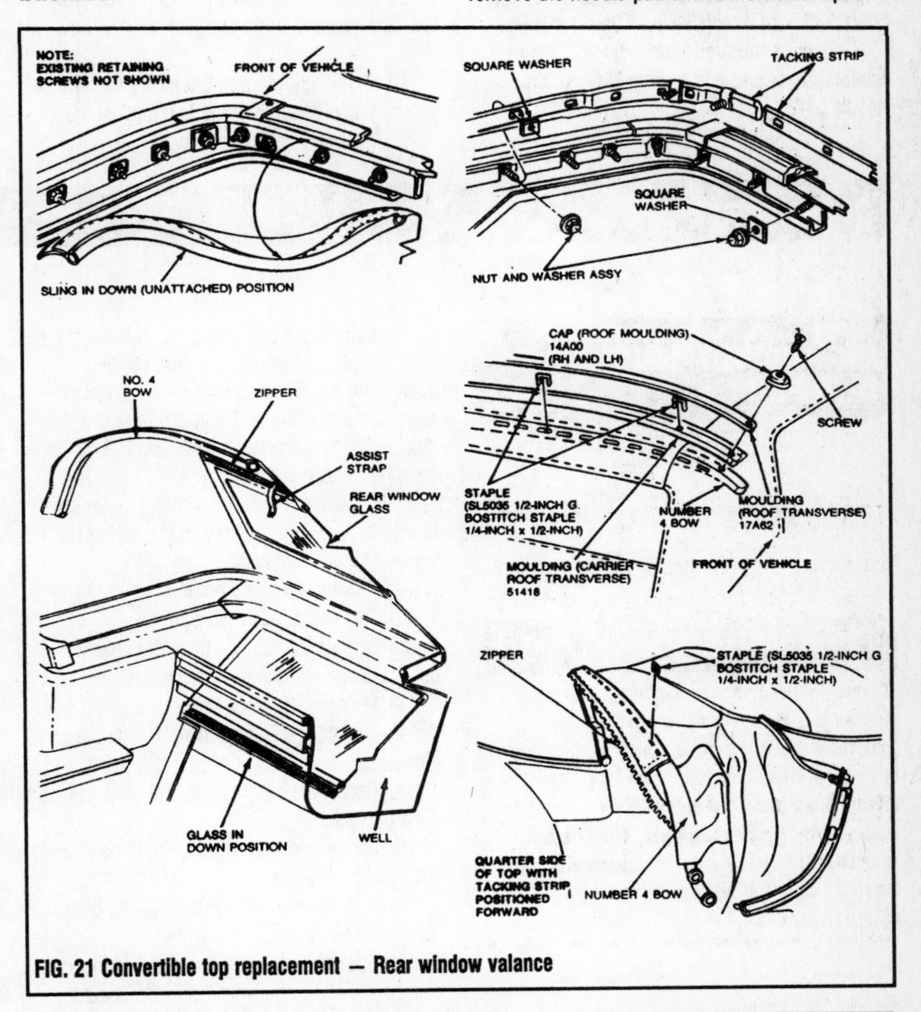

FIG. 21 Convertible top replacement — Rear window valance

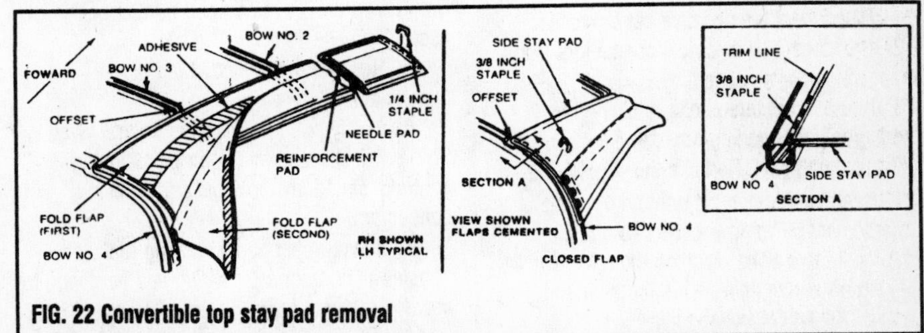

FIG. 22 Convertible top stay pad removal

4. Remove the staples attaching the stay pad at each corner and to the bows 2, 3, and 4 and remove the stay pad.

To Install:

5. Remove the needle and reinforcement pads from the stay pad. Fold the stay pad closed and position the pad against the offset of the bows and staple each corner with 1/4 in. (6mm) staples.

6. Open the stay pad and staple at bows 2, 3, and 4 with 3/8 in. (10mm) staples.

7. Position the reinforcement in the stay pad and staple at bow No. 1 with 1/4 in. (6mm) staples. Stretch the reinforcement and staple at bows 2, 3, and 4 with 3/8 in. (10mm) staples.

8. Apply adhesive to the exposed side of the reinforcement pad and to both sides of the needle pad.

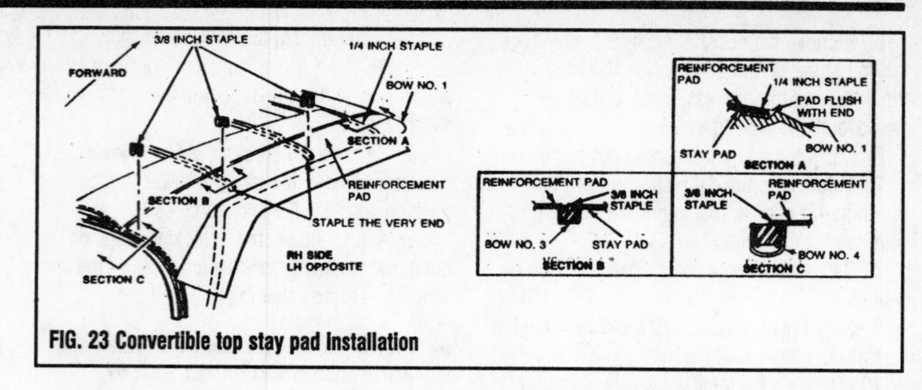

FIG. 23 Convertible top stay pad installation

9. Position the needle pad on the stay pad at the edge of bow No. 4 and just short of bow No. 1.

10. Apply adhesive to the inside surface of the stay pad flaps. Fold the flap against the needle pad and apply adhesive to the edge. Fold the outside flap over the inside flap and staple at bow No. 1 with 1/4 in. (6mm) staples and at bow No. 4 with 3/8 in. (10mm) staples.

11. Install the vinyl cover.

INTERIOR

Instrument Panel and Pad

REMOVAL & INSTALLATION

CAUTION

Some vehicles are equipped with air bags. Before attempting to service air bag equipped vehicles be sure that the system is properly disarmed and all safety precautions are taken. Serious personal injury and vehicle damage could result if this note is disregarded.

➧ SEE FIGS. 24–25

1. Disconnect the negative battery cable.

2. Disconnect all electrical connectors from the steering column assembly.

3. Remove the three bolts retaining the steering column opening cover and reinforcement panel. Remove the panel.

4. Remove the steering column opening reinforcement by removing the two bolts. Remove the two bolts retaining the lower steering column opening reinforcement and remove the reinforcement.

5. Remove the floor console. Remove the six nuts retaining the steering column to the instrument panel (two are retaining the hood release mechanism and four are retaining the column to lower brake pedal support). Lower the steering column to the floor.

6. Remove the steering column upper and lower shrouds and disconnect the wiring from the combination switch.

7. Remove the brake pedal support nut.

8. Snap out the defroster grille.

9. Remove the screws from the speaker covers. Snap out the speaker covers.

10. Remove the front screws retaining the right hand and left hand scuff plates at the cowl trim panel.

11. Remove the right hand and left hand side cowl trim panels.

12. Disconnect the wiring at right hand and left hand cowl sides.

13. Remove the cowl side retaining bolts (one each side).

14. Open the glove compartment door and flex the glove compartment bin tabs inward. Drop down the glove compartment door assembly.

15. Remove the five cowl top screw attachments.

16. Gently pull the instrument panel away from cowl. Disconnect the air conditioning controls, speedometer cable and wire connectors.

17. If instrument panel is being replaced, transfer all components, wiring and hardware to new instrument panel.

To Install:

18. Position the instrument panel in place. Connect the electrical connections. Install the instrument panel upper screws. Install the instrument panel lower screws.

19. Install the lower brace and tighten the bolts. Install the radio speaker grilles.

20. Using all openings install all the instrument panel electrical connections, air conditioning outlets, air conditioning control cables, antenna wires and anything else that may have been removed.

21. Continue the installation in the reverse order of the removal procedure.

22. Check for proper operation of the air bag indicator. Check for proper operation of all components.

Console

REMOVAL & INSTALLATION

➧ SEE FIG. 26

1. Disconnect the negative battery cable.

2. Remove the two access covers at the rear of the console assembly, in order to expose the console mounting screws. Remove the screws.

3. Remove the gearshift opening panel and console floor bracket retaining screws.

4. Remove the rear access panel and the three console to floor bracket retaining screws.

5. Move the parking brake lever in the UP

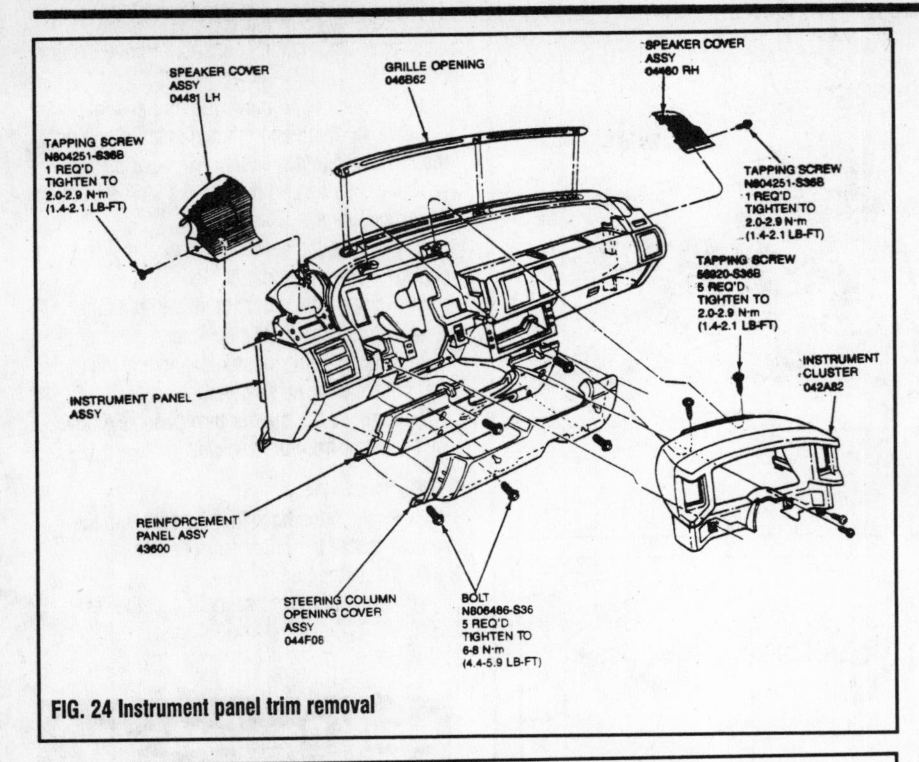

FIG. 24 Instrument panel trim removal

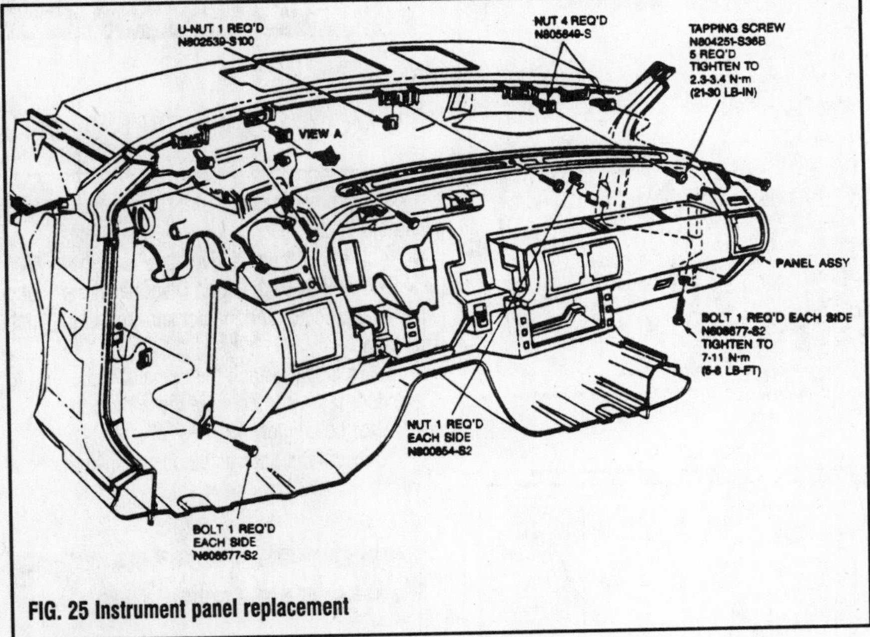

FIG. 25 Instrument panel replacement

position. Remove the four retaining screws and lift the panel up. Disconnect any necessary wire connections.

6. Remove the two console-to-rear floor bracket retaining screws. Insert a small screwdriver into the two notches at the bottom front upper finish panel and snap it out.

7. Open the glove compartment door and drop the glove compartment assembly down. Remove the two console-to-instrument panel retaining screws.

8. Remove the console-to-bracket retaining screws. Remove the console from the vehicle.

To install:

9. Position the console assembly in the vehicle.

10. Install the console-to-bracket retaining screws.

11. Install the two console-to-instrument panel retaining screws. Install the glove compartment assembly.

12. Snap in the front upper finish panel. Install the two console-to-rear floor bracket retaining screws..

13. Connect any necessary wire connections. Install the four panel retaining screws..

14. Install the rear access panel and the three console to floor bracket retaining screws.

15. Install the gearshift opening panel and console floor bracket retaining screws.

16. Install the console mounting screws. Install the two access covers at the rear of the console assembly.

17. Connect the negative battery cable.

Door Trim Panels

REMOVAL & INSTALLATION

➧ SEE FIG. 27

1. Remove the window handle retaining screw and remove the handle.

2. Remove the door latch handle retaining screw and remove the handle.

3. Remove the screws from the armrest door pull cup area.

4. Remove the retaining screws from the armrest.

5. On cars with power door locks and/or power windows, remove the retaining screws and power switch cover assembly. Remove the screws holding the switch housing.

6. Remove the mirror remote control bezel nut.

7. Remove the door trim panel retaining screws.

8. With a flat, wood spatula, pry the trim retaining clips from the door panel. These clips can be easily torn from the trim panel, so be very careful to pry as closely as possible to the clips.

9. Pull the panel out slightly and disconnect all wiring.

10. If a new panel is being installed, transfer all necessary parts.

11. Installation is the reverse of removal.

Headliner

REMOVAL & INSTALLATION

➧ SEE FIG. 28

1. Disconnect the negative battery cable. Remove the front seats. Remove the rear seats.

2. Remove the sun visors. Remove the sun

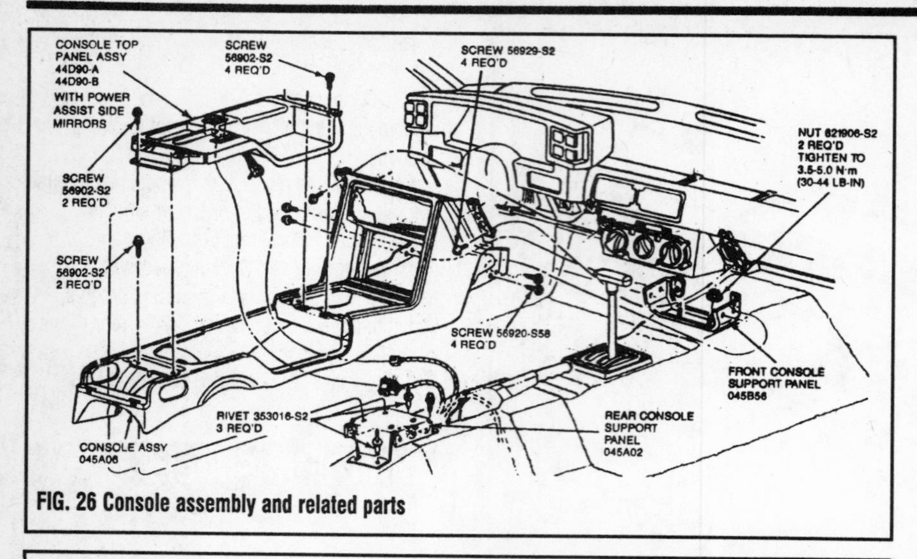

FIG. 26 Console assembly and related parts

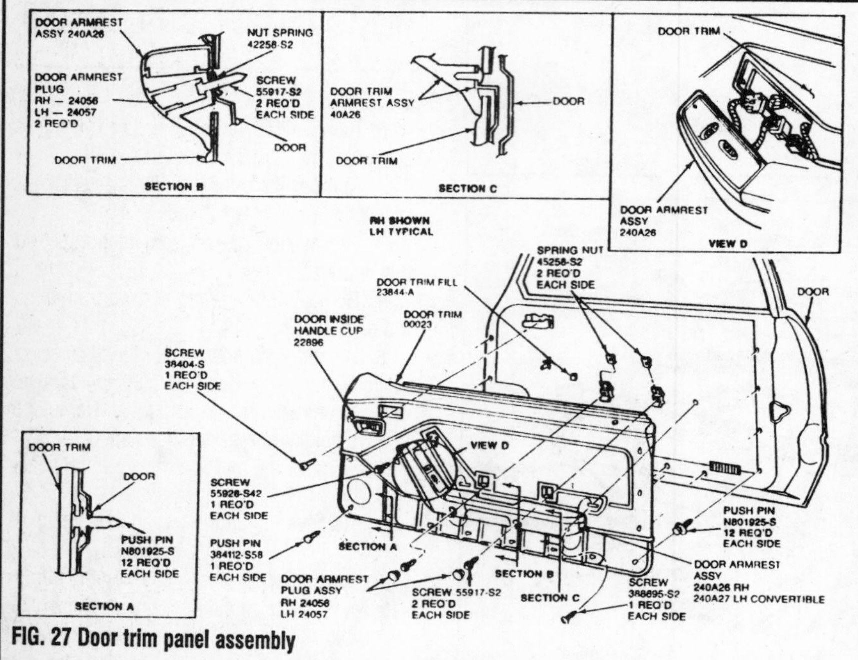

FIG. 27 Door trim panel assembly

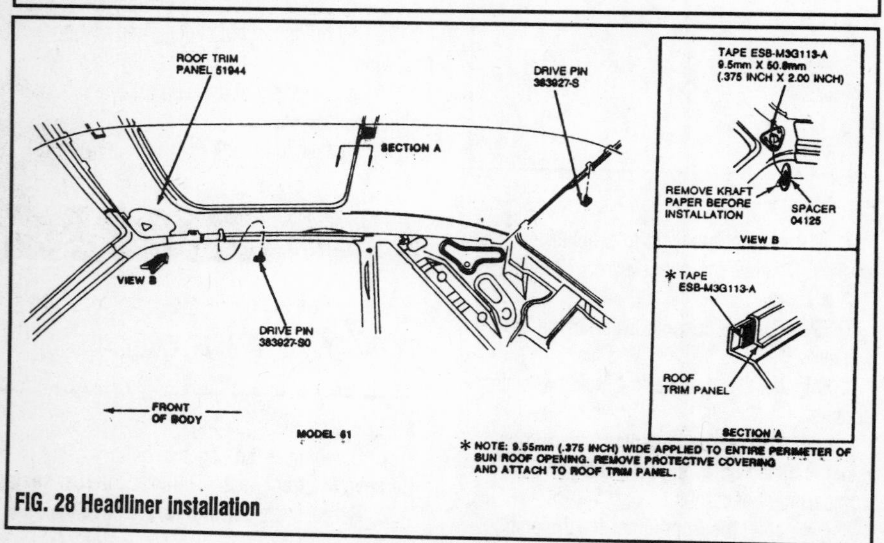

FIG. 28 Headliner installation

visor arm clip retaining screws, remove the arm clip.

3. If equipped, remove the roof console. Remove all dome and reading lights. Snap out the assist strap trim covers. Remove the retaining screws and remove the straps from their mountings.

4. Remove the center body pillar inside finish panel. Remove the coat hooks.

5. Remove the rear roof side trim panel.

6. Remove the roof side inner molding, the liftgate header rail garnish molding and the upper rear corner pillar finish panel.

7. Remove the quarter trim panel. Remove the headliner from the vehicle.

To install:

8. Position the headliner assembly in the vehicle. Install the proper trim panels.

9. Install the sun visors.

10. Install the front seats. Install the rear seats.

Power Door Lock Actuator

REMOVAL & INSTALLATION

1. Remove the door trim panel and watershield.

2. Drill out the pop-rivet attaching the actuator motor to the door. Disconnect the wiring at the connector and the actuator rod at the latch assembly.

3. To install, attach the actuator motor rod to the door latch and connect the wire to the actuator connector.

4. Install the door actuator motor to the door with a pop-rivet or equivalent.

Front Door Latch

REMOVAL & INSTALLATION

➧ SEE FIG. 29

1. Remove the door trim panel and the watershield.

2. Check all the connections of the remote control link and the rod and service if necessary.

3. Remove the remote control assembly and the link clip.

4. Remove the clip attaching the control assembly and the link clip.

5. Remove the clip from the actuator motor, if so equipped.

6. Remove the clip attaching the push-button rod to the latch.

7. Remove the clip attaching the outside door handle rod to the latch assembly.

8. Remove the three screws attaching the latch assembly to the door.

9. Remove the latch assembly (with the remote control link lock cylinder rod) and anti-theft shield from the door cavity.

To install:

10. Install the new bushings and clips onto the new latch assembly. Install the anti-theft shield, remote control link and the lock cylinder rod onto the latch assembly levers.

11. Position the latch (with the link and rod) onto the door cavity, aligning the screw holes in the latch and door. Install the three screws and tighten to 36–72 inch lbs.

12. Attach the outside door handle rod to the latch with a clip.

13. Attach the push-button rod to the latch assembly with clip.

14. Remove the clip from the actuator motor (if so equipped).

15. Attach the lock cylinder rod to the lock cylinder with clip.

16. Install the remote control assembly (and the link clip).

17. Open and close the door to check the latch assembly operation.

18. Install the watershield and the door trim panel.

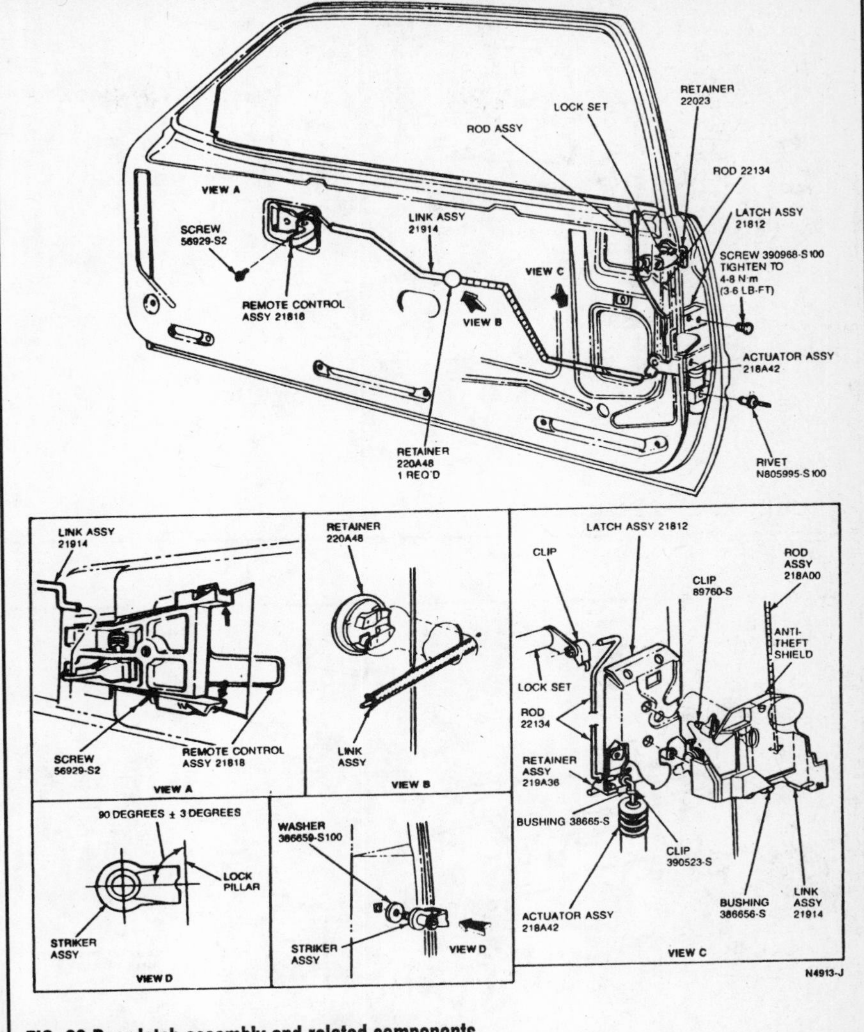

FIG. 29 Door latch assembly and related components

Door Lock Assembly

REMOVAL & INSTALLATION

➡ **When a lock cylinder must be replaced, replace both sides in a set to avoid carrying an extra set of keys.**

1. Remove the door trim panel and watershield.

2. Remove the clip attaching the lock cylinder rod-to-lock cylinder.

3. Pry the lock cylinder out of the slot in the door.

To install:

4. Work the lock cylinder assembly into the outer door panel.

5. Install the cylinder retainer into the slot and push the retainer onto the lock cylinder.

6. Connect the lock cylinder rod to the lock cylinder and install the clip. Lock and unlock the door to check for proper operation.

7. Install the watershield and door trim panel.

Hatchback Lock

REMOVAL & INSTALLATION

➧ SEE FIG. 30

1. Open the hatchback door.

2. Remove the screws retaining the latch assembly.

3. Remove the pop-rivet and retainer. The lock cylinder must be removed with the retainer.

4. Installation is the reverse of the removal procedure. Tighten the latch retaining screws to 7–10 inch lbs. Adjust the striker plate and tighten to 20–28 ft. lbs.

Trunk Lid Lock

REMOVAL & INSTALLATION

➧ SEE FIG. 31

1. Remove the latch retaining screws. Remove the latch.

2. Remove the retainer clip by drilling out the rivet with a 1/4 in. (6mm) drill. Remove the lock support.

3. Remove the lock cylinder retainer as you remove the lock cylinder.

4. Installation is the reverse of the removal procedure. Torque the retaining screws 6–10 ft. lbs.

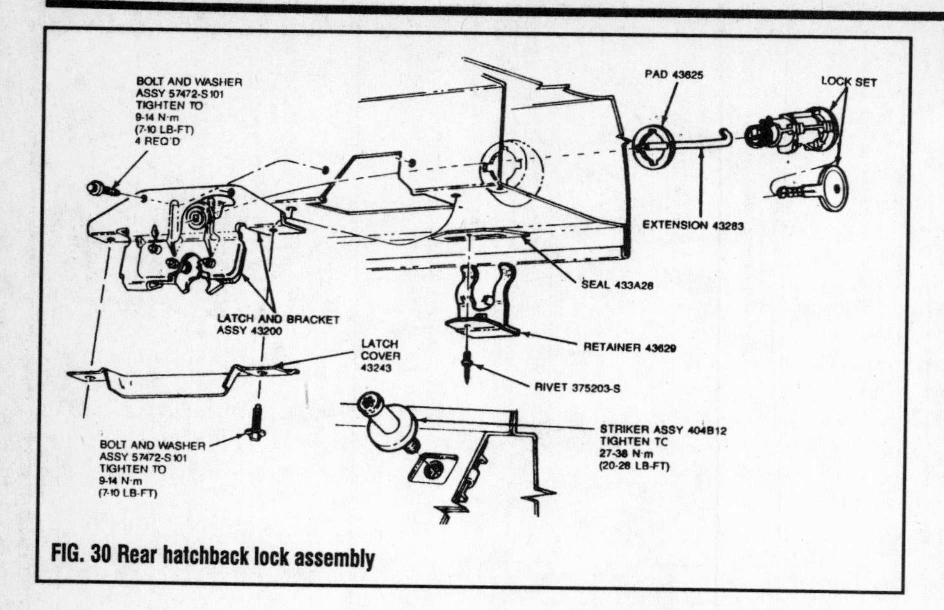

FIG. 30 Rear hatchback lock assembly

FIG. 31 Rear trunk lock assembly

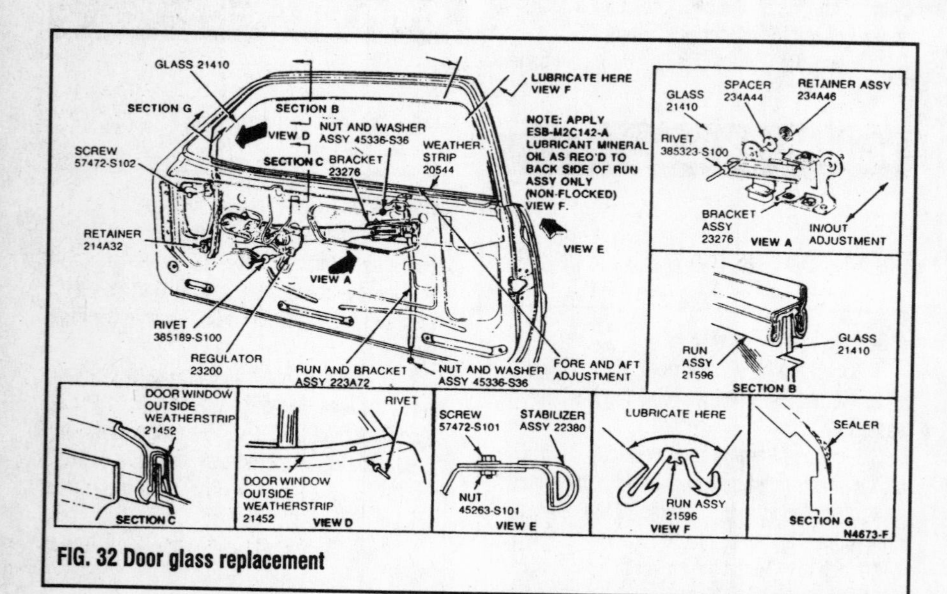

FIG. 32 Door glass replacement

Door Glass

REMOVAL & INSTALLATION

➧ SEE FIG. 32

1. Remove the door trim panel and watershield.
2. Remove the crew attaching each glass rear stabilizer to the inner panel, and remove the stabilizer.
3. Loosen the 2 screws that attach the door glass front run retainer to the inner door panel.
4. Lower the glass to gain access to the glass bracket rivets.
5. Drill out the glass bracket attaching rivets and push out the rivets.

❊❊ WARNING

Before removing the rivets you should insert a suitable block support between the door outer panel and the glass bracket, to stabilize the glass during rivet removal.

6. Remove the glass.
7. Installation is the reverse of removal. Replace the rivets with 1/4-20 × 1 in. nuts and bolts. When the glass is operating properly, tighten the bolts to 20 in. lbs.

Door Glass Regulator

REMOVAL & INSTALLATION

➧ SEE FIG. 33

1. Remove the trim panel and watershield.
2. Prop the window glass in the full up position.
3. Disconnect the window motor wiring if so equipped.
4. Drill out the 3 rivets (manual windows) or 4 rivets (electric windows), attaching the regulator to the inner door panel.
5. Remove the upper screw and washer and the lower nut and washer, attaching the run and bracket to the inner door panel. Slide the run tube up between the door belt and glass. It's a good idea to cover the glass with a protective cloth.
6. Remove the regulator slide from the glass

bracket and remove the regulator through the door access hole.

7. Installation is the reverse of removal. Replace the rivets with 1/4-20 × 1 in. bolts and nuts.

Electric Window Motor

REMOVAL & INSTALLATION

1. Raise the window to the full up position, if possible. If glass cannot be raised and is in a partially down or in the full down position, it must be supported so that it will not fall into door well during the motor removal.
2. Disconnect the negative (−) battery cable.
3. Remove the door trim panel and watershield.
4. Remove the three motor mounting screws and disengage the motor and drive drive assembly from the regulator quadrant gear.

To install:

5. Install the new motor and drive assembly. Tighten the three motor mounting screws to 50–85 inch lbs.
6. Connect the window motor wiring leads.
7. Connect the negative (−) battery cable.
8. Check the power window for proper operation.
9. Install the door trim panel and the watershield.

➡ **Verify that all the drain holes at bottom of doors are open to prevent water accumulation over the motor.**

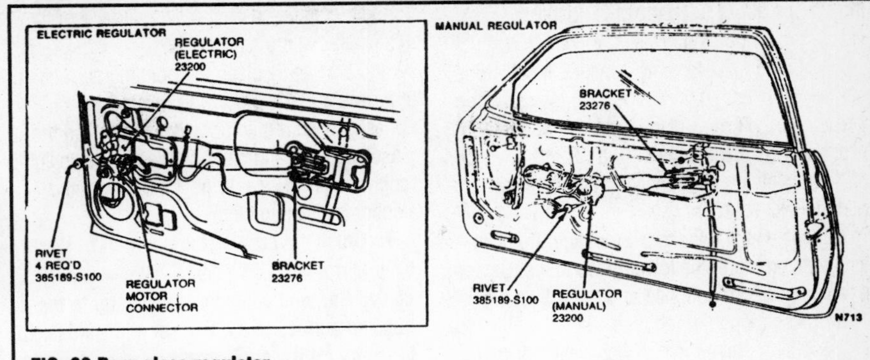

FIG. 33 Door glass regulator

FIG. 34 Windshield glass and related trim

Windshield Glass

REMOVAL & INSTALLATION

➧ SEE FIG. 34

1. Disconnect the negative battery cable. Remove the windshield wiper arms and blades.
2. Remove all windshield trim moldings. Remove the leaf screen. Remove the interior mirror.
3. Using a three foot length of single strand music wire, smallest diameter available. Cut the urethane seal and rubber seal around the entire edge of the windshield.
4. Be sure that you are wearing safety glasses. Force the music wire through the seal at the bottom of the windshield. With someone holding the wire inside the vehicle and the other person holding the wire outside the vehicle move the wire along the bottom and then along the sides and top of the windshield to cut the seal.
5. Using tool D81T–33610–H or a glass holding tool, remove the windshield from the vehicle.

To install:

6. If the existing urethane remains on the windshield opening flange, the new urethane can be applied over it, but at no time should the thickness of the material be above 0.10 in. (2.5mm).
7. Using a clean brush apply urethane metal primer ESB–M2G234–A or equivalent to any sheet metal that has been exposed along the windshield.
8. Apply vinyl foam tape C6AZ–19627–A or equivalent that meets Ford Motor Company's specification ESB–M3G77–A along the cowl and lower A pillars about 4 in. (102mm).
9. Allow the primer to dry for a minimum of about 30 minutes.
10. Be sure that the windshield is clean and

free of any dirt or used material. Install the rear view mirror mounting bracket, as required.

11. Using a lint free rag wipe the inside edge of the windshield, 0.80 in. (20mm) along the top and 2.75 in. (70mm) along the sides and bottom with urethane glass wipe ESB–M5B280–A. Wipe off immediately after application because this material will flash dry.

12. Install the windshield molding. Position the glass on top of the lower glass stops. Center it top and bottom and side to side. Adjust the lower glass stops, as required.

13. Using crayon make alignment marks at points on four sides of both the glass and the window opening.

14. Remove the window glass and the molding assemblies from the vehicle.

15. Using a clean brush apply urethane primer to the edge of the windshield, 0.80 in. (20mm) along the top and 2.75 in. (70mm) along the sides and bottom.

16. Apply an even bead of urethane ESB–M2G316–A around the entire sheet metal flange using an air pressure cartridge gun, air pressure should be about 40 psi. The bead should be triangular in shape 0.55 in. (14mm) high, and 0.33 in. (8mm) at the base.

17. Apply a double bead of urethane along the cowl top and bottom of the opening. Install the windshield taking care to align the glass with the alignment marks. This must be done within 15 minutes of applying the urethane.

18. Install the wiper arms, wiper blades and leaf screen. Install the rear view mirror.

Inside Rear View Mirror

REPLACEMENT

➧ SEE FIG. 35

1. Loosen the mirror assembly-to-mounting bracket setscrew.

2. Remove the mirror assembly by sliding it upward and away from the mounting bracket.

3. If the bracket vinyl pad remains on windshield, apply low heat from an electric heat gun until the vinyl softens. Peel the vinyl off the windshield and discard.

To Install:

4. Make the sure glass, bracket, and adhesive kit, (Rear view Mirror Repair Kit D9AZ–19554–B or equivalent) are at least at room temperature of 65–75° F (18–24° C).

5. Locate and mark the mirror mounting bracket location on the outside surface of the windshield with a wax pencil.

6. Thoroughly clean the bonding surfaces of the glass and the bracket to remove the old adhesive. Use a mild abrasive cleaner on the glass and fine sandpaper on the bracket to lightly roughen the surface. Wipe it clean with the alcohol-moistened cloth.

7. Crush the accelerator vial (part of Rear view Mirror Repair Kit D9AZ–19554–B or equivalent), and apply the accelerator to the bonding surface of the bracket and windshield. Let it dry for three minutes.

8. Apply two drops of adhesive (Rear view Mirror Repair Kit D9AZ–19554–B or equivalent) to the mounting surface of the bracket. Using a clean toothpick or wooden match, quickly spread the adhesive evenly over the mounting surface of the bracket.

9. Quickly position the mounting bracket on the windshield. The 3/8 in. (10mm) circular depression in the bracket must be toward the inside of the passenger compartment. Press the bracket firmly against the windshield for one minute.

10. Allow the bond to set for five minutes. Remove any excess bonding material from the windshield with an alcohol dampened cloth.

11. Attach the mirror to the mounting bracket and tighten the setscrew to 10–20 inch lbs.

Front Bucket Seats

REMOVAL & INSTALLATION

➧ SEE FIG. 36

1. Remove the plastic shield retaining screws and remove the shield.

2. Remove the bolts and nut and washer assemblies retaining the seat tracks to the floor.

3. Remove the seat and track assembly from the vehicle and place on a clean working area.

➡ **Use care when handling seat and track assembly. Dropping the assembly or sitting on the seat not secured in the vehicle may result in damaged components.**

4. Remove the seat track-to-seat cushion attaching screws. Remove the seat cushion and assist spring from the tracks.

5. If the seat tracks are being replaced, transfer the assist springs and spacers, if so equipped, to the new track assembly.

To Install:

6. Mount the seat tracks to the seat cushion.

7. Install the seat track-to-seat cushion retaining screws.

8. Place the seat assembly into vehicle and ensure proper alignment.

9. Install the screws, studs, plastic shields, and nut and washer assemblies.

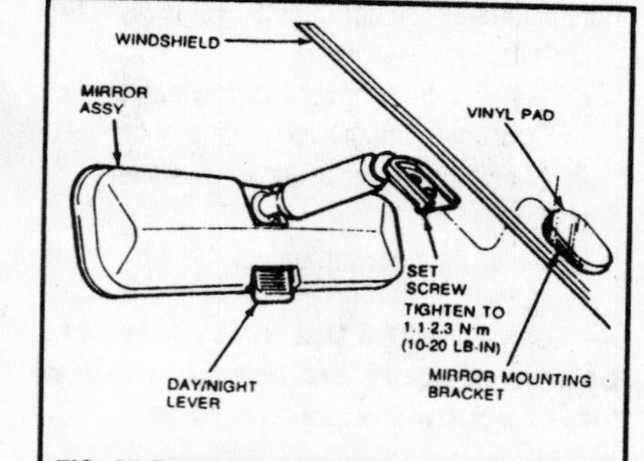

FIG. 35 Rear view mirror replacement

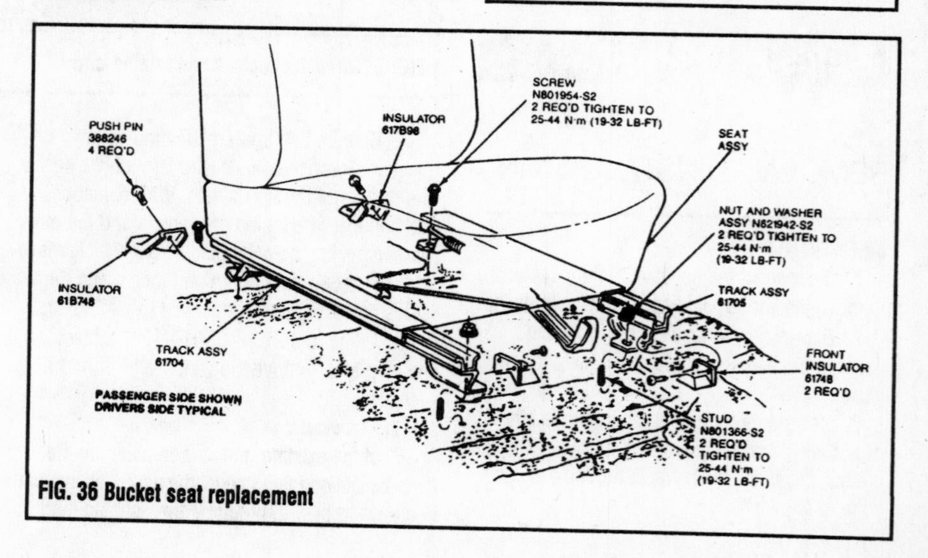

FIG. 36 Bucket seat replacement

Rear Seats

REMOVAL & INSTALLATION

Seat Cushion

➧ SEE FIG. 37

1. Apply knee pressure to the lower portion of the rear seat cushion. Push rearward to disengage the seat cushion from the retainer brackets.

To install:

2. Position the seat cushion assembly into the vehicle.
3. Place the seat belts on top of the cushion.
4. Apply knee pressure to the lower portion of the seat cushion assembly. Push rearward and down to lock the seat cushion into position.
5. Pull the rear seat cushion forward to be certain it is secured into its floor retainer.

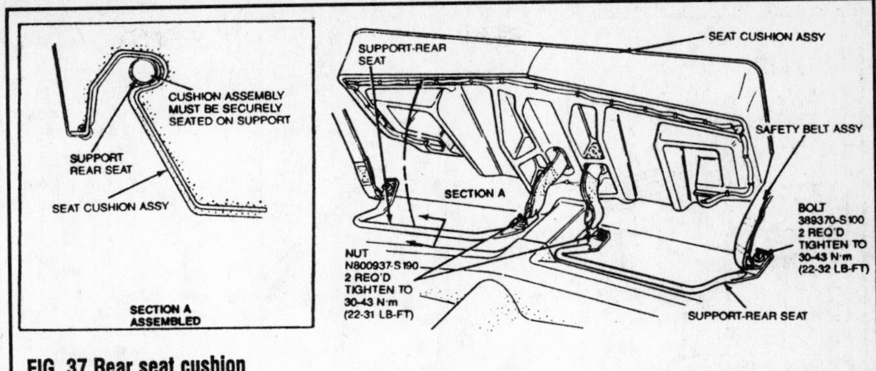

FIG. 37 Rear seat cushion

Seat Back Rest

➧ SEE FIG. 38

1. Remove the rear seat cushion.
2. Remove the seat back bracket attaching bolts.
3. Grasp the seat back assembly at the bottom and lift it up to disengage the hanger wire from the retainer brackets.

To install:

4. Position the seat back in the vehicle so that the hanger wires are engaged with the retaining brackets.
5. Install the seat back bolts and tighten to 5–7 ft. lbs.
6. Install the rear seat cushion.

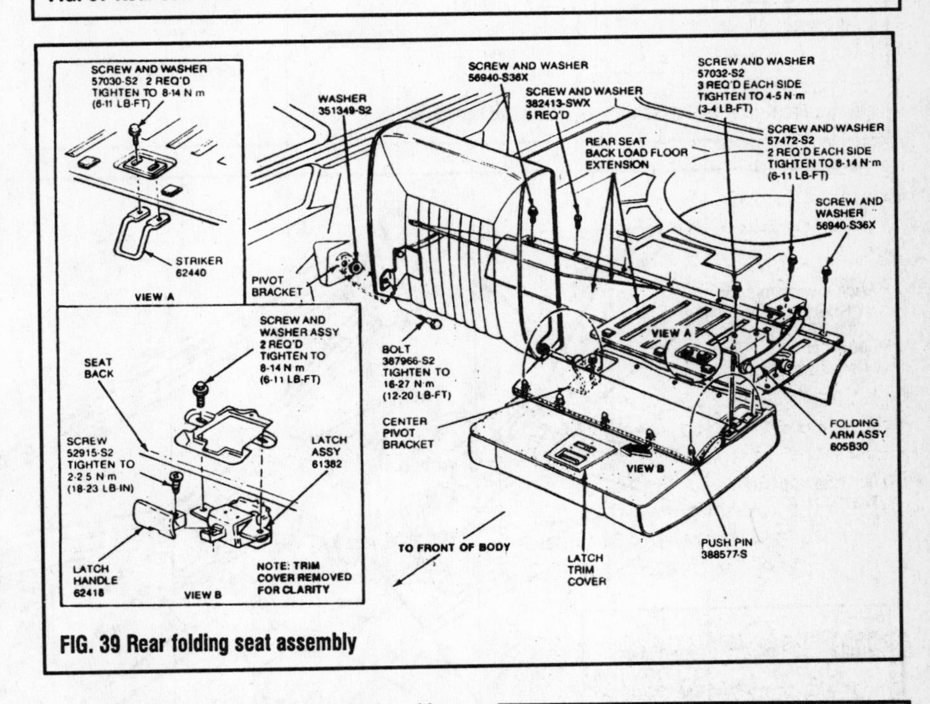

FIG. 39 Rear folding seat assembly

Split Folding Rear Seat Back

➧ SEE FIG. 39

1. Fold the seat back down.
2. Carefully pry up the five push pins to disengage the carpet from the seat back.
3. Remove the three screws attaching the folding arm to the seat back.
4. Remove the seat back from the inboard pivot pin by sliding the seat back toward the outboard side of the vehicle.

To install:

5. Position the seat back onto the inboard pivot pin.
6. Install the three screws attaching the folding arm to the seat back. Tighten to 3–4 ft. lbs.
7. Position the carpet on the seat back and install the push pins.
8. Raise the seat back to the latched position. Adjust the striker plate as necessary.

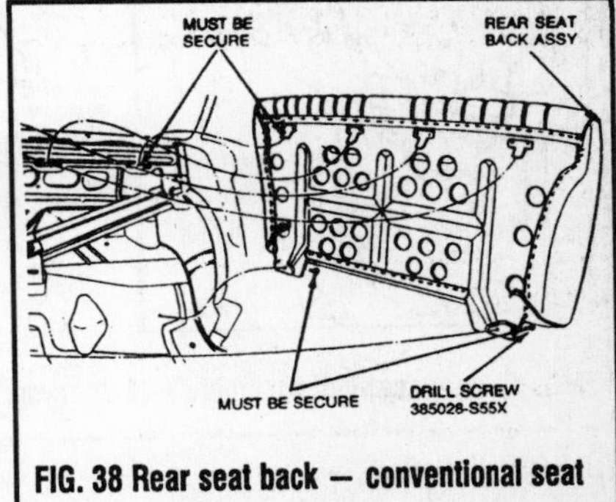

FIG. 38 Rear seat back — conventional seat

Seat Belt Systems

REMOVAL & INSTALLATION

Front

➧ SEE FIGS. 40–41

1. Remove the D-ring cover. Using the bit, tool T77L-2100–A or equivalent, remove the belt bolt. Remove the B–pillar upper trim panel.
2. Remove the scuff plate retaining screws and panel. Remove the belt through the slot in the upper center trim panel.
3. Remove the belt anchor to sill bolt and rubber washer. Remove the belt retractor bolt. Remove the web guide retaining screw and slide the guide rearward to remove it from the B-pillar.
4. Remove the outboard safety belt assembly from the vehicle. Remove the nut from the inboard buckle assembly. On the left side disconnect the buzzer wire and pry off the locator.
5. Pull the buckle upward and remove it from the seat.

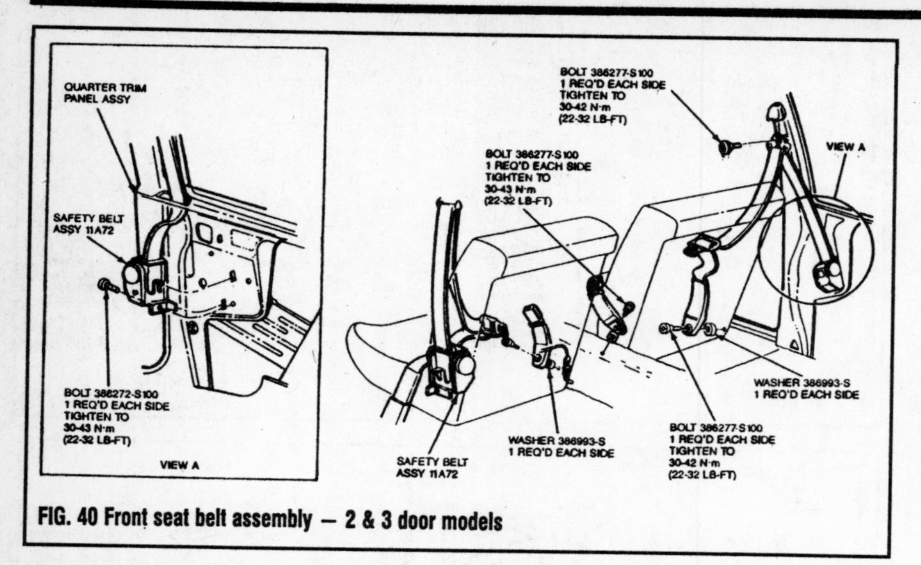

FIG. 40 Front seat belt assembly — 2 & 3 door models

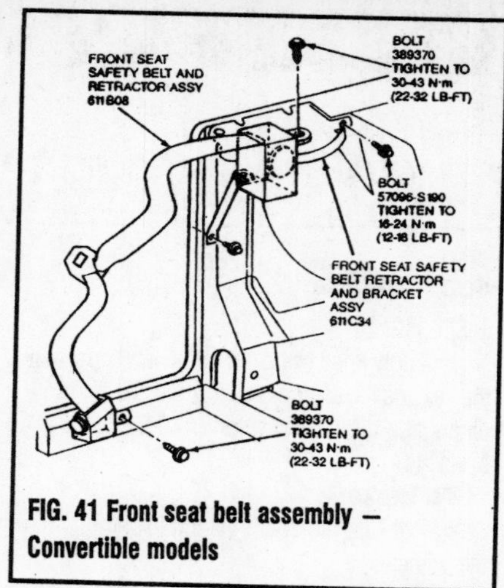

FIG. 41 Front seat belt assembly — Convertible models

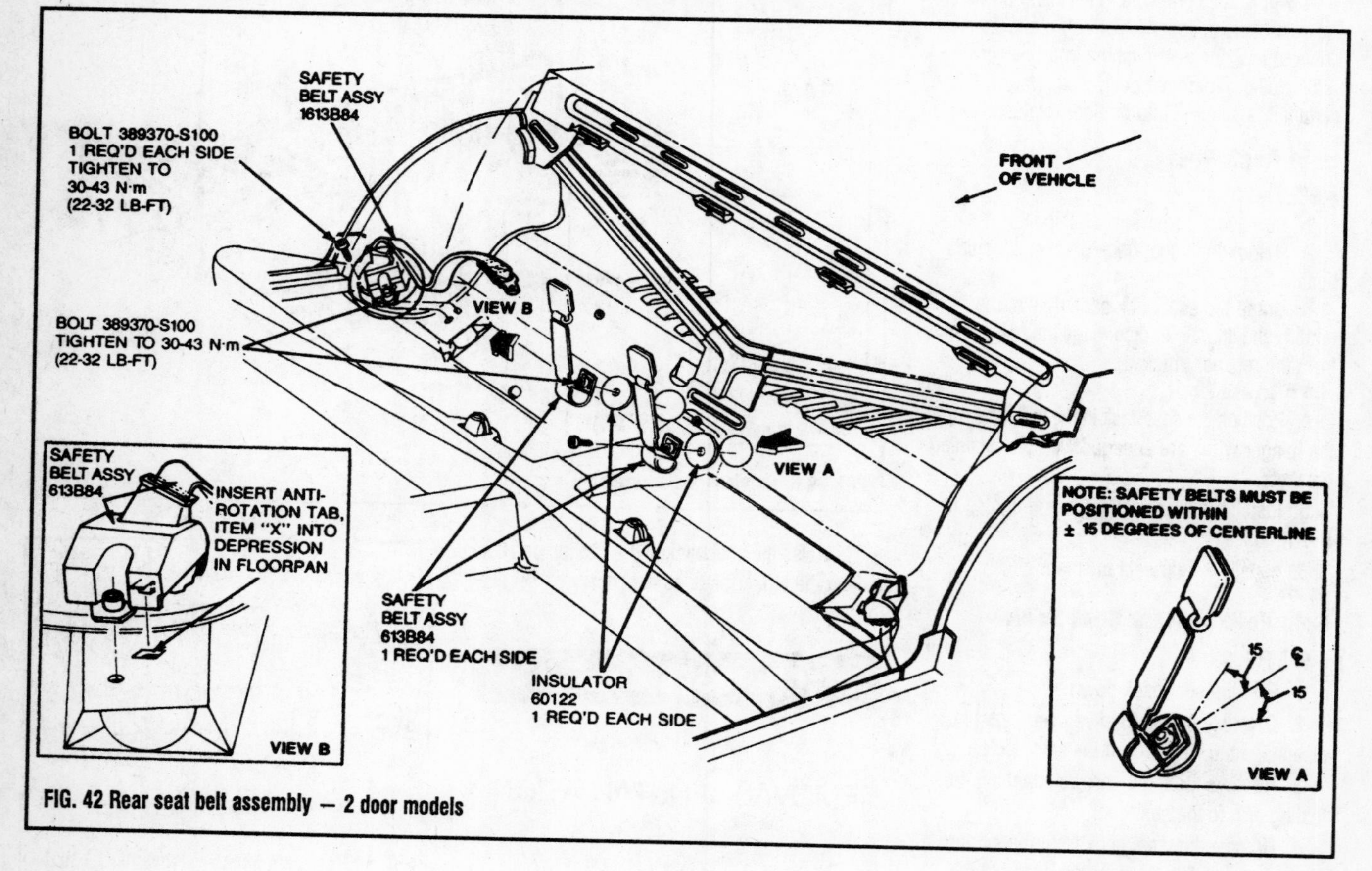

FIG. 42 Rear seat belt assembly — 2 door models

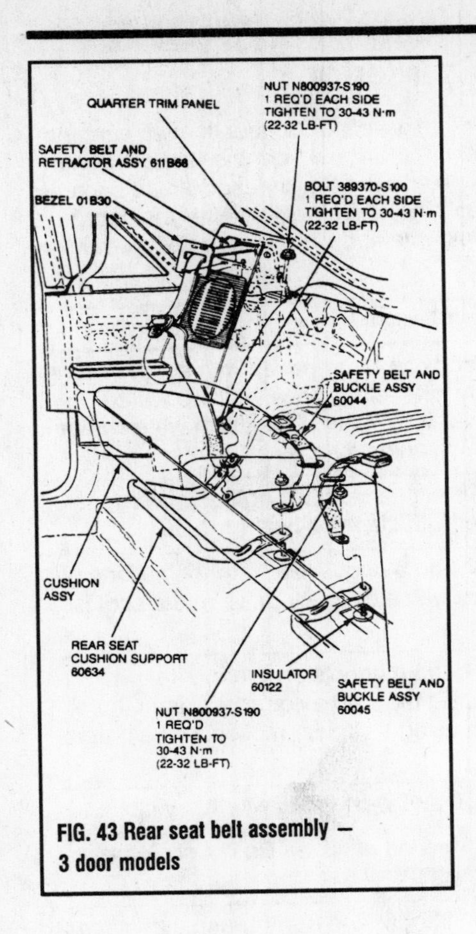

FIG. 43 Rear seat belt assembly — 3 door models

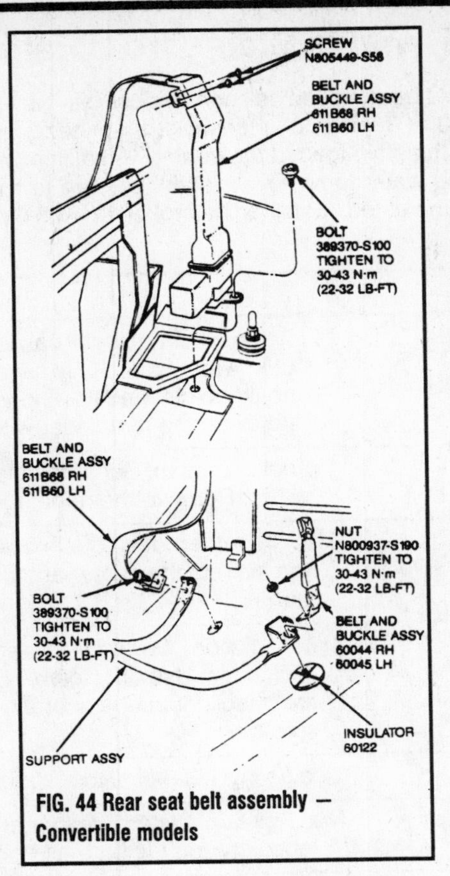

FIG. 44 Rear seat belt assembly — Convertible models

6. Installation is the reverse of the removal procedure.

Rear

➧ SEE FIGS. 42–44

1. Remove the rear seat back and cushion.
2. Remove the angel wing trim and package tray trim.
3. Remove the buckle end anchor nuts. Remove the buckle end belts.
4. Remove the retaining bolt to both rear seat retractors. Remove the retractors.
5. Installation is the reverse of the removal procedure.

How to Remove Stains from Fabric Interior

For best results, spots and stains should be removed as soon as possible. Never use gasoline, lacquer thinner, acetone, nail polish remover or bleach. Use a 3′ x 3″ piece of cheesecloth. Squeeze most of the liquid from the fabric and wipe the stained fabric from the outside of the stain toward the center with a lifting motion. Turn the cheesecloth as soon as one side becomes soiled. When using water to remove a stain, be sure to wash the entire section after the spot has been removed to avoid water stains. Encrusted spots can be broken up with a dull knife and vacuumed before removing the stain.

Type of Stain	How to Remove It
Surface spots	Brush the spots out with a small hand brush or use a commercial preparation such as K2R to lift the stain.
Mildew	Clean around the mildew with warm suds. Rinse in cold water and soak the mildew area in a solution of 1 part table salt and 2 parts water. Wash with upholstery cleaner.
Water stains	Water stains in fabric materials can be removed with a solution made from 1 cup of table salt dissolved in 1 quart of water. Vigorously scrub the solution into the stain and rinse with clear water. Water stains in nylon or other synthetic fabrics should be removed with a commercial type spot remover.
Chewing gum, tar, crayons, shoe polish (greasy stains)	Do not use a cleaner that will soften gum or tar. Harden the deposit with an ice cube and scrape away as much as possible with a dull knife. Moisten the remainder with cleaning fluid and scrub clean.

How to Remove Stains from Fabric Interior

For best results, spots and stains should be removed as soon as possible. Never use gasoline, lacquer thinner, acetone, nail polish remover or bleach. Use a 3′ x 3″ piece of cheesecloth. Squeeze most of the liquid from the fabric and wipe the stained fabric from the outside of the stain toward the center with a lifting motion. Turn the cheesecloth as soon as one side becomes soiled. When using water to remove a stain, be sure to wash the entire section after the spot has been removed to avoid water stains. Encrusted spots can be broken up with a dull knife and vacuumed before removing the stain.

Type of Stain	How to Remove It
Ice cream, candy	Most candy has a sugar base and can be removed with a cloth wrung out in warm water. Oily candy, after cleaning with warm water, should be cleaned with upholstery cleaner. Rinse with warm water and clean the remainder with cleaning fluid.
Wine, alcohol, egg, milk, soft drink (non-greasy stains)	Do not use soap. Scrub the stain with a cloth wrung out in warm water. Remove the remainder with cleaning fluid.
Grease, oil, lipstick, butter and related stains	Use a spot remover to avoid leaving a ring. Work from the outisde of the stain to the center and dry with a clean cloth when the spot is gone.
Headliners (cloth)	Mix a solution of warm water and foam upholstery cleaner to give thick suds. Use only foam—liquid may streak or spot. Clean the entire headliner in one operation using a circular motion with a natural sponge.
Headliner (vinyl)	Use a vinyl cleaner with a sponge and wipe clean with a dry cloth.
Seats and door panels	Mix 1 pint upholstery cleaner in 1 gallon of water. Do not soak the fabric around the buttons.
Leather or vinyl fabric	Use a multi-purpose cleaner full strength and a stiff brush. Let stand 2 minutes and scrub thoroughly. Wipe with a clean, soft rag.
Nylon or synthetic fabrics	For normal stains, use the same procedures you would for washing cloth upholstery. If the fabric is extremely dirty, use a multi-purpose cleaner full strength with a stiff scrub brush. Scrub thoroughly in all directions and wipe with a cotton towel or soft rag.

GLOSSARY

AIR/FUEL RATIO: The ratio of air to gasoline by weight in the fuel mixture drawn into the engine.

AIR INJECTION: One method of reducing harmful exhaust emissions by injecting air into each of the exhaust ports of an engine. The fresh air entering the hot exhaust manifold causes any remaining fuel to be burned before it can exit the tailpipe.

ALTERNATOR: A device used for converting mechanical energy into electrical energy.

AMMETER: An instrument, calibrated in amperes, used to measure the flow of an electrical current in a circuit. Ammeters are always connected in series with the circuit being tested.

AMPERE: The rate of flow of electrical current present when one volt of electrical pressure is applied against one ohm of electrical resistance.

ANALOG COMPUTER: Any microprocessor that uses similar (analogous) electrical signals to make its calculations.

ARMATURE: A laminated, soft iron core wrapped by a wire that converts electrical energy to mechanical energy as in a motor or relay. When rotated in a magnetic field, it changes mechanical energy into electrical energy as in a generator.

ATMOSPHERIC PRESSURE: The pressure on the Earth's surface caused by the weight of the air in the atmosphere. At sea level, this pressure is 14.7 psi at 32°F (101 kPa at 0°C).

ATOMIZATION: The breaking down of a liquid into a fine mist that can be suspended in air.

AXIAL PLAY: Movement parallel to a shaft or bearing bore.

BACKFIRE: The sudden combustion of gases in the intake or exhaust system that results in a loud explosion.

BACKLASH: The clearance or play between two parts, such as meshed gears.

BACKPRESSURE: Restrictions in the exhaust system that slow the exit of exhaust gases from the combustion chamber.

BAKELITE: A heat resistant, plastic insulator material commonly used in printed circuit boards and transistorized components.

BALL BEARING: A bearing made up of hardened inner and outer races between which hardened steel balls roll.

BALLAST RESISTOR: A resistor in the primary ignition circuit that lowers voltage after the engine is started to reduce wear on ignition components.

BEARING: A friction reducing, supportive device usually located between a stationary part and a moving part.

BIMETAL TEMPERATURE SENSOR: Any sensor or switch made of two dissimilar types of metal that bend when heated or cooled due to the different expansion rates of the alloys. These types of sensors usually function as an on/off switch.

BLOWBY: Combustion gases, composed of water vapor and unburned fuel, that leak past the piston rings into the crankcase during normal engine operation. These gases are removed by the PCV system to prevent the buildup of harmful acids in the crankcase.

BRAKE PAD: A brake shoe and lining assembly used with disc brakes.

BRAKE SHOE: The backing for the brake lining. The term is, however, usually applied to the assembly of the brake backing and lining.

BUSHING: A liner, usually removable, for a bearing; an anti-friction liner used in place of a bearing.

BYPASS: System used to bypass ballast resistor during engine cranking to increase voltage supplied to the coil.

CALIPER: A hydraulically activated device in a disc brake system, which is mounted straddling the brake rotor (disc). The caliper contains at least one piston and two brake pads. Hydraulic pressure on the piston(s) forces the pads against the rotor.

CAMSHAFT: A shaft in the engine on which are the lobes (cams) which operate the valves. The camshaft is driven by the crankshaft, via a belt, chain or gears, at one half the crankshaft speed.

CAPACITOR: A device which stores an electrical charge.

CARBON MONOXIDE (CO): A colorless, odorless gas given off as a normal byproduct of combustion. It is poisonous and extremely dangerous in confined areas, building up slowly to toxic levels without warning if adequate ventilation is not available.

CARBURETOR: A device, usually mounted on the intake manifold of an engine, which mixes the air and fuel in the proper proportion to allow even combustion.

CATALYTIC CONVERTER: A device installed in the exhaust system, like a muffler, that converts harmful byproducts of combustion into carbon dioxide and water vapor by means of a heat-producing chemical reaction.

CENTRIFUGAL ADVANCE: A mechanical method of advancing the spark timing by using fly weights in the distributor that react to centrifugal force generated by the distributor shaft rotation.

CHECK VALVE: Any one-way valve installed to permit the flow of air, fuel or vacuum in one direction only.

CHOKE: A device, usually a movable valve, placed in the intake path of a carburetor to restrict the flow of air.

CIRCUIT: Any unbroken path through which an electrical current can flow. Also used to describe fuel flow in some instances.

CIRCUIT BREAKER: A switch which protects an electrical circuit from overload by opening the circuit when the current flow exceeds a predetermined level. Some circuit breakers must be reset manually, while most reset automatically

COIL (IGNITION): A transformer in the ignition circuit which steps up the voltage provided to the spark plugs.

COMBINATION MANIFOLD: An assembly which includes both the intake and exhaust manifolds in one casting.

COMBINATION VALVE: A device used in some fuel systems that routes fuel vapors to a charcoal storage canister instead of venting them into the atmosphere. The valve relieves fuel tank pressure and allows fresh air into the tank as the fuel level drops to prevent a vapor lock situation.

COMPRESSION RATIO: The comparison of the total volume of the cylinder and combustion chamber with the piston at BDC and the piston at TDC.

CONDENSER: 1. An electrical device which acts to store an electrical charge, preventing voltage surges.
2. A radiator-like device in the air conditioning system in which refrigerant gas condenses into a liquid, giving off heat.

CONDUCTOR: Any material through which an electrical current can be transmitted easily.

CONTINUITY: Continuous or complete circuit. Can be checked with an ohmmeter.

COUNTERSHAFT: An intermediate shaft which is rotated by a mainshaft and transmits, in turn, that rotation to a working part.

CRANKCASE: The lower part of an engine in which the crankshaft and related parts operate.

CRANKSHAFT: The main driving shaft of an engine which receives reciprocating motion from the pistons and converts it to rotary motion.

CYLINDER: In an engine, the round hole in the engine block in which the piston(s) ride.

CYLINDER BLOCK: The main structural member of an engine in which is found the cylinders, crankshaft and other principal parts.

CYLINDER HEAD: The detachable portion of the engine, fastened, usually, to the top of the cylinder block, containing all or most of the combustion chambers. On overhead valve engines, it contains the valves and their operating parts. On overhead cam engines, it contains the camshaft as well.

DEAD CENTER: The extreme top or bottom of the piston stroke.

DETONATION: An unwanted explosion of the air/fuel mixture in the combustion chamber caused by excess heat and compression, advanced timing, or an overly lean mixture. Also referred to as "ping".

DIAPHRAGM: A thin, flexible wall separating two cavities, such as in a vacuum advance unit.

DIESELING: A condition in which hot spots in the combustion chamber cause the engine to run on after the key is turned off.

DIFFERENTIAL: A geared assembly which allows the transmission of motion between drive axles, giving one axle the ability to turn faster than the other.

DIODE: An electrical device that will allow current to flow in one direction only.

DISC BRAKE: A hydraulic braking assembly consisting of a brake disc, or rotor, mounted on an axle, and a caliper assembly containing, usually two brake pads which are activated by hydraulic pressure. The pads are forced against the sides of the disc, creating friction which slows the vehicle.

DISTRIBUTOR: A mechanically driven device on an engine which is responsible for electrically firing the spark plug at a predetermined point of the piston stroke.

DOWEL PIN: A pin, inserted in mating holes in two different parts allowing those parts to maintain a fixed relationship.

DRUM BRAKE: A braking system which consists of two brake shoes and one or two wheel cylinders, mounted on a fixed backing plate, and a brake drum, mounted on an axle, which revolves around the assembly. Hydraulic action applied to the wheel cylinders forces the shoes outward against the drum, creating friction, slowing the vehicle.

DWELL: The rate, measured in degrees of shaft rotation, at which an electrical circuit cycles on and off.

ELECTRONIC CONTROL UNIT (ECU): Ignition module, amplifier or igniter. See Module for definition.

ELECTRONIC IGNITION: A system in which the timing and firing of the spark plugs is controlled by an electronic control unit, usually called a module. These systems have no points or condenser.

ENDPLAY: The measured amount of axial movement in a shaft.

ENGINE: A device that converts heat into mechanical energy.

EXHAUST MANIFOLD: A set of cast passages or pipes which conduct exhaust gases from the engine.

FEELER GAUGE: A blade, usually metal, of precisely predetermined thickness, used to measure the clearance between two parts. These blades usually are available in sets of assorted thicknesses.

F-HEAD: An engine configuration in which the intake valves are in the cylinder head, while the camshaft and exhaust valves are located in the cylinder block. The camshaft operates the intake valves via lifters and pushrods, while it operates the exhaust valves directly.

FIRING ORDER: The order in which combustion occurs in the cylinders of an engine. Also the order in which spark is distributed to the plugs by the distributor.

FLATHEAD: An engine configuration in which the camshaft and all the valves are located in the cylinder block.

FLOODING: The presence of too much fuel in the intake manifold and combustion chamber which prevents the air/fuel mixture from firing, thereby causing a no-start situation.

FLYWHEEL: A disc shaped part bolted to the rear end of the crankshaft. Around the outer perimeter is affixed the ring gear. The starter drive engages the ring gear, turning the flywheel, which rotates the crankshaft, imparting the initial starting motion to the engine.

FOOT POUND (ft.lb. or sometimes, ft. lbs.): The amount of energy or work needed to raise an item weighing one pound, a distance of one foot.

FUSE: A protective device in a circuit which prevents circuit overload by breaking the circuit when a specific amperage is present. The device is constructed around a strip or wire of a lower amperage rating than the circuit it is designed to protect. When an amperage higher than that stamped on the fuse is present in the circuit, the strip or wire melts, opening the circuit.

GEAR RATIO: The ratio between the number of teeth on meshing gears.

GENERATOR: A device which converts mechanical energy into electrical energy.

HEAT RANGE: The measure of a spark plug's ability to dissipate heat from its firing end. The higher the heat range, the hotter the plug fires.

HUB: The center part of a wheel or gear.

HYDROCARBON (HC): Any chemical compound made up of hydrogen and carbon. A major pollutant formed by the engine as a byproduct of combustion.

HYDROMETER: An instrument used to measure the specific gravity of a solution.

INCH POUND (in.lb. or sometimes, in. lbs.): One twelfth of a foot pound.

INDUCTION: A means of transferring electrical energy in the form of a magnetic field. Principle used in the ignition coil to increase voltage.

INJECTION PUMP: A device, usually mechanically operated, which meters and delivers fuel under pressure to the fuel injector.

INJECTOR: A device which receives metered fuel under relatively low pressure and is activated to inject the fuel into the engine under relatively high pressure at a predetermined time.

INPUT SHAFT: The shaft to which torque is applied, usually carrying the driving gear or gears.

INTAKE MANIFOLD: A casting of passages or pipes used to conduct air or a fuel/air mixture to the cylinders.

JOURNAL: The bearing surface within which a shaft operates.

KEY: A small block usually fitted in a notch between a shaft and a hub to prevent slippage of the two parts.

MANIFOLD: A casting of passages or set of pipes which connect the cylinders to an inlet or outlet source.

MANIFOLD VACUUM: Low pressure in an engine intake manifold formed just below the throttle plates. Manifold vacuum is highest at idle and drops under acceleration.

MASTER CYLINDER: The primary fluid pressurizing device in a hydraulic system. In automotive use, it is found in brake and hydraulic clutch systems and is pedal activated, either directly or, in a power brake system, through the power booster.

MODULE: Electronic control unit, amplifier or igniter of solid state or integrated design which controls the current flow in the ignition primary circuit based on input from the pick- up coil. When the module opens the primary circuit, the high secondary voltage is induced in the coil.

NEEDLE BEARING: A bearing which consists of a number (usually a large number) of long, thin rollers.

OHM:(Ω) The unit used to measure the resistance of conductor to electrical flow. One ohm is the amount of resistance that limits current flow to one ampere in a circuit with one volt of pressure.

OHMMETER: An instrument used for measuring the resistance, in ohms, in an electrical circuit.

OUTPUT SHAFT: The shaft which transmits torque from a device, such as a transmission.

OVERDRIVE: A gear assembly which produces more shaft revolutions than that transmitted to it.

OVERHEAD CAMSHAFT (OHC): An engine configuration in which the camshaft is mounted on top of the cylinder head and operates the valves either directly or by means of rocker arms.

OVERHEAD VALVE (OHV): An engine configuration in which all of the valves are located in the cylinder head and the camshaft is located in the cylinder block. The camshaft operates the valves via lifters and pushrods.

OXIDES OF NITROGEN (NOx): Chemical compounds of nitrogen produced as a byproduct of combustion. They combine with hydrocarbons to produce smog.

OXYGEN SENSOR: Used with the feedback system to sense the presence of oxygen in the exhaust gas and signal the computer which can reference the voltage signal to an air/fuel ratio.

PINION: The smaller of two meshing gears.

PISTON RING: An open ended ring which fits into a groove on the outer diameter of the piston. Its chief function is to form a seal between the piston and cylinder wall. Most automotive pistons have three rings: two for compression sealing; one for oil sealing.

PRELOAD: A predetermined load placed on a bearing during assembly or by adjustment.

PRIMARY CIRCUIT: Is the low voltage side of the ignition system which consists of the ignition switch, ballast resistor or resistance wire, bypass, coil, electronic control unit and pick-up coil as well as the connecting wires and harnesses.

PRESS FIT: The mating of two parts under pressure, due to the inner diameter of one being smaller than the outer diameter of the other, or vice versa; an interference fit.

RACE: The surface on the inner or outer ring of a bearing on which the balls, needles or rollers move.

REGULATOR: A device which maintains the amperage and/or voltage levels of a circuit at predetermined values.

RELAY: A switch which automatically opens and/or closes a circuit.

RESISTANCE: The opposition to the flow of current through a circuit or electrical device, and is measured in ohms. Resistance is equal to the voltage divided by the amperage.

RESISTOR: A device, usually made of wire, which offers a preset amount of resistance in an electrical circuit.

RING GEAR: The name given to a ring-shaped gear attached to a differential case,or affixed to a flywheel or as part a planetary gear set.

ROLLER BEARING: A bearing made up of hardened inner and outer races between which hardened steel rollers move.

ROTOR: 1. The disc-shaped part of a disc brake assembly, upon which the brake pads bear; also called, brake disc.
2. The device mounted atop the distributor shaft, which passes current to the distributor cap tower contacts.

SECONDARY CIRCUIT: The high voltage side of the ignition system, usually above 20,000 volts. The secondary includes the ignition coil, coil wire, distributor cap and rotor, spark plug wires and spark plugs.

SENDING UNIT: A mechanical, electrical, hydraulic or electromagnetic device which transmits information to a gauge.

SENSOR: Any device designed to measure engine operating conditions or ambient pressures and temperatures. Usually electronic in nature and designed to send a voltage signal to an on-board computer, some sensors may operate as a simple on/off switch or they may provide a variable voltage signal (like a potentiometer) as conditions or measured parameters change.

SHIM: Spacers of precise, predetermined thickness used between parts to establish a proper working relationship.

SLAVE CYLINDER: In automotive use, a device in the hydraulic clutch system which is activated by hydraulic force, disengaging the clutch.

SOLENOID: A coil used to produce a magnetic field, the effect of which is to produce work.

SPARK PLUG: A device screwed into the combustion chamber of a spark ignition engine. The basic construction is a conductive core inside of a ceramic insulator, mounted in an outer conductive base. An electrical charge from the spark plug wire travels along the conductive core and jumps a preset air gap to a grounding point or points at the end of the conductive base. The resultant spark ignites the fuel/air mixture in the combustion chamber.

SPLINES: Ridges machined or cast onto the outer diameter of a shaft or inner diameter of a bore to enable parts to mate without rotation.

TACHOMETER: A device used to measure the rotary speed of an engine, shaft, gear, etc., usually in rotations per minute.

THERMOSTAT: A valve, located in the cooling system of an engine, which is closed when cold and opens gradually in response to engine heating, controlling the temperature of the coolant and rate of coolant flow.

TOP DEAD CENTER (TDC): The point at which the piston reaches the top of its travel on the compression stroke.

TORQUE: The twisting force applied to an object.

TORQUE CONVERTER: A turbine used to transmit power from a driving member to a driven member via hydraulic action, providing changes in drive ratio and torque. In automotive use, it links the driveplate at the rear of the engine to the automatic transmission.

TRANSDUCER: A device used to change a force into an electrical signal.

TRANSISTOR: A semi-conductor component which can be actuated by a small voltage to perform an electrical switching function.

TUNE-UP: A regular maintenance function, usually associated with the replacement and adjustment of parts and components in the electrical and fuel systems of a vehicle for the purpose of attaining optimum performance.

TURBOCHARGER: An exhaust driven pump which compresses intake air and forces it into the combustion chambers at higher than atmospheric pressures. The increased air pressure allows more fuel to be burned and results in increased horsepower being produced.

VACUUM ADVANCE: A device which advances the ignition timing in response to increased engine vacuum.

VACUUM GAUGE: An instrument used to measure the presence of vacuum in a chamber.

VALVE: A device which control the pressure, direction of flow or rate of flow of a liquid or gas.

VALVE CLEARANCE: The measured gap between the end of the valve stem and the rocker arm, cam lobe or follower that activates the valve.

VISCOSITY: The rating of a liquid's internal resistance to flow.

VOLTMETER: An instrument used for measuring electrical force in units called volts. Voltmeters are always connected parallel with the circuit being tested.

WHEEL CYLINDER: Found in the automotive drum brake assembly, it is a device, actuated by hydraulic pressure, which, through internal pistons, pushes the brake shoes outward against the drums.

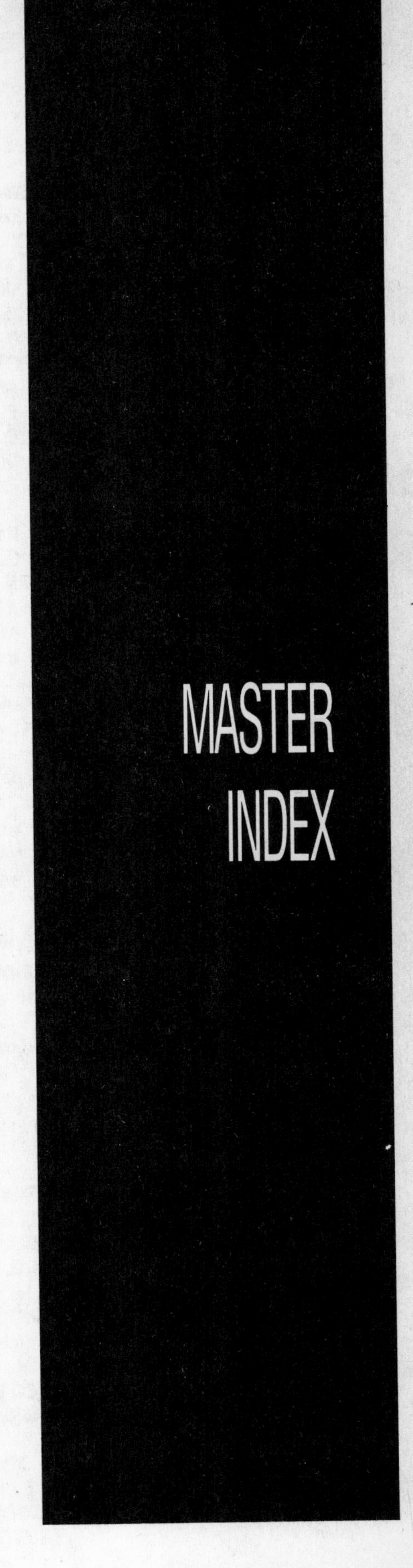
MASTER
INDEX